Introduction to Linear and Matrix Algebra

Nathaniel Johnston

Introduction to Linear and Matrix Algebra

 Springer

Nathaniel Johnston
Department of Mathematics and
Computer Science
Mount Allison University
Sackville, NB, Canada

ISBN 978-3-030-52813-3   ISBN 978-3-030-52811-9   (eBook)
https://doi.org/10.1007/978-3-030-52811-9

Mathematics Subject Classification: 15Axx, 97H60, 00-01

This Springer imprint is published by the registered company Springer Nature Switzerland AG
The registered company address is: Gewerbestrasse 11, 6330 Cham, Switzerland

For Cora

I hope that one day you're interested enough to read this book,
and I hope it's successful enough that you have to.

Preface

The Purpose of this Book

Linear algebra, more so than any other mathematical subject, can be approached in numerous ways. Many textbooks present the subject in a very concrete and numerical manner, spending much of their time solving systems of linear equations and having students perform laborious row-reductions on matrices. Many other books instead focus very heavily on linear transformations and other basis-independent properties, almost to the point that their connection to matrices is considered an inconvenient afterthought that students should avoid using at all costs.

This book is written from the perspective that both linear transformations and matrices are useful objects in their own right, but it is the connection between the two that really unlocks the magic of linear algebra. Sometimes, when we want to know something about a linear transformation, the easiest way to get an answer is to grab onto a basis and look at the corresponding matrix. Conversely, there are many interesting families of matrices and matrix operations that seemingly have nothing to do with linear transformations, yet can nonetheless illuminate how some basis-independent objects behave.

For this reason, we introduce both matrices and linear transformations early, in Chapter 1, and frequently switch back and forth between these two perspectives. For example, we motivate matrix multiplication in the standard way via the composition of linear transformations, but are also careful to say that this is not the *only* useful way of looking at matrix multiplication—for example, multiplying the adjacency matrix of a graph with itself gives useful information about walks on that graph (see Section 1.B), despite there not being a linear transformation in sight.

We spend much of the first chapter discussing the geometry of vectors, and we emphasize the geometric nature of matrices and linear transformations repeatedly throughout the rest of the book. For example, the invertibility of matrices (see Section 2.2) is not just presented as an algebraic concept that we determine via Gaussian elimination, but its geometric interpretation as linear transformations that do not "squash" space is also emphasized. Even more dramatically, the determinant, which is notoriously difficult to motivate algebraically, is first introduced geometrically as the factor by which a linear transformation stretches space (see Section 3.2).

We believe that repeatedly emphasizing this interplay between algebra and geometry (i.e., between matrices and linear transformations) leads to a deeper understanding of the topics presented in this book. It also better prepares students for future studies in linear algebra, where linear transformations take center stage.

Features of this Book

This book makes use of numerous features to make it as easy to read and understand as possible. Here, we highlight some of these features and discuss how to best make use of them.

Focus

Linear algebra has no shortage of fields in which it is applicable, and this book presents many of them when appropriate. However, these applications are presented first and foremost to illustrate the mathematical theory being introduced, and for how *mathematically* interesting they are, rather than for how important they are in other fields of study. For example, some games that can be analyzed and solved via linear algebra are presented in Section 2.A—not because they are "useful", but rather because

- they let us make use of all of the tools that we developed earlier in that chapter,
- they give us a reason to introduce and explore a new topic (finite fields), and
- (most importantly) they are *interesting*.

We similarly look at some other mathematical applications of linear algebra in Sections 1.B (introductory graph theory), 2.1.5 (solving real-world problems via linear systems), 3.B (power iteration and Google's PageRank algorithm), and 3.D (solving linear recurrence relations to, for example, find an explicit formula for the Fibonacci numbers).

This book takes a rather theoretical approach and thus tries to keep computations clean whenever possible. The examples that we work through in the text to illustrate computational methods like Gaussian elimination are carefully constructed to avoid large fractions (or even fractions at all, when possible), as are the exercises.

It is also worth noting that we do not discuss the history of linear algebra, such as when Gaussian elimination was invented, who first studied eigenvalues, and how the various hideous formulas for the determinant were originally derived. On a very related note, this book is extremely anachronistic— topics are presented in an order that makes them easy to learn, not in the order that they were studied or discovered historically.

Notes in the Margin

This text makes heavy use of notes in the margin, which are used to introduce some additional terminology or provide reminders that would be distracting in the main text. They are most commonly used to try to address potential points of confusion for the reader, so it is best not to skip them.

For example, if we make use of the fact that $\cos(\pi/6) = \sqrt{3}/2$ in the middle of a long calculation, we just make note of that fact in the margin rather than dragging out that calculation even longer to make it explicit in-line. Similarly, if we start discussing a concept that we have not made use of in the past 3 or 4 sections, we provide a reminder in the margin of what that concept is.

Exercises

Several exercises can be found at the end of every section in this book, and whenever possible there are four types of them:

- There are **computational exercises** that ask the reader to implement some algorithm or make use of the tools presented in that section to solve a numerical problem by hand.

- There are **computer software exercises**, denoted by a computer icon (🖥), that ask the reader to use mathematical software like MATLAB, Octave (gnu.org/software/octave), Julia (julialang.org), or SciPy (scipy.org) to solve a numerical problem that is larger or uglier than could reasonably be solved by hand. The latter three of these software packages are free and open source.

- There are **true/false exercises** that test the reader's critical thinking skills and reading comprehension by asking them whether some statements are true or false.

- There are **proof exercises** that ask the reader to prove a general statement. These typically are either routine proofs that follow straight from the definition (and thus were omitted from the main text itself), or proofs that can be tackled via some technique that we saw in that section. For example, after proving the triangle inequality in Section 1.2, Exercise 1.2.2.1 asks the reader to prove the "reverse" triangle inequality, which can be done simply by moving terms around in the original proof of the triangle inequality.

Roughly half of the exercises are marked with an asterisk (∗), which means that they have a solution provided in Appendix C. Exercises marked with *two* asterisks (∗∗) are referenced in the main text and are thus particularly important (and also have solutions in Appendix C).

There are also 150 exercises freely available for this course online as part of the Open Problem Library for WeBWorK (github.com/openwebwork/webwork-open-problem-library, in the "MountAllison" directory). These exercises are typically computational in nature and feature randomization so as to create an essentially endless set of problems for students to work through. All 150 of these exercises are also available on Edfinity (edfinity.com).

To the Instructor and Independent Reader

This book is intended to accompany an introductory proof-based linear algebra course, typically targeted at students who have already completed one or two university-level mathematics courses (which are typically calculus courses, but need not be). It is expected that this is one of the first proof-based courses that the student will be taking, so proof techniques are kept as conceptually simple as possible (for example, techniques like proof by induction are completely avoided in the main text). A brief introduction to proofs and proof techniques can be found in Appendix A.3.

Sectioning

The sectioning of the book is designed to make it as simple to teach from as possible. The author spends approximately the following amount of time on each chunk of this book:

- **Subsection:** 1 hour lecture
- **Section:** 1 week (3 subsections per section)
- **Chapter:** 4 weeks (4 sections per chapter)
- **Book:** 12-week course (3 chapters)

Of course, this is just a rough guideline, as some sections are longer than others (in particular, Sections 1.1 and 1.2 are quite short compared to most later sections). Furthermore, there are numerous in-depth "Extra Topic" sections that can be included in addition to, or instead of, some of its main sections. Alternatively, the additional topics covered in those sections can serve as independent study topics for students.

Extra Topic Sections

Almost half of this book's sections are called "Extra Topic" sections. The purpose of the book being arranged in this way is that it provides a clear main path through the book (Sections 1.1–1.4, 2.1–2.4, and 3.1–3.4) that can be supplemented by the Extra Topic sections at the reader's/instructor's discretion.

We want to emphasize that the Extra Topic sections are not labeled as such because they are less important than the main sections, but only because they are not prerequisites to any of the main sections. For example, linear programming (Section 2.B) is one of the most important topics in modern mathematics and is a tool that is used in almost every science, but it is presented in an Extra Topic section since none of the other sections of this book depend on it.

For a graph that depicts the various dependencies of the sections of this book on each other, see Figure ★.

Lead-in to *Advanced Linear and Matrix Algebra*

This book is the first part of a two-book series, with the follow-up book titled *Advanced Linear and Matrix Algebra* [Joh20]. While most students will only take one linear algebra course and thus only need this first book, these books are designed to provide a natural transition for those students who do go on to a second course in linear algebra.

Because these books aim to not overlap with each other or repeat content, some topics that instructors might expect to find in an introductory linear algebra textbook are not present here. Most notably, this book barely makes any mention of orthonormal bases or the Gram–Schmidt process, and orthogonal projections are only discussed in the 1-dimensional case (i.e., projections onto a line). Furthermore, this book considers the concrete vector spaces $\mathbb{R}^n$ and $\mathbb{C}^n$ exclusively (and briefly $\mathbb{F}^n$, where $\mathbb{F}$ is a finite field, in Section 2.A)—abstract vector spaces make no appearances here.

The reason for these omissions is simply that they are covered early in [Joh20]. In particular, that book starts in Section 1.1 with abstract vector spaces, introduces inner products by Section 1.3, and then explores applications of inner products like orthonormal bases, the Gram–Schmidt process, and orthogonal projections in Section 1.4.

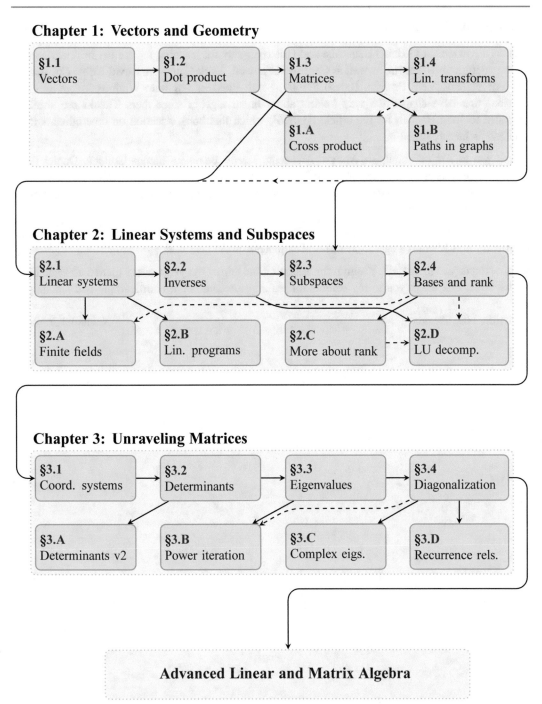

Figure ★: A graph depicting the dependencies of the sections of this book on each other. Solid arrows indicate that the section is required before proceeding to the section that it points to, while dashed arrows indicate recommended (but not required) prior reading. The "main path" through the book consists of Sections 1–4 of each chapter. The extra sections A–D are optional and can be explored at the reader's discretion, as none of the main sections depend on them.

Acknowledgments

Thanks are extended to Heinz Bauschke and Hristo Sendov for teaching me the joy of linear algebra as an undergraduate student, as well as Geoffrey Cruttwell, Mark Hamilton, David Kribs, Chi-Kwong Li, Neil McKay, Vern Paulsen, Rajesh Pereira, Sarah Plosker, and John Watrous for various discussions that have shaped the way I think about linear algebra since then. Thanks are similarly extended to John Hannah for the article [Han96], which this book's section on determinants (Section 3.2) is largely based on.

Thank you to Amira Abouleish, Maryse Arseneau, Jeremi Beaulieu, Sienna Collette, Daniel Gold, Patrice Pagulayan, Everett Patterson, Noah Warner, Ethan Wright, and countless other students in my linear algebra classes at Mount Allison University for drawing my attention to typos and parts of the book that could be improved.

Parts of the layout of this book were inspired by the *Legrand Orange Book* template by Velimir Gayevskiy and Mathias Legrand at LaTeXTemplates.com.

Finally, thank you to my wife Kathryn for tolerating me during the years of my mental absence glued to this book, and thank you to my parents for making me care about both learning and teaching.

Sackville, NB, Canada   Nathaniel Johnston

Table of Contents

1. Vectors and Geometry

> The power of mathematics is often to change one thing
> into another, to change geometry into language.
>
> Marcus du Sautoy

This chapter serves as an introduction to the various objects—vectors, matrices, and linear transformations—that are the central focus of linear algebra. Instead of investigating what we can do with these objects, for now we simply focus on understanding their basic properties, how they interact with each other, and their geometric intuition.

1.1 Vectors and Vector Operations

In earlier math courses, focus was on how to manipulate expressions involving a single variable. For example, we learned how to solve equations like $4x - 3 = 7$ and we learned about properties of functions like $f(x) = 3x + 8$, where in each case the one variable was called "x". One way of looking at linear algebra is the natural extension of these ideas to the situation where we have two or more variables. For example, we might try solving an equation like $3x + 2y = 1$, or we might want to investigate the properties of a function that takes in two independent variables and outputs two dependent variables.

The notation $a \in S$ means that the object a is in the set S, so $\mathbf{v} \in \mathbb{R}^n$ means that the vector $\mathbf{v}$ is in the set $\mathbb{R}^n$ of n-dimensional space.

To make expressions involving several variables easier to deal with, we use **vectors**, which are ordered lists of numbers or variables. We say that the number of entries in the vector is its **dimension**, and if a vector has n entries, we say that it "lives in" or "is an element of" $\mathbb{R}^n$. We denote vectors themselves by lowercase bold letters like $\mathbf{v}$ and $\mathbf{w}$, and we write their entries within parentheses. For example, $\mathbf{v} = (2, 3) \in \mathbb{R}^2$ is a 2-dimensional vector and $\mathbf{w} = (1, 3, 2) \in \mathbb{R}^3$ is a 3-dimensional vector (just like $4 \in \mathbb{R}$ is a real number).

In the 2- and 3-dimensional cases, we can visualize vectors as arrows that indicate displacement in different directions by the amount specified in their entries. The vector's first entry represents displacement in the x-direction, its second entry represents displacement in the y-direction, and in the 3-dimensional case its third entry represents displacement in the z-direction, as in Figure 1.1.

The front of a vector, where the tip of the arrow is located, is called its **head**, and the opposite end is called its **tail**. One way to compute the entries of a vector is to subtract the coordinates of its tail from the corresponding coordinates of its head. For example, the vector that goes from the point

© Springer Nature Switzerland AG 2021
N. Johnston, *Introduction to Linear and Matrix Algebra*,
https://doi.org/10.1007/978-3-030-52811-9_1

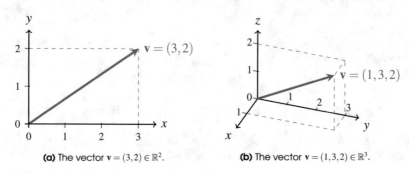

(a) The vector $\mathbf{v} = (3,2) \in \mathbb{R}^2$.   **(b)** The vector $\mathbf{v} = (1,3,2) \in \mathbb{R}^3$.

Figure 1.1: Vectors can be visualized as arrows in (a) 2 and (b) 3 dimensions.

Some other books denote vectors with arrows like $\vec{v}$, or $\overrightarrow{AB}$ if they wish to specify that its tail is located at point A and its head is located at point B.

$(-1,1)$ to the point $(2,2)$ is $(2,2) - (-1,1) = (3,1)$. However, this is also the same as the vector that points from $(1,0)$ to $(4,1)$, since $(4,1) - (1,0) = (3,1)$ as well.

It is thus important to keep in mind that the coordinates of a vector specify its length and direction, but *not* its location in space; we can move vectors around in space without actually changing the vector itself, as in Figure 1.2. To remove this ambiguity when discussing vectors, we often choose to display them with their tail located at the origin—this is called the **standard position** of the vector.

When a vector is in standard position, the coordinates of the point at its head are exactly the same as the entries of the vector.

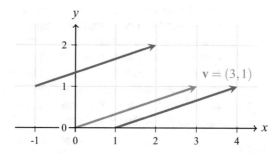

Figure 1.2: Three copies of the vector $\mathbf{v} = (3,1)$ located at different positions in the plane. The vector highlighted in orange is in standard position, since its tail is located at the origin.

1.1.1   Vector Addition

Even though we can represent vectors in 2 and 3 dimensions via arrows, we emphasize that one of our goals is to keep vectors (and all of our linear algebra tools) as dimension-independent as possible. Our visualizations involving arrows can thus help us build intuition for how vectors behave, but our definitions and theorems themselves should work just as well in $\mathbb{R}^7$ (even though we cannot really visualize this space) as they do in $\mathbb{R}^3$. For this reason, we typically introduce new concepts by first giving the algebraic, dimension-independent definition, followed by some examples to illustrate the geometric significance of the new concept. We start with vector addition, the simplest vector operation that there is.

Definition 1.1.1

Vector Addition

Suppose $\mathbf{v} = (v_1, v_2, \ldots, v_n) \in \mathbb{R}^n$ and $\mathbf{w} = (w_1, w_2, \ldots, w_n) \in \mathbb{R}^n$ are vectors. Then their **sum**, denoted by $\mathbf{v} + \mathbf{w}$, is the vector

$$\mathbf{v} + \mathbf{w} \stackrel{\text{def}}{=} (v_1 + w_1, v_2 + w_2, \ldots, v_n + w_n).$$

Vector addition can be motivated in at least two different ways. On the one hand, it is algebraically the simplest operation that could reasonably be considered a way of adding up two vectors: most students, if asked to add up two vectors, would add them up entry-by-entry even if they had not seen Definition 1.1.1. On the other hand, vector addition also has a simple geometric picture in terms of arrows: If $\mathbf{v}$ and $\mathbf{w}$ are positioned so that the tail of $\mathbf{w}$ is located at the same point as the head of $\mathbf{v}$ (in which case we say that $\mathbf{v}$ and $\mathbf{w}$ are positioned **head-to-tail**), then $\mathbf{v} + \mathbf{w}$ is the vector pointing from the tail of $\mathbf{v}$ to the head of $\mathbf{w}$, as in Figure 1.3(a). In other words, $\mathbf{v} + \mathbf{w}$ represents the total displacement accrued by following $\mathbf{v}$ and then following $\mathbf{w}$.

If we instead work entirely with vectors in standard position, then $\mathbf{v} + \mathbf{w}$ is the vector that points along the diagonal between sides $\mathbf{v}$ and $\mathbf{w}$ of a parallelogram, as in Figure 1.3(b).

Despite the triangle and parallelogram pictures looking different, the vector $\mathbf{v} + \mathbf{w}$ is the same in each.

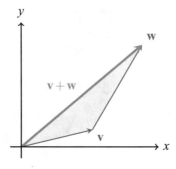

(a) Adding vectors head-to-tail.

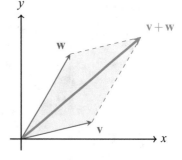

(b) Adding vectors in standard position.

Figure 1.3: How to visualize the addition of two vectors. If $\mathbf{v}$ and $\mathbf{w}$ are (a) positioned head-to-tail then $\mathbf{v} + \mathbf{w}$ forms the third side of the triangle with sides $\mathbf{v}$ and $\mathbf{w}$, but if $\mathbf{v}$ and $\mathbf{w}$ are (b) in standard position, then $\mathbf{v} + \mathbf{w}$ is the diagonal of the parallelogram with sides $\mathbf{v}$ and $\mathbf{w}$.

Before actually making use of vector addition, it will be useful to know some of the basic properties that it satisfies. We list two of the most important such properties in the following theorem for easy reference.

Theorem 1.1.1

Vector Addition Properties

Suppose $\mathbf{v}, \mathbf{w}, \mathbf{x} \in \mathbb{R}^n$ are vectors. Then the following properties hold:

  a) $\mathbf{v} + \mathbf{w} = \mathbf{w} + \mathbf{v}$, and   (commutativity)

  b) $(\mathbf{v} + \mathbf{w}) + \mathbf{x} = \mathbf{v} + (\mathbf{w} + \mathbf{x})$.   (associativity)

Proof. Both parts of this theorem can be proved directly by making use of the relevant definitions. To prove part (a), we use the definition of vector addition together with the fact that the addition of real numbers is commutative (i.e., $x + y = y + x$ for all $x, y \in \mathbb{R}$):

$$\mathbf{v} + \mathbf{w} = (v_1 + w_1, v_2 + w_2, \ldots, v_n + w_n)$$
$$= (w_1 + v_1, w_2 + v_2, \ldots, w_n + v_n) = \mathbf{w} + \mathbf{v}.$$

The proof of part (b) of the theorem similarly follows fairly quickly from the definition of vector addition, and the corresponding property of real numbers, so we leave its proof to Exercise 1.1.14. ∎

The two properties of vector addition that are described by Theorem 1.1.1 are called **commutativity** and **associativity**, respectively, and they basically say that we can unambiguously talk about the sum of any set of vectors without having to worry about the order in which we perform the addition. For example, this theorem shows that expressions like $\mathbf{v} + \mathbf{w} + \mathbf{x}$ make sense, since there is no need to question whether it means $(\mathbf{v} + \mathbf{w}) + \mathbf{x}$ or $\mathbf{v} + (\mathbf{w} + \mathbf{x})$.

While neither of these properties are surprising, it is still important to carefully think about which properties each vector operation satisfies as we introduce it. Later in this chapter, we will introduce two operations (matrix multiplication in Section 1.3.2 and the cross product in Section 1.A) that are *not* commutative (i.e., the order of "multiplication" matters since $\mathbf{v} \times \mathbf{w} \neq \mathbf{w} \times \mathbf{v}$), so it is important to be careful not to assume that basic properties like these hold without actually checking them first.

Example 1.1.1
Numerical Examples of Vector Addition

Compute the following vector sums:

a) $(2,5,-1)+(1,-1,2)$,

b) $(1,2)+(3,1)+(2,-1)$, and

c) the sum of the 8 vectors that point from the origin to the corners of a cube with opposite corners at $(0,0,0)$ and $(1,1,1)$, as shown:

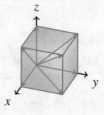

Even though we are adding 8 vectors, we can only see 7 vectors in the image. The missing vector that we cannot see is $(0,0,0)$.

Solutions:

a) $(2,5,-1)+(1,-1,2) = (2+1,5-1,-1+2) = (3,4,1)$.

b) $(1,2)+(3,1)+(2,-1) = (1+3+2,2+1-1) = (6,2)$. Note that this sum can be visualized by placing all three vectors head-to-tail, as shown below. This same procedure works for any number of vectors.

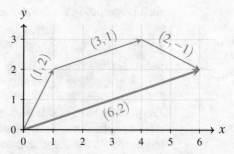

Sums with lots of terms are often easier to evaluate if we can exploit some form of symmetry, as we do here in example (c).

c) We could list all 8 vectors and explicitly compute the sum, but a quicker method is to notice that the 8 vectors we are adding are exactly those that have any combination of 0's and 1's in their 3 entries (i.e., $(0,0,1)$, $(1,0,1)$, and so on). When we add them, in any given entry, exactly half (i.e., 4) of the vectors have a 0 in that entry, and the other half have a 1 there. We thus conclude that the sum of these vectors is $(4,4,4)$.

1.1.2 Scalar Multiplication

The other basic operation on vectors that we introduce at this point is one that changes a vector's length and/or reverses its direction, but does not otherwise change the direction in which it points.

Definition 1.1.2

Scalar Multiplication

"Scalar" just means "number".

Suppose $\mathbf{v} = (v_1, v_2, \ldots, v_n) \in \mathbb{R}^n$ is a vector and $c \in \mathbb{R}$ is a scalar. Then their **scalar multiplication**, denoted by $c\mathbf{v}$, is the vector

$$c\mathbf{v} \overset{\text{def}}{=} (cv_1, cv_2, \ldots, cv_n).$$

We remark that, once again, algebraically this is exactly the definition that someone would likely expect the quantity $c\mathbf{v}$ to have. Multiplying each entry of $\mathbf{v}$ by c seems like a rather natural operation, and it has the simple geometric interpretation of stretching $\mathbf{v}$ by a factor of c, as in Figure 1.4. In particular, if $|c| > 1$ then scalar multiplication stretches $\mathbf{v}$, but if $|c| < 1$ then it shrinks $\mathbf{v}$. When $c < 0$ then this operation also reverses the direction of $\mathbf{v}$, in addition to any stretching or shrinking that it does if $|c| \neq 1$.

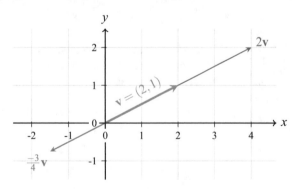

Figure 1.4: Scalar multiplication can be used to stretch, shrink, and/or reverse the direction of a vector.

Two special cases of scalar multiplication are worth pointing out:

- If $c = 0$ then $c\mathbf{v}$ is the **zero vector**, all of whose entries are 0, which we denote by $\mathbf{0}$.
- If $c = -1$ then $c\mathbf{v}$ is the vector whose entries are the negatives of $\mathbf{v}$'s entries, which we denote by $-\mathbf{v}$.

In other words, vector subtraction is also performed in the "obvious" entrywise way.

We also define **vector subtraction** via $\mathbf{v} - \mathbf{w} \overset{\text{def}}{=} \mathbf{v} + (-\mathbf{w})$, and we note that it has the geometric interpretation that $\mathbf{v} - \mathbf{w}$ is the vector pointing from the head of $\mathbf{w}$ to the head of $\mathbf{v}$ when $\mathbf{v}$ and $\mathbf{w}$ are in standard position. It is perhaps easiest to keep this geometric picture straight ("it points from the head of *which* vector to the head of the other one?") if we just think of $\mathbf{v} - \mathbf{w}$ as the vector that must be added to $\mathbf{w}$ to get $\mathbf{v}$ (so it points from $\mathbf{w}$ to $\mathbf{v}$). Alternatively, $\mathbf{v} - \mathbf{w}$ is the *other* diagonal (besides $\mathbf{v} + \mathbf{w}$) in the parallelogram with sides $\mathbf{v}$ and $\mathbf{w}$, as in Figure 1.5.

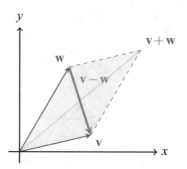

Figure 1.5: How to visualize the subtraction of two vectors. If **v** and **w** are in standard position then **v** − **w** is one of the diagonals of the parallelogram defined by **v** and **w** (and **v** + **w** is the other diagonal, as in Figure 1.3(b)).

It is straightforward to verify some simple properties of the zero vector, such as the facts that $\mathbf{v} - \mathbf{v} = \mathbf{0}$ and $\mathbf{v} + \mathbf{0} = \mathbf{v}$ for every vector $\mathbf{v} \in \mathbb{R}^n$, by working entry-by-entry with the vector operations. There are also quite a few other simple ways in which scalar multiplication interacts with vector addition, some of which we now list explicitly for easy reference.

Theorem 1.1.2

Scalar Multiplication Properties

Suppose $\mathbf{v}, \mathbf{w} \in \mathbb{R}^n$ are vectors and $c, d \in \mathbb{R}$ are scalars. Then the following properties hold:

    a) $c(\mathbf{v} + \mathbf{w}) = c\mathbf{v} + c\mathbf{w}$,
    b) $(c + d)\mathbf{v} = c\mathbf{v} + d\mathbf{v}$, and
    c) $c(d\mathbf{v}) = (cd)\mathbf{v}$.

Property (a) says that scalar multiplication **distributes** over vector addition, and property (b) says that scalar multiplication distributes over real number addition.

Proof. All three parts of this theorem can be proved directly by making use of the relevant definitions. To prove part (a), we use the corresponding properties of real numbers in each entry of the vector:

$$
\begin{aligned}
c(\mathbf{v} + \mathbf{w}) &= c(v_1 + w_1, v_2 + w_2, \dots, v_n + w_n) & \text{(vector addition)} \\
&= (c(v_1 + w_1), c(v_2 + w_2), \dots, c(v_n + w_n)) & \text{(scalar mult.)} \\
&= (cv_1 + cw_1, cv_2 + cw_2, \dots, cv_n + cw_n) & \text{(property of } \mathbb{R}) \\
&= (cv_1, cv_2, \dots, cv_n) + (cw_1, cw_2, \dots, cw_n) & \text{(vector addition)} \\
&= c(v_1, v_2, \dots, v_n) + c(w_1, w_2, \dots, w_n) & \text{(scalar mult.)} \\
&= c\mathbf{v} + c\mathbf{w}.
\end{aligned}
$$

The proofs of parts (b) and (c) of the theorem similarly follow fairly quickly from the definitions of vector addition and scalar multiplication, and the corresponding properties of real numbers, so we leave their proofs to Exercise 1.1.15. ∎

Example 1.1.2

Numerical Examples of Vector Operations

Compute the indicated vectors:

    a) $3\mathbf{v} - 2\mathbf{w}$, where $\mathbf{v} = (2, 1, -1)$ and $\mathbf{w} = (-1, 0, 3)$, and
    b) the sum of the 6 vectors that point from the center $(0,0)$ of a regular hexagon to its corners, one of which is located at $(1,0)$, as shown:

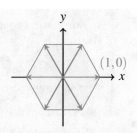

Solutions:

a) $3\mathbf{v} - 2\mathbf{w} = (6,3,-3) - (-2,0,6) = (8,3,-9)$.

b) We could use trigonometry to find the entries of all 6 vectors explicitly, but an easier way to compute this sum is to label the vectors, in counter-clockwise order starting at an arbitrary location, as $\mathbf{v}$, $\mathbf{w}$, $\mathbf{x}$, $-\mathbf{v}$, $-\mathbf{w}$, $-\mathbf{x}$ (since the final 3 vectors point in the opposite directions of the first 3 vectors). It follows that the sum is $\mathbf{v} + \mathbf{w} + \mathbf{x} - \mathbf{v} - \mathbf{w} - \mathbf{x} = \mathbf{0}$.

> This method of solving (b) has the nice feature that it still works even if we rotate the hexagon or change the number of sides.

By making use of these properties of vector addition and scalar multiplication, we can solve vector equations in much the same way that we solve equations involving real numbers: we can add and subtract vectors on both sides of an equation, and multiply and divide by scalars on both sides of the equation, until the unknown vector is isolated. We illustrate this procedure with some examples.

Example 1.1.3
Vector Algebra

Solve the following equations for the vector $\mathbf{x}$:

a) $\mathbf{x} - (3,2,1) = (1,2,3) - 3\mathbf{x}$, and

b) $\mathbf{x} + 2(\mathbf{v} + \mathbf{w}) = -\mathbf{v} - 3(\mathbf{x} - \mathbf{w})$.

Solutions:

> The "$\Longrightarrow$" symbol here is an **implication arrow** and is read as "implies". It means that the upcoming statement (e.g., $\mathbf{x} = (1,1,1)$) follows logically from the one before it (e.g., $4\mathbf{x} = (4,4,4)$).

a) We solve this equation as follows:

$$\mathbf{x} - (3,2,1) = (1,2,3) - 3\mathbf{x}$$
$$\Longrightarrow \quad \mathbf{x} = (4,4,4) - 3\mathbf{x} \quad \text{(add } (3,2,1) \text{ to both sides)}$$
$$\Longrightarrow \quad 4\mathbf{x} = (4,4,4) \quad \text{(add } 3\mathbf{x} \text{ to both sides)}$$
$$\Longrightarrow \quad \mathbf{x} = (1,1,1). \quad \text{(divide both sides by 4)}$$

b) The method of solving this equation is the same as in part (a), but this time the best we can do is express $\mathbf{x}$ in terms of $\mathbf{v}$ and $\mathbf{w}$:

$$\mathbf{x} + 2(\mathbf{v} + \mathbf{w}) = -\mathbf{v} - 3(\mathbf{x} - \mathbf{w})$$
$$\Longrightarrow \quad \mathbf{x} + 2\mathbf{v} + 2\mathbf{w} = -\mathbf{v} - 3\mathbf{x} + 3\mathbf{w} \quad \text{(expand parentheses)}$$
$$\Longrightarrow \quad 4\mathbf{x} = -3\mathbf{v} + \mathbf{w} \quad \text{(add } 3\mathbf{x}, \text{ subtract } 2\mathbf{v} + 2\mathbf{w})$$
$$\Longrightarrow \quad \mathbf{x} = \tfrac{1}{4}(\mathbf{w} - 3\mathbf{v}). \quad \text{(divide both sides by 4)}$$

1.1.3 Linear Combinations

One common task in linear algebra is to start out with some given collection of vectors $\mathbf{v}_1, \mathbf{v}_2, \ldots, \mathbf{v}_k$ and then use vector addition and scalar multiplication to construct new vectors out of them. The following definition gives a name to this concept.

Definition 1.1.3

Linear Combinations

A **linear combination** of the vectors $\mathbf{v}_1, \mathbf{v}_2, \ldots, \mathbf{v}_k \in \mathbb{R}^n$ is any vector of the form

$$c_1\mathbf{v}_1 + c_2\mathbf{v}_2 + \cdots + c_k\mathbf{v}_k,$$

where $c_1, c_2, \ldots, c_k \in \mathbb{R}$.

We will see how to determine whether or not a vector is a linear combination of a given set of vectors in Section 2.1.

For example, $(1,2,3)$ is a linear combination of the vectors $(1,1,1)$ and $(-1,0,1)$ since $(1,2,3) = 2(1,1,1) + (-1,0,1)$. On the other hand, $(1,2,3)$ is *not* a linear combination of the vectors $(1,1,0)$ and $(2,1,0)$ since every vector of the form $c_1(1,1,0) + c_2(2,1,0)$ has a 0 in its third entry, and thus cannot possibly equal $(1,2,3)$.

When working with linear combinations, some particularly important vectors are those with all entries equal to 0, except for a single entry that equals 1. Specifically, for each $j = 1, 2, \ldots, n$, we define the vector $\mathbf{e}_j \in \mathbb{R}^n$ by

$$\mathbf{e}_j \overset{\text{def}}{=} (0, 0, \ldots, 0, 1, 0, \ldots, 0).$$
$$\uparrow j\text{-th entry}$$

Whenever we use these vectors, the dimension of $\mathbf{e}_j$ will be clear from context or by saying things like $\mathbf{e}_3 \in \mathbb{R}^7$.

For example, in $\mathbb{R}^2$ there are two such vectors: $\mathbf{e}_1 = (1,0)$ and $\mathbf{e}_2 = (0,1)$. Similarly, in $\mathbb{R}^3$ there are three such vectors: $\mathbf{e}_1 = (1,0,0)$, $\mathbf{e}_2 = (0,1,0)$, and $\mathbf{e}_3 = (0,0,1)$. In general, in $\mathbb{R}^n$ there are n of these vectors, $\mathbf{e}_1, \mathbf{e}_2, \ldots, \mathbf{e}_n$, and we call them the **standard basis vectors** (for reasons that we discuss in the next chapter). Notice that in $\mathbb{R}^2$ and $\mathbb{R}^3$, these are the vectors that point a distance of 1 in the direction of the x-, y-, and z-axes, as in Figure 1.6.

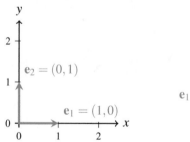

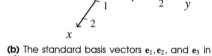

(a) The standard basis vectors $\mathbf{e}_1$ and $\mathbf{e}_2$ in $\mathbb{R}^2$.

(b) The standard basis vectors $\mathbf{e}_1, \mathbf{e}_2$, and $\mathbf{e}_3$ in $\mathbb{R}^3$.

Figure 1.6: The standard basis vectors point a distance of 1 along the x-, y-, and z-axes.

For now, the reason for our interest in these standard basis vectors is that every vector $\mathbf{v} \in \mathbb{R}^n$ can be written as a linear combination of them. In particular, if $\mathbf{v} = (v_1, v_2, \ldots, v_n)$ then

When we see expressions like this, it is useful to remind ourselves of the "type" of each object: $v_1, v_2, \ldots, v_n$ are scalars and $\mathbf{e}_1, \mathbf{e}_2, \ldots, \mathbf{e}_n$ are vectors.

$$\mathbf{v} = v_1\mathbf{e}_1 + v_2\mathbf{e}_2 + \cdots + v_n\mathbf{e}_n,$$

which can be verified just by computing each of the entries of the linear combination on the right. This idea of writing vectors in terms of the standard basis vectors (or other distinguished sets of vectors that we introduce later) is one of the most useful techniques that we make use of in linear algebra: in many situations, if we can prove that some property holds for the standard basis vectors, then we can use linear combinations to show that it must hold for *all* vectors.

Example 1.1.4

Numerical Examples of Linear Combinations

Compute the indicated linear combinations of standard basis vectors:

a) Compute $3\mathbf{e}_1 - 2\mathbf{e}_2 + \mathbf{e}_3 \in \mathbb{R}^3$, and

b) Write $(3, 5, -2, -1)$ as a linear combination of $\mathbf{e}_1, \mathbf{e}_2, \mathbf{e}_3, \mathbf{e}_4 \in \mathbb{R}^4$.

Solutions:

a) $3\mathbf{e}_1 - 2\mathbf{e}_2 + \mathbf{e}_3 = 3(1,0,0) - 2(0,1,0) + (0,0,1) = (3,-2,1)$. In general, when adding multiples of the standard basis vectors, the resulting vector has the coefficient of $\mathbf{e}_1$ in its first entry, the coefficient of $\mathbf{e}_2$ in its second entry, and so on.

b) Just like in part (a), the entries of the vectors are the scalars in the linear combination: $(3,5,-2,-1) = 3\mathbf{e}_1 + 5\mathbf{e}_2 - 2\mathbf{e}_3 - \mathbf{e}_4$.

Remark 1.1.1

No Vector Multiplication

At this point, it seems natural to ask why we have defined vector addition $\mathbf{v} + \mathbf{w}$ and scalar multiplication $c\mathbf{v}$ in the "obvious" entrywise ways, but we have not similarly defined the entrywise product of two vectors:

$$\mathbf{vw} \overset{\text{def}}{=} (v_1 w_1, v_2 w_2, \ldots, v_n w_n).$$

The answer is simply that entrywise vector multiplication is not particularly useful—it does not often come up in real-world problems or play a role in more advanced mathematical structures, nor does it have a simple geometric interpretation. There are some other more useful ways of "multiplying" vectors together, called the dot product and the cross product, which we explore in Sections 1.2 and 1.A, respectively.

Exercises

solutions to starred exercises on page 435

1.1.1 Draw each of the following vectors in standard position in $\mathbb{R}^2$:

*(a) $\mathbf{v} = (3,2)$
*(c) $\mathbf{x} = (1,-3)$
(b) $\mathbf{w} = (-0.5, 3)$
(d) $\mathbf{y} = (-2,-1)$

*1.1.2 Draw each of the vectors from Exercise 1.1.1, but with their tail located at the point $(1,2)$.

*1.1.3 If each of the vectors from Exercise 1.1.1 are positioned so that their heads are located at the point $(3,3)$, find the location of their tails.

1.1.4 Draw each of the following vectors in standard position in $\mathbb{R}^3$:

*(a) $\mathbf{v} = (0,0,2)$
*(c) $\mathbf{x} = (1,2,0)$
(b) $\mathbf{w} = (-1,2,1)$
(d) $\mathbf{y} = (3,2,-1)$

1.1.5 If the vectors $\mathbf{v}, \mathbf{w}, \mathbf{x},$ and $\mathbf{y}$ are as in Exercise 1.1.1, then compute

*(a) $\mathbf{v} + \mathbf{w}$
*(c) $\mathbf{y} - 2\mathbf{x}$
(b) $\mathbf{v} + \mathbf{w} + \mathbf{y}$
(d) $\mathbf{v} + 2\mathbf{w} + 2\mathbf{x} + 2\mathbf{y}$

1.1.6 If the vectors $\mathbf{v}, \mathbf{w}, \mathbf{x},$ and $\mathbf{y}$ are as in Exercise 1.1.4, then compute

*(a) $\mathbf{v} + \mathbf{y}$
*(c) $4\mathbf{x} - 2\mathbf{w}$
(b) $4\mathbf{w} + 3\mathbf{w} - (2\mathbf{w} + 6\mathbf{w})$
(d) $2\mathbf{x} - \mathbf{w} - \mathbf{y}$

* **1.1.7** Write each of the vectors $\mathbf{v}, \mathbf{w}, \mathbf{x},$ and $\mathbf{y}$ from Exercise 1.1.4 as a linear combination of the standard basis vectors $\mathbf{e}_1, \mathbf{e}_2, \mathbf{e}_3 \in \mathbb{R}^3$.

1.1.8 Suppose that the side vectors of a parallelogram are $\mathbf{v} = (1,4)$ and $\mathbf{w} = (-2,1)$. Find vectors describing both of the parallelogram's diagonals.

* **1.1.9** Suppose that the diagonal vectors of a parallelogram are $\mathbf{x} = (3,-2)$ and $\mathbf{y} = (1,4)$. Find vectors describing the parallelogram's sides.

1.1.10 Solve the following vector equations for $\mathbf{x}$:

*(a) $(1,2) - \mathbf{x} = (3,4) - 2\mathbf{x}$
(b) $3((1,-1) + \mathbf{x}) = 2\mathbf{x}$
*(c) $2(\mathbf{x} + 2(\mathbf{x} + 2\mathbf{x})) = 3(\mathbf{x} + 3(\mathbf{x} + 3\mathbf{x}))$
(d) $-2(\mathbf{x} - (1,-2)) = \mathbf{x} + 2(\mathbf{x} + (1,1))$

1.1.11 Write the vector $\mathbf{x}$ in terms of the vectors $\mathbf{v}$ and $\mathbf{w}$:

*(a) $\mathbf{v} - \mathbf{x} = \mathbf{w} + \mathbf{x}$
(b) $2\mathbf{v} - 3\mathbf{x} = 4\mathbf{x} - 5\mathbf{w}$
*(c) $4(\mathbf{x} + \mathbf{v}) - \mathbf{x} = 2(\mathbf{w} + \mathbf{x})$
(d) $2(\mathbf{x} + 2(\mathbf{x} + 2\mathbf{x})) = 2(\mathbf{v} + 2\mathbf{v})$

*1.1.12 Does there exist a scalar $c \in \mathbb{R}$ such that $c(1,2) = (3,4)$? Justify your answer both algebraically and geometrically.

1.1.13 Let $n \geq 3$ be an integer and consider the set of n vectors that point from the center of the regular n-gon in $\mathbb{R}^2$, to its corners.

(a) Show that if n is even then the sum of these n vectors is $\mathbf{0}$. [Hint: We solved the $n = 6$ case in Example 1.1.2(b).]
(b) Show that if n is odd then the sum of these n vectors is $\mathbf{0}$. [Hint: This is more difficult. Try working with the x- and y-entries of the sum individually.]

**1.1.14 Prove part (b) of Theorem 1.1.1.

**1.1.15 Recall Theorem 1.1.2, which established some of the basic properties of scalar multiplication.

(a) Prove part (b) of the theorem.
(b) Prove part (c) of the theorem.

1.2   Lengths, Angles, and the Dot Product

When discussing geometric properties of vectors, like their length or the angle between them, we would like our definitions to be as dimension-independent as possible, so that it is just as easy to discuss the length of a vector in $\mathbb{R}^7$ as it is to discuss the length of one in $\mathbb{R}^2$. At first it might be somewhat surprising that discussing the length of a vector in high-dimensional spaces is something that we can do at all—after all, we cannot really visualize anything past 3 dimensions. We thus stress that the dimension-independent definitions of length and angle that we introduce in this section are not theorems that we prove, but rather are *definitions* that we adopt so that they satisfy the basic geometric properties that lengths and angles "should" satisfy.

1.2.1   The Dot Product

The main tool that helps us extend geometric notions from $\mathbb{R}^2$ and $\mathbb{R}^3$ to arbitrary dimensions is the dot product, which is a way of combining two vectors so as to create a single number:

Definition 1.2.1
Dot Product

> Suppose $\mathbf{v} = (v_1, v_2, \ldots, v_n) \in \mathbb{R}^n$ and $\mathbf{w} = (w_1, w_2, \ldots, w_n) \in \mathbb{R}^n$ are vectors. Then their **dot product**, denoted by $\mathbf{v} \cdot \mathbf{w}$, is the quantity
>
> $$\mathbf{v} \cdot \mathbf{w} \stackrel{\text{def}}{=} v_1 w_1 + v_2 w_2 + \cdots + v_n w_n.$$

It is important to keep in mind that the output of the dot product is a *number*, not a vector. So, for example, the expression $\mathbf{v} \cdot (\mathbf{w} \cdot \mathbf{x})$ does not make sense, since $\mathbf{w} \cdot \mathbf{x}$ is a number, and so we cannot take its dot product with $\mathbf{v}$. On the other hand, the expression $\mathbf{v}/(\mathbf{w} \cdot \mathbf{x})$ *does* make sense, since dividing a vector by a number is a valid mathematical operation. As we introduce more operations between different types of objects, it will become increasingly important to keep in mind the type of object that we are working with at all times.

Example 1.2.1
Numerical
Examples
of the Dot
Product

> Compute (or state why it's impossible to compute) the following dot products:
>
> a) $(1,2,3) \cdot (4,-3,2)$,
> b) $(3,6,2) \cdot (-1,5,2,1)$, and

Recall that $\mathbf{e}_j$ is the vector with a 1 in its j-th entry and 0s elsewhere.

c) $(v_1, v_2, \ldots, v_n) \cdot \mathbf{e}_j$, where $1 \leq j \leq n$.

Solutions:

a) $(1,2,3) \cdot (4,-3,2) = 1 \cdot 4 + 2 \cdot (-3) + 3 \cdot 2 = 4 - 6 + 6 = 4$.

b) $(3,6,2) \cdot (-1,5,2,1)$ does not exist, since these vectors do not have the same number of entries.

c) For this dot product to make sense, we have to assume that the vector $\mathbf{e}_j$ has n entries (the same number of entries as $(v_1, v_2, \ldots, v_n)$). Then

$$(v_1, v_2, \ldots, v_n) \cdot \mathbf{e}_j = 0v_1 + \cdots + 0v_{j-1} + 1v_j + 0v_{j+1} + \cdots + 0v_n$$
$$= v_j.$$

The dot product can be interpreted geometrically as roughly measuring the amount of overlap between $\mathbf{v}$ and $\mathbf{w}$. For example, if $\mathbf{v} = \mathbf{w} = (1,0)$ then $\mathbf{v} \cdot \mathbf{w} = 1$, but as we rotate $\mathbf{w}$ away from $\mathbf{v}$, their dot product decreases down to 0 when $\mathbf{v}$ and $\mathbf{w}$ are perpendicular (i.e., when $\mathbf{w} = (0,1)$ or $\mathbf{w} = (0,-1)$), as illustrated in Figure 1.7. It then decreases even farther down to -1 when $\mathbf{w}$ points in the opposite direction of $\mathbf{v}$ (i.e., when $\mathbf{w} = (-1,0)$).

More specifically, if we rotate $\mathbf{w}$ counter-clockwise from $\mathbf{v}$ by an angle of θ then its coordinates become $\mathbf{w} = (\cos(\theta), \sin(\theta))$. The dot product between $\mathbf{v}$ and $\mathbf{w}$ is then $\mathbf{v} \cdot \mathbf{w} = 1\cos(\theta) + 0\sin(\theta) = \cos(\theta)$, which is largest when θ is small (i.e., when $\mathbf{w}$ points in almost the same direction as $\mathbf{v}$).

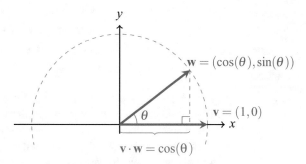

Figure 1.7: The dot product of two vectors decreases as we rotate them away from each other. Here, the dot product between $\mathbf{v}$ and $\mathbf{w}$ is $\mathbf{v} \cdot \mathbf{w} = 1\cos(\theta) + 0\sin(\theta) = \cos(\theta)$, which is largest when θ is small.

Before we can make use of the dot product, we should make ourselves aware of the mathematical properties that it satisfies. The following theorem catalogs the most important of these properties, none of which are particularly surprising or difficult to prove.

Theorem 1.2.1

Dot Product Properties

Suppose $\mathbf{v}, \mathbf{w}, \mathbf{x} \in \mathbb{R}^n$ are vectors and $c \in \mathbb{R}$ is a scalar. Then the following properties hold:

a) $\mathbf{v} \cdot \mathbf{w} = \mathbf{w} \cdot \mathbf{v}$,   (commutativity)
b) $\mathbf{v} \cdot (\mathbf{w} + \mathbf{x}) = \mathbf{v} \cdot \mathbf{w} + \mathbf{v} \cdot \mathbf{x}$, and   (distributivity)
c) $\mathbf{v} \cdot (c\mathbf{w}) = c(\mathbf{v} \cdot \mathbf{w})$.

Proof. To prove part (a) of the theorem, we use the definition of the dot product together with the fact that the multiplication of real numbers is commutative:

$$\mathbf{v} \cdot \mathbf{w} = v_1 w_1 + v_2 w_2 + \cdots + v_n w_n$$
$$= w_1 v_1 + w_2 v_2 + \cdots + w_n v_n = \mathbf{w} \cdot \mathbf{v}.$$

The proofs of parts (b) and (c) of the theorem similarly follow fairly quickly from the definition of the dot product and the corresponding properties of real numbers, so we leave their proofs to Exercise 1.2.13.   ∎

The properties described by Theorem 1.2.1 can be combined to generate new properties of the dot product as well. For example, property (c) of that theorem tells us that we can pull scalars out of the second vector in a dot product, but by combining properties (a) and (c), we can show that we can also pull scalars out of the first vector in a dot product:

property (a)

$$(c\mathbf{v}) \cdot \mathbf{w} = \mathbf{w} \cdot (c\mathbf{v}) = c(\mathbf{w} \cdot \mathbf{v}) = c(\mathbf{v} \cdot \mathbf{w}).$$

property (c)

Similarly, by using properties (a) and (b) together, we see that we can "multiply out" parenthesized dot products much like we multiply out real numbers:

$$
\begin{aligned}
(\mathbf{v}+\mathbf{w}) \cdot (\mathbf{x}+\mathbf{y}) &= (\mathbf{v}+\mathbf{w}) \cdot \mathbf{x} + (\mathbf{v}+\mathbf{w}) \cdot \mathbf{y} && \text{(property (b))} \\
&= \mathbf{x} \cdot (\mathbf{v}+\mathbf{w}) + \mathbf{y} \cdot (\mathbf{v}+\mathbf{w}) && \text{(property (a))} \\
&= \mathbf{x} \cdot \mathbf{v} + \mathbf{x} \cdot \mathbf{w} + \mathbf{y} \cdot \mathbf{v} + \mathbf{y} \cdot \mathbf{w} && \text{(property (b))} \\
&= \mathbf{v} \cdot \mathbf{x} + \mathbf{w} \cdot \mathbf{x} + \mathbf{v} \cdot \mathbf{y} + \mathbf{w} \cdot \mathbf{y}. && \text{(property (a))}
\end{aligned}
$$

In particular, if you have used the acronym "FOIL" to help you multiply out real expressions like $(x+2)(x^2+3x)$, the exact same method works with the dot product.

All of this is just to say that the dot product behaves similarly to the multiplication of real numbers, and has all of the nice properties that we might hope that something we call a "product" might have. The reason that the dot product is actually *useful* though is that it can help us discuss the length of vectors and the angle between vectors, as in the next two subsections.

1.2.2 Vector Length

In 2 or 3 dimensions, we can use geometric techniques to compute the length of a vector $\mathbf{v}$, which we represent by $\|\mathbf{v}\|$. The length of a vector $\mathbf{v} = (v_1, v_2) \in \mathbb{R}^2$ can be computed by noticing that $\mathbf{v} = (v_1, 0) + (0, v_2)$, so $\mathbf{v}$ forms the hypotenuse of a right-angled triangle with shorter sides given by the vectors $(v_1, 0)$ and $(0, v_2)$, as illustrated in Figure 1.8(a). Since the length of $(v_1, 0)$ is $|v_1|$ and the length of $(0, v_2)$ is $|v_2|$, the Pythagorean theorem tells us that

The Pythagorean theorem says that if a right-angled triangle has longest side (hypotenuse) of length c and other sides of length a and b, then $c^2 = a^2 + b^2$ (so $c = \sqrt{a^2+b^2}$).

$$\|\mathbf{v}\| = \sqrt{\big\|(v_1,0)\big\|^2 + \big\|(0,v_2)\big\|^2} = \sqrt{|v_1|^2 + |v_2|^2} = \sqrt{v_1^2 + v_2^2} = \sqrt{\mathbf{v} \cdot \mathbf{v}}.$$

This argument still works, but is slightly trickier, for 3-dimensional vectors $\mathbf{v} = (v_1, v_2, v_3) \in \mathbb{R}^3$. In this case, we instead write $\mathbf{v} = (v_1, v_2, 0) + (0, 0, v_3)$, so that $\mathbf{v}$ forms the hypotenuse of a right-angled triangle with shorter sides given by the vectors $(v_1, v_2, 0)$ and $(0, 0, v_3)$, as in Figure 1.8(b). Since the length of $(v_1, v_2, 0)$ is $\sqrt{v_1^2 + v_2^2}$ (it is just a vector in $\mathbb{R}^2$ with an extra "0" entry tacked on) and the length of $(0,0,v_3)$ is $|v_3|$, the Pythagorean theorem tells us that

$$\|\mathbf{v}\| = \sqrt{\|(v_1,v_2,0)\|^2 + \|(0,0,v_3)\|^2}$$
$$= \sqrt{\left(\sqrt{v_1{}^2+v_2{}^2}\right)^2 + |v_3|^2} = \sqrt{v_1^2+v_2^2+v_3^2} = \sqrt{\mathbf{v}\cdot\mathbf{v}}.$$

The length of a vector is also called its **Euclidean norm** or simply its **norm**.

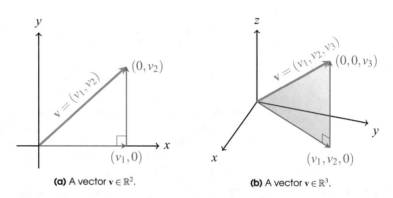

(a) A vector $\mathbf{v} \in \mathbb{R}^2$.   **(b)** A vector $\mathbf{v} \in \mathbb{R}^3$.

Figure 1.8: A breakdown of how the Pythagorean theorem can be used to determine the length of vectors in (a) $\mathbb{R}^2$ and (b) $\mathbb{R}^3$.

When considering higher-dimensional vectors, we can no longer visualize them quite as easily as we could in the 2- and 3-dimensional cases, so it's not necessarily obvious what we even mean by the "length" of a vector in, for example, $\mathbb{R}^7$. In these cases, we simply *define* the length of a vector so as to continue the pattern that we observed above.

Definition 1.2.2

Length of a Vector

The **length** of a vector $\mathbf{v} = (v_1,v_2,\ldots,v_n) \in \mathbb{R}^n$, denoted by $\|\mathbf{v}\|$, is the quantity

$$\|\mathbf{v}\| \stackrel{\text{def}}{=} \sqrt{\mathbf{v}\cdot\mathbf{v}} = \sqrt{v_1^2 + v_2^2 + \cdots + v_n^2}.$$

It is worth noting that this definition does indeed make sense, since the quantity $\mathbf{v}\cdot\mathbf{v} = v_1^2 + v_2^2 + \cdots + v_n^2$ is non-negative, so we can take its square root. To get a feeling for how the length of a vector works, we compute the length of a few example vectors.

Example 1.2.2

Numerical Examples of Vector Length

Compute the lengths of the following vectors:

a) $(2,-5,4,6)$,

b) $(\cos(\theta),\sin(\theta))$, and

c) the main diagonal of a cube in $\mathbb{R}^3$ with side length 1.

Solutions:

a) $\|(2,-5,4,6)\| = \sqrt{2^2+(-5)^2+4^2+6^2} = \sqrt{81} = 9$.

b) $\|(\cos(\theta),\sin(\theta))\| = \sqrt{\cos^2(\theta)+\sin^2(\theta)} = \sqrt{1} = 1$.

c) The cube with side length 1 can be positioned so that it has one vertex at $(0,0,0)$ and its opposite vertex at $(1,1,1)$, as shown below:

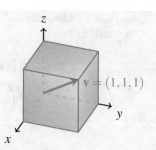

The main diagonal of this cube is the vector $\mathbf{v} = (1,1,1)$, which has length $\|\mathbf{v}\| = \sqrt{1^2 + 1^2 + 1^2} = \sqrt{3}$.

We now start describing the basic properties of the length of a vector. Our first theorem just presents two very simple properties that should be expected geometrically: if we multiply a vector by a scalar, then its length is multiplied by the absolute value of amount, and the zero vector is the only vector with length equal to 0 (all other vectors have positive length).

Theorem 1.2.2

Length Properties

Suppose $\mathbf{v} \in \mathbb{R}^n$ is a vector and $c \in \mathbb{R}$ is a scalar. Then the following properties hold:

a) $\|c\mathbf{v}\| = |c|\|\mathbf{v}\|$, and

b) $\|\mathbf{v}\| \geq 0$, with equality if and only if $\mathbf{v} = \mathbf{0}$.

Proof. Both of these properties follow fairly quickly from the definition of vector length. For property (a), we compute

$$\|c\mathbf{v}\| = \sqrt{(cv_1)^2 + (cv_2)^2 + \cdots + (cv_n)^2}$$
$$= \sqrt{c^2(v_1^2 + v_2^2 + \cdots + v_n^2)}$$
$$= \sqrt{c^2}\sqrt{v_1^2 + v_2^2 + \cdots + v_n^2} = |c|\|\mathbf{v}\|.$$

In the final equality here, we use the fact that $\sqrt{c^2} = |c|$.

For property (b), the fact that $\|\mathbf{v}\| \geq 0$ follows from the fact that the square root function is defined to return the *non-negative* square root of its input. It is straightforward to show that $\|\mathbf{0}\| = 0$, so to complete the proof we just need to show that if $\|\mathbf{v}\| = 0$ then $\mathbf{v} = \mathbf{0}$. Well, if $\|\mathbf{v}\| = 0$ then $v_1^2 + v_2^2 + \cdots + v_n^2 = 0$, and since $v_j^2 \geq 0$ for each $1 \leq j \leq n$, with equality if and only if $v_j = 0$, we see that it must be the case that $v_1 = v_2 = \cdots = v_n = 0$ (i.e., $\mathbf{v} = \mathbf{0}$). ∎

If one of the terms in the sum $v_1^2 + v_2^2 + \cdots + v_n^2$ were strictly positive, the sum would be strictly positive too.

It is often particularly useful to focus attention on **unit vectors**: vectors with length equal to 1. Unit vectors often arise in situations where the vector's direction is important, but its length is not. Importantly, Theorem 1.2.2(a) tells us that we can always rescale any vector to have length 1 just by dividing the vector by its length, as in Figure 1.9(a):

$$\left\|\frac{\mathbf{v}}{\|\mathbf{v}\|}\right\| = \frac{1}{\|\mathbf{v}\|}\|\mathbf{v}\| = 1.$$

If $\mathbf{v} = \mathbf{0}$ then we can still write $\mathbf{v} = \|\mathbf{v}\|\mathbf{u}$ where $\mathbf{u}$ is a unit vector, but $\mathbf{u}$ is no longer unique (in fact, it can be *any* unit vector).

Rescaling a vector like this so that it has length 1 is called **normalization**.

As a result of the fact that we can rescale vectors like this, there is exactly one unit vector that points in each direction, and we can think of the set of all unit vectors in $\mathbb{R}^2$ as the unit circle, in $\mathbb{R}^3$ as the unit sphere, and so on, as in Figure 1.9(b). Furthermore, we can always decompose vectors into the product of their length and direction. That is, we can write every non-zero vector $\mathbf{v} \in \mathbb{R}^n$

The **unit circle** is the circle in $\mathbb{R}^2$ of radius 1 centered at the origin. The **unit sphere** is the sphere in $\mathbb{R}^3$ of radius 1 centered at the origin.

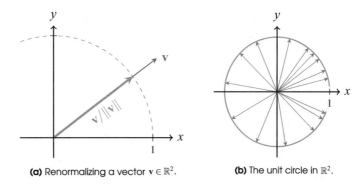

(a) Renormalizing a vector $\mathbf{v} \in \mathbb{R}^2$.

(b) The unit circle in $\mathbb{R}^2$.

Figure 1.9: By normalizing vectors, we find that (a) there is exactly one unit vector pointing in each direction, and (b) the set of unit vectors in $\mathbb{R}^2$ makes up the unit circle.

in the form $\mathbf{v} = \|\mathbf{v}\|\mathbf{u}$, where $\mathbf{u} = \mathbf{v}/\|\mathbf{v}\|$ is the unique unit vector pointing in the same direction as $\mathbf{v}$.

The next property that we look at is an inequality that relates the lengths of two vectors to their dot product. The intuition for this theorem comes from Figure 1.7, where we noticed that the dot product of the vector $\mathbf{v} = (1,0)$ with any other vector of length 1 was always between -1 and 1. In general, the dot product of two vectors cannot be "too large" compared to the lengths of the vectors.

Theorem 1.2.3

Cauchy–Schwarz Inequality

Suppose that $\mathbf{v}, \mathbf{w} \in \mathbb{R}^n$ are vectors. Then $|\mathbf{v} \cdot \mathbf{w}| \leq \|\mathbf{v}\|\|\mathbf{w}\|$.

Proof. The proof works by computing the length of an arbitrary linear combination of $\mathbf{v}$ and $\mathbf{w}$. Specifically, if $c, d \in \mathbb{R}$ are any real numbers then $\|c\mathbf{v} + d\mathbf{w}\|^2$ is the square of a length, so it must be non-negative. By expanding the length in terms of the dot product, we see that

This is the first theorem in this book whose proof does not follow immediately from the definitions, but rather requires a clever insight.

$$\begin{aligned} 0 \leq \|c\mathbf{v} + d\mathbf{w}\|^2 &= (c\mathbf{v} + d\mathbf{w}) \cdot (c\mathbf{v} + d\mathbf{w}) \\ &= c^2(\mathbf{v} \cdot \mathbf{v}) + 2cd(\mathbf{v} \cdot \mathbf{w}) + d^2(\mathbf{w} \cdot \mathbf{w}) \\ &= c^2\|\mathbf{v}\|^2 + 2cd(\mathbf{v} \cdot \mathbf{w}) + d^2\|\mathbf{w}\|^2 \end{aligned}$$

for all real numbers c and d. Well, if $\mathbf{w} = \mathbf{0}$ then the Cauchy–Schwarz inequality follows trivially since it just says that $0 \leq 0$, and otherwise we can choose $c = \|\mathbf{w}\|$ and $d = -(\mathbf{v} \cdot \mathbf{w})/\|\mathbf{w}\|$ in the above inequality to see that

$$\begin{aligned} 0 &\leq \|\mathbf{v}\|^2\|\mathbf{w}\|^2 - 2\|\mathbf{w}\|(\mathbf{v} \cdot \mathbf{w})^2/\|\mathbf{w}\| + (\mathbf{v} \cdot \mathbf{w})^2\|\mathbf{w}\|^2/\|\mathbf{w}\|^2 \\ &= \|\mathbf{v}\|^2\|\mathbf{w}\|^2 - (\mathbf{v} \cdot \mathbf{w})^2. \end{aligned}$$

Rearranging and taking the square root of both sides of this inequality gives us $|\mathbf{v} \cdot \mathbf{w}| \leq \|\mathbf{v}\|\|\mathbf{w}\|$, which is exactly what we wanted to prove. ∎

While we will repeatedly make use of the Cauchy–Schwarz inequality as we progress through this book, for now it has two immediate and important applications. The first is that it lets us prove one final property of vector lengths—the fact that $\|\mathbf{v} + \mathbf{w}\|$ is never larger than $\|\mathbf{v}\| + \|\mathbf{w}\|$. To get some intuition for why this is the case, simply recall that in $\mathbb{R}^2$ and $\mathbb{R}^3$, the vectors $\mathbf{v}$, $\mathbf{w}$, and $\mathbf{v} + \mathbf{w}$ can be arranged to form the sides of a triangle, as in Figure 1.10. The inequality $\|\mathbf{v} + \mathbf{w}\| \leq \|\mathbf{v}\| + \|\mathbf{w}\|$ thus simply says that the length of one side of a triangle is never larger than the sum of the lengths of the other two sides.

The triangle inequality is sometimes expressed via the statement "stopping for coffee on your way to class cannot be a shortcut".

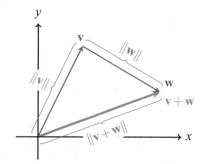

Figure 1.10: The shortest path between two points is a straight line, so $\|\mathbf{v}+\mathbf{w}\|$ is never larger than $\|\mathbf{v}\|+\|\mathbf{w}\|$. This fact is called the triangle inequality, and it is proved in Theorem 1.2.4.

Theorem 1.2.4
Triangle Inequality

> Suppose that $\mathbf{v}, \mathbf{w} \in \mathbb{R}^n$ are vectors. Then $\|\mathbf{v}+\mathbf{w}\| \le \|\mathbf{v}\| + \|\mathbf{w}\|$.

Proof. We start by expanding $\|\mathbf{v}+\mathbf{w}\|^2$ in terms of the dot product:

$$
\begin{aligned}
\|\mathbf{v}+\mathbf{w}\|^2 &= (\mathbf{v}+\mathbf{w})\cdot(\mathbf{v}+\mathbf{w}) && \text{(definition of length)}\\
&= (\mathbf{v}\cdot\mathbf{v}) + 2(\mathbf{v}\cdot\mathbf{w}) + (\mathbf{w}\cdot\mathbf{w}) && \text{(dot product properties (FOIL))}\\
&= \|\mathbf{v}\|^2 + 2(\mathbf{v}\cdot\mathbf{w}) + \|\mathbf{w}\|^2 && \text{(definition of length)}\\
&\le \|\mathbf{v}\|^2 + 2\|\mathbf{v}\|\|\mathbf{w}\| + \|\mathbf{w}\|^2 && \text{(Cauchy–Schwarz inequality)}\\
&= (\|\mathbf{v}\| + \|\mathbf{w}\|)^2. && \text{(factor cleverly)}
\end{aligned}
$$

We can then take the square root of both sides of the above inequality to see $\|\mathbf{v}+\mathbf{w}\| \le \|\mathbf{v}\| + \|\mathbf{w}\|$, as desired. ∎

The other immediate application of the Cauchy–Schwarz inequality is that it gives us a way to discuss the angle between vectors, which is the topic of the next subsection.

1.2.3 The Angle Between Vectors

In order to get a bit of an idea of how to discuss the angle between vectors in terms of things like the dot product, we first focus on vectors in $\mathbb{R}^2$ or $\mathbb{R}^3$. In these lower-dimensional cases, we can use geometric techniques to determine the angle between two vectors $\mathbf{v}$ and $\mathbf{w}$. If $\mathbf{v}, \mathbf{w} \in \mathbb{R}^2$ then we can place $\mathbf{v}$ and $\mathbf{w}$ in standard position, so that the vectors $\mathbf{v}$, $\mathbf{w}$, and $\mathbf{v}-\mathbf{w}$ form the sides of a triangle, as in Figure 1.11(a).

The law of cosines says that if the side lengths of a triangle are a, b and c, and the angle between the sides with lengths a and b is θ, then

$$c^2 = a^2 + b^2$$
$$- 2ab\cos(\theta).$$

We can then use the law of cosines to relate $\|\mathbf{v}\|$, $\|\mathbf{w}\|$, $\|\mathbf{v}-\mathbf{w}\|$, and the angle θ between $\mathbf{v}$ and $\mathbf{w}$. Specifically, we find that

$$\|\mathbf{v}-\mathbf{w}\|^2 = \|\mathbf{v}\|^2 + \|\mathbf{w}\|^2 - 2\|\mathbf{v}\|\|\mathbf{w}\|\cos(\theta).$$

On the other hand, the basic properties of the dot product that we saw back in Theorem 1.2.1 tell us that

$$
\begin{aligned}
\|\mathbf{v}-\mathbf{w}\|^2 &= (\mathbf{v}-\mathbf{w})\cdot(\mathbf{v}-\mathbf{w})\\
&= \mathbf{v}\cdot\mathbf{v} - \mathbf{v}\cdot\mathbf{w} - \mathbf{w}\cdot\mathbf{v} + \mathbf{w}\cdot\mathbf{w} = \|\mathbf{v}\|^2 - 2(\mathbf{v}\cdot\mathbf{w}) + \|\mathbf{w}\|^2.
\end{aligned}
$$

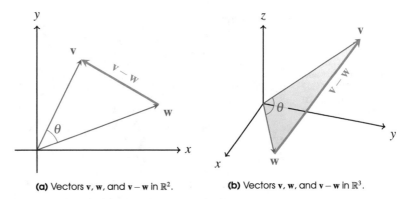

(a) Vectors $\mathbf{v}$, $\mathbf{w}$, and $\mathbf{v} - \mathbf{w}$ in $\mathbb{R}^2$. **(b)** Vectors $\mathbf{v}$, $\mathbf{w}$, and $\mathbf{v} - \mathbf{w}$ in $\mathbb{R}^3$.

Figure 1.11: The vectors $\mathbf{v}$, $\mathbf{w}$, and $\mathbf{v} - \mathbf{w}$ can be arranged to form a triangle in (a) $\mathbb{R}^2$ and (b) $\mathbb{R}^3$. The angle θ between $\mathbf{v}$ and $\mathbf{w}$ can then be expressed in terms of $\|\mathbf{v}\|$, $\|\mathbf{w}\|$, and $\|\mathbf{v} - \mathbf{w}\|$ via the law of cosines.

By setting these two expressions for $\|\mathbf{v} - \mathbf{w}\|^2$ equal to each other, we see that

$$\|\mathbf{v}\|^2 + \|\mathbf{w}\|^2 - 2\|\mathbf{v}\|\|\mathbf{w}\|\cos(\theta) = \|\mathbf{v}\|^2 - 2(\mathbf{v} \cdot \mathbf{w}) + \|\mathbf{w}\|^2.$$

Simplifying and rearranging this equation then gives a formula for θ in terms of the lengths of $\mathbf{v}$ and $\mathbf{w}$ and their dot product:

arccos is the inverse function of cos: if $0 \le \theta \le \pi$, then $\arccos(x) = \theta$ is equivalent to $\cos(\theta) = x$. It is sometimes written as $\cos^{-1}$ or acos.	$$\cos(\theta) = \frac{\mathbf{v} \cdot \mathbf{w}}{\|\mathbf{v}\|\|\mathbf{w}\|}, \quad \text{so} \quad \theta = \arccos\left(\frac{\mathbf{v} \cdot \mathbf{w}}{\|\mathbf{v}\|\|\mathbf{w}\|}\right).$$

This argument still works, but is slightly trickier to visualize, when working with vector $\mathbf{v}, \mathbf{w} \in \mathbb{R}^3$ that are 3-dimensional. In this case, we can still arrange $\mathbf{v}$, $\mathbf{w}$, and $\mathbf{v} - \mathbf{w}$ to form a triangle, and the calculation that we did in $\mathbb{R}^2$ is the exact same—the only change is that the triangle is embedded in 3-dimensional space, as in Figure 1.11(b).

When considering vectors in higher-dimensional spaces, we no longer have a visual guide for what the angle between two vectors means, so instead we simply *define* the angle so as to be consistent with the formula that we derived above:

Definition 1.2.3

Angle Between Vectors

The **angle** θ between two non-zero vectors $\mathbf{v}, \mathbf{w} \in \mathbb{R}^n$ is the quantity

$$\theta = \arccos\left(\frac{\mathbf{v} \cdot \mathbf{w}}{\|\mathbf{v}\|\|\mathbf{w}\|}\right).$$

It is worth noting that we typically measure angles in radians, not degrees. Also, the Cauchy–Schwarz inequality is very important when defining the angle between vectors in this way, since it ensures that the fraction $(\mathbf{v} \cdot \mathbf{w})/(\|\mathbf{v}\|\|\mathbf{w}\|)$ is between -1 and 1, which is what we require for its arccosine to exist in the first place.

Example 1.2.3

Numerical Examples of Vector Angles

Compute the angle between the following pairs of vectors:
   a) $\mathbf{v} = (1, 2)$ and $\mathbf{w} = (3, 4)$,
   b) $\mathbf{v} = (1, 2, -1, -2)$ and $\mathbf{w} = (1, -1, 1, -1)$, and
   c) the diagonals of two adjacent faces of a cube.

Solutions:

a) $\mathbf{v} \cdot \mathbf{w} = 3 + 8 = 11$, $\|\mathbf{v}\| = \sqrt{5}$, and $\|\mathbf{w}\| = 5$, so the angle between $\mathbf{v}$ and $\mathbf{w}$ is

$$\theta = \arccos\left(\frac{11}{5\sqrt{5}}\right) \approx 0.1799 \text{ radians (or } \approx 10.30 \text{ degrees)}.$$

b) $\mathbf{v} \cdot \mathbf{w} = 1 - 2 - 1 + 2 = 0$, so the angle between $\mathbf{v}$ and $\mathbf{w}$ is

$$\theta = \arccos(0) = \pi/2 \text{ (i.e., 90 degrees)}.$$

Notice that, in this case, we were able to compute the angle between $\mathbf{v}$ and $\mathbf{w}$ without even computing $\|\mathbf{v}\|$ or $\|\mathbf{w}\|$. For this reason, it is a good idea to compute $\mathbf{v} \cdot \mathbf{w}$ first (as we did here)—if $\mathbf{v} \cdot \mathbf{w} = 0$ then we know right away that the angle is $\theta = \pi/2$.

c) The cube with side length 1 can be positioned so that it has one vertex at $(0,0,0)$ and its opposite vertex at $(1,1,1)$. There are lots of pairs of face diagonals that we could choose, so we (arbitrarily) choose the face diagonals $\mathbf{v} = (1,0,1) - (1,1,0) = (0,-1,1)$ and $\mathbf{w} = (0,1,1) - (1,1,0) = (-1,0,1)$, as shown below.

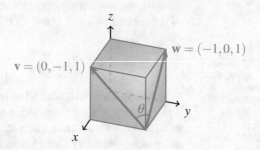

Then $\mathbf{v} \cdot \mathbf{w} = 0 + 0 + 1 = 1$, $\|\mathbf{v}\| = \sqrt{2}$, and $\|\mathbf{w}\| = \sqrt{2}$, so the angle between $\mathbf{v}$ and $\mathbf{w}$ is

$$\theta = \arccos\left(\frac{1}{\sqrt{2} \cdot \sqrt{2}}\right) = \arccos\left(\frac{1}{2}\right) = \pi/3 \text{ (i.e., 60 degrees)}.$$

In part (b) of the above example, we were able to conclude that the angle between $\mathbf{v}$ and $\mathbf{w}$ was $\pi/2$ based only on the fact that $\mathbf{v} \cdot \mathbf{w} = 0$ (since $\arccos(0) = \pi/2$). This implication goes both ways (i.e., if the angle between two vectors is $\theta = \pi/2$ then their dot product equals 0) and is an important enough special case that it gets its own name.

Definition 1.2.4

Orthogonality

Two vectors $\mathbf{v}, \mathbf{w} \in \mathbb{R}^n$ are said to be **orthogonal** if $\mathbf{v} \cdot \mathbf{w} = 0$.

We think of the word "orthogonal" as a synonym for "perpendicular" in small dimensions, as this is exactly what it means in $\mathbb{R}^2$ and $\mathbb{R}^3$ (recall that an angle of $\pi/2$ radians is 90 degrees)—see Figure 1.12. However, orthogonality also applies to higher-dimensional situations (e.g., the two vectors $\mathbf{v}, \mathbf{w} \in \mathbb{R}^4$ from Example 1.2.3(b) are orthogonal), despite us not being able to visualize them.

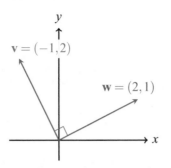

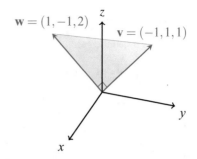

(a) The vectors $(-1,2)$ and $(2,1)$ in $\mathbb{R}^2$ are orthogonal since $(-1,2)\cdot(2,1) = -2+2 = 0$.

(b) The vectors $(-1,1,1)$ and $(1,-1,2)$ in $\mathbb{R}^3$ are orthogonal since $(-1,1,1)\cdot(1,-1,2) = -1-1+2 = 0$.

Figure 1.12: In 2 or 3 dimensions, two vectors being orthogonal (i.e., having dot product equal to 0) means that they are perpendicular to each other.

Exercises

solutions to starred exercises on page 436

1.2.1 Compute the dot product $\mathbf{v} \cdot \mathbf{w}$ of each of the following pairs of vectors.

* *(a)* $\mathbf{v} = (-2,4)$, $\mathbf{w} = (2,1)$
* (b) $\mathbf{v} = (1,2,3)$, $\mathbf{w} = (-3,2,-1)$
* *(c)* $\mathbf{v} = (3,-1,0,1)$, $\mathbf{w} = (0,2,1,3)$
* (d) $\mathbf{v} = (\sqrt{2}, \sqrt{3}, \sqrt{5})$, $\mathbf{w} = (\sqrt{2}, \sqrt{3}, \sqrt{5})$
* *(e)* $\mathbf{v} = \mathbf{0} \in \mathbb{R}^9$, $\mathbf{w} = (8,1,5,-7,3,9,1,-3,2)$

1.2.2 Compute the length $\|\mathbf{v}\|$ of each of the following vectors $\mathbf{v}$, and also give a unit vector $\mathbf{u}$ pointing in the same direction as $\mathbf{v}$.

* *(a)* $\mathbf{v} = (3,4)$
* (b) $\mathbf{v} = (2,1,-2)$
* *(c)* $\mathbf{v} = (-2\sqrt{2},-3,\sqrt{10},3)$
* (d) $\mathbf{v} = (\cos(\theta),\sin(\theta))$

1.2.3 Compute the angle between each of the following pairs of vectors.

* *(a)* $\mathbf{v} = (1,\sqrt{3})$, $\mathbf{w} = (\sqrt{3},1)$
* (b) $\mathbf{v} = (0,-2,2)$, $\mathbf{w} = (1,0,1)$
* *(c)* $\mathbf{v} = (1,1,1,1)$, $\mathbf{w} = (-2,0,-2,0)$
* (d) $\mathbf{v} = (2,1,-3)$, $\mathbf{w} = (-1,2,3)$
* *(e)* $\mathbf{v} = (\cos(\theta),\sin(\theta))$, $\mathbf{w} = (-\sin(\theta),\cos(\theta))$

1.2.4 Determine which of the following statements are true and which are false.

* *(a)* If $\mathbf{v},\mathbf{w},\mathbf{x} \in \mathbb{R}^n$ are vectors with $\mathbf{v} \cdot \mathbf{w} = \mathbf{v} \cdot \mathbf{x}$, then $\mathbf{w} = \mathbf{x}$.
* (b) If $\mathbf{v},\mathbf{w},\mathbf{x} \in \mathbb{R}^n$ are vectors with $\mathbf{v} \cdot \mathbf{w} = 0$ and $\mathbf{w} \cdot \mathbf{x} = 0$, then $\mathbf{v} \cdot \mathbf{x} = 0$ too.
* * (c) If $\mathbf{v},\mathbf{w} \in \mathbb{R}^n$ are vectors with $\|\mathbf{v}\| + \|\mathbf{w}\| \le 2$, then $\|\mathbf{v} + \mathbf{w}\| \le 2$ too.
* (d) There exist vectors $\mathbf{v},\mathbf{w} \in \mathbb{R}^5$ such that $\|\mathbf{v}\| = 2$, $\|\mathbf{w}\| = 4$, and $\|\mathbf{v} - \mathbf{w}\| = 1$.
* * (e) There exist vectors $\mathbf{v},\mathbf{w} \in \mathbb{R}^3$ such that $\|\mathbf{v}\| = 1$, $\|\mathbf{w}\| = 2$, and $\mathbf{v} \cdot \mathbf{w} = -1$.
* (f) If $\mathbf{v},\mathbf{w} \in \mathbb{R}^n$ are unit vectors, then $|\mathbf{v} \cdot \mathbf{w}| \le 1$.
* * (g) If $\mathbf{v},\mathbf{w} \in \mathbb{R}^n$ are vectors with $|\mathbf{v} \cdot \mathbf{w}| \le 1$, then $\|\mathbf{v}\| \le 1$ or $\|\mathbf{w}\| \le 1$ (or both).

1.2.5 How can we use the quantity $\mathbf{v} \cdot \mathbf{w}$ to determine whether the angle between $\mathbf{v}$ and $\mathbf{w}$ is acute, a right angle, or obtuse **without** computing $\|\mathbf{v}\|$, $\|\mathbf{w}\|$, or using any trigonometric functions?

[Hint: Definition 1.2.1 solves the right angle case.]

* **1.2.6** Suppose $\mathbf{v} = (3,\sqrt{3})$. Find all vectors $\mathbf{w} \in \mathbb{R}^2$ such that $\|\mathbf{w}\| = 2$ and the angle between $\mathbf{v}$ and $\mathbf{w}$ is $\theta = \pi/3$.

1.2.7 Let $\mathbf{v},\mathbf{w},\mathbf{x} \in \mathbb{R}^n$ be non-zero vectors. Determine which of the following expressions do and do not make sense.

* *(a)* $\mathbf{v} \cdot (\mathbf{w} - \mathbf{x})$
* (b) $(\mathbf{v} \cdot \mathbf{w})\mathbf{x}$
* *(c)* $\mathbf{v} + (\mathbf{w} \cdot \mathbf{x})$
* (d) $\mathbf{v}/\|\mathbf{v}\|$
* *(e)* $\mathbf{v}^2$
* (f) $(\mathbf{v} + \mathbf{w})/\mathbf{x}$

1.2.8 Let $\mathbf{e}_j \in \mathbb{R}^n$ be the standard basis vector with 1 in its j-th entry and 0 in all other entries.

* (a) Compute $\|\mathbf{e}_j\|$.
* (b) Suppose $1 \le i, j \le n$. Compute $\mathbf{e}_i \cdot \mathbf{e}_j$. [Side note: You will get a different answer depending on whether $i = j$ or $i \ne j$.]

* **1.2.9** Let $\mathbf{v} = (1,2) \in \mathbb{R}^2$.

* (a) Find a non-zero vector that is orthogonal to $\mathbf{v}$.
* (b) Is it possible to find a non-zero vector that is orthogonal to $\mathbf{v}$ as well as the vectors that you found in part (a)? Justify your answer.

1.2.10 Let $\mathbf{v} = (1,2,3) \in \mathbb{R}^3$.

* (a) Find a non-zero vector that is orthogonal to $\mathbf{v}$.
* (b) Find a non-zero vector that is orthogonal to $\mathbf{v}$ and is *also* orthogonal to the vector that you found in part (a).
* (c) Is it possible to find a non-zero vector that is orthogonal to $\mathbf{v}$ as well as both of the vectors that you found in parts (a) and (b)? Justify your answer.

*1.2.11   A rectangle in $\mathbb{R}^3$ has three of its corners at the points $(1,0,-1)$, $(2,2,2)$, and $(-1,2,3)$. What are the coordinates of its fourth corner?

1.2.12   Let $\mathbf{v} \in \mathbb{R}^n$. Show that

$$\left(\frac{v_1 + v_2 + \cdots + v_n}{n} \right)^2 \leq \frac{1}{n} \left(v_1^2 + v_2^2 + \cdots + v_n^2 \right).$$

[Side note: In words, this says that the square of the average of a set of numbers is never larger than the average of their squares.]

**1.2.13   Recall Theorem 1.2.1, which established some of the basic properties of the dot product.

  (a) Prove part (b) of the theorem.
  (b) Prove part (c) of the theorem.

1.2.14   Find the coordinates of the vectors that point from the center of a regular hexagon to its corners if it is centered at $(0,0)$ and has one corner located at $(1,0)$.

*1.2.15   If $\mathbf{v}$ and $\mathbf{w}$ are n-dimensional vectors with complex (instead of real) entries, we write $\mathbf{v}, \mathbf{w} \in \mathbb{C}^n$ and we define their dot product by

$$\mathbf{v} \cdot \mathbf{w} \stackrel{\text{def}}{=} \overline{v_1} w_1 + \overline{v_2} w_2 + \cdots + \overline{v_n} w_n,$$

where $\overline{a+ib} = a - ib$ is the complex conjugate (see Appendix A.1).

  (a) Show that $\mathbf{v} \cdot \mathbf{w} = \overline{\mathbf{w} \cdot \mathbf{v}}$.
  (b) Show that $\mathbf{v} \cdot (c\mathbf{w}) = c(\mathbf{v} \cdot \mathbf{w})$ for all complex scalars $c \in \mathbb{C}$, but $(c\mathbf{v}) \cdot \mathbf{w} = \overline{c}(\mathbf{v} \cdot \mathbf{w})$.

1.2.16   Suppose that $\mathbf{x} \in \mathbb{R}^n$. Show that $\mathbf{x} \cdot \mathbf{y} = 0$ for all $\mathbf{y} \in \mathbb{R}^n$ if and only if $\mathbf{x} = \mathbf{0}$.

**1.2.17   In this exercise, we determine when equality holds in the Cauchy–Schwarz and triangle inequalities.

  (a) Prove that $|\mathbf{v} \cdot \mathbf{w}| = \|\mathbf{v}\| \|\mathbf{w}\|$ if and only if either $\mathbf{w} = \mathbf{0}$ or there exists a scalar $c \in \mathbb{R}$ such that $\mathbf{v} = c\mathbf{w}$.
  (b) Prove that $\|\mathbf{v} + \mathbf{w}\| = \|\mathbf{v}\| + \|\mathbf{w}\|$ if and only if either $\mathbf{w} = \mathbf{0}$ or there exists a scalar $0 \leq c \in \mathbb{R}$ such that $\mathbf{v} = c\mathbf{w}$.

1.2.18   Let $\mathbf{v}, \mathbf{w} \in \mathbb{R}^n$ be vectors that are orthogonal to each other. Prove that $\|\mathbf{v} + \mathbf{w}\|^2 = \|\mathbf{v}\|^2 + \|\mathbf{w}\|^2$.
[Side note: This is the **Pythagorean theorem** in $\mathbb{R}^n$.]

*1.2.19   Let $\mathbf{v}, \mathbf{w} \in \mathbb{R}^n$ be vectors.

  (a) Show that $\|\mathbf{v} + \mathbf{w}\|^2 + \|\mathbf{v} - \mathbf{w}\|^2 = 2\|\mathbf{v}\|^2 + 2\|\mathbf{w}\|^2$.
       [Side note: This is called the **parallelogram law**.]
  (b) Draw a parallelogram with sides $\mathbf{v}$ and $\mathbf{w}$ and explain geometrically what the result of part (a) says.

1.2.20   Let $\mathbf{v}, \mathbf{w} \in \mathbb{R}^n$ be vectors. Prove that $\mathbf{v} \cdot \mathbf{w} = \frac{1}{4}\left(\|\mathbf{v} + \mathbf{w}\|^2 - \|\mathbf{v} - \mathbf{w}\|^2 \right)$.
[Side note: This is called the **polarization identity**.]

*1.2.21   Let $\mathbf{v}, \mathbf{w} \in \mathbb{R}^n$ be vectors. Prove that $\|\mathbf{v} - \mathbf{w}\| \geq \|\mathbf{v}\| - \|\mathbf{w}\|$.
[Side note: This is called the **reverse triangle inequality**.]

1.2.22   In this exercise, we tweak the proof of Theorem 1.2.3 slightly to get another proof of the Cauchy–Schwarz inequality.

  (a) What inequality results from choosing $c = \|\mathbf{w}\|$ and $d = \|\mathbf{v}\|$ in the proof?
  (b) What inequality results from choosing $c = \|\mathbf{w}\|$ and $d = -\|\mathbf{v}\|$ in the proof?
  (c) Combine the inequalities from parts (a) and (b) to prove the Cauchy–Schwarz inequality.

*1.2.23   This exercise guides you through another proof of the Cauchy–Schwarz inequality. Let $\mathbf{v}, \mathbf{w} \in \mathbb{R}^n$ be vectors, and consider the function $f(x) = \|\mathbf{v} - x\mathbf{w}\|^2$.

  (a) Show that this is a quadratic function in x.
  (b) What is the discriminant of this quadratic? Recall that the discriminant of the quadratic $ax^2 + bx + c$ is $b^2 - 4ac$.
  (c) Why must the discriminant from part (b) be ≤ 0? [Hint: How many roots does f have?]
  (d) Use parts (b) and (c) to prove the Cauchy–Schwarz inequality.

1.2.24   This exercise guides you through yet another proof of the Cauchy–Schwarz inequality.

  (a) Show that if $x, y \in \mathbb{R}$ then $|x + y| \leq |x| + |y|$. Do *not* use Theorem 1.2.4.
  (b) Show that if $x_1, \ldots, x_n \in \mathbb{R}$ then $|x_1 + \cdots + x_n| \leq |x_1| + \cdots + |x_n|$.
  (c) Show that if $x, y \in \mathbb{R}$ then $xy \leq \frac{1}{2}(x^2 + y^2)$.
  (d) Show that if $\mathbf{v}, \mathbf{w} \in \mathbb{R}^n$ are non-zero vectors then $|\mathbf{v} \cdot \mathbf{w}| / (\|\mathbf{v}\| \|\mathbf{w}\|) \leq 1$. [Hint: Start by writing out $\mathbf{v} \cdot \mathbf{w} = v_1 w_1 + \cdots + v_n w_n$ in the fraction on the left. Then use part (b), and finally use part (c).]

1.3   Matrices and Matrix Operations

One more concept that we will need before we start running with linear algebra is that of a **matrix**, which is a 2D array of numbers like

$$A = \begin{bmatrix} 1 & 3 \\ 2 & -1 \end{bmatrix} \quad \text{and} \quad B = \begin{bmatrix} 3 & 0 & 2 \\ 0 & -1 & 1 \end{bmatrix}.$$

The plural of "matrix" is "matrices".

Note that every row of a matrix must have the same number of entries as every other row, and similarly every column must have the same number of entries as every other column. The rows and columns must line up with each other, and every spot in the matrix (i.e., every intersection of a row and column) must contain an entry. For example, the following are *not* valid matrices:

$$\begin{bmatrix} 1 & 2 & 3 \\ & 4 & 5 \end{bmatrix} \quad \text{and} \quad \begin{bmatrix} 2 & -3 & & 4 \\ & 1 & 2 & 0 \end{bmatrix}.$$

In almost any kind of matrix notation, rows come before columns: $a_{i,j}$ means the entry in the i-th row and j-th column of A, a 3×4 matrix is one with 3 rows and 4 columns, etc.

We typically denote matrices by uppercase letters like $A, B, C, \ldots$, and we write their entries via the corresponding lowercase letters, together with subscripts that indicate which row and column (in that order) the entry comes from. For example, if A and B are as above, then $a_{1,2} = 3$, since the entry of A in the 1st row and 2nd column is 3. Similarly, $a_{2,1} = 2$ and $b_{2,3} = 1$. The entry in the i-th row and j-th column is also called its "(i, j)-entry", and we sometimes alternatively denote it using square brackets like $[A]_{i,j}$. For example, the $(2,1)$-entry of B is $b_{2,1} = [B]_{2,1} = 0$.

The number of rows and columns that a matrix has are collectively referred to as its **size**, and we always list the number of rows first. For example, the matrix A above has size 2×2, whereas B has size 2×3. The set of all real matrices with m rows and n columns is denoted by $\mathcal{M}_{m,n}$, or simply by $\mathcal{M}_n$ if $m = n$ (in which case the matrix is called **square**). So if A and B again refer to the matrices displayed above, then $A \in \mathcal{M}_2$ and $B \in \mathcal{M}_{2,3}$.

1.3.1 Matrix Addition and Scalar Multiplication

We do not yet have a nice geometric interpretation of matrices like we did for vectors (we will develop a geometric understanding of matrices in the next section), but for now we note that we can define addition and scalar multiplication for matrices in the exact same entrywise manner that we did for vectors.

Definition 1.3.1

Matrix Addition and Scalar Multiplication

Suppose $A, B \in \mathcal{M}_{m,n}$ are matrices and $c \in \mathbb{R}$ is a scalar. Then the **sum** $A + B$ and **scalar multiplication** cA are the $m \times n$ matrices whose (i, j)-entries, for each $1 \leq i \leq m$ and $1 \leq j \leq n$, are

$$[A+B]_{i,j} = a_{i,j} + b_{i,j} \quad \text{and} \quad [cA]_{i,j} = ca_{i,j},$$

respectively.

We also use O to denote the **zero matrix** whose entries all equal 0 (or $O_{m,n}$ if we wish to emphasize or clarify that it is $m \times n$, or O_n if it is $n \times n$). Similarly, we define matrix subtraction $(A - B = A + (-1)B)$ and the negative of a matrix $(-A = (-1)A)$ in the obvious entrywise ways. There is nothing fancy or surprising about how these matrix operations work, but it is worthwhile to work through a couple of quick examples to make sure that we are comfortable with them.

Example 1.3.1

Numerical Examples of Matrix Operations

Suppose $A = \begin{bmatrix} 1 & 3 \\ 2 & -1 \end{bmatrix}$, $B = \begin{bmatrix} 2 & 1 \\ 0 & 1 \end{bmatrix}$, and $C = \begin{bmatrix} 1 & 0 & 1 \\ 0 & -1 & 1 \end{bmatrix}$. Compute

a) $A + B$,

b) $2A - 3B$, and

c) $A + 2C$.

Solutions:

a) $A + B = \begin{bmatrix} 1 & 3 \\ 2 & -1 \end{bmatrix} + \begin{bmatrix} 2 & 1 \\ 0 & 1 \end{bmatrix} = \begin{bmatrix} 3 & 4 \\ 2 & 0 \end{bmatrix}$.

b) $2A - 3B = \begin{bmatrix} 2 & 6 \\ 4 & -2 \end{bmatrix} - \begin{bmatrix} 6 & 3 \\ 0 & 3 \end{bmatrix} = \begin{bmatrix} -4 & 3 \\ 4 & -5 \end{bmatrix}$.

c) This expression does not make sense; we cannot add or subtract matrices that have different sizes (A is a 2×2 matrix, but $2C$ is a 2×3 matrix).

As we might expect, matrix addition and scalar multiplication satisfy the same algebraic properties like commutativity and associativity that we saw for vectors back in Theorems 1.1.1 and 1.1.2. For completeness, we state these properties explicitly, but we emphasize that none of these are meant to be surprising.

Theorem 1.3.1

Properties of Matrix Operations

Suppose $A, B, C \in \mathcal{M}_{m,n}$ are $m \times n$ matrices and $c, d \in \mathbb{R}$ are scalars. Then
a) $A + B = B + A$,   (commutativity)
b) $(A + B) + C = A + (B + C)$,   (associativity)
c) $c(A + B) = cA + cB$,   (distributivity)
d) $(c + d)A = cA + dA$, and   (distributivity)
e) $c(dA) = (cd)A$.

Proof. All five of these properties can be proved fairly quickly by using the relevant definitions. To see that part (a) holds, we compute

We already know that $a + b = b + a$ when $a, b \in \mathbb{R}$. We use this fact in each of the mn entries of this matrix sum.

$$A + B = \begin{bmatrix} a_{1,1} + b_{1,1} & a_{1,2} + b_{1,2} & \cdots & a_{1,n} + b_{1,n} \\ a_{2,1} + b_{2,1} & a_{2,2} + b_{2,2} & \cdots & a_{2,n} + b_{2,n} \\ \vdots & \vdots & \ddots & \vdots \\ a_{m,1} + b_{m,1} & a_{m,2} + b_{m,2} & \cdots & a_{m,n} + b_{m,n} \end{bmatrix}$$

$$= \begin{bmatrix} b_{1,1} + a_{1,1} & b_{1,2} + a_{1,2} & \cdots & b_{1,n} + a_{1,n} \\ b_{2,1} + a_{2,1} & b_{2,2} + a_{2,2} & \cdots & b_{2,n} + a_{2,n} \\ \vdots & \vdots & \ddots & \vdots \\ b_{m,1} + a_{m,1} & b_{m,2} + a_{m,2} & \cdots & b_{m,n} + a_{m,n} \end{bmatrix} = B + A,$$

where we used commutativity of addition of real numbers within each entry of the matrix.

The remaining parts of the theorem can be proved similarly—just use the definitions of matrix addition and scalar multiplication together with the fact that all of these properties hold for addition and multiplication of real numbers (see Exercise 1.3.19). ∎

Some additional properties of matrix addition and scalar multiplication that we did not explicitly mention in this theorem, but which should be fairly clear, are the facts that $A + O = A$ and $A - A = O$ for all matrices $A \in \mathcal{M}_{m,n}$.

1.3.2 Matrix Multiplication

While matrix addition and scalar multiplication are in a sense nothing new, we now introduce the standard method for multiplying matrices, which is *very* new and seems quite unintuitive at first.

Definition 1.3.2

Matrix Multiplication

If $A \in \mathcal{M}_{m,n}$ and $B \in \mathcal{M}_{n,p}$ are matrices, then their **product** AB is the $m \times p$ matrix whose (i,j)-entry, for each $1 \leq i \leq m$ and $1 \leq j \leq p$, is

$$[AB]_{i,j} \overset{\text{def}}{=} a_{i,1}b_{1,j} + a_{i,2}b_{2,j} + \cdots + a_{i,n}b_{n,j}.$$

In other words, the product AB is the matrix whose entries are all of the possible dot products of the rows of A with the columns of B, as illustrated in Figure 1.13. Before proceeding with some examples, we emphasize that the matrix product AB only makes sense if A has the same number of columns as B has rows. For example, it does not make sense to multiply a 2×3 matrix by another 2×3 matrix, but it does make sense to multiply a 2×3 matrix by a 3×7 matrix.

In other words, the inner dimensions in a matrix product must agree: we can multiply an $m \times n$ matrix by an $r \times p$ matrix if and only if $n = r$.

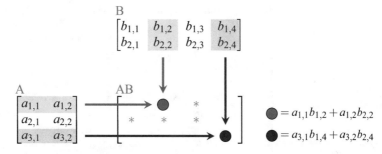

Figure 1.13: A visualization of the multiplication of a 3×2 matrix A and a 2×4 matrix B. Each entry in the 3×4 matrix AB is the dot product of the corresponding row of A and column of B.

Example 1.3.2

Numerical Examples of Matrix Multiplication

Compute each of the matrix products indicated below (if possible) if

$$A = \begin{bmatrix} 1 & 2 \\ 3 & 4 \end{bmatrix}, \quad B = \begin{bmatrix} 5 & 6 & 7 \\ 8 & 9 & 10 \end{bmatrix}, \quad \text{and} \quad C = \begin{bmatrix} 1 & 0 \\ 0 & -1 \\ 2 & -1 \end{bmatrix}.$$

a) AB,
b) AC,
c) BA, and
d) BC.

Solutions:

a) We carefully compute each entry of AB one at a time. For example, the $(1,1)$-entry of AB is the dot product of the first row of A with the first column of B: $(1,2) \cdot (5,8) = 1 \cdot 5 + 2 \cdot 8 = 21$. Similarly computing the other entries gives

$$AB = \begin{bmatrix} 1 & 2 \\ 3 & 4 \end{bmatrix} \begin{bmatrix} 5 & 6 & 7 \\ 8 & 9 & 10 \end{bmatrix}$$

$$= \begin{bmatrix} (1,2) \cdot (5,8) & (1,2) \cdot (6,9) & (1,2) \cdot (7,10) \\ (3,4) \cdot (5,8) & (3,4) \cdot (6,9) & (3,4) \cdot (7,10) \end{bmatrix}$$

$$= \begin{bmatrix} 1 \times 5 + 2 \times 8 & 1 \times 6 + 2 \times 9 & 1 \times 7 + 2 \times 10 \\ 3 \times 5 + 4 \times 8 & 3 \times 6 + 4 \times 9 & 3 \times 7 + 4 \times 10 \end{bmatrix}$$

$$= \begin{bmatrix} 21 & 24 & 27 \\ 47 & 54 & 61 \end{bmatrix}.$$

b) *AC* does not exist since *A* has 2 columns but *C* has 3 rows, and those numbers would have to match for the matrix product to exist.

c) *BA* does not exist since *B* has 3 columns but *A* has 2 rows, and those numbers would have to match for the matrix product to exist.

Keep in mind that *AB*'s size consists of the outer dimensions of *A* and *B*: if *A* is $m \times n$ and *B* is $n \times p$ then *AB* is $m \times p$.

d) Again, we compute each entry of *BC* one at a time. For example, the $(1,1)$-entry of *BC* is the dot product of the first row of *B* with the first column of *C*: $(5,6,7) \cdot (1,0,2) = 5 \cdot 1 + 6 \cdot 0 + 7 \cdot 2 = 19$. Similarly computing the other entries gives

$$BC = \begin{bmatrix} 5 & 6 & 7 \\ 8 & 9 & 10 \end{bmatrix} \begin{bmatrix} 1 & 0 \\ 0 & -1 \\ 2 & -1 \end{bmatrix}$$

$$= \begin{bmatrix} (5,6,7) \cdot (1,0,2) & (5,6,7) \cdot (0,-1,-1) \\ (8,9,10) \cdot (1,0,2) & (8,9,10) \cdot (0,-1,-1) \end{bmatrix}$$

$$= \begin{bmatrix} 5 \times 1 + 6 \times 0 + 7 \times 2 & 5 \times 0 + 6 \times -1 + 7 \times -1 \\ 8 \times 1 + 9 \times 0 + 10 \times 2 & 8 \times 0 + 9 \times -1 + 10 \times -1 \end{bmatrix}$$

$$= \begin{bmatrix} 19 & -13 \\ 28 & -19 \end{bmatrix}.$$

When performing matrix multiplication, it is a good idea to frequently double-check that the sizes of the matrices we are working with actually make sense. In particular, the inner dimensions of the matrices must be equal, and the outer dimensions of the matrices will be the dimensions of the matrix product:

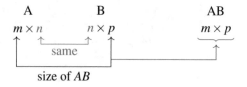

Since matrix multiplication is so much less straightforward than the other vector and matrix operations we have introduced, it is not immediately clear what properties it satisfies. Keeping track of these properties is thus somewhat more important than it was for, say, vector addition:

Theorem 1.3.2

Properties of Matrix Multiplication

Let A, B, and C be matrices (with sizes such that the multiplications and additions below make sense), and let $c \in \mathbb{R}$ be a scalar. Then

a) $(AB)C = A(BC)$,   (associativity)

b) $A(B+C) = AB + AC$,   (left distributivity)

c) $(A+B)C = AC + BC$, and   (right distributivity)

d) $c(AB) = (cA)B$.

Proof. The proofs of all of these statements are quite similar to each other, so we only explicitly prove part (b)—the remaining parts of the theorem are left to Exercise 1.3.20.

To this end we note that, for each i and j, the (i, j)-entry of $A(B+C)$ is

$$a_{i,1}(b_{1,j} + c_{1,j}) + a_{i,2}(b_{2,j} + c_{2,j}) + \cdots + a_{i,n}(b_{n,j} + c_{n,j}),$$

whereas the (i, j)-entry of $AB + AC$ is

$$(a_{i,1}b_{1,j} + a_{i,2}b_{2,j} + \cdots + a_{i,n}b_{n,j}) + (a_{i,1}c_{1,j} + a_{i,2}c_{2,j} + \cdots + a_{i,n}c_{n,j}).$$

It is straightforward to see that these two quantities are equal just by expanding out parentheses and regrouping terms. Each entry of $A(B+C)$ thus equals the corresponding entry of $AB + AC$, so the matrices themselves are the same. ∎

Remark 1.3.1

Matrix Multiplication is Not Commutative

An operation *not* being commutative should be our default assumption, since the order in which events occur matters. We would prefer to lose all of our money and then get a million dollars, rather than get a million dollars and then lose all of our money.

It is worth stressing the fact that we did *not* list commutativity (i.e., $AB = BA$) as one of the properties of matrix multiplication in Theorem 1.3.2. The reason for this omission is simple: it's not true. We saw in Example 1.3.2 that it's entirely possible that BA does not even exist when AB does.

Even if both AB and BA exist, they may or may not have the same size as each other (see Exercise 1.3.15), and even if they are of the same size, they still might not equal each other. For example, if

$$A = \begin{bmatrix} 1 & 1 \\ 0 & 1 \end{bmatrix} \quad \text{and} \quad B = \begin{bmatrix} 1 & 0 \\ 1 & 1 \end{bmatrix}$$

then

$$AB = \begin{bmatrix} 2 & 1 \\ 1 & 1 \end{bmatrix} \quad \text{and} \quad BA = \begin{bmatrix} 1 & 1 \\ 1 & 2 \end{bmatrix}.$$

While we will see that AB and BA share *some* properties with each other in later chapters of this book, for now it is best to think of them as completely different matrices.

Matrix Powers

One particularly important square matrix is the one that consists entirely of 0 entries, except with 1s on its diagonal (i.e., its $(1,1)$-entry, $(2,2)$-entry, $(3,3)$-entry, and so on, all equal 1). This is called the **identity matrix**, and if it has size $n \times n$ then it is denoted by I_n (or if its size is clear from context or irrelevant, we denote it just by I). For example, the identity matrices in $\mathcal{M}_2$ and $\mathcal{M}_3$ are

$$I_2 = \begin{bmatrix} 1 & 0 \\ 0 & 1 \end{bmatrix} \quad \text{and} \quad I_3 = \begin{bmatrix} 1 & 0 & 0 \\ 0 & 1 & 0 \\ 0 & 0 & 1 \end{bmatrix}.$$

The reason that the identity matrix is so important is that multiplying it by another matrix does not change that other matrix (similar to how the zero matrix "does nothing" under matrix addition, the identity matrix "does nothing" under matrix multiplication):

Theorem 1.3.3

Multiplication by the Identity Matrix

If $A \in \mathcal{M}_{m,n}$ then $AI_n = A = I_m A$.

Proof. For the first equality, we recall that the (i,j)-entry of AI_n is the dot product of the i-th row of A with the j-th column of I_n. However, the j-th column of I_n is $\mathbf{e}_j$, so it follows from Example 1.2.1(c) that the (i,j)-entry of AI_n is the j-th entry of the i-th row of A: $a_{i,j}$. It follows that all entries of AI_n and A coincide, so $AI_n = A$.

Because the second equality is proved in such a similar manner, we leave it as Exercise 1.3.21. ∎

Recall that a matrix is square if it has the same number of rows as columns.

When a matrix A is square, we are able to multiply it by itself and thus obtain **powers** of A. That is, we define $A^2 = AA$, $A^3 = AAA$, and in general

$$A^k \overset{\text{def}}{=} \underbrace{AA \cdots A}_{k \text{ copies}}.$$

Recall that if $x \in \mathbb{R}$ then $x^0 = 1$. The definition $A^0 = I$ is analogous.

Matrix powers satisfy many of the nice properties that we would expect them to, like $A^{k+\ell} = A^k A^\ell$ and $(A^k)^\ell = A^{k\ell}$ for all $k, \ell \geq 1$ (see Exercise 1.3.22). We also define $A^0 = I$ so that these same properties even hold if $k = 0$ or $\ell = 0$.

Example 1.3.3

Numerical Examples of Matrix Powers

Compute each of the indicated matrix powers if $A = \begin{bmatrix} 2 & 1 \\ -1 & 3 \end{bmatrix}$.

a) A^2,

b) A^4, and

c) I^7.

Solutions:

a) We just multiply A by itself:

$$A^2 = \begin{bmatrix} 2 & 1 \\ -1 & 3 \end{bmatrix} \begin{bmatrix} 2 & 1 \\ -1 & 3 \end{bmatrix} = \begin{bmatrix} 3 & 5 \\ -5 & 8 \end{bmatrix}.$$

For now, matrix powers are only defined if the exponent is a non-negative integer. We explore how to extend this definition to negative integers in Section 2.2, and to arbitrary real numbers in Section 3.4.

b) We could multiply A by itself 2 more times, or we could use the fact that $A^4 = (A^2)^2$, so we just need to square our answer from part (a):

$$A^4 = (A^2)^2 = \begin{bmatrix} 3 & 5 \\ -5 & 8 \end{bmatrix} \begin{bmatrix} 3 & 5 \\ -5 & 8 \end{bmatrix} = \begin{bmatrix} -16 & 55 \\ -55 & 39 \end{bmatrix}.$$

c) We can repeatedly use Theorem 1.3.3 to see that $I^k = I$ for all non-negative integers k, so $I^7 = I$.

As a result of the fact that matrix multiplication is not commutative, we have to be somewhat careful when doing algebra with matrices. For example, when expanding the expression $(A+B)^2$, the reader might be tempted to write $(A+B)^2 = A^2 + 2AB + B^2$. However, if we go through the calculation a bit more carefully then we find that the correct expression is

$$(A+B)^2 = (A+B)(A+B)$$
$$= A(A+B) + B(A+B)$$
$$= AA + AB + BA + BB = A^2 + AB + BA + B^2,$$

and in general we cannot simplify any further by combining the AB and BA terms in the final line above.

Example 1.3.4

Introduction to Matrix Algebra

Suppose $A, B \in \mathcal{M}_n$. Expand and simplify the following expressions:

a) $(A+I)^2(A-I)$, and
b) $(A-A)(A+B-I)^7$.

Solutions:

Even though it is not the case that $(A+B)^2 = A^2 + 2AB + B^2$ in general, it *is* the case that $(A+I)^2 = A^2 + 2A + I$, since A and I commute.

a) We first square the $A+I$ and then multiply it by $A-I$:

$$(A+I)^2(A-I) = (A^2 + AI + IA + I)(A-I)$$
$$= (A^2 + 2A + I)(A-I)$$
$$= (A^3 + 2A^2 + A) - (A^2 + 2A + I)$$
$$= A^3 + A^2 - A - I.$$

b) We could expand $(A+B-I)^7$ (which would be ugly!) and then multiply by $A - A = O$ (the zero matrix), or we could simply notice that O times *any* matrix is equal to O. It follows that $(A-A)(A+B-I)^7 = O(A+B-I)^7 = O$.

Row and Column Vectors

One of the most useful features of matrix multiplication is that it can be used to move vectors around. In particular, if $A \in \mathcal{M}_{m,n}$ and $\mathbf{v} \in \mathcal{M}_{n,1}$ then $A\mathbf{v} \in \mathcal{M}_{m,1}$, and since each of $\mathbf{v}$ and $A\mathbf{v}$ have just one column, we can think of them as vectors (and we think of A as transforming $\mathbf{v}$ into $A\mathbf{v}$). In fact, we call these matrices with just one column **column vectors**, and similarly we refer to matrices with just one row as **row vectors**. For example,

We can think of numbers as 1×1 matrices or 1-dimensional vectors, and we can think of vectors as $1 \times n$ or $m \times 1$ matrices. Numbers are thus special cases of vectors, which are special cases of matrices.

$$\text{if} \quad A = \begin{bmatrix} 1 & 2 \\ 3 & 4 \end{bmatrix} \quad \text{and} \quad \mathbf{v} = \begin{bmatrix} 2 \\ -1 \end{bmatrix} \quad \text{then} \quad A\mathbf{v} = \begin{bmatrix} 2-2 \\ 6-4 \end{bmatrix} = \begin{bmatrix} 0 \\ 2 \end{bmatrix},$$

so we think of the matrix A as transforming $\mathbf{v} = (2, -1)$ into $A\mathbf{v} = (0, 2)$, but we have to write $\mathbf{v}$ and $A\mathbf{v}$ as columns for the matrix multiplication to actually work out.

When performing matrix multiplication (as well as a few other tasks that we will investigate later in this book), the difference between row vectors and column vectors is important, since the product $A\mathbf{v}$ does not make sense if $\mathbf{v}$ is a row vector—the inner dimensions of the matrix multiplication do not match. However, when thinking of vectors geometrically as we did in the previous sections, this difference is often unimportant.

To help make it clearer whether or not we actually care about the shape (i.e., row or column) of a vector, we use square brackets when thinking of a vector as a row or a column (as we have been doing throughout this section for matrices), and we use round parentheses if its shape is unimportant to us

It is not uncommon to abuse notation slightly and say things like $[1,2,3] \in \mathbb{R}^3$ (even though $[1,2,3]$ is a row vector and thus technically lives in $\mathcal{M}_{1,3}$).

(as we did in the previous two sections). For example, $\mathbf{v} = [1,2,3] \in \mathcal{M}_{1,3}$ is a row vector,

$$\mathbf{w} = \begin{bmatrix} 1 \\ 2 \\ 3 \end{bmatrix} \in \mathcal{M}_{3,1}$$

is a column vector, and $\mathbf{x} = (1,2,3) \in \mathbb{R}^3$ is a vector for which we do not care if it is a row or column. If the shape of a vector matters (e.g., if we multiply it by a matrix) and we give no indication otherwise, it is assumed to be a *column* vector from now on.

Remark 1.3.2

Why is Matrix Multiplication So Weird?

At this point, it seems natural to wonder why matrix multiplication is defined in such a seemingly bizarre way. After all, it is not even commutative, so why not just multiply matrices entrywise, similar to how we add them? Such an operation would certainly be much simpler.

The usual answer to this question is that matrix multiplication can be used as a function $\mathbf{v} \mapsto A\mathbf{v}$ that moves vectors around $\mathbb{R}^n$. We introduce this idea in more depth in Section 1.4, and this connection between matrices and functions of vectors is one of the central themes of linear algebra.

1.3.3   The Transpose

We now introduce an operation on matrices that changes the shape of a matrix, but not its contents. Specifically, it swaps the role of the rows and columns of a matrix:

Definition 1.3.3

The Transpose

Suppose $A \in \mathcal{M}_{m,n}$ is an $m \times n$ matrix. Then its **transpose**, which we denote by A^T, is the $n \times m$ matrix whose (i, j)-entry is $a_{j,i}$.

Another way of thinking about the transpose is as reflecting the entries of A across its main diagonal:

$$A = \begin{bmatrix} 1 & 2 & 3 \\ 4 & 5 & 6 \end{bmatrix} \quad \Longrightarrow \quad A^T = \begin{bmatrix} 1 & 4 \\ 2 & 5 \\ 3 & 6 \end{bmatrix}.$$

Example 1.3.5

Numerical Examples of the Transpose

Compute each of the indicated matrices if

$$A = \begin{bmatrix} 1 & 2 \\ 3 & 4 \end{bmatrix} \quad \text{and} \quad B = \begin{bmatrix} -1 & 1 & 1 \\ 0 & 1 & 0 \end{bmatrix}.$$

a) A^T,
b) B^T,
c) $(AB)^T$, and
d) $B^T A^T$.

Solutions:

a) $A^T = \begin{bmatrix} 1 & 3 \\ 2 & 4 \end{bmatrix}.$

b) $B^T = \begin{bmatrix} -1 & 0 \\ 1 & 1 \\ 1 & 0 \end{bmatrix}$.

c) $(AB)^T = \left(\begin{bmatrix} 1 & 2 \\ 3 & 4 \end{bmatrix} \begin{bmatrix} -1 & 1 & 1 \\ 0 & 1 & 0 \end{bmatrix} \right)^T = \begin{bmatrix} -1 & 3 & 1 \\ -3 & 7 & 3 \end{bmatrix}^T = \begin{bmatrix} -1 & -3 \\ 3 & 7 \\ 1 & 3 \end{bmatrix}$.

d) $B^T A^T = \begin{bmatrix} -1 & 0 \\ 1 & 1 \\ 1 & 0 \end{bmatrix} \begin{bmatrix} 1 & 3 \\ 2 & 4 \end{bmatrix} = \begin{bmatrix} -1 & -3 \\ 3 & 7 \\ 1 & 3 \end{bmatrix}$.

Parts (c) and (d) of the previous example suggest that $(AB)^T = B^T A^T$ might be a general rule about matrix transposes. The following theorem shows that this is indeed the case (and also establishes a few other less surprising properties of the transpose).

Theorem 1.3.4

Properties of the Transpose

Let A and B be matrices with sizes such that the operations below make sense and let $c \in \mathbb{R}$ be a scalar. Then

a) $(A^T)^T = A$,
b) $(A + B)^T = A^T + B^T$,
c) $(AB)^T = B^T A^T$, and
d) $(cA)^T = cA^T$.

Proof. Properties (a), (b), and (d) are all fairly intuitive, so we leave their proofs to Exercise 1.3.23. To prove that property (c) holds, we compute the (i,j)-entry of $(AB)^T$ and $B^T A^T$:

Recall that $[(AB)^T]_{i,j}$ means the (i,j)-entry of $(AB)^T$.

$$[(AB)^T]_{i,j} = [AB]_{j,i} = a_{j,1}b_{1,i} + a_{j,2}b_{2,i} + \cdots + a_{j,n}b_{n,i},$$

whereas

$$[B^T A^T]_{i,j} = [B^T]_{i,1}[A^T]_{1,j} + [B^T]_{i,2}[A^T]_{2,j} + \cdots + [B^T]_{i,n}[A^T]_{n,j}$$
$$= b_{1,i}a_{j,1} + b_{2,i}a_{j,2} + \cdots + b_{n,i}a_{j,n}.$$

Since these two quantities are the same, we conclude that $[(AB)^T]_{i,j} = [B^T A^T]_{i,j}$ for all i and j, so $(AB)^T = B^T A^T$, as desired. $\blacksquare$

The transpose also has the useful property that it converts a column vector into the corresponding row vector, and vice-versa. Furthermore, if $\mathbf{v}, \mathbf{w} \in \mathbb{R}^n$ are column vectors then we can use our usual matrix multiplication rule to see that

Strictly speaking, $\mathbf{v}^T \mathbf{w}$ is a 1×1 matrix, whereas $\mathbf{v} \cdot \mathbf{w}$ is a scalar, but these are "essentially" the same thing, so we do not care about the distinction.

$$\mathbf{v}^T \mathbf{w} = \begin{bmatrix} v_1 & v_2 & \cdots & v_n \end{bmatrix} \begin{bmatrix} w_1 \\ w_2 \\ \vdots \\ w_n \end{bmatrix} = v_1 w_1 + v_2 w_2 + \cdots + v_n w_n = \mathbf{v} \cdot \mathbf{w}.$$

In other words, we can use matrix multiplication to recover the dot product (which is not surprising, since we defined matrix multiplication via taking all possible dot products of rows and columns of the matrices being multiplied).

1.3.4 Block Matrices

There are often patterns in the entries of a large matrix, and it might be useful to break that large matrix down into smaller chunks based on some partition of its rows and columns. For example, if

$$A = \begin{bmatrix} 1 & 0 & 0 & 1 & 0 & 0 \\ 0 & 1 & 0 & 0 & 1 & 0 \\ 0 & 0 & 1 & 0 & 0 & 1 \\ 0 & 0 & 0 & 2 & 1 & -1 \\ 0 & 0 & 0 & 0 & -2 & 3 \end{bmatrix} \quad \text{and} \quad B = \begin{bmatrix} 1 & 2 & 0 & 0 \\ 2 & 1 & 0 & 0 \\ -1 & 1 & 0 & 0 \\ 0 & 0 & 1 & 2 \\ 0 & 0 & 2 & 1 \\ 0 & 0 & -1 & 1 \end{bmatrix}$$

Blocks are sometimes called **submatrices**.

then we can break A and B down into **blocks** that are individually rather simple as follows:

The vertical and horizontal bars in A and B do not have any mathematical meaning—they just help us visualize the different matrix blocks.

$$A = \left[\begin{array}{ccc|ccc} 1 & 0 & 0 & 1 & 0 & 0 \\ 0 & 1 & 0 & 0 & 1 & 0 \\ 0 & 0 & 1 & 0 & 0 & 1 \\ \hline 0 & 0 & 0 & 2 & 1 & -1 \\ 0 & 0 & 0 & 0 & -2 & 3 \end{array} \right] \quad \text{and} \quad B = \left[\begin{array}{cc|cc} 1 & 2 & 0 & 0 \\ 2 & 1 & 0 & 0 \\ -1 & 1 & 0 & 0 \\ \hline 0 & 0 & 1 & 2 \\ 0 & 0 & 2 & 1 \\ 0 & 0 & -1 & 1 \end{array} \right].$$

We then recognize two of the blocks of A as I_3 (the 3×3 identity matrix) and some of the blocks of A and B as the zero matrix O. If we call the remaining blocks

$$C = \begin{bmatrix} 2 & 1 & -1 \\ 0 & -2 & 3 \end{bmatrix} \quad \text{and} \quad D = \begin{bmatrix} 1 & 2 \\ 2 & 1 \\ -1 & 1 \end{bmatrix},$$

We do not need to specify the sizes of the zero matrices O here, since they are determined by the sizes of the other blocks.

then we can write A and B in terms of C and D as follows:

$$A = \begin{bmatrix} I_3 & I_3 \\ O & C \end{bmatrix} \quad \text{and} \quad B = \begin{bmatrix} D & O \\ O & D \end{bmatrix}.$$

When A and B are written in this way, as matrices whose entries are themselves matrices, they are called **block matrices**. The remarkable thing about block matrices is that we can multiply them just like regular matrices. For example, if we want to compute the matrix AB, instead of computing the product the long way (i.e., computing $4 \cdot 5 = 20$ dot products of 6-dimensional vectors), we can simply multiply them as 2×2 block matrices:

Be careful: order matters when multiplying block matrices. The bottom-right block is CD, not DC.

$$\begin{aligned} AB &= \begin{bmatrix} I_3 & I_3 \\ O & C \end{bmatrix} \begin{bmatrix} D & O \\ O & D \end{bmatrix} \\ &= \begin{bmatrix} I_3 D + I_3 O & I_3 O + I_3 D \\ O D + C O & O O + C D \end{bmatrix} = \begin{bmatrix} D & D \\ O & CD \end{bmatrix}. \end{aligned}$$

It is not difficult to compute $CD = \begin{bmatrix} 5 & 4 \\ -7 & 1 \end{bmatrix}$, so it follows that

$$AB = \begin{bmatrix} D & D \\ O & CD \end{bmatrix} = \left[\begin{array}{cc|cc} 1 & 2 & 1 & 2 \\ 2 & 1 & 2 & 1 \\ -1 & 1 & -1 & 1 \\ \hline 0 & 0 & 5 & 4 \\ 0 & 0 & -7 & 1 \end{array}\right],$$

which is the same answer that we would have gotten if we multiplied A and B directly. In general, multiplying block matrices like this is valid (a fact that we prove explicitly in Appendix B.1) as long as we choose the sizes of the blocks so that each and every matrix multiplication being performed makes sense, in which case we say that the block matrices have **conformable partitions**.

Example 1.3.6

Numerical Examples (and Non-Examples) of Block Matrix Multiplication

Suppose that

$$A = \begin{bmatrix} 1 & 2 \\ 3 & 4 \end{bmatrix} \quad \text{and} \quad B = \begin{bmatrix} -1 & 1 & 1 \\ 0 & 1 & 0 \end{bmatrix}.$$

Either compute each of the following block matrix products, or explain why they do not make sense:

a) $\begin{bmatrix} A & B \\ B & A \end{bmatrix}^2$,

b) $\begin{bmatrix} A & B \\ O & I_3 \end{bmatrix} \begin{bmatrix} A & A \\ O & A \\ I_2 & O \end{bmatrix}$,

Recall that I_2 and I_3 refer to the 2×2 and 3×3 identity matrices, respectively. For example,
$$I_2 = \begin{bmatrix} 1 & 0 \\ 0 & 1 \end{bmatrix}.$$

c) $\begin{bmatrix} A & B \\ O & I_3 \end{bmatrix} \begin{bmatrix} A & A \\ O & A \end{bmatrix}$, and

d) $\begin{bmatrix} A & B \\ O & I_3 \end{bmatrix} \begin{bmatrix} B & O \\ I_3 & I_3 \end{bmatrix}$.

Solutions:

a) The block matrix $\begin{bmatrix} A & B \\ B & A \end{bmatrix}$ itself makes no sense, since A has two columns but B has 3 columns. Written out explicitly, it would have the form

$$\begin{bmatrix} A & B \\ B & A \end{bmatrix} = \left[\begin{array}{ccc|ccc} 1 & 2 & & -1 & 1 & 1 \\ 3 & 4 & & 0 & 1 & 0 \\ \hline -1 & 1 & 1 & & 1 & 2 \\ 0 & 1 & 0 & & 3 & 4 \end{array}\right],$$

which is not a valid matrix since the columns do not match up.

b) The individual block matrices exist, but we cannot multiply them, since the first matrix has 2 block columns and the second matrix has 3 block rows (the inner dimensions must be the same).

We can also add block matrices in the obvious way (i.e., just add up the blocks in the same positions), as long as the blocks being added have the same sizes.

c) The individual block matrices exist and they have sizes appropriate for multiplying (the inner dimensions are both 2), but there is another problem this time: if we perform the block multiplication, we get

$$\begin{bmatrix} A & B \\ O & I_3 \end{bmatrix} \begin{bmatrix} A & A \\ O & A \end{bmatrix} = \begin{bmatrix} A^2 & A^2 + BA \\ O & A \end{bmatrix},$$

which makes no sense since the product BA is not defined (B has 3 columns, but A has 2 rows).

d) Finally, this is a block matrix multiplication that actually makes sense:

$$\begin{bmatrix} A & B \\ O & I_3 \end{bmatrix} \begin{bmatrix} B & O \\ I_3 & I_3 \end{bmatrix} = \begin{bmatrix} AB+B & B \\ I_3 & I_3 \end{bmatrix} = \left[\begin{array}{ccc|ccc} -2 & 4 & 2 & -1 & 1 & 1 \\ -3 & 8 & 3 & 0 & 1 & 0 \\ 1 & 0 & 0 & 1 & 0 & 0 \\ 0 & 1 & 0 & 0 & 1 & 0 \\ 0 & 0 & 1 & 0 & 0 & 1 \end{array} \right].$$

The computation of the top-left block $AB+B$ did take some effort, which we leave to the reader.

Sometimes, partitioning matrices in multiple different ways can lead to new insights about how matrix multiplication works. For example, although we defined matrix multiplication via every entry of AB being the dot product of the corresponding row of A and column of B, we can use block matrix multiplication to come up with some additional (equivalent) characterizations of matrix multiplication. For example, our first result in this direction tells us that the matrix-vector product $A\mathbf{v}$ is a linear combination of the columns of A:

Theorem 1.3.5

Matrix-Vector Multiplication

Suppose $A \in \mathcal{M}_{m,n}$ has columns $\mathbf{a}_1, \mathbf{a}_2, \ldots, \mathbf{a}_n$ (in that order) and $\mathbf{v} \in \mathbb{R}^n$ is a column vector. Then

$$A\mathbf{v} = v_1\mathbf{a}_1 + v_2\mathbf{a}_2 + \cdots + v_n\mathbf{a}_n.$$

Proof. We simply perform block matrix multiplication:

$$A\mathbf{v} = \begin{bmatrix} \mathbf{a}_1 \mid \mathbf{a}_2 \mid \cdots \mid \mathbf{a}_n \end{bmatrix} \begin{bmatrix} v_1 \\ v_2 \\ \vdots \\ v_n \end{bmatrix} = v_1\mathbf{a}_1 + v_2\mathbf{a}_2 + \cdots + v_n\mathbf{a}_n,$$

where the final equality comes from thinking of A as a $1 \times n$ block matrix and $\mathbf{v}$ as an $n \times 1$ matrix. ∎

Of course, we could just directly compute $A\mathbf{v}$ from the definition of matrix multiplication, but it is convenient to have multiple different ways of thinking about and computing the same thing. In a similar vein, we can use block matrix multiplication to see that the matrix multiplication AB acts column-wise on the matrix B:

Theorem 1.3.6

Column-Wise Form of Matrix Multiplication

Suppose $A \in \mathcal{M}_{m,n}$ and $B \in \mathcal{M}_{n,p}$ are matrices. If B has columns $\mathbf{b}_1, \mathbf{b}_2, \ldots, \mathbf{b}_p$ (in that order), then

$$AB = \begin{bmatrix} A\mathbf{b}_1 \mid A\mathbf{b}_2 \mid \cdots \mid A\mathbf{b}_p \end{bmatrix}.$$

Proof. Again, we simply perform block matrix multiplication:

$$AB = A\begin{bmatrix} \mathbf{b}_1 \mid \mathbf{b}_2 \mid \cdots \mid \mathbf{b}_p \end{bmatrix} = \begin{bmatrix} A\mathbf{b}_1 \mid A\mathbf{b}_2 \mid \cdots \mid A\mathbf{b}_p \end{bmatrix},$$

where the final equality comes from thinking of A as a 1×1 block matrix and B as a $1 \times p$ block matrix. ∎

In other words, the matrix AB is exactly the matrix that is obtained by multiplying each of the columns of B on the left by A. Similarly, the matrix multiplication AB can also be thought of as multiplying each of the rows of A on the right by B (see Exercise 1.3.26). Furthermore, combining Theorems 1.3.5 and 1.3.6 shows that the columns of AB are linear combinations of the columns of A (and similarly, the rows of AB are linear combinations of the rows of B—see Exercise 1.3.27).

Example 1.3.7

Multiple Methods of Matrix Multiplication

Suppose $A = \begin{bmatrix} 1 & 2 \\ 3 & 4 \end{bmatrix}$ and $B = \begin{bmatrix} -1 & 1 & 1 \\ 0 & 1 & 0 \end{bmatrix}$. Compute AB via

a) the definition of matrix multiplication, and
b) Theorem 1.3.6.

Solutions:

For part (a), we compute the matrix product using the same method as Example 1.3.2.

a) If we perform matrix multiplication in the usual way, we get

$$AB = \begin{bmatrix} 1 & 2 \\ 3 & 4 \end{bmatrix} \begin{bmatrix} -1 & 1 & 1 \\ 0 & 1 & 0 \end{bmatrix}$$

$$= \begin{bmatrix} (1,2) \cdot (-1,0) & (1,2) \cdot (1,1) & (1,2) \cdot (1,0) \\ (3,4) \cdot (-1,0) & (3,4) \cdot (1,1) & (3,4) \cdot (1,0) \end{bmatrix}$$

$$= \begin{bmatrix} -1 & 3 & 1 \\ -3 & 7 & 3 \end{bmatrix}.$$

b) The columns of B are

$$\mathbf{b}_1 = \begin{bmatrix} -1 \\ 0 \end{bmatrix}, \quad \mathbf{b}_2 = \begin{bmatrix} 1 \\ 1 \end{bmatrix}, \quad \text{and} \quad \mathbf{b}_3 = \begin{bmatrix} 1 \\ 0 \end{bmatrix}.$$

Multiplying each of these columns by A gives

$$A\mathbf{b}_1 = \begin{bmatrix} -1 \\ -3 \end{bmatrix}, \quad A\mathbf{b}_2 = \begin{bmatrix} 3 \\ 7 \end{bmatrix}, \quad \text{and} \quad A\mathbf{b}_3 = \begin{bmatrix} 1 \\ 3 \end{bmatrix}.$$

Finally, placing these column into a matrix gives us exactly AB:

$$AB = \begin{bmatrix} -1 & 3 & 1 \\ -3 & 7 & 3 \end{bmatrix}.$$

Exercises

solutions to starred exercises on page 437

1.3.1 Suppose

$$A = \begin{bmatrix} 2 & -1 \\ 0 & 3 \end{bmatrix} \quad \text{and} \quad B = \begin{bmatrix} -1 & 1 \\ -2 & 0 \end{bmatrix}.$$

Compute the following matrices:

*(a) $A + B$

(b) $2B - 3A$

*(c) $A - B$

(d) $2(A+B) - A$

1.3.2 Let A and B be as in Exercise 1.3.1, and let

$$C = \begin{bmatrix} 1 & 2 & -1 \\ 1 & 0 & 1 \end{bmatrix} \quad \text{and} \quad D = \begin{bmatrix} 0 & -2 & 1 \\ 2 & 2 & 0 \end{bmatrix}.$$

Compute the following matrices:

*(a) AB

(b) CD^T

*(c) AC

(d) $C^T D$

*(e) A^2

(f) $(C^T D)^2$

*(g) BD

(h) $(C-D)^T (A - B^T)$

*(i) $(A+B)(C+D)$

(j) $D^T (A^T + B)^T C$

1.3.3 Determine which of the following statements are true and which are false.

*(a) If A and B are matrices such that AB is square, then A and B must be square.

(b) The matrices A^TA and AA^T are always defined, regardless of the size of A.

*(c) The matrices A^TA and AA^T are always square, regardless of the size of A.

(d) If A and B are square matrices of the same size then $(A+B)^2 = A^2 + 2AB + B^2$.

*(e) If A and B are square matrices of the same size then $(AB)^2 = A^2B^2$.

(f) If A and B are matrices such that $AB = O$ and $A \neq O$, then $B = O$.

*(g) If A is a square matrix for which $A^2 = I$, then $A = I$ or $A = -I$.

(h) If A is a square matrix for which $A^2 = A$, then either $A = I$ or $A = O$.

*(i) If $A \in \mathcal{M}_{n,m}$ and $\mathbf{b} \in \mathbb{R}^m$ is a column vector, then $A\mathbf{b}$ is a linear combination of the rows of A.

1.3.4 Suppose $A \in \mathcal{M}_{2,2}, B \in \mathcal{M}_{2,5}$, and $C \in \mathcal{M}_{5,2}$ are matrices. Determine which of the following expressions do and do not make sense. If they do make sense, what is the size of the resulting matrix?

*(a) AB (b) A^7
*(c) AC (d) B^2
*(e) $A - B$ (f) BB^T
*(g) ABC (h) $B^TB + C^TC$
*(i) $A + BC$ (j) $(A + C^TC)(B + C^T)$

1.3.5 Let $A = \begin{bmatrix} 0 & 1 \\ -1 & 0 \end{bmatrix}$. Compute A^{1000}.

[Hint: Start by computing $A^2, A^3, \ldots$, and see if you notice a pattern.]

1.3.6 Let $B = \begin{bmatrix} 1 & 2 \\ 2 & -1 \end{bmatrix}$. Compute B^{1000}.

[Hint: Start by computing $B^2, B^3, \ldots$, and see if you notice a pattern.]

*1.3.7 Let $A = \begin{bmatrix} 1 & 1 \\ 0 & 1 \end{bmatrix}$.

(a) Compute A^2.
(b) Compute A^3.
(c) Find a general formula for A^k (where $k \geq 0$ is an integer).

1.3.8 Let $B = \begin{bmatrix} 1 & 1 & 1 \\ 0 & 1 & 1 \\ 0 & 0 & 1 \end{bmatrix}$.

(a) Compute B^2.
(b) Compute B^3.
(c) Find a general formula for B^k (where $k \geq 0$ is an integer).

1.3.9 Let $C = \begin{bmatrix} 1 & -1 \\ -1 & 1 \end{bmatrix}$.

(a) Compute C^2.
(b) Compute C^3.
(c) Find a general formula for C^k (where $k \geq 0$ is an integer) and prove that this formula is correct. [Hint: Use induction.]

1.3.10 Let $h \in \mathbb{R}$ and
$$A = \begin{bmatrix} 0.8 & 0.6 \\ 0.2 & h \end{bmatrix}.$$

Use computer software to compute A^{2500} when $h = 0.39$, $h = 0.40$, and $h = 0.41$. What do you think happens as you take larger and larger powers of A in each of these three cases? What do you think is special about $h = 0.40$ that leads to this change in behavior?

*1.3.11 Suppose J_n is the $n \times n$ matrix all of whose entries are 1. For example,
$$J_2 = \begin{bmatrix} 1 & 1 \\ 1 & 1 \end{bmatrix} \quad \text{and} \quad J_3 = \begin{bmatrix} 1 & 1 & 1 \\ 1 & 1 & 1 \\ 1 & 1 & 1 \end{bmatrix}.$$

Compute J_n^2.

1.3.12 Suppose $A \in \mathcal{M}_{2,2}, B \in \mathcal{M}_{2,3}$, and $C \in \mathcal{M}_{3,4}$ are matrices. Determine which of the following block matrix multiplications do and do not make sense.

*(a) $\begin{bmatrix} A & A \\ A & A \end{bmatrix}^2$

(b) $\begin{bmatrix} C & I_3 \\ I_4 & O \end{bmatrix} \begin{bmatrix} O & I_3 \\ I_4 & C^T \end{bmatrix}$

*(c) $\begin{bmatrix} A & B \\ B^T & I_3 \end{bmatrix} \begin{bmatrix} O & C^T \\ B^T & I_3 \end{bmatrix}$

(d) $\begin{bmatrix} A & B \\ O & C^T \end{bmatrix} \begin{bmatrix} I_2 & B \\ B^T & O \end{bmatrix}$

1.3.13 Compute the product AB using block matrix multiplication, using the partitioning of A and B indicated.

*(a) $A = \begin{bmatrix} 1 & 0 & 1 & 0 \\ 0 & 1 & 0 & 1 \end{bmatrix}$, $B = \begin{bmatrix} 1 & 2 & 2 & 1 \\ 2 & 1 & 1 & 2 \\ 2 & 1 & 1 & 2 \\ 1 & 2 & 2 & 1 \end{bmatrix}$

(b) $A = \begin{bmatrix} 1 & 2 & 0 \\ 3 & 1 & 0 \\ 0 & 0 & 2 \end{bmatrix}$, $B = \begin{bmatrix} -1 & 1 & 0 & 0 \\ 2 & 2 & 0 & 0 \\ 0 & 0 & 1 & 1 \end{bmatrix}$

*(c) $A = \begin{bmatrix} 1 & 0 & 0 & 1 \\ 0 & 1 & 0 & 1 \\ 0 & 0 & 1 & 1 \\ 0 & 0 & 0 & 2 \end{bmatrix}$, $B = \begin{bmatrix} 1 & 2 & 3 \\ 2 & 3 & 4 \\ 3 & 4 & 5 \\ 1 & 2 & 3 \end{bmatrix}$

1.3.14 Consider the 7×7 matrix

$$A = \begin{bmatrix} 1 & 0 & 0 & 0 & 1 & 1 & 1 \\ 0 & 1 & 0 & 0 & 1 & 1 & 1 \\ 0 & 0 & 1 & 0 & 1 & 1 & 1 \\ 0 & 0 & 0 & 1 & 1 & 1 & 1 \\ 1 & 1 & 1 & 1 & 0 & 0 & 0 \\ 1 & 1 & 1 & 1 & 0 & 0 & 0 \\ 1 & 1 & 1 & 1 & 0 & 0 & 0 \end{bmatrix}.$$

Compute A^2. [Hint: Partition A as a block matrix to save yourself a lot of work.]

⁎⁎1.3.15 Suppose $A \in \mathcal{M}_{m,n}$ and $B \in \mathcal{M}_{r,p}$ are matrices.

(a) What restrictions must be placed on m, n, r, and p to ensure that both of the products AB and BA exist?

(b) What additional restrictions must be placed on m, n, r, and p to ensure that AB and BA have the same size as each other?

1.3.16 Let $A, B \in \mathcal{M}_{m,n}$. Show that $\mathbf{v}^T A \mathbf{w} = \mathbf{v}^T B \mathbf{w}$ for all $\mathbf{v} \in \mathbb{R}^m$ and $\mathbf{w} \in \mathbb{R}^n$ if and only if $A = B$.

[Hint: What is $\mathbf{e}_i^T A \mathbf{e}_j$?]

⁎⁎1.3.17 Suppose that $A \in \mathcal{M}_{m,n}$.

(a) Show that $\mathbf{x} \cdot (A\mathbf{y}) = (A^T \mathbf{x}) \cdot \mathbf{y}$ for all $\mathbf{x} \in \mathbb{R}^m$ and $\mathbf{y} \in \mathbb{R}^n$. [Side note: This is actually why we care about the transpose in the first place—it gives us a way of moving a matrix around in a dot product.]

(b) Show that if $\mathbf{x} \cdot (A\mathbf{y}) = (B\mathbf{x}) \cdot \mathbf{y}$ for all $\mathbf{x} \in \mathbb{R}^m$ and $\mathbf{y} \in \mathbb{R}^n$ then $B = A^T$.

⁎⁎1.3.18 Let $\mathbf{e}_i$ be the standard basis vector with 1 in its i-th entry and 0 in all other entries, and let $A, B \in \mathcal{M}_{m,n}$.

(a) Show that $A\mathbf{e}_i$ is the i-th column of A.

(b) Show that $\mathbf{e}_i^T A$ is the i-th row of A.

(c) Use parts (a) and (b), together with Theorem 1.3.6, to give an alternate proof of Theorem 1.3.3.

(d) Use part (a) to show that $A\mathbf{v} = B\mathbf{v}$ for all $\mathbf{v} \in \mathbb{R}^n$ if and only if $A = B$.

⁎⁎1.3.19 Recall Theorem 1.3.1, which established some of the basic properties of matrix addition and scalar multiplication.

(a) Prove part (b) of the theorem.

(b) Prove part (c) of the theorem.

(c) Prove part (d) of the theorem.

(d) Prove part (e) of the theorem.

⁎⁎1.3.20 Recall Theorem 1.3.2, which established some of the basic properties of matrix multiplication.

(a) Prove part (a) of the theorem.

(b) Prove part (c) of the theorem.

(c) Prove part (d) of the theorem.

⁎⁎1.3.21 Prove the second equality of Theorem 1.3.3. That is, show that if $A \in \mathcal{M}_{m,n}$ then $I_m A = A$.

⁎⁎1.3.22 Suppose $A \in \mathcal{M}_n$ and k and ℓ are non-negative integers.

(a) Show that $A^{k+\ell} = A^k A^\ell$.

(b) Show that $\left(A^k\right)^\ell = A^{k\ell}$.

⁎⁎1.3.23 Recall Theorem 1.3.4, which established some of the basic properties of the transpose.

(a) Prove part (a) of the theorem.

(b) Prove part (b) of the theorem.

(c) Prove part (d) of the theorem.

1.3.24 Suppose that $A_1, A_2, \ldots, A_k$ are matrices whose sizes are such that the product $A_1 A_2 \ldots A_k$ makes sense. Prove that $(A_1 A_2 \cdots A_k)^T = A_k^T \cdots A_2^T A_1^T$.

[Hint: Use induction.]

1.3.25 Suppose that $A \in \mathcal{M}_n$. Show that $(A^n)^T = (A^T)^n$ for all integers $n \geq 0$.

⁎⁎1.3.26 Suppose $A \in \mathcal{M}_{m,n}$ and $B \in \mathcal{M}_{n,p}$ are matrices, and let $\mathbf{a}_j^T$ be the j-th row of A. Show that

$$AB = \begin{bmatrix} \mathbf{a}_1^T B \\ \hline \mathbf{a}_2^T B \\ \hline \vdots \\ \hline \mathbf{a}_m^T B \end{bmatrix}.$$

[Hint: Try mimicking the proof of Theorem 1.3.6.]

⁎⁎1.3.27 Suppose $A \in \mathcal{M}_{m,n}$ and $B \in \mathcal{M}_{n,p}$ are matrices.

(a) Show that the columns of AB are linear combinations of the columns of A.

(b) Show that the rows of AB are linear combinations of the rows of B.

1.4 Linear Transformations

The final main ingredient of linear algebra, after vectors and matrices, are linear transformations: functions that act on vectors and that do not "mess up" vector addition and scalar multiplication. Despite being the most abstract and difficult object to grasp in this chapter, they are of paramount importance and permeate all of linear algebra, so the reader is encouraged to explore this section thoroughly.

Definition 1.4.1

Linear
Transformations

A **linear transformation** is a function $T : \mathbb{R}^n \to \mathbb{R}^m$ that satisfies the following two properties:

    a) $T(\mathbf{v}+\mathbf{w}) = T(\mathbf{v}) + T(\mathbf{w})$ for all vectors $\mathbf{v}, \mathbf{w} \in \mathbb{R}^n$, and

    b) $T(c\mathbf{v}) = cT(\mathbf{v})$ for all vectors $\mathbf{v} \in \mathbb{R}^n$ and all scalars $c \in \mathbb{R}$.

The notation $T : \mathbb{R}^n \to \mathbb{R}^m$ means that T is a function that sends vectors in $\mathbb{R}^n$ to vectors in $\mathbb{R}^m$. More generally, $f : X \to Y$ means that the function f sends members of the set X to the set Y.

For example, it follows fairly quickly from Theorem 1.3.2 that matrix multiplication is a linear transformation—if $A \in M_{m,n}$ then the function that sends $\mathbf{v} \in \mathbb{R}^n$ to $A\mathbf{v} \in \mathbb{R}^m$ preserves vector addition (i.e., $A(\mathbf{v}+\mathbf{w}) = A\mathbf{v}+A\mathbf{w}$) and scalar multiplication (i.e., $A(c\mathbf{v}) = c(A\mathbf{v})$).

Geometrically, linear transformations can be thought of as the functions that rotate, stretch, shrink, and/or reflect $\mathbb{R}^n$, but do so somewhat uniformly. For example, if we draw a square grid on $\mathbb{R}^2$ as in Figure 1.14, then a linear transformation $T : \mathbb{R}^2 \to \mathbb{R}^2$ can rotate, stretch, shrink, and/or reflect that grid, but it will still be made up of cells of the same size and shape (which in general will be parallelograms rather than squares). Furthermore, if a vector $\mathbf{v}$ is located in the x-th cell in the direction of $\mathbf{e}_1$ and the y-th cell in the direction of $\mathbf{e}_2$, then $T(\mathbf{v})$ is located in the x-th cell in the direction of $T(\mathbf{e}_1)$ and the y-th cell in the direction of $T(\mathbf{e}_2)$ (again, see Figure 1.14).

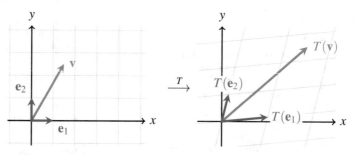

Figure 1.14: A linear transformation $T : \mathbb{R}^2 \to \mathbb{R}^2$ transforms a square grid into a grid made up of parallelograms, and it preserves which cell of the grid each vector is in (in this case, $\mathbf{v}$ is in the 2nd square to the right, 3rd up, and $T(\mathbf{v})$ is similarly in the 2nd parallelogram in the direction of $T(\mathbf{e}_1)$ and 3rd in the direction of $T(\mathbf{e}_2)$).

Alternatively, we can also think of linear transformations as the functions that preserve linear combinations. That is, a function $T : \mathbb{R}^n \to \mathbb{R}^m$ is a linear transformation if and only if

$$T(c_1\mathbf{v}_1 + c_2\mathbf{v}_2 + \cdots + c_k\mathbf{v}_k) = c_1 T(\mathbf{v}_1) + c_2 T(\mathbf{v}_2) + \cdots + c_k T(\mathbf{v}_k)$$

for all $\mathbf{v}_1, \mathbf{v}_2, \ldots, \mathbf{v}_k \in \mathbb{R}^n$ and all $c_1, c_2, \ldots, c_k \in \mathbb{R}$ (see Exercise 1.4.9).

Example 1.4.1

Determining if a
Transformation
is Linear

Determine whether or not the following functions $T : \mathbb{R}^2 \to \mathbb{R}^2$ are linear transformations. Also illustrate their effect on $\mathbb{R}^2$ geometrically.

    a) $T(v_1, v_2) = (1+v_1, 2+v_2)$,

    b) $T(v_1, v_2) = (v_1 - v_2, v_1 v_2)$, and

    c) $T(v_1, v_2) = (v_1 - v_2, v_1 + v_2)$.

In general, if T is a linear transformation then it must be the case that $T(\mathbf{0}) = \mathbf{0}$ (see Exercise 1.4.8).

Solutions:

    a) This transformation is *not* linear. One way to see this is to notice that $2T(0,0) = 2(1,2) = (2,4)$, but $T(2(0,0)) = T(0,0) = (1,2)$. Since these are not the same, T is not linear. The fact that T is not linear is illustrated by the fact that it translates $\mathbb{R}^2$, rather than just rotation, stretching, shrinking, and/or reflecting it:

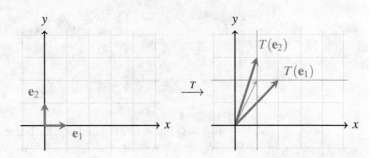

b) This transformation is *not* linear. One way to see this is to notice that $2T(1,1) = 2(0,1) = (0,2)$, but $T(2(1,1)) = T(2,2) = (0,4)$. Since these are not the same, T is not linear. Geometrically, we can see that T is not linear by the fact that it stretches and shrinks different parts of $\mathbb{R}^2$ by different amounts:

Any function that multiplies the entries of the input vector by each other is not linear. Functions like sin, cos, ln, and the absolute value also cannot be applied to entries of the vector in linear transformations.

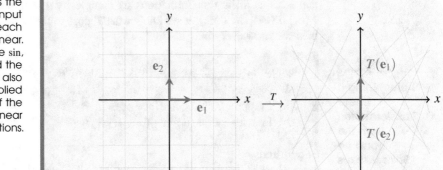

c) This transformation *is* linear. To see this, we check the two defining properties of linear transformations:

$$T(\mathbf{v}+\mathbf{w}) = T(v_1 + w_1, v_2 + w_2)$$
$$= ((v_1 + w_1) - (v_2 + w_2), (v_1 + w_1) + (v_2 + w_2))$$
$$= (v_1 - v_2, v_1 + v_2) + (w_1 - w_2, w_1 + w_2) = T(\mathbf{v}) + T(\mathbf{w}),$$

and

$$T(c\mathbf{v}) = T(cv_1, cv_2) = (cv_1 - cv_2, cv_1 + cv_2)$$
$$= c(v_1 - v_2, v_1 + v_2) = cT(\mathbf{v}).$$

Geometrically, we can see that this is a linear transformation since it just rotates and stretches $\mathbb{R}^2$ by a uniform amount. Specifically, T rotates $\mathbb{R}^2$ counter-clockwise by 45 degrees and stretches it by a factor of $\sqrt{2}$:

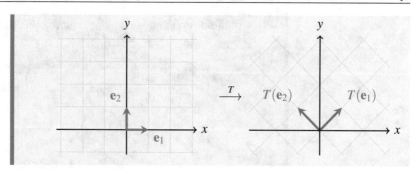

To justify our geometric intuition for linear transformations, first recall that for every vector $\mathbf{v} = (v_1, v_2, \ldots, v_n) \in \mathbb{R}^n$, we can write $\mathbf{v} = v_1\mathbf{e}_1 + v_2\mathbf{e}_2 + \cdots + v_n\mathbf{e}_n$, where $\mathbf{e}_1, \mathbf{e}_2, \ldots, \mathbf{e}_n$ are the standard basis vectors. By using the fact that linear transformations preserve linear combinations, we see that

> Recall that $\mathbf{e}_j$ is the vector with 1 in its j-th entry and 0 elsewhere.

$$T(\mathbf{v}) = T(v_1\mathbf{e}_1 + v_2\mathbf{e}_2 + \cdots + v_n\mathbf{e}_n) = v_1 T(\mathbf{e}_1) + v_2 T(\mathbf{e}_2) + \cdots + v_n T(\mathbf{e}_n),$$

which is exactly what we said before: if $\mathbf{v} \in \mathbb{R}^2$ extends a distance of v_1 in the direction of $\mathbf{e}_1$ and a distance of v_2 in the direction of $\mathbf{e}_2$, then $T(\mathbf{v})$ extends the same amounts in the directions of $T(\mathbf{e}_1)$ and $T(\mathbf{e}_2)$, respectively (and similarly for $\mathbf{e}_3, \ldots, \mathbf{e}_n$ in higher dimensions). We thus learn the very important fact that linear transformations are completely determined by the vectors $T(\mathbf{e}_1), T(\mathbf{e}_2), \ldots, T(\mathbf{e}_n)$: if we know what T does to the standard basis vectors, then we know everything about T.

Example 1.4.2

Determining a Linear Transformation from the Standard Basis Vectors

Suppose $T : \mathbb{R}^2 \to \mathbb{R}^2$ is a linear transformation for which $T(\mathbf{e}_1) = (1,1)$ and $T(\mathbf{e}_2) = (-1,1)$.

a) Compute $T(2,3)$, and
b) Find a general formula for $T(v_1, v_2)$.

Solutions:
a) Since $(2,3) = 2\mathbf{e}_1 + 3\mathbf{e}_2$, we know that $T(2,3) = T(2\mathbf{e}_1 + 3\mathbf{e}_2) = 2T(\mathbf{e}_1) + 3T(\mathbf{e}_2) = 2(1,1) + 3(-1,1) = (-1,5)$.

b) We can mimic the computation from part (a): $(v_1, v_2) = v_1\mathbf{e}_1 + v_2\mathbf{e}_2$, so

$$\begin{aligned}
T(v_1, v_2) &= T(v_1\mathbf{e}_1 + v_2\mathbf{e}_2) \\
&= v_1 T(\mathbf{e}_1) + v_2 T(\mathbf{e}_2) \\
&= v_1(1,1) + v_2(-1,1) = (v_1 - v_2, v_1 + v_2).
\end{aligned}$$

In other words, this is exactly the same as the linear transformation from Example 1.4.1(c).

1.4.1 Linear Transformations as Matrices

> If $A \in \mathcal{M}_{m,n}$ is a matrix, then a function that sends $\mathbf{v}$ to $A\mathbf{v}$ is sometimes called a **matrix transformation**.

One of the easiest ways to see that a function $T : \mathbb{R}^n \to \mathbb{R}^m$ is indeed a linear transformation is to find a matrix whose multiplication has the same effect as T. For example,

$$\text{if} \quad A = \begin{bmatrix} 1 & -1 \\ 1 & 1 \end{bmatrix} \quad \text{then} \quad A\mathbf{v} = \begin{bmatrix} 1 & -1 \\ 1 & 1 \end{bmatrix} \begin{bmatrix} v_1 \\ v_2 \end{bmatrix} = \begin{bmatrix} v_1 - v_2 \\ v_1 + v_2 \end{bmatrix},$$

which shows that multiplying a column vector by the matrix A has the same effect as the linear transformation $T : \mathbb{R}^2 \to \mathbb{R}^2$ defined by $T(v_1, v_2) = (v_1 - v_2, v_1 + v_2)$ (i.e., the linear transformation from Examples 1.4.1(c) and 1.4.2(b)). One of the most remarkable facts about linear transformations is that this procedure can always be carried out—every linear transformation can be represented via matrix multiplication, and there is a straightforward method for constructing a matrix that does the job:

Theorem 1.4.1 **Standard Matrix of a Linear Transformation**	A function $T : \mathbb{R}^n \to \mathbb{R}^m$ is a linear transformation if and only if there exists a matrix $[T] \in \mathcal{M}_{m,n}$ such that $$T(\mathbf{v}) = [T]\mathbf{v} \quad \text{for all} \quad \mathbf{v} \in \mathbb{R}^n.$$ Furthermore, the unique matrix $[T]$ with this property is called the **standard matrix** of T, and it is $$[T] \overset{\text{def}}{=} \big[\, T(\mathbf{e}_1) \mid T(\mathbf{e}_2) \mid \cdots \mid T(\mathbf{e}_n) \,\big].$$

In other words, $[T]$ is the matrix with the vectors $T(\mathbf{e}_1)$, $T(\mathbf{e}_2)$, ..., $T(\mathbf{e}_n)$ as its columns.

Proof. It follows immediately from Theorem 1.3.2 that if $[T] \in \mathcal{M}_{m,n}$ then the function that sends $\mathbf{v}$ to $[T]\mathbf{v}$ is a linear transformation. We thus only have to prove that for every linear transformation $T : \mathbb{R}^n \to \mathbb{R}^m$, the matrix

$$[T] = \big[T(\mathbf{e}_1) \mid T(\mathbf{e}_2) \mid \cdots \mid T(\mathbf{e}_n)\big]$$

satisfies $[T]\mathbf{v} = T(\mathbf{v})$, and no other matrix has this property.

To see that $[T]\mathbf{v} = T(\mathbf{v})$, we use the block matrix multiplication techniques that we learned in Section 1.3.4:

This theorem says that linear transformations and matrix transformations are the same thing.

$$[T]\mathbf{v} = \big[T(\mathbf{e}_1) \mid T(\mathbf{e}_2) \mid \cdots \mid T(\mathbf{e}_n)\big] \begin{bmatrix} v_1 \\ v_2 \\ \vdots \\ v_n \end{bmatrix} \quad \text{(definition of } [T])$$

$$= v_1 T(\mathbf{e}_1) + v_2 T(\mathbf{e}_2) + \cdots + v_n T(\mathbf{e}_n) \quad \text{(block matrix multiplication)}$$

$$= T(v_1\mathbf{e}_1 + v_2\mathbf{e}_2 + \cdots + v_n\mathbf{e}_n) \quad \text{(since } T \text{ is linear)}$$

$$= T(\mathbf{v}). \quad (\mathbf{v} = v_1\mathbf{e}_1 + v_2\mathbf{e}_2 + \cdots + v_n\mathbf{e}_n)$$

To verify that $[T]$ is unique, suppose that $A \in \mathcal{M}_{m,n}$ is *any* matrix with the property that $T(\mathbf{v}) = A\mathbf{v}$ for all $\mathbf{v} \in \mathbb{R}^n$. Then $T(\mathbf{v}) = [T]\mathbf{v}$ and $T(\mathbf{v}) = A\mathbf{v}$, so $[T]\mathbf{v} = A\mathbf{v}$ for all $\mathbf{v} \in \mathbb{R}^n$ as well. It follows from Exercise 1.3.18(d) that $A = [T]$, which completes the proof. ∎

Because of this one-to-one correspondence between matrices and linear transformations, they are often thought of as the same thing.

The above theorem says that there is a one-to-one correspondence between matrices and linear transformations—every matrix $A \in \mathcal{M}_{m,n}$ can be used to create a linear transformation $T : \mathbb{R}^n \to \mathbb{R}^m$ via $T(\mathbf{v}) = A\mathbf{v}$, and conversely every linear transformation $T : \mathbb{R}^n \to \mathbb{R}^m$ can be used to create its standard matrix $[T] \in \mathcal{M}_{m,n}$. This bridge between matrices and linear transformations is one of the most useful tools in all of linear algebra, as it lets us use geometric techniques (based on linear transformations) for dealing with matrices, and it lets us use algebraic techniques (based on matrices) for dealing with linear transformations.

Example 1.4.3

**Representing
Linear
Transformations
Via Matrices**

Find the standard matrix of the following linear transformations.
a) $T(v_1, v_2) = (v_1 + 2v_2, 3v_1 + 4v_2)$, and
b) $T(v_1, v_2, v_3) = (3v_1 - v_2 + v_3, 2v_1 + 4v_2 - 2v_3)$.

Solutions:
a) We use Theorem 1.4.1, which tells us to compute $T(\mathbf{e}_1) = (1, 3)$
   and $T(\mathbf{e}_2) = (2, 4)$ and place these as columns into a matrix, in that
   order:
   $$[T] = \begin{bmatrix} 1 & 2 \\ 3 & 4 \end{bmatrix}.$$

   Notice that the entries of $[T]$ are just the coefficients in the definition
   of T, read row-by-row. This always happens.

Be careful: if
$T : \mathbb{R}^n \to \mathbb{R}^m$ then its
standard matrix is
an $m \times n$ matrix; the
dimensions are the
opposite of what
we might expect
at first.

b) We could explicitly compute $T(\mathbf{e}_1)$, $T(\mathbf{e}_2)$, and $T(\mathbf{e}_3)$ and place
   them as columns in a matrix like we did in part (a), or we could
   simply place the coefficients of v_1, v_2, and v_3 in the output of T, in
   order, in the *rows* of a matrix, as we suggested at the end of part (a):
   $$[T] = \begin{bmatrix} 3 & -1 & 1 \\ 2 & 4 & -2 \end{bmatrix}.$$

   Notice that T maps from $\mathbb{R}^3$ to $\mathbb{R}^2$, so $[T]$ is a 2×3 matrix.

Example 1.4.4

**Representing
Matrices
Geometrically**

Interpret the following matrices as linear transformations and represent
them geometrically (in the same way that we represented linear transformations in Figure 1.14).

a) $A = \begin{bmatrix} 2.1 & 0.3 \\ 0.2 & 1.2 \end{bmatrix}$
b) $B = \begin{bmatrix} 2 & 1 \\ 1 & 2 \\ 1 & 1 \end{bmatrix}$

Solutions:
a) Since $A\mathbf{e}_1$ (i.e., the first column of A) equals $(2.1, 0.2)$ and $A\mathbf{e}_2$ (i.e.,
   its second column) equals $(0.3, 1.2)$, we can think of this matrix
   as the linear transformation that sends $\mathbf{e}_1$ and $\mathbf{e}_2$ to $(2.1, 0.2)$ and
   $(0.3, 1.2)$, respectively:

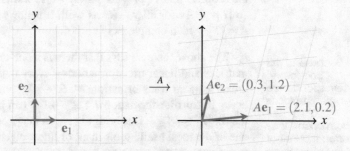

b) Since $B\mathbf{e}_1$ (i.e., the first column of B) equals $(2, 1, 1)$ and $B\mathbf{e}_2$ (i.e.,
   its second column) equals $(1, 2, 1)$, we can think of this matrix as
   the linear transformation that sends $\mathbf{e}_1$ and $\mathbf{e}_2$ from $\mathbb{R}^2$ to $(2, 1, 1)$
   and $(1, 2, 1)$, respectively, in $\mathbb{R}^3$:

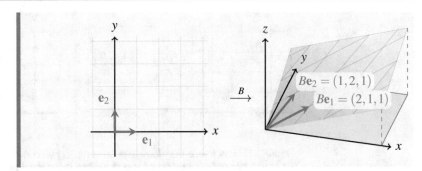

Be careful: B is 3×2, so it acts as a linear transformation that sends $\mathbb{R}^2$ to $\mathbb{R}^3$.

Remark 1.4.1

Modifying Linear Transformations

Just like we can add and subtract matrices and multiply them by scalars, we can also add and subtract linear transformations and multiply them by scalars. These operations are all defined in the obvious ways: if $S, T : \mathbb{R}^n \to \mathbb{R}^m$ and $c \in \mathbb{R}$ then $S + T : \mathbb{R}^n \to \mathbb{R}^m$ and $cT : \mathbb{R}^n \to \mathbb{R}^m$ are the linear transformations defined by

$$(S+T)(\mathbf{v}) = S(\mathbf{v}) + T(\mathbf{v}) \quad \text{and} \quad (cT)(\mathbf{v}) = cT(\mathbf{v}) \quad \text{for all} \quad \mathbf{v} \in \mathbb{R}^n,$$

respectively.

We do not spend any significant time discussing the properties of these operations because they are all fairly intuitive and completely analogous to the properties of addition and scalar multiplication of matrices. In particular, the standard matrix of $S + T$ is $[S + T] = [S] + [T]$ and similarly the standard matrix of cT is $[cT] = c[T]$ (see Exercise 1.4.11), so we can think of addition and scalar multiplication of linear transformations as the addition and scalar multiplication of their standard matrices.

1.4.2 A Catalog of Linear Transformations

In order to get more comfortable with the relationship between linear transformations and matrices, we construct the standard matrices of a few very geometrically-motivated linear transformations that come up frequently.

Geometrically, the identity transformation leaves $\mathbb{R}^n$ unchanged, while the zero transformation squishes it down to a single point at the origin.

The two simplest linear transformations that exist are the **zero transformation** $O : \mathbb{R}^n \to \mathbb{R}^m$, defined by $O(\mathbf{v}) = \mathbf{0}$ for all $\mathbf{v} \in \mathbb{R}^n$, and the **identity transformation** $I : \mathbb{R}^n \to \mathbb{R}^n$, defined by $I(\mathbf{v}) = \mathbf{v}$ for all $\mathbf{v} \in \mathbb{R}^n$. It is perhaps not surprising that the standard matrices of these transformations are the zero matrix and the identity matrix, respectively. To verify this claim, just notice that if $O \in \mathcal{M}_{m,n}$ and $I \in \mathcal{M}_n$ are the zero matrix and the identity matrix, then $O\mathbf{v} = \mathbf{0}$ and $I\mathbf{v} = \mathbf{v}$ for all $\mathbf{v} \in \mathbb{R}^n$ too.

Diagonal Matrices

The next simplest type of linear transformation $T : \mathbb{R}^n \to \mathbb{R}^n$ is one that does not change the direction of the standard basis vectors, but just stretches them by certain (possibly different) amounts, as in Figure 1.15. These linear transformations are the ones for which there exist scalars $c_1, c_2, \ldots, c_n \in \mathbb{R}^n$ such that $T(v_1, v_2, \ldots, v_n) = (c_1 v_1, c_2 v_2, \ldots, c_n v_n)$.

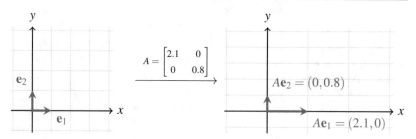

Figure 1.15: The linear transformation associated with a diagonal matrix stretches $\mathbb{R}^n$ in the direction of each of the standard basis vectors, but does not skew or rotate space. The transformation displayed here stretches the x-direction by a factor of 2.1 and stretches the y-direction by a factor of 0.8.

The standard matrix of this linear transformation is

$$[T] = \left[\, c_1\mathbf{e}_1 \mid c_2\mathbf{e}_2 \mid \cdots \mid c_n\mathbf{e}_n \,\right] = \begin{bmatrix} c_1 & 0 & \cdots & 0 \\ 0 & c_2 & \cdots & 0 \\ \vdots & \vdots & \ddots & \vdots \\ 0 & 0 & \cdots & c_n \end{bmatrix}.$$

An **off-diagonal** entry of a matrix A is an entry $a_{i,j}$ with $i \neq j$.

Matrices of this form (with all off-diagonal entries equal to 0) are called **diagonal matrices**, and they are useful because they are so much easier to work with than other matrices. For example, it is straightforward to verify that the product of two diagonal matrices is another diagonal matrix, and their multiplication just happens entrywise:

Similarly, a **diagonal linear transformation** is one whose standard matrix is diagonal.

$$\begin{bmatrix} c_1 & 0 & \cdots & 0 \\ 0 & c_2 & \cdots & 0 \\ \vdots & \vdots & \ddots & \vdots \\ 0 & 0 & \cdots & c_n \end{bmatrix} \begin{bmatrix} d_1 & 0 & \cdots & 0 \\ 0 & d_2 & \cdots & 0 \\ \vdots & \vdots & \ddots & \vdots \\ 0 & 0 & \cdots & d_n \end{bmatrix} = \begin{bmatrix} c_1d_1 & 0 & \cdots & 0 \\ 0 & c_2d_2 & \cdots & 0 \\ \vdots & \vdots & \ddots & \vdots \\ 0 & 0 & \cdots & c_nd_n \end{bmatrix}.$$

Slightly more generally, if D is a diagonal matrix and A is an arbitrary matrix, then the matrix multiplication DA simply multiplies each row of A by the corresponding diagonal entry of D. Similarly, the product AD multiplies each *column* of A by the corresponding diagonal entry of D (see Exercise 1.4.10). For example,

The identity and zero matrices are both diagonal matrices.

$$\begin{bmatrix} 1 & 0 & 0 \\ 0 & 2 & 0 \\ 0 & 0 & 3 \end{bmatrix} \begin{bmatrix} 1 & 1 & 1 \\ 1 & 1 & 1 \\ 1 & 1 & 1 \end{bmatrix} = \begin{bmatrix} 1 & 1 & 1 \\ 2 & 2 & 2 \\ 3 & 3 & 3 \end{bmatrix} \quad \text{and}$$

$$\begin{bmatrix} 1 & 1 & 1 \\ 1 & 1 & 1 \\ 1 & 1 & 1 \end{bmatrix} \begin{bmatrix} 1 & 0 & 0 \\ 0 & 2 & 0 \\ 0 & 0 & 3 \end{bmatrix} = \begin{bmatrix} 1 & 2 & 3 \\ 1 & 2 & 3 \\ 1 & 2 & 3 \end{bmatrix}.$$

In $\mathbb{R}^2$, we can imagine shining a light down perpendicular to the given line: the projection of $\mathbf{v}$ onto that line is the shadow cast by $\mathbf{v}$.

Projections

The next type of linear transformation that we look at is a **projection onto a line**. If we fix a particular line in $\mathbb{R}^n$ then the projection onto it sends each vector to a description of how much it points in the direction of that line. For example, the **projection onto the x-axis** sends every vector $\mathbf{v} = (v_1, v_2) \in \mathbb{R}^2$ to $(v_1, 0)$ and the **projection onto the y-axis** sends them to $(0, v_2)$, as in Figure 1.16(a).

More generally, given any vector $\mathbf{v}$, we can draw a right-angled triangle with $\mathbf{v}$ as the hypotenuse and one of its legs on the given line. Then the projection onto that line is the function that sends $\mathbf{v}$ to this leg of the triangle, as in Figure 1.16(b). To make this projection easier to talk about mathematically, we let $\mathbf{u}$ be a unit vector on the line that we are projecting onto (recall that we think of unit vectors as specifying directions) and let $P_{\mathbf{u}}$ denote the projection onto the line in the direction of $\mathbf{u}$.

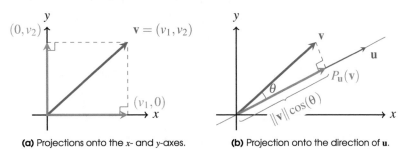

(a) Projections onto the x- and y-axes. **(b)** Projection onto the direction of $\mathbf{u}$.

Figure 1.16: The projection of a vector $\mathbf{v}$ onto the line in the direction of $\mathbf{u}$ is denoted by $P_{\mathbf{u}}(\mathbf{v})$, and it points a distance of $\|\mathbf{v}\|\cos(\theta)$ in the direction of $\mathbf{u}$, where θ is the angle between $\mathbf{v}$ and $\mathbf{u}$.

To see that $P_{\mathbf{u}}$ is indeed a linear transformation, and to find a formula for computing $P_{\mathbf{u}}(\mathbf{v})$, we use many of the same geometric techniques that we used in previous sections. Since $\mathbf{v}$ and $P_{\mathbf{u}}(\mathbf{v})$ form the hypotenuse and leg of a right-angled triangle, respectively, we know that $P_{\mathbf{u}}(\mathbf{v})$ points a distance of $\|\mathbf{v}\|\cos(\theta)$ in the direction of $\mathbf{u}$, where θ is the angle between $\mathbf{v}$ and $\mathbf{u}$. In other words, $P_{\mathbf{u}}(\mathbf{v}) = \mathbf{u}(\|\mathbf{v}\|\cos(\theta))$.

We then recall from Definition 1.2.3 that $\cos(\theta) = \mathbf{u}\cdot\mathbf{v}/(\|\mathbf{u}\|\|\mathbf{v}\|)$. By plugging this into our formula for $P_{\mathbf{u}}(\mathbf{v})$ and using the fact that $\|\mathbf{u}\|=1$, we see that

$$P_{\mathbf{u}}(\mathbf{v}) = \mathbf{u}\big(\|\mathbf{v}\|(\mathbf{u}\cdot\mathbf{v})/\|\mathbf{v}\|\big) = \mathbf{u}(\mathbf{u}\cdot\mathbf{v}).$$

Be careful: $\mathbf{u}^T\mathbf{u}$ is a number (in particular, it equals $\mathbf{u}\cdot\mathbf{u} = \|\mathbf{u}\|^2 = 1$), but $\mathbf{u}\mathbf{u}^T$ is a matrix! Since $\mathbf{u}$ is $n\times 1$ and $\mathbf{u}^T$ is $1\times n$, $\mathbf{u}\mathbf{u}^T$ is $n\times n$.

Finally, if we recall from Section 1.3.3 that $\mathbf{u}\cdot\mathbf{v} = \mathbf{u}^T\mathbf{v}$, it follows that $P_{\mathbf{u}}(\mathbf{v}) = \mathbf{u}(\mathbf{u}^T\mathbf{v}) = (\mathbf{u}\mathbf{u}^T)\mathbf{v}$. In other words, $P_{\mathbf{u}}$ is indeed a linear transformation, and it has standard matrix

$$\big[P_{\mathbf{u}}\big] = \mathbf{u}\mathbf{u}^T.$$

It is worth noting that even though Figure 1.16 illustrates how projections work in $\mathbb{R}^2$, this calculation of $P_{\mathbf{u}}$'s standard matrix works the exact same way in $\mathbb{R}^n$, regardless of the dimension.

Example 1.4.5

Standard Matrices of Projections

Find the standard matrices of the linear transformations that project onto the lines in the direction of the following vectors, and depict these projections geometrically.
  a) $\mathbf{u} = (1,0) \in \mathbb{R}^2$, and
  b) $\mathbf{w} = (1,2,3) \in \mathbb{R}^3$.

Solutions:
  a) Since $\mathbf{u}$ is a unit vector, the standard matrix of $P_{\mathbf{u}}$ is simply

$$\big[P_{\mathbf{u}}\big] = \mathbf{u}\mathbf{u}^T = \begin{bmatrix} 1 \\ 0 \end{bmatrix}\begin{bmatrix} 1 & 0 \end{bmatrix} = \begin{bmatrix} 1 & 0 \\ 0 & 0 \end{bmatrix}.$$

We note that this agrees with observations that we made earlier, since $\mathbf{u}$ points in the direction of the x-axis, and multiplication by this matrix indeed has the same effect as the projection onto the x-axis:

$$[P_{\mathbf{u}}]\mathbf{v} = \begin{bmatrix} 1 & 0 \\ 0 & 0 \end{bmatrix} \begin{bmatrix} v_1 \\ v_2 \end{bmatrix} = \begin{bmatrix} v_1 \\ 0 \end{bmatrix}.$$

We can visualize this projection as just squashing everything down onto the x-axis (in particular, $[P_{\mathbf{u}}]\mathbf{e}_1 = \mathbf{e}_1$ and $[P_{\mathbf{u}}]\mathbf{e}_2 = \mathbf{0}$):

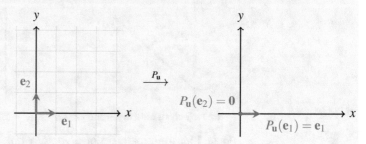

b) Since $\mathbf{w}$ is not a unit vector, we have to first normalize it (i.e., divide it by its length) to turn it into a unit vector pointing in the same direction. Well, $\|\mathbf{w}\| = \sqrt{1^2 + 2^2 + 3^2} = \sqrt{14}$, so we let $\mathbf{u} = \mathbf{w}/\|\mathbf{w}\| = (1,2,3)/\sqrt{14}$. The standard matrix of $P_{\mathbf{u}}$ is then

$$[P_{\mathbf{u}}] = \mathbf{u}\mathbf{u}^T = \frac{1}{14} \begin{bmatrix} 1 \\ 2 \\ 3 \end{bmatrix} \begin{bmatrix} 1 & 2 & 3 \end{bmatrix} = \frac{1}{14} \begin{bmatrix} 1 & 2 & 3 \\ 2 & 4 & 6 \\ 3 & 6 & 9 \end{bmatrix}.$$

This projection sends each of $\mathbf{e}_1$, $\mathbf{e}_2$, and $\mathbf{e}_3$ to different multiples of $\mathbf{w}$. In fact, it squishes all of $\mathbb{R}^3$ down onto the line in the direction of $\mathbf{w}$:

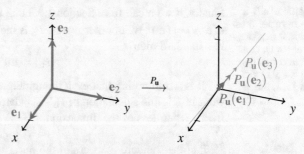

As a bit of a sanity check, we note that this matrix leaves $\mathbf{w}$ (and thus $\mathbf{u}$) unchanged, as the projection $P_{\mathbf{u}}$ should:

$$[P_{\mathbf{u}}]\mathbf{w} = \frac{1}{14} \begin{bmatrix} 1 & 2 & 3 \\ 2 & 4 & 6 \\ 3 & 6 & 9 \end{bmatrix} \begin{bmatrix} 1 \\ 2 \\ 3 \end{bmatrix} = \begin{bmatrix} 1 \\ 2 \\ 3 \end{bmatrix} = \mathbf{w}.$$

> In general, if $\mathbf{u} \in \mathbb{R}^n$ is a unit vector then $P_{\mathbf{u}}(\mathbf{u}) = \mathbf{u}$: $P_{\mathbf{u}}$ leaves everything on the line in the direction of $\mathbf{u}$ alone.

> The "F" in $F_{\mathbf{u}}$ stands for flip or reflection.

Reflections

The function $F_{\mathbf{u}} : \mathbb{R}^n \to \mathbb{R}^n$ that reflects space through the line in the direction of the unit vector $\mathbf{u}$ is also a linear transformation. To verify this claim and determine its standard matrix, we note that $F_{\mathbf{u}}(\mathbf{v}) = \mathbf{v} + 2(P_{\mathbf{u}}(\mathbf{v}) - \mathbf{v})$, as demonstrated in Figure 1.17.

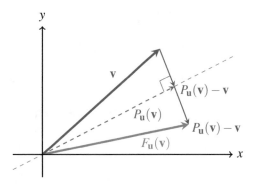

Figure 1.17: The reflection of a vector **v** across the line in the direction of **u** is denoted by $F_{\mathbf{u}}(\mathbf{v})$ and is equal to $\mathbf{v} + 2(P_{\mathbf{u}}(\mathbf{v}) - \mathbf{v})$.

Just like before, even though this picture is in $\mathbb{R}^2$, the calculation of $F_{\mathbf{u}}$'s standard matrix works the same way in all dimensions.

By simplifying, we see that

Be careful when factoring in the final step here. It is tempting to write

$$2(\mathbf{u}\mathbf{u}^T)\mathbf{v} - \mathbf{v}$$
$$= (2\mathbf{u}\mathbf{u}^T - 1)\mathbf{v},$$

but this is wrong since $\mathbf{u}\mathbf{u}^T$ is a matrix, so we cannot subtract 1 from it.

$$
\begin{aligned}
F_{\mathbf{u}}(\mathbf{v}) &= \mathbf{v} + 2(P_{\mathbf{u}}(\mathbf{v}) - \mathbf{v}) && \text{(by Figure 1.17)}\\
&= 2P_{\mathbf{u}}(\mathbf{v}) - \mathbf{v} && \text{(expand parentheses)}\\
&= 2(\mathbf{u}\mathbf{u}^T)\mathbf{v} - \mathbf{v} && \text{(since } P_{\mathbf{u}}(\mathbf{v}) = (\mathbf{u}\mathbf{u}^T)\mathbf{v})\\
&= (2\mathbf{u}\mathbf{u}^T - I)\mathbf{v}. && \text{(factor carefully)}
\end{aligned}
$$

The standard matrix of the reflection $F_{\mathbf{u}}$ is thus

$$[F_{\mathbf{u}}] = 2\mathbf{u}\mathbf{u}^T - I.$$

Example 1.4.6

Standard Matrices of Reflections

Find the standard matrices of the linear transformations that reflect through the lines in the direction of the following vectors, and depict these reflections geometrically.

a) $\mathbf{u} = (0,1) \in \mathbb{R}^2$, and
b) $\mathbf{w} = (1,1,1) \in \mathbb{R}^3$.

Solutions:

a) Since **u** is a unit vector, the standard matrix of $F_{\mathbf{u}}$ is simply

$$[F_{\mathbf{u}}] = 2\mathbf{u}\mathbf{u}^T - I = 2\begin{bmatrix} 0 \\ 1 \end{bmatrix}\begin{bmatrix} 0 & 1 \end{bmatrix} - \begin{bmatrix} 1 & 0 \\ 0 & 1 \end{bmatrix} = \begin{bmatrix} -1 & 0 \\ 0 & 1 \end{bmatrix}.$$

Geometrically, this matrix acts as a reflection through the y-axis:

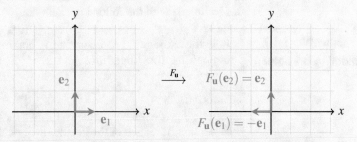

We can verify our computation of $[F_{\mathbf{u}}]$ by noting that multiplication by it indeed just flips the sign of the x-entry of the input vector:

$$[F_{\mathbf{u}}]\mathbf{v} = \begin{bmatrix} -1 & 0 \\ 0 & 1 \end{bmatrix} \begin{bmatrix} v_1 \\ v_2 \end{bmatrix} = \begin{bmatrix} -v_1 \\ v_2 \end{bmatrix}.$$

b) Since $\mathbf{w}$ is not a unit vector, we have to first normalize it (i.e., divide it by its length) to turn it into a unit vector pointing in the same direction. Well, $\|\mathbf{w}\| = \sqrt{1^2 + 1^2 + 1^2} = \sqrt{3}$, so we let $\mathbf{u} = \mathbf{w}/\|\mathbf{w}\| = (1,1,1)/\sqrt{3}$. The standard matrix of $F_{\mathbf{u}}$ is then

$$[F_{\mathbf{u}}] = 2\mathbf{u}\mathbf{u}^T - I = 2 \begin{bmatrix} 1 \\ 1 \\ 1 \end{bmatrix} \begin{bmatrix} 1 & 1 & 1 \end{bmatrix} / 3 - \begin{bmatrix} 1 & 0 & 0 \\ 0 & 1 & 0 \\ 0 & 0 & 1 \end{bmatrix}$$

$$= \frac{1}{3} \begin{bmatrix} -1 & 2 & 2 \\ 2 & -1 & 2 \\ 2 & 2 & -1 \end{bmatrix}.$$

This reflection is a bit more difficult to visualize since it acts on $\mathbb{R}^3$ instead of $\mathbb{R}^2$, but we can at least try:

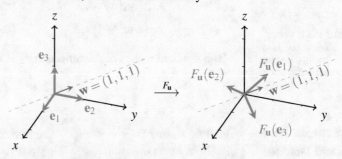

As a bit of a sanity check, we can verify that this matrix leaves $\mathbf{w}$ (and thus $\mathbf{u}$) unchanged, as the reflection $F_{\mathbf{u}}$ should:

$$[F_{\mathbf{u}}]\mathbf{w} = \frac{1}{3} \begin{bmatrix} -1 & 2 & 2 \\ 2 & -1 & 2 \\ 2 & 2 & -1 \end{bmatrix} \begin{bmatrix} 1 \\ 1 \\ 1 \end{bmatrix} = \begin{bmatrix} 1 \\ 1 \\ 1 \end{bmatrix} = \mathbf{w}.$$

In general, to project or reflect a vector we should start by finding the standard matrix of the corresponding linear transformation. Once we have that matrix, all we have to do is multiply it by the starting vector in order to find where it ends up after the linear transformation is applied to it. We illustrate this technique with the following example.

Example 1.4.7

Reflecting a Vector

Find the entries of the vector that is obtained by reflecting $\mathbf{v} = (-1, 3)$ through the line going through the origin at an angle of $\pi/3$ counterclockwise from the x-axis.

Solution:

Our first goal is to compute $[F_{\mathbf{u}}]$, where $\mathbf{u}$ is a unit vector that points in the same direction of the line that we want to reflect through. One such unit vector is $\mathbf{u} = (\cos(\pi/3), \sin(\pi/3)) = (1, \sqrt{3})/2$, so

$$[F_{\mathbf{u}}] = 2\mathbf{u}\mathbf{u}^T - I = \frac{1}{2}\begin{bmatrix} 1 \\ \sqrt{3} \end{bmatrix}\begin{bmatrix} 1 & \sqrt{3} \end{bmatrix} - \begin{bmatrix} 1 & 0 \\ 0 & 1 \end{bmatrix} = \frac{1}{2}\begin{bmatrix} -1 & \sqrt{3} \\ \sqrt{3} & 1 \end{bmatrix}.$$

The reflected vector that we want is $F_{\mathbf{u}}(\mathbf{v})$, which we can compute simply by multiplying $\mathbf{v}$ by the standard matrix $[F_{\mathbf{u}}]$:

$$[F_{\mathbf{u}}]\mathbf{v} = \frac{1}{2}\begin{bmatrix} -1 & \sqrt{3} \\ \sqrt{3} & 1 \end{bmatrix}\begin{bmatrix} -1 \\ 3 \end{bmatrix} = \frac{1}{2}\begin{bmatrix} 3\sqrt{3}+1 \\ 3-\sqrt{3} \end{bmatrix} \approx \begin{bmatrix} 3.0981 \\ 0.6340 \end{bmatrix}.$$

We should have probably reflected the square grid on the right through the line in the direction of $\mathbf{u}$ too, but we find $F_{\mathbf{u}}$ easier to visualize as is.

We can visualize this reflection geometrically as follows:

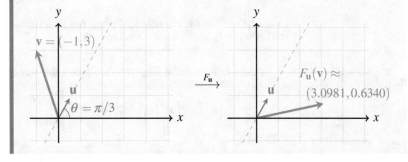

Rotations in Two Dimensions

As one final category of frequently-used linear transformations, we consider (counter-clockwise) **rotations by an angle** θ, which we denote by $R^\theta : \mathbb{R}^2 \to \mathbb{R}^2$. In order for these transformations to be linear, they must be centered at the origin so that $R^\theta(\mathbf{0}) = \mathbf{0}$. To see that R^θ then is indeed linear, we simply note that it is geometrically clear that $R^\theta(\mathbf{v}+\mathbf{w}) = R^\theta(\mathbf{v}) + R^\theta(\mathbf{w})$ for all $\mathbf{v}, \mathbf{w} \in \mathbb{R}^2$, since this just means that adding two vectors and rotating the result is the same as rotating the two vectors and *then* adding them (see Figure 1.18(a)). Similarly, the requirement that $R^\theta(c\mathbf{v}) = cR^\theta(\mathbf{v})$ for all $\mathbf{v} \in \mathbb{R}^2$ and all $c \in \mathbb{R}$ follows from the fact that if we scale a vector and then rotate it, we get the same result as if we rotate and *then* scale, as in Figure 1.18(b).

Using a superscript (rather than a subscript) for the angle θ in R^θ might seem strange at first—we will explain why it makes sense in Remark 1.4.2.

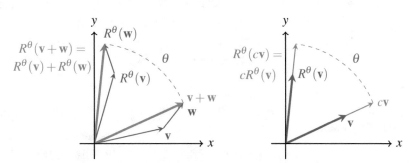

(a) Rotating $\mathbf{v}$ and $\mathbf{w}$ and then adding them has the same effect as adding and then rotating them, so $R^\theta(\mathbf{v}+\mathbf{w}) = R^\theta(\mathbf{v}) + R^\theta(\mathbf{w})$.

(b) Rotating $\mathbf{v}$ and then multiplying it by c has the same effect as multiplying by c and then rotating, so $R^\theta(c\mathbf{v}) = cR^\theta(\mathbf{v})$.

Figure 1.18: A visualization of the linearity of the rotation function R^θ.

Since R^θ is linear, to find its standard matrix (and thus an algebraic description of how it acts on arbitrary vectors), it is enough to compute $R^\theta(\mathbf{e}_1)$ and

$R^\theta(\mathbf{e}_2)$. By regarding $R^\theta(\mathbf{e}_1)$ and $R^\theta(\mathbf{e}_2)$ as the hypotenuses (with length 1) of right-angled triangles as in Figure 1.19, we see that $R^\theta(\mathbf{e}_1) = (\cos(\theta), \sin(\theta))$ and $R^\theta(\mathbf{e}_2) = (-\sin(\theta), \cos(\theta))$.

Recall that linear transformations are determined by what they do to the standard basis vectors.

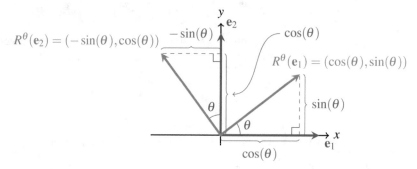

Figure 1.19: R^θ rotates the standard basis vectors $\mathbf{e}_1$ and $\mathbf{e}_2$ to $R^\theta(\mathbf{e}_1) = (\cos(\theta), \sin(\theta))$ and $R^\theta(\mathbf{e}_2) = (-\sin(\theta), \cos(\theta))$, respectively.

It then follows from Theorem 1.4.1 that the standard matrix of R^θ is

$$\left[R^\theta\right] = \left[\, R^\theta(\mathbf{e}_1) \mid R^\theta(\mathbf{e}_2)\, \right] = \begin{bmatrix} \cos(\theta) & -\sin(\theta) \\ \sin(\theta) & \cos(\theta) \end{bmatrix}.$$

Example 1.4.8

Standard Matrices of Rotations

Find the standard matrix of the linear transformation that rotates $\mathbb{R}^2$ by

a) $\pi/4$ radians counter-clockwise, and

b) $\pi/6$ radians clockwise.

Solutions:

a) We compute

$$\left[R^{\pi/4}\right] = \begin{bmatrix} \cos(\pi/4) & -\sin(\pi/4) \\ \sin(\pi/4) & \cos(\pi/4) \end{bmatrix} = \frac{1}{\sqrt{2}} \begin{bmatrix} 1 & -1 \\ 1 & 1 \end{bmatrix}.$$

b) A rotation clockwise by $\pi/6$ radians is equivalent to a counter-clockwise rotation by $-\pi/6$ radians, so we compute

Recall that $\sin(-x) = -\sin(x)$ and $\cos(-x) = \cos(x)$ for all $x \in \mathbb{R}$.

$$\left[R^{-\pi/6}\right] = \begin{bmatrix} \cos(-\pi/6) & -\sin(-\pi/6) \\ \sin(-\pi/6) & \cos(-\pi/6) \end{bmatrix} = \frac{1}{2} \begin{bmatrix} \sqrt{3} & 1 \\ -1 & \sqrt{3} \end{bmatrix}.$$

Example 1.4.9

Rotating Vectors

Rotate each of the following vectors by the indicated amount.

a) Rotate $\mathbf{v} = (1, 3)$ by $\pi/4$ radians counter-clockwise, and

b) rotate $\mathbf{w} = (\sqrt{3}, 3)$ by $\pi/6$ radians clockwise.

Solutions:

a) We saw in Example 1.4.8(a) that the standard matrix of this rotation is

$$\left[R^{\pi/4}\right] = \frac{1}{\sqrt{2}}\begin{bmatrix} 1 & -1 \\ 1 & 1 \end{bmatrix}, \quad \text{so}$$

$$\left[R^{\pi/4}\right]\mathbf{v} = \frac{1}{\sqrt{2}}\begin{bmatrix} 1 & -1 \\ 1 & 1 \end{bmatrix}\begin{bmatrix} 1 \\ 3 \end{bmatrix} = \begin{bmatrix} -\sqrt{2} \\ 2\sqrt{2} \end{bmatrix}.$$

Recall that $\pi/4$ radians and $\pi/6$ radians equal 45 degrees and 30 degrees, respectively.

We can visualize this rotation as follows:

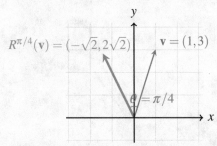

b) Similarly, we computed the standard matrix of this rotation in Example 1.4.8(b), so we can simply compute $R^{\pi/6}(\mathbf{w})$ via matrix multiplication:

$$\left[R^{-\pi/6}\right]\mathbf{w} = \frac{1}{2}\begin{bmatrix} \sqrt{3} & 1 \\ -1 & \sqrt{3} \end{bmatrix}\begin{bmatrix} \sqrt{3} \\ 3 \end{bmatrix} = \begin{bmatrix} 3 \\ \sqrt{3} \end{bmatrix}.$$

We can visualize this rotation as follows:

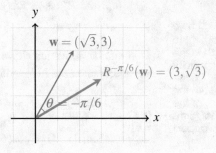

Rotations in Higher Dimensions

While projections and reflections work regardless of the dimension of the space that they act on, extending rotations to dimensions higher than 2 is somewhat delicate. For example, in $\mathbb{R}^3$ it does not make sense to say something like "rotate $\mathbf{v} = (1,2,3)$ counter-clockwise by an angle of $\pi/4$"—this instruction is under-specified, since one angle alone does not tell us in which direction we should rotate.

We show how to construct a rotation around a particular line in $\mathbb{R}^3$ in Exercise 1.4.24.
In $\mathbb{R}^3$, we can get around this problem by specifying an axis of rotation via a unit vector (much like we used a unit vector to specify which line to project onto or reflect through earlier). However, we still have the problem that a rotation that looks clockwise from one side looks counter-clockwise from the other. Furthermore, there is a slightly simpler method that extends more straightforwardly to even higher dimensions—repeatedly rotate in a plane containing two of the coordinate axes.

To illustrate how to do this, consider the linear transformation R_{yz}^{θ} that rotates $\mathbb{R}^3$ by an angle of θ around the x-axis, from the positive y-axis toward the positive z-axis. It is straightforward to see that rotating $\mathbf{e}_1$ in this way has no effect at all (i.e., $R_{yz}^{\theta}(\mathbf{e}_1) = \mathbf{e}_1$), as shown in Figure 1.20. Similarly, to rotate $\mathbf{e}_2$ or $\mathbf{e}_3$ we can just treat the yz-plane as if it were $\mathbb{R}^2$ and repeat the derivation from Figure 1.18 to see that $R_{yz}^{\theta}(\mathbf{e}_2) = (0, \cos(\theta), \sin(\theta))$ and $R_{yz}^{\theta}(\mathbf{e}_3) = (0, -\sin(\theta), \cos(\theta))$.

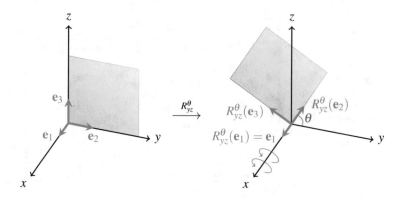

Figure 1.20: The linear transformation R_{yz}^{θ} rotates vectors by an angle of θ around the x-axis in the direction from the positive y-axis to the positive z-axis. In particular, it leaves $\mathbf{e}_1$ unchanged and it rotates $\mathbf{e}_2$ and $\mathbf{e}_3$ (and any vector in the yz-plane) just as rotations in $\mathbb{R}^2$ do.

It follows that the standard matrix of R_{yz}^{θ} is

$$\left[R_{yz}^{\theta}\right] = \left[\, R_{yz}^{\theta}(\mathbf{e}_1) \mid R_{yz}^{\theta}(\mathbf{e}_2) \mid R_{yz}^{\theta}(\mathbf{e}_3) \,\right] = \begin{bmatrix} 1 & 0 & 0 \\ 0 & \cos(\theta) & -\sin(\theta) \\ 0 & \sin(\theta) & \cos(\theta) \end{bmatrix}.$$

We list $[R_{zx}^{\theta}]$ here instead of $[R_{xz}^{\theta}]$ to keep $-\sin(\theta)$ in the upper triangular part. $[R_{xz}^{\theta}]$ is the same but with the signs of the off-diagonal terms swapped.

A similar derivation shows that we can analogously rotate around the other two coordinate axes in $\mathbb{R}^3$ via the standard matrices

$$\left[R_{zx}^{\theta}\right] = \begin{bmatrix} \cos(\theta) & 0 & -\sin(\theta) \\ 0 & 1 & 0 \\ \sin(\theta) & 0 & \cos(\theta) \end{bmatrix} \quad \text{and} \quad \left[R_{xy}^{\theta}\right] = \begin{bmatrix} \cos(\theta) & -\sin(\theta) & 0 \\ \sin(\theta) & \cos(\theta) & 0 \\ 0 & 0 & 1 \end{bmatrix}.$$

Example 1.4.10

Rotating Vectors in 3 Dimensions

Rotate $\mathbf{v} = (3, -1, 2)$ around the z-axis by an angle of $\theta = 2\pi/3$ in the direction from the positive x-axis to the positive y-axis.

Solutions:

Since our goal is to compute $R_{xy}^{2\pi/3}(\mathbf{v})$, we start by constructing the standard matrix of $R_{xy}^{2\pi/3}$:

$$\left[R_{xy}^{2\pi/3}\right] = \begin{bmatrix} \cos(2\pi/3) & -\sin(2\pi/3) & 0 \\ \sin(2\pi/3) & \cos(2\pi/3) & 0 \\ 0 & 0 & 1 \end{bmatrix} = \frac{1}{2}\begin{bmatrix} -1 & -\sqrt{3} & 0 \\ \sqrt{3} & -1 & 0 \\ 0 & 0 & 2 \end{bmatrix}.$$

All that remains is to multiply this matrix by $\mathbf{v} = (3, -1, 2)$:

$$[R_{xy}^{2\pi/3}]\mathbf{v} = \frac{1}{2}\begin{bmatrix} -1 & -\sqrt{3} & 0 \\ \sqrt{3} & -1 & 0 \\ 0 & 0 & 2 \end{bmatrix}\begin{bmatrix} 3 \\ -1 \\ 2 \end{bmatrix} = \frac{1}{2}\begin{bmatrix} \sqrt{3}-3 \\ 3\sqrt{3}+1 \\ 4 \end{bmatrix} \approx \begin{bmatrix} -0.6340 \\ 3.0981 \\ 2.000 \end{bmatrix}.$$

We can visualize this rotation as follows:

The z-entry of $R_{xy}^{2\pi/3}(\mathbf{v})$ is of course the same as that of $\mathbf{v}$, since we are rotating around the z-axis.

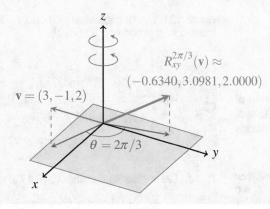

By applying rotations around different coordinate axes one after another, we can rotate vectors in $\mathbb{R}^3$ (and higher dimensions) however we like. Applying linear transformations in a sequence like this is called **composition**, and is the subject of the next subsection.

1.4.3 Composition of Linear Transformations

As we mentioned earlier, there are some simple ways of combining linear transformations to create new ones. In particular, we noted that if $S, T : \mathbb{R}^n \to \mathbb{R}^m$ are linear transformations and $c \in \mathbb{R}$ then $S + cT : \mathbb{R}^n \to \mathbb{R}^m$ is also a linear transformation.

We now introduce another method of combining linear transformations that is slightly more exotic, and has the interpretation of applying one linear transformation after another:

Definition 1.4.2

Composition of Linear Transformations

Suppose $T : \mathbb{R}^n \to \mathbb{R}^m$ and $S : \mathbb{R}^m \to \mathbb{R}^p$ are linear transformations. Then their **composition** is the function $S \circ T : \mathbb{R}^n \to \mathbb{R}^p$ defined by

$$(S \circ T)(\mathbf{v}) = S(T(\mathbf{v})) \quad \text{for all} \quad \mathbf{v} \in \mathbb{R}^n.$$

That is, the composition $S \circ T$ of two linear transformations is the function that has the same effect on vectors as first applying T to them and then applying S. In other words, while T sends $\mathbb{R}^n$ to $\mathbb{R}^m$ and S sends $\mathbb{R}^m$ to $\mathbb{R}^p$, the composition $S \circ T$ skips the intermediate step and sends $\mathbb{R}^n$ directly to $\mathbb{R}^p$, as illustrated in Figure 1.21.

Be careful: $S \circ T$ is read right-to-left: T is applied first and *then* S is applied. This convention was chosen so that S and T appear in the same order in the defining equation $(S \circ T)(\mathbf{v}) = S(T(\mathbf{v}))$.

Even though S and T are linear transformations, at first it is not particularly obvious whether or not $S \circ T$ is also linear, and if so, what its standard matrix is. The following theorem shows that $S \circ T$ is indeed linear, and its standard matrix is simply the product of the standard matrices of S and T. In fact, this theorem is the main reason that matrix multiplication was defined in the seemingly bizarre way that it was.

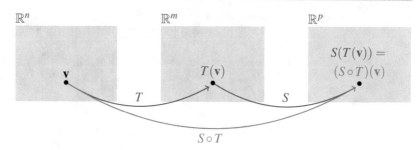

Figure 1.21: The composition of S and T, denoted by $S \circ T$, is the linear transformation that sends $\mathbf{v}$ to $S(T(\mathbf{v}))$.

Theorem 1.4.2

Composition of Linear Transformations

Suppose $T : \mathbb{R}^n \to \mathbb{R}^m$ and $S : \mathbb{R}^m \to \mathbb{R}^p$ are linear transformations with standard matrices $[T] \in \mathcal{M}_{m,n}$ and $[S] \in \mathcal{M}_{p,m}$, respectively. Then $S \circ T : \mathbb{R}^n \to \mathbb{R}^p$ is a linear transformation, and its standard matrix is

$$[S \circ T] = [S][T].$$

Notice that $[S \circ T] = [S][T]$ is a $p \times n$ matrix.

Proof. Let $\mathbf{v} \in \mathbb{R}^n$ and compute $(S \circ T)(\mathbf{v})$:

$$(S \circ T)(\mathbf{v}) = S(T(\mathbf{v})) = S([T]\mathbf{v}) = [S]([T]\mathbf{v}) = ([S][T])\mathbf{v}.$$

In other words, $S \circ T$ is a function that acts on $\mathbf{v}$ in the exact same way as matrix multiplication by the matrix $[S][T]$. It thus follows from Theorem 1.4.1 that $S \circ T$ is a linear transformation and its standard matrix is $[S][T]$, as claimed. ∎

By applying linear transformations one after another like this, we can construct simple algebraic descriptions of complicated geometric transformations. For example, it might be difficult to visualize exactly what happens to $\mathbb{R}^2$ if we rotate it, reflect it, and then rotate it again, but this theorems tells us that we can unravel exactly what happens just by multiplying together the standard matrices of the three individual linear transformations.

Example 1.4.11

Composition of Linear Transformations

Find the standard matrix of the linear transformation T that reflects $\mathbb{R}^2$ through the line $y = \frac{4}{3}x$ and then stretches it in the x-direction by a factor of 2 and in the y-direction by a factor of 3.

Solution:
We compute the two standard matrices individually and then multiply them together. A unit vector on the line $y = \frac{4}{3}x$ is $\mathbf{u} = (3/5, 4/5)$, and the reflection $F_{\mathbf{u}}$ has standard matrix

$$[F_{\mathbf{u}}] = 2 \begin{bmatrix} 3/5 \\ 4/5 \end{bmatrix} \begin{bmatrix} 3/5 & 4/5 \end{bmatrix} - \begin{bmatrix} 1 & 0 \\ 0 & 1 \end{bmatrix} = \frac{1}{25} \begin{bmatrix} -7 & 24 \\ 24 & 7 \end{bmatrix}.$$

Recall that when composing linear transformations, the rightmost transformation is applied first, so the standard matrix we are after is $[D][F_{\mathbf{u}}]$, not $[F_{\mathbf{u}}][D]$.

The diagonal stretch D has standard matrix

$$[D] = \begin{bmatrix} 2 & 0 \\ 0 & 3 \end{bmatrix}.$$

The standard matrix of the composite linear transformation $T = D \circ F_{\mathbf{u}}$ is thus the product of these two individual standard matrices:

$$[T] = [D][F_{\mathbf{u}}] = \frac{1}{25} \begin{bmatrix} 2 & 0 \\ 0 & 3 \end{bmatrix} \begin{bmatrix} -7 & 24 \\ 24 & 7 \end{bmatrix} = \frac{1}{25} \begin{bmatrix} -14 & 48 \\ 72 & 21 \end{bmatrix}.$$

We can visualize this composite linear transformation T as a sequence of two separate distortions of $\mathbb{R}^2$—first a reflection and then a diagonal scaling—or we can think of it just as a single linear transformation that has the same effect:

Keep in mind that the diagonal scaling D scales in the direction of the x- and y-axes, not in the direction of the reflected square grid, so the grid at the bottom-right is made up of parallelograms instead of rectangles.

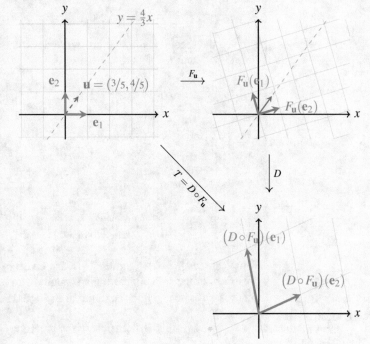

Example 1.4.12

Composition of Rotations

Explain why, if θ and ϕ are any two angles, then $R^\theta \circ R^\phi = R^{\theta+\phi}$. Use this fact to derive the angle-sum trigonometric identities

$$\sin(\theta + \phi) = \sin(\theta)\cos(\phi) + \cos(\theta)\sin(\phi)$$

and

$$\cos(\theta + \phi) = \cos(\theta)\cos(\phi) - \sin(\theta)\sin(\phi).$$

Solution:

The fact that $R^\theta \circ R^\phi = R^{\theta+\phi}$ is clear geometrically: rotating a vector $\mathbf{v} \in \mathbb{R}^2$ by ϕ radians and then by θ more radians is the same as rotating it by $\theta + \phi$ radians:

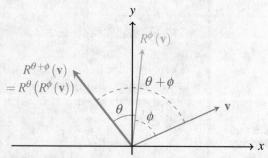

To see the second claim, we just write down the standard matrix of each of $R^\theta \circ R^\phi$ and $R^{\theta+\phi}$:

$$\left[R^\theta \circ R^\phi\right] = \left[R^\theta\right]\left[R^\phi\right] = \begin{bmatrix} \cos(\theta) & -\sin(\theta) \\ \sin(\theta) & \cos(\theta) \end{bmatrix} \begin{bmatrix} \cos(\phi) & -\sin(\phi) \\ \sin(\phi) & \cos(\phi) \end{bmatrix}$$

$$= \begin{bmatrix} \cos(\theta)\cos(\phi) - \sin(\theta)\sin(\phi) & -\cos(\theta)\sin(\phi) - \sin(\theta)\cos(\phi) \\ \sin(\theta)\cos(\phi) + \cos(\theta)\sin(\phi) & -\sin(\theta)\sin(\phi) + \cos(\theta)\cos(\phi) \end{bmatrix}.$$

On the other hand,

$$\left[R^{\theta+\phi}\right] = \begin{bmatrix} \cos(\theta+\phi) & -\sin(\theta+\phi) \\ \sin(\theta+\phi) & \cos(\theta+\phi) \end{bmatrix}.$$

The matrices $\left[R^\theta \circ R^\phi\right]$ and $\left[R^{\theta+\phi}\right]$ must be equal (since they are standard matrices of the same linear transformation), so it follows that all of their entries must be equal. By looking at their $(2,1)$-entries, we see that

$$\sin(\theta+\phi) = \sin(\theta)\cos(\phi) + \cos(\theta)\sin(\phi),$$

and their $(1,1)$-entries similarly tell us that

$$\cos(\theta+\phi) = \cos(\theta)\cos(\phi) - \sin(\theta)\sin(\phi).$$

The $(1,2)$- and $(2,2)$-entries of these matrices give us no additional information; they just tell us again that these same identities hold.

If we choose $\phi = \theta$ then these identities tell us that $\sin(2\theta) = 2\sin(\theta)\cos(\theta)$ and $\cos(2\theta) = \cos^2(\theta) - \sin^2(\theta)$.

Remark 1.4.2

Rotation Notation

Example 1.4.12 actually explains why we use the notation R^θ, despite it seeming to conflict with the fact that we also use superscripts for powers of matrices. It turns out that there is no conflict between these notations at all—$[R^\theta]$ really is the θ-th power of the matrix

$$R = \begin{bmatrix} \cos(1) & -\sin(1) \\ \sin(1) & \cos(1) \end{bmatrix} \approx \begin{bmatrix} 0.5403 & -0.8415 \\ 0.8415 & 0.5403 \end{bmatrix}$$

that rotates counter-clockwise by 1 radian. Geometrically, we defined R^θ as rotating counter-clockwise by an angle of θ, but we can equivalently think of it as rotating counter-clockwise by 1 radian θ times.

For now, this interpretation of R^θ as the θ-th power of R only makes sense when θ is a non-negative integer, but it also works for non-integer powers of R, which we introduce in Section 3.4.2.

We close this section by noting that the composition of three or more linear transformations works exactly how we might expect based on how it works for two of them: if R, S, and T are linear transformations then their composition $R \circ S \circ T$ is the linear transformation that applies T, then S, then R, and its standard matrix is $[R \circ S \circ T] = [R][S][T]$.

Exercises

solutions to starred exercises on page 439

1.4.1 Find the standard matrix of the following linear transformations:

* (a) $T(v_1, v_2) = (v_1 + 2v_2, 3v_1 - v_2)$.
  (b) $T(v_1, v_2) = (v_1 + v_2, 2v_1 - v_2, -v_1 + 3v_2)$.
* (c) $T(v_1, v_2, v_3) = (v_1 + v_2, v_1 + v_2 - v_3)$.
  (d) $T(v_1, v_2, v_3) = (v_2, 2v_1 + v_3, v_2 - v_3)$.

1.4.2 Find the standard matrices of the linear transformations T that act as follows:

* (a) $T(1,0) = (3,-1)$ and $T(0,1) = (1,2)$.
  (b) $T(1,0) = (1,3)$ and $T(1,1) = (3,7)$.
* (c) $T(1,0) = (-1,0,1)$ and $T(0,1) = (2,3,0)$.
  (d) $T(1,0,0) = (2,1)$, $T(0,1,0) = (-1,1)$, and $T(0,0,1) = (0,3)$.
* (e) $T(1,0,0) = (1,2,3)$, $T(1,1,0) = (0,1,2)$, and $T(1,1,1) = (0,0,1)$.

1.4.3 Determine which of the following statements are true and which are false.

*(a) Linear transformations are functions.
 (b) Every matrix transformation (i.e., function that sends **v** to $A\mathbf{v}$ for some fixed matrix $A \in \mathcal{M}_{m,n}$) is a linear transformation.
*(c) Every linear transformation $T : \mathbb{R}^2 \to \mathbb{R}^4$ is completely determined by the vectors $T(\mathbf{e}_1)$ and $T(\mathbf{e}_2)$.
 (d) The standard matrix of a linear transformation $T : \mathbb{R}^4 \to \mathbb{R}^3$ has size 4×3.
*(e) There exists a linear transformation $T : \mathbb{R}^2 \to \mathbb{R}^2$ such that $T(\mathbf{e}_1) = (2,1)$, $T(\mathbf{e}_2) = (1,3)$, and $T(1,1) = (3,3)$.
 (f) If $\mathbf{u} \in \mathbb{R}^n$ is a vector then the standard matrix of the projection onto the line in the direction of $\mathbf{u}$ is $\mathbf{u}\mathbf{u}^T$.
*(g) If R_{xy}^θ, R_{yz}^θ, and R_{xz}^θ are rotations around the coordinate axes in $\mathbb{R}^3$ as described in the text, then $R_{xy}^\theta \circ R_{yz}^\theta = R_{xz}^\theta$.

1.4.4 Determine which of the following functions $T : \mathbb{R}^2 \to \mathbb{R}^2$ are and are not linear transformations. If T is a linear transformation, find its standard matrix. If it is not a linear transformation, justify your answer (i.e., show a property of linear transformations that it fails).

*(a) $T(v_1, v_2) = (v_1^2, v_2)$.
 (b) $T(v_1, v_2) = (v_1 + 2v_2, v_2 - v_1)$.
*(c) $T(v_1, v_2) = (\sin(v_1) + v_2, v_1 - \cos(v_2))$.
 (d) $T(v_1, v_2) = (\sin(3)v_1 + v_2, v_1 - \cos(2)v_2)$.
*(e) $T(v_1, v_2) = (\sqrt{v_1} + \sqrt{v_2}, \sqrt{v_1 + v_2})$.
 (f) $T(v_1, v_2) = (\sqrt{3}v_1, \sqrt{2}v_2 - \sqrt{5}v_1)$.
*(g) $T(v_1, v_2) = (|v_1|, |v_2|)$.

1.4.5 Determine which of the following geometric transformations of $\mathbb{R}^2$ are and are not linear transformations. If it is a linear transformation, find its standard matrix. If it is not a linear transformation, justify your answer (i.e., show a property of linear transformations that it fails).

* (a) The projection onto the line $y = 3x$.
  (b) The projection onto the line $y = 3x - 2$.
*(c) The reflection across the line $y = 2x$.
  (d) The reflection across the line $y = 2x + 1$.
*(e) The counter-clockwise rotation around the point $(0,0)$ by $\pi/5$ radians.
  (f) The counter-clockwise rotation around the point $(0,1)$ by $\pi/5$ radians.

1.4.6 Find the standard matrix of the composite linear transformation $S \circ T$, when S and T are defined as follows:

*(a) $S(v_1, v_2) = (2v_2, v_1 + v_2)$,
     $T(v_1, v_2) = (v_1 + 2v_2, 3v_1 - v_2)$.
 (b) $S(v_1, v_2) = (v_1 - 2v_2, 3v_1 + v_2)$,
     $T(v_1, v_2, v_3) = (v_1 + v_2, v_1 + v_2 - v_3)$.
*(c) $S(v_1, v_2, v_3) = (v_1, v_1 + v_2, v_1 + v_2 + v_3)$,
     $T(v_1, v_2) = (v_1 + v_2, 2v_1 - v_2, -v_1 + 3v_2)$.

1.4.7 Find the standard matrix of the composite linear transformation that acts on $\mathbb{R}^2$ as follows:

 (a) Rotates $\mathbb{R}^2$ counter-clockwise about the origin by an angle of $\pi/3$ radians, and then stretches in the x- and y-directions by factors of 2 and 3, respectively.
*(b) Projects $\mathbb{R}^2$ onto the line $y = x$ and then rotates $\mathbb{R}^2$ clockwise about the origin by an angle of $\pi/4$ radians.
 (c) Reflects $\mathbb{R}^2$ through the line $y = 2x$ and then reflects $\mathbb{R}^2$ through the line $y = x/2$.

∗∗1.4.8 Show that if $T : \mathbb{R}^n \to \mathbb{R}^m$ is a linear transformation, then $T(\mathbf{0}) = \mathbf{0}$.

∗∗1.4.9 Show that a function $T : \mathbb{R}^n \to \mathbb{R}^m$ is a linear transformation if and only if

$$T(c_1 \mathbf{v}_1 + \cdots + c_k \mathbf{v}_k) = c_1 T(\mathbf{v}_1) + \cdots + c_k T(\mathbf{v}_k)$$

for all $\mathbf{v}_1, \ldots, \mathbf{v}_k \in \mathbb{R}^n$ and all $c_1, \ldots, c_k \in \mathbb{R}$.

∗∗1.4.10 Let $D \in \mathcal{M}_n$ be a diagonal matrix and let $A \in \mathcal{M}_n$ be an arbitrary matrix. Use block matrix multiplication to show that, for all $1 \leq i \leq n$, ...

 (a) the product DA multiplies the i-th row of A by $d_{i,i}$, and

(b) the product AD multiplies the i-th column of A by $d_{i,i}$.

∗∗1.4.11 Let $S, T : \mathbb{R}^n \to \mathbb{R}^m$ be linear transformations and let $c \in \mathbb{R}$ be a scalar. Show that the following relationships between various standard matrices hold. [Hint: Recall the definitions of $S + T$ and cT from Remark 1.4.1.]

(a) $[S + T] = [S] + [T]$, and
(b) $[cT] = c[T]$.

1.4.12 Let $Q, S, T : \mathbb{R}^n \to \mathbb{R}^m$ be linear transformations. Prove the following. [Hint: Use Exercise 1.4.11(a) and properties that we already know about matrix addition.]

(a) $S + T = T + S$, and
(b) $Q + (S + T) = (Q + S) + T$.

1.4.13 Let $S, T : \mathbb{R}^n \to \mathbb{R}^m$ be linear transformations and let $c, d \in \mathbb{R}$ be scalars. Prove the following.

[Hint: Use Exercise 1.4.11 and properties that we already know about scalar multiplication for matrices.]

(a) $c(S + T) = cS + cT$,
(b) $(c + d)T = cT + dT$, and
(c) $c(dT) = (cd)T$.

∗1.4.14 Suppose

$$A = \begin{bmatrix} \cos(\pi/4) & -\sin(\pi/4) \\ \sin(\pi/4) & \cos(\pi/4) \end{bmatrix}.$$

Compute the matrix A^{160}.

[Hint: Think about this problem geometrically.]

∗∗1.4.15 Suppose

$$A = \begin{bmatrix} \cos(\theta) & -\sin(\theta) \\ \sin(\theta) & \cos(\theta) \end{bmatrix}$$

for some $\theta \in \mathbb{R}$. Show that $A^T A = I$.

[Side note: Matrices with this property are called **unitary matrices**, and we will study them in depth in Section 1.4 of [Joh20].]

1.4.16 Suppose $\mathbf{u} \in \mathbb{R}^n$ is a unit vector and $A = \mathbf{u}\mathbf{u}^T \in \mathcal{M}_n$. Show that $A^2 = A$. Provide a geometric interpretation of this fact.

∗∗1.4.17 Suppose $\mathbf{u} \in \mathbb{R}^n$ is a unit vector and $A = 2\mathbf{u}\mathbf{u}^T - I \in \mathcal{M}_n$. Show that $A^2 = I$. Provide a geometric interpretation of this fact.

1.4.18 Show that if $P_{\mathbf{u}} : \mathbb{R}^n \to \mathbb{R}^n$ is a projection onto a line and $\mathbf{v} \in \mathbb{R}^n$ then $\|P_{\mathbf{u}}(\mathbf{v})\| \leq \|\mathbf{v}\|$.

∗1.4.19 Show that the linear transformation that reflects $\mathbb{R}^2$ through the line $y = mx$ has standard matrix

$$\frac{1}{1 + m^2} \begin{bmatrix} 1 - m^2 & 2m \\ 2m & m^2 - 1 \end{bmatrix}.$$

1.4.20 Show that the linear transformation that reflects $\mathbb{R}^2$ through the line that points at an angle θ counter-clockwise from the x-axis has standard matrix

$$\begin{bmatrix} \cos(2\theta) & \sin(2\theta) \\ \sin(2\theta) & -\cos(2\theta) \end{bmatrix}.$$

1.4.21 Suppose $\mathbf{u}, \mathbf{v} \in \mathbb{R}^2$ are vectors with an angle of θ between them. Show that $F_{\mathbf{u}} \circ F_{\mathbf{v}} = R^{2\theta}$.

[Hint: Use Exercise 1.4.20.]

1.4.22 Give an example of a function $T : \mathbb{R}^2 \to \mathbb{R}^2$ that is *not* a linear transformation, but satisfies the property $T(c\mathbf{v}) = cT(\mathbf{v})$ for all $\mathbf{v} \in \mathbb{R}^2$ and $c \in \mathbb{R}$.

[Hint: You probably cannot find a "nice" function that works. Try a piecewise/branch function.]

∗∗1.4.23 A **shear matrix** is a square matrix with every diagonal entry equal to 1, and exactly one non-zero off-diagonal entry. For example, the 2×2 shear matrices have the form

$$\begin{bmatrix} 1 & c \\ 0 & 1 \end{bmatrix} \quad \text{or} \quad \begin{bmatrix} 1 & 0 \\ c & 1 \end{bmatrix}$$

for some non-zero scalar $c \in \mathbb{R}$. Let $S_{i,j}^c$ denote the shear matrix with a c in the (i, j)-entry (where $i \neq j$).

(a) Illustrate the geometric effect of the 2×2 shear matrices $S_{1,2}^c$ and $S_{2,1}^c$ shown above as linear transformations.
    [Side note: This picture is the reason for the name "shear" matrix.]
(b) Show that $(S_{i,j}^c)^n = S_{i,j}^{nc}$ for all integers $n \geq 1$ (and hence the superscript notation that we use here for shear matrices agrees with exponent notation, just like it did for rotation matrices in Remark 1.4.2).
    [Hint: We can write every shear matrix in the form $S_{i,j}^c = I + cE_{i,j}$, where $E_{i,j}$ is the matrix with a 1 in its (i, j)-entry and zeros elsewhere. What is $E_{i,j}^2$?]

∗∗1.4.24 Let $R_{\mathbf{u}}^\theta$ be the linear transformation that rotates $\mathbb{R}^3$ around the unit vector $\mathbf{u}$ clockwise (when looking in the direction that $\mathbf{u}$ points) by an angle of θ. Its standard matrix is

$$[R_{\mathbf{u}}^\theta] = \cos(\theta)I + (1 - \cos(\theta))\mathbf{u}\mathbf{u}^T$$

$$+ \sin(\theta) \begin{bmatrix} 0 & -u_3 & u_2 \\ u_3 & 0 & -u_1 \\ -u_2 & u_1 & 0 \end{bmatrix}.$$

(a) Compute $[R_{\mathbf{u}}^\theta]$ if $\mathbf{u} = (1, 0, 0)$. Which matrix that we already saw in the text does it equal?
(b) Compute $R_{\mathbf{u}}^{\pi/4}(\mathbf{v})$ if $\mathbf{u} = (1, 2, 2)/3$ and $\mathbf{v} = (3, 2, 1)$.
(c) Show that $R_{yz}^{\pi/3} \circ R_{xy}^{\pi/3}$ is a rotation matrix. Furthermore, find a unit vector $\mathbf{u}$ and an angle θ such that $R_{\mathbf{u}}^\theta = R_{yz}^{\pi/3} \circ R_{xy}^{\pi/3}$.
    [Side note: the composition of *any* two rotations is always a rotation (even in dimensions higher than 3), but proving this fact is outside of the scope of this book.]

1.5 Summary and Review

In this chapter, we introduced the central objects that are studied in linear algebra: vectors, matrices, and linear transformations. We developed some basic ways of manipulating and combining these objects, such as vector addition and scalar multiplication, and we saw that these operations satisfy the basic properties, like distributivity and associativity, that we would expect them to based on our familiarity with properties of real numbers.

On the other hand, the formula for matrix multiplication was seemingly quite bizarre at first, but was later justified by the fact that it implements the action of linear transformations. That is, we can think of a linear transformation $T : \mathbb{R}^n \to \mathbb{R}^m$ as being "essentially the same" as its standard matrix

Recall that $\mathbf{e}_1, \mathbf{e}_2, \ldots, \mathbf{e}_n$ are the standard basis vectors.

$$[T] = \big[\, T(\mathbf{e}_1) \mid T(\mathbf{e}_2) \mid \cdots \mid T(\mathbf{e}_n) \,\big]$$

in the following two senses:

- Applying T to $\mathbf{v}$ is equivalent to performing matrix-vector multiplication with $[T]$. That is, $T(\mathbf{v}) = [T]\mathbf{v}$.
- Composing two linear transformations S and T is equivalent to multiplying their standard matrices. That is, $[S \circ T] = [S][T]$.

For these reasons, we often do not even differentiate between matrices and linear transformations in the later sections of this book. Instead, we just talk about matrices, with the understanding that a matrix is no longer "just" a 2D array of numbers for us, but is also a function that moves vectors around $\mathbb{R}^n$ in a linear way (i.e., it is a linear transformation). Furthermore, the columns of the matrix tell us exactly where the linear transformation sends the standard basis vectors $\mathbf{e}_1, \mathbf{e}_2, \ldots, \mathbf{e}_n$ (see Figure 1.22).

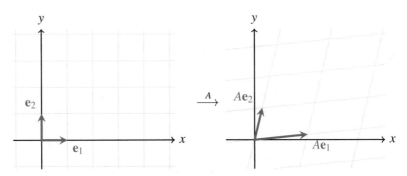

Figure 1.22: A 2×2 matrix A transforms the unit square (with sides $\mathbf{e}_1$ and $\mathbf{e}_2$) into a parallelogram with sides $A\mathbf{e}_1$ and $A\mathbf{e}_2$, which are the columns of A.

This interpretation of matrices reinforces the idea that most linear algebraic objects have both an algebraic interpretation as well as a geometric one. Most importantly, we have the following interpretations of vectors and matrices:

- Algebraically, vectors are lists of numbers. Geometrically, they are arrows in space that represent movement or displacement.
- Algebraically, matrices are arrays of numbers. Geometrically, they are linear transformations—functions that deform a square grid in $\mathbb{R}^n$ into a parallelogram grid.

Exercises

solutions to starred exercises on page 442

1.5.1 Determine which of the following statements are true and which are false.

*(a) If A and B are matrices such that AB is square, then BA must also be square.

(b) If A and B are square matrices of the same size, then $(AB)^T = A^T B^T$.

*(c) If A and B are square matrices of the same size, then $(AB)^3 = A^3 B^3$.

(d) If A and B are square matrices of the same size, then $(A+B)(A-B) = A^2 - B^2$.

*(e) If A is a square matrix for which $A^2 = O$, then $A = O$.

(f) There exists a matrix $A \in \mathcal{M}_2$ such that

$$A\begin{bmatrix} x \\ y \end{bmatrix} = \begin{bmatrix} 2x - y \\ y - x \end{bmatrix}$$

for all $x, y \in \mathbb{R}$.

*(g) If $R, S, T : \mathbb{R}^n \to \mathbb{R}^n$ are linear transformations, then $R \circ (S \circ T) = (R \circ S) \circ T$.

1.5.2 Suppose

$$A = \frac{1}{\sqrt{2}} \begin{bmatrix} 1 & -1 \\ 1 & 1 \end{bmatrix}.$$

(a) Show (algebraically) that $\|A\mathbf{v}\| = \|\mathbf{v}\|$ for all $\mathbf{v} \in \mathbb{R}^2$.

(b) Provide a geometric explanation for the result of part (a).

1.5.3 For each of the following matrices, determine whether they are the standard matrix of a projection onto a line, a reflection through a line, a rotation, or none of these. In each case, find a unit vector being projected onto or reflected through, or the angle being rotated by.

*(a) $\begin{bmatrix} 1 & 0 \\ 0 & 0 \end{bmatrix}$

(b) $\begin{bmatrix} -1 & 0 \\ 0 & 1 \end{bmatrix}$

*(c) $\dfrac{1}{\sqrt{2}} \begin{bmatrix} 1 & -1 \\ 1 & 1 \end{bmatrix}$

(d) $\begin{bmatrix} 1 & 0 \\ 0 & 1 \end{bmatrix}$

*(e) $\begin{bmatrix} 1 & 2 \\ 3 & 4 \end{bmatrix}$

(f) $\dfrac{1}{2} \begin{bmatrix} 1 & \sqrt{3} \\ -\sqrt{3} & 1 \end{bmatrix}$

*(g) $\dfrac{1}{9} \begin{bmatrix} 1 & 2 & 2 \\ 2 & 4 & 4 \\ 2 & 4 & 4 \end{bmatrix}$

(h) $\dfrac{1}{9} \begin{bmatrix} -1 & 4 & 8 \\ 4 & -7 & 4 \\ 8 & 4 & -1 \end{bmatrix}$

1.5.4 Suppose

$$A = \begin{bmatrix} -1 & 2 \\ 0 & 3 \end{bmatrix} \quad \text{and} \quad \mathbf{v} = \begin{bmatrix} 1 \\ 0 \end{bmatrix}.$$

Compute $A^{500}\mathbf{v}$.

*1.5.5 Explain why the following matrix power identity holds for all $\theta \in \mathbb{R}$ and all positive integers n.

$$\begin{bmatrix} \cos(\theta) & -\sin(\theta) \\ \sin(\theta) & \cos(\theta) \end{bmatrix}^n = \begin{bmatrix} \cos(n\theta) & -\sin(n\theta) \\ \sin(n\theta) & \cos(n\theta) \end{bmatrix}.$$

[Hint: Think geometrically.]

1.A   Extra Topic: Areas, Volumes, and the Cross Product

There is one more operation on vectors that we have not yet introduced, called the cross product. To help motivate it, consider the problem of finding a vector that is orthogonal to $\mathbf{v} = (v_1, v_2) \in \mathbb{R}^2$. It is clear from inspection that one vector that works is $\mathbf{w} = (v_2, -v_1)$, since then $\mathbf{v} \cdot \mathbf{w} = v_1 v_2 - v_2 v_1 = 0$ (see Figure 1.23(a)).

If we ramp this type of problem up slightly to 3 dimensions, we can instead ask for a vector $\mathbf{x} \in \mathbb{R}^3$ that is orthogonal to *two* vectors $\mathbf{v} = (v_1, v_2, v_3) \in \mathbb{R}^3$ and $\mathbf{w} = (w_1, w_2, w_3) \in \mathbb{R}^3$. It is much more difficult to eyeball a solution in this case, but we will verify momentarily that the following vector works:

Definition 1.A.1

Cross Product

We just write the cross product as a column vector (instead of a row vector) here to make it easier to read.

If $\mathbf{v} = (v_1, v_2, v_3) \in \mathbb{R}^3$ and $\mathbf{w} = (w_1, w_2, w_3) \in \mathbb{R}^3$ are vectors then their **cross product**, denoted by $\mathbf{v} \times \mathbf{w}$, is defined by

$$\mathbf{v} \times \mathbf{w} \overset{\text{def}}{=} \begin{pmatrix} v_2 w_3 - v_3 w_2 \\ v_3 w_1 - v_1 w_3 \\ v_1 w_2 - v_2 w_1 \end{pmatrix}.$$

To see that the cross product is orthogonal to each of $\mathbf{v}$ and $\mathbf{w}$, we simply compute the relevant dot products:

$$\mathbf{v} \cdot (\mathbf{v} \times \mathbf{w}) = v_1(v_2w_3 - v_3w_2) + v_2(v_3w_1 - v_1w_3) + v_3(v_1w_2 - v_2w_1)$$
$$= v_1v_2w_3 - v_1v_3w_2 + v_2v_3w_1 - v_2v_1w_3 + v_3v_1w_2 - v_3v_2w_1$$
$$= 0.$$

The dot product $\mathbf{w} \cdot (\mathbf{v} \times \mathbf{w})$ can similarly be shown to equal 0 (see Exercise 1.A.10), so we conclude that $\mathbf{v} \times \mathbf{w}$ is orthogonal to each of $\mathbf{v}$ and $\mathbf{w}$, as illustrated in Figure 1.23(b).

Keep in mind that the cross product is only defined in $\mathbb{R}^3$; it is one of the very few dimension-dependent operations that we investigate.

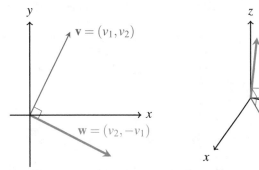

(a) A vector $\mathbf{w}$ that is orthogonal to a given vector $\mathbf{v} \in \mathbb{R}^2$.

(b) The cross product $\mathbf{v} \times \mathbf{w}$ is orthogonal to both $\mathbf{v}$ and $\mathbf{w}$ in $\mathbb{R}^3$.

Figure 1.23: How to construct a vector that is orthogonal to (a) one vector in $\mathbb{R}^2$ and (b) two vectors in $\mathbb{R}^3$.

It helps to remember that, in the formula for $\mathbf{v} \times \mathbf{w}$, the first entry has no 1s in subscripts, the second has no 2s, and the third has no 3s.

The formula for the cross product is a bit messy, so it can help to have a mnemonic that keeps track of which entries of $\mathbf{v}$ and $\mathbf{w}$ are multiplied and subtracted in which entries of $\mathbf{v} \times \mathbf{w}$. One way of remembering the formula is to write the entries of $\mathbf{v}$ in order twice, like $v_1, v_2, v_3, v_1, v_2, v_3$, and then similarly write the entries of $\mathbf{w}$ twice underneath, creating a 2×6 array. We completely ignore the leftmost and rightmost columns of this array, and draw the three "X"s possible between the 2nd and 3rd, 3rd and 4th, and 4th and 5th columns of the array, as in Figure 1.24.

This mnemonic also suggests where the name "cross" product comes from.

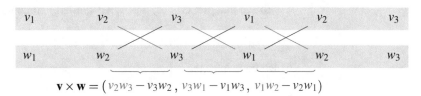

$$\mathbf{v} \times \mathbf{w} = (v_2w_3 - v_3w_2, \ v_3w_1 - v_1w_3, \ v_1w_2 - v_2w_1)$$

Figure 1.24: A mnemonic for computing the cross product: we add along the forward (green) diagonals and subtract along the backward (purple) diagonals.

The first "X" tells us the value of the first entry of $\mathbf{v} \times \mathbf{w}$, the second "X" tells us the second entry, and the third "X" tells us the third entry: we multiply along the diagonals of the "X"s, adding up the forward diagonals and subtracting the backward diagonals.

Example 1.A.1

Numerical Examples of the Cross Product

Find a vector orthogonal to the following vectors:

a) $(1,2)$, and

b) $(1,2,3)$ and $(-1,0,1)$.

Solutions:

a) Just reverse the order of the entries in the vector and take the negative of one of them: $(2,-1)$ works, since $(1,2) \cdot (2,-1) = 2 - 2 = 0$. Alternatively, we could think of $(1,2)$ as living in $\mathbb{R}^3$ (with z-entry equal to 0) and take the cross product with a vector that points in the direction of the z-axis: $(1,2,0) \times (0,0,1) = (2,-1,0)$, which is the same as the answer we found before, but embedded in $\mathbb{R}^3$.

b) $(1,2,3) \times (-1,0,1) = (2,-4,2)$ is orthogonal to both of these vectors. As a sanity check, it is straightforward to verify that $(1,2,3) \cdot (2,-4,2) = 0$ and $(-1,0,1) \cdot (2,-4,2) = 0$.

As always, since we have defined a new mathematical operation, we would like to know what basic properties it satisfies. The following theorem lists these "obvious" properties, which we note are quite a bit less obvious than the properties of the dot product were:

Theorem 1.A.1

Properties of the Cross Product

Let $\mathbf{v}, \mathbf{w}, \mathbf{x} \in \mathbb{R}^3$ be vectors and let $c \in \mathbb{R}$ be a scalar. Then

a) $\mathbf{v} \times \mathbf{w} = -(\mathbf{w} \times \mathbf{v})$,   (anticommutativity)

b) $\mathbf{v} \times (\mathbf{w} + \mathbf{x}) = \mathbf{v} \times \mathbf{w} + \mathbf{v} \times \mathbf{x}$,   (distributivity)

c) $\mathbf{v} \times \mathbf{v} = \mathbf{0}$, and

d) $(c\mathbf{v}) \times \mathbf{w} = c(\mathbf{v} \times \mathbf{w})$.

Proof. We only explicitly prove properties (a) and (c), since these are the more surprising ones. The proofs of properties (b) and (d) are similar, so we leave them to Exercise 1.A.11.

For property (a), we just compute $\mathbf{v} \times \mathbf{w}$ and $\mathbf{w} \times \mathbf{v}$ directly from the definition of the cross product:

$$(v_1, v_2, v_3) \times (w_1, w_2, w_3) = (v_2 w_3 - v_3 w_2, v_3 w_1 - v_1 w_3, v_1 w_2 - v_2 w_1)$$
$$(w_1, w_2, w_3) \times (v_1, v_2, v_3) = (w_2 v_3 - w_3 v_2, w_3 v_1 - w_1 v_3, w_1 v_2 - w_2 v_1).$$

It is straightforward to verify that the two vectors above are negatives of each other, as desired.

To see that property (c) holds, we similarly compute

$$(v_1, v_2, v_3) \times (v_1, v_2, v_3) = (v_2 v_3 - v_2 v_3, v_1 v_3 - v_1 v_3, v_1 v_2 - v_1 v_2) = (0, 0, 0),$$

which completes the proof. ∎

1.A.1   Areas

We have already seen that the direction of the cross product encodes an important geometric property—it points in the direction orthogonal to each of $\mathbf{v}$ and $\mathbf{w}$. It turns out that its length also has a nice geometric interpretation as well, as demonstrated by the following theorem.

Theorem 1.A.2

Area of a Parallelogram

Quantities (b), (c), and (d) are equal to each other in any dimension, not just $\mathbb{R}^3$.

Let $\mathbf{v}, \mathbf{w} \in \mathbb{R}^3$ be vectors and let θ be the angle between them. Then the following four quantities are all equal to each other:

a) $\|\mathbf{v} \times \mathbf{w}\|$,
b) $\sqrt{\|\mathbf{v}\|^2 \|\mathbf{w}\|^2 - (\mathbf{v} \cdot \mathbf{w})^2}$,
c) $\|\mathbf{v}\| \|\mathbf{w}\| \sin(\theta)$, and
d) the area of the parallelogram whose sides are $\mathbf{v}$ and $\mathbf{w}$.

Proof. We start by showing that the quantities (c) and (d) are equal to each other. That is, we show that the area of a parallelogram with sides $\mathbf{v}$ and $\mathbf{w}$ is $\|\mathbf{v}\| \|\mathbf{w}\| \sin(\theta)$. To this end, recall that the area of a parallelogram is equal to its base times its height. The length of its base is straightforward to compute—it is $\|\mathbf{v}\|$. To determine its height, we drop a line from the head of $\mathbf{w}$ perpendicularly down to $\mathbf{v}$, as in Figure 1.25(a).

The choice of $\mathbf{v}$ as the base here is arbitrary—we could instead choose $\mathbf{w}$ as the base and get the same final answer.

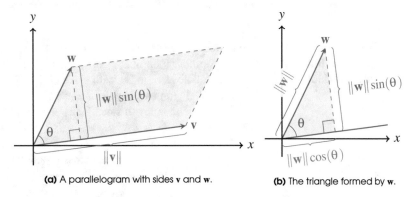

(a) A parallelogram with sides $\mathbf{v}$ and $\mathbf{w}$.   **(b)** The triangle formed by $\mathbf{w}$.

Figure 1.25: A parallelogram with sides $\mathbf{v}$ and $\mathbf{w}$ has base length $\|\mathbf{v}\|$ and height $\|\mathbf{w}\| \sin(\theta)$, and thus area equal to $\|\mathbf{v}\| \|\mathbf{w}\| \sin(\theta)$.

This process creates a right-angled triangle whose legs have lengths $\|\mathbf{w}\| \cos(\theta)$ and $\|\mathbf{w}\| \sin(\theta)$, as in Figure 1.25(b). In particular, the side perpendicular to $\mathbf{v}$ has length $\|\mathbf{w}\| \sin(\theta)$, which is thus the height of the parallelogram with sides $\mathbf{v}$ and $\mathbf{w}$. The area of that parallelogram is then $\|\mathbf{v}\| \|\mathbf{w}\| \sin(\theta)$, as claimed.

The equivalence of (a) and (b) in this theorem is sometimes called **Lagrange's identity**.

Next, we show that the quantities (a) and (b) are the same by verifying that $\|\mathbf{v} \times \mathbf{w}\|^2 = \|\mathbf{v}\|^2 \|\mathbf{w}\|^2 - (\mathbf{v} \cdot \mathbf{w})^2$ and then taking the square root of both sides of the equation. To this end, we simply compute

$$\|\mathbf{v} \times \mathbf{w}\|^2 = \left\| (v_2 w_3 - v_3 w_2, v_3 w_1 - v_1 w_3, v_1 w_2 - v_2 w_1) \right\|^2$$
$$= (v_2 w_3 - v_3 w_2)^2 + (v_3 w_1 - v_1 w_3)^2 + (v_1 w_2 - v_2 w_1)^2$$
$$= v_2^2 w_3^2 + v_3^2 w_2^2 + v_3^2 w_1^2 + v_1^2 w_3^2 + v_1^2 w_2^2 + v_2^2 w_1^2$$
$$- 2(v_2 v_3 w_2 w_3 + v_1 v_3 w_1 w_3 + v_1 v_2 w_1 w_2).$$

On the other hand, we can also expand $\|\mathbf{v}\|^2 \|\mathbf{w}\|^2 - (\mathbf{v} \cdot \mathbf{w})^2$ to get

The terms $v_1^2 w_1^2$, $v_2^2 w_2^2$, and $v_3^2 w_3^2$ were both added and subtracted here and thus canceled out.

$$\|\mathbf{v}\|^2 \|\mathbf{w}\|^2 - (\mathbf{v} \cdot \mathbf{w})^2 = (v_1^2 + v_2^2 + v_3^2)(w_1^2 + w_2^2 + w_3^2) - (v_1 w_1 + v_2 w_2 + v_3 w_3)^2$$
$$= v_1^2 w_2^2 + v_1^2 w_3^2 + v_2^2 w_1^2 + v_2^2 w_3^2 + v_3^2 w_1^2 + v_3^2 w_2^2$$
$$- 2(v_2 v_3 w_2 w_3 + v_1 v_3 w_1 w_3 + v_1 v_2 w_1 w_2),$$

which is the exact same as the expanded form of $\|\mathbf{v} \times \mathbf{w}\|^2$ that we computed previously. We thus conclude that $\|\mathbf{v} \times \mathbf{w}\|^2 = \|\mathbf{v}\|^2 \|\mathbf{w}\|^2 - (\mathbf{v} \cdot \mathbf{w})^2$, as desired.

Finally, to see that the quantities (b) and (c) are the same, we use the fact that the angle between $\mathbf{v}$ and $\mathbf{w}$ is $\theta = \arccos\left(\mathbf{v}\cdot\mathbf{w}/(\|\mathbf{v}\|\|\mathbf{w}\|)\right)$ (recall Definition 1.2.3), which can be rearranged into the form $\mathbf{v}\cdot\mathbf{w} = \|\mathbf{v}\|\|\mathbf{w}\|\cos(\theta)$. Plugging this formula for $\mathbf{v}\cdot\mathbf{w}$ into (b) shows that

> Remember that $\sin^2(\theta) + \cos^2(\theta) = 1$, so $1 - \cos^2(\theta) = \sin^2(\theta)$.

$$\sqrt{\|\mathbf{v}\|^2\|\mathbf{w}\|^2 - (\mathbf{v}\cdot\mathbf{w})^2} = \sqrt{\|\mathbf{v}\|^2\|\mathbf{w}\|^2 - \|\mathbf{v}\|^2\|\mathbf{w}\|^2\cos^2(\theta)}$$
$$= \sqrt{\|\mathbf{v}\|^2\|\mathbf{w}\|^2(1 - \cos^2(\theta))}$$
$$= \sqrt{\|\mathbf{v}\|^2\|\mathbf{w}\|^2\sin^2(\theta)}$$
$$= \|\mathbf{v}\|\|\mathbf{w}\|\sin(\theta),$$

where the final equality makes use of the fact that $\sqrt{\sin^2(\theta)} = |\sin(\theta)| = \sin(\theta)$ since $0 \le \theta \le \pi$, so $\sin(\theta) \ge 0$. This completes the proof. ∎

Example 1.A.2

Areas Via the Cross Product

Compute the areas of the following regions:
 a) The parallelogram in $\mathbb{R}^3$ with sides $(1,2,-1)$ and $(2,1,2)$,
 b) the parallelogram in $\mathbb{R}^2$ with sides $(1,2)$ and $(3,1)$,
 c) the parallelogram in $\mathbb{R}^4$ with sides $(2,2,-1,1)$ and $(-1,1,2,1)$, and
 d) the triangle in $\mathbb{R}^2$ with sides $(1,2)$ and $(3,1)$.

Solutions:
 a) We compute the length of the cross product: $(1,2,-1) \times (2,1,2) = (5,-4,-3)$, which has length $\sqrt{5^2 + (-4)^2 + (-3)^2} = 5\sqrt{2}$.

 b) To compute this area, we can embed these vectors into $\mathbb{R}^3$ by adding a z-entry equal to 0 and then using the cross product: $(1,2,0) \times (3,1,0) = (0,0,-5)$, which has length 5.

 c) We cannot use the cross product here since these vectors live in $\mathbb{R}^4$, but we can still use formula (b) from Theorem 1.A.2 to see that the area of this parallelogram is

$$\sqrt{\|\mathbf{v}\|^2\|\mathbf{w}\|^2 - (\mathbf{v}\cdot\mathbf{w})^2} = \sqrt{10\cdot 7 - (-1)^2} = \sqrt{69}.$$

 d) The triangle with sides $\mathbf{v}$ and $\mathbf{w}$ has exactly half of the area of the parallelogram with the same sides, as illustrated below, so the area of this triangle is $5/2$ (recall that we found that this parallelogram has area 5 in part (b)).

> A parallelogram with sides $\mathbf{v}$ and $\mathbf{w}$ is made up of two copies of the triangle with sides $\mathbf{v}$ and $\mathbf{w}$, and thus has double the area.

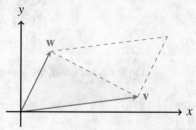

1.A.2 Volumes

We can also extend these ideas slightly to learn about the volume of parallelepipeds in $\mathbb{R}^3$, but we need to make use of both the cross product and the dot product in order to make it work:

Theorem 1.A.3

Volume of a Parallelepiped

A parallelepiped is a slanted and stretched cube in $\mathbb{R}^3$, just like a parallelogram is a slanted and stretched square in $\mathbb{R}^2$.

Let $\mathbf{v}, \mathbf{w}, \mathbf{x} \in \mathbb{R}^3$ be vectors. Then the volume of the parallelepiped with sides $\mathbf{v}$, $\mathbf{w}$, and $\mathbf{x}$ is $|\mathbf{v} \cdot (\mathbf{w} \times \mathbf{x})|$.

Proof. We first expand the expression $|\mathbf{v} \cdot (\mathbf{w} \times \mathbf{x})|$ in terms of the lengths of $\mathbf{v}$ and $\mathbf{w} \times \mathbf{x}$. If θ is the angle between $\mathbf{v}$ and $\mathbf{w} \times \mathbf{x}$, then Definition 1.2.3 tells us that $|\mathbf{v} \cdot (\mathbf{w} \times \mathbf{x})| = \|\mathbf{v}\| \|\mathbf{w} \times \mathbf{x}\| |\cos(\theta)|$.

Our goal now becomes showing that this quantity equals the volume of the parallelepiped with sides $\mathbf{v}$, $\mathbf{w}$, and $\mathbf{x}$. This parallelepiped's volume is equal to the area of its base times its height. However, its base (when oriented as in Figure 1.26) is a parallelogram with sides $\mathbf{w}$ and $\mathbf{x}$, and thus has area $\|\mathbf{w} \times \mathbf{x}\|$ according to Theorem 1.A.2.

We must use $|\cos(\theta)|$ rather than just $\cos(\theta)$ since $\mathbf{w} \times \mathbf{x}$ might point up or down.

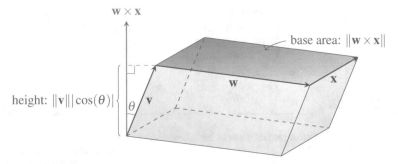

Figure 1.26: The volume of a parallelepiped with sides $\mathbf{v}$, $\mathbf{w}$, and $\mathbf{x}$ is the area of its base $\|\mathbf{w} \times \mathbf{x}\|$ times its height $\|\mathbf{v}\| |\cos(\theta)|$, where θ is the angle between $\mathbf{w} \times \mathbf{x}$ and $\mathbf{v}$.

The quantity $\mathbf{v} \cdot (\mathbf{w} \times \mathbf{x})$ is called the **scalar triple product** of $\mathbf{v}$, $\mathbf{w}$, and $\mathbf{x}$—see Exercise 1.A.16.

It follows that all we have left to do is show that the height of this parallelepiped is $\|\mathbf{v}\| |\cos(\theta)|$. To verify this claim, recall that $\mathbf{w} \times \mathbf{x}$ is perpendicular to each of $\mathbf{w}$ and $\mathbf{x}$, and is thus perpendicular to the base parallelogram. The height of the parallelepiped is then the amount that $\mathbf{v}$ points in the direction of $\mathbf{w} \times \mathbf{x}$, which we can see is indeed $\|\mathbf{v}\| |\cos(\theta)|$ by drawing a right-angled triangle with hypotenuse $\mathbf{v}$, as in Figure 1.26. ∎

Example 1.A.3

Volume Via the Cross Product

Compute the volume of the parallelepiped with sides $(1,0,1)$, $(-1,2,2)$, and $(3,2,1)$.

Solution:

We use the formula provided by Theorem 1.A.3:

$$|(1,0,1) \cdot ((-1,2,2) \times (3,2,1))| = |(1,0,1) \cdot (-2,7,-8)|$$
$$= |-2-8| = 10.$$

Remark 1.A.1

Why Parallelograms and Parallelepipeds?

At this point, we might wonder why we care about the areas of parallelograms and volumes of parallelepipeds in the first place—they perhaps seem like somewhat arbitrary shapes to focus so much attention on.

The answer is that they are exactly the shapes that a square or a cube can be transformed into by a linear transformation. For example, a linear transformation $T : \mathbb{R}^2 \to \mathbb{R}^2$ moves the sides $\mathbf{e}_1$ and $\mathbf{e}_2$ of the unit square to $T(\mathbf{e}_1)$ and $T(\mathbf{e}_2)$, which are the sides of a parallelogram. The area of this parallelogram (which we can compute via Theorem 1.A.2) describes how much T has stretched or shrunk the unit square and thus $\mathbb{R}^2$ as a whole:

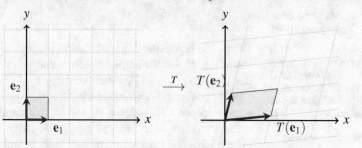

This measure of how much a linear transformation stretches space is called its **determinant**, and the techniques that we presented in this section let us compute it for linear transformations acting on $\mathbb{R}^2$ or $\mathbb{R}^3$. We will investigate determinants of general linear transformations on $\mathbb{R}^n$ in Section 3.2.

Exercises

solutions to starred exercises on page 442

1.A.1 Compute the cross product $\mathbf{v} \times \mathbf{w}$ of the following pairs of vectors:

* *(a) $\mathbf{v} = (1,2,3)$ and $\mathbf{w} = (3,2,1)$
* (b) $\mathbf{v} = (87,17,-43)$ and $\mathbf{w} = (87,17,-43)$
* *(c) $\mathbf{v} = (-1,2,0)$ and $\mathbf{w} = (-4,1,-2)$
* (d) $\mathbf{v} = (1,0,0)$ and $\mathbf{w} = (0,1,0)$

1.A.2 Compute the area of the parallelograms with the following pairs of vectors as their sides:

* *(a) $\mathbf{v} = (2,1)$ and $\mathbf{w} = (-2,3)$
* (b) $\mathbf{v} = (3,0,0)$ and $\mathbf{w} = (0,4,0)$
* *(c) $\mathbf{v} = (1,2,3)$ and $\mathbf{w} = (3,-1,2)$
* (d) $\mathbf{v} = (1,0,1,-2)$ and $\mathbf{w} = (-2,1,3,1)$

1.A.3 Compute the area of the triangles with the following pairs of vectors as their sides:

* * (a) $\mathbf{v} = (0,4)$ and $\mathbf{w} = (1,1)$
* (b) $\mathbf{v} = (0,2,2)$ and $\mathbf{w} = (1,-2,1)$
* * (c) $\mathbf{v} = (-1,1,-1)$ and $\mathbf{w} = (3,2,1)$
* (d) $\mathbf{v} = (1,2,1,2)$ and $\mathbf{w} = (0,-1,2,1)$

1.A.4 Compute the volume of the parallelepipeds with the following sets of vectors as their sides:

* * (a) $\mathbf{v} = (1,0,0)$, $\mathbf{w} = (0,2,0)$, and $\mathbf{x} = (0,0,3)$
* (b) $\mathbf{v} = (0,4,1)$, $\mathbf{w} = (1,1,0)$, and $\mathbf{x} = (2,0,-1)$
* * (c) $\mathbf{v} = (1,1,1)$, $\mathbf{w} = (2,-1,1)$, and $\mathbf{x} = (2,-2,-1)$
* (d) $\mathbf{v} = (-2,1,3)$, $\mathbf{w} = (1,-4,2)$, and $\mathbf{x} = (3,3,-2)$

1.A.5 Find the area of the parallelogram or parallelepiped that the unit square or cube is sent to by the linear transformations with the following standard matrices. [Hint: Refer back to Remark 1.A.1.]

* (a) $\begin{bmatrix} 1 & 0 \\ 0 & 1 \end{bmatrix}$ 　　(b) $\begin{bmatrix} 1 & 7 \\ 0 & 1 \end{bmatrix}$

* (c) $\begin{bmatrix} 0 & -2 \\ 2 & 3 \end{bmatrix}$ 　　(b) $\begin{bmatrix} 1 & 2 \\ 3 & 4 \end{bmatrix}$

* (e) $\begin{bmatrix} 1 & -3 & 2 \\ 0 & 2 & -7 \\ 0 & 0 & 3 \end{bmatrix}$ 　(b) $\begin{bmatrix} 1 & 2 & 3 \\ 4 & 5 & 6 \\ 7 & 8 & 9 \end{bmatrix}$

1.A.6 Determine which of the following statements are true and which are false.

* (a) The cross product of two unit vectors is a unit vector.
* (b) If $\mathbf{v}, \mathbf{w} \in \mathbb{R}^3$ then $\|\mathbf{v} \times \mathbf{w}\| = \|\mathbf{w} \times \mathbf{v}\|$.
* (c) If $\mathbf{v} \in \mathbb{R}^3$ then $\mathbf{v} \times \mathbf{v} = \mathbf{v}^2$.
* (d) If $\mathbf{v}, \mathbf{w} \in \mathbb{R}^3$ then $(\mathbf{v}+\mathbf{v}) \times \mathbf{w} = 2(\mathbf{v} \times \mathbf{w})$.
* *(e) If $\mathbf{v}, \mathbf{w} \in \mathbb{R}^4$ then $\mathbf{v} \times \mathbf{w} = -(\mathbf{w} \times \mathbf{v})$.
* (f) There exist vectors $\mathbf{v}, \mathbf{w} \in \mathbb{R}^3$ with $\|\mathbf{v}\| = 1$, $\|\mathbf{w}\| = 1$, and $\mathbf{v} \times \mathbf{w} = (1,1,1)$.

1.A.7 Recall that $\mathbf{e}_i$ is the vector whose i-th entry is 1 and all other entries are 0. Compute the following cross products: $\mathbf{e}_1 \times \mathbf{e}_2$, $\mathbf{e}_2 \times \mathbf{e}_3$, and $\mathbf{e}_3 \times \mathbf{e}_1$.

* **1.A.8** Suppose that $\mathbf{v}$ and $\mathbf{w}$ are unit vectors in $\mathbb{R}^3$ with $\mathbf{v} \times \mathbf{w} = (1/3, 2/3, 2/3)$. What is the angle between $\mathbf{v}$ and $\mathbf{w}$?

1.A.9 Suppose that $\mathbf{v}, \mathbf{w}, \mathbf{x} \in \mathbb{R}^3$ are vectors. Show that $\|\mathbf{v} \times \mathbf{w}\| = \|\mathbf{v}\|\|\mathbf{w}\|$ if and only if $\mathbf{v}$ and $\mathbf{w}$ are orthogonal.

∗∗1.A.10 We showed that if $\mathbf{v}, \mathbf{w} \in \mathbb{R}^3$ then $\mathbf{v} \cdot (\mathbf{v} \times \mathbf{w}) = 0$. Show that $\mathbf{w} \cdot (\mathbf{v} \times \mathbf{w}) = 0$ too.

∗∗1.A.11 Recall Theorem 1.A.1, which established some of the basic properties of the cross product.

(a) Prove part (b) of the theorem.
(b) Prove part (d) of the theorem.

1.A.12 Suppose that $\mathbf{v}, \mathbf{w} \in \mathbb{R}^3$ are vectors and $c \in \mathbb{R}$ is a scalar. Show that $\mathbf{v} \times (c\mathbf{w}) = c(\mathbf{v} \times \mathbf{w})$.

∗1.A.13 Is the cross product associative? In other words, is it true that $(\mathbf{v} \times \mathbf{w}) \times \mathbf{x} = \mathbf{v} \times (\mathbf{w} \times \mathbf{x})$ for all $\mathbf{v}, \mathbf{w}, \mathbf{x} \in \mathbb{R}^3$?

1.A.14 In Theorem 1.A.1 we showed that $\mathbf{v} \times \mathbf{v} = \mathbf{0}$. Prove the stronger statement that $\mathbf{v} \times \mathbf{w} = \mathbf{0}$ if and only if $\mathbf{w} = \mathbf{0}$ or $\mathbf{v} = c\mathbf{w}$ for some $c \in \mathbb{R}$.

1.A.15 Suppose that $\mathbf{v}, \mathbf{w}, \mathbf{x} \in \mathbb{R}^3$ are non-zero vectors.

(a) Give an example to show that it is possible that $\mathbf{v} \times \mathbf{w} = \mathbf{v} \times \mathbf{x}$ even if $\mathbf{w} \neq \mathbf{x}$.
(b) Show that if $\mathbf{v} \times \mathbf{w} = \mathbf{v} \times \mathbf{x}$ *and* $\mathbf{v} \cdot \mathbf{w} = \mathbf{v} \cdot \mathbf{x}$ then it must be the case that $\mathbf{w} = \mathbf{x}$.

∗∗1.A.16 Suppose that $\mathbf{v}, \mathbf{w}, \mathbf{x} \in \mathbb{R}^3$ are vectors.

(a) Show that $\mathbf{v} \cdot (\mathbf{w} \times \mathbf{x}) = \mathbf{w} \cdot (\mathbf{x} \times \mathbf{v}) = \mathbf{x} \cdot (\mathbf{v} \times \mathbf{w})$. [Side note: In other words, the scalar triple product is unchanged under cyclic shifts of $\mathbf{v}$, $\mathbf{w}$, and $\mathbf{x}$.]
(b) Show that

$$\mathbf{v} \cdot (\mathbf{w} \times \mathbf{x}) = -\mathbf{v} \cdot (\mathbf{x} \times \mathbf{w})$$
$$= -\mathbf{w} \cdot (\mathbf{v} \times \mathbf{x})$$
$$= -\mathbf{x} \cdot (\mathbf{w} \times \mathbf{v}).$$

[Side note: In other words, the scalar triple product is negated if we swap any two of the three vectors.]

1.A.17 Suppose that $\mathbf{v}, \mathbf{w}, \mathbf{x} \in \mathbb{R}^3$ are vectors. Show that $\mathbf{v} \times (\mathbf{w} \times \mathbf{x}) = (\mathbf{v} \cdot \mathbf{x})\mathbf{w} - (\mathbf{v} \cdot \mathbf{w})\mathbf{x}$. [Side note: This is called the **vector triple product** of $\mathbf{v}$, $\mathbf{w}$, and $\mathbf{x}$.]

∗1.A.18 Suppose that $\mathbf{v}, \mathbf{w}, \mathbf{x} \in \mathbb{R}^3$ are vectors. Show that

$$\mathbf{v} \times (\mathbf{w} \times \mathbf{x}) + \mathbf{w} \times (\mathbf{x} \times \mathbf{v}) + \mathbf{x} \times (\mathbf{v} \times \mathbf{w}) = \mathbf{0}.$$

[Side note: This is called the **Jacobi identity**.]

1.A.19 Suppose that $\mathbf{v}, \mathbf{w} \in \mathbb{R}^3$ are vectors. Show that

$$\mathbf{v} \times \mathbf{w} = \frac{1}{2}(\mathbf{v} - \mathbf{w}) \times (\mathbf{v} + \mathbf{w}).$$

1.B   Extra Topic: Paths in Graphs

While we already saw one use of matrices (they give us a method of representing linear transformations), we stress that this is just one particular application of them. As another example to illustrate the utility of matrices and matrix multiplication, we now explore how they can help us uncover properties of graphs.

1.B.1   Undirected Graphs

A **graph** is a finite set of vertices, together with a set of edges connecting those vertices. Vertices are typically drawn as dots (sometimes with labels like $A, B, C, \ldots$) and edges are typically drawn as (not necessarily straight) lines connecting those dots, as in Figure 1.27.

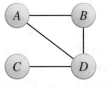

(a) A graph with 4 vertices.

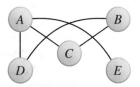

(b) A graph with 5 vertices.

Figure 1.27: Some small graphs. Notice that edges need not be straight, and they are allowed to cross each other.

For now, the graphs we consider are **undirected**: edges connect two vertices, but there is no difference between an edge that connects vertex i to j versus one that connects j to i.

We think of the vertices as objects of some type, and edges as representing the existence of a relationship between those objects. For example, a graph might represent

- A collection of cities (vertices) and the roads that connect them (edges),
- People (vertices) and the friendships that they have with other people on a social networking website (edges), or
- Satellites (vertices) and communication links between them (edges).

We emphasize that a graph is determined only by which vertices and edges between vertices are present—the particular locations of the vertices and methods of drawing the edges are unimportant. For example, the two graphs displayed in Figure 1.28 are in fact the exact same graph, despite looking quite different on the surface.

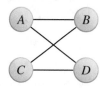

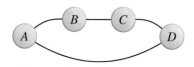

(a) A graph with 4 vertices.

(b) The same graph as the one on the left, but represented in a different way.

Figure 1.28: Two different representations of the same graph. The geometric arrangement of the vertices and edges does not matter—only the number of vertices and presence or absence of edges between those vertices is relevant.

A problem that comes up fairly frequently is how to count the number of paths of a certain length between different vertices on a graph. For example, this could tell us how many ways there are to drive from Toronto to Montréal, passing through exactly three other cities on the way, or the number of friends of friends that we have on a social networking website. For small graphs, it is straightforward enough to count the paths of small lengths by hand.

Example 1.B.1

Counting Short Paths by Hand

Count the number of paths of the indicated type in the graph displayed in Figure 1.27(a):

a) Of length 2 from A to B, and

b) of length 3 from A to D.

Solutions:

a) We simply examine the graph and notice that there is only one such path: A–D–B.

By "path of length 3", we mean a path of length *exactly* 3; a path of length 2 does not count.

b) Paths of length 3 are a bit more difficult to eyeball, but some examination reveals that there are 4 such paths:

$$A-B-A-D, \quad A-D-A-D, \quad A-D-B-D, \quad \text{and} \quad A-D-C-D.$$

As the size of the graph or the length of the paths increases, counting these paths by hand becomes increasingly impractical. Even for paths just of length 3 or 4, it's often difficult to be sure that we have found *all* paths of the indicated length. To lead us toward a better way of solving these types of problems, we construct a matrix that describes the graph:

Definition 1.B.1	The **adjacency matrix** of a graph with n vertices is the matrix $A \in \mathcal{M}_n$
Adjacency Matrix	whose entries are

$$a_{i,j} = \begin{cases} 1, & \text{if there is an edge between the } i-\text{th and } j\text{-th vertices} \\ 0, & \text{otherwise.} \end{cases}$$

The adjacency matrix of a graph is always **symmetric**: $a_{i,j} = a_{j,i}$ for all i, j (equivalently, $A = A^T$).

The adjacency matrix of a graph encodes all of the information about the graph—its size is the number of vertices and its entries indicate which pairs of vertices do and do not have edges between them. We can think of the adjacency matrix as a compact representation of a graph that eliminates the (irrelevant) details of how the vertices are oriented in space and how the edges between them are drawn. For example, the graphs in Figure 1.28 have the same adjacency matrices (since they are the same graph).

Example 1.B.2

Constructing Adjacency Matrices

Construct the adjacency matrix of the graph from each of the following figures:
a) Figure 1.27(a),
b) Figure 1.27(b), and
c) Figure 1.28.

Solutions:
a) Since vertex A is connected to vertices B and D via edges, we place a 1 in the 2nd and 4th entries of the first row of the adjacency matrix. Using similar reasoning for the other rows results in the following adjacency matrix:

$$\begin{bmatrix} 0 & 1 & 0 & 1 \\ 1 & 0 & 0 & 1 \\ 0 & 0 & 0 & 1 \\ 1 & 1 & 1 & 0 \end{bmatrix}.$$

b) Since there are 5 vertices, this adjacency matrix is 5×5:

$$\begin{bmatrix} 0 & 0 & 1 & 1 & 1 \\ 0 & 0 & 1 & 1 & 0 \\ 1 & 1 & 0 & 0 & 0 \\ 1 & 1 & 0 & 0 & 0 \\ 1 & 0 & 0 & 0 & 0 \end{bmatrix}.$$

If the vertices of a graph are not labeled then there is some ambiguity in the definition of its adjacency matrix, since it is not clear which vertices are the first, second, third, and so on. In practice, this does not matter much—the vertices can be ordered in any arbitrary way without changing the adjacency matrix's important properties.

c) Both of the graphs in this figure really are the same (only their geometric representation differs), so we get the same answer regardless of which of them we use:

$$\begin{bmatrix} 0 & 1 & 0 & 1 \\ 1 & 0 & 1 & 0 \\ 0 & 1 & 0 & 1 \\ 1 & 0 & 1 & 0 \end{bmatrix}.$$

One way of thinking of the adjacency matrix A is that the entry $a_{i,j}$ counts the number of paths of length 1 between vertices i and j (since $a_{i,j} = 1$ if there is such a path and $a_{i,j} = 0$ otherwise). Remarkably, the entries of higher powers of the adjacency matrix similarly count the number of paths of longer lengths between vertices.

Theorem 1.B.1

**Counting Paths
in Graphs**

Suppose $A \in \mathcal{M}_n$ is the adjacency matrix of a graph, and let $k \geq 1$ be an integer. Then for each $1 \leq i, j \leq n$, the number of paths of length k between vertices i and j is equal to $\left[A^k\right]_{i,j}$.

Proof. We already noted why this theorem is true when $k = 1$. To show that it is true when $k = 2$, recall from the definition of matrix multiplication that

Recall that the notation $[A^k]_{i,j}$ means the (i,j)-entry of the matrix A^k.

$$\left[A^2\right]_{i,j} = a_{i,1}a_{1,j} + a_{i,2}a_{2,j} + \cdots + a_{i,n}a_{n,j}.$$

The first term in the sum above $(a_{i,1}a_{1,j})$ equals 1 exactly if there is an edge between vertices i and 1 and also an edge between vertices 1 and j. In other words, the quantity $a_{i,1}a_{1,j}$ counts the number of paths of length 2 between vertices i and j, with vertex 1 as the intermediate vertex.

Similarly, $a_{i,2}a_{2,j}$ counts the number of paths of length 2 between vertices i and j with vertex 2 as the intermediate vertex, and so on. By adding these terms up, we see that $[A^2]_{i,j}$ counts the number of paths of length 2 between vertices i and j, with *any* vertex as the intermediate vertex.

The general result for arbitrary $k \geq 1$ can be proved in a very similar manner via induction, but is left as Exercise 1.B.11.   ∎

Example 1.B.3

**Counting Paths
via the
Adjacency
Matrix**

Count the number of paths of the indicated type in the following graph:

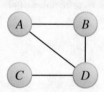

Be careful: we use A to refer to a vertex in the graph as well as the graph's adjacency matrix. It should be clear from context which one we mean.

a) Of length 3 from A to D, and
b) of length 6 from B to C.

Solutions:

a) We computed the adjacency matrix of this graph in Example 1.B.2(a):

$$A = \begin{bmatrix} 0 & 1 & 0 & 1 \\ 1 & 0 & 0 & 1 \\ 0 & 0 & 0 & 1 \\ 1 & 1 & 1 & 0 \end{bmatrix}.$$

To compute the number of paths of length 3, we compute A^3:

We will see an extremely fast way to compute large powers of matrices (and thus the number of long paths in graphs) in Section 3.4.

$$A^3 = \begin{bmatrix} 0 & 1 & 0 & 1 \\ 1 & 0 & 0 & 1 \\ 0 & 0 & 0 & 1 \\ 1 & 1 & 1 & 0 \end{bmatrix} \begin{bmatrix} 0 & 1 & 0 & 1 \\ 1 & 0 & 0 & 1 \\ 0 & 0 & 0 & 1 \\ 1 & 1 & 1 & 0 \end{bmatrix} \begin{bmatrix} 0 & 1 & 0 & 1 \\ 1 & 0 & 0 & 1 \\ 0 & 0 & 0 & 1 \\ 1 & 1 & 1 & 0 \end{bmatrix}$$

$$= \begin{bmatrix} 2 & 1 & 1 & 1 \\ 1 & 2 & 1 & 1 \\ 1 & 1 & 1 & 0 \\ 1 & 1 & 0 & 3 \end{bmatrix} \begin{bmatrix} 0 & 1 & 0 & 1 \\ 1 & 0 & 0 & 1 \\ 0 & 0 & 0 & 1 \\ 1 & 1 & 1 & 0 \end{bmatrix} = \begin{bmatrix} 2 & 3 & 1 & 4 \\ 3 & 2 & 1 & 4 \\ 1 & 1 & 0 & 3 \\ 4 & 4 & 3 & 2 \end{bmatrix}.$$

Since the $(1,4)$-entry of this matrix is 4, there are 4 such paths from the first vertex (A) to the fourth vertex (D). Notice that this is exactly the same answer that we computed for this problem in Example 1.B.1(b).

b) To count paths of length 6, we need to compute A^6. While we could laboriously multiply five times to get $A^6 = AAAAAA$, it is quicker to recall that $A^6 = (A^3)^2$, so we can just square the matrix that we computed in part (a):

We did not need to compute the entire matrix A^6 here—we could have just computed its $(2,3)$-entry to save time and effort.

$$A^6 = (A^3)^2 = \begin{bmatrix} 2 & 3 & 1 & 4 \\ 3 & 2 & 1 & 4 \\ 1 & 1 & 0 & 3 \\ 4 & 4 & 3 & 2 \end{bmatrix}^2 = \begin{bmatrix} 30 & 29 & 17 & 31 \\ 29 & 30 & 17 & 31 \\ 17 & 17 & 11 & 14 \\ 31 & 31 & 14 & 45 \end{bmatrix}.$$

Since the $(2,3)$-entry of this matrix is 17, there are 17 paths of length 6 from the second vertex (B) to the third vertex (C).

Example 1.B.4

Counting Airline Routes

The following graph represents nonstop airline routes between some cities in Eastern Canada:

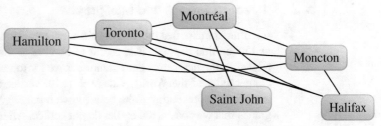

a) How many routes are there from Moncton to Hamilton with exactly one stopover in another city in between?
b) How many routes are there from Saint John to Halifax with no more than two stopovers in between?

Solutions:

Choosing a different ordering of the cities might change the adjacency matrix, but not the final answer.

a) We start by constructing the adjacency matrix of this graph. Note that we must order the vertices in some way, so we (arbitrarily) order the vertices Hamilton–Toronto–Montréal–Saint John–Moncton–

Halifax, which results in the adjacency matrix

$$A = \begin{bmatrix} 0 & 1 & 0 & 0 & 1 & 1 \\ 1 & 0 & 1 & 1 & 1 & 1 \\ 0 & 1 & 0 & 1 & 1 & 1 \\ 0 & 1 & 1 & 0 & 0 & 0 \\ 1 & 1 & 1 & 0 & 0 & 1 \\ 1 & 1 & 1 & 0 & 1 & 0 \end{bmatrix}.$$

The number of routes from Moncton to Hamilton with exactly one stopover in between is then $[A^2]_{5,1}$, so we compute A^2:

A route with one stopover is a path of length 2, a route with two stopovers is a path of length 3, and so on.

$$A^2 = \begin{bmatrix} 3 & 2 & 3 & 1 & 2 & 2 \\ 2 & 5 & 3 & 1 & 3 & 3 \\ 3 & 3 & 4 & 1 & 2 & 2 \\ 1 & 1 & 1 & 2 & 2 & 2 \\ 2 & 3 & 2 & 2 & 4 & 3 \\ 2 & 3 & 2 & 2 & 3 & 4 \end{bmatrix}.$$

Since the $(5,1)$-entry of this matrix is 2, there are 2 such routes.

b) The number of routes with *no more than* two stopovers equals $[A^3]_{4,6} + [A^2]_{4,6} + a_{4,6}$, so we start by computing $[A^3]_{4,6}$, which is the dot product of the 4th row of A with the 6th column of A^2:

$$[A^3]_{4,6} = (0, 1, 1, 0, 0, 0) \cdot (2, 3, 2, 2, 3, 4) = 3 + 2 = 5.$$

The number of such routes is thus

$$[A^3]_{4,6} + [A^2]_{4,6} + a_{4,6} = 5 + 2 + 0 = 7.$$

1.B.2 Directed Graphs and Multigraphs

It is often the case that the relationship between objects is not entirely symmetric, and when this happens it is useful to be able to distinguish which object is related to the other one. For example, if we try to use a graph to represent the connectivity of the World Wide Web (with vertices representing web pages and edges representing links between web pages), we find that the types of graphs that we considered earlier do not suffice. After all, it is entirely possible that Web Page A links to Web Page B, but not vice-versa—should there be an edge between the two vertices corresponding to those web pages?

Directed graphs are sometimes called **digraphs** for short.

To address situations like this, we use **directed graphs**, which consist of vertices and edges just like before, except the edges point *from* one vertex to another (whereas they just *connected* two vertices in our previous undirected setup). Some examples are displayed in Figure 1.29.

Fortunately, counting paths in directed graphs is barely any different from doing so in undirected graphs—we construct the adjacency matrix of the graph and then look at the entries of its powers. The only difference is that the adjacency matrix of a directed graph has a 1 in its (i, j)-entry if the graph has an edge going *from* vertex i to vertex j, so it is no longer necessarily the case that $a_{i,j} = a_{j,i}$.

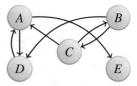

(a) A directed graph with 4 vertices.   **(b)** A directed graph with 5 vertices.

Figure 1.29: Some small directed graphs. Notice that vertices can now potentially have a pair of edges between them—one going in each direction.

Example 1.B.5

Counting Paths in Directed Graphs

Count the number of paths of the indicated type in the graph displayed in Figure 1.29(b):

a) Of length 2 from B to A,

b) of length 4 from C to E, and

c) of length 87 from A to C.

Solutions:

Unlike in the case of undirected graphs, the adjacency matrix of a directed graph is not necessarily symmetric.

a) We start by computing the adjacency matrix of this graph:

$$A = \begin{bmatrix} 0 & 0 & 0 & 1 & 1 \\ 0 & 0 & 1 & 1 & 0 \\ 1 & 1 & 0 & 0 & 0 \\ 1 & 0 & 0 & 0 & 0 \\ 0 & 0 & 0 & 0 & 0 \end{bmatrix}.$$

To count the number of paths of length 2, we compute A^2:

$$A^2 = \begin{bmatrix} 1 & 0 & 0 & 0 & 0 \\ 2 & 1 & 0 & 0 & 0 \\ 0 & 0 & 1 & 2 & 1 \\ 0 & 0 & 0 & 1 & 1 \\ 0 & 0 & 0 & 0 & 0 \end{bmatrix}.$$

Since the $(2,1)$-entry of this matrix is 2, there are 2 such paths from B to A.

We could explicitly compute the entire matrix A^4, but that would be more work than we have to do to solve this problem.

b) We need to compute $[A^4]_{3,5}$, which is the dot product of the 3rd row of A^2 with the 5th column of A^2:

$$[A^4]_{3,5} = (0,0,1,2,1) \cdot (0,0,1,1,0) = 0+0+1+2+0 = 3.$$

There are thus 3 paths of length 4 from C to E.

c) We could compute A^{87} and look at its $(1,3)$-entry (ugh!), but an easier way to solve this problem is to notice that the only edge that points to C comes from B, and the only edge that points to B comes from C. There are thus no paths whatsoever, of any length, to C from any vertex other than B.

In fact, this argument shows that the third column of A^k alternates back and forth between $(0,1,0,0,0)$ when k is odd (and there is 1 path of length k from B to C) and $(0,0,1,0,0)$ when k is even (and there is 1 path of length k from C back to itself).

As an even further generalization of graphs, we can consider **multigraphs**, which are graphs that allow multiple edges between the same pair of vertices, and even allow edges that connect a vertex to itself. Multigraphs could be used to represent roads connecting cities, for example—after all, a pair of cities often has more than just one road connecting them.

Multigraphs can be either directed or undirected, as illustrated in Figure 1.30. In either case, the method of using the adjacency matrix to count paths between vertices still works. The only difference with multigraphs is that the adjacency matrix no longer consists entirely of 0s and 1s, but rather its entries $a_{i,j}$ describe *how many* edges there are from vertex i to vertex j.

> An edge from a vertex to itself is called a **loop**.

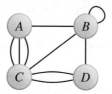 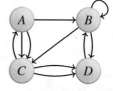

(a) An undirected multigraph.   **(b)** A directed multigraph.

Figure 1.30: Some small multigraphs, which allow for multiple edges between the same pairs of vertices, and even edges from a vertex to itself. Multigraphs can either be (a) undirected or (b) directed.

Example 1.B.6

Counting Paths in Multigraphs

Count the number of paths of the indicated type:
  a) Of length 2 from A to D in Figure 1.30(a), and
  b) of length 3 from B to C in Figure 1.30(b).

Solutions:
  a) The adjacency matrix of this graph is

$$A = \begin{bmatrix} 0 & 1 & 3 & 0 \\ 1 & 1 & 1 & 1 \\ 3 & 1 & 0 & 2 \\ 0 & 1 & 2 & 0 \end{bmatrix}.$$

To compute the number of paths of length 2 between vertices A and D, we compute $[A^2]_{1,4}$, which is the dot product of the 1st row of the adjacency matrix with its fourth column:

$$[A^2]_{1,4} = (0,1,3,0) \cdot (0,1,2,0) = 0+1+6+0 = 7.$$

  b) The adjacency matrix of this graph is

$$A = \begin{bmatrix} 0 & 1 & 2 & 0 \\ 0 & 1 & 1 & 1 \\ 1 & 0 & 0 & 2 \\ 0 & 1 & 0 & 0 \end{bmatrix}.$$

To compute the number of paths of length 3 from B to C, we need the quantity $[A^3]_{2,3}$. Before we can compute this though, we need A^2:

$$A^2 = \begin{bmatrix} 2 & 1 & 1 & 5 \\ 1 & 2 & 1 & 3 \\ 0 & 3 & 2 & 0 \\ 0 & 1 & 1 & 1 \end{bmatrix}.$$

Then $[A^3]_{2,3}$ is the dot product of the 2nd row of A^2 with the 3rd column of A:

$$[A^3]_{2,3} = (1,2,1,3) \cdot (2,1,0,0) = 2+2+0+0 = 4.$$

$[A^3]_{2,3}$ is also the dot product of the 2nd row of A with the 3rd column of A^2:
$(0,1,1,1) \cdot (1,1,2,1) = 4$ too.

Remark 1.B.1

Your Friends are More Popular than You Are

The friendship paradox was first observed by Scott L. Feld in 1991 (Fel91).

The notation $\sum_{j=1}^{n} w_j$ means $w_1 + w_2 + \cdots + w_n$.

Linear algebra is actually used for much more in graph theory than just as a tool for being able to count paths in graphs. As another application of the tools that we have learned so far, we now demonstrate the interesting (but slightly disheartening) fact that, on average, a person has fewer friends than their friends have. This counter-intuitive fact is sometimes called the **friendship paradox**.

To pin down this fact mathematically, let n be the number of people in the world ($n \approx 7.5$ billion) and let $\mathbf{w} = (w_1, w_2, \ldots, w_n) \in \mathbb{R}^n$ be such that, for each $1 \leq j \leq n$, w_j is the number of friends that the j-th person has. Then the average number of friends that people have is

$$\frac{1}{n} \sum_{j=1}^{n} w_j.$$ (average number of friendships = total friendships divided by number of people)

Counting the average number of friends of friends is somewhat more difficult, so it helps to draw a small graph to illustrate how it works. For now, we consider a graph that illustrates the friendships between just 4 people:

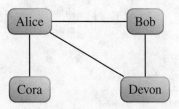

Now we imagine that each person in this graph creates one list for each of their friends, and those lists contain the names of the friends of that friend, as follows:

Alice's lists:				Bob's lists:	
Bob	**Cora**	**Devon**		**Alice**	**Devon**
Alice	Alice	Alice		Bob	Alice
Devon		Bob		Cora	Bob
				Devon	

Cora's list:		Devon's lists:	
Alice		**Alice**	**Bob**
Bob		Bob	Alice
Cora		Cora	Devon
Devon		Devon	

In this example, the average number of friends is $(3+2+1+2)/4 = 2$. Since $2 \leq 2.25$, we see that the friendship paradox indeed holds in this case.

The average number of friends of friends is the average length of these lists (excluding the bold headers), which is $(2+1+2+3+2+3+3+2)/8 = 2.25$ in this example.

In general, the j-th person creates w_j lists—one for each of their friends—so there are a total of $\sum_{j=1}^{n} w_j$ lists created. Furthermore, the j-th person will have a list made for them by each of their w_j friends, and each of those lists contain w_j entries. It follows that person j contributes w_j^2 entries to these lists, so there are $\sum_{j=1}^{n} w_j^2$ total friends of friends listed. The average number of friends that friends have is thus

Alice is friends with 3 people, each of whom make a list for her of length 3, so she contributes 9 total names to the lists. Similarly, Bob and Devon each contribute $2 \times 2 = 4$ names, and Cora contributes $1 \times 1 = 1$.

$$\frac{\sum_{j=1}^{n} w_j^2}{\sum_{j=1}^{n} w_j}. \qquad \text{(average number of friends of friends =}$$
$$\text{total names on lists divided by number of lists)}$$

If we let $\mathbf{v} = (1,1,\ldots,1) \in \mathbb{R}^n$ and then apply the Cauchy–Schwarz inequality to $\mathbf{v}$ and $\mathbf{w}$ then see that

$$\sum_{j=1}^{n} w_j = |\mathbf{v} \cdot \mathbf{w}| \leq \|\mathbf{v}\| \|\mathbf{w}\| = \sqrt{n} \sqrt{\sum_{j=1}^{n} w_j^2}.$$

Squaring both sides of the above inequality, and then dividing both sides by $n \sum_{j=1}^{n} w_j$, results in the inequality

$$\frac{1}{n} \sum_{j=1}^{n} w_j \leq \frac{\sum_{j=1}^{n} w_j^2}{\sum_{j=1}^{n} w_j},$$

which is exactly the friendship paradox: the average number of friends is no larger than the average number of friends that friends have.

We close this section by noting that the friendship paradox also applies to many things other than friendships. For example, our social media followers typically have more followers than we do, our mathematical co-authors are typically more prolific than we are, and our dance partners have typically danced with more people than we have (and the same goes for partners in... other activities).

Exercises

solutions to starred exercises on page 443

1.B.1 Write the adjacency matrix of each of the following (undirected) graphs:

*(a)

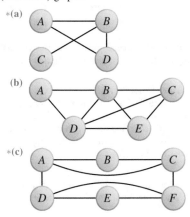

(b)

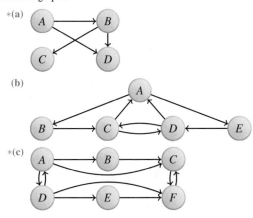

*(c)

1.B.2 Write the adjacency matrix of each of the following directed graphs:

*(a)

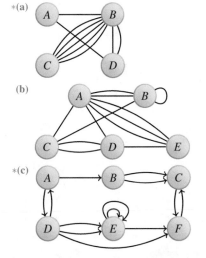

(b)

*(c)

1.B.3 Write the adjacency matrix of each of the following multigraphs:

*(a)

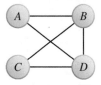

(b)

*(c)

1.B.4 Determine which of the following statements are true and which are false.

*(a) The adjacency matrix A of every undirected graph is symmetric (i.e., has the property $A = A^T$).
(b) The adjacency matrix of a graph with 7 edges is of size 7×7.
*(c) If there is a path of length two between two specific vertices in an undirected graph, then there must be a path of length three between those same two vertices.
(d) If a graph has no loops then the diagonal entries of its adjacency matrix must all equal 0.

1.B.5 Draw a (multi)graph with the given adjacency matrix. Choose your (multi)graph to be undirected or directed as appropriate.

*(a) $\begin{bmatrix} 0 & 1 & 1 & 0 \\ 1 & 0 & 0 & 1 \\ 1 & 0 & 0 & 1 \\ 0 & 1 & 1 & 0 \end{bmatrix}$

(b) $\begin{bmatrix} 0 & 1 & 1 & 1 & 1 \\ 0 & 0 & 1 & 1 & 1 \\ 0 & 0 & 0 & 1 & 1 \\ 0 & 0 & 0 & 0 & 1 \\ 0 & 0 & 0 & 0 & 0 \end{bmatrix}$

*(c) $\begin{bmatrix} 0 & 1 & 1 & 1 & 1 \\ 1 & 0 & 1 & 1 & 1 \\ 1 & 1 & 0 & 1 & 1 \\ 1 & 1 & 1 & 0 & 1 \\ 1 & 1 & 1 & 1 & 0 \end{bmatrix}$

(d) $\begin{bmatrix} 1 & 2 & 2 & 1 \\ 2 & 0 & 0 & 1 \\ 2 & 0 & 0 & 0 \\ 1 & 1 & 0 & 1 \end{bmatrix}$

*(e) $\begin{bmatrix} 0 & 1 & 1 & 1 \\ 0 & 0 & 0 & 0 \\ 0 & 0 & 0 & 0 \\ 1 & 0 & 0 & 0 \end{bmatrix}$

(f) $\begin{bmatrix} 2 & 0 & 0 & 1 \\ 0 & 1 & 3 & 0 \\ 0 & 0 & 1 & 1 \\ 3 & 0 & 1 & 1 \end{bmatrix}$

1.B.6 Compute the number of paths of the given length in the graph from the indicated exercise.

*(a) Of length 2, from A to C, Exercise 1.B.1(a).
(b) Of length 3, from A to E, Exercise 1.B.1(b).
*(c) Of length 2, from D to E, Exercise 1.B.1(c).
(d) Of length 2, from A to D, Exercise 1.B.2(a).
*(e) Of length 3, from A back to A, Exercise 1.B.2(b).
(f) Of length 3, from A to F, Exercise 1.B.2(c).
*(g) Of length 2, from D to C, Exercise 1.B.3(a).
(h) Of length 3, from A to E, Exercise 1.B.3(b).
*(i) Of length 3, from D to F, Exercise 1.B.3(c).
⌨(j) Of length 5, from A to D, Exercise 1.B.1(a).
*⌨(k) Of length 6, from B to E, Exercise 1.B.1(b).
⌨(l) Of length 7, from C to D, Exercise 1.B.1(c).

1.B.7 Compute the number of paths of the indicated type in the following (undirected) graph:

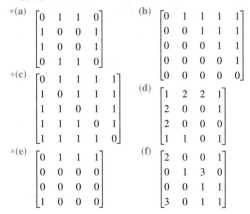

*(a) Of length 2 from A to B.
(b) Of length 3 from A to C.
*(c) Of length 4 from A to D.
(d) Of length 8 from B to C.

1.B.8 Suppose that the friendships between six people on a social network are represented by the following graph:

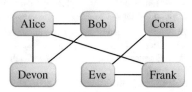

Use the adjacency matrix of this graph to answer the following questions:

(a) How many friends of friends does Bob have?
(b) In how many ways are Alice and Frank friends of friends of friends? (In other words, how many paths of length three are there between Alice and Frank?)

∗**1.B.9** Sometimes, we can use graphs to help us learn things about matrices (rather than the other way around).

(a) Draw a directed graph that has adjacency matrix

$$A = \begin{bmatrix} 1 & 1 \\ 0 & 1 \end{bmatrix}.$$

(b) Determine (based on the graph from part (a)) how many paths of length n there are between each pair of vertices in the graph.
(c) Use part (b) to find a formula for A^n. [Hint: Your answer should be the same as it was for Exercise 1.3.7(c).]

1.B.10 Powers of the adjacency matrix can also be used to find the length of the shortest path between two vertices in a graph.

(a) Describe how to use the adjacency matrix of a graph to find the length of the shortest path between two vertices in that graph.
(b) Use this method to find the length of the shortest path from vertex B to vertex E in the graph from Exercise 1.B.2(b).

∗∗**1.B.11** Complete the proof of Theorem 1.B.1. [Hint: Use induction and mimic the proof given in the $k = 2$ case.]

2. Linear Systems and Subspaces

Linear algebra is the central subject of mathematics.
You cannot learn too much linear algebra.

Benedict Gross

In this chapter, we start introducing some more interesting and useful properties of matrices and linear transformations. While we begin the chapter by motivating these various properties via systems of linear equations, it is important to keep in mind that systems of linear equations are just *one* of the many uses of matrices.

As we make our way through to the end of this chapter (and indeed, throughout the rest of this book), we will see that the various concepts that arise from systems of linear equations are also useful when investigating seemingly unrelated objects like linear transformations, graphs, and integer sequences.

2.1 Systems of Linear Equations

Much of linear algebra revolves around solving and manipulating the simplest types of equations that exist—linear equations:

Definition 2.1.1
Linear Equations

A **linear equation** in n variables $x_1, x_2, \ldots, x_n$ is an equation that can be written in the form

$$a_1 x_1 + a_2 x_2 + \cdots + a_n x_n = b,$$

where $a_1, a_2, \ldots, a_n$ and b are constants called the **coefficients** of the linear equation.

The names of the variables do not matter. We typically use x, y, and z if there are only 2 or 3 variables, and we use $x_1, x_2, \ldots, x_n$ if there are more.

For example, the following equations are all linear:

$$x + 3y = 4, \qquad 2x - \pi y = 3, \qquad 4x + 3 = 6y,$$
$$\sqrt{3}x - y = \sqrt{5}, \qquad \cos(1)x + \sin(1)y = 2, \quad \text{and} \quad x + y - 2z = 7.$$

Note that even though the top-right equation above is not quite in the form described by Definition 2.1.1, it can be rearranged so as to be in that form, so it is linear. In particular, it is equivalent to the equation $4x - 6y = -3$, which is in the desired form. Also, the bottom-left and bottom-middle equations are indeed linear since the square root and trigonometric functions are only applied to

© Springer Nature Switzerland AG 2021
N. Johnston, *Introduction to Linear and Matrix Algebra*,
https://doi.org/10.1007/978-3-030-52811-9_2

the coefficients (not the variables) in the equations. By contrast, the following equations are all *not* linear:

$$\sqrt{x} + 3y = 4, \qquad 2x - 7y^2 = 3, \qquad 4y + 2xz = 1,$$
$$2^x - 2^y = 3, \qquad \cos(x) + \sin(y) = 0, \quad \text{and} \quad \ln(x) - y/z = 2.$$

In general, an equation is linear if each variable is only multiplied by a constant: variables cannot be multiplied by other variables, they cannot be raised to an exponent other than 1, and they cannot have other functions applied to them.

Geometrically, we can think of linear equations as representing lines and planes (and higher-dimensional flat shapes that we cannot quite picture). For example, the general equation of a line is $ax + by = c$, and the general equation of a plane is $ax + by + cz = d$, both of which are linear equations (see Figure 2.1).

> The equation of a line is sometimes instead written in the form $y = mx + b$, where m and b are constants. This is also a linear equation.

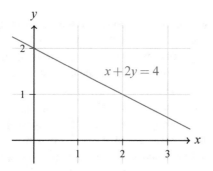

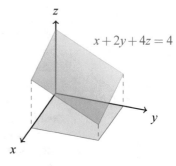

(a) A linear equation in 2 variables represents a line in $\mathbb{R}^2$. The equation of this line is perhaps more usually written in the (equivalent) form $y = -x/2 + 2$.

(b) A linear equation in 3 variables represents a plane in $\mathbb{R}^3$. This plane actually extends arbitrarily far in all directions—it is just truncated here so it is easier to visualize.

Figure 2.1: Linear equations represent flat objects: lines, planes, and higher-dimensional hyperplanes.

Oftentimes, we want to solve multiple linear equations at the same time. That is, we want to find values for the variables $x_1, x_2, \ldots, x_n$ such that multiple different linear equations are all satisfied simultaneously. This leads us naturally to consider *systems* of linear equations:

> **Definition 2.1.2**
>
> **Systems of Linear Equations**

A **system of linear equations** (or a **linear system**) is a finite set of linear equations, each with the same variables $x_1, x_2, \ldots, x_n$. Also,

- a **solution** of a system of linear equations is a vector $\mathbf{x} = (x_1, x_2, \ldots, x_n)$ whose entries satisfy all of the linear equations in the system, and

- the **solution set** of a system of linear equations is the set of *all* solutions of the system.

Geometrically, a solution of a system of linear equations is a point at the intersection of all of the lines, planes, or hyperplanes defined by the linear equations in the system. For example, consider the following linear system that consists of two equations:

> Try plugging $x = 2$, $y = 1$ back into the linear equations to verify that they are both true at this point.

$$x + 2y = 4$$
$$-x + \ y = -1$$

The lines defined by these equations are plotted in Figure 2.2. Based on this graph, it appears that these lines have a unique point of intersection, and it is located at the point $(2,1)$. The vector $\mathbf{x} = (2,1)$ thus seems to be the unique solution of this system of linear equations. To instead find this solution algebraically, we could add the two equations in the linear system to get the new equation $3y = 3$, which tells us that $y = 1$. Plugging $y = 1$ back into the original equation $x + 2y = 4$ then tells us that $x = 2$.

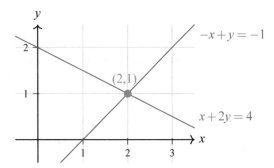

Figure 2.2: The system of linear equations $x+2y=4$, $-x+y=-1$ is represented graphically by two lines that intersect at a single point $(2,1)$. This point of intersection is the unique solution of the linear system.

However, systems of linear equations do not always have a unique solution like in the previous example. To illustrate the other possibilities, consider the following two systems of linear equations:

$$\begin{aligned} x + 2y &= 4 \\ 2x + 4y &= 8 \end{aligned} \qquad\qquad \begin{aligned} x + 2y &= 4 \\ x + 2y &= 3 \end{aligned}$$

The first linear system is strange because the second equation is simply a multiple of the first equation, and thus tells us nothing new—any pair of x, y values that satisfy the first equation also satisfy the second equation. In other words, both equations describe the same line, so there are infinitely many solutions, as displayed in Figure 2.3(a).

On the other hand, the second linear system has no solutions at all because $x + 2y$ cannot simultaneously equal 4 and 3. Geometrically, these two equations represent parallel lines, as illustrated in Figure 2.3(b). The fact that this linear system has no solutions corresponds to the fact that parallel lines do not intersect.

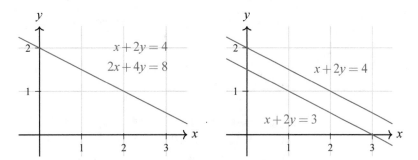

(a) A linear system with infinitely many solutions, represented by two equivalent (and thus always intersecting) lines.

(b) A linear system with no solutions, represented by two parallel (and thus non-intersecting) lines.

Figure 2.3: A visual demonstration of the fact that systems of linear equations can have (a) infinitely many solutions or (b) no solutions.

These examples show that systems of linear equations can have no solutions, exactly one solution, or infinitely many solutions. We will show shortly (in the upcoming Theorem 2.1.1) that these are the only possibilities. That is, there does not exist a system of linear equations with exactly 2 solutions, or exactly 3 solutions, or exactly 4 solutions, and so on.

2.1.1 Matrix Equations

One of the primary uses of matrices is that they give us a way of working with linear systems much more compactly and cleanly. In particular, any system of linear equations

Go back to Definition 1.3.2 or Theorem 1.3.5 to remind yourself that this ugly mess is just what a matrix times a vector looks like.

$$a_{1,1}x_1 + a_{1,2}x_2 + \cdots + a_{1,n}x_n = b_1$$
$$a_{2,1}x_1 + a_{2,2}x_2 + \cdots + a_{2,n}x_n = b_2$$
$$\vdots$$
$$a_{m,1}x_1 + a_{m,2}x_2 + \cdots + a_{m,n}x_n = b_m$$

can be rewritten as the single matrix equation $A\mathbf{x} = \mathbf{b}$, where $A \in \mathcal{M}_{m,n}$ is the **coefficient matrix** whose (i,j)-entry is $a_{i,j}$, $\mathbf{b} = (b_1, b_2, \ldots, b_m) \in \mathbb{R}^m$ is a vector containing the constants from the right-hand side, and $\mathbf{x} = (x_1, x_2, \ldots, x_n) \in \mathbb{R}^n$ is a vector containing the variables.

Example 2.1.1

Writing Linear Systems as Matrix Equations

Write the following linear systems as matrix equations:

a) $x + 2y = 4$
    $3x + 4y = 6$

b) $3x - 2y + z = -3$
    $2x + 3y - 2z = 5$

Solutions:

a) We place the coefficients of the linear system in a matrix A, the variables in a vector $\mathbf{x}$, and the numbers from the right-hand side in a vector $\mathbf{b}$, obtaining the following matrix equation $A\mathbf{x} = \mathbf{b}$:

$$\begin{bmatrix} 1 & 2 \\ 3 & 4 \end{bmatrix} \begin{bmatrix} x \\ y \end{bmatrix} = \begin{bmatrix} 4 \\ 6 \end{bmatrix}.$$

Indeed, if we were to perform the matrix multiplication on the left then we would get exactly the linear system that we started with.

The number of rows of A equals the number of linear equations. The number of columns of A equals the number of variables.

b) We proceed similarly to before, being careful to note that the coefficient matrix now has 3 columns and the vector $\mathbf{x}$ now contains 3 variables:

$$\begin{bmatrix} 3 & -2 & 1 \\ 2 & 3 & -2 \end{bmatrix} \begin{bmatrix} x \\ y \\ z \end{bmatrix} = \begin{bmatrix} -3 \\ 5 \end{bmatrix}.$$

As a sanity check, we can we perform the matrix multiplication on the left to see that we get back the original linear system:

$$\begin{bmatrix} 3x - 2y + z \\ 2x + 3y - 2z \end{bmatrix} = \begin{bmatrix} -3 \\ 5 \end{bmatrix}.$$

The advantage of writing linear systems in this way (beyond the fact that it requires less writing) is that we can now make use of the various properties of matrices and matrix multiplication that we already know to help us understand linear systems a bit better. For example, we can now prove the observation that we made earlier: every linear system has either zero, one, or infinitely many solutions.

Theorem 2.1.1

Trichotomy for Linear Systems

Every system of linear equations has either

   a) no solutions,

   b) exactly one solution, or

   c) infinitely many solutions.

Refer back to Figures 2.2 and 2.3 for geometric interpretations of the three possibilities described by this theorem.

Proof. Another way of phrasing this theorem is as follows: if a system of linear equations has at least two solutions then it must have infinitely many solutions. With this in mind, we start by assuming that there are two distinct solutions to the linear system.

If $A\mathbf{x} = \mathbf{b}$ is the matrix form of the linear system (where $A \in \mathcal{M}_{m,n}$, $\mathbf{x} \in \mathbb{R}^n$, and $\mathbf{b} \in \mathbb{R}^m$), then there existing two distinct solutions of the linear system means that there exist vectors $\mathbf{x}_1 \neq \mathbf{x}_2 \in \mathbb{R}^n$ such that $A\mathbf{x}_1 = \mathbf{b}$ and $A\mathbf{x}_2 = \mathbf{b}$. Then for any scalar $c \in \mathbb{R}$, it is the case that

$$A\big((1-c)\mathbf{x}_1 + c\mathbf{x}_2\big) = (1-c)A\mathbf{x}_1 + cA\mathbf{x}_2 = (1-c)\mathbf{b} + c\mathbf{b} = \mathbf{b},$$

so every vector of the form $(1-c)\mathbf{x}_1 + c\mathbf{x}_2$ is a solution of the linear system. Since there are infinitely many such vectors (one for each choice of $c \in \mathbb{R}$), the proof is complete. ∎

Geometrically, this proof shows that every vector whose head is on the line going through the heads of $\mathbf{x}_1$ and $\mathbf{x}_2$ is a solution of the linear system too.

When a system of linear equations has at least one solution (i.e., in cases (b) and (c) of the above theorem), it is called **consistent**. If it has no solutions (i.e., in case (a) of the theorem), it is called **inconsistent**. For example, the linear systems depicted in Figures 2.2 and 2.3(a) are consistent, whereas the system in Figure 2.3(b) is inconsistent. Similarly, we can visualize linear systems in three variables via intersecting planes as in Figure 2.4. A linear system in three variables is consistent if all of the planes have at least one common point of intersection (as in Figures 2.4(a), (b), and (c)), whereas it is inconsistent otherwise (as in Figure 2.4(d), where each pair of planes intersect, but there is no point where all three of them intersect).

2.1.2 Row Echelon Form

We now turn our attention to the problem of actually finding the solutions of a system of linear equations. If the linear system has a "triangular" form, then solving it is fairly intuitive. For example, consider the following system of linear equations:

$$\begin{aligned} x + 3y - 2z &= 5 \\ 2y - 6z &= 4 \\ 3z &= 6 \end{aligned}$$

The final equation in this system tells us that $3z = 6$, so $z = 2$. Plugging $z = 2$ into the other two equations (and moving all constants over to the right-hand side) transforms them into the form

$$\begin{aligned} x + 3y &= 9 \\ 2y &= 16 \end{aligned}$$

It is also possible for the intersection of several planes to be a plane (if all of the planes are really the same plane in disguise), in which case the associated linear system has infinitely many solutions.

It is also possible for several planes to all be parallel to each other, in which case the associated linear system has no solutions.

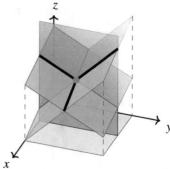

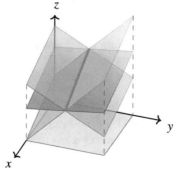

(a) Two planes typically intersect in a line, and thus a system of two linear equations in three variables usually has infinitely many solutions.

(b) Three planes typically intersect in a point, and thus a system of three linear equations in three variables usually has exactly one solution.

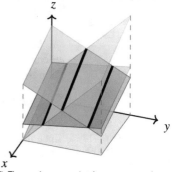

(c) Three (or more) planes can also intersect in a line, and thus a system of three or more linear equations in three variables can have infinitely many solutions.

(d) Three (or more) planes can also not have a common intersection at all, and thus a system of three or more linear equations in three variables can have no solutions.

Figure 2.4: Planes can intersect at either infinitely many points (as in (a) and (c)), at a single point (as in (b)), or not at all (as in (d)). There are also some other ways for planes to intersect at infinitely many points or no points that are not displayed here.

We can again then just read off the value of one of the variables from the final equation: $2y = 16$, so $y = 8$. Finally, plugging $y = 8$ into the first equation then tells us that $x + 24 = 9$, so $x = -15$. This system of linear equations thus has a unique solution, and it is $(x, y, z) = (-15, 8, 2)$.

The procedure that we used to solve the previous example is called **back substitution**, and it worked because of the triangular shape of the linear system. We were able to easily solve for z, which we then could plug into the second equation and easily solve for y, which we could plug into the first equation and easily solve for x. With this in mind, our goal now is to put *every* system of equations into this triangular form. We first start by eliminating the variable x from the second and third equations, and then we eliminate the variable y from the third equation (and then to actually *solve* the system, we do back substitution like before).

Example 2.1.2

Solving a Linear System

Put the following system of linear equations into a triangular form that is easy to solve via back substitution:

$$x + 3y - 2z = 5$$
$$x + 5y - 8z = 9$$
$$2x + 4y + 5z = 12$$

Solution:

To put this linear system into the desired triangular form, we start by subtracting the first equation from the second equation, so that the new second equation is

(equation 2)	$x + 5y - 8z = 9$
$-$(equation 1)	$-x - 3y + 2z = -5$
(new equation 2)	$2y - 6z = 4$

Since every term in our new equation 2 is a multiple of 2, we could cancel out this common factor and instead write it in the simpler form $y - 3z = 2$ if we wanted to.

The reason for performing this calculation is that our new equation 2 does not have an "x" term. To similarly eliminate the "x" term from equation 3, we subtract 2 copies of the first equation from the third equation:

(equation 3)	$2x + 4y + 5z = 12$
-2(equation 1)	$-2x - 6y + 4z = -10$
(new equation 3)	$-2y + 9z = 2$

Our system of linear equations is now in the form

$$x + 3y - 2z = 5$$
$$2y - 6z = 4$$
$$-2y + 9z = 2$$

To complete the process of putting this system into triangular form, we just need to eliminate the "y" term in the third equation, which can be accomplished by adding the second equation to the third equation:

(equation 3)	$-2y + 9z = 2$
$+$(equation 2)	$2y - 6z = 4$
(new equation 3)	$3z = 6$

Finally, our linear system is now in the triangular form

$$x + 3y - 2z = 5$$
$$2y - 6z = 4$$
$$3z = 6$$

which is exactly the linear system that we solved earlier via back substitution. We thus conclude that the (unique) solution to this linear system is $(x, y, z) = (-15, 8, 2)$, just like we computed earlier.

To reduce the amount of writing we have to do when solving the linear system $A\mathbf{x} = \mathbf{b}$, we typically write it more compactly as an **augmented matrix** $[A \mid \mathbf{b}]$, so that we only have to write down the coefficients in the linear system at every step of the computation, rather than also having to write down the variable names (e.g., x, y, and z) and other details that do not change. For example, the augmented matrix of the linear system from Example 2.1.2 is

$$\left[\begin{array}{ccc|c} 1 & 3 & -2 & 5 \\ 1 & 5 & -8 & 9 \\ 2 & 4 & 5 & 12 \end{array}\right].$$

Then all of the equation manipulations that we performed earlier correspond to operations where we modify the rows of this augmented matrix. For example, our first step toward solving Example 2.1.2 was to replace the second equation by (equation 2) − (equation 1), which is equivalent to replacing the second row of the augmented matrix by (row 2) − (row 1). In order to even further reduce the amount of writing we have to do when solving linear systems, we now introduce some standard notation that we can use to indicate which row modifications we are performing at each step of our computation, so that we do not have to repeatedly write out expressions like "(equation 3) − 2(equation 1)":

Multiplication. Multiplying row j by a non-zero scalar $c \in \mathbb{R}$ is denoted by cR_j.
    Swap. Swapping rows i and j is denoted by $R_i \leftrightarrow R_j$.
  Addition. For any scalar $c \in \mathbb{R}$, replacing row i by (row i) + c(row j) is denoted by $R_i + cR_j$.

The three operations described above are called **elementary row operations**, and two matrices are called **row equivalent** if one can be converted to the other via elementary row operations. To get comfortable with row operations and their associated notation, we now use them to re-solve the system of linear equations from Example 2.1.2.

Example 2.1.3

Solving a Linear System via Matrix Notation

Use matrix notation to solve this system of linear equations:

$$x + 3y - 2z = 5$$
$$x + 5y - 8z = 9$$
$$2x + 4y + 5z = 12$$

Solution:

At each step, we highlight which row was just modified to make the operations a bit easier to see.

We already displayed the augmented matrix of this linear system earlier—we start with that matrix and perform row operations so as to get it into an upper triangular form, just like earlier:

$$\left[\begin{array}{ccc|c} 1 & 3 & -2 & 5 \\ 1 & 5 & -8 & 9 \\ 2 & 4 & 5 & 12 \end{array}\right] \xrightarrow{R_2 - R_1} \left[\begin{array}{ccc|c} 1 & 3 & -2 & 5 \\ 0 & 2 & -6 & 4 \\ 2 & 4 & 5 & 12 \end{array}\right]$$

Be careful when grouping multiple row operations into one step like we did here. The operations are performed one after another, not at the same time. Exercise 2.1.16 illustrates how things can go wrong if row operations are performed simultaneously.

$$\xrightarrow{R_3 - 2R_1} \left[\begin{array}{ccc|c} 1 & 3 & -2 & 5 \\ 0 & 2 & -6 & 4 \\ 0 & -2 & 9 & 2 \end{array}\right] \xrightarrow{R_3 + R_2} \left[\begin{array}{ccc|c} 1 & 3 & -2 & 5 \\ 0 & 2 & -6 & 4 \\ 0 & 0 & 3 & 6 \end{array}\right].$$

The row operations illustrated above carry out the exact same calculation as the entirety of Example 2.1.2. At this point, the system of equations is in a form that can be solved by back substitution. However, we can also completely solve the system of equations by performing a few extra row operations, as we now illustrate:

$$\begin{bmatrix} 1 & 3 & -2 & 5 \\ 0 & 2 & -6 & 4 \\ 0 & 0 & 3 & 6 \end{bmatrix} \xrightarrow{\frac{1}{3}R_3} \begin{bmatrix} 1 & 3 & -2 & 5 \\ 0 & 2 & -6 & 4 \\ 0 & 0 & 1 & 2 \end{bmatrix}$$

$$\xrightarrow[R_2+6R_3]{R_1+2R_3} \begin{bmatrix} 1 & 3 & 0 & 9 \\ 0 & 2 & 0 & 16 \\ 0 & 0 & 1 & 2 \end{bmatrix} \xrightarrow{\frac{1}{2}R_2} \begin{bmatrix} 1 & 3 & 0 & 9 \\ 0 & 1 & 0 & 8 \\ 0 & 0 & 1 & 2 \end{bmatrix}$$

$$\xrightarrow{R_1-3R_2} \begin{bmatrix} 1 & 0 & 0 & -15 \\ 0 & 1 & 0 & 8 \\ 0 & 0 & 1 & 2 \end{bmatrix}.$$

At this point, we can simply read the solution to the system of equations off from the right-hand side of the bottom augmented matrix: $(x, y, z) = (-15, 8, 2)$, just like we computed earlier.

Remark 2.1.1

Elementary Row Operations are Reversible

It is not difficult to see that if $\mathbf{x}$ is a solution of a linear system $[\,A \mid \mathbf{b}\,]$ then it is also a solution of any linear system that is obtained by applying elementary row operations to it. For example, swapping the order of two equations in a linear system does not alter its solutions, and multiplying an equation through by a scalar does not affect whether or not it is true.

Furthermore, the elementary row operations are all reversible—they can be undone by other elementary row operations of the same type. In particular, the multiplication operation cR_j is reversed by $\frac{1}{c}R_j$, the swap operation $R_i \leftrightarrow R_j$ is undone by itself, and the addition operation $R_i + cR_j$ is undone by $R_i - cR_j$.

This reversibility of row operations ensures that, not only are solutions to the linear system not lost when row-reducing (as we showed above), but no new solutions are introduced either (since no solutions are lost when undoing the row operations). In fact, this is exactly why we require $c \neq 0$ in the multiplication row operation cR_j; if $c = 0$ then this row operation is not reversible, so it may introduce additional solutions to the linear system. For example, in the linear systems

$$\begin{bmatrix} 1 & 2 & 3 \\ 3 & 4 & 7 \end{bmatrix} \xrightarrow{0R_2} \begin{bmatrix} 1 & 2 & 3 \\ 0 & 0 & 0 \end{bmatrix},$$

the system on the left has a unique solution $\mathbf{x} = (1,1)$, but the system on the right has infinitely many others, such as $\mathbf{x} = (3,0)$.

The previous examples illustrate the most commonly-used method for solving systems of linear equations: use row operations to first make the matrix "triangular", and then either solve the system by back substitution or by performing additional row operations. To make this procedure more precise, and to help us solve some trickier linear systems where a triangular form seems difficult to obtain, we now define exactly what our triangular form should look like and discuss how to use row operations to get there.

Definition 2.1.3

(Reduced) Row Echelon Form

A matrix is said to be in **row echelon form** (**REF**) if:

a) all rows consisting entirely of zeros are below the non-zero rows, and

b) in each non-zero row, the first non-zero entry (called the **leading entry**) is to the left of any leading entries below it.

If the matrix also satisfies the following additional property, then it is in **reduced row echelon form** (**RREF**):

c) each leading entry equals 1 and is the only non-zero entry in its column.

The word "echelon" means level or rank, and it refers to the fact that the rows of a matrix in this form are arranged somewhat like a staircase.

For example, of the two augmented matrices

$$\left[\begin{array}{ccc|c} 1 & 3 & -2 & 5 \\ 0 & 2 & -6 & 4 \\ 0 & 0 & 3 & 6 \end{array}\right] \quad \text{and} \quad \left[\begin{array}{ccc|c} 1 & 0 & 0 & -15 \\ 0 & 1 & 0 & 8 \\ 0 & 0 & 1 & 2 \end{array}\right]$$

that appeared in Example 2.1.3, both are in row echelon form, but only the matrix on the right is in *reduced* row echelon form.

Slightly more generally, if we use ★ to represent non-zero leading entries and ∗ to represent arbitrary (potentially zero) non-leading entries, then the following matrices are in row echelon form:

$$\begin{bmatrix} \bigstar & * & * & * \\ 0 & 0 & \bigstar & * \\ 0 & 0 & 0 & \bigstar \\ 0 & 0 & 0 & 0 \end{bmatrix} \quad \text{and} \quad \begin{bmatrix} 0 & 0 & \bigstar & * & * & * & * & * & * & * \\ 0 & 0 & 0 & \bigstar & * & * & * & * & * & * \\ 0 & 0 & 0 & 0 & 0 & \bigstar & * & * & * & * \\ 0 & 0 & 0 & 0 & 0 & 0 & 0 & 0 & \bigstar & * \end{bmatrix}.$$

*A **leading column** of a matrix in row echelon form is a column containing a leading entry. For example, the leading columns of the matrix on the left here are its first, third, and fourth.*

If we modify these matrices so that their leading entries are 1 and they also have 0's *above* those leading 1's, then they will be in *reduced* row echelon form:

$$\begin{bmatrix} 1 & * & 0 & 0 \\ 0 & 0 & 1 & 0 \\ 0 & 0 & 0 & 1 \\ 0 & 0 & 0 & 0 \end{bmatrix} \quad \text{and} \quad \begin{bmatrix} 0 & 0 & 1 & 0 & * & 0 & * & * & 0 & * \\ 0 & 0 & 0 & 1 & * & 0 & * & * & 0 & * \\ 0 & 0 & 0 & 0 & 0 & 1 & * & * & 0 & * \\ 0 & 0 & 0 & 0 & 0 & 0 & 0 & 0 & 1 & * \end{bmatrix}.$$

Roughly speaking, a matrix is in row echelon form if we can solve the associated linear system via back substitution, whereas it is in *reduced* row echelon form if we can just read the solution to the linear system directly from the entries of the matrix.

Example 2.1.4

(Reduced) Row Echelon Forms

Determine which of the following matrices are and are not in (reduced) row echelon form:

a) $\begin{bmatrix} 1 & 2 & 3 \\ 0 & 1 & 0 \end{bmatrix}$

b) $\begin{bmatrix} 1 & 0 & 2 \\ 0 & 1 & 4 \end{bmatrix}$

c) $\begin{bmatrix} 2 & 0 & -1 & 5 \\ 0 & 0 & 3 & 0 \\ 0 & 0 & 0 & 0 \end{bmatrix}$

d) $\begin{bmatrix} 1 & 2 & 3 & 4 \\ 0 & 0 & 2 & -1 \\ 0 & 0 & 1 & 0 \end{bmatrix}$

Solutions:

a) This matrix is in row echelon form. However, it is not in reduced row echelon form, because the leading 1 in the second row has a non-zero entry above it.

b) This matrix is in reduced row echelon form.

Every matrix that is in reduced row echelon form is also in row echelon form (but not vice-versa).

c) This matrix is also in row echelon form. However, it is not in reduced row echelon form, because its leading entries are 2 and 3 (not 1).

d) This matrix is not in row echelon form since the leading entry in the second row has a non-zero entry below it.

2.1.3 Gaussian Elimination

There are numerous different sequences of row operations that can be used to put a matrix (or augmented matrix) into row echelon form (or reduced row echelon form), and the process of doing so is called **row-reduction**. However, it is useful to have a standard method of choosing which row operations to apply when. With this in mind, we now present one method, called **Gaussian elimination**, which can be used to put any matrix into (not necessarily reduced) row echelon form and thus solve the associated linear system. In particular, we illustrate this algorithm with the following matrix:

$$\begin{bmatrix} 0 & 0 & -1 & 1 & 0 \\ 0 & -2 & 1 & -5 & 2 \\ 0 & 2 & -2 & 6 & -3 \\ 0 & -4 & 2 & -10 & 5 \end{bmatrix}.$$

Step 1: Position a leading entry. Locate the leftmost non-zero column of the matrix, and swap rows (if necessary) so that the topmost entry of this column is non-zero. This top-left entry becomes the leading entry of the top row:

Alternatively, we could have instead swapped rows 1 and 2 or rows 1 and 4 here.

leading entry

$$\begin{bmatrix} 0 & 0 & -1 & 1 & 0 \\ 0 & -2 & 1 & -5 & 2 \\ 0 & 2 & -2 & 6 & -3 \\ 0 & -4 & 2 & -10 & 5 \end{bmatrix} \xrightarrow{R_1 \leftrightarrow R_3} \begin{bmatrix} 0 & 2 & -2 & 6 & -3 \\ 0 & -2 & 1 & -5 & 2 \\ 0 & 0 & -1 & 1 & 0 \\ 0 & -4 & 2 & -10 & 5 \end{bmatrix}$$

This step is sometimes called **pivoting** on the leading entry (and the leading entry is sometimes called the **pivot**).

Step 2: Zero out the leading entry's column. Use the "addition" row operation to create zeros in all entries below the leading entry from step 1. Optionally, the arithmetic can be made somewhat easier by first dividing the top row by the leading entry (2 in this case), but we do not do so here:

new zeros

Be careful with your arithmetic! Double negatives are quite common when performing row operations, and lots of them appear here when we subtract row 2 from the others.

$$\begin{bmatrix} 0 & 2 & -2 & 6 & -3 \\ 0 & -2 & 1 & -5 & 2 \\ 0 & 0 & -1 & 1 & 0 \\ 0 & -4 & 2 & -10 & 5 \end{bmatrix} \begin{matrix} {} \\ R_2+R_1 \\ R_4+2R_1 \\ \longrightarrow \end{matrix} \begin{bmatrix} 0 & 2 & -2 & 6 & -3 \\ 0 & 0 & -1 & 1 & -1 \\ 0 & 0 & -1 & 1 & 0 \\ 0 & 0 & -2 & 2 & -1 \end{bmatrix}$$

Step 3: Repeat until we cannot. Partition the matrix into a block matrix whose top block consists of the row with a leading entry from Step 1, and whose bottom block consists of all lower rows. Repeat Steps 1 and 2 on the bottom block.

$$\begin{matrix} \text{ignore this top row} \\ \end{matrix}$$

$$\left[\begin{array}{ccccc} 0 & 2 & -2 & 6 & -3 \\ 0 & 0 & -1 & 1 & -1 \\ 0 & 0 & -1 & 1 & 0 \\ 0 & 0 & -2 & 2 & -1 \end{array}\right] \xrightarrow[\;R_4-2R_2\;]{R_3-R_2} \left[\begin{array}{ccc|cc} 0 & 2 & -2 & 6 & -3 \\ 0 & 0 & -1 & 1 & -1 \\ 0 & 0 & 0 & 0 & 1 \\ 0 & 0 & 0 & 0 & 1 \end{array}\right]$$

new zeros — new leading entry

Now that we have completed Steps 1 and 2 again, we repeat this process, now ignoring the top *two* rows (instead of just the top row). In general, we continue in this way until all of the bottom rows consist of nothing but zeros, or until we reach the bottom of the matrix (whichever comes first):

ignore these top rows

$$\left[\begin{array}{ccccc} 0 & 2 & -2 & 6 & -3 \\ 0 & 0 & -1 & 1 & -1 \\ 0 & 0 & 0 & 0 & 1 \\ 0 & 0 & 0 & 0 & 1 \end{array}\right] \xrightarrow{\;R_4-R_3\;} \left[\begin{array}{ccccc} 0 & 2 & -2 & 6 & -3 \\ 0 & 0 & -1 & 1 & -1 \\ 0 & 0 & 0 & 0 & 1 \\ 0 & 0 & 0 & 0 & 0 \end{array}\right]$$

new leading entry —   new zero —

At this point, all of the remaining rows contain nothing but zeros, so we are done—the matrix is now in row echelon form. If this matrix represented a linear system then we could solve it via back substitution at this point. However, it is sometimes convenient to go slightly farther and put the matrix into *reduced* row echelon form. We can do so via an extension of Gaussian elimination called **Gauss–Jordan elimination**, which consists of just one additional step.

Step 4: **Reduce even more, from right-to-left.** Starting with the rightmost leading entry, and moving from bottom-right to top-left, use the "multiplication" row operation to change each leading entry to 1 and use the "addition" row operation to create zeros in all entries above the leading entries.

Keep in mind that Step 4 is only necessary if we want the matrix in *reduced* row echelon form.

new zeros

$$\left[\begin{array}{ccccc} 0 & 2 & -2 & 6 & -3 \\ 0 & 0 & -1 & 1 & -1 \\ 0 & 0 & 0 & 0 & 1 \\ 0 & 0 & 0 & 0 & 0 \end{array}\right] \xrightarrow[\;R_2+R_3\;]{R_1+3R_3} \left[\begin{array}{ccccc} 0 & 2 & -2 & 6 & 0 \\ 0 & 0 & -1 & 1 & 0 \\ 0 & 0 & 0 & 0 & 1 \\ 0 & 0 & 0 & 0 & 0 \end{array}\right]$$

rightmost leading entry

The next rightmost leading entry is the $(2,3)$-entry, so we now use row operations to change it to 1 and create zeros in the remaining entries of this column.

$$\left[\begin{array}{ccccc} 0 & 2 & -2 & 6 & 0 \\ 0 & 0 & -1 & 1 & 0 \\ 0 & 0 & 0 & 0 & 1 \\ 0 & 0 & 0 & 0 & 0 \end{array}\right] \xrightarrow{\;-R_2\;} \left[\begin{array}{ccccc} 0 & 2 & -2 & 6 & 0 \\ 0 & 0 & 1 & -1 & 0 \\ 0 & 0 & 0 & 0 & 1 \\ 0 & 0 & 0 & 0 & 0 \end{array}\right]$$

next leading entry   scaled to 1

new zero

$$\xrightarrow{\;R_1+2R_2\;} \left[\begin{array}{ccccc} 0 & 2 & 0 & 4 & 0 \\ 0 & 0 & 1 & -1 & 0 \\ 0 & 0 & 0 & 0 & 1 \\ 0 & 0 & 0 & 0 & 0 \end{array}\right]$$

Finally, the only remaining leading entry is the $(1,2)$-entry, which we scale to equal 1.

This example was chosen very carefully so that the arithmetic worked out nicely. In general, we should be very careful when performing row operations, as there are often lots of ugly fractions to deal with.

$$\underset{\text{next leading entry}}{\downarrow} \begin{bmatrix} 0 & \boxed{2} & 0 & 4 & 0 \\ 0 & 0 & 1 & -1 & 0 \\ 0 & 0 & 0 & 0 & 1 \\ 0 & 0 & 0 & 0 & 0 \end{bmatrix} \xrightarrow{\frac{1}{2}R_1} \underset{\text{scaled to 1}}{\downarrow} \begin{bmatrix} 0 & \boxed{1} & 0 & 2 & 0 \\ 0 & 0 & 1 & -1 & 0 \\ 0 & 0 & 0 & 0 & 1 \\ 0 & 0 & 0 & 0 & 0 \end{bmatrix}$$

Finally, the matrix on the right above is the reduced row echelon form of the matrix that we started with.

Neither Gaussian elimination nor Gauss–Jordan elimination is "better" than the other for solving linear systems—which one we use largely depends on personal preference. However, one advantage of the reduced row echelon form of a matrix is that it is unique (i.e., every matrix can be row-reduced to one, and only one, matrix in reduced row echelon form), whereas non-reduced row echelon forms are not. That is, applying a different sequence of row operations than the one specified by Gauss–Jordan elimination might get us to a different row echelon form along the way, but we always end up at the same *reduced* row echelon form. This fact is hopefully somewhat intuitive—reduced row echelon form was defined specifically to have a form that row operations are powerless to simplify further—but it is proved explicitly in Appendix B.2.

2.1.4 Solving Linear Systems

After we have used row operations to put an augmented matrix into (reduced) row echelon form, it is straightforward to find the solution(s) of the corresponding linear system just by writing down the linear system associated with that (reduced) row echelon form. However, the details can differ somewhat depending on whether the linear system has no solutions, one solution, or infinitely many solutions, so we consider each of these possibilities one at a time.

Linear Systems with No Solutions

If we interpret the RREF that we computed via Gauss–Jordan elimination in the previous subsection as an augmented matrix, then the associated system of linear equations has the following form:

Remember that each row corresponds to an equation, and each column corresponds to a variable or the right-hand side of the equation. We assigned the 2nd, 3rd, and 4th columns the variables $x, y,$ and z, respectively.

$$\left[\begin{array}{cccc|c} 0 & 1 & 0 & 2 & 0 \\ 0 & 0 & 1 & -1 & 0 \\ 0 & 0 & 0 & 0 & 1 \\ 0 & 0 & 0 & 0 & 0 \end{array} \right] \qquad \begin{aligned} x + 2z &= 0 \\ y - z &= 0 \\ 0 &= 1 \\ 0 &= 0 \end{aligned}$$

Of particular note is the third equation, which says $0 = 1$. This tells us that the linear system has no solutions, since there is no way to choose $x, y,$ and z so as to make that equation true. In fact, this is precisely how we can identify when a linear system is inconsistent (i.e., has no solutions) in general: there is a row in its row echelon forms consisting of zeros on the left-hand side and a non-zero entry on the right-hand side.

> (!) A linear system has no solutions if and only if its row echelon forms have a row that looks like $\begin{bmatrix} 0 & 0 & \cdots & 0 \mid b \end{bmatrix}$ for some scalar $b \neq 0$.

Linear Systems with a Unique Solution

If the RREF of an augmented matrix and the associated system of linear equations were to instead have the form

$$\left[\begin{array}{ccc|c} 1 & 0 & 0 & 4 \\ 0 & 1 & 0 & -3 \\ 0 & 0 & 1 & 2 \end{array}\right] \qquad\qquad \begin{aligned} x &= 4 \\ y &= -3 \\ z &= 2 \end{aligned}$$

then we could directly see that the system has a unique solution: $(x,y,z) = (4,-3,2)$. In general, a linear system has a unique solution exactly when (a) it is consistent (i.e., there is at least one solution, so no row of the reduced row echelon form corresponds to the unsolvable equation $0 = 1$) and (b) every column in the left-hand block of the row echelon forms has a leading entry in it.

> (!) A linear system has a unique solution if and only if it is consistent and the number of leading entries in its row echelon forms equals the number of variables.

Linear Systems with Infinitely Many Solutions

Finally, if the RREF of an augmented matrix and the associated system of linear equations were to have the form

$$\left[\begin{array}{ccc|c} 1 & 4 & 0 & 3 \\ 0 & 0 & 1 & 2 \\ 0 & 0 & 0 & 0 \end{array}\right] \qquad\qquad \begin{aligned} x+4y &= 3 \\ z &= 2 \\ 0 &= 0 \end{aligned}$$

then the linear system would have infinitely many solutions. The reason for this is that we could choose y to have any value that we like and then solve for $x = 3 - 4y$. In general, we call the variables corresponding to columns containing a leading entry a **leading variable**, and we call the other variables **free variables**. In the system above, y is the free variable, while x and z are leading variables.

The solution set of a system with infinitely many solutions can always be described by solving for the leading variables in terms of the free variables, and each free variable corresponds to one "dimension" or "degree of freedom" in the solution set. For example, if there is one free variable then the solution set is a line, if there are two then it is a plane, and so on.

To get a bit of a handle on what the solution set looks like geometrically, it is often useful to write it in terms of vectors. For example, if we return to the example above, the solutions are the vectors of the form

Recall from earlier that $x = 3 - 4y$ and $z = 2$. We just leave the free variable y alone and write $y = y$.

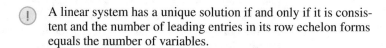

$$\begin{pmatrix} x \\ y \\ z \end{pmatrix} = \begin{pmatrix} 3 - 4y \\ y \\ 2 \end{pmatrix} = \begin{pmatrix} 3 \\ 0 \\ 2 \end{pmatrix} + y \begin{pmatrix} -4 \\ 1 \\ 0 \end{pmatrix}.$$

This set of vectors has the geometric interpretation of being the line (in $\mathbb{R}^3$) through the point $(3,0,2)$ in the direction of the vector $(-4,1,0)$, as in Figure 2.5. We investigate this geometric interpretation of the solution set in much more depth in Section 2.3.

A system of linear equations has infinitely many solutions exactly when (a) there is at least one solution and (b) there is at least one column in the

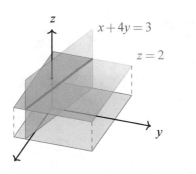

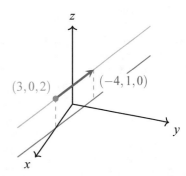

(a) The solution set is the intersection of the two planes $x+4y=3$ and $z=2$.

(b) The solution set is the line going through the point $(3,0,2)$ in the direction of the vector $(-4,1,0)$.

Figure 2.5: Geometrically, the solution set to the system of linear equations $x+4y=3$, $z=2$ can be thought of as the intersection of those two planes, or as the line through the point $(3,0,2)$ in the direction of the vector $(-4,1,0)$.

left-hand block of the row echelon forms without a leading entry (i.e., there is at least one free variable). In this case, all of the solutions can be described by letting the free variables take on any value (hence the term "free") and then solving for the leading variables in terms of those free variables.

> (!) A linear system has infinitely many solutions if and only if it is consistent and the number of leading entries in its row echelon forms is less than the number of variables.

The method that we have developed here for determining how many solutions a linear system has once it has been put into row echelon form is summarized by the flowchart in Figure 2.6.

Keep in mind that we can use *any* row echelon form here to determine how many solutions there are. Computing the reduced row echelon form is often overkill, especially if there are no solutions.

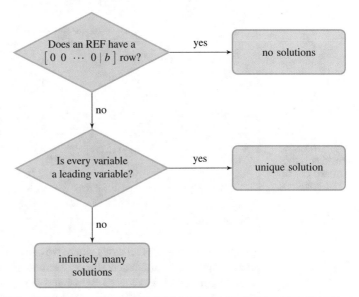

Figure 2.6: A flowchart that describes how to use a row echelon form of an augmented matrix to determine how many solutions the associated linear system has.

Example 2.1.5

Solving Systems of Linear Equations in Row Echelon Form

Find all solutions of the linear systems associated with the following augmented matrices in row echelon form:

a) $\begin{bmatrix} 1 & -1 & 0 & 1 & | & 2 \\ 0 & 0 & 1 & -1 & | & 1 \\ 0 & 0 & 0 & 0 & | & 0 \end{bmatrix}$ b) $\begin{bmatrix} 1 & 2 & -4 & | & -4 \\ 0 & 3 & -1 & | & 2 \\ 0 & 0 & 8 & | & 8 \end{bmatrix}$

Solutions:

Since there are two free variables in the solution of the linear system in part (a), its solution set is 2-dimensional (i.e., a plane).

a) This matrix is already in reduced row echelon form, so we can directly read the solutions from the matrix. If we label the variables corresponding to the columns from left-to-right as w, x, y, z, then w and y are the leading variables and x and z are the free variables. The first equation says (after moving the free variables to the right-hand-side) $w = 2 + x - z$, and the second equation says $y = 1 + z$ (and x and z can be anything).

b) This matrix is in non-reduced row echelon form. Since each column on the left-hand-side has a leading entry, and there are no rows of the form $\begin{bmatrix} 0 & 0 & 0 & | & b \end{bmatrix}$, we know the system has a unique solution. To find it, we back substitute. If we label the variables from left-to-right as x, y, z, then the third equation says $8z = 8$, so $z = 1$. Plugging that into the second equation tells us $3y - 1 = 2$, so $y = 1$. Plugging those values into the first equation tells us $x + 2 - 4 = -4$, so $x = -2$. The unique solution is thus $(x, y, z) = (-2, 1, 1)$.

Example 2.1.6

Solving a System of Linear Equations

Find all solutions of the linear system associated with the augmented matrix

$$\begin{bmatrix} 1 & 2 & -2 & | & -4 \\ 2 & 4 & 1 & | & 0 \\ 1 & 2 & 7 & | & 2 \end{bmatrix}.$$

Solution:

This matrix is not in row echelon form, so our first step is to use row operations to get it there:

Notice that we only needed a (non-reduced) row echelon form of the matrix to see that there are no solutions.

$\begin{bmatrix} 1 & 2 & -2 & | & -4 \\ 2 & 4 & 1 & | & 0 \\ 1 & 2 & 7 & | & 2 \end{bmatrix} \xrightarrow[R_3 - R_1]{R_2 - 2R_1} \begin{bmatrix} 1 & 2 & -2 & | & -4 \\ 0 & 0 & 5 & | & 8 \\ 0 & 0 & 9 & | & 6 \end{bmatrix}$

$\xrightarrow{R_3 - \frac{9}{5}R_2} \begin{bmatrix} 1 & 2 & -2 & | & -4 \\ 0 & 0 & 5 & | & 8 \\ 0 & 0 & 0 & | & -42/5 \end{bmatrix}$

Since the bottom row of this row echelon form corresponds to the equation $0x + 0y + 0z = -42/5$, we see that this system of equations has no solution.

There are some special cases where we can determine how many solutions a linear system has without actually performing any computations. For example, if the linear system has a zero right-hand side (i.e., it is of the form $A\mathbf{x} = \mathbf{0}$) then we know that it must have at least one solution, since $\mathbf{x} = \mathbf{0}$ solves any such system. Linear systems of this form are called **homogeneous systems**, and we now state our previous observation about them a bit more formally:

Theorem 2.1.2	Suppose $A \in \mathcal{M}_{m,n}$. Then the linear system $A\mathbf{x} = \mathbf{0}$ is consistent.
Homogeneous Systems	

The shape of a matrix can also tell us a great deal about how many solutions a linear system has, as demonstrated by the following theorem:

Theorem 2.1.3	Suppose $A \in \mathcal{M}_{m,n}$ and $m < n$. Then the linear system $A\mathbf{x} = \mathbf{b}$ has either
Short and Fat Systems	no solutions or infinitely many solutions.

We often say that a matrix is "short and fat" if it has more columns than rows, and similarly that it is "tall and skinny" if it has more rows than columns.

To convince ourselves that the above theorem is true, consider what happens if we row-reduce a matrix with more columns than rows. Since each row has at most 1 leading entry, it is not possible for each column to have a leading entry in it (see Figure 2.7), so we see from the flowchart in Figure 2.6 that the associated linear system cannot have a unique solution.

$$
\begin{bmatrix}
\bigstar & * & * & * & * & \cdots & * & * \\
0 & \bigstar & * & * & * & \cdots & * & * \\
\vdots & \vdots & \ddots & \vdots & \vdots & \ddots & \vdots & \vdots \\
0 & 0 & \cdots & \bigstar & * & \cdots & * & *
\end{bmatrix}
$$

Figure 2.7: A short and fat linear system (i.e., one with more variables than equations) cannot have a unique solution, since the coefficient matrix does not have enough rows for each column to have a leading entry.

By combining the previous two theorems, we immediately get the following corollary:

Corollary 2.1.4	Suppose $A \in \mathcal{M}_{m,n}$ and $m < n$. Then the linear system $A\mathbf{x} = \mathbf{0}$ has infinitely
Short and Fat Homogeneous Systems	many solutions.

2.1.5 Applications of Linear Systems

To start demonstrating the utility of solving systems of linear equations, we now look at how they can be used to solve some problems that we introduced earlier, as well as various other scientific problems of interest. Our first such example shows how they can be used to find vectors that are orthogonal to other given vectors.

Example 2.1.7	Find a non-zero vector that is orthogonal to...
Finding Orthogonal Vectors	a) $(1,2,3)$ and $(0,1,-1)$ b) $(1,2,2,2)$, $(2,1,-1,0)$, and $(1,0,2,1)$

Remember that we want to write v_1 in terms only of the free variable, so do not leave it in the form $v_1 = -2v_2 - 3v_3$. Use the fact that $v_2 = v_3$ to replace the leading variable v_2 by the free variable v_3.

Solutions:

a) We could use the cross product (see Section 1.A) to find such a vector, or we can set this up as a linear system. We want to find a vector $\mathbf{v} = (v_1, v_2, v_3)$ such that

$$v_1 + 2v_2 + 3v_3 = 0$$
$$v_2 - v_3 = 0$$

This linear system is already in row echelon form, with v_3 as a free variable and v_1, v_2 as leading variables. Its solutions are thus the vectors (v_1, v_2, v_3) with $v_2 = v_3$ and $v_1 = -2v_2 - 3v_3 = -5v_3$, and v_3 arbitrary. To find a single explicit vector that works, we can choose $v_3 = 1$ to get $\mathbf{v} = (-5, 1, 1)$.

b) Once again, we set up the system of linear equations that is described by the three orthogonality requirements:

$$
\begin{aligned}
v_1 + 2v_2 + 2v_3 + 2v_4 &= 0 \\
2v_1 + v_2 - v_3 &= 0 \\
v_1 + 2v_3 + v_4 &= 0
\end{aligned}
$$

The reduced row echelon form of the associated augmented matrix is

$$
\left[\begin{array}{cccc|c}
1 & 0 & 0 & 0 & 0 \\
0 & 1 & 0 & 1/2 & 0 \\
0 & 0 & 1 & 1/2 & 0
\end{array}\right].
$$

Try to compute this RREF on your own.

We thus see that v_4 is a free variable, and v_1, v_2, v_3 are all leading variables. Solutions (v_1, v_2, v_3, v_4) to this system have the form $v_1 = 0$, $v_2 = -v_4/2$, and $v_3 = -v_4/2$, with v_4 arbitrary. If we choose $v_4 = 2$ then we find the specific vector $\mathbf{v} = (0, -1, -1, 2)$, which is indeed orthogonal to all three of the given vectors.

Really though, we can choose v_4 to be any non-zero value that we want.

Next, we use linear systems to determine whether or not a vector is a linear combination of some given collection of vectors (recall that we saw how to do this in some very limited special cases in Section 1.1.3, but we did not learn how to do it in general).

Example 2.1.8

Finding a Linear Combination

Determine whether or not the following vectors are linear combinations of the vectors $(-1, 3, 2)$ and $(3, 1, -1)$:

a) $(1, 1, 3)$
b) $(1, 3, 1)$

Solutions:

Refer back to Section 1.1.3 if you need a refresher on linear combinations.

a) More explicitly, we are being asked whether or not there exist scalars $c_1, c_2 \in \mathbb{R}$ such that

$$
(1, 1, 3) = c_1(-1, 3, 2) + c_2(3, 1, -1).
$$

If we compare the entries of the vectors on the left- and right-hand sides of that equation, we see that this is actually a system of linear equations:

$$
\begin{aligned}
-c_1 + 3c_2 &= 1 \\
3c_1 + c_2 &= 1 \\
2c_1 - c_2 &= 3
\end{aligned}
$$

We can solve this linear system by using row operations to put it in row echelon form:

$$\begin{bmatrix} -1 & 3 & | & 1 \\ 3 & 1 & | & 1 \\ 2 & -1 & | & 3 \end{bmatrix} \xrightarrow[R_3+2R_1]{R_2+3R_1} \begin{bmatrix} -1 & 3 & | & 1 \\ 0 & 10 & | & 4 \\ 0 & 5 & | & 5 \end{bmatrix}$$

$$\xrightarrow{R_3-\frac{1}{2}R_2} \begin{bmatrix} -1 & 3 & | & 1 \\ 0 & 10 & | & 4 \\ 0 & 0 & | & 3 \end{bmatrix}.$$

> Notice that the columns on the left are the vectors in the linear combination and the augmented column is the vector that we want to write as a linear combination of the others.

From here we see that the linear system has no solution, since the final row in this row echelon form says that $0 = 3$. It follows that there do not exist scalars c_1, c_2 with the desired properties, so $(1, 1, 3)$ is *not* a linear combination of $(-1, 3, 2)$ and $(3, 1, -1)$.

b) Similarly, we now want to know whether or not there exist scalars $c_1, c_2 \in \mathbb{R}$ such that

$$(1, 3, 1) = c_1(-1, 3, 2) + c_2(3, 1, -1).$$

If we set this up as a linear system and solve it just as we did in part (a), we find the following row echelon form:

> Notice that this augmented matrix is the same as the one from part (a), except for its right-hand-side column.

$$\begin{bmatrix} -1 & 3 & | & 1 \\ 3 & 1 & | & 3 \\ 2 & -1 & | & 1 \end{bmatrix} \xrightarrow[R_3+2R_1]{R_2+3R_1} \begin{bmatrix} -1 & 3 & | & 1 \\ 0 & 10 & | & 6 \\ 0 & 5 & | & 3 \end{bmatrix}$$

$$\xrightarrow{R_3-\frac{1}{2}R_2} \begin{bmatrix} -1 & 3 & | & 1 \\ 0 & 10 & | & 6 \\ 0 & 0 & | & 0 \end{bmatrix}.$$

> After the coefficients c_1, c_2 are found, it is easy to verify that $(1, 3, 1) = \frac{4}{5}(-1, 3, 2) + \frac{3}{5}(3, 1, -1)$. The hard part is finding c_1 and c_2 in the first place.

This linear system can now be solved via back substitution: $c_2 = 3/5$, so $-c_1 + 3c_2 = 1$ implies $c_1 = 4/5$. We thus conclude that $(1, 3, 1)$ *is* a linear combination of $(-1, 3, 2)$ and $(3, 1, -1)$, and in particular

$$(1, 3, 1) = \tfrac{4}{5}(-1, 3, 2) + \tfrac{3}{5}(3, 1, -1).$$

The above example has a natural geometric interpretation as well. There is some plane that contains both of the vectors $(-1, 3, 2)$ and $(3, 1, -1)$—this plane also contains the vector $(1, 3, 1)$ from part (b) of the example, but not the vector $(1, 1, 3)$ from part (a), as illustrated in Figure 2.8.

Example 2.1.9

Solving Mixing Problems

> We start exploring this idea of whether or not one vector lies in the same (hyper)plane as some other vectors in depth in Section 2.3.

Suppose a farmer has two types of milk: one that is 3.5% butterfat and another that is 1% butterfat. How much of each should they mix together to create 500 liters of milk that is 2% butterfat?

Solution:

If we let x and y denote the number of liters of 3.5% butterfat milk and 1% butterfat milk that the farmer will mix together, respectively, then the fact that they want a total of 500 liters of milk tells us that

$$x + y = 500.$$

We similarly want there to be a total of $0.02 \times 500 = 10$ liters of butterfat

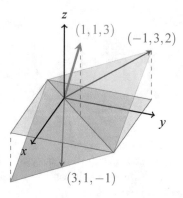 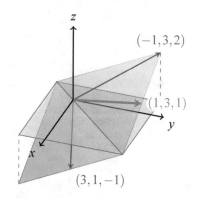

(a) $(1,1,3)$ is not a linear combination of $(-1,3,2)$ and $(3,1,-1)$.

(b) $(1,3,1)$ is a linear combination of $(-1,3,2)$ and $(3,1,-1)$.

Figure 2.8: Geometrically, any two vectors lie on a common plane. A third vector is a linear combination of those two vectors if and only if it also lies on the same plane (and the two original vectors are not multiples of each other).

in the resulting mixture, so we want

$$0.035x + 0.01y = 10.$$

We thus have a linear system with two variables and two equations, and it is straightforward to solve it via Gauss–Jordan elimination:

We could have instead used Gaussian elimination and back substitution. Either method works.

$$\left[\begin{array}{cc|c} 1 & 1 & 500 \\ 0.035 & 0.01 & 10 \end{array}\right] \xrightarrow{R_2 - 0.035R_1} \left[\begin{array}{cc|c} 1 & 1 & 500 \\ 0 & -0.025 & -7.5 \end{array}\right]$$

$$\xrightarrow{-40R_2} \left[\begin{array}{cc|c} 1 & 1 & 500 \\ 0 & 1 & 300 \end{array}\right] \xrightarrow{R_1 - R_2} \left[\begin{array}{cc|c} 1 & 0 & 200 \\ 0 & 1 & 300 \end{array}\right].$$

We thus see that the farmer should mix together $x = 200$ liters of the 3.5% butterfat milk and $y = 300$ liters of the 1% butterfat milk.

Remark 2.1.2

Numerical Instability of Linear Systems

Great care must be taken when constructing a linear system in practice, as changing its coefficients even slightly can change the resulting solution(s) considerably. For example, if we were to change the bottom-right entry in the linear system from Example 2.1.9 from 0.01 to 0.015 then the solution changes considerably, from $(x, y) = (200, 300)$ to $(x, y) = (125, 375)$. For another example of this phenomenon, see Exercise 2.1.10.

Changing the coefficients just slightly can even change a linear system from having a unique solution to having no solution or infinitely many solutions (or vice-versa), which can be particularly problematic when the linear system was constructed by performing various (necessarily not perfect) measurements. However, identifying these problems and dealing with them is outside of the scope of this book—there are entire textbooks devoted to numerical linear algebra where these problems are explored.

The following example illustrates how we can use linear systems to determine how many molecules of various chemical compounds are required as input to a chemical reaction, as well as how many molecules of different compounds are produced as output.

Example 2.1.10

Balancing Chemical Equations

In molecular formulas, subscripts say how many copies of the preceding element are in the molecule. For example, the formula H_2O for water means that it consists of 2 atoms of hydrogen (H) and 1 atom of oxygen (O).

When butane (C_4H_{10}) burns in the presence of oxygen gas (O_2), it produces carbon dioxide (CO_2) and water vapor (H_2O). We thus say that the "unbalanced" equation for burning butane is

$$C_4H_{10} + O_2 \rightarrow CO_2 + H_2O.$$

Find the corresponding "balanced" equation. That is, find integers $w, x, y,$ and z (which represent how many copies of each molecule are present in the reaction) such that the chemical equation

$$wC_4H_{10} + xO_2 \rightarrow yCO_2 + zH_2O$$

has the same amount of each element on the left- and right-hand sides.

Solution:

On the left-hand side of this chemical equation, we have $2x$ atoms of oxygen (O) and on the right-hand side we have $2y + z$ atoms of oxygen, so we require

$$2x = 2y + z, \quad \text{or equivalently} \quad 2x - 2y - z = 0.$$

Similarly matching up the number of atoms of carbon (C) and hydrogen (H) on the left- and right-hand sides reveals that

$$4w - y = 0 \quad \text{and} \quad 10w - 2z = 0.$$

We thus have a linear system with 4 variables and 3 equations, which we can solve using the techniques of this section:

Try to compute this RREF on your own.

$$\left[\begin{array}{cccc|c} 0 & 2 & -2 & -1 & 0 \\ 4 & 0 & -1 & 0 & 0 \\ 10 & 0 & 0 & -2 & 0 \end{array}\right] \xrightarrow{\text{row-reduce}} \left[\begin{array}{cccc|c} 1 & 0 & 0 & -1/5 & 0 \\ 0 & 1 & 0 & -13/10 & 0 \\ 0 & 0 & 1 & -4/5 & 0 \end{array}\right].$$

Since z is a free variable, we are free to choose it however we like, and we should choose it so that all of the variables end up having integer values. In particular, if we choose $z = 10$ then we get $w = 2$, $x = 13$, and $y = 8$, so that the balanced chemical equation is

Any positive integer multiple of this equation is also balanced.

$$2C_4H_{10} + 13O_2 \rightarrow 8CO_2 + 10H_2O.$$

That is, 2 molecules of butane can be burned in the presence of 13 molecules of oxygen gas to produce 8 molecules of carbon dioxide and 10 molecules of water vapor.

Some other problems that we can solve with linear systems include characterizing which vectors go to a certain place via a given linear transformation (Exercise 2.1.24), finding matrices that commute with each other (Exercise 2.1.28), and finding polynomials whose graphs go through a specific set of points in the plane (Exercise 2.1.29). In general, most new computational problems that we see from here on in this book can be solved by rephrasing the problem as a linear system. Being able to solve linear systems is our mathematical sledgehammer that breaks down the majority of problems that we encounter from here.

Exercises

solutions to starred exercises on page 444

2.1.1 Which of the following matrices are in row echelon form? Reduced row echelon form?

*(a) $\begin{bmatrix} 1 & 0 \\ 0 & 1 \end{bmatrix}$

(b) $\begin{bmatrix} 1 & 2 & 1 \\ 0 & 1 & 2 \end{bmatrix}$

*(c) $\begin{bmatrix} 0 & 0 & 0 \\ 0 & 1 & 2 \\ 0 & 0 & 0 \end{bmatrix}$

(d) $\begin{bmatrix} 2 & 0 & 1 \\ 0 & 2 & 0 \\ 0 & 0 & 0 \end{bmatrix}$

*(e) $\begin{bmatrix} 1 & 0 \\ 0 & 2 \\ 0 & 1 \end{bmatrix}$

(f) $\begin{bmatrix} 1 & 2 & 7 \\ 0 & 0 & 0 \\ 0 & 0 & 0 \end{bmatrix}$

*(g) $\begin{bmatrix} 1 & 2 & 0 & 3 & 0 \\ 0 & 0 & 1 & -1 & 0 \\ 0 & 0 & 0 & 0 & 1 \end{bmatrix}$

(h) $\begin{bmatrix} 0 & 2 & 3 & -1 \\ 0 & 0 & 1 & 1 \\ 0 & 0 & 0 & 0 \end{bmatrix}$

2.1.2 Use Gauss–Jordan elimination to compute the reduced row echelon form of each of the following matrices.

*(a) $\begin{bmatrix} 1 & 2 \\ 3 & 4 \end{bmatrix}$

(b) $\begin{bmatrix} 2 & 3 & -1 \\ 4 & 1 & 3 \end{bmatrix}$

*(c) $\begin{bmatrix} 1 & 2 \\ 3 & 6 \\ -1 & -2 \end{bmatrix}$

(d) $\begin{bmatrix} 3 & 9 & 3 \\ 1 & 3 & 0 \\ 1 & 3 & 2 \end{bmatrix}$

*(e) $\begin{bmatrix} -1 & 2 & 2 \\ 4 & -1 & 1 \\ 2 & 4 & 2 \end{bmatrix}$

(f) $\begin{bmatrix} -1 & 2 & 7 & 2 \\ 4 & -1 & -7 & 1 \\ 2 & 4 & 10 & 2 \end{bmatrix}$

*(g) $\begin{bmatrix} 4 & 0 & 8 \\ 4 & 1 & 11 \\ 1 & 2 & 8 \\ -1 & 0 & -2 \end{bmatrix}$

(h) $\begin{bmatrix} 1 & -2 & 0 & 1 \\ 4 & -8 & 1 & 7 \\ 2 & -4 & 1 & 5 \\ 1 & -2 & -1 & -2 \end{bmatrix}$

💻 2.1.3 Use computer software to compute the reduced row echelon form of each of the following matrices.

*(a) $\begin{bmatrix} 3 & -2 & 0 & -2 \\ 0 & -2 & -2 & 2 \\ 0 & 2 & 0 & -1 \\ 1 & 0 & 2 & -2 \end{bmatrix}$

(b) $\begin{bmatrix} 4 & 2 & -1 & 2 & 1 \\ 1 & 2 & -1 & 0 & 4 \\ 5 & 1 & 2 & 6 & -1 \\ -3 & 4 & 2 & 2 & 5 \end{bmatrix}$

*(c) $\begin{bmatrix} 0 & -2 & -2 & 3 & 4 & 1 \\ 1 & 0 & -2 & 5 & 4 & 0 \\ 5 & 0 & 2 & 1 & 0 & 3 \\ -1 & 1 & 0 & 2 & 4 & -1 \end{bmatrix}$

(d) $\begin{bmatrix} 1 & -1 & -3 & -1 & -3 & -3 & -1 \\ 0 & -1 & -2 & 1 & -2 & 3 & 2 \\ 3 & -1 & -5 & 0 & 0 & -5 & 3 \\ 2 & -3 & -8 & -2 & -9 & -5 & 1 \\ -2 & 3 & 8 & -2 & 5 & -3 & -1 \end{bmatrix}$

∗2.1.4 If we interpret each of the matrices from Exercise 2.1.1 as augmented matrices, which of the corresponding systems of equations have no solutions, a unique solution, and infinitely many solutions? For example, the matrix in Exercise 2.1.1(h) corresponds to the linear system

$$2y + 3z = -1$$
$$z = 1$$

2.1.5 Find all solutions of the following systems of linear equations.

*(a) $\begin{aligned} x + 2y &= 3 \\ 2x + y &= 3 \end{aligned}$

(b) $\begin{aligned} x + y + z &= 4 \\ x - y + z &= 0 \end{aligned}$

*(c) $\begin{aligned} x - y &= 2 \\ x + 2y &= 4 \\ 2x - y &= 5 \end{aligned}$

(d) $\begin{aligned} 2x + y - z &= 1 \\ x - 3y + z &= -2 \\ 2x - 2y + 3z &= 7 \end{aligned}$

*(e) $\begin{aligned} x + y + z &= 1 \\ -x \quad\; + z &= 2 \\ 2x + y \quad\; &= 0 \end{aligned}$

(f) $\begin{aligned} w + x + y - z &= 0 \\ 2x + 3y + z &= -1 \\ y - z &= -3 \\ 3z &= 3 \end{aligned}$

*(g) $\begin{aligned} v - 2w - x - 2y - z &= 1 \\ 2v \quad\; - 2x - 6y - 4z &= 2 \\ 4w \quad\; + y - 2z &= 3 \end{aligned}$

💻 2.1.6 Use computer software to find all solutions of the following systems of linear equations.

*(a) $\begin{aligned} 6v + 5w + 3x - 2y - 2z &= 1 \\ 3v - w + x + 5y + 4z &= 2 \\ 3v + 4w + x + 3y + 4z &= 3 \\ 2v + 6w + x - 2y - z &= 4 \end{aligned}$

(b) $\begin{aligned} v - w + 2x + 6y + 6z &= 3 \\ 4v + 3w \quad\; - y + 4z &= 0 \\ 5v + w - 2x - 2y - 2z &= 2 \\ w - 2x + 3y - 2z &= -1 \\ v + 5w - x \quad\; + 5z &= 3 \end{aligned}$

2.1.7 Determine which of the following statements are true and which are false.

*(a) The equation $4x - \sin(1)y + 2z = \sqrt[3]{5}$ is linear.

(b) The equation $xy + 4z = 4$ is linear.

*(c) In $\mathbb{R}^3$, if two lines are not parallel then they must intersect at a point.

(d) In $\mathbb{R}^3$, if two planes are not parallel then they must intersect in a line.

*(e) A system of linear equations can have exactly 2 distinct solutions.

(f) Every homogeneous system of linear equations is consistent.

*(g) If a linear system has more equations than variables, then it must have no solution.

(h) If a linear system has fewer equations than variables, then it cannot have a unique solution.

∗(i) If a linear system has fewer equations than variables, then it must have infinitely many solutions.

∗**2.1.8** Let $t \in \mathbb{R}$ and consider the linear system described by the following augmented matrix:

$$\left[\begin{array}{ccc|c} 1 & -1 & & 0 \\ 1 & 1 & & 1 \\ 0 & 1 & & t \end{array} \right].$$

For which values of t does this system have (i) no solutions, (ii) a unique solution, and (iii) infinitely many solutions?

2.1.9 Let $h, k \in \mathbb{R}$ and consider the following system of linear equations in the variables x, y, and z:

$$x + y + hz = 1$$
$$y - z = k$$
$$x - y + 2z = 3$$

For which values of h and k does this system have (i) no solutions, (ii) a unique solution, and (iii) infinitely many solutions?

∗∗▢ **2.1.10** Consider the following system of linear equations in the variables w, x, y, and z:

$$w + x/2 + y/3 + z/4 = 1$$
$$w/2 + x/3 + y/4 + z/5 = 1$$
$$w/3 + x/4 + y/5 + z/6 = 1$$
$$w/4 + x/5 + y/6 + z/7 = h$$

Use computer software to solve this linear system when $h = 0.95$, $h = 1.00$, and $h = 1.05$. Do you notice anything surprising about how the solution changes as h changes? Explain what causes this surprising change in the solution.

∗**2.1.11** Let $a, b, c, d \in \mathbb{R}$ be scalars and consider the system of linear equations described by the augmented matrix

$$\left[\begin{array}{cc|c} a & b & 1 \\ c & d & 1 \end{array} \right].$$

Show that this system of linear equations has a unique solution whenever $ad - bc \neq 0$.

2.1.12 Show how to swap the two rows of the matrix

$$\begin{bmatrix} a & b \\ c & d \end{bmatrix}$$

using only the "addition" and "multiplication" row operations (i.e., do not directly use the "swap" row operation). [Hint: Start by subtracting row 1 from row 2.]

∗∗**2.1.13** Suppose $A, B, R \in \mathcal{M}_{m,n}$.

(a) Suppose A and R are row equivalent, and so are B and R. Explain why A and B are row equivalent.
(b) Show that A and B are row equivalent if and only if they have the same reduced row echelon form.

2.1.14 Let $A \in \mathcal{M}_{m,n}$ and $\mathbf{b} \in \mathbb{R}^m$. Show that the system of linear equations $A\mathbf{x} = \mathbf{b}$ is consistent if and only if $\mathbf{b}$ is a linear combination of the columns of A.

∗**2.1.15** Use a system of linear equations to find a vector that is orthogonal to each of the listed vectors.

(a) $(1, 2, 3)$ and $(3, 2, 1)$
(b) $(1, 2, 0, 1)$, $(-2, 1, 1, 3)$, and $(-1, -1, 2, 1)$

∗∗**2.1.16** Care must be taken when performing multiple row operations simultaneously—even though it is OK to write multiple row operations in one step, this exercise illustrates why they must be performed sequentially (i.e., one after another) rather than simultaneously.

Consider the linear system represented by the augmented matrix

$$\left[\begin{array}{cc|c} 1 & 1 & 3 \\ 1 & 2 & 5 \end{array} \right].$$

(a) Perform the row operation $R_2 - R_1$, followed by the row operation $R_1 - R_2$, on this augmented matrix. What is the resulting augmented matrix, and what is the (unique) solution of the associated linear system?
(b) Perform the row operations $R_2 - R_1$ and $R_1 - R_2$ simultaneously on this augmented matrix (i.e., in the operation $R_1 - R_2$, subtract the *original* row 2 from row 1, not the newly-computed row 2). What is the resulting augmented matrix, and what are the solutions of the associated linear system?

2.1.17 Determine whether or not $\mathbf{b}$ is a linear combination of the other vectors. [Hint: Mimic Example 2.1.8.]

∗(a) $\mathbf{b} = (3, 4)$, $\mathbf{v}_1 = (6, 8)$
(b) $\mathbf{b} = (2, -3)$, $\mathbf{v}_1 = (1, 2)$, $\mathbf{v}_2 = (4, -1)$
∗(c) $\mathbf{b} = (1, 5, -6)$, $\mathbf{v}_1 = (1, 0, 0)$, $\mathbf{v}_2 = (0, 1, 0)$, $\mathbf{v}_3 = (0, 0, 1)$
(d) $\mathbf{b} = (2, 1, 2)$, $\mathbf{v}_1 = (-2, 2, 1)$, $\mathbf{v}_2 = (1, 2, 3)$
∗(e) $\mathbf{b} = (2, 1, 2)$, $\mathbf{v}_1 = (-4, 4, -1)$, $\mathbf{v}_2 = (2, -1, 1)$
(f) $\mathbf{b} = (2, 1, 2, 3)$, $\mathbf{v}_1 = (1, 2, 3, 4)$, $\mathbf{v}_2 = (4, 3, 2, 1)$, $\mathbf{v}_3 = (1, -1, 1, -1)$
∗(g) $\mathbf{b} = (1, 3, -3, -1)$, $\mathbf{v}_1 = (1, 2, 3, 4)$, $\mathbf{v}_2 = (4, 3, 2, 1)$, $\mathbf{v}_3 = (1, -1, 1, -1)$

∗**2.1.18** Let $\mathbf{v} = (v_1, v_2, v_3)$ and $\mathbf{w} = (w_1, w_2, w_3)$ be non-zero vectors. Use a system of linear equations to show that every vector that is orthogonal to both $\mathbf{v}$ and $\mathbf{w}$ is a multiple of

$$(v_2 w_3 - v_3 w_2, v_3 w_1 - v_1 w_3, v_1 w_2 - v_2 w_1)$$

[Side note: This vector is the cross product, which we explored in Section 1.A.]

∗**2.1.19** Suppose we have two mixtures of salt water: one that is 13% salt (by weight) and another that is 5% salt. How much of each should be mixed together to get 120 kilograms of a salt water mixture that is 8% salt?

[Hint: Mimic Example 2.1.9.]

2.1.20 Suppose a chemist has two acid solutions: one that is 30% acid (and the rest is water) and another that is 8% acid. How much of each should be mixed together to get 110 liters of a solution that is 25% acid?

*2.1.21 When zinc sulfide (ZnS) is heated in the presence of oxygen gas (O_2), it produces zinc oxide (ZnO) and sulfur dioxide (SO_2). Determine how many of each of these molecules are involved in the reaction. That is, balance the following unbalanced chemical equation:

$$ZnS + O_2 \rightarrow ZnO + SO_2.$$

[Hint: Mimic Example 2.1.10.]

2.1.22 When ethane (C_2H_6) burns in the presence of oxygen gas (O_2), it produces carbon dioxide (CO_2) and water vapor (H_2O). Determine how many of each of these molecules are involved in the reaction. That is, balance the following unbalanced chemical equation:

$$C_2H_6 + O_2 \rightarrow CO_2 + H_2O.$$

2.1.23 Find an equation of the plane in $\mathbb{R}^3$ with the points $(1,1,1)$, $(2,3,4)$, and $(-1,-1,0)$ on it.

[Hint: Recall that the general equation of a plane is of the form $ax + by + cz = d$. Plug in the given points to set up a system of linear equations.]

**2.1.24 Suppose $\mathbf{u} = (1,2,2)/3$ and let $P_{\mathbf{u}}$ be the linear transformation that projects $\mathbb{R}^3$ onto the line through the origin in the direction of $\mathbf{u}$. Find all vectors $\mathbf{v} \in \mathbb{R}^3$ with the property that $P_{\mathbf{u}}(\mathbf{v}) = (2,4,4)$.

[Hint: Find the standard matrix of $P_{\mathbf{u}}$ and use it to set up a system of linear equations.]

2.1.25 Systems of linear equations can be used to determine whether or not a vector is a linear combination of another set of vectors.

*(a) Is $(3,-2,1)$ a linear combination of the vectors $(1,4,2)$ and $(2,-1,1)$?
[Hint: Rewrite the equation $(3,-2,1) = c_1(1,4,2) + c_2(2,-1,1)$ as a system of linear equations.]
(b) Is $(2,7,-3,3)$ a linear combination of the vectors $(1,2,0,1)$, $(-2,1,1,3)$, and $(-1,-1,2,1)$?

2.1.26 Find the standard matrix of the linear transformation T that acts as described. [Hint: Set up a system of linear equations to determine $T(\mathbf{e}_1), T(\mathbf{e}_2), \ldots, T(\mathbf{e}_n)$.]

*(a) $T(1,1) = (3,7)$, $T(1,-1) = (-1,-1)$
(b) $T(1,2) = (5,3,4)$, $T(2,-1) = (0,1,3)$
*(c) $T(1,1,1) = (4,6,1)$, $T(2,-1,1) = (1,1,-4)$, $T(0,0,1) = (1,2,0)$

2.1.27 We can sometimes solve systems of non-linear equations by cleverly converting them into systems of linear equations. Some particular tricks that are useful are multiplying equations by variables to eliminate fractions, and performing a change of variables.

Use these tricks to find all solutions of the following systems of non-linear equations.

*(a) $1/x + 2/y = 3/xy$
$x + y = 2$

(b) $y/x + 2x/y = 6/xy$
$x^2 + y^2 = 5$

*(c) $1/x \quad\quad - 2/z = -4$
$1/x + 1/y + 1/z = 4$
$2/x - 6/y - 2/z = 4$

(d) $\sin(x) + 2\cos(y) - \cos(z) = 3$
$2\sin(x) + \cos(y) + \cos(z) = 0$
$-\sin(x) - \cos(y) + \cos(z) = -2$

**2.1.28 Recall that matrix multiplication is not commutative in general. Nonetheless, it is *sometimes* the case that $AB = BA$, and we can find matrices with this property by solving linear systems.

(a) Suppose
$$A = \begin{bmatrix} 0 & 1 \\ 1 & 0 \end{bmatrix}.$$
Find all $B \in \mathcal{M}_2$ with the property that $AB = BA$. [Hint: Write
$$B = \begin{bmatrix} a & b \\ c & d \end{bmatrix}$$
and use the matrix multiplication to set up a system of linear equations.]
(b) Suppose
$$C = \begin{bmatrix} 1 & 1 \\ 1 & 0 \end{bmatrix}.$$
Find all $D \in \mathcal{M}_2$ with the property that $CD = DC$.
(c) Use parts (a) and (b) to show that the only matrices that commute with everything in $\mathcal{M}_2$ are the matrices of the form cI, where $c \in \mathbb{R}$. [Side note: These are sometimes called **scalar matrices**.]

2.1.29 An **interpolating polynomial is a polynomial that passes through a given set of points in $\mathbb{R}^2$ (often these points are found via some experiment, and we want some curve that can be used to estimate other data points).

(a) Find constants b and m such that the line $y = mx + b$ goes through the points $(1,2)$ and $(3,8)$.
[Hint: The fact that the line goes through $(1,2)$ tells us that $2 = m + b$ and the fact that the line goes through $(3,8)$ tells us that $8 = 3m + b$.]
(b) Find constants a, b, and c such that the parabola $y = ax^2 + bx + c$ goes through the points $(1,3)$, $(2,6)$, $(3,13)$.
[Hint: The fact that the parabola goes through $(2,6)$ tells us that $6 = 4a + 2b + c$, and so on.]
(c) What happens in part (b) if you try to find a line $y = mx + b$ that goes through all 3 points?
(d) What happens in part (b) if you try to find a cubic $y = ax^3 + bx^2 + cx + d$ that goes through the 3 points?

□ **2.1.30** The world's population (in billions) over the past 50 years is summarized in the following table:

Year:	1970	1980	1990	2000	2010	2020
Pop.:	3.70	4.46	5.33	6.14	6.96	7.79

(a) Use computer software and the method of Exercise 2.1.29 to construct a degree-5 interpolating polynomial for this data. That is, find scalars a, b, c, d, e, and f such that if x is the number of decades that have passed since 1970 then

$$p(x) = ax^5 + bx^4 + cx^3 + dx^2 + ex + f$$

is the world's population (in billions) at the start of the decade (i.e., $p(0) = 3.70$, $p(1) = 4.46$, and so on).

(b) Use this polynomial to estimate what the world's population was in 1960, and compare with the actual population in that year (3.03 billion). How accurate was your estimate?

(c) Use this polynomial to estimate what the world's population will be in 2060. Do you think that this estimate is reasonable?

2.2 Elementary Matrices and Matrix Inverses

In the previous section, we demonstrated how we can use matrices and the three elementary row operations to solve systems of linear equations. In this section, we rephrase these elementary row operations purely in terms of matrices and matrix multiplication, so that we can more easily make use of linear systems and Gaussian elimination when proving theorems and making connections with other aspects of linear algebra.

2.2.1 Elementary Matrices

Suppose we are given the following 3×4 matrix, which we would like to put into row echelon form via Gaussian elimination (or reduced row echelon form via Gauss–Jordan elimination):

$$\begin{bmatrix} 0 & 2 & 4 & 0 \\ 1 & 1 & 0 & -1 \\ 3 & 4 & 2 & 1 \end{bmatrix}.$$

We could also swap rows 1 and 3, but doing so makes the numbers a bit uglier.

We should start by swapping rows 1 and 2, since we want to get a non-zero leading entry in the top-left corner of the matrix. We of course could do this operation "directly" like we did in the previous section, but another way of doing it is to multiply on the left by a certain matrix as follows (for now, do not worry about *why* we would do the row operation this way):

$$\begin{bmatrix} 0 & 1 & 0 \\ 1 & 0 & 0 \\ 0 & 0 & 1 \end{bmatrix} \begin{bmatrix} 0 & 2 & 4 & 0 \\ 1 & 1 & 0 & -1 \\ 3 & 4 & 2 & 1 \end{bmatrix} = \begin{bmatrix} 1 & 1 & 0 & -1 \\ 0 & 2 & 4 & 0 \\ 3 & 4 & 2 & 1 \end{bmatrix}. \tag{2.2.1}$$

Next, we want to subtract 3 times row 1 from row 3, which we can again carry out by multiplying on the left by a certain cleverly-chosen matrix:

$$\begin{bmatrix} 1 & 0 & 0 \\ 0 & 1 & 0 \\ -3 & 0 & 1 \end{bmatrix} \begin{bmatrix} 1 & 1 & 0 & -1 \\ 0 & 2 & 4 & 0 \\ 3 & 4 & 2 & 1 \end{bmatrix} = \begin{bmatrix} 1 & 1 & 0 & -1 \\ 0 & 2 & 4 & 0 \\ 0 & 1 & 2 & 4 \end{bmatrix}, \tag{2.2.2}$$

and we then probably want to divide row 2 by 2, which we can (yet again) implement by multiplying on the left by a certain matrix:

$$\begin{bmatrix} 1 & 0 & 0 \\ 0 & 1/2 & 0 \\ 0 & 0 & 1 \end{bmatrix} \begin{bmatrix} 1 & 1 & 0 & -1 \\ 0 & 2 & 4 & 0 \\ 0 & 1 & 2 & 4 \end{bmatrix} = \begin{bmatrix} 1 & 1 & 0 & -1 \\ 0 & 1 & 2 & 0 \\ 0 & 1 & 2 & 4 \end{bmatrix}. \tag{2.2.3}$$

We could continue in this way if we wanted to (see the upcoming Example 2.2.1), but the point is that we have now shown how to perform all 3 of the elementary row operations (first "swap", then "addition", and then "multiplication") by multiplying on the left by certain matrices. In fact, the matrices that we multiplied by on the left are the matrices that we get by applying the desired row operations to the identity matrix.

For example, we carried out the "swap" row operation $R_1 \leftrightarrow R_2$ in Equation (2.2.1) by multiplying on the left by

$$\begin{bmatrix} 0 & 1 & 0 \\ 1 & 0 & 0 \\ 0 & 0 & 1 \end{bmatrix},$$

which is exactly the matrix that we get if we swap rows 1 and 2 of the identity matrix. Similarly, in Equations (2.2.2) and (2.2.3) we multiplied on the left by the matrices

$$\begin{bmatrix} 1 & 0 & 0 \\ 0 & 1 & 0 \\ -3 & 0 & 1 \end{bmatrix} \quad \text{and} \quad \begin{bmatrix} 1 & 0 & 0 \\ 0 & 1/2 & 0 \\ 0 & 0 & 1 \end{bmatrix},$$

respectively, which are the matrices that are obtained by applying the desired row operations $R_3 - 3R_1$ and $\frac{1}{2}R_2$ to the identity matrix. We will make extensive use of these special matrices that implement the elementary row operations throughout the rest of this section (and some of the later sections of this book), so we give them a name.

Definition 2.2.1

**Elementary
Matrices**

A square matrix $A \in \mathcal{M}_n$ is called an **elementary matrix** if it can be obtained from the identity matrix via a single elementary row operation.

We have already seen a few specific examples of elementary matrices, but it is worthwhile to briefly discuss what they look like in general. Just like there are three types of elementary row operations ("swap", "addition", and "multiplication"), there are also three types of elementary matrices.

The elementary matrices corresponding to the "swap" row operation $R_i \leftrightarrow R_j$ look like

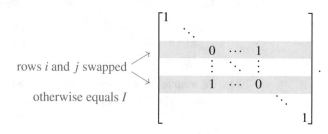

rows i and j swapped

otherwise equals I

Geometrically, these "swap" matrices act as linear transformations that do not stretch or deform $\mathbb{R}^n$, but rather only reflect space so as to swap two of the coordinate axes. In particular, the unique "swap" matrix acting on $\mathbb{R}^2$ is the reflection through the line $y = x$ that interchanges the x- and y-axes, as illustrated in Figure 2.9.

Similarly, the elementary matrices corresponding to the "addition" row operation $R_i + cR_j$ and the "multiplication" row operation cR_j look like

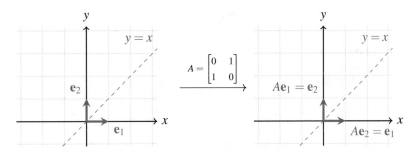

Figure 2.9: The "swap" matrix $A = \begin{bmatrix} 0 & 1 \\ 1 & 0 \end{bmatrix}$ acts as a reflection through the line $y = x$.

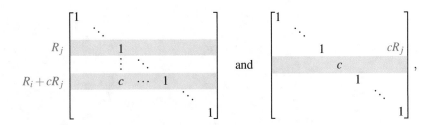

respectively. Geometrically, the "multiplication" matrices implement a diagonal scaling and thus just stretch $\mathbb{R}^n$ by a factor of c in the j-th coordinate direction. The "addition" matrices, however, are slightly more exotic—they shear space in the direction of one of the coordinate axes, as illustrated in Figure 2.10, and are thus sometimes called **shear matrices**.

We first saw shear matrices back in Exercise 1.4.23.

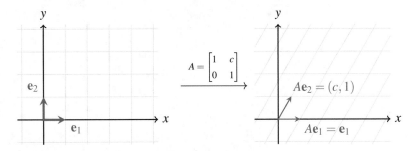

Figure 2.10: The "addition" matrix $A = \begin{bmatrix} 1 & c \\ 0 & 1 \end{bmatrix}$ shears space in the x-direction. A larger value of c corresponds to a more extreme shear.

While elementary matrices themselves are not terribly exciting, they are useful because we can use them to build up to more interesting things. For example, since multiplying on the left by an elementary matrix is equivalent to performing a *single* elementary row operation, multiplying on the left by several elementary matrices, one after another, must be equivalent to performing a *sequence* of elementary row operations. In particular, this means that we can encode the row operations applied by Gaussian elimination or Gauss–Jordan elimination as matrix multiplication. We illustrate what we mean by this with an example.

Example 2.2.1

A Matrix
Decomposition
Based on
the RREF

Let

$$A = \begin{bmatrix} 0 & 2 & 4 & 0 \\ 1 & 1 & 0 & -1 \\ 3 & 4 & 2 & 1 \end{bmatrix}.$$

Find a 3×3 matrix E such that $EA = R$, where R is the reduced row echelon form of A.

Solution:

We actually already started to do this earlier in this section, in Equations (2.2.1)–(2.2.3), when we started row-reducing A. In particular, we showed that if

$$E_1 = \begin{bmatrix} 0 & 1 & 0 \\ 1 & 0 & 0 \\ 0 & 0 & 1 \end{bmatrix}, \quad E_2 = \begin{bmatrix} 1 & 0 & 0 \\ 0 & 1 & 0 \\ -3 & 0 & 1 \end{bmatrix}, \quad \text{and} \quad E_3 = \begin{bmatrix} 1 & 0 & 0 \\ 0 & 1/2 & 0 \\ 0 & 0 & 1 \end{bmatrix},$$

then

$$E_3 E_2 E_1 A = \begin{bmatrix} 1 & 1 & 0 & -1 \\ 0 & 1 & 2 & 0 \\ 0 & 1 & 2 & 4 \end{bmatrix},$$

which is a partially row-reduced form of A. To find E, we just need to continue this process until we reach the RREF of A. The next two elementary row operations we perform are $R_1 - R_2$ and $R_3 - R_2$, which correspond to the elementary matrices

$$E_4 = \begin{bmatrix} 1 & -1 & 0 \\ 0 & 1 & 0 \\ 0 & 0 & 1 \end{bmatrix} \quad \text{and} \quad E_5 = \begin{bmatrix} 1 & 0 & 0 \\ 0 & 1 & 0 \\ 0 & -1 & 1 \end{bmatrix},$$

so that

$$E_5 E_4 E_3 E_2 E_1 A = \begin{bmatrix} 1 & 0 & -2 & -1 \\ 0 & 1 & 2 & 0 \\ 0 & 0 & 0 & 4 \end{bmatrix}.$$

Finally, we apply the elementary row operations $\frac{1}{4}R_3$ and then $R_1 + R_3$, which correspond to multiplying by the elementary matrices

$$E_6 = \begin{bmatrix} 1 & 0 & 0 \\ 0 & 1 & 0 \\ 0 & 0 & 1/4 \end{bmatrix} \quad \text{and} \quad E_7 = \begin{bmatrix} 1 & 0 & 1 \\ 0 & 1 & 0 \\ 0 & 0 & 1 \end{bmatrix},$$

respectively. It follows that

$$E_7 E_6 E_5 E_4 E_3 E_2 E_1 A = \begin{bmatrix} 1 & 0 & -2 & 0 \\ 0 & 1 & 2 & 0 \\ 0 & 0 & 0 & 1 \end{bmatrix},$$

which is the reduced row echelon form of A. So to get $EA = R$, we can set $E = E_7 E_6 E_5 E_4 E_3 E_2 E_1$, which can be computed by straightforward (albeit

Computing E in this way requires *six* matrix multiplications! We will see a better way soon.

tedious) calculation to be

$$E = E_7 E_6 E_5 E_4 E_3 E_2 E_1 = \frac{1}{8} \begin{bmatrix} -5 & 2 & 2 \\ 4 & 0 & 0 \\ -1 & -6 & 2 \end{bmatrix}.$$

The matrices $E_1, E_2, \ldots, E_7$ in the previous example can be thought of like a log that keeps track of which row operations were used to transform A into its reduced row echelon form R—first the row operation corresponding to E_1 was performed, then the one corresponding to E_2, and so on. Similarly, the matrix $E = E_7 E_6 E_5 E_4 E_3 E_2 E_1$ can be thought of as a condensed version of that log—the entries of the rows of E tell us which linear combinations of the rows of A give us its reduced row echelon form. For example, in Example 2.2.1 the first row of E tells us that to get the first row of the RREF of A, we should add $-5/8$ of A's first row, $2/8 = 1/4$ of its second row, and $1/4$ of its third row.

Recall from Exercise 1.3.27 that the rows of $EA = R$ are linear combinations of the rows of A.

To actually compute the product of all of those elementary matrices $E = E_7 E_6 E_5 E_4 E_3 E_2 E_1$, we could perform six matrix multiplications (ugh!), but a perhaps slightly less unpleasant way is to just apply the same row operations to the identity matrix that we apply to A in its row-reduction. The following example illustrates what we mean by this.

Example 2.2.2

A Matrix Decomposition via Row Operations

Let

$$A = \begin{bmatrix} 0 & 2 & 4 & 0 \\ 1 & 1 & 0 & -1 \\ 3 & 4 & 2 & 1 \end{bmatrix}.$$

Compute the reduced row echelon form of the 1×2 block matrix $[A \mid I]$.

Solution:

We just apply Gauss–Jordan elimination to this block matrix and see what happens:

$$\begin{bmatrix} 0 & 2 & 4 & 0 & 1 & 0 & 0 \\ 1 & 1 & 0 & -1 & 0 & 1 & 0 \\ 3 & 4 & 2 & 1 & 0 & 0 & 1 \end{bmatrix} \xrightarrow{R_1 \leftrightarrow R_2} \begin{bmatrix} 1 & 1 & 0 & -1 & 0 & 1 & 0 \\ 0 & 2 & 4 & 0 & 1 & 0 & 0 \\ 3 & 4 & 2 & 1 & 0 & 0 & 1 \end{bmatrix}$$

$$\xrightarrow{R_3 - 3R_1} \begin{bmatrix} 1 & 1 & 0 & -1 & 0 & 1 & 0 \\ 0 & 2 & 4 & 0 & 1 & 0 & 0 \\ 0 & 1 & 2 & 4 & 0 & -3 & 1 \end{bmatrix}$$

$$\xrightarrow{\frac{1}{2}R_2} \begin{bmatrix} 1 & 1 & 0 & -1 & 0 & 1 & 0 \\ 0 & 1 & 2 & 0 & 1/2 & 0 & 0 \\ 0 & 1 & 2 & 4 & 0 & -3 & 1 \end{bmatrix}$$

$$\xrightarrow[R_3 - R_2]{R_1 - R_2} \begin{bmatrix} 1 & 0 & -2 & -1 & -1/2 & 1 & 0 \\ 0 & 1 & 2 & 0 & 1/2 & 0 & 0 \\ 0 & 0 & 0 & 4 & -1/2 & -3 & 1 \end{bmatrix}.$$

At this point, the matrix is in row echelon form, and we now begin the backwards row-reduction step to put it into *reduced* row echelon form:

$$\left[\begin{array}{cccc|ccc} 1 & 0 & -2 & -1 & -1/2 & 1 & 0 \\ 0 & 1 & 2 & 0 & 1/2 & 0 & 0 \\ 0 & 0 & 0 & 4 & -1/2 & -3 & 1 \end{array}\right]$$

$$\xrightarrow{\frac{1}{4}R_3} \left[\begin{array}{cccc|ccc} 1 & 0 & -2 & -1 & -1/2 & 1 & 0 \\ 0 & 1 & 2 & 0 & 1/2 & 0 & 0 \\ 0 & 0 & 0 & 1 & -1/8 & -3/4 & 1/4 \end{array}\right]$$

$$\xrightarrow{R_1+R_3} \left[\begin{array}{cccc|ccc} 1 & 0 & -2 & 0 & -5/8 & 1/4 & 1/4 \\ 0 & 1 & 2 & 0 & 1/2 & 0 & 0 \\ 0 & 0 & 0 & 1 & -1/8 & -3/4 & 1/4 \end{array}\right].$$

In particular, the left block in this reduced row echelon form is simply the RREF of A, while the right block is exactly the same as the matrix E that we computed in Example 2.2.1.

In the following theorem, R is often chosen to be a row echelon form of A, but it does not have to be.

The previous example suggests that if we row-reduce a block matrix $[\,A\mid I\,]$ to some other form $[\,R\mid E\,]$ then these blocks must satisfy the property $EA = R$. We now state and prove this observation formally.

Theorem 2.2.1

Row-Reduction is Multiplication on the Left

If $A, R \in \mathcal{M}_{m,n}$ and $E \in \mathcal{M}_m$ are matrices such that the block matrix $[\,A\mid I\,]$ can be row-reduced to $[\,R\mid E\,]$, then $R = EA$.

Proof. We make use of some block matrix multiplication trickery along with the fact that performing an elementary row operation is equivalent to multiplication on the left by the corresponding elementary matrix. In particular, if row-reducing $[\,A\mid I\,]$ to $[\,R\mid E\,]$ makes use of the elementary row operations corresponding to elementary matrices $E_1, E_2, \ldots, E_k$, in that order, then

$$[\,R\mid E\,] = E_k \cdots E_2 E_1 [\,A\mid I\,] = [\,E_k \cdots E_2 E_1 A \mid E_k \cdots E_2 E_1\,].$$

This means (by looking at the right half of the above block matrix) that $E = E_k \cdots E_2 E_1$, which then implies (by looking at the left half of the block matrix) that $R = EA$. ∎

The above theorem says that, not only is performing a single row operation equivalent to multiplication on the left by an elementary matrix, but performing a *sequence* of row operations is also equivalent to multiplication on the left by some (potentially non-elementary) matrix.

2.2.2   The Inverse of a Matrix

One of the nice features of the elementary matrices is that they are **invertible**: multiplication by them can be undone by multiplying by some other matrix. For example, if

Refer back to Remark 2.1.1, which noted that every elementary row operation is reversible.

$$E_1 = \left[\begin{array}{ccc} 1 & 0 & 0 \\ 0 & 1 & 0 \\ -3 & 0 & 1 \end{array}\right] \quad \text{and} \quad E_2 = \left[\begin{array}{ccc} 1 & 0 & 0 \\ 0 & 1 & 0 \\ 3 & 0 & 1 \end{array}\right],$$

then it is straightforward to check that $E_1 E_2 = I$ and $E_2 E_1 = I$. We thus say that E_1 and E_2 are **inverses** of each other, which makes sense in this case because

E_1 is the elementary matrix corresponding to the elementary row operation $R_3 - 3R_1$, and E_2 is the elementary matrix corresponding to the elementary row operation $R_3 + 3R_1$, and performing these row operations one after another has the same effect as not doing anything at all.

It is also useful to talk about invertibility of (not necessarily elementary) matrices in general, and the idea is exactly the same—two matrices are inverses of each other if multiplying by one of them "undoes" the multiplication by the other.

Definition 2.2.2

Invertible Matrices

A square matrix $A \in \mathcal{M}_n$ is called **invertible** if there exists a matrix, which we denote by A^{-1} and call the **inverse** of A, such that

$$AA^{-1} = A^{-1}A = I.$$

Some books use the terms **singular** and **non-singular** to mean non-invertible and invertible, respectively.

In this definition, we referred to A^{-1} as *the* inverse of A (as opposed to *an* inverse of A). To justify this terminology, we should show that every matrix has at most one inverse. To see this, suppose for a moment that a matrix $A \in \mathcal{M}_n$ had *two* inverses $B, C \in \mathcal{M}_n$ (i.e., $AB = BA = I$ and $AC = CA = I$). It would then follow that

$$B = IB = (CA)B = C(AB) = CI = C,$$

so in fact these two inverses must be the same as each other. It follows that inverses (when they exist) are indeed unique.

Given a pair of matrices, it is straightforward to check whether or not they are inverses of each other—just multiply them together and see if we get the identity matrix.

Example 2.2.3

Verifying Inverses

Determine whether or not the following pairs of matrices are inverses of each other:

a) $\begin{bmatrix} 1 & 2 \\ 3 & 4 \end{bmatrix}$ and $\dfrac{1}{2} \begin{bmatrix} -4 & 2 \\ 3 & -1 \end{bmatrix}$,

b) $\begin{bmatrix} 1 & 2 \\ 2 & 4 \end{bmatrix}$ and $\begin{bmatrix} -4 & 2 \\ 2 & -1 \end{bmatrix}$, and

c) $\begin{bmatrix} 0 & 1 & 1 \\ 1 & 0 & 1 \\ 1 & 1 & 0 \end{bmatrix}$ and $\dfrac{1}{2} \begin{bmatrix} -1 & 1 & 1 \\ 1 & -1 & 1 \\ 1 & 1 & -1 \end{bmatrix}$.

Solutions:

a) We just compute the product of these two matrices in both ways:

$$\begin{bmatrix} 1 & 2 \\ 3 & 4 \end{bmatrix} \left(\frac{1}{2} \begin{bmatrix} -4 & 2 \\ 3 & -1 \end{bmatrix} \right) = \begin{bmatrix} 1 & 0 \\ 0 & 1 \end{bmatrix} \quad \text{and}$$

$$\left(\frac{1}{2} \begin{bmatrix} -4 & 2 \\ 3 & -1 \end{bmatrix} \right) \begin{bmatrix} 1 & 2 \\ 3 & 4 \end{bmatrix} = \begin{bmatrix} 1 & 0 \\ 0 & 1 \end{bmatrix}.$$

Since both of these products equal the identity matrix, these two matrices are inverses of each other.

In fact, we will show
in Example 2.2.7
that the matrix

$$\begin{bmatrix} 1 & 2 \\ 2 & 4 \end{bmatrix}$$

does not have an
inverse at all.

b) Again, we compute the product of these two matrices:

$$\begin{bmatrix} 1 & 2 \\ 2 & 4 \end{bmatrix} \begin{bmatrix} -4 & 2 \\ 2 & -1 \end{bmatrix} = \begin{bmatrix} 0 & 0 \\ 0 & 0 \end{bmatrix}.$$

Since this does not equal the identity matrix, we conclude that these matrices are *not* inverses of each other (note that we do not need to compute the product of these two matrices in the opposite order in this case—as long *at least one* of those products is not the identity, they are not inverses of each other).

c) Yet again, we compute the product of these two matrices in both ways:

$$\begin{bmatrix} 0 & 1 & 1 \\ 1 & 0 & 1 \\ 1 & 1 & 0 \end{bmatrix} \left(\frac{1}{2} \begin{bmatrix} -1 & 1 & 1 \\ 1 & -1 & 1 \\ 1 & 1 & -1 \end{bmatrix} \right) = \begin{bmatrix} 1 & 0 & 0 \\ 0 & 1 & 0 \\ 0 & 0 & 1 \end{bmatrix} \quad \text{and}$$

$$\left(\frac{1}{2} \begin{bmatrix} -1 & 1 & 1 \\ 1 & -1 & 1 \\ 1 & 1 & -1 \end{bmatrix} \right) \begin{bmatrix} 0 & 1 & 1 \\ 1 & 0 & 1 \\ 1 & 1 & 0 \end{bmatrix} = \begin{bmatrix} 1 & 0 & 0 \\ 0 & 1 & 0 \\ 0 & 0 & 1 \end{bmatrix}.$$

Since both of these products equal the identity matrix, these two matrices are inverses of each other.

We can also think about inverses geometrically in terms of linear transformations. Since matrices and linear transformations are essentially the same thing, it should not be surprising that we say that a linear transformation $T : \mathbb{R}^n \to \mathbb{R}^n$ is invertible if there exists a linear transformation, which we denote by T^{-1}, such that $T \circ T^{-1} = T^{-1} \circ T = I$. In other words, T^{-1} is the linear transformation that undoes the action of T: it sends each vector $T(\mathbf{v})$ back to $\mathbf{v}$, as in Figure 2.11.

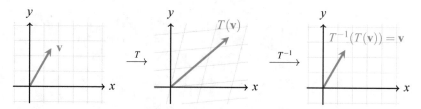

Figure 2.11: T^{-1} is the linear transformation that undoes the action of T. That is, $T^{-1}(T(\mathbf{v})) = \mathbf{v}$ for all $\mathbf{v} \in \mathbb{R}^n$ (and $T(T^{-1}(\mathbf{v})) = \mathbf{v}$ too).

However, not every linear transformation has an inverse (i.e., not every linear transformation is invertible), which we can again make sense of geometrically. For example, recall from Section 1.4.2 that a projection $P_{\mathbf{u}}$ onto a line in the direction of a unit vector $\mathbf{u}$ is a linear transformation. The problem that gets in the way of invertibility of $P_{\mathbf{u}}$ is that multiple vectors get projected onto the same vector on that line, so how could we ever undo the projection? Given just $P_{\mathbf{u}}(\mathbf{v})$, we have no way of determining what $\mathbf{v}$ is (see Figure 2.12).

In other words, the projection $P_{\mathbf{u}}$ "squashes" space – it takes in all of $\mathbb{R}^2$ but only outputs a (1-dimensional) line, and there is no way to recover the information that was thrown away in the squashing process. This is completely analogous to the fact that we cannot divide real numbers by 0: given the value

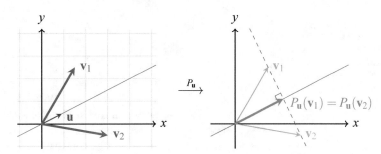

Figure 2.12: Not all matrices or linear transformations are invertible. For example, the projection $P_{\mathbf{u}}$ onto the line in the direction of a unit vector $\mathbf{u}$ is not invertible, since multiple input vectors are sent to the same output vector. In particular, every single vector on the dashed line on the right is sent to $P_{\mathbf{u}}(\mathbf{v}_1) = P_{\mathbf{u}}(\mathbf{v}_2)$.

$0x$, we cannot determine x: it could be absolutely anything. Multiplication by 0 throws away information, so we cannot undo/invert it (i.e., we cannot divide by 0). In fact, 0 is the unique non-invertible 1×1 matrix.

As always, it is worthwhile to spend some time thinking about what properties our new mathematical operation (matrix inversion) has.

Theorem 2.2.2 **Properties of Matrix Inverses** T is not an exponent here. Recall that A^T is the transpose of A.	Let $A \in \mathcal{M}_n$ be an invertible matrix, let c be a non-zero real number, and let k be a positive integer. Then a) A^{-1} is invertible and $(A^{-1})^{-1} = A$. b) cA is invertible and $(cA)^{-1} = \frac{1}{c}A^{-1}$. c) A^T is invertible and $(A^T)^{-1} = (A^{-1})^T$. d) A^k is invertible and $(A^k)^{-1} = (A^{-1})^k$.

Proof. Most parts of this theorem are intuitive enough, so we just prove part (b), and we leave the others to Exercise 2.2.17.

To see why part (b) holds, we just compute the product of cA and its proposed inverse in both ways:

$$(cA)\left(\frac{1}{c}A^{-1}\right) = \frac{c}{c}AA^{-1} = I \quad \text{and} \quad \left(\frac{1}{c}A^{-1}\right)(cA) = \frac{c}{c}A^{-1}A = I,$$

which shows that cA has $\frac{1}{c}A^{-1}$ as its inverse, as desired. ■

In particular, part (d) of the above theorem tells us that we can unambiguously extend our definition of matrix powers (which we have already defined for non-negative exponents) to negative exponents. Specifically, if we define

$$A^{-k} \stackrel{\text{def}}{=} (A^{-1})^k \quad \text{for all integers} \quad k \geq 1,$$

then matrix powers still have the "nice" properties like $A^{k+\ell} = A^k A^\ell$ and $A^{k\ell} = (A^k)^\ell$ even when k and ℓ are allowed to be negative (see Exercise 2.2.18).

Example 2.2.4 **Computing Negative Matrix Powers**	Compute A^{-2} if $A = \begin{bmatrix} 1 & 2 \\ 3 & 4 \end{bmatrix}$. **Solution:** We saw in Example 2.2.3(a) that the inverse of A is $$A^{-1} = \frac{1}{2}\begin{bmatrix} -4 & 2 \\ 3 & -1 \end{bmatrix}.$$

It follows that

$$A^{-2} = (A^{-1})^2 = \left(\frac{1}{2} \begin{bmatrix} -4 & 2 \\ 3 & -1 \end{bmatrix} \right) \left(\frac{1}{2} \begin{bmatrix} -4 & 2 \\ 3 & -1 \end{bmatrix} \right) = \frac{1}{4} \begin{bmatrix} 22 & -10 \\ -15 & 7 \end{bmatrix}.$$

Example 2.2.3 showed that it is straightforward to check whether or not a given matrix is an inverse of another matrix. However, actually *finding* the inverse of a matrix (or showing that none exists) is a bit more involved. To get started toward solving this problem, we first think about how to find the inverse of an elementary matrix.

Recall from Remark 2.1.1 that elementary row operations are reversible— we can undo any of the three elementary row operations by applying another elementary row operation:

- The row operation cR_i is undone by the row operation $\frac{1}{c}R_j$.
- The row operation $R_i \leftrightarrow R_j$ is undone by applying itself a second time.
- The row operation $R_i + cR_j$ is undone by the row operation $R_i - cR_j$.

It follows that every elementary matrix is invertible, and that its inverse is also an elementary matrix. Furthermore, pairs of elementary matrices are inverses of each other exactly when they correspond to elementary row operations that undo each other. We illustrate this idea with some examples.

Example 2.2.5

Inverses of Elementary Matrices

Find the inverse of each of the following elementary matrices:

a) $\begin{bmatrix} 1 & 0 \\ 0 & 3 \end{bmatrix}$

b) $\begin{bmatrix} 1 & 2 \\ 0 & 1 \end{bmatrix}$

c) $\begin{bmatrix} 1 & 0 & 0 \\ 0 & 0 & 1 \\ 0 & 1 & 0 \end{bmatrix}$

d) $\begin{bmatrix} 1 & 0 & 0 \\ -5 & 1 & 0 \\ 0 & 0 & 1 \end{bmatrix}$

Solutions:

a) This is the elementary matrix corresponding to the row operation $3R_2$, so its inverse is the elementary matrix for the row operation $\frac{1}{3}R_2$:

$$\begin{bmatrix} 1 & 0 \\ 0 & 1/3 \end{bmatrix}.$$

b) This is the elementary matrix for the row operation $R_1 + 2R_2$, so its inverse is the elementary matrix for the row operation $R_1 - 2R_2$:

$$\begin{bmatrix} 1 & -2 \\ 0 & 1 \end{bmatrix}.$$

c) This is the elementary matrix for the row operation $R_2 \leftrightarrow R_3$, so its inverse is just itself:

$$\begin{bmatrix} 1 & 0 & 0 \\ 0 & 0 & 1 \\ 0 & 1 & 0 \end{bmatrix}.$$

d) This is the elementary matrix for to the row operation $R_2 - 5R_1$, so its inverse is the elementary matrix for the row operation $R_2 + 5R_1$:

$$\begin{bmatrix} 1 & 0 & 0 \\ 5 & 1 & 0 \\ 0 & 0 & 1 \end{bmatrix}.$$

To ramp this technique up into a method for finding the inverse of an arbitrary matrix, we need the following theorem, which tells us that if two matrices are invertible, then so is their product (and furthermore, we can compute the inverse of the product directly from the inverses of the matrices in the product):

Theorem 2.2.3

Product of Invertible Matrices

Suppose $A, B \in \mathcal{M}_n$ are invertible. Then AB is also invertible, and

$$(AB)^{-1} = B^{-1}A^{-1}.$$

Compare this theorem with Theorem 1.3.4(c), where the transpose of a product was similarly the product of the transposes in the opposite order.

Proof. Since we are given a formula for the proposed inverse, we just need to multiply it by AB and see that we do indeed get the identity matrix:

$$(AB)(B^{-1}A^{-1}) = A(BB^{-1})A^{-1} = AIA^{-1} = AA^{-1} = I, \quad \text{and}$$
$$(B^{-1}A^{-1})(AB) = B^{-1}(A^{-1}A)B = B^{-1}IB = B^{-1}B = I.$$

Since both products result in the identity matrix, we conclude that AB is indeed invertible, and its inverse is $B^{-1}A^{-1}$, as claimed. ∎

The fact that the order of multiplication is switched in the above theorem perhaps seems strange at first, but actually makes a great deal of sense—when inverting (or "undoing") operations in real life, we often must undo them in the order opposite to which we originally did them. For example, in the morning we put on our socks and then put on our shoes, but at night we take off our shoes and *then* take off our socks. For this reason, Theorem 2.2.3 is sometimes light-heartedly referred to as the "socks-and-shoes rule".

Example 2.2.6

Finding an Inverse of a Product

Find the inverse of the matrix $\begin{bmatrix} 1 & 2 \\ 0 & 1 \end{bmatrix} \begin{bmatrix} 1 & 0 \\ 0 & 3 \end{bmatrix} = \begin{bmatrix} 1 & 6 \\ 0 & 3 \end{bmatrix}.$

Solution:

We saw in Example 2.2.5 that

$$\begin{bmatrix} 1 & 2 \\ 0 & 1 \end{bmatrix}^{-1} = \begin{bmatrix} 1 & -2 \\ 0 & 1 \end{bmatrix} \quad \text{and} \quad \begin{bmatrix} 1 & 0 \\ 0 & 3 \end{bmatrix}^{-1} = \begin{bmatrix} 1 & 0 \\ 0 & 1/3 \end{bmatrix}.$$

To find the inverse of their product, we compute the product of their inverses in the opposite order:

$$\begin{bmatrix} 1 & 6 \\ 0 & 3 \end{bmatrix}^{-1} = \left(\begin{bmatrix} 1 & 2 \\ 0 & 1 \end{bmatrix} \begin{bmatrix} 1 & 0 \\ 0 & 3 \end{bmatrix} \right)^{-1} = \begin{bmatrix} 1 & 0 \\ 0 & 3 \end{bmatrix}^{-1} \begin{bmatrix} 1 & 2 \\ 0 & 1 \end{bmatrix}^{-1}$$

$$= \begin{bmatrix} 1 & 0 \\ 0 & 1/3 \end{bmatrix} \begin{bmatrix} 1 & -2 \\ 0 & 1 \end{bmatrix} = \begin{bmatrix} 1 & -2 \\ 0 & 1/3 \end{bmatrix}.$$

Theorem 2.2.3 also extends to the product of more than two matrices in a straightforward way. For example, if A, B, and C are invertible matrices of the same size, then applying the theorem twice shows that

$$(ABC)^{-1} = ((AB)C)^{-1} = C^{-1}(AB)^{-1} = C^{-1}B^{-1}A^{-1}.$$

In general, the inverse of a product of invertible matrices is again invertible, regardless of how many matrices there are in the product, and its inverse is the product of the individual inverses in the opposite order.

2.2.3 A Characterization of Invertible Matrices

Since we now know how to find the inverse of a product of invertible matrices, it follows that if we can write a matrix as a product of elementary matrices, then we can find its inverse by inverting each of those elementary matrices and multiplying together those inverses in opposite order (just like we did in Example 2.2.6).

However, this observation is not yet useful, since we do not know how to break down a general invertible matrix into a product of elementary matrices, nor do we know whether or not this is always possible. The following theorem and its proof solve these problems, and also introduces an important connection between invertible matrices and systems of linear equations.

Theorem 2.2.4

Characterization of Invertible Matrices

Suppose $A \in \mathcal{M}_n$. The following are equivalent:

   a) A is invertible.
   b) The linear system $A\mathbf{x} = \mathbf{b}$ has a solution for all $\mathbf{b} \in \mathbb{R}^n$.
   c) The linear system $A\mathbf{x} = \mathbf{b}$ has a unique solution for all $\mathbf{b} \in \mathbb{R}^n$.
   d) The linear system $A\mathbf{x} = \mathbf{0}$ has a unique solution.
   e   The reduced row echelon form of A is I (the identity matrix).
   f A can be written as a product of elementary matrices.

To prove this theorem, we show that the 6 properties imply each other as follows:

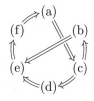

Proof. To start, we show that (a) $\implies$ (c) $\implies$ (d) $\implies$ (e) $\implies$ (f) $\implies$ (a), which means that any one of these five statements implies any of the other five (we will take care of condition (b) later).

To see that (a) $\implies$ (c), we note that if A is invertible then $\mathbf{x} = A^{-1}\mathbf{b}$ is a solution of the linear system $A\mathbf{x} = \mathbf{b}$, since

$$A(A^{-1}\mathbf{b}) = (AA^{-1})\mathbf{b} = I\mathbf{b} = \mathbf{b}.$$

To see that this solution is unique, suppose that there were two solutions $\mathbf{x}, \mathbf{y} \in \mathbb{R}^n$. Then $A\mathbf{x} = \mathbf{b}$ and $A\mathbf{y} = \mathbf{b}$, so subtracting gives us $A(\mathbf{x} - \mathbf{y}) = \mathbf{0}$. It then follows that

$$\mathbf{x} - \mathbf{y} = (A^{-1}A)(\mathbf{x} - \mathbf{y}) = A^{-1}(A(\mathbf{x} - \mathbf{y})) = A^{-1}\mathbf{0} = \mathbf{0},$$

so $\mathbf{x} = \mathbf{y}$ (i.e., the solution is unique).

The implication (c) $\implies$ (d) follows simply by choosing $\mathbf{b} = \mathbf{0}$, so we jump straight to proving that (d) $\implies$ (e). If we represent the linear system $A\mathbf{x} = \mathbf{0}$ in augmented matrix form $[A \mid \mathbf{0}]$ and then apply Gauss–Jordan elimination, we get the augmented matrix $[R \mid \mathbf{0}]$, where R is the reduced row echelon form of A. Since R is square, if $R \neq I$ then it must have a row of zeros at the bottom, so this linear system must have at least one free variable and thus cannot have a unique solution. We thus conclude that if the linear system $A\mathbf{x} = \mathbf{0}$ *does* have a unique solution (i.e., (d) holds) then $R = I$ (i.e., (e) holds).

Next, we show that (e) $\implies$ (f). If the RREF of A is I, then there exists a sequence of row operations that transforms A into I. Equivalently, there exists a sequence of elementary matrices $E_1, E_2, \ldots, E_k$ such that $E_k \cdots E_2 E_1 A = I$. By multiplying on the left by E_k^{-1}, E_{k-1}^{-1}, and so on to E_1^{-1}, we see that

$$A = E_1^{-1} E_2^{-1} \cdots E_k^{-1}.$$

Since the inverse of an elementary matrix is again an elementary matrix, A can be written as a product of elementary matrices, so (f) follows.

To complete this portion of the proof, we note that the implication (f) $\implies$ (a) follows from the facts that elementary matrices are invertible, and the product of invertible matrices is invertible (Theorem 2.2.3).

All that remains is to show that property (b) is equivalent to the other properties. Property (c) trivially implies property (b), so we just need to show that (b) implies any of the five other properties. We show that (b) implies (e).

To find a **b** such that row-reducing $[\,A \mid \mathbf{b}\,]$ produces $[\,R \mid \mathbf{e}_n\,]$, just apply the inverse sequence of row operations to $\mathbf{e}_n$.

To this end, choose $\mathbf{b} \in \mathbb{R}^n$ so that applying Gauss–Jordan elimination to $[\,A \mid \mathbf{b}\,]$ produces $[\,R \mid \mathbf{e}_n\,]$, where R is the reduced row echelon form of A. If R had a zero row at the bottom, then this linear system would not have a solution. However, since we are assuming (b) holds, this cannot be the case. It follows that R must have no zero rows, so it equals I, which (finally!) completes the proof. ■

Remark 2.2.1

How to Think About Elementary Matrices

We can think of the relationship between elementary matrices and invertible matrices much like we think of the relationship between prime numbers and positive integers—Theorem 2.2.4 tells us that every invertible matrix can be written as a product of elementary matrices, just like every positive integer can be written as a product of prime numbers.

One important difference between these two settings is that the method of writing an invertible matrix as a product of elementary matrices is very non-unique, whereas the prime decomposition of a positive integer is unique. However, we can still think of elementary matrices as the basic building blocks of invertible matrices, and oftentimes we can prove general facts and properties of invertible matrices just by considering what those properties look like for elementary matrices.

Example 2.2.7

Determining Whether or Not a Matrix is Invertible

Determine whether or not the following matrices are invertible:

a) $\begin{bmatrix} 1 & 2 \\ 2 & 4 \end{bmatrix}$

b) $\begin{bmatrix} 1 & 2 & -2 \\ 0 & 1 & -2 \\ 1 & 1 & 1 \end{bmatrix}$

Solutions:

a) By the previous theorem, we know that a matrix is invertible if and only if its reduced row echelon form is the identity matrix. With that in mind, we compute the RREF of this matrix:

$$\begin{bmatrix} 1 & 2 \\ 2 & 4 \end{bmatrix} \xrightarrow{R_2 - 2R_1} \begin{bmatrix} 1 & 2 \\ 0 & 0 \end{bmatrix}.$$

Since this RREF is not the identity matrix, we conclude that this matrix is not invertible.

b) Again, we determine invertibility by computing the RREF of this matrix:

$$\begin{bmatrix} 1 & 2 & -2 \\ 0 & 1 & -2 \\ 1 & 1 & 1 \end{bmatrix} \xrightarrow{R_3 - R_1} \begin{bmatrix} 1 & 2 & -2 \\ 0 & 1 & -2 \\ 0 & -1 & 3 \end{bmatrix}$$

$$\xrightarrow[R_3 + R_2]{R_1 - 2R_2} \begin{bmatrix} 1 & 0 & 2 \\ 0 & 1 & -2 \\ 0 & 0 & 1 \end{bmatrix} \xrightarrow[R_2 + 2R_3]{R_1 - 2R_3} \begin{bmatrix} 1 & 0 & 0 \\ 0 & 1 & 0 \\ 0 & 0 & 1 \end{bmatrix}.$$

Since this RREF is the identity matrix, it follows that the original matrix is invertible.

Fortunately, we can also use the above theorem to actually *compute* the inverse of a matrix when it exists. If A is invertible then block matrix multiplication shows that $A^{-1}[A \mid I] = [I \mid A^{-1}]$. However, multiplication on the left by A^{-1} is equivalent to performing a sequence of row operations on $[A \mid I]$, so we conclude that $[A \mid I]$ can be row-reduced to $[I \mid A^{-1}]$. This leads immediately to the following theorem, which tells us how to simultaneously check if a matrix is invertible, and find its inverse if it exists:

Theorem 2.2.5

Finding Matrix Inverses

Suppose $A \in \mathcal{M}_n$. Then A is invertible if and only if there exists a matrix $E \in \mathcal{M}_n$ such that the RREF of the block matrix $[A \mid I]$ is $[I \mid E]$. Furthermore, if A is invertible then it is necessarily the case that $A^{-1} = E$.

Before working through some explicit examples, it is worth comparing this result with Theorem 2.2.1, which says that if we can row-reduce $[A \mid I]$ to $[R \mid E]$ then $R = EA$. The above theorem clarifies the special case when the reduced row echelon form of A is $R = I$, so the equation $R = EA$ simplifies to $I = EA$ (and thus $E = A^{-1}$).

Example 2.2.8

Computing Inverses

For each of the following matrices, either compute their inverse or show that none exists:

We generalize part (c) of this example a bit later, in Exercise 2.3.14.

a) $\begin{bmatrix} 2 & 2 \\ 4 & 5 \end{bmatrix}$

b) $\begin{bmatrix} 1 & -2 \\ -3 & 6 \end{bmatrix}$

c) $\begin{bmatrix} 1 & 2 & 3 \\ 4 & 5 & 6 \\ 7 & 8 & 9 \end{bmatrix}$

d) $\begin{bmatrix} 1 & 1 & 1 \\ 1 & 2 & 4 \\ 1 & 3 & 9 \end{bmatrix}$

Solutions:

a) To determine whether or not this matrix is invertible, and find its inverse if it is, we augment it with the identity and row-reduce:

$$\left[\begin{array}{cc|cc} 2 & 2 & 1 & 0 \\ 4 & 5 & 0 & 1 \end{array}\right] \xrightarrow{R_2 - 2R_1} \left[\begin{array}{cc|cc} 2 & 2 & 1 & 0 \\ 0 & 1 & -2 & 1 \end{array}\right]$$

$$\xrightarrow{R_1 - 2R_2} \left[\begin{array}{cc|cc} 2 & 0 & 5 & -2 \\ 0 & 1 & -2 & 1 \end{array}\right] \xrightarrow{\frac{1}{2}R_1} \left[\begin{array}{cc|cc} 1 & 0 & 5/2 & -1 \\ 0 & 1 & -2 & 1 \end{array}\right].$$

Since we were able to row-reduce this matrix so that the block on the left is the identity, it is invertible, and its inverse is the block on the right:

$$\begin{bmatrix} 5/2 & -1 \\ -2 & 1 \end{bmatrix}.$$

b) Again, we augment with the identity matrix and row-reduce:

$$\left[\begin{array}{cc|cc} 1 & -2 & 1 & 0 \\ -3 & 6 & 0 & 1 \end{array}\right] \xrightarrow{R_2+3R_1} \left[\begin{array}{cc|cc} 1 & -2 & 1 & 0 \\ 0 & 0 & 3 & 1 \end{array}\right].$$

Since the matrix on the left has a zero row, the original matrix is not invertible.

If we ever get a zero row on the left at any point while row-reducing, we can immediately stop and conclude the matrix is not invertible—the reduced row echelon form will also have a zero row.

c) Yet again, we augment with the identity matrix and row-reduce:

$$\left[\begin{array}{ccc|ccc} 1 & 2 & 3 & 1 & 0 & 0 \\ 4 & 5 & 6 & 0 & 1 & 0 \\ 7 & 8 & 9 & 0 & 0 & 1 \end{array}\right] \begin{array}{c} R_2-4R_1 \\ R_3-7R_1 \\ \xrightarrow{\hspace{1cm}} \end{array} \left[\begin{array}{ccc|ccc} 1 & 2 & 3 & 1 & 0 & 0 \\ 0 & -3 & -6 & -4 & 1 & 0 \\ 0 & -6 & -12 & -7 & 0 & 1 \end{array}\right]$$

$$\xrightarrow{R_3-2R_2} \left[\begin{array}{ccc|ccc} 1 & 2 & 3 & 1 & 0 & 0 \\ 0 & -3 & -6 & -4 & 1 & 0 \\ 0 & 0 & 0 & 1 & -2 & 1 \end{array}\right].$$

Since we have row-reduced the matrix on the left so that it has a zero row, we conclude that it is not invertible.

d) As always, we augment with the identity matrix and row-reduce:

$$\left[\begin{array}{ccc|ccc} 1 & 1 & 1 & 1 & 0 & 0 \\ 1 & 2 & 4 & 0 & 1 & 0 \\ 1 & 3 & 9 & 0 & 0 & 1 \end{array}\right] \begin{array}{c} R_2-R_1 \\ R_3-R_1 \\ \xrightarrow{\hspace{1cm}} \end{array} \left[\begin{array}{ccc|ccc} 1 & 1 & 1 & 1 & 0 & 0 \\ 0 & 1 & 3 & -1 & 1 & 0 \\ 0 & 2 & 8 & -1 & 0 & 1 \end{array}\right]$$

$$\begin{array}{c} R_1-R_2 \\ R_3-2R_2 \\ \xrightarrow{\hspace{1cm}} \end{array} \left[\begin{array}{ccc|ccc} 1 & 0 & -2 & 2 & -1 & 0 \\ 0 & 1 & 3 & -1 & 1 & 0 \\ 0 & 0 & 2 & 1 & -2 & 1 \end{array}\right]$$

$$\xrightarrow{\frac{1}{2}R_3} \left[\begin{array}{ccc|ccc} 1 & 0 & -2 & 2 & -1 & 0 \\ 0 & 1 & 3 & -1 & 1 & 0 \\ 0 & 0 & 1 & 1/2 & -1 & 1/2 \end{array}\right]$$

$$\begin{array}{c} R_1+2R_3 \\ R_2-3R_3 \\ \xrightarrow{\hspace{1cm}} \end{array} \left[\begin{array}{ccc|ccc} 1 & 0 & 0 & 3 & -3 & 1 \\ 0 & 1 & 0 & -5/2 & 4 & -3/2 \\ 0 & 0 & 1 & 1/2 & -1 & 1/2 \end{array}\right].$$

We will see a generalization of the fact that the matrix from part (d) is invertible a bit later, in Example 2.3.8.

Since we were able to row-reduce this block matrix so that the block on the left is the identity, we conclude that the original matrix is invertible and its inverse is the block on the right:

$$\begin{bmatrix} 3 & -3 & 1 \\ -5/2 & 4 & -3/2 \\ 1/2 & -1 & 1/2 \end{bmatrix}.$$

While our general method of computing inverses works for matrices of any size, when the matrix is 2×2 we do not actually have to row-reduce at all, as the following theorem provides an explicit formula for the inverse.

Theorem 2.2.6

**Inverse of a
2 × 2 Matrix**

The quantity $ad - bc$
is called the
determinant of A.
We will explore it in
depth in Section 3.2.

The matrix $A = \begin{bmatrix} a & b \\ c & d \end{bmatrix}$ is invertible if and only if $ad - bc \neq 0$, and

$$A^{-1} = \frac{1}{ad - bc} \begin{bmatrix} d & -b \\ -c & a \end{bmatrix}.$$

Proof. If $ad - bc \neq 0$ then we can show that the inverse of A is as claimed just by multiplying it by A:

$$\left(\frac{1}{ad - bc} \begin{bmatrix} d & -b \\ -c & a \end{bmatrix} \right) \begin{bmatrix} a & b \\ c & d \end{bmatrix} = \frac{1}{ad - bc} \begin{bmatrix} ad - bc & 0 \\ 0 & ad - bc \end{bmatrix} = \begin{bmatrix} 1 & 0 \\ 0 & 1 \end{bmatrix},$$

and similarly when computing the product in the opposite order. Since this product is the identity matrix, A is invertible and its inverse is as claimed.

On the other hand, if $ad - bc = 0$ then $ad = bc$. From here we split into two cases:

Case 1: If either $a = 0$ or $b = 0$ then $ad = bc$ implies that two of the entries of A in the same row or column are both 0. It follows that A can be row-reduced so as to have a zero row, and is thus not invertible.

Case 2: If $a, b \neq 0$ then $ad = bc$ implies $d/b = c/a$, so the second row of A is a multiple of its first row. It again follows that A can be row-reduced so as to have a zero row, and thus is not invertible.

In either case, the equation $ad - bc = 0$ implies that A is not invertible, which completes the proof. ∎

There is actually an explicit formula for the inverse of a matrix of any size, which we derive in Section 3.A.1. However, it is hideous for matrices of size 3×3 and larger and thus the method of computing the inverse based on Gauss–Jordan elimination is typically much easier to use.

Example 2.2.9

**Inverse of a
2 × 2 Matrix**

The inverses of all
four of these
matrices were also
investigated earlier
in Examples 2.2.3
and 2.2.8.

For each of the following matrices, either compute their inverse or show that none exists:

a) $\begin{bmatrix} 1 & 2 \\ 3 & 4 \end{bmatrix}$ b) $\begin{bmatrix} 1 & 2 \\ 2 & 4 \end{bmatrix}$

c) $\begin{bmatrix} 2 & 2 \\ 4 & 5 \end{bmatrix}$ d) $\begin{bmatrix} 1 & -2 \\ -3 & 6 \end{bmatrix}$

Solutions:

a) We first compute $ad - bc = 1 \times 4 - 2 \times 3 = 4 - 6 = -2 \neq 0$, so this matrix is invertible. Its inverse is

$$\frac{1}{-2} \begin{bmatrix} 4 & -2 \\ -3 & 1 \end{bmatrix} = \frac{1}{2} \begin{bmatrix} -4 & 2 \\ 3 & -1 \end{bmatrix}.$$

b) Again, we start by computing $ad - bc = 1 \times 4 - 2 \times 2 = 4 - 4 = 0$. Since this quantity equals 0, we conclude that the matrix is not invertible.

c) Yet again, we first compute $ad - bc = 2 \times 5 - 2 \times 4 = 10 - 8 = 2 \neq 0$, so this matrix is invertible. Its inverse is

$$\frac{1}{2}\begin{bmatrix} 5 & -4 \\ -2 & 2 \end{bmatrix}.$$

d) Once again, computing $ad - bc = 1 \times 6 - (-2) \times (-3) = 6 - 6 = 0$ shows that this matrix is not invertible.

At this point, it is perhaps worthwhile to return to the fact that a linear system $A\mathbf{x} = \mathbf{b}$ with an invertible coefficient matrix has a unique solution (i.e., condition (c) of Theorem 2.2.4). In fact, we can write this solution explicitly in terms of the inverse matrix: $\mathbf{x} = A^{-1}\mathbf{b}$ is the (necessarily unique) solution, since

$$A\mathbf{x} = A(A^{-1}\mathbf{b}) = (A^{-1}A)\mathbf{b} = I\mathbf{b} = \mathbf{b}.$$

Solving a linear system by finding the inverse of the coefficient matrix is a useful technique because it allows us to solve linear systems with multiple different right-hand-side vectors $\mathbf{b}$ after performing Gauss–Jordan elimination just once (to compute A^{-1}), rather than having to do it for every single right-hand-side vector.

To illustrate what we mean by this, consider the linear system $A\mathbf{x} = \mathbf{b}$, where

$$A = \begin{bmatrix} 1 & 1 & 1 \\ 1 & 2 & 4 \\ 1 & 3 & 9 \end{bmatrix} \quad \text{and} \quad \mathbf{b} = \begin{bmatrix} 2 \\ 1 \\ 2 \end{bmatrix}.$$

We computed the inverse of this coefficient matrix in Example 2.2.8(d), and it is

$$A^{-1} = \begin{bmatrix} 3 & -3 & 1 \\ -5/2 & 4 & -3/2 \\ 1/2 & -1 & 1/2 \end{bmatrix}.$$

The unique solution of this linear system is thus

$$\mathbf{x} = A^{-1}\mathbf{b} = \begin{bmatrix} 3 & -3 & 1 \\ -5/2 & 4 & -3/2 \\ 1/2 & -1 & 1/2 \end{bmatrix}\begin{bmatrix} 2 \\ 1 \\ 2 \end{bmatrix} = \begin{bmatrix} 5 \\ -4 \\ 1 \end{bmatrix}.$$

There is an even better method of solving multiple linear systems with the same coefficient matrices based on something called the "LU decomposition", which we explore in Section 2.D.2.

If the right-hand-side $\mathbf{b}$ were to then change to something different then we could again compute the unique solution simply via matrix multiplication, rather than having to row-reduce all over again. For example, if

$$\mathbf{b} = \begin{bmatrix} 1 \\ 0 \\ -1 \end{bmatrix} \quad \text{then} \quad \mathbf{x} = A^{-1}\mathbf{b} = \begin{bmatrix} 3 & -3 & 1 \\ -5/2 & 4 & -3/2 \\ 1/2 & -1 & 1/2 \end{bmatrix}\begin{bmatrix} 1 \\ 0 \\ -1 \end{bmatrix} = \begin{bmatrix} 2 \\ -1 \\ 0 \end{bmatrix}.$$

Again, it is perhaps useful to keep in mind the idea that elementary matrices act as a log that keeps track of row operations. Since A^{-1} is the product of all of the elementary matrices that were used to row-reduce A to I, we can think of multiplication by A^{-1} as performing all of the row operations needed to solve the linear system $A\mathbf{x} = \mathbf{b}$ simultaneously.

Finally, we close this section with one final result that shows that, up until now, we have been doing twice as much work as is necessary to show that two matrices are inverses of each other. Even though we defined matrices A and A^{-1} as being inverses of each other if $AA^{-1} = A^{-1}A = I$, it is actually enough just to check one of these two equations:

Theorem 2.2.7 **One-Sided Inverses**	Suppose $A \in \mathcal{M}_n$. If there exists a matrix $B \in \mathcal{M}_n$ such that $AB = I$ or $BA = I$ then A is invertible and $A^{-1} = B$.

Proof. If $BA = I$ then multiplying the linear system $A\mathbf{x} = \mathbf{0}$ on the left by B shows that $BA\mathbf{x} = B\mathbf{0} = \mathbf{0}$, but also $BA\mathbf{x} = I\mathbf{x} = \mathbf{x}$, so $\mathbf{x} = \mathbf{0}$. The linear system $A\mathbf{x} = \mathbf{0}$ thus has $\mathbf{x} = \mathbf{0}$ as its unique solution, so it follows from Theorem 2.2.4 that A is invertible. Multiplying the equation $BA = I$ on the right by A^{-1} shows that $B = A^{-1}$, as desired.

The proof of the case when $AB = I$ is similar, and left as Exercise 2.2.19. ∎

Remark 2.2.2

One-Sided and Non-Square Inverses

When we defined inverses in Definition 2.2.2, we required that the matrix A was square. Some of the ideas from this section can be extended to the non-square case, but the details are more delicate. For example, consider the matrix

$$A = \begin{bmatrix} 1 & 0 \\ 0 & 2 \\ 0 & 0 \end{bmatrix}.$$

This matrix is not square, and thus not invertible, but it does have a "left" inverse—there is a matrix B such that $BA = I$. For example,

In fact, A has infinitely many left inverses—we can replace the entries of the third column of B by whatever we like.

$$BA = \begin{bmatrix} 1 & 0 & 0 \\ 0 & 1/2 & 0 \end{bmatrix} \begin{bmatrix} 1 & 0 \\ 0 & 2 \\ 0 & 0 \end{bmatrix} = \begin{bmatrix} 1 & 0 \\ 0 & 1 \end{bmatrix}.$$

However, it does not have a "right" inverse: there is no matrix C such that $AC = I$, since the only way that AC could be square is if C is 2×3, in which case the third row of AC must be the zero vector.

Geometrically, this makes sense since A acts as a linear transformation that sends $\mathbb{R}^2$ to the xy-plane in $\mathbb{R}^3$, and there is another linear transformation (a left inverse) that sends that plane back to $\mathbb{R}^2$:

The fact that A has a left inverse but not a right inverse does not violate Theorem 2.2.7, since that theorem only applies to square matrices.

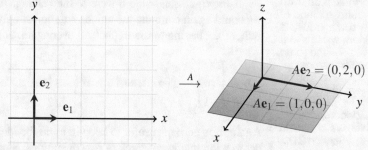

However, there does not exist a linear transformation (a right inverse) that squashes all of $\mathbb{R}^3$ into $\mathbb{R}^2$ in such a way that A then blows it back up to all of $\mathbb{R}^3$.

Exercises

solutions to starred exercises on page 447

2.2.1 Write down the elementary matrix $E \in \mathcal{M}_3$ corresponding to each of the following row operations.

*(a) $R_1 \leftrightarrow R_3$ (b) $R_2 \leftrightarrow R_3$
*(c) $R_1 + 3R_2$ (d) $R_3 - 4R_1$
*(e) $3R_1$ (f) $6R_3$
*(g) $R_2 - 2R_3$ (h) $R_2 + 3R_1$

2.2.2 For each of the following matrices, either find the inverse of the matrix, or show that it is not invertible.

*(a) $\begin{bmatrix} 1 & 0 \\ 0 & 1 \end{bmatrix}$ (b) $\begin{bmatrix} 6 & 3 \\ 2 & 1 \end{bmatrix}$

*(c) $\begin{bmatrix} 2 & 3 \\ 3 & 2 \end{bmatrix}$ (d) $\begin{bmatrix} 1 & 2 & 1 \\ 0 & 1 & 2 \end{bmatrix}$

*(e) $\begin{bmatrix} 2 & 4 & 0 \\ 1 & -2 & 0 \\ 2 & 0 & -1 \end{bmatrix}$ (f) $\begin{bmatrix} 2 & 6 & 1 \\ 0 & 0 & 0 \\ 3 & -2 & 7 \end{bmatrix}$

*(g) $\begin{bmatrix} 1 & 0 \\ 0 & 2 \\ 0 & 1 \end{bmatrix}$ (h) $\begin{bmatrix} 1 & -2 & -2 \\ -2 & 1 & -2 \\ -2 & -2 & 1 \end{bmatrix}$

🖥 **2.2.3** For each of the following matrices, use computer software to either find the inverse of the matrix, or to show that it is not invertible.

*(a) $\begin{bmatrix} 0 & 1 & 1 & 2 \\ 4 & 2 & 3 & 0 \\ 3 & 1 & 5 & 5 \\ 5 & 0 & 5 & 2 \end{bmatrix}$ (b) $\begin{bmatrix} -1 & -2 & 2 & 2 \\ -1 & 1 & 1 & 0 \\ 3 & 3 & 5 & 1 \\ 5 & 2 & 2 & 0 \end{bmatrix}$

*(c) $\begin{bmatrix} -1 & -3 & 3 & 5 & 4 \\ 1 & 5 & -2 & 4 & 5 \\ 5 & -3 & 0 & 1 & 0 \\ 4 & 1 & -2 & 1 & 1 \\ 4 & 4 & 5 & 3 & 5 \end{bmatrix}$

(d) $\begin{bmatrix} -1 & -2 & -1 & 1 & 2 & 4 \\ -1 & -1 & 0 & 0 & -1 & 0 \\ 2 & 4 & 1 & 1 & -4 & -6 \\ 1 & 2 & 1 & 0 & -2 & -5 \\ -3 & -6 & -1 & -4 & 7 & 9 \\ 0 & -1 & 1 & -5 & 3 & 1 \end{bmatrix}$

2.2.4 For each of the following matrices $A \in \mathcal{M}_{m,n}$, find a matrix $E \in \mathcal{M}_m$ such that EA equals the reduced row echelon form of A.

*(a) $\begin{bmatrix} 1 & 2 \\ 3 & 4 \end{bmatrix}$ (b) $\begin{bmatrix} 2 & -1 \\ -4 & 2 \end{bmatrix}$

*(c) $\begin{bmatrix} 1 & 2 & 3 \\ 4 & 5 & 6 \end{bmatrix}$ (d) $\begin{bmatrix} 0 & -1 & 3 \\ 0 & 2 & 1 \end{bmatrix}$

*(e) $\begin{bmatrix} 2 & 1 \\ -1 & 2 \\ 1 & 1 \end{bmatrix}$ (f) $\begin{bmatrix} 1 & 2 & 1 \\ -1 & 0 & -1 \\ 0 & 1 & 2 \end{bmatrix}$

*(g) $\begin{bmatrix} 0 & 1 & 3 \\ 3 & -2 & 3 \\ 1 & 0 & 3 \end{bmatrix}$ (h) $\begin{bmatrix} 4 & 0 & 2 & 4 \\ -1 & -1 & 1 & 2 \\ 5 & 0 & 0 & 2 \end{bmatrix}$

2.2.5 Determine which of the following statements are true and which are false.

*(a) The inverse of an elementary matrix is an elementary matrix.
 (b) The transpose of an elementary matrix is an elementary matrix.
*(c) Every square matrix can be written as a product of elementary matrices.
 (d) Every invertible matrix can be written as a product of elementary matrices.
*(e) The $n \times n$ identity matrix is invertible.
 (f) If A and B are invertible matrices, then so is AB.
*(g) If A and B are invertible matrices, then so is $A + B$.
 (h) If $A^6 = I$ then A is invertible.
*(i) If $A^7 = O$ then A is not invertible.
 (j) If A and B are matrices such that $AB = O$ and A is invertible, then $B = O$.
*(k) If A, B, and X are invertible matrices such that $XA = B$, then $X = A^{-1}B$.

2.2.6 Let $a \neq 0$ be a real number. Show that the following matrix is invertible, and find its inverse:

$$\begin{bmatrix} a & 1 & 1 \\ 0 & a & 1 \\ 0 & 0 & a \end{bmatrix}.$$

* **2.2.7** Let a and b be real numbers, and consider the matrix

$$A = \begin{bmatrix} a & b & b & \cdots & b \\ b & a & b & \cdots & b \\ b & b & a & \cdots & b \\ \vdots & \vdots & \vdots & \ddots & \vdots \\ b & b & b & \cdots & a \end{bmatrix} \in \mathcal{M}_n.$$

(a) Show that A is not invertible if $a = b$.
(b) Find a formula for A^{-1} (and thus show that A is invertible) if $a \neq b$. [Hint: Try some small examples and guess a form for A^{-1}.]

2.2.8 Suppose $A \in \mathcal{M}_n$.

(a) Show that if A has a row consisting entirely of zeros then it is not invertible.
[Hint: What can we say about the rows of AB in this case, where $B \in \mathcal{M}_n$ is any other matrix?]
(b) Show that if A has a column consisting entirely of zeros then it is not invertible.

2.2.9 Compute the indicated powers of the matrix

$$A = \begin{bmatrix} 2 & 1 \\ 3 & 2 \end{bmatrix}.$$

*(a) A^2 (b) A^{-1}
*(c) A^{-2} (d) A^{-4}

2.2.10 A matrix $A \in \mathcal{M}_n$ is called **nilpotent** if there is some positive integer k such that $A^k = O$. In this exercise, we will compute some formulas for matrix inverses involving nilpotent matrices.

(a) Suppose $A^2 = O$. Prove that $I - A$ is invertible, and its inverse is $I + A$.

(b) Suppose $A^3 = O$. Prove that $I - A$ is invertible, and its inverse is $I + A + A^2$.

(c) Suppose k is a positive integer and $A^k = O$. Prove that $I - A$ is invertible, and find its inverse.

*2.2.11 Suppose $A \in \mathcal{M}_{m,n}$ is such that $A^T A$ is invertible, and let $P = A(A^T A)^{-1} A^T$.

(a) What is the size of P?

(b) Show that $P = P^T$. [Side note: Matrices with this property are called **symmetric**.]

(c) Show that $P^2 = P$. [Side note: Matrices such that $P^T = P^2 = P$ are called **orthogonal projections**.]

*2.2.12 Suppose that $P, Q \in \mathcal{M}_n$ and P is invertible. Show that row-reducing the augmented matrix $[\, P \mid Q \,]$ produces the matrix $[\, I \mid P^{-1} Q \,]$.

2.2.13 Suppose that A, B, and X are square matrices of the same size. Given the following matrix equations, find a formula for X in terms of A and B (you may assume all matrices are invertible as necessary).

[Hint: Multiply on the left and/or right by matrix inverses.]

*(a) $AX = B$ (b) $AXB = O$

*(c) $AXB = I$ (d) $AX + BX = A - B$

2.2.14 Suppose $A \in \mathcal{M}_n$ is invertible and $B \in \mathcal{M}_n$ is obtained by swapping two of the rows of A. Explain why B^{-1} can be obtained by swapping those same two *columns* of A^{-1}.

2.2.15 Show that a diagonal matrix is invertible if and only if all of its diagonal entries are non-zero, and find a formula for its inverse.

**2.2.16 Suppose that $A \in \mathcal{M}_n$ is an upper triangular matrix.

(a) Show that A is invertible if and only if all of its diagonal entries are non-zero.
[Hint: Consider the linear system $A\mathbf{x} = \mathbf{0}$.]

(b) Show that if A is invertible then A^{-1} is also upper triangular. [Hint: First show that if $\mathbf{b}$ has its last k entries equal to 0 (for some k) then the solution $\mathbf{x}$ to $A\mathbf{x} = \mathbf{b}$ also has its last k entries equal to 0.]

(c) Show that if A is invertible then the diagonal entries of A^{-1} are the reciprocals of the diagonal entries of A, in the same order.

(d) What changes in parts (a)–(c) if A is *lower* triangular instead of upper triangular?

**2.2.17 Recall Theorem 2.2.2, which established some of the basic properties of matrix inverses.

(a) Prove part (a) of the theorem.
(b) Prove part (c) of the theorem.
(c) Prove part (d) of the theorem.

**2.2.18 Suppose $A \in \mathcal{M}_n$ and k and ℓ are integers (which may potentially be negative).

(a) Show that $A^{k+\ell} = A^k A^\ell$.

(b) Show that $\left(A^k\right)^\ell = A^{k\ell}$.

**2.2.19 Complete the proof of Theorem 2.2.7 by showing that if $A, B \in \mathcal{M}_n$ are such that $AB = I$ then A is invertible and $A^{-1} = B$.

**2.2.20 In this exercise, we generalize Exercise 2.2.15 to *block* diagonal matrices.

(a) Show that

$$\begin{bmatrix} A & O \\ O & D \end{bmatrix}^{-1} = \begin{bmatrix} A^{-1} & O \\ O & D^{-1} \end{bmatrix},$$

assuming that A and D are invertible.

(b) Show that the block diagonal matrix

$$\begin{bmatrix} A_1 & O & \cdots & O \\ O & A_2 & \cdots & O \\ \vdots & \vdots & \ddots & \vdots \\ O & O & \cdots & A_n \end{bmatrix}$$

is invertible if and only if each of $A_1, A_2, \ldots, A_n$ are invertible, and find a formula for its inverse.

2.2.21 Show that the following block matrix inverse formulas hold, as long as the matrix blocks are of sizes so that the given expressions make sense and all of the indicated inverses exist.

*(a) $\begin{bmatrix} A & I \\ I & O \end{bmatrix}^{-1} = \begin{bmatrix} O & I \\ I & -A \end{bmatrix}$

(b) $\begin{bmatrix} I & B \\ O & I \end{bmatrix}^{-1} = \begin{bmatrix} I & -B \\ O & I \end{bmatrix}$

*(c) $\begin{bmatrix} A & B \\ O & D \end{bmatrix}^{-1} = \begin{bmatrix} A^{-1} & -A^{-1}BD^{-1} \\ O & D^{-1} \end{bmatrix}$

(d) If we define $S = D - CA^{-1}B$, then

$$\begin{bmatrix} A & B \\ C & D \end{bmatrix}^{-1} =$$
$$\begin{bmatrix} A^{-1} + A^{-1}BS^{-1}CA^{-1} & -A^{-1}BS^{-1} \\ -S^{-1}CA^{-1} & S^{-1} \end{bmatrix}.$$

[Side note: The matrix S in part (d) is called the **Schur complement**.]

2.2.22 Find the inverse of each of the following matrices. [Hint: Partition the matrices as block matrices and use one of the formulas from Exercise 2.2.21.]

*(a) $\begin{bmatrix} 1 & 2 & 1 & 0 \\ 2 & 3 & 0 & 1 \\ 1 & 0 & 0 & 0 \\ 0 & 1 & 0 & 0 \end{bmatrix}$ (b) $\begin{bmatrix} 1 & 0 & 4 & 6 \\ 0 & 1 & 1 & 5 \\ 0 & 0 & 1 & 0 \\ 0 & 0 & 0 & 1 \end{bmatrix}$

*(c) $\begin{bmatrix} 1 & 1 & 2 & -1 & 0 \\ 1 & 2 & 2 & 0 & 1 \\ 0 & 0 & 1 & 0 & 0 \\ 0 & 0 & 0 & 2 & 0 \\ 0 & 0 & 0 & 0 & 3 \end{bmatrix}$ (d) $\begin{bmatrix} 2 & 1 & 1 & 0 \\ 1 & 2 & 1 & 1 \\ 1 & 1 & 2 & 1 \\ 0 & 1 & 1 & 2 \end{bmatrix}$

2.2.23 The **Sherman-Morrison formula** is a result that says that if $A \in \mathcal{M}_n$ is invertible and $\mathbf{v}, \mathbf{w} \in \mathbb{R}^n$, then

$$(A + \mathbf{v}\mathbf{w}^T)^{-1} = A^{-1} - \frac{A^{-1}\mathbf{v}\mathbf{w}^T A^{-1}}{1 + \mathbf{w}^T A^{-1}\mathbf{v}},$$

as long as the denominator on the right is non-zero. Prove that this formula holds by multiplying $A + \mathbf{v}\mathbf{w}^T$ by its proposed inverse.

[Side note: This formula is useful in situations where we know the inverse of A and want to use that information to obtain the inverse of a slight perturbation of A without having to compute it from scratch.]

▢ **2.2.24** Let A_n be the $n \times n$ matrix

$$A_n = \begin{bmatrix} 1 & 1 & 1 & \cdots & 1 \\ 1 & 2 & 1 & \cdots & 1 \\ 1 & 1 & 3 & \cdots & 1 \\ \vdots & \vdots & \vdots & \ddots & \vdots \\ 1 & 1 & 1 & \cdots & n \end{bmatrix}.$$

(a) Use computer software to compute A_n^{-1} when $n = 2, 3, 4, 5$.
(b) Based on the computations from part (a), guess a formula for A_n^{-1} that works for all n.

2.3 Subspaces, Spans, and Linear Independence

In the previous two sections, we considered some topics (linear systems and matrix inverses) that were motivated very algebraically. We now switch gears a bit and think about how we can make use of geometric concepts to deepen our understanding of these topics.

2.3.1 Subspaces

As a starting point, recall that linear systems can be interpreted geometrically as asking for the point(s) of intersection of a collection of lines or planes (depending on the number of variables involved). The following definition introduces **subspaces**, which can be thought of as any-dimensional analogues of lines and planes.

> **Definition 2.3.1**
> **Subspaces**
>
> A **subspace** of $\mathbb{R}^n$ is a non-empty set $\mathcal{S}$ of vectors in $\mathbb{R}^n$ with the properties that
>
> a) if $\mathbf{v}, \mathbf{w} \in \mathcal{S}$ then $\mathbf{v} + \mathbf{w} \in \mathcal{S}$, and
> b) if $\mathbf{v} \in \mathcal{S}$ and $c \in \mathbb{R}$ then $c\mathbf{v} \in \mathcal{S}$.

*Properties (a) and (b) are sometimes called **closure under vector addition** and **closure under scalar multiplication**, respectively.*

The idea behind this definition is that property (a) ensures that subspaces are "flat", and property (b) makes it so that they are "infinitely long" (just like lines and planes). Subspaces do not have any holes or edges, and if they extend even a little bit in a given direction, then they must extend forever in that direction.

The defining properties of subspaces mimic the properties of lines and planes, but with one caveat—every subspace contains $\mathbf{0}$ (the zero vector). The reason for this is simply that if we choose $\mathbf{v} \in \mathcal{S}$ arbitrarily and let $c = 0$ in property (b) of Definition 2.3.1 then

$$\mathbf{0} = 0\mathbf{v} = c\mathbf{v} \in \mathcal{S}.$$

This implies, for example, that a line through the origin is indeed a subspace, but a line in $\mathbb{R}^2$ with y-intercept equal to anything other than 0 is not a subspace (see Figure 2.13).

Before working with subspaces algebraically, we look at couple of quick examples geometrically to try to build some intuition for how they work.

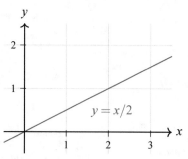

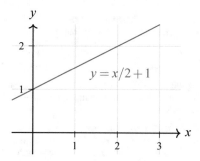

(a) The line $y = x/2$ is a subspace, since lines through the origin are subspaces.

(b) The line $y = x/2 + 1$ is not a subspace, since it does not go through the origin.

Figure 2.13: Lines and planes are subspaces if and only if they go through the origin, so the line (a) $y = x/2$ is a subspace but the line (b) $y = x/2 + 1$ is not.

Example 2.3.1

Geometric Examples of Subspaces

Determine whether or not the following sets of vectors are subspaces.

a) The line in $\mathbb{R}^2$ through the points $(2, 1)$ and $(-4, -2)$.
b) The line in $\mathbb{R}^2$ with slope $1/2$ and y-intercept -1.
c) The plane in $\mathbb{R}^3$ with equation $x + y - 3z = 0$.

Solutions:

a) This line is sketched below, and it does indeed go through the origin. It should be quite believable that if we add any two vectors on this line, the result is also on the line. Similarly, multiplying any vector **v** on this line by a scalar c just stretches it but does not change its direction, so $c\mathbf{v}$ is also on the line. It follows that this *is* a subspace.

> Every line through the origin (in any number of dimensions) is always a subspace.

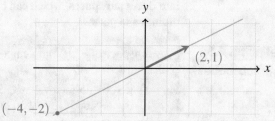

b) This line is sketched below, but notice that it does not contain **0** (i.e., it does not go through the origin). This immediately tells us that it *is not* a subspace. Algebraically, we can prove that it is not a subspace by noticing that $(2, 0)$ is on the line, but $2(2, 0) = (4, 0)$ is not, so it is not closed under scalar multiplication.

> Sets like this line, which *would* be subspaces if they went through the origin, are sometimes called **affine spaces**.

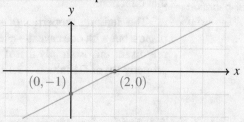

c) This plane *is* a subspace, and is illustrated below. It should be quite believable that if **v** lies on this plane and $c \in \mathbb{R}$ is a scalar, then $c\mathbf{v}$ also lies on the plane. The fact that $\mathbf{v} + \mathbf{w}$ lies on the plane whenever

> We will show that this plane is subspace algebraically in Example 2.3.2.

v and **w** do can be seen from the parallelogram law for adding vectors—the entire parallelogram with **v** and **w** as its sides lies on the same plane.

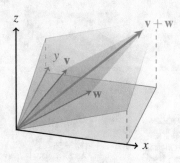

Every *plane* through the origin (in any number of dimensions) is also always a subspace.

It is worth observing that the definition of a subspace is very reminiscent of the definition of a linear transformation (Definition 1.4.1). In both cases, we require that vector addition and scalar multiplication are "respected" in some sense: linear transformations are the functions for which we can apply the function either before or after performing vector addition or scalar multiplication without affecting the result, and subspaces are the sets for which we can perform vector addition and scalar multiplication without leaving the set.

Also, just like we did with linear transformations, we can rephrase the defining properties of subspaces in terms of linear combinations. In particular, a non-empty set S is a subspace of $\mathbb{R}^n$ if and only if

Be careful: $\mathbf{v}_1, \mathbf{v}_2, \ldots,$ $\mathbf{v}_k$ are vectors, not entries of a vector, which are instead denoted like $v_1, v_2,$ $\ldots, v_n$.

$$c_1\mathbf{v}_1 + c_2\mathbf{v}_2 + \cdots + c_k\mathbf{v}_k \in S$$

for all $\mathbf{v}_1, \mathbf{v}_2, \ldots, \mathbf{v}_k \in S$ and all $c_1, c_2, \ldots, c_k \in \mathbb{R}$ (see Exercise 2.3.15).

Even though we cannot visualize subspaces in higher than 3 dimensions, we keep the line/plane intuition in mind—a subspace of $\mathbb{R}^n$ looks like a copy of $\mathbb{R}^m$ (for some $m \leq n$) going through the origin. For example, lines look like copies or $\mathbb{R}^1$, planes look like copies of $\mathbb{R}^2$, and so on. The following example clarifies this intuition a bit and demonstrates how to show algebraically that a set is or is not a subspace.

Example 2.3.2

Algebraic Examples of Subspaces

Determine whether or not the following sets of vectors are subspaces.

a) The graph of the function $y = x^2$ in $\mathbb{R}^2$.
b) The set $\{(x, y) \in \mathbb{R}^2 : x \geq 0, y \geq 0\}$.
c) The plane in $\mathbb{R}^3$ with equation $x + y - 3z = 0$.

The notation $\{(x, y) \in \mathbb{R}^2 : \text{some property}\}$ means the set of all vectors in $\mathbb{R}^2$ that satisfy the conditions described by "some property".

Solutions:

a) This set is *not* a subspace, since (for example) the vectors $\mathbf{v} = (-1, 1)$ and $\mathbf{w} = (1, 1)$ are in the set, but $\mathbf{v} + \mathbf{w} = (0, 2)$ is not, since it is not the case that $2 = 0^2$:

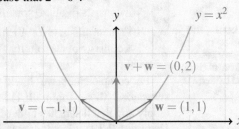

b) This set is *not* a subspace, since (for example) the vector $\mathbf{v} = (1,1)$ is in the set, but $-\mathbf{v} = (-1,-1)$ is not, since the set only contains vectors with non-negative entries:

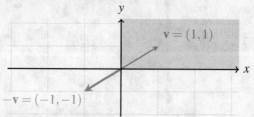

c) This plane *is* a subspace, as we illustrated geometrically in Example 2.3.1(c). To prove it algebraically, we must show that both of the properties described in the definition of a subspace hold.

For property (a), suppose that $\mathbf{v} = (v_1, v_2, v_3)$ and $\mathbf{w} = (w_1, w_2, w_3)$ are both on the plane, so $v_1 + v_2 - 3v_3 = 0$ and $w_1 + w_2 - 3w_3 = 0$. Then by adding these two equations, we see that

$$(v_1 + w_1) + (v_2 + w_2) - 3(v_3 + w_3) = 0,$$

so $\mathbf{v} + \mathbf{w}$ lies on the plane too.

For property (b), suppose that $\mathbf{v} = (v_1, v_2, v_3)$ is on the plane, so $v_1 + v_2 - 3v_3 = 0$. Then for any scalar c, multiplying this equation by c shows that

$$(cv_1) + (cv_2) - 3(cv_3) = 0,$$

so $c\mathbf{v}$ lies on the plane too.

Many of the most useful subspaces that we work with come from interpreting matrices as linear systems or linear transformations. For example, one very natural subspace arises from the fact that matrices represent linear systems of equations: for every matrix $A \in \mathcal{M}_{m,n}$, there is an associated linear system $A\mathbf{x} = \mathbf{0}$, and the set of vectors $\mathbf{x}$ satisfying this system is a subspace (we will verify this claim shortly).

Similarly, another subspace appears naturally if we think about matrices as linear transformations: every matrix $A \in \mathcal{M}_{m,n}$ can be thought of as a linear transformation that sends $\mathbf{x} \in \mathbb{R}^n$ to $A\mathbf{x} \in \mathbb{R}^m$. The set of possible outputs of this linear transformation—that is, the set of vectors of the form $A\mathbf{x}$—is also a subspace. We summarize these observations in Definition 2.3.2.

Definition 2.3.2

Matrix Subspaces

Suppose $A \in \mathcal{M}_{m,n}$.

a) The **range** of A is the subspace of $\mathbb{R}^m$, denoted by $\mathrm{range}(A)$, that consists of all vectors of the form $A\mathbf{x}$.

b) The **null space** of A is the subspace of $\mathbb{R}^n$, denoted by $\mathrm{null}(A)$, that consists of all solutions $\mathbf{x}$ of the linear system $A\mathbf{x} = \mathbf{0}$.

For completeness, we should verify that the range and null space are indeed subspaces. To see that $\mathrm{null}(A)$ is a subspace, we have to show that it is non-empty and that it satisfies both of the defining properties of subspaces given in Definition 2.3.1. It is non-empty because $A\mathbf{0} = \mathbf{0}$, so $\mathbf{0} \in \mathrm{null}(A)$. To see that it satisfies property (a), let $\mathbf{v}, \mathbf{w} \in \text{null}(A)$, so that $A\mathbf{v} = \mathbf{0}$ and $A\mathbf{w} = \mathbf{0}$. Then

$$A(\mathbf{v} + \mathbf{w}) = A\mathbf{v} + A\mathbf{w} = \mathbf{0} + \mathbf{0} = \mathbf{0},$$

so $\mathbf{v} + \mathbf{w} \in \text{null}(A)$ too. Similarly, for property (b) we note that if $\mathbf{v} \in \text{null}(A)$ and $c \in \mathbb{R}$ then
$$A(c\mathbf{v}) = c(A\mathbf{v}) = c\mathbf{0} = \mathbf{0},$$

so $c\mathbf{v} \in \text{null}(A)$ too. It follows that $\text{null}(A)$ is indeed a subspace.

Similarly, to see that $\text{range}(A)$ is a subspace, we first notice that it is non-empty since $A\mathbf{0} = \mathbf{0} \in \text{range}(A)$. Furthermore, if $A\mathbf{x}, A\mathbf{y} \in \text{range}(A)$ and $c \in \mathbb{R}$ then $A\mathbf{x} + A\mathbf{y} = A(\mathbf{x} + \mathbf{y}) \in \text{range}(A)$ as well, as is $A(c\mathbf{x}) = c(A\mathbf{x})$.

Example 2.3.3

Determining Range and Null Space

Describe and plot the range and null space of the matrix $A = \begin{bmatrix} 2 & -2 \\ 1 & -1 \end{bmatrix}$.

Solution:

To get our hands on the range of A, we let $\mathbf{x} = (x, y)$ and compute

$$A\mathbf{x} = \begin{bmatrix} 2 & -2 \\ 1 & -1 \end{bmatrix} \begin{bmatrix} x \\ y \end{bmatrix} = \begin{bmatrix} 2(x-y) \\ x-y \end{bmatrix}.$$

We will develop more systematic ways of finding the range and null space of a matrix in the next several subsections.

Notice that $A\mathbf{x}$ only depends on $x - y$, not x and y individually. We can thus define a new variable $z = x - y$ and see that $A\mathbf{x} = (2z, z) = z(2, 1)$, so the range of A is the set of all multiples of the vector $(2, 1)$. In other words, it is the line going through the origin and the point $(2, 1)$:

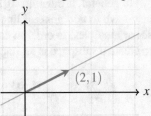

On the other hand, $\text{null}(A)$ is the set of vectors $\mathbf{x} = (x, y)$ such that $A\mathbf{x} = \mathbf{0}$. This is a linear system that we can solve as follows:

$$\begin{bmatrix} 2 & -2 & | & 0 \\ 1 & -1 & | & 0 \end{bmatrix} \xrightarrow{R_1 \leftrightarrow R_2} \begin{bmatrix} 1 & -1 & | & 0 \\ 2 & -2 & | & 0 \end{bmatrix} \xrightarrow{R_2 - 2R_1} \begin{bmatrix} 1 & -1 & | & 0 \\ 0 & 0 & | & 0 \end{bmatrix},$$

In this example, the range and null space were both lines. We will see later (in Theorem 2.4.10) that the only other possibility for 2×2 matrices is that one of the range or null space is all of $\mathbb{R}^2$ and the other one is $\{\mathbf{0}\}$.

which shows that y is a free variable and x is a leading variable with $x = y$. It follows that $\mathbf{x} = (x, x) = x(1, 1)$, so the null space is the line going through the origin and the point $(1, 1)$:

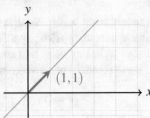

2.3.2 The Span of a Set of Vectors

Some of the previous examples of sets that are *not* subspaces suggest a method by which we could turn them into subspaces: we could add all linear combinations of members of the set to it. For example, the set containing only the vector $(2,1)$ is not a subspace of $\mathbb{R}^2$ because, for example, $2(2,1) = (4,2)$ is not in the set. To fix this problem, we could simply add $(4,2)$ to the set and ask whether or not $\{(2,1),(4,2)\}$ is a subspace. This larger also fails to be a subspace for a very similar reason—there are other scalar multiples of $(2,1)$ still missing from it. However, if we add *all* of the scalar multiples of $(2,1)$ to the set, we get the line through the origin and the point $(2,1)$, which *is* a subspace, as shown Figure 2.14.

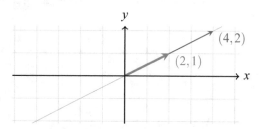

Figure 2.14: To turn the set $\{(2,1)\}$ into a subspace, we have to add all of the scalar multiples of $(2,1)$ to it, creating a line through the origin.

> Recall that a linear combination of the vectors $\mathbf{v}_1, \ldots, \mathbf{v}_k$ is any vector of the form $c_1 \mathbf{v}_1 + \cdots + c_k \mathbf{v}_k$, where $c_1, \ldots, c_k \in \mathbb{R}$.

In general, if our starting set contains more than just one vector, we might also have to add general linear combinations of those vectors (not just their scalar multiples) in order to create a subspace. This idea of enlarging a set so as to create a subspace is an important one that we now give a name and explore.

> **Definition 2.3.3**
>
> **Span**
>
> If $B = \{\mathbf{v}_1, \mathbf{v}_2, \ldots, \mathbf{v}_k\}$ is a set of vectors in $\mathbb{R}^n$, then the set of all linear combinations of those vectors is called their **span**, and it is denoted by
>
> $$\text{span}(B) \quad \text{or} \quad \text{span}(\mathbf{v}_1, \mathbf{v}_2, \ldots, \mathbf{v}_k).$$

> Recall that in $\mathbb{R}^2$, $\mathbf{e}_1 = (1,0)$ and $\mathbf{e}_2 = (0,1)$.

For example, $\text{span}((2,1))$ is the line through the origin and the point $(2,1)$, which was depicted in Figure 2.14 and discussed earlier. As a more useful example, we note that if $\mathbf{e}_1$ and $\mathbf{e}_2$ are the standard basis vectors in $\mathbb{R}^2$ then $\text{span}(\mathbf{e}_1, \mathbf{e}_2) = \mathbb{R}^2$. To verify this claim, we just need to show that every vector in $\mathbb{R}^2$ can be written as a linear combination of $\mathbf{e}_1$ and $\mathbf{e}_2$, which we learned how to do back in Section 1.1.3:

$$(x,y) = x\mathbf{e}_1 + y\mathbf{e}_2.$$

It should seem believable that the natural generalization of this fact to arbitrary dimensions also holds: $\text{span}(\mathbf{e}_1, \mathbf{e}_2, \ldots, \mathbf{e}_n) = \mathbb{R}^n$ for all $n \geq 1$, since we can always write any vector, in any number of dimensions as a linear combination of these standard basis vectors:

$$\mathbf{v} = (v_1, v_2, \ldots, v_n) = v_1\mathbf{e}_1 + v_2\mathbf{e}_2 + \cdots + v_n\mathbf{e}_n.$$

Example 2.3.4
Determining Spans

Describe the following spans both geometrically and algebraically.

a) $\mathrm{span}\big((1,2),(2,1)\big)$

b) $\mathrm{span}\big((1,2,1),(2,1,1)\big)$

Solutions:

a) Geometrically, this span seems like it "should" be all of $\mathbb{R}^2$, since we can lay a parallelogram grid with sides $(1,2)$ and $(2,1)$ down on top of $\mathbb{R}^2$. This grid tells us how to create a linear combination of those two vectors that equals any other vector (we just follow the lines on the grid and keep track of how far we had to go in each direction):

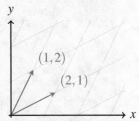

To verify that $\mathrm{span}\big((1,2),(2,1)\big) = \mathbb{R}^2$ algebraically, we show that for every $(x,y) \in \mathbb{R}^2$ we can find coefficients $c_1, c_2 \in \mathbb{R}$ such that

$$(x,y) = c_1(1,2) + c_2(2,1).$$

If we compare the entries of the vectors on the left- and right-hand sides of that equation, we see that this is actually a system of linear equations:

> Be careful: the variables in this linear system are c_1 and c_2, not x and y. We are trying to solve for c_1 and c_2, while we should just think of x and y as unknown constants that are given ahead of time.

$$c_1 + 2c_2 = x$$
$$2c_1 + c_2 = y$$

We can solve this linear system by using row operations to put it in row echelon form:

$$\begin{bmatrix} 1 & 2 & x \\ 2 & 1 & y \end{bmatrix} \xrightarrow{R_2 - 2R_1} \begin{bmatrix} 1 & 2 & x \\ 0 & -3 & y - 2x \end{bmatrix}.$$

From here, we could explicitly solve the system to find values for c_1 and c_2, but it is enough to notice that this row echelon form has no row of the form $\begin{bmatrix} 0 & 0 & | & b \end{bmatrix}$, so it has a solution regardless of the values of x and y. That is, we can always write (x,y) as a linear combination of $(1,2)$ and $(2,1)$.

> The span of any two non-zero vectors is always either a line (if the vectors are collinear) or a plane (if they are not collinear).

b) This set should be the plane in $\mathbb{R}^3$ containing the vectors $(1,2,1)$ and $(2,1,1)$, which is displayed below. Just like in the previous 2D example, we can lay out a parallelogram grid using these two vectors that tells us how to use a linear combination to reach any vector on this plane.

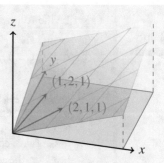

To come up with an algebraic description of this plane, we use the same procedure that we did in part (a). We start with a vector $(x,y,z) \in \mathbb{R}^3$ and we try to find $c_1, c_2 \in \mathbb{R}$ such that

$$(x,y,z) = c_1(1,2,1) + c_2(2,1,1).$$

By setting the entries of the vectors on the left- and right-hand-sides equal to each other, we can interpret this as a system of linear equations, which we can solve as follows:

Again, the variables that we are solving for in this linear system are c_1 and c_2, not x, y, and z.

$$\begin{bmatrix} 1 & 2 & x \\ 2 & 1 & y \\ 1 & 1 & z \end{bmatrix} \xrightarrow[R_3 - R_1]{R_2 - 2R_1} \begin{bmatrix} 1 & 2 & x \\ 0 & -3 & y - 2x \\ 0 & -1 & z - x \end{bmatrix}$$

$$\xrightarrow{R_3 - \frac{1}{3}R_2} \begin{bmatrix} 1 & 2 & x \\ 0 & -3 & y - 2x \\ 0 & 0 & z - \frac{1}{3}x - \frac{1}{3}y \end{bmatrix}.$$

Now that this linear system is in row echelon form, we see that it has a solution if and only if the bottom-right entry, $z = \frac{1}{3}x + \frac{1}{3}y$, equals zero. After rearranging and simplifying, this equation becomes $x + y - 3z = 0$, so the vector (x,y,z) is a linear combination of $(1,2,1)$ and $(2,1,1)$ if and only if $x + y - 3z = 0$.

We motivated the span of a set of vectors as a way of turning that set into a subspace. We now show that the span of a set of vectors is indeed always a subspace, as we would hope.

Theorem 2.3.1

Spans are Subspaces

In fact, we can think of $\text{span}(\mathbf{v}_1, \mathbf{v}_2, \ldots, \mathbf{v}_k)$ as the *smallest* subspace containing $\mathbf{v}_1, \mathbf{v}_2, \ldots, \mathbf{v}_k$.

Let $\mathbf{v}_1, \mathbf{v}_2, \ldots, \mathbf{v}_k \in \mathbb{R}^n$. Then $\text{span}(\mathbf{v}_1, \mathbf{v}_2, \ldots, \mathbf{v}_k)$ is a subspace of $\mathbb{R}^n$.

Proof. We first note that $\text{span}(\mathbf{v}_1, \mathbf{v}_2, \ldots, \mathbf{v}_k)$ is non-empty (for example, it contains each of $\mathbf{v}_1, \mathbf{v}_2, \ldots, \mathbf{v}_k$). To show that property (a) in the definition of subspaces (Definition 2.3.1) holds, suppose that $\mathbf{v}, \mathbf{w} \in \text{span}(\mathbf{v}_1, \mathbf{v}_2, \ldots, \mathbf{v}_k)$. Then there exist $c_1, c_2, \ldots, c_k, d_1, d_2, \ldots, d_k \in \mathbb{R}$ such that

$$\mathbf{v} = c_1\mathbf{v}_1 + c_2\mathbf{v}_2 + \cdots + c_k\mathbf{v}_k$$
$$\mathbf{w} = d_1\mathbf{v}_1 + d_2\mathbf{v}_2 + \cdots + d_k\mathbf{v}_k.$$

By adding these two equations, we see that

$$\mathbf{v} + \mathbf{w} = (c_1 + d_1)\mathbf{v}_1 + (c_2 + d_2)\mathbf{v}_2 + \cdots + (c_k + d_k)\mathbf{v}_k,$$

so $\mathbf{v} + \mathbf{w}$ is also a linear combination of $\mathbf{v}_1, \mathbf{v}_2, \ldots, \mathbf{v}_k$, which implies that $\mathbf{v} + \mathbf{w} \in \text{span}(\mathbf{v}_1, \mathbf{v}_2, \ldots, \mathbf{v}_k)$ too.

Similarly, to show that property (b) holds, note that if $c \in \mathbb{R}$ then

$$c\mathbf{v} = (cc_1)\mathbf{v}_1 + (cc_2)\mathbf{v}_2 + \cdots + (cc_k)\mathbf{v}_k,$$

so $c\mathbf{v} \in \text{span}(\mathbf{v}_1, \mathbf{v}_2, \ldots, \mathbf{v}_k)$, which shows that $\text{span}(\mathbf{v}_1, \mathbf{v}_2, \ldots, \mathbf{v}_k)$ is indeed a subspace. ■

Spans provide us with an any-dimensional way of describing a subspace in terms of some of the vectors that it contains. For example, instead of saying things like "the line going through the origin and $(2,3)$" or "the plane going through the origin, $(1,2,3)$, and $(3,-1,2)$", we can now say $\text{span}((2,3))$ or $\text{span}((1,2,3),(3,-1,2))$, respectively. Being able to describe subspaces like this is especially useful in higher-dimensional situations where our intuition about (hyper-)planes breaks down.

> If $S = \text{span}(\mathbf{v}_1, \mathbf{v}_2, \ldots, \mathbf{v}_k)$ then we sometimes say that $\{\mathbf{v}_1, \mathbf{v}_2, \ldots, \mathbf{v}_k\}$ **spans** S or that S is **spanned** by $\{\mathbf{v}_1, \mathbf{v}_2, \ldots, \mathbf{v}_k\}$.

It turns out that the range of a matrix can be expressed very conveniently as the span of a set of vectors in a way that requires no calculation whatsoever—it can be eyeballed directly from the entries of the matrix.

Theorem 2.3.2

Range Equals the Span of Columns

Suppose $A \in \mathcal{M}_{m,n}$ has columns $\mathbf{a}_1, \mathbf{a}_2, \ldots, \mathbf{a}_n$. Then

$$\text{range}(A) = \text{span}(\mathbf{a}_1, \mathbf{a}_2, \ldots, \mathbf{a}_n).$$

Proof. Recall from Theorem 1.3.5 that $A\mathbf{x}$ can be written as a linear combination of the columns $\mathbf{a}_1, \mathbf{a}_2, \ldots, \mathbf{a}_n$ of A:

$$A\mathbf{x} = x_1\mathbf{a}_1 + x_2\mathbf{a}_2 + \cdots + x_n\mathbf{a}_n.$$

> For this reason, the range of A is sometimes called its **column space** and denoted by $\text{col}(A)$.

The range of A is the set of all vectors of the form $A\mathbf{x}$, which thus equals the set of all linear combinations of $\mathbf{a}_1, \mathbf{a}_2, \ldots, \mathbf{a}_n$, which is $\text{span}(\mathbf{a}_1, \mathbf{a}_2, \ldots, \mathbf{a}_n)$. ■

The above theorem provides an alternate proof of the fact that $\text{range}(A)$ is a subspace, since it equals the span of a set of vectors, and we just showed in Theorem 2.3.1 that the span of *any* set of vectors is a subspace. Furthermore, it is useful for giving us a way to actually get our hands on and compute the range of a matrix, as we now demonstrate.

Example 2.3.5

Determining Range via Columns

Describe the range of $A = \begin{bmatrix} 2 & -2 \\ 1 & -1 \end{bmatrix}$ as the span of a set of vectors.

Solution:

Theorem 2.3.2 says that the range of A is the span of its columns, so

$$\text{range}(A) = \text{span}((2,1),(-2,-1)).$$

This is a fine answer on its own, but it is worthwhile to simplify it somewhat—the vector $(-2,-1)$ does not actually contribute anything to the span since it is a multiple of $(2,1)$. We could thus just as well say that

$$\text{range}(A) = \text{span}((2,1)).$$

Furthermore, both of these spans are the line through the origin and $(2,1)$, which agrees with the answer that we found in Example 2.3.3.

While we like to describe subspaces as spans of sets of vectors, the previous example demonstrates that doing so might be somewhat redundant—we *can* describe a line as a span of 2 (or more) vectors, but it seems somewhat silly to do so when we could just use 1 vector instead. Similarly, we could describe a plane as the span of 93 vectors, but why would we? Only 2 vectors are needed. We start looking at this idea of whether or not a set of vectors contains "redundancies" like these in the next subsection.

We close this subsection by describing how the span of the rows or columns of a matrix can tell us about whether or not it is invertible.

Theorem 2.3.3

Spanning Sets and Invertible Matrices

Suppose $A \in \mathcal{M}_n$. The following are equivalent:

a) A is invertible.
b) The columns of A span $\mathbb{R}^n$.
c) The rows of A span $\mathbb{R}^n$.

Proof. The fact that properties (a) and (b) are equivalent follows from combining two of our recent previous results: Theorem 2.2.4 tells us that A is invertible if and only if the linear system $A\mathbf{x} = \mathbf{b}$ has a solution for all $\mathbf{b} \in \mathbb{R}^n$, which means that every $\mathbf{b} \in \mathbb{R}^n$ can be written as a linear combination of the columns of A (by Theorem 1.3.5). This is exactly what it means for the columns of A to span $\mathbb{R}^n$.

> There are actually numerous other conditions that are equivalent to invertibility as well. We will collect these conditions at the end of the chapter, in Theorem 2.5.1.

The equivalence of properties (b) and (c) follows from the fact that A is invertible if and only if A^T is invertible (Theorem 2.2.2(c)), and the rows of A are the columns of A^T. ∎

Since the range of a matrix equals the span of its columns, the equivalence of properties (a) and (b) in the above theorem can be rephrased as follows:

> (!) A matrix $A \in \mathcal{M}_n$ is invertible if and only if its range is all of $\mathbb{R}^n$.

The geometric interpretation of this fact is the same as the geometric interpretation of invertibility that we saw back in Section 2.2.2: if the range of a matrix is smaller than $\mathbb{R}^n$ (for example, a plane in $\mathbb{R}^3$), then that matrix "squashes" space, and that squashing cannot be undone (see Figure 2.15). However, if its range is all of $\mathbb{R}^n$ then it just shuffles vectors around in space, and its inverse shuffles them back.

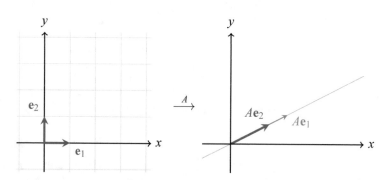

Figure 2.15: Non-invertible matrices $A \in \mathcal{M}_n$ have range that is strictly smaller than $\mathbb{R}^n$, so space is "squashed" by the corresponding linear transformation. In this figure, $\mathbb{R}^2$ is sent to a (1-dimensional) line, which cannot be "unsquashed" back into all of $\mathbb{R}^2$.

2.3.3   Linear Dependence and Independence

Recall from earlier that a row echelon form of a matrix can have entire rows of zeros at the bottom of it. For example, a row echelon form of

$$\left[\begin{array}{cc|c} 1 & -1 & 2 \\ -1 & 1 & -2 \end{array}\right] \quad \text{is} \quad \left[\begin{array}{cc|c} 1 & -1 & 2 \\ 0 & 0 & 0 \end{array}\right].$$

This happens when there is some linear combination of the rows of the matrix that equals the zero row, and we interpret this roughly as saying that one of the rows of the matrix (i.e., one of the equations in the associated linear system) does not "contribute anything new" to the linear system. In the example above, the second equation of the associated linear system says $-x+y=-2$, which is redundant since the first equation already said that $x-y=2$, and these equations can be obtained from each other just by multiplying one another by -1.

The following definition captures this idea that a redundancy among vectors or linear equations can be identified by whether or not some linear combination of them equals zero.

Definition 2.3.4

Linear Dependence and Independence

A set of vectors $B = \{\mathbf{v}_1, \mathbf{v}_2, \ldots, \mathbf{v}_k\}$ is **linearly dependent** if there exist scalars $c_1, c_2, \ldots, c_k \in \mathbb{R}$, at least one of which is not zero, such that

$$c_1\mathbf{v}_1 + c_2\mathbf{v}_2 + \cdots + c_k\mathbf{v}_k = \mathbf{0}.$$

If B is not linearly dependent then it is called **linearly independent**.

For example, the set of vectors $\{(2,3),(1,0),(0,1)\}$ is linearly dependent because

$$(2,3) - 2(1,0) - 3(0,1) = (0,0).$$

On the other hand, the set of vectors $\{(1,0,0),(0,1,0),(0,0,1)\}$ is linearly independent, since the only linear combination of those three vectors that equals $(0,0,0)$ is the "trivial" linear combination:

More generally, the set of standard basis vectors $\{\mathbf{e}_1,\mathbf{e}_2,\ldots,\mathbf{e}_n\} \subseteq \mathbb{R}^n$ is always linearly independent, regardless of n.

$$0(1,0,0) + 0(0,1,0) + 0(0,0,1) = (0,0,0).$$

In general, to check whether or not a set of vectors $\{\mathbf{v}_1, \mathbf{v}_2, \ldots, \mathbf{v}_k\}$ is linearly independent, we set

$$c_1\mathbf{v}_1 + c_2\mathbf{v}_2 + \cdots + c_k\mathbf{v}_k = \mathbf{0}$$

and then try to solve for the scalars $c_1, c_2, \ldots, c_k$. If they must all equal 0 (i.e., if this homogeneous linear system has a unique solution), then the set is linearly independent, and otherwise (i.e., if the linear system has infinitely many solutions) it is linearly dependent.

Example 2.3.6

Determining Linear Independence

Determine whether the following sets of vectors are linearly dependent or linearly independent:

a) $\{(1,-1,0),(-2,1,2),(1,1,-4)\}$
b) $\{(1,2,3),(1,0,1),(0,-1,2)\}$

Solutions:
a) More explicitly, we are being asked whether or not there exist (not

all zero) scalars $c_1, c_2, c_3 \in \mathbb{R}$ such that

$$c_1(1,-1,0) + c_2(-2,1,2) + c_3(1,1,-4) = (0,0,0).$$

If we compare the entries of the vectors on the left- and right-hand sides of this equation, we see that it is actually a linear system:

$$
\begin{aligned}
c_1 - 2c_2 + c_3 &= 0 \\
-c_1 + c_2 + c_3 &= 0 \\
2c_2 - 4c_3 &= 0
\end{aligned}
$$

We can solve this linear system by using row operations to put it in row echelon form:

Notice that the columns of this coefficient matrix are exactly the columns whose linear (in)dependence we are trying to determine.

$$
\left[\begin{array}{ccc|c}
1 & -2 & 1 & 0 \\
-1 & 1 & 1 & 0 \\
0 & 2 & -4 & 0
\end{array}\right]
\xrightarrow{R_2+R_1}
\left[\begin{array}{ccc|c}
1 & -2 & 1 & 0 \\
0 & -1 & 2 & 0 \\
0 & 2 & -4 & 0
\end{array}\right]
$$

$$
\xrightarrow{R_3+2R_2}
\left[\begin{array}{ccc|c}
1 & -2 & 1 & 0 \\
0 & -1 & 2 & 0 \\
0 & 0 & 0 & 0
\end{array}\right].
$$

From here, we see that the linear system has infinitely many solutions (since it is consistent and c_3 is a free variable). If we choose $c_3 = 1$ and then solve by back substitution, we find that $c_2 = 2$ and $c_1 = 3$, which tells us that

We could have also chosen c_3 to be any other non-zero value.

$$3(1,-1,0) + 2(-2,1,2) + (1,1,-4) = (0,0,0).$$

It follows that $\{(1,-1,0),(-2,1,2),(1,1,-4)\}$ is linearly dependent.

b) Similarly, we now want to know whether or not there exist scalars $c_1, c_2, c_3 \in \mathbb{R}$, not all equal to 0, such that

$$c_1(1,2,3) + c_2(1,0,1) + c_3(0,-1,2) = (0,0,0).$$

If we set this up as a linear system and solve it just as we did in part (a), we find the following row echelon form:

Recall from Theorem 2.1.2 that linear systems with zeros on the right-hand side are called **homogeneous** and always have either infinitely many solutions like in part (a), or unique solution (the zero vector) like in part (b).

$$
\left[\begin{array}{ccc|c}
1 & 1 & 0 & 0 \\
2 & 0 & -1 & 0 \\
3 & 1 & 2 & 0
\end{array}\right]
\xrightarrow[R_3-3R_1]{R_2-2R_1}
\left[\begin{array}{ccc|c}
1 & 1 & 0 & 0 \\
0 & -2 & -1 & 0 \\
0 & -2 & 2 & 0
\end{array}\right]
$$

$$
\xrightarrow{R_3-R_2}
\left[\begin{array}{ccc|c}
1 & 1 & 0 & 0 \\
0 & -2 & -1 & 0 \\
0 & 0 & 3 & 0
\end{array}\right]
$$

From here we see that the linear system has a unique solution, since it is consistent and each column has a leading entry. Since $c_1 = c_2 = c_3 = 0$ is a solution, it must be the only solution, so we conclude that the set $\{(1,2,3),(1,0,1),(0,-1,2)\}$ is linearly independent.

We saw in the previous example that we can check linear (in)dependence of a set of vectors by placing those vectors as columns in a matrix and augmenting with a $\mathbf{0}$ right-hand side. It is worth stating this observation as a theorem:

Theorem 2.3.4

Checking Linear Dependence

Suppose $A \in \mathcal{M}_{m,n}$ has columns $\mathbf{a}_1, \mathbf{a}_2, \ldots, \mathbf{a}_n$. The following are equivalent:

a) The set $\{\mathbf{a}_1, \mathbf{a}_2, \ldots, \mathbf{a}_n\}$ is linearly dependent.

b) The linear system $A\mathbf{x} = \mathbf{0}$ has a non-zero solution.

Proof. We use block matrix multiplication. If $A = \begin{bmatrix} \mathbf{a}_1 \mid \mathbf{a}_2 \mid \cdots \mid \mathbf{a}_n \end{bmatrix}$ then

$$A\mathbf{x} = \begin{bmatrix} \mathbf{a}_1 \mid \mathbf{a}_2 \mid \cdots \mid \mathbf{a}_n \end{bmatrix} \begin{bmatrix} x_1 \\ x_2 \\ \vdots \\ x_n \end{bmatrix} = x_1 \mathbf{a}_1 + x_2 \mathbf{a}_2 + \cdots + x_n \mathbf{a}_n.$$

It follows that there exists a non-zero $\mathbf{x} \in \mathbb{R}^n$ such that $A\mathbf{x} = \mathbf{0}$ if and only if there exist constants $x_1, x_2, \ldots, x_n$ (not all 0) such that $x_1 \mathbf{a}_1 + x_2 \mathbf{a}_2 + \cdots + x_n \mathbf{a}_n = \mathbf{0}$ (i.e., $\{\mathbf{a}_1, \mathbf{a}_2, \ldots, \mathbf{a}_n\}$ is a linearly dependent set). ■

While it is typically most convenient mathematically to work with linear dependence in the form described by Definition 2.3.4, we can gain some intuition for it by rephrasing it in a slightly different way. The idea is to notice that we can rearrange the equation

$$c_1 \mathbf{v}_1 + c_2 \mathbf{v}_2 + \cdots + c_k \mathbf{v}_k = \mathbf{0}$$

so as to solve for one of the vectors (say $\mathbf{v}_1$) in terms of the others. In other words, we have the following equivalent characterization of linear (in)dependence:

This fact is proved in Exercise 2.3.21.

! A set of vectors is linearly dependent if and only if at least one of the vectors in the set is a linear combination of the others.

For example, $\{(1, -1, 0), (-2, 1, 2), (1, 1, -4)\}$ is linearly dependent, since

$$3(1, -1, 0) + 2(-2, 1, 2) + (1, 1, -4) = (0, 0, 0).$$

By rearranging this equation, we can solve for any one of the vectors in terms of the other two as follows:

$$(1, -1, 0) = -\tfrac{2}{3}(-2, 1, 2) - \tfrac{1}{3}(1, 1, -4),$$
$$(-2, 1, 2) = -\tfrac{3}{2}(1, -1, 0) - \tfrac{1}{2}(1, 1, -4), \quad \text{and}$$
$$(1, 1, -4) = -3(1, -1, 0) - 2(-2, 1, 2).$$

Geometrically, this means that $(1, -1, 0)$, $(-2, 1, 2)$, and $(1, 1, -4)$ all lie on a common plane. In other words, they span a plane instead of all of $\mathbb{R}^3$, and in this case we could discard any one of these vectors and the span would still be a plane (see Figure 2.16(a)).

It is important to be a bit careful when rephrasing linear (in)dependence in terms of one vector being a linear combination of the other vectors in the set. To illustrate why, consider the set $\{(1, 2, 1), (2, 4, 2), (2, 1, 1)\}$, which is linearly dependent since

Remember that for linear dependence, we only need *at least one* of the coefficients to be non-zero. This set is still linearly dependent even though the third coefficient is 0.

$$2(1, 2, 1) - (2, 4, 2) + 0(2, 1, 1) = (0, 0, 0).$$

However, it is not possible to write $(2,1,1)$ as a linear combination of $(1,2,1)$ and $(2,4,2)$, since those two vectors point in the same direction (see Figure 2.16(b)) and so any linear combination of them also lies on the line in that same direction. This illustrates why we say that a set is linearly dependent if *at least one* of the vectors from the set is a linear combination of the others. In this case, we can write $(2,4,2)$ as a linear combination of $(1,2,1)$ and $(2,1,1)$: $(2,4,2) = 2(1,2,1) + 0(2,1,1)$.

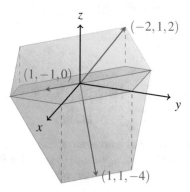

(a) A linearly dependent set of 3 vectors that spans a plane in $\mathbb{R}^3$.

(b) The set $\{(1,2,1),(2,4,2),(2,1,1)\}$ is linearly dependent, but removing $(2,1,1)$ from it changes its span from a plane to a line.

Figure 2.16: If a set of 3 non-zero vectors in $\mathbb{R}^3$ is linearly dependent then it spans a line or a plane, and there is at least one vector in the set with the property that we can remove it without affecting the span.

One special case of linear (in)dependence that is worth pointing out is the case when there are only two vectors. A set of two vectors is linearly dependent if and only if one of the vectors is a multiple of the other one. This is straightforward enough to prove using the definition of linear dependence (see Exercise 2.3.20), but geometrically it says that two vectors form a linearly dependent set if and only if they lie on the same line.

Example 2.3.7

Linear Independence of a Set of Two Vectors

We could also check linear (in)dependence using the method of Example 2.3.6, but that is overkill when there are only two vectors.

Determine whether the following sets of vectors are linearly dependent or linearly independent:

a) $\{(1,2),(3,1)\}$

b) $\{(1,2,3),(2,4,6)\}$

Solutions:

a) Since $(1,2)$ and $(3,1)$ are not multiples of each other (i.e., there is no $c \in \mathbb{R}$ such that $(3,1) = c(1,2)$), we conclude that this set is linearly independent. Geometrically, this means that these two vectors do not lie on a common line:

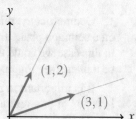

b) Since $(2,4,6) = 2(1,2,3)$, these two vectors are multiples of each other and thus the set is linearly dependent. Geometrically, this means that these two vectors lie on the same line:

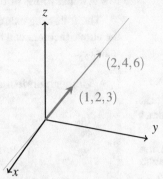

Similarly, a set of 3 vectors is linearly dependent if and only if they lie on a common plane, and in general a set of n vectors is linearly dependent if and only if they lie on a common $(n-1)$-dimensional hyperplane. More generally, linear dependence means that a set of vectors is bigger than the dimension of the hyperplane that they span—an idea that we clarify and make more precise in the next section.

We close this subsection by introducing a connection between linear independence and invertible matrices that is analogous to the connection between spanning sets and invertible matrices that was provided by Theorem 2.3.3.

Theorem 2.3.5

Independence and Invertible Matrices

Suppose $A \in \mathcal{M}_n$. The following are equivalent:

a) A is invertible.

b) The columns of A form a linearly independent set.

c) The rows of A form a linearly independent set.

Proof. The fact that properties (a) and (b) are equivalent follows from combining two of our recent previous results: Theorem 2.2.4 tells us that A is invertible if and only if the linear system $A\mathbf{x} = \mathbf{0}$ has a unique solution (which is necessarily $\mathbf{x} = \mathbf{0}$), and then Theorem 2.3.4 says that is equivalent to the columns of A forming a linearly independent set.

The equivalence of properties (b) and (c) follows from the facts that A is invertible if and only if A^T is invertible (Theorem 2.2.2(c)), and the rows of A are the columns of A^T. ∎

The final line of this proof is identical to the final line of the proof of Theorem 2.3.3.

For example, if we go back to the sets of vectors that we considered in Example 2.3.6, we see that

$$\begin{bmatrix} 1 & -2 & 1 \\ -1 & 1 & 1 \\ 0 & 2 & -4 \end{bmatrix}$$

is not invertible, since its columns are linearly dependent, while the matrix

$$\begin{bmatrix} 1 & 1 & 0 \\ 2 & 0 & -1 \\ 3 & 1 & 2 \end{bmatrix}$$

is invertible, since its columns are linearly independent. Note that this conclusion agrees with the fact that a row echelon form of the former matrix has a zero row, while a row echelon form of the latter matrix does not (we computed row echelon forms of these matrices back Example 2.3.6).

The following example illustrates a somewhat more sophisticated way that the above theorem can be used to establish invertibility of a matrix or family of matrices.

Example 2.3.8

Vandermonde Matrices

For example, the matrix

$$\begin{bmatrix} 1 & 1 & 1 \\ 1 & 2 & 4 \\ 1 & 3 & 9 \end{bmatrix}$$

from Example 2.2.8(d) is a Vandermonde matrix with $a_0 = 1$, $a_1 = 2$, and $a_2 = 3$.

A **Vandermonde matrix** is a square matrix of the form

$$V = \begin{bmatrix} 1 & a_0 & a_0^2 & \cdots & a_0^n \\ 1 & a_1 & a_1^2 & \cdots & a_1^n \\ \vdots & \vdots & \vdots & \ddots & \vdots \\ 1 & a_n & a_n^2 & \cdots & a_n^n \end{bmatrix},$$

where $a_0, a_1, \ldots, a_n$ are real numbers. Show that V is invertible if and only if $a_0, a_1, \ldots, a_n$ are distinct (i.e., $a_i \neq a_j$ whenever $i \neq j$).

Solution:

Suppose for now that $a_0, a_1, \ldots, a_n$ are distinct. We just learned from Theorem 2.3.5 that V is invertible if and only if its columns are linearly independent, which is the case if and only if

Be careful: V is an $(n+1) \times (n+1)$ matrix, not an $n \times n$ matrix.

$$c_0 \begin{bmatrix} 1 \\ 1 \\ \vdots \\ 1 \end{bmatrix} + c_1 \begin{bmatrix} a_0 \\ a_1 \\ \vdots \\ a_n \end{bmatrix} + c_2 \begin{bmatrix} a_0^2 \\ a_1^2 \\ \vdots \\ a_n^2 \end{bmatrix} + \cdots + c_n \begin{bmatrix} a_0^n \\ a_1^n \\ \vdots \\ a_n^n \end{bmatrix} = \begin{bmatrix} 0 \\ 0 \\ \vdots \\ 0 \end{bmatrix} \qquad (2.3.1)$$

implies $c_0 = c_1 = \cdots = c_n = 0$.

If we define the polynomial $p(x) = c_0 + c_1 x + c_2 x^2 + \cdots + c_n x^n$, then the first entry of Equation (2.3.1) says that $p(a_0) = 0$, the next entry says that $p(a_1) = 0$, and so on to $p(a_n) = 0$. Since p is a degree-n polynomial, and we just showed that it has $n + 1$ distinct roots (since $a_0, a_1, \ldots, a_n$ are distinct), the factor theorem (Theorem A.2.2) shows that it must be the zero polynomial. That is, its coefficients must all equal 0, so $c_0 = c_1 = \cdots = c_n = 0$, so the columns of A form a linearly independent set, which implies that A is invertible via Theorem 2.3.5.

The factor theorem, as well as other useful polynomial-related facts, are covered in Appendix A.2.1.

On the other hand, if $a_0, a_1, \ldots, a_n$ are *not* distinct then two of the rows of V are identical, so the rows of V do not form a linearly independent set, so V is not invertible.

Remark 2.3.1

Polynomial Interpolation

We assume throughout this remark that $x_0, x_1, \ldots, x_n$ are distinct.

Suppose we are given a set of points (x_0, y_0), (x_1, y_1), ..., (x_n, y_n) in $\mathbb{R}^2$ and we would like to find a polynomial whose graph goes through all of those points (in other words, we want to find a polynomial p for which $p(x_0) = y_0$, $p(x_1) = y_1$, ..., $p(x_n) = y_n$).

Since we can find a line going through any two points, the $n = 1$ case of this problem can be solved by a polynomial p of degree 1:

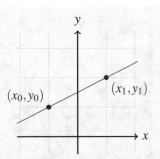

More generally, we can use the invertibility of Vandermonde matrices to show that there exists a unique degree-n polynomial going through any given set of $n+1$ points—this is called the **interpolating polynomial** of those points. So just like there is a unique line going through any 2 points, there is a unique parabola going through any 3 points, a unique cubic going through any 4 points, and so on:

Actually, the interpolating polynomial might have leading coefficient 0 and thus degree less than n.

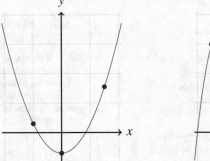
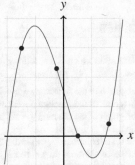

To see why such a polynomial exists (and is unique), recall that a degree-n polynomial p can be written in the form

$$p(x) = c_n x^n + \cdots + c_2 x^2 + c_1 x + c_0$$

for some scalars $c_0, c_1, \ldots, c_n$. The fact that we are requiring $p(x_0) = y_0$, $p(x_1) = y_1$, ..., $p(x_n) = y_n$, can be written explicitly as the linear system

For example, $p(x_0) = y_0$ can be written more explicitly as $c_n x_0^n + \cdots + c_0 = y_0$, which is the first equation in this linear system.

$$\begin{bmatrix} 1 & x_0 & x_0^2 & \cdots & x_0^n \\ 1 & x_1 & x_1^2 & \cdots & x_1^n \\ \vdots & \vdots & \vdots & \ddots & \vdots \\ 1 & x_n & x_n^2 & \cdots & x_n^n \end{bmatrix} \begin{bmatrix} c_0 \\ c_1 \\ \vdots \\ c_n \end{bmatrix} = \begin{bmatrix} y_0 \\ y_1 \\ \vdots \\ y_n \end{bmatrix},$$

where we recognize the coefficient matrix on the left as a Vandermonde matrix. We showed in Example 2.3.8 that this matrix is invertible whenever $x_0, x_1, \ldots, x_n$ are distinct. It follows that this linear system has a unique solution, so there is a unique polynomial p of degree n whose graph goes through the desired points.

For example, to find the quadratic (i.e., degree-2) polynomial whose graph goes through the points $(-2, 13)$, $(1, -2)$, and $(3, 8)$, we construct

This linear system is constructed by placing powers of the x-coordinates in the coefficient matrix and the y-coordinates in the right-hand-side vector.

the linear system

$$\begin{bmatrix} 1 & -2 & 4 \\ 1 & 1 & 1 \\ 1 & 3 & 9 \end{bmatrix} \begin{bmatrix} c_0 \\ c_1 \\ c_2 \end{bmatrix} = \begin{bmatrix} 13 \\ -2 \\ 8 \end{bmatrix},$$

which has unique solution $(c_0, c_1, c_2) = (-1, -3, 2)$. The desired quadratic (i.e., the interpolating polynomial of the 3 given points) is thus

$$p(x) = c_2 x^2 + c_1 x + c_0 = 2x^2 - 3x - 1.$$

To double-check that this polynomial does indeed interpolate those points, we could plug values into p to see that it is indeed the case that $p(-2) = 13$, $p(1) = -2$, and $p(3) = 8$.

Exercises

solutions to starred exercises on page 451

2.3.1 Determine which of the following sets are and are not subspaces.

 *(a) The graph of the function $y = x + 1$ in $\mathbb{R}^2$.
 (b) The graph of the function $y = 3x$ in $\mathbb{R}^2$.
 *(c) The graph of the function $y = \sin(x)$ in $\mathbb{R}^2$.
 (d) The set of unit vectors in $\mathbb{R}^3$.
 *(e) The set of solutions (x, y, z) to $x + 2y + 3z = 4$.
 (f) The set of solutions (x, y, z) to $x - y + 8z = 0$.
 *(g) $\{(x, y) \in \mathbb{R}^2 : x + 2y = 0\}$
 (h) $\{(x, y) \in \mathbb{R}^2 : x + y \geq 0\}$
 *(i) $\{(x, y) \in \mathbb{R}^2 : xy \geq 0\}$
 (j) $\{(x, y, z) \in \mathbb{R}^3 : xy + yz = 0\}$

2.3.2 For each of the following sets of vectors in $\mathbb{R}^2$, determine whether its span is a line or all of $\mathbb{R}^2$. If it is a line then give an equation of that line.

 *(a) $\{(1, 1), (2, 2), (3, 3), (-1, -1)\}$
 (b) $\{(2, 3), (0, 0)\}$
 *(c) $\{(1, 2), (2, 1)\}$

2.3.3 For each of the following sets of vectors in $\mathbb{R}^3$, determine whether its span is a line, a plane, or all of $\mathbb{R}^3$. If it is a line then say which direction it points in, and if it is a plane then give an equation of that plane (recall that an equation of a plane has the form $ax + by + cz = d$).

 *(a) $\{(1, 1, 1), (0, 0, 0), (-2, -2, -2)\}$
 (b) $\{(0, 1, -1), (1, 2, 1), (3, -1, 4)\}$
 *(c) $\{(1, 2, 1), (0, 1, -1), (2, 5, 1)\}$
 (d) $\{(1, 1, 0), (1, 0, -1), (0, 1, 1), (1, 2, 1)\}$

2.3.4 Describe the range and null space of the following matrices geometrically. That is, determine whether they are $\{\mathbf{0}\}$, a line, or all of $\mathbb{R}^2$. If they are a line then give an equation of that line.

 *(a) $\begin{bmatrix} 1 & 1 \\ 1 & 1 \end{bmatrix}$
 (b) $\begin{bmatrix} 1 & 2 \\ 2 & 4 \end{bmatrix}$
 *(c) $\begin{bmatrix} 0 & 1 \\ 2 & 3 \end{bmatrix}$
 (d) $\begin{bmatrix} 0 & 0 \\ 0 & 0 \end{bmatrix}$

2.3.5 Describe the range and null space of the following matrices geometrically. That is, determine whether they are $\{\mathbf{0}\}$, a line, a plane, or all of $\mathbb{R}^3$. If they are a line then say which direction they point in, and if they are a plane then give an equation of that plane (recall that an equation of a plane has the form $ax + by + cz = d$).

 *(a) $\begin{bmatrix} 1 & 1 & 2 \\ 1 & 1 & 2 \\ 2 & 2 & 4 \end{bmatrix}$
 (b) $\begin{bmatrix} 1 & 0 & 0 \\ 0 & 1 & 1 \\ 0 & 1 & 1 \end{bmatrix}$
 *(c) $\begin{bmatrix} 1 & 1 & 1 \\ 1 & 1 & 0 \\ 1 & 0 & 0 \end{bmatrix}$
 (d) $\begin{bmatrix} 1 & 2 & 3 \\ 2 & 3 & 4 \\ 3 & 4 & 6 \end{bmatrix}$

2.3.6 Determine which of the following sets are and are not linearly independent.

 *(a) $\{(1, 2), (3, 4)\}$
 (b) $\{(1, 0, 1), (1, 1, 1)\}$
 *(c) $\{(1, 0, -1), (1, 1, 1), (1, 2, -1)\}$
 (d) $\{(1, 2, 3), (4, 5, 6), (7, 8, 9)\}$
 *(e) $\{(1, 1), (2, 1), (3, -2)\}$
 (f) $\{(2, 1, 0), (0, 0, 0), (1, 1, 2)\}$
 *(g) $\{(1, 2, 4, 1), (2, 4, -1, 3), (-1, 1, 1, -1)\}$
 (h) $\{(0, 1, 1, 1), (1, 0, 1, 1), (1, 1, 0, 1), (1, 1, 1, 0)\}$

💻 **2.3.7** Use computer software to determine which of the following sets of vectors span all of $\mathbb{R}^4$.

 *(a) $\{(1, 2, 3, 4), (3, 1, 4, 2), (2, 4, 1, 3), (4, 3, 2, 1)\}$
 (b) $\{(4, 2, 5, 2), (3, 1, 2, 4), (1, 4, 2, 3), (3, 1, 4, 2)\}$
 *(c) $\{(4, 4, 4, 3), (3, 3, -1, 1), (-1, 2, 1, 2),$
    $(1, 0, 1, -1), (3, 3, 2, 2)\}$
 (d) $\{(2, -1, 4, -1), (3, 1, 2, 1), (0, 1, 3, 4),$
    $(-1, -1, 5, 2), (1, 3, 1, 6)\}$

💻 **2.3.8** Use computer software to determine which of the following sets are and are not linearly independent.

 *(a) $\{(1, 2, 3, 4), (3, 1, 4, 2), (2, 4, 1, 3), (4, 3, 2, 1)\}$
 (b) $\{(3, 5, 1, 4), (4, 4, 5, 5), (5, 0, 4, 3), (1, 1, 5, -1)\}$
 *(c) $\{(5, 4, 5, 1, 5), (4, 3, 3, 0, 4), (-1, 0, 3, -1, 4)\}$
 (d) $\{(5, -1, 2, 4, 3), (4, -8, 1, -4, -9), (2, 2, 1, 4, 5)\}$

2.3.9 Determine which of the following statements are true and which are false.

*(a) If vectors $\mathbf{v}_1, \mathbf{v}_2, \ldots, \mathbf{v}_k \in \mathbb{R}^n$ are such that no two of these vectors are scalar multiples of each other then they must be linearly independent.

(b) Suppose the vectors $\mathbf{v}_1, \mathbf{v}_2, \ldots, \mathbf{v}_k \in \mathbb{R}^n$ are drawn, head-to-tail (i.e., the head of $\mathbf{v}_1$ at the tail of $\mathbf{v}_2$, the head of $\mathbf{v}_2$ at the tail of $\mathbf{v}_3$, and so on). If the vectors form a closed loop (i.e., the head of $\mathbf{v}_k$ is at the tail of $\mathbf{v}_1$), then they must be linearly dependent.

*(c) The set containing just the zero vector, $\{\mathbf{0}\}$, is a subspace of $\mathbb{R}^n$.

(d) $\mathbb{R}^n$ is a subspace of $\mathbb{R}^n$.

*(e) If $\mathbf{v}, \mathbf{w} \in \mathbb{R}^3$ then $\mathrm{span}(\mathbf{v}, \mathbf{w})$ is a plane through the origin.

(f) Let $\mathbf{v}_1, \mathbf{v}_2, \ldots, \mathbf{v}_k$ be vectors in $\mathbb{R}^n$. If $k \geq n$ then it must be the case that $\mathrm{span}(\mathbf{v}_1, \mathbf{v}_2, \ldots, \mathbf{v}_k) = \mathbb{R}^n$.

2.3.10 For what values of k is the following set of vectors linearly independent?

$$\{(1,2,3), (-1,k,1), (1,1,0)\}$$

2.3.11 For which values of k, if any, do the given vectors span (i) a line, (ii) a plane, or (iii) all of $\mathbb{R}^3$?

*(a) $(0,1,-1), (1,2,1), (k,-1,4)$

(b) $(1,2,3), (3,k,k+3), (2,4,k)$

2.3.12 Use the method of Remark 2.3.1 to find a polynomial that goes through the given points.

*(a) $(2,1), (5,7)$

(b) $(1,1), (2,2), (4,4)$

*(c) $(1,1), (2,2), (4,10)$

(d) $(-1,-3), (0,1), (1,-1), (2,3)$

2.3.13 Suppose $A \in \mathcal{M}_{m,n}$ is a matrix. Show that if $\mathbf{b} \neq \mathbf{0}$ is a column vector then the solution set of the linear system $A\mathbf{x} = \mathbf{b}$ is not a subspace of $\mathbb{R}^n$. [Side note: Recall that it *is* a subspace (the null space) if $\mathbf{b} = \mathbf{0}$.]

2.3.14 Let A_n be the $n \times n$ matrix with entries $1, 2, \ldots, n^2$ written left-to-right, top-to-bottom. For example,

$$A_2 = \begin{bmatrix} 1 & 2 \\ 3 & 4 \end{bmatrix} \quad \text{and} \quad A_3 = \begin{bmatrix} 1 & 2 & 3 \\ 4 & 5 & 6 \\ 7 & 8 & 9 \end{bmatrix}.$$

Show that A_n is invertible if and only if $n \leq 2$.

[Hint: Write the third row of A_n as a linear combination of its first two rows.]

2.3.15 Show that a non-empty set $\mathcal{S}$ is a subspace of $\mathbb{R}^n$ if and only if

$$c_1\mathbf{v}_1 + \cdots + c_k\mathbf{v}_k \in \mathcal{S}$$

for all $\mathbf{v}_1, \ldots, \mathbf{v}_k \in \mathcal{S}$ and all $c_1, \ldots, c_k \in \mathbb{R}$.

2.3.16 Suppose $A = \begin{bmatrix} 0 & 1 \\ 1 & 0 \end{bmatrix}$.

(a) Find all vectors $\mathbf{v} \in \mathbb{R}^2$ with the property that $\{\mathbf{v}, A\mathbf{v}\}$ is a linearly dependent set.

(b) Interpret your answer to part (a) geometrically. In particular, use the fact that A is the standard matrix of some well-known linear transformation.

2.3.17 Use computer software to determine whether or not $\mathbf{v} = (1,2,3,4,5)$ is in the range of the given matrix.

*(a) $\begin{bmatrix} 3 & -1 & 1 & 0 & -1 \\ 2 & 1 & -1 & 4 & 0 \\ 1 & 1 & -1 & 1 & 2 \\ 2 & 1 & 2 & 0 & 0 \\ -1 & 3 & -1 & 3 & 2 \end{bmatrix}$

(b) $\begin{bmatrix} 4 & 1 & 2 & 3 & 2 & 2 & -1 \\ 0 & -1 & 4 & 2 & 0 & 4 & 2 \\ -1 & 2 & 0 & 2 & 4 & 3 & 3 \\ 3 & 1 & -1 & 0 & 2 & 2 & 3 \\ -1 & 0 & -1 & 4 & 3 & 0 & 3 \end{bmatrix}$

*2.3.18 Let $A \in \mathcal{M}_n$. Show that the set of vectors $\mathbf{v} \in \mathbb{R}^n$ satisfying $A\mathbf{v} = \mathbf{v}$ is a subspace of $\mathbb{R}^n$.

[Side note: This subspace is sometimes called the **fixed-point subspace** of A.]

2.3.19 Suppose $\mathbf{v}, \mathbf{w} \in \mathbb{R}^n$. Show that

$$\mathrm{span}(\mathbf{v}, \mathbf{w}) = \mathrm{span}(\mathbf{v}, \mathbf{v} + \mathbf{w}).$$

2.3.20 Prove that a set of two vectors $\{\mathbf{v}, \mathbf{w}\} \subset \mathbb{R}^n$ is linearly dependent if and only if $\mathbf{v}$ and $\mathbf{w}$ lie on the same line (i.e., $\mathbf{v} = c\mathbf{w}$ or $\mathbf{w} = \mathbf{0}$).

2.3.21 Prove that a non-empty set of vectors is linearly dependent if and only if at least one of the vectors in the set is a linear combination of the others.

2.3.22 Suppose that the set of vectors $\{\mathbf{v}, \mathbf{w}, \mathbf{x}\} \subseteq \mathbb{R}^n$ is linearly independent.

(a) Show that the set $\{\mathbf{v} + \mathbf{w}, \mathbf{v} + \mathbf{x}, \mathbf{w} - \mathbf{x}\}$ is linearly dependent.

(b) Show that the set $\{\mathbf{v} + \mathbf{w}, \mathbf{v} + \mathbf{x}, \mathbf{w} + \mathbf{x}\}$ is linearly independent.

(c) Determine whether or not $\{\mathbf{v}, \mathbf{v} + \mathbf{w}, \mathbf{v} + \mathbf{w} + \mathbf{x}\}$ is linearly independent, and justify your answer.

2.3.23 Suppose that $B \subseteq \mathbb{R}^n$ is a finite set of vectors. Show that if $\mathbf{0} \in B$ then B is linearly dependent.

2.3.24 Suppose that $\mathbf{w}, \mathbf{v}_1, \mathbf{v}_2, \ldots, \mathbf{v}_k \in \mathbb{R}^n$ are vectors such that $\mathbf{w}$ is a linear combination of $\mathbf{v}_1, \mathbf{v}_2, \ldots, \mathbf{v}_k$. Show that

$$\mathrm{span}(\mathbf{w}, \mathbf{v}_1, \mathbf{v}_2, \ldots, \mathbf{v}_k) = \mathrm{span}(\mathbf{v}_1, \mathbf{v}_2, \ldots, \mathbf{v}_k).$$

*2.3.25 Suppose that $B \subseteq C \subseteq \mathbb{R}^n$ are finite sets of vectors.

(a) Show that if B is linearly dependent, then so is C.

(b) Show that if C is linearly independent, then so is B.

2.3.26 Prove that a set of n vectors in $\mathbb{R}^m$ is linearly dependent whenever $n > m$.

∗∗2.3.27 Suppose that $A \in \mathcal{M}_{m,n}$ and $B \in \mathcal{M}_{n,p}$.

  (a) Show that range$(AB) \subseteq$ range(A).

  (b) Show that null$(B) \subseteq$ null(AB).

  (c) Provide an example to show that range(AB) might not be contained in range(B), and null(A) might not be contained in null(AB).

∗∗ 2.3.28 Suppose $\mathbf{v}_1, \mathbf{v}_2, \ldots, \mathbf{v}_n \in \mathbb{R}^n$ are vectors and $c_1, c_2, \ldots, c_n \in \mathbb{R}$ are non-zero scalars. Show that $\{\mathbf{v}_1, \mathbf{v}_2, \ldots, \mathbf{v}_n\}$ is linearly independent if and only if $\{c_1\mathbf{v}_1, c_2\mathbf{v}_2, \ldots, c_n\mathbf{v}_n\}$ is linearly independent.

▢ 2.3.29 Let V_n be the $n \times n$ Vandermonde matrix (see Example 2.3.8)

$$V_n = \begin{bmatrix} 1 & 1 & 1 & \cdots & 1 \\ 1 & 2 & 4 & \cdots & 2^{n-1} \\ 1 & 3 & 9 & \cdots & 3^{n-1} \\ \vdots & \vdots & \vdots & \ddots & \vdots \\ 1 & n & n^2 & \cdots & n^{n-1} \end{bmatrix}.$$

  (a) Use computer software to compute V_n^{-1} when $n = 2, 3, 4, 5$.

  (b) Based on the computations from part (a), guess a formula for the entries in the first row of V_n^{-1} that works for all n.

     [Side note: There is an explicit formula for the inverse of arbitrary Vandermonde matrices, but it is quite nasty.]

2.4   Bases and Rank

When describing a subspace as the span of a set of vectors, it is desirable to avoid redundancies. For example, it seems a bit silly to describe the plane $x + y - 3z = 0$ in $\mathbb{R}^3$ (see Figure 2.17) as span$\big((1,2,1),(2,1,1),(3,3,2)\big)$, since the vector $(3,3,2)$ is a linear combination of $(1,2,1)$ and $(2,1,1)$ and thus contributes nothing to the span—we could instead just say that the plane is span$\big((1,2,1),(2,1,1)\big)$. In this section, we explore this idea of finding a smallest possible description of a subspace.

> In particular, $(3,3,2) = (1,2,1)+(2,1,1)$.

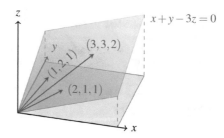

> We typically prefer to write subspaces as a span of as few vectors as possible.

Figure 2.17: The plane $x + y - 3z = 0$ is the span of the three vectors $(1,2,1)$, $(2,1,1)$, and $(3,3,2)$. However, it is also the span of any two of these three vectors, since they are all linear combinations of each other.

2.4.1   Bases and the Dimension of Subspaces

The key tool for making spanning sets as small as possible is linear independence. For example, a plane in $\mathbb{R}^3$ is spanned by any linearly independent set of two vectors on the plane (i.e., two vectors that are not parallel to each other). More than two vectors could be used to span the plane, but they would necessarily be linearly dependent. On the other hand, there is no way to use *fewer* than two vectors to span a plane, since the span of just one vector is a line (see Figure 2.18).

Two or more vectors can also span a line, if they all point in the same or opposite directions.

(a) A single vector spans a line.

(b) Two non-parallel vectors span a plane.

(c) Several vectors on a plane span that plane.

Figure 2.18: The most efficient way to express a plane as a span of vectors on that plane is to use two non-parallel vectors, since one vector is too few (it spans just a line) and three or more non-parallel vectors are redundant (all vectors after the second one tell us nothing new about the plane).

For planes in $\mathbb{R}^3$, it is thus the case that every linearly independent set contains at most 2 vectors, whereas every spanning set contains at least 2 vectors. A similar size trade-off occurs in general for all subspaces (as we will prove later in this section), which leads to the following definition:

Definition 2.4.1

Bases

A **basis** of a subspace $\mathcal{S} \subseteq \mathbb{R}^n$ is a set of vectors in $\mathcal{S}$ that

a) spans $\mathcal{S}$, and

b) is linearly independent.

A basis is not too big and not too small—it's just right.

The idea of a basis is that it is a set that is "big enough" to span the subspace, but it is not "so big" that it contains redundancies. That is, it is "just" big enough to span the subspace.

We actually already saw some bases in previous sections of this book. For example, the set $\{\mathbf{e}_1, \mathbf{e}_2, \ldots, \mathbf{e}_n\} \subset \mathbb{R}^n$ of standard basis vectors is (as their name suggests) indeed a basis of $\mathbb{R}^n$, and it is called the **standard basis**. To verify that it is a basis, we need to show (a) that its span is $\mathbb{R}^n$ (which we already showed immediately after introducing spans in Section 2.3.2), and (b) that $\{\mathbf{e}_1, \mathbf{e}_2, \ldots, \mathbf{e}_n\}$ is linearly independent. To see linear independence, suppose that

$$c_1\mathbf{e}_1 + c_2\mathbf{e}_2 + \cdots + c_n\mathbf{e}_n = \mathbf{0}.$$

The vector on the left is just $(c_1, c_2, \ldots, c_n)$, and if that vector equals $\mathbf{0}$ then $c_1 = c_2 = \cdots = c_n = 0$, which means exactly that $\{\mathbf{e}_1, \mathbf{e}_2, \ldots, \mathbf{e}_n\}$ is linearly independent. We now look at some less trivial examples of bases of subspaces.

Example 2.4.1

Showing That a Set is a Basis

Show that the following sets B are bases of the indicated subspace $\mathcal{S}$.

a) $B = \{(1,2,1),(2,1,1)\}$, $\mathcal{S}$ is the plane $x+y-3z=0$ in $\mathbb{R}^3$.

b) $B = \{(1,1,2),(1,2,1),(2,1,1)\}$, $\mathcal{S} = \mathbb{R}^3$.

Solutions:

Recall that a set $\{\mathbf{v}, \mathbf{w}\}$ with two vectors is linearly dependent if and only if $\mathbf{v}$ and $\mathbf{w}$ are multiples of each other.

a) We showed in Example 2.3.4(b) that $\mathrm{span}\big((1,2,1),(2,1,1)\big)$ is exactly this plane, so we just need to show linear independence. Since there are only two vectors in this set, linear independence follows from the fact that $(1,2,1)$ and $(2,1,1)$ are not multiples of each other.

b) We have to show that $\mathrm{span}(B) = \mathbb{R}^3$ and that B is linearly independent. To see that B spans $\mathbb{R}^3$, we find which vectors (x,y,z) are in the span using the same method of Example 2.3.4:

$$\begin{bmatrix} 1 & 1 & 2 & | & x \\ 1 & 2 & 1 & | & y \\ 2 & 1 & 1 & | & z \end{bmatrix} \xrightarrow[R_3-2R_1]{R_2-R_1} \begin{bmatrix} 1 & 1 & 2 & | & x \\ 0 & 1 & -1 & | & y-x \\ 0 & -1 & -3 & | & z-2x \end{bmatrix}$$

$$\xrightarrow{R_3+R_2} \begin{bmatrix} 1 & 1 & 2 & | & x \\ 0 & 1 & -1 & | & y-x \\ 0 & 0 & -4 & | & z+y-3x \end{bmatrix}.$$

Now that this linear system is in row echelon form, we see that it has a solution no matter what x, y, and z are (their values only change what the solution *is*, not whether or not a solution *exists*), so $\text{span}(B) = \mathbb{R}^3$.

Next, we have to show that B is linearly independent. To do so, we place the vectors from B as columns into a matrix, augment with a **0** right-hand side, and solve (i.e., we use Theorem 2.3.4):

Notice that the linear independence calculation is identical to the span calculation that we did above, except for the augmented right-hand side.

$$\begin{bmatrix} 1 & 1 & 2 & | & 0 \\ 1 & 2 & 1 & | & 0 \\ 2 & 1 & 1 & | & 0 \end{bmatrix} \xrightarrow[R_3-2R_1]{R_2-R_1} \begin{bmatrix} 1 & 1 & 2 & | & 0 \\ 0 & 1 & -1 & | & 0 \\ 0 & -1 & -3 & | & 0 \end{bmatrix}$$

$$\xrightarrow{R_3+R_2} \begin{bmatrix} 1 & 1 & 2 & | & 0 \\ 0 & 1 & -1 & | & 0 \\ 0 & 0 & -4 & | & 0 \end{bmatrix}.$$

From here we can see that this linear system has a unique solution, which must be the zero vector, so B is linearly independent. Since it spans $\mathbb{R}^3$ and is linearly independent, we conclude that B is a basis of $\mathbb{R}^3$.

Remark 2.4.1

The Zero Subspace and the Empty Basis

We similarly call subspaces other than $\{0\}$ **non-zero subspaces**.

Bases give us a standardized way of describing subspaces of $\mathbb{R}^n$, so we would like every subspace to have at least one. We will see in the upcoming Theorem 2.4.5 that subspaces do indeed all have bases, but there is one subspace that is worth devoting some special attention to in this regard—the **zero subspace** $\{0\}$.

The only subsets of $\{0\}$ are $\{\}$ and $\{0\}$ itself, so these are the only sets that could possibly be bases of it. However, we showed in Exercise 2.3.23 that $\{0\}$ is linearly dependent, so the only possible basis of the zero subspace $\{0\}$ is the empty set $\{\}$.

Fortunately, $\{\}$ is indeed a basis of $\{0\}$. To see that it is linearly independent, notice that if we choose $B = \{\}$ (i.e., $k = 0$) in Definition 2.3.4 then there is no non-zero linear combination of the members of B adding up to **0**, so it is not linearly dependent. Similarly, the span of the empty set $\{\}$ is indeed $\{0\}$ (not $\{\}$), since every linear combination of the members of $\{\}$ is an "empty sum" (i.e., a sum with no terms in it). Empty sums are chosen to equal **0** simply by definition.

In general, bases are very non-unique—every subspace other than $\{0\}$ has infinitely many different bases. For example, we already showed that $\mathbb{R}^3$ has at least two bases: $\{e_1, e_2, e_3\}$ and $\{(1,1,2),(1,2,1),(2,1,1)\}$ (from Example 2.4.1(b)). To obtain even more bases of $\mathbb{R}^3$, we could just rotate any one vector from an existing basis, being careful not to rotate it into the plane spanned by the other two basis vectors (see Figure 2.19).

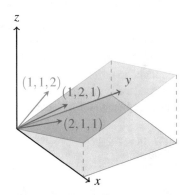

Figure 2.19: The set $\{(1,1,2),(1,2,1),(2,1,1)\}$ is a basis of $\mathbb{R}^3$, and it will remain a basis of $\mathbb{R}^3$ even if we change $(1,1,2)$ to any vector that is not a linear combination of (i.e., not in the planed spanned by) the other two vectors.

However, one thing that never changes is the *number* of vectors in a basis. For example, the fact that all of the bases of $\mathbb{R}^3$ that we have seen consist of exactly 3 vectors is not a coincidence—they all have exactly that many vectors. To prove this claim, we first need the following theorem that pins down the idea that linearly independent sets are "small", while spanning sets are "big".

Theorem 2.4.1

Linearly Independent Sets Versus Spanning Sets

Let S be a subspace of $\mathbb{R}^n$, and suppose that $B, C \subseteq S$ are finite sets with the properties that B is linearly independent and $\text{span}(C) = S$. Then

$$|B| \leq |C|.$$

Proof. We begin by giving names to the vectors in B and C: we write $B = \{\mathbf{v}_1, \mathbf{v}_2, \ldots, \mathbf{v}_k\}$ and $C = \{\mathbf{w}_1, \mathbf{w}_2, \ldots, \mathbf{w}_\ell\}$, where $k = |B|$ and $\ell = |C|$.

The notation $|B|$ means the number of vectors in the set B.

We prove the result by showing that if $\ell < k$ and C spans S, then B must in fact be linearly *dependent*. To prove this claim, we observe that since C spans S, and $\mathbf{v}_i \in B \subseteq S$ for all $1 \leq i \leq k$, we can write each $\mathbf{v}_i$ as a linear combination of $\mathbf{w}_1, \mathbf{w}_2, \ldots, \mathbf{w}_\ell$:

$$\mathbf{v}_i = a_{i,1}\mathbf{w}_1 + a_{i,2}\mathbf{w}_2 + \cdots + a_{i,\ell}\mathbf{w}_\ell \quad \text{for} \quad 1 \leq i \leq k, \tag{2.4.1}$$

where each of the $a_{i,j}$'s $(1 \leq i \leq k, 1 \leq j \leq \ell)$ is a scalar.

To make the rest of the proof simpler, we place these scalars into vectors and matrices in the "usual" way—we define $A \in \mathcal{M}_{k,\ell}$ to be the matrix whose (i,j)-entry is $a_{i,j}$, $V \in \mathcal{M}_{n,k}$ to be the matrix whose columns are $\mathbf{v}_1, \mathbf{v}_2, \ldots, \mathbf{v}_k$,

In words, this theorem says that in any subspace, every linearly independent set is smaller than every spanning set.

and $W \in \mathcal{M}_{n,\ell}$ to be the matrix whose columns are $\mathbf{w}_1, \mathbf{w}_2, \ldots, \mathbf{w}_\ell$. We can then succinctly rewrite the linear system (2.4.1) as the single matrix equation $V = WA^T$, since block matrix multiplication shows that the i-th column of WA^T equals

$$a_{i,1}\mathbf{w}_1 + a_{i,2}\mathbf{w}_2 + \cdots + a_{i,\ell}\mathbf{w}_\ell,$$

which equals $\mathbf{v}_i$ (the i-th column of V).

Next, Corollary 2.1.4 tells us that there is a non-zero solution $\mathbf{x} \in \mathbb{R}^k$ to the linear system $A^T\mathbf{x} = \mathbf{0}$, since A^T has more columns than rows. It follows that

$$V\mathbf{x} = (WA^T)\mathbf{x} = W(A^T\mathbf{x}) = W\mathbf{0} = \mathbf{0}.$$

Finally, since the linear system $V\mathbf{x} = \mathbf{0}$ has a non-zero solution, Theorem 2.3.4 tells us that the set $B = \{\mathbf{v}_1, \mathbf{v}_2, \ldots, \mathbf{v}_k\}$ consisting of the columns of V is linearly dependent, which completes the proof. ∎

The previous theorem roughly says that there is a sort of tug-of-war that happens between linear independent and spanning sets—a set has to be "small enough" in order for it to be linearly independent, while it has to be "big enough" in order for it to span the subspace it lives in. Since bases require that *both* of these properties hold, their size is actually completely determined by the subspace that they span.

Corollary 2.4.2	Suppose S is a subspace of $\mathbb{R}^n$. Every basis of S has the same number of vectors.
Uniqueness of Size of Bases	

Proof. This result follows immediately from Theorem 2.4.1: if $B \subseteq S$ and $C \subseteq S$ are bases of S then $|B| \leq |C|$ since B is linearly independent and C spans S. On the other hand, $|C| \leq |B|$ since C is linearly independent and B spans S. It follows that $|B| = |C|$. ∎

Until now, we have never actually defined exactly what we mean by the "dimension" of a subspace of $\mathbb{R}^n$. While this concept is intuitive enough in the cases that we can visualize (i.e., dimensions 1, 2, and 3), one of the main uses of the above corollary is that it lets us unambiguously extend this concept to larger subspaces that we cannot draw pictures of.

Definition 2.4.2	Suppose S is a subspace of $\mathbb{R}^n$. The number of vectors in a basis of S is called the **dimension** of S and is denoted by $\dim(S)$.
Dimension of a Subspace	

While the above definition of dimension might seem a bit strange at first, it actually matches our intuitive notion of dimension very well. For example, $\dim(\mathbb{R}^n) = n$ since the standard basis of $\mathbb{R}^n$ (i.e., the set $\{\mathbf{e}_1, \mathbf{e}_2, \ldots, \mathbf{e}_n\}$) contains exactly n vectors. Similarly, lines are 1-dimensional since a single vector acts as a basis of a line, planes in $\mathbb{R}^3$ are 2-dimensional since two non-parallel vectors form a basis of a plane (see Figure 2.20), and the zero subspace $\{\mathbf{0}\}$ is 0-dimensional as a result of Remark 2.4.1 (which agrees with our intuition that a single point is 0-dimensional).

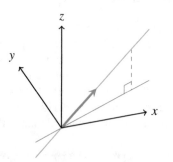

(a) A basis of a line contains just 1 vector, so lines are 1-dimensional.

(b) A basis of a plane contains 2 vectors, so planes are 2-dimensional.

Figure 2.20: Our new definition of dimension agrees with our intuitive understanding of dimension: lines and planes are 1- and 2-dimensional, respectively, as we would expect.

Furthermore, by combining Theorem 2.4.1 with Definition 2.4.2, we see that not only are linearly independent sets "small" and spanning sets "large", but the point that separates "small" from "large" is exactly the dimension of the subspace that is being worked in. That is, for every subspace S of $\mathbb{R}^n$, the following fact holds:

(!)   size of a linearly independent set $\leq \dim(S) \leq$ size of a spanning set

So, for example, any linearly independent set that is contained within a plane has at *most* 2 vectors, and any set that spans a plane has at *least* 2 vectors.

Example 2.4.2

Finding a Basis and Dimension

Find a basis of the following subspaces S of $\mathbb{R}^3$ and thus compute their dimension.

  a) $S = \operatorname{span}\big((1,1,1),(1,2,3),(3,2,1)\big)$.

  b) The set S of vectors (x,y,z) satisfying the equation $2x - y + 3z = 0$.

Solutions:

  a) It seems natural to first guess that $B = \big\{(1,1,1),(1,2,3),(3,2,1)\big\}$ might be a basis of this subspace. By the definition of S, we already know that $S = \operatorname{span}(B)$, so we just need to check linear independence of B, which we do in the usual way by placing the vectors from B into a matrix as columns, and augmenting with a $\mathbf{0}$ right-hand side:

$$\left[\begin{array}{ccc|c} 1 & 1 & 3 & 0 \\ 1 & 2 & 2 & 0 \\ 1 & 3 & 1 & 0 \end{array}\right] \xrightarrow[R_3 - R_1]{R_2 - R_1} \left[\begin{array}{ccc|c} 1 & 1 & 3 & 0 \\ 0 & 1 & -1 & 0 \\ 0 & 2 & -2 & 0 \end{array}\right]$$

$$\xrightarrow{R_3 - 2R_2} \left[\begin{array}{ccc|c} 1 & 1 & 3 & 0 \\ 0 & 1 & -1 & 0 \\ 0 & 0 & 0 & 0 \end{array}\right].$$

Be slightly careful here: the discarded vector must be a linear combination of the other vectors in the set. In this case, any of the vectors can be discarded, but sometimes we have to be careful (see Exercise 2.4.23).

We see from the above row echelon form that this system of equations has infinitely many solutions, so B is a linearly *dependent* set. We can thus discard one of the vectors from B without affecting $\operatorname{span}(B)$. If we (arbitrarily) decide to discard $(1,1,1)$ then we arrive at the set $C = \{(1,2,3),(3,2,1)\}$, which has the same span as B. Furthermore, C is linearly independent since $(1,2,3)$ and $(3,2,1)$ are not multiples of each other. It follows that C is a basis of S, and S is 2-dimensional (i.e., a plane) since C contains 2 vectors.

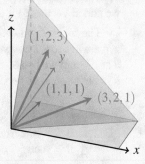

b) Unlike in part (a), we do not have an "obvious" choice of basis to use as our starting point here. Instead, we start by picking (arbitrarily) the vector $(1, -1, -1)$, which we can see is in $\mathcal{S}$ since $2(1) - (-1) + 3(-1) = 0$. However, this vector by itself does not span $\mathcal{S}$ since we can find a vector in $\mathcal{S}$ but not in $\text{span}((1, -1, -1))$—in particular, $(3, 3, -1)$ is one such vector. If we set $B = \{(1, -1, -1), (3, 3, -1)\}$, then we can see that B is linearly independent since the two vectors in B are not multiples of each other. To check whether or not $\text{span}(B) = \mathcal{S}$, we check which vectors (x, y, z) are linear combinations of the members of B:

To see that $(3, 3, -1) \in \mathcal{S}$ we compute $2(3) - (3) + 3(-1) = 0$.

$$\begin{bmatrix} 1 & 3 & x \\ -1 & 3 & y \\ -1 & -1 & z \end{bmatrix} \xrightarrow[R_3+R_1]{R_2+R_1} \begin{bmatrix} 1 & 3 & x \\ 0 & 6 & x+y \\ 0 & 2 & x+z \end{bmatrix}$$

$$\xrightarrow{R_3-\frac{1}{3}R_2} \begin{bmatrix} 1 & 3 & x \\ 0 & 6 & x+y \\ 0 & 0 & \frac{2}{3}x - \frac{1}{3}y + z \end{bmatrix}.$$

The above system of linear equations has a solution (and thus (x, y, z) is in $\text{span}(B)$) if and only if $\frac{2}{3}x - \frac{1}{3}y + z = 0$. Multiplying this equation by 3 shows that $2x - y + 3z = 0$, so $\text{span}(B)$ is exactly the subspace $\mathcal{S}$ and B is a basis of $\mathcal{S}$. Furthermore, $\dim(\mathcal{S}) = 2$ since B contains 2 vectors.

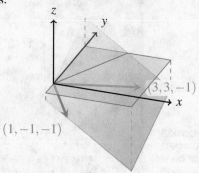

In general, any two vectors on a plane that are not multiples of each other form a basis of that plane.

When constructing bases of subspaces, it is very useful to be able to tweak linearly independent or spanning sets that we have already found. For example, in Example 2.4.2(a) we started with a set B that spanned a subspace $\mathcal{S}$, and we discarded vectors from B until it was small enough to be a basis of $\mathcal{S}$. On the other hand, in Example 2.4.2(b) we started with a linearly independent set and we added vectors to it until it was large enough to be a basis of $\mathcal{S}$.

The following theorem says that both of these procedures always work—we can always toss away vectors from a spanning set until it is small enough to be a basis, and we can always add new vectors to a linearly independent set until it is big enough to be a basis.

Theorem 2.4.3

Creating a Basis from a Set of Vectors

Suppose $\mathcal{S}$ is a subspace of $\mathbb{R}^n$ and $B \subseteq \mathcal{S}$ is a finite set of vectors.

a) If B is linearly independent then there is a basis C of $\mathcal{S}$ with $B \subseteq C$.

b) If B spans $\mathcal{S}$ then there is a basis C of $\mathcal{S}$ with $C \subseteq B$.

Proof. Suppose $B = \{\mathbf{v}_1, \mathbf{v}_2, \ldots, \mathbf{v}_k\}$. To prove part (a), notice that if $\text{span}(B) = \mathcal{S}$ then we are done: B is linearly independent and spans $\mathcal{S}$, and is thus a basis

Part (a) of this
theorem is
sometimes phrased
as "any linearly
independent set
can be extended to
a basis".

of it. On the other hand, if it does not span S then there exists a vector $\mathbf{w} \in S$ that is not a linear combination of $\mathbf{v}_1, \mathbf{v}_2, \ldots, \mathbf{v}_k$.

Our next step is to show that the set $\{\mathbf{w}, \mathbf{v}_1, \mathbf{v}_2, \ldots, \mathbf{v}_k\}$ is linearly independent. To show this, we suppose that

$$d\mathbf{w} + c_1\mathbf{v}_1 + c_2\mathbf{v}_2 + \cdots + c_k\mathbf{v}_k = \mathbf{0},$$

with the goal of showing that $d = c_1 = c_2 = \cdots = c_k = 0$. Well, $d = 0$ since otherwise we could rearrange the equation to show that $\mathbf{w}$ is a linear combination of $\mathbf{v}_1, \mathbf{v}_2, \ldots, \mathbf{v}_k$. It is thus the case that

$$c_1\mathbf{v}_1 + c_2\mathbf{v}_2 + \cdots + c_k\mathbf{v}_k = \mathbf{0}.$$

However, since $\{\mathbf{v}_1, \mathbf{v}_2, \ldots, \mathbf{v}_k\}$ is a linearly independent set, this means that $c_1 = c_2 = \cdots = c_k = 0$ too, so $\{\mathbf{w}, \mathbf{v}_1, \mathbf{v}_2, \ldots, \mathbf{v}_k\}$ is indeed linearly independent.

We then repeat: if $\{\mathbf{w}, \mathbf{v}_1, \mathbf{v}_2, \ldots, \mathbf{v}_k\}$ spans S then it must be a basis of S, whereas if it does not span S we can add another vector to it while preserving linear independence. Since S is contained in $\mathbb{R}^n$, we cannot find a linearly independent subset of it containing more than n vectors, so this process eventually terminates—we must arrive at a linearly independent set $C \subseteq S$ (with no more than n vectors) that spans S, and is thus a basis of it.

The proof of part (b) of the theorem is quite similar, and is thus left as Exercise 2.4.22. ∎

If we know the dimension of a subspace S in advance (as is often the case with lines, planes, and other "familiar" subspaces), we can save ourselves quite a bit of work when trying to determine whether or not a particular set is a basis of S. The following theorem summarizes the simplest way to solve this problem. In particular, it shows that if the size of the proposed basis is right then we only have to check one of the two defining properties of bases (spanning or linear independence)—the other property comes for free.

Theorem 2.4.4

Determining if a Set is a Basis

> Suppose S is a subspace of $\mathbb{R}^n$ and $B \subseteq S$ is a set containing k vectors.
>
> a) If $k \neq \dim(S)$ then B is not a basis of S.
> b) If $k = \dim(S)$ then the following are equivalent:
>   i) B spans S,
>   ii) B is linearly independent, and
>   iii) B is a basis of S.

Proof. Part (a) of the theorem follows immediately from the definition of dimension (Definition 2.4.2) and the fact that all bases have the same size (Corollary 2.4.2).

For part (b), we note that condition (iii) immediately implies conditions (i) and (ii), since all bases (by definition) are linearly independent and span S.

A lot of the proofs
involving bases just
involve definition
chasing. We do not
do anything
particularly clever in
these proofs, but
rather just string
together definitions
as appropriate.

To see that condition (ii) implies condition (iii), we note that Theorem 2.4.3(a) says that, since B is linearly independent, we can add 0 or more vectors to B to create a basis of S. However, B already has k vectors and $\dim(S) = k$, so all bases of S have exactly k vectors. It follows that the only possibility is that B becomes a basis when we add 0 vectors to it—i.e., B itself is already a basis of S.

The proof that condition (i) implies condition (iii) is similar, and left as Exercise 2.4.24. ∎

Example 2.4.3

Determining if a Set is a Basis

Determine whether or not the given set B is a basis of $\mathbb{R}^3$.

a) $B = \{(1,2,3),(4,1,2)\}$

b) $B = \{(1,-1,-1),(2,2,-1),(1,1,0)\}$

Solutions:

a) B is not a basis of $\mathbb{R}^3$ since $\mathbb{R}^3$ is 3-dimensional, but B contains only 2 vectors.

b) Since $\mathbb{R}^3$ is 3-dimensional and B contains 3 vectors, we conclude that B *might* be a basis—we just have to make sure that it is linearly independent or spans all of $\mathbb{R}^3$. We decide to check that it is linearly independent, which we do by placing the vectors from B as columns into a matrix and augmenting with a $\mathbf{0}$ right-hand side as usual:

Checking linear independence is a bit easier than checking that a set is spanning, since the right-hand side of the linear system then has the zero vector $\mathbf{0}$ rather than an arbitrary vector $\mathbf{x}$.

$$\left[\begin{array}{ccc|c} 1 & 2 & 1 & 0 \\ -1 & 2 & 1 & 0 \\ -1 & -1 & 0 & 0 \end{array}\right] \xrightarrow[R_3+R_1]{R_2+R_1} \left[\begin{array}{ccc|c} 1 & 2 & 1 & 0 \\ 0 & 4 & 2 & 0 \\ 0 & 1 & 1 & 0 \end{array}\right]$$

$$\xrightarrow{R_3-\frac{1}{4}R_2} \left[\begin{array}{ccc|c} 1 & 2 & 1 & 0 \\ 0 & 4 & 2 & 0 \\ 0 & 0 & 1/2 & 0 \end{array}\right].$$

Since the above linear system has a unique solution, we conclude that B is indeed linearly independent, so it is a basis of $\mathbb{R}^3$.

Example 2.4.4

Determining if a Set is a Basis... Again

Determine whether or not $B = \{(1,-1,-1),(2,2,-1)\}$ is a basis of the plane $S \subset \mathbb{R}^3$ defined by the equation $3x - y + 4z = 0$.

Solution:

This plane is 2-dimensional and B contains 2 vectors, so B *might* be a basis of S. To confirm that it is a basis, we first double-check that its members are actually in S in the first place, which we do by verifying that they both satisfy the given equation of the plane:

$$3(1) - (-1) + 4(-1) = 0 \quad \text{and} \quad 3(2) - (2) + 4(-1) = 0.$$

With that technicality out of the way, we now just need to check that B either spans all of S or that B is linearly independent. Checking linear independence is much easier in this case, since we can just notice that $(1,-1,-1)$ and $(2,2,-1)$ are not multiples of each other, so B is indeed a basis of S.

Up until now, we have not actually showed that all subspaces *have* bases. We now fill this gap and show that we can indeed construct a basis of any subspace. Furthermore, a basis can be found via the procedure that we used to find one in Example 2.4.2(b): we just pick some vector in the subspace, then pick another one that is not a linear combination of the vector we already picked, and so on until we have enough that they span the whole subspace.

Theorem 2.4.5

Existence of Bases

Every subspace of $\mathbb{R}^n$ has a basis.

Proof. We start by choosing a largest possible linearly independent set of vectors in the subspace $\mathcal{S}$, and calling it $B = \{\mathbf{v}_1, \mathbf{v}_2, \ldots, \mathbf{v}_m\}$ (a largest such set indeed exists, and has $m \leq n$, since no linearly independent set in $\mathbb{R}^n$ can have more than n vectors). We claim that B is in fact a basis of $\mathcal{S}$, so our goal is to show that $\mathrm{span}(B) = \mathcal{S}$.

> If $\mathcal{S} = \{\mathbf{0}\}$ is the zero subspace then the largest linearly independent set of vectors in it is the empty set $\{\}$. The rest of the proof falls apart in this case, but that's okay because we already know from Remark 2.4.1 that $\mathcal{S}$ has a basis: $\{\}$.

To verify this claim, let $\mathbf{w}$ be any vector in $\mathcal{S}$—we want to show that $\mathbf{w} \in \mathrm{span}(B)$. We know that the set $\{\mathbf{w}, \mathbf{v}_1, \mathbf{v}_2, \ldots, \mathbf{v}_m\}$ is linearly dependent since it contains $m + 1$ vectors, and the largest linearly independent set in $\mathcal{S}$ contains m vectors. We can thus find scalars $d, c_1, c_2, \ldots, c_m$ (not all equal to zero) such that

$$d\mathbf{w} + c_1\mathbf{v}_1 + c_2\mathbf{v}_2 + \cdots + c_m\mathbf{v}_m = \mathbf{0}.$$

Since B is linearly independent, we know that if $d = 0$ then $c_1 = c_2 = \cdots = c_m = 0$ too, which contradicts our assumption that not all of these scalars equal 0. We thus conclude that $d \neq 0$, so we can rearrange this equation into the form

$$\mathbf{w} = -\frac{c_1}{d}\mathbf{v}_1 - \frac{c_2}{d}\mathbf{v}_2 - \cdots - \frac{c_m}{d}\mathbf{v}_m.$$

It follows that $\mathbf{w} \in \mathrm{span}(B)$, so $\mathrm{span}(B) = \mathcal{S}$, as claimed. ∎

The above theorem can be thought of as the converse of Theorem 2.3.1: not only is every span a subspace, but also every subspace can be written as the span of a finite set of vectors. Furthermore, that set of vectors can even be chosen to be linearly independent (i.e., a basis).

2.4.2 The Fundamental Matrix Subspaces

Recall from Theorem 2.3.2 that the range of a matrix is equal to the span of its columns. However, its columns in general do not form a *basis* of its range, since they might be linearly dependent. We thus now spend some time reconsidering the range and null space of a matrix, particularly with the goal of constructing bases (not just any spanning sets) of these subspaces.

Example 2.4.5

Determining Range and Null Space

Find bases of the range and null space of $A = \begin{bmatrix} 1 & 1 & 2 \\ 2 & -1 & 1 \\ 1 & 0 & 1 \end{bmatrix}$.

Solution:

The range of A is the span of its columns:

$$\mathrm{range}(A) = \mathrm{span}\big((1,2,1),(1,-1,0),(2,1,1)\big).$$

To find a basis of $\mathrm{range}(A)$, we need to find a linearly independent subset of these columns.

> Recall that linear combinations like this one can be found by solving the linear system $(2,1,1) = c_1(1,2,1) + c_2(1,-1,0)$.

To this end, we note that $(2,1,1) = (1,2,1) + (1,-1,0)$, so we can discard $(2,1,1)$ from this set without affecting linear independence. And indeed, it is straightforward to see that $\{(1,2,1),(1,-1,0)\}$ is linearly independent (and thus a basis of $\mathrm{range}(A)$) since the two vectors in that set are not multiples of each other. If follows that $\mathrm{range}(A)$ is a plane.

On the other hand, to find a basis of $\mathrm{null}(A)$, we just solve the system of linear equations $A\mathbf{x} = \mathbf{0}$:

$$\begin{bmatrix} 1 & 1 & 2 & | & 0 \\ 2 & -1 & 1 & | & 0 \\ 1 & 0 & 1 & | & 0 \end{bmatrix} \xrightarrow[R_3 - R_1]{R_2 - 2R_1} \begin{bmatrix} 1 & 1 & 2 & | & 0 \\ 0 & -3 & -3 & | & 0 \\ 0 & -1 & -1 & | & 0 \end{bmatrix}$$

$$\xrightarrow{\frac{-1}{3}R_2} \begin{bmatrix} 1 & 1 & 2 & | & 0 \\ 0 & 1 & 1 & | & 0 \\ 0 & -1 & -1 & | & 0 \end{bmatrix} \xrightarrow[R_3 + R_2]{R_1 - R_2} \begin{bmatrix} 1 & 0 & 1 & | & 0 \\ 0 & 1 & 1 & | & 0 \\ 0 & 0 & 0 & | & 0 \end{bmatrix}.$$

From here we see that x_3 is a free variable, and we can solve for x_1 and x_2 (the leading variables) in terms of x_3: the first row tells us that $x_1 + x_3 = 0$ (i.e., $x_1 = -x_3$) and the second row tells us that $x_2 + x_3 = 0$ (i.e., $x_2 = -x_3$).

It follows that null(A) consists of the vectors of the form $(x_1, x_2, x_3) = (-x_3, -x_3, x_3) = x_3(-1, -1, 1)$, so $\{(-1, -1, 1)\}$ is a basis of null(A). These subspaces are displayed below (range(A) on the left and null(A) on the right):

We will see later (in Theorem 2.4.10) that the dimensions of the range and null space of an $n \times n$ matrix always add up to n (just like here the dimensions are $2 + 1 = 3$ for a 3×3 matrix).

To compute a basis of null(A) in the previous example, we had to compute a row echelon form of A. It turns out that we can directly find a basis of range(A) from a row echelon form as well, just by taking the columns of A that become leading columns in the REF. For example, we showed that the RREF of

$$A = \begin{bmatrix} 1 & 1 & 2 \\ 2 & -1 & 1 \\ 1 & 0 & 1 \end{bmatrix} \quad \text{is} \quad \begin{bmatrix} 1 & 0 & 1 \\ 0 & 1 & 1 \\ 0 & 0 & 0 \end{bmatrix},$$

which has leading entries in its first and second columns. Taking the first and second columns of A itself results in the set $\{(1, 2, 1), (1, -1, 0)\}$, which is a basis of range(A).

To see why this method works in general, recall that if a matrix A has row echelon form R then $\mathbf{x}$ solves the linear system $A\mathbf{x} = \mathbf{0}$ if and only if it solves the linear system $R\mathbf{x} = \mathbf{0}$ (this is why we started looking at row echelon forms in the first place—they have the same solution set but are easier to work with). In other words, a linear combination of the columns of A equals $\mathbf{0}$ if and only if the same linear combination of the columns of R equals $\mathbf{0}$.

By "the same linear combination", we mean a linear combination with the same coefficients.

It follows that a particular subset of the columns of A is a basis of range(A) if and only if that same subset of columns of R is a basis of range(R). Since the leading columns of R form a basis of range(R) (after all, the non-leading columns of R are linear combinations of its leading columns), we arrive at the following fact, which we illustrate in Figure 2.21:

(!) The columns of a matrix A that have a leading entry in one of its row echelon forms make up a basis of range(A).

For example, the three columns highlighted in blue in Figure 2.21 form a basis of range(A).

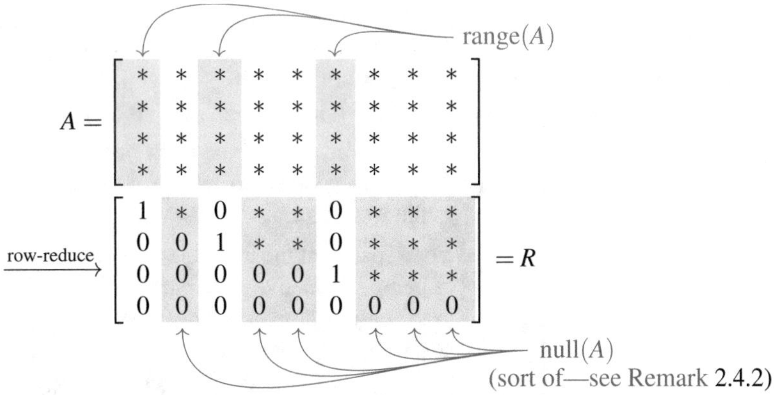

Figure 2.21: After we row-reduce A to its reduced row echelon form R, we can immediately read off a basis of range(A) from the columns of A (actually, any row echelon form of A suffices). A basis of null(A) can also be determined from R, but it is not quite as straightforward (see Remark 2.4.2).

Example 2.4.6

Determining Range and Null Space (More Efficiently)

Find bases of the range and null space of the following matrix:

$$A = \begin{bmatrix} 1 & 0 & 1 & 0 & -1 \\ 1 & 1 & 0 & 0 & 1 \\ -1 & 0 & -1 & 1 & 4 \\ 2 & 1 & 1 & -1 & -3 \end{bmatrix}$$

Solution:

We start by finding the reduced row echelon form of A:

We prefer this method of determining a basis of the range, rather than that of Example 2.4.5, since it just requires one calculation (finding the RREF) to get bases of both the range and the null space.

$$\begin{bmatrix} 1 & 0 & 1 & 0 & -1 \\ 1 & 1 & 0 & 0 & 1 \\ -1 & 0 & -1 & 1 & 4 \\ 2 & 1 & 1 & -1 & -3 \end{bmatrix} \xrightarrow[\substack{R_2-R_1 \\ R_3+R_1 \\ R_4-2R_1}]{} \begin{bmatrix} 1 & 0 & 1 & 0 & -1 \\ 0 & 1 & -1 & 0 & 2 \\ 0 & 0 & 0 & 1 & 3 \\ 0 & 1 & -1 & -1 & -1 \end{bmatrix}$$

$$\xrightarrow[R_4-R_2]{} \begin{bmatrix} 1 & 0 & 1 & 0 & -1 \\ 0 & 1 & -1 & 0 & 2 \\ 0 & 0 & 0 & 1 & 3 \\ 0 & 0 & 0 & -1 & -3 \end{bmatrix} \xrightarrow[R_4+R_3]{} \begin{bmatrix} 1 & 0 & 1 & 0 & -1 \\ 0 & 1 & -1 & 0 & 2 \\ 0 & 0 & 0 & 1 & 3 \\ 0 & 0 & 0 & 0 & 0 \end{bmatrix}.$$

Since this row echelon form has leading entries in columns 1, 2, and 4, we know that the set consisting of columns 1, 2, and 4 of A (i.e., $\{(1,1,-1,2),(0,1,0,1),(0,0,1,-1)\}$) is a basis of range$(A)$.

Be careful to use columns of the *original* matrix A when constructing a basis of range(A), not columns of a row echelon form.

To find a basis of null(A), we proceed as we did in Example 2.4.5. We think of A as the coefficient matrix in the linear system $A\mathbf{x} = \mathbf{0}$ and note that the RREF tells us that x_3 and x_5 are free variables, so we solve for x_1, x_2, and x_4 (the leading variables) in terms of them. Explicitly, we get

$$(x_1,x_2,x_3,x_4,x_5) = x_3(-1,1,1,0,0) + x_5(1,-2,0,-3,1).$$

We thus conclude that $\{(-1,1,1,0,0),(1,-2,0,-3,1)\}$ is a basis of $\text{null}(A)$.

It is actually possible to construct a basis of $\text{null}(A)$ pretty much directly from the entries of the RREF of A, without having to explicitly write down a solution of the linear system $A\mathbf{x} = \mathbf{0}$ first. The trick is as follows:

- First, extend or truncate each of the non-leading columns of a reduced row echelon form so as to have the same number of entries as the *rows* of A.

- Next, space out the non-zero entries of these vectors so that, instead of being grouped together at the top of the vector, they are located in the entries corresponding to the leading columns.

- Finally, put a -1 in the "diagonal" entry—if came from the j-th column of the RREF, put a -1 in its j-entry.

For example, if the reduced row echelon form of a matrix A is

$$\begin{bmatrix} 1 & 2 & 0 & 3 & 4 & 0 & 5 \\ 0 & 0 & 1 & 6 & 7 & 0 & 8 \\ 0 & 0 & 0 & 0 & 0 & 1 & 9 \end{bmatrix}$$

then each of the four non-leading columns correspond to one vector in a basis of $\text{null}(A)$ as follows:

In particular, we place the 3 potentially non-zero entries of each of these new vectors in their first, third, and sixth entries, since the leading columns of the RREF are its first, third, and sixth columns.

$$\begin{bmatrix} 2 \\ 0 \\ 0 \end{bmatrix} \rightarrow \begin{bmatrix} 2 \\ -1 \\ 0 \\ 0 \\ 0 \\ 0 \\ 0 \end{bmatrix}, \quad \begin{bmatrix} 3 \\ 6 \\ 0 \end{bmatrix} \rightarrow \begin{bmatrix} 3 \\ 0 \\ 6 \\ -1 \\ 0 \\ 0 \\ 0 \end{bmatrix}, \quad \begin{bmatrix} 4 \\ 7 \\ 0 \end{bmatrix} \rightarrow \begin{bmatrix} 4 \\ 0 \\ 7 \\ 0 \\ -1 \\ 0 \\ 0 \end{bmatrix}, \quad \begin{bmatrix} 5 \\ 8 \\ 9 \end{bmatrix} \rightarrow \begin{bmatrix} 5 \\ 0 \\ 8 \\ 0 \\ 0 \\ 9 \\ -1 \end{bmatrix}$$

One basis of $\text{null}(A)$ is thus

$$\{(2,-1,0,0,0,0,0),(3,0,6,-1,0,0,0),$$
$$(4,0,7,0,-1,0,0),(5,0,8,0,0,9,-1)\}.$$

It turns out that the subspaces $\text{range}(A^T)$ and $\text{null}(A^T)$ play just as important of a role as $\text{range}(A)$ and $\text{null}(A)$ themselves, so we give these four subspaces a common name:

Suppose A is a matrix. The four subspaces

$$\text{range}(A), \ \text{null}(A), \ \text{range}(A^T), \ \text{and} \ \text{null}(A^T)$$

are called the **four fundamental subspaces** associated with A.

One method of finding bases of $\text{range}(A^T)$ and $\text{null}(A^T)$ would be to simply compute the RREF of A^T and use the method for finding bases of the range and null space discussed in the previous two examples. We demonstrate this method in the following example, but we will see a more efficient way shortly:

Example 2.4.7

Determining Range and Null Space of a Transposed Matrix

Find bases of $\operatorname{range}(A^T)$ and $\operatorname{null}(A^T)$, where

$$A = \begin{bmatrix} 1 & 0 & 1 & 0 & -1 \\ 1 & 1 & 0 & 0 & 1 \\ -1 & 0 & -1 & 1 & 4 \\ 2 & 1 & 1 & -1 & -3 \end{bmatrix}.$$

Solution:

We start by finding the reduced row echelon form of A^T:

$$\begin{bmatrix} 1 & 1 & -1 & 2 \\ 0 & 1 & 0 & 1 \\ 1 & 0 & -1 & 1 \\ 0 & 0 & 1 & -1 \\ -1 & 1 & 4 & -3 \end{bmatrix} \xrightarrow[\ R_5+R_1\ ]{R_3-R_1} \begin{bmatrix} 1 & 1 & -1 & 2 \\ 0 & 1 & 0 & 1 \\ 0 & -1 & 0 & -1 \\ 0 & 0 & 1 & -1 \\ 0 & 2 & 3 & -1 \end{bmatrix}$$

$$\xrightarrow[\substack{R_3+R_2 \\ R_5-2R_2}]{R_1-R_2} \begin{bmatrix} 1 & 0 & -1 & 1 \\ 0 & 1 & 0 & 1 \\ 0 & 0 & 0 & 0 \\ 0 & 0 & 1 & -1 \\ 0 & 0 & 3 & -3 \end{bmatrix} \xrightarrow{R_3 \leftrightarrow R_4} \begin{bmatrix} 1 & 0 & -1 & 1 \\ 0 & 1 & 0 & 1 \\ 0 & 0 & 1 & -1 \\ 0 & 0 & 0 & 0 \\ 0 & 0 & 3 & -3 \end{bmatrix}$$

$$\xrightarrow[\ R_5-3R_3\ ]{R_1+R_3} \begin{bmatrix} 1 & 0 & 0 & 0 \\ 0 & 1 & 0 & 1 \\ 0 & 0 & 1 & -1 \\ 0 & 0 & 0 & 0 \\ 0 & 0 & 0 & 0 \end{bmatrix}.$$

One of the reasons why $\operatorname{range}(A^T)$ and $\operatorname{null}(A^T)$ are interesting is provided by Exercise 2.4.26. We investigate these two subspaces more thoroughly in (Joh20).

Since this row echelon form has leading entries in columns 1, 2, and 3, we know that columns 1, 2, and 3 of A^T (i.e., rows 1, 2, and 3 of A) form a basis of $\operatorname{range}(A^T)$. Explicitly, this basis is

$$\{(1,0,1,0,-1),(1,1,0,0,1),(-1,0,-1,1,4)\}.$$

To find a basis of $\operatorname{null}(A^T)$, we proceed as we did in the previous examples: we think of A as the coefficient matrix in the linear system $A\mathbf{x} = \mathbf{0}$ and note that the RREF tells us that x_4 is a free variable, so we solve for x_1, x_2, and x_3 (the leading variables) in terms of it. Explicitly, we get

$$(x_1, x_2, x_3, x_4) = x_4(0, -1, 1, 1),$$

so $\{(0,-1,1,1)\}$ is a basis of $\operatorname{null}(A^T)$.

However, if we already have the RREF of A, we can compute a basis of $\operatorname{range}(A^T)$ without going through the additional trouble of row-reducing A^T. To see how why, we start with the following theorem:

Theorem 2.4.6

Range of Transposed Matrix Equals the Span of Rows

Suppose $A \in \mathcal{M}_{m,n}$ has rows $\mathbf{a}_1, \mathbf{a}_2, \ldots, \mathbf{a}_m$. Then

$$\text{range}(A^T) = \text{span}(\mathbf{a}_1, \mathbf{a}_2, \ldots, \mathbf{a}_m).$$

The above theorem follows immediately from the corresponding observation (Theorem 2.3.2) about the range of a matrix equaling the span of its columns, and the fact that the rows of A are the (transpose of the) columns of A^T.

For this reason, $\text{range}(A^T)$ is sometimes called the **row space** of A and denoted by $\text{row}(A)$.

The way we can make use of this theorem is by noticing that if R is a row echelon form of A then the rows of R are linear combinations of the rows of A, and vice-versa (after all, the three elementary row operations that we use to row-reduce matrices are just very specific ways of taking linear combinations). It follows that $\text{range}(A^T) = \text{range}(R^T)$, which is the span of the rows of R. Since the non-zero rows of R are linearly independent, we arrive at the following fact:

> ! The non-zero rows in a row echelon form of A form a basis of $\text{range}(A^T)$.

We thus now know how to compute bases of three of the four fundamental subspaces of a matrix directly from its reduced row echelon form. To see how we can similarly handle the remaining fundamental subspace, $\text{null}(A^T)$, we recall that row-reducing $[\,A \mid I\,]$ to $[\,R \mid E\,]$, where R is the reduced row echelon form of A, results in the equation $R = EA$ holding (refer back to Theorem 2.2.1). Taking the transpose of both sides of that equation gives $A^T E^T = R^T$. If we write E^T in terms of its columns $\mathbf{v}_1, \ldots, \mathbf{v}_m$ (i.e., the rows of E), then we see that

$$A^T E^T = \left[\, A^T \mathbf{v}_1 \mid A^T \mathbf{v}_2 \mid \cdots \mid A^T \mathbf{v}_m \,\right] = R^T.$$

If the bottom k rows of R consist entirely of zeros, it follows that the rightmost k columns of R^T are $\mathbf{0}$ as well, so $A^T \mathbf{v}_{m-k+1} = A^T \mathbf{v}_{m-k+2} = \cdots = A^T \mathbf{v}_m = \mathbf{0}$. In other words, the rightmost k columns of E^T (i.e., the bottom k rows of E) are in $\text{null}(A^T)$.

The fact that these k rows are linearly independent follows from Theorem 2.3.5, which tells us that rows of invertible matrices (like E) are always linearly independent. To see that they span $\text{null}(A^T)$ (and thus form a basis of that subspace), we just note that if we could find another vector to add to this set (while preserving linear independence), then running the argument above backwards would show that R has an additional zero row, which violates uniqueness of the reduced row echelon form (Theorem B.2.1). This leads to the following observation:

> ! The block matrix $[\,A \mid I\,]$ can be row-reduced to $[\,R \mid E\,]$, where R is a row echelon form of A. If R has k zero rows, then the bottom k rows of E form a basis of $\text{null}(A^T)$.

To summarize, we can find bases of each of the four fundamental subspaces by row-reducing $[\,A \mid I\,]$ to its reduced row echelon form $[\,R \mid E\,]$. The non-zero rows of R then form a basis of $\text{range}(A^T)$, the rows of E beside the zero rows of R form a basis of $\text{null}(A^T)$ (these two facts are summarized in Figure 2.22), the columns of A corresponding to leading columns in R form a basis of $\text{range}(A)$, and a basis of $\text{null}(A)$ can be found by solving for the leading variables in

The *columns* of A and R can be used to construct bases on range(A) and null(A), as in Figure 2.21.

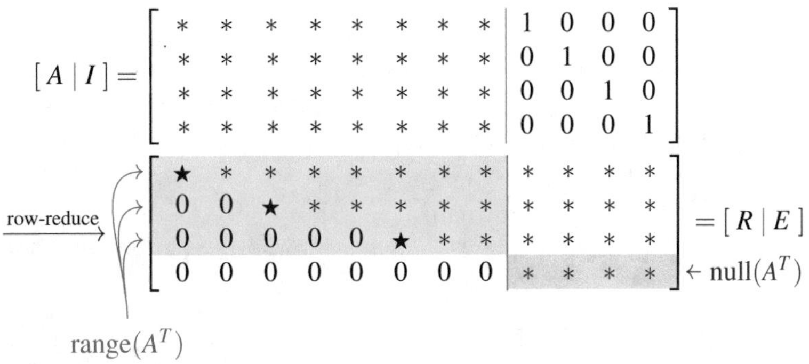

$$[A \mid I] = \begin{bmatrix} * & * & * & * & * & * & * & * & 1 & 0 & 0 & 0 \\ * & * & * & * & * & * & * & * & 0 & 1 & 0 & 0 \\ * & * & * & * & * & * & * & * & 0 & 0 & 1 & 0 \\ * & * & * & * & * & * & * & * & 0 & 0 & 0 & 1 \end{bmatrix}$$

$$\xrightarrow{\text{row-reduce}} \begin{bmatrix} \bigstar & * & * & * & * & * & * & * & * & * & * & * \\ 0 & 0 & \bigstar & * & * & * & * & * & * & * & * & * \\ 0 & 0 & 0 & 0 & 0 & \bigstar & * & * & * & * & * & * \\ 0 & 0 & 0 & 0 & 0 & 0 & 0 & 0 & * & * & * & * \end{bmatrix} = [R \mid E]$$

$\leftarrow \text{null}(A^T)$

$\text{range}(A^T)$

Figure 2.22: After we row-reduce $[A \mid I]$ to a row echelon form $[R \mid E]$, we can immediately read off bases of range(A^T) and null(A^T) from the rows of that matrix.

terms of the free variables in the linear system $R\mathbf{x} = \mathbf{0}$ (these two facts were summarized in Figure 2.21).

Example 2.4.8

Computing Bases of the Four Fundamental Subspaces

Find bases of the four fundamental subspaces of $A = \begin{bmatrix} 1 & 1 & 1 & -1 \\ 0 & 1 & 1 & 0 \\ -1 & 1 & 1 & 1 \end{bmatrix}$.

Solution:

We start by augmenting this matrix with a 3×3 identity matrix and row-reducing to find its reduced row echelon form:

$$\begin{bmatrix} 1 & 1 & 1 & -1 & 1 & 0 & 0 \\ 0 & 1 & 1 & 0 & 0 & 1 & 0 \\ -1 & 1 & 1 & 1 & 0 & 0 & 1 \end{bmatrix} \xrightarrow{R_3 + R_1} \begin{bmatrix} 1 & 1 & 1 & -1 & 1 & 0 & 0 \\ 0 & 1 & 1 & 0 & 0 & 1 & 0 \\ 0 & 2 & 2 & 0 & 1 & 0 & 1 \end{bmatrix}$$

$$\xrightarrow[R_3 - 2R_2]{R_1 - R_2} \begin{bmatrix} 1 & 0 & 0 & -1 & 1 & -1 & 0 \\ 0 & 1 & 1 & 0 & 0 & 1 & 0 \\ 0 & 0 & 0 & 0 & 1 & -2 & 1 \end{bmatrix}.$$

From here, we can immediately identify bases of three of the four fundamental subspaces as rows and columns of these matrices (using Figures 2.21 and 2.22 as a guide):

- range(A): $\{(1, 0, -1), (1, 1, 1)\}$,
- null(A^T): $\{(1, -2, 1)\}$, and
- range(A^T): $\{(1, 0, 0, -1), (0, 1, 1, 0)\}$.

The first two columns of A form a basis of range(A), the first two rows of the RREF form a basis of range(A^T), and the final row of the augmented matrix to the right of the RREF forms a basis of null(A^T).

To find a basis of null(A), we solve the linear system $R\mathbf{x} = \mathbf{0}$, where R is the reduced row echelon form that we computed above. In this linear system, x_1 and x_2 are leading variables and x_3 and x_4 are free variables. Furthermore, $x_1 - x_4 = 0$ and $x_2 + x_3 = 0$, so writing the solution in vector notation gives

$$(x_1, x_2, x_3, x_4) = x_3(0, -1, 1, 0) + x_4(1, 0, 0, 1).$$

It follows that one basis of null(A) is $\{(0, -1, 1, 0), (1, 0, 0, 1)\}$.

These ideas lead naturally to yet another characterization of invertibility of a matrix. Most of these properties follow immediately from the relevant definitions or our previous characterizations of invertibility (e.g., Theorems 2.3.3

and 2.3.5), so we leave its proof to Exercise 2.4.19.

Theorem 2.4.7

Bases and Invertible Matrices

Suppose $A \in \mathcal{M}_n$. The following are equivalent:

a) A is invertible.
b) $\text{range}(A) = \mathbb{R}^n$.
c) $\text{null}(A) = \{\mathbf{0}\}$.
d) The columns of A form a basis of $\mathbb{R}^n$.
e) The rows of A form a basis of $\mathbb{R}^n$.

We could also add "$\text{range}(A^T) = \mathbb{R}^n$" and "$\text{null}(A^T) = \{\mathbf{0}\}$" as additional equivalent conditions in the above theorem, but perhaps we have shown that invertibility is equivalent to enough other properties at this point that they are starting to feel redundant.

2.4.3 The Rank of a Matrix

In Section 2.2.2, we looked at invertible matrices $A \in \mathcal{M}_n$, which (among their many other equivalent characterizations) were the matrices whose columns span $\mathbb{R}^n$. In other words, their range is all of $\mathbb{R}^n$—they do not "squash" space down into a smaller-dimensional subspace. We now look at the rank of a matrix, which can be thought of as a measure of how non-invertible it is (i.e., how small is the subspace it squashes $\mathbb{R}^n$ down to).

Definition 2.4.4

Rank of a Matrix

Suppose A is a matrix. Its **rank**, denoted by $\text{rank}(A)$, is the dimension of its range.

Projections were introduced in Section 1.4.2.

For example, if $\mathbf{u} \in \mathbb{R}^n$ is a unit vector and $A = \mathbf{u}\mathbf{u}^T$ is the standard matrix of the projection onto the line in the direction of $\mathbf{u}$, then the range of A is 1-dimensional, so $\text{rank}(A) = 1$. Similarly, we showed that the matrix

$$A = \begin{bmatrix} 1 & 1 & 2 \\ 2 & -1 & 1 \\ 1 & 0 & 1 \end{bmatrix}$$

from Example 2.4.5 has range equal to $\text{span}\big((1,2,1),(1,-1,0)\big)$. Since this range is a (2-dimensional) plane, its rank equals 2 (see Figure 2.23).

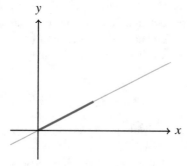

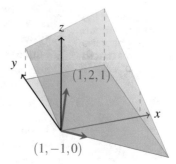

(a) A projection onto a line has rank 1, since its range is 1-dimensional.

(b) A 3×3 matrix with rank 2 has a plane as its range.

Figure 2.23: The rank of a matrix is the dimension of its range.

By recalling the fact that $A \in \mathcal{M}_n$ is invertible if and only if its range is $\mathbb{R}^n$, we conclude that A is invertible if and only if its rank is n. Since $\text{rank}(A) \leq n$ for all $A \in \mathcal{M}_n$, invertible matrices are exactly those with the largest possible rank, so we say that they have **full rank**. Roughly speaking, we think of higher-rank matrices as being "closer to" invertible than lower-rank matrices, and we will present several different ways of justifying this intuition as we investigate the basic properties of rank throughout this subsection.

More generally, if $A \in \mathcal{M}_{m,n}$ then $\text{rank}(A) \leq \min\{m,n\}$ so we say that A has **full rank** when $\text{rank}(A) = \min\{m,n\}$.

One of the reasons why the rank of a matrix is so useful is that it can be interpreted in so many different ways. While it equals the dimension of the range of a matrix (which is a very geometrically-motivated quantity), it can also be phrased in purely algebraic ways coming from linear systems.

Theorem 2.4.8

Characterization of Rank

Suppose A is a matrix. The following quantities are equal to each other:

a) $\text{rank}(A)$
b) $\text{rank}(A^T)$
c) The number of non-zero rows in any row echelon form of A.
d) The number of leading columns in any row echelon form of A.

The equivalence of (a) and (b) says that for every matrix, the dimension of the column space equals the dimension of the row space. We thus sometimes say "column rank equals row rank".

Proof. We start by noting that the quantities (c) and (d) are equal to each other since every non-zero row in a row echelon form has exactly one leading entry, as does every leading column. The remainder of the proof is devoted to showing that (a) and (d) are equal to each other, as are (b) and (c).

The equivalence of (a) and (d) comes from our method of constructing a basis of $\text{range}(A)$ from Section 2.4.2. Specifically, the basis of $\text{range}(A)$ that we constructed there consists of the columns of A corresponding to leading columns in its row echelon forms. The number of these leading columns thus equals the number of vectors in the basis, which (by definition) equals the dimension of $\text{range}(A)$, and that dimension (again, by definition) equals $\text{rank}(A)$.

The fact that (b) and (c) are equal to each other follows from a similar argument: the non-zero rows in a row echelon form of A form a basis of $\text{range}(A^T)$, and the number of vectors in this basis equals $\text{rank}(A^T)$. This completes the proof. ∎

In fact, we could give a characterization of rank that is analogous to almost any one of the characterizations of invertible matrices that we have provided throughout this chapter (Theorems 2.2.4, 2.3.3, and 2.3.5). For example, A is invertible if and only if its columns form a linearly independent set, and more generally the rank of A is equal to the maximal size of a linearly independent subset of its columns (see Exercise 2.4.29).

Example 2.4.9

Computing the Rank

Compute the rank of the following matrices:

a) The standard matrix $A \in \mathcal{M}_n$ of a reflection across a line.
b) The standard matrix $B \in \mathcal{M}_2$ of a rotation.
c) $C = \begin{bmatrix} 0 & 0 & -2 & 2 & -2 \\ 2 & -2 & -1 & 3 & 3 \\ -1 & 1 & -1 & 0 & -3 \end{bmatrix}$

We explored reflections and rotations in Section 1.4.2.

Solutions:

a) The rank of this matrix is n since its every vector in $\mathbb{R}^n$ can be reached by the reflection, so its range is all of $\mathbb{R}^n$, which is n-

dimensional. Equivalently, A is invertible since reflecting across a line twice is equivalent to doing nothing at all, so $A^2 = I$ (i.e., A is its own inverse). Since invertible matrices have rank n, it follows that $\text{rank}(A) = n$.

b) Similar to the matrix A from part (a), every vector in $\mathbb{R}^2$ can be reached by a rotation, so its range is all of $\mathbb{R}^2$, which is 2-dimensional. It follows that $\text{rank}(B) = 2$.

c) We can find the rank of this matrix by row-reducing to row echelon form:

$$\begin{bmatrix} 0 & 0 & -2 & 2 & -2 \\ 2 & -2 & -1 & 3 & 3 \\ -1 & 1 & -1 & 0 & -3 \end{bmatrix} \xrightarrow{R_1 \leftrightarrow R_3} \begin{bmatrix} -1 & 1 & -1 & 0 & -3 \\ 2 & -2 & -1 & 3 & 3 \\ 0 & 0 & -2 & 2 & -2 \end{bmatrix}$$

$$\xrightarrow{R_2 + 2R_1} \begin{bmatrix} -1 & 1 & -1 & 0 & -3 \\ 0 & 0 & -3 & 3 & -3 \\ 0 & 0 & -2 & 2 & -2 \end{bmatrix} \xrightarrow{R_3 - \frac{2}{3}R_2} \begin{bmatrix} -1 & 1 & -1 & 0 & -3 \\ 0 & 0 & -3 & 3 & -3 \\ 0 & 0 & 0 & 0 & 0 \end{bmatrix}$$

Since this row echelon form has 2 non-zero rows, we conclude that $\text{rank}(C) = 2$.

Somewhat complementary to the rank of a matrix is its **nullity**, which is the dimension of its null space (we denote the nullity of A simply by $\text{nullity}(A)$). We saw that a matrix $A \in \mathcal{M}_n$ is invertible if and only if the linear system $A\mathbf{x} = \mathbf{0}$ has a unique solution (i.e., its null space is $\{\mathbf{0}\}$), which is equivalent to it having nullity 0. This leads immediately to the following theorem, which follows immediately from Theorem 2.4.7 and the definitions of rank and nullity.

Theorem 2.4.9

Rank and Invertible Matrices

Let $A \in \mathcal{M}_n$. The following are equivalent:

a) A is invertible.
b) $\text{rank}(A) = n$.
c) $\text{nullity}(A) = 0$.

More generally, a natural trade-off occurs between the rank and nullity of a matrix: the smaller the rank of a matrix is (i.e., the "less invertible" it is), the smaller the span of its columns is, so the more ways we can construct linear combinations of those columns to get $\mathbf{0}$, and thus the larger its nullity is. The following theorem makes this observation precise.

Theorem 2.4.10

Rank–Nullity

In a linear system $A\mathbf{x} = \mathbf{b}$, the rank of A is the number of leading variables and its nullity is the number of free variables. This theorem thus just says that if we add up the number of leading and free variables, we get the total number of variables.

Suppose $A \in \mathcal{M}_{m,n}$ is a matrix. Then $\text{rank}(A) + \text{nullity}(A) = n$.

Proof. We use the equivalence of the quantities (a) and (d) from Theorem 2.4.8. Let $r = \text{rank}(A)$ and consider the linear system $A\mathbf{x} = \mathbf{0}$, which has r leading variables. Since this linear system has n variables total, it must have $n - r$ free variables. But we saw in Section 2.4.2 that each free variable gives one member of a basis of $\text{null}(A)$, so $\text{nullity}(A) = \dim(\text{null}(A)) = n - r$. It follows that

$$\text{rank}(A) + \text{nullity}(A) = r + (n - r) = n,$$

as claimed.   ∎

In terms of systems of linear equations, the rank of a coefficient matrix tells us how many leading variables it has, or equivalently how many of the equations in the system actually matter (i.e., cannot be removed without changing the solution set). On the other hand, the nullity tells us how many free variables it has (see Figure 2.24).

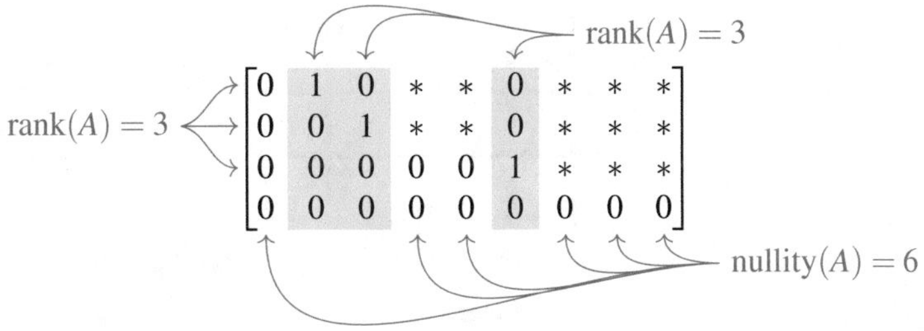

$$\text{rank}(A) = 3 \longleftrightarrow \begin{bmatrix} 0 & 1 & 0 & * & * & 0 & * & * & * \\ 0 & 0 & 1 & * & * & 0 & * & * & * \\ 0 & 0 & 0 & 0 & 0 & 1 & * & * & * \\ 0 & 0 & 0 & 0 & 0 & 0 & 0 & 0 & 0 \end{bmatrix} \qquad \text{rank}(A) = 3$$

$$\text{nullity}(A) = 6$$

Figure 2.24: The rank of a matrix equals the number of leading columns that its RREF has, which equals the number of non-zero rows that its RREF (or any of its row echelon forms) has. Its nullity is the number of non-leading columns in its RREF, which equals the number of free variables in the associated linear system.

Example 2.4.10

Computing the Rank and Nullity

Compute the rank and nullity of the matrix

$$A = \begin{bmatrix} 1 & 0 & -1 & 0 & 0 \\ 2 & 1 & -1 & 2 & 2 \\ -1 & -1 & 0 & -2 & -2 \\ 1 & 0 & -1 & 0 & 1 \end{bmatrix}.$$

Solution:

We can find the rank and nullity of this matrix by row-reducing to row echelon form:

To find the rank and nullity, we just need a row echelon form of the matrix, not necessarily its *reduced* row echelon form.

$$\begin{bmatrix} 1 & 0 & -1 & 0 & 0 \\ 2 & 1 & -1 & 2 & 2 \\ 1 & 0 & -1 & 0 & 1 \\ -1 & -1 & 0 & -2 & -2 \end{bmatrix} \xrightarrow[\substack{R_2-2R_1 \\ R_3-R_1 \\ R_4+R_1}]{} \begin{bmatrix} 1 & 0 & -1 & 0 & 0 \\ 0 & 1 & 1 & 2 & 2 \\ 0 & 0 & 0 & 0 & 1 \\ 0 & -1 & -1 & -2 & -2 \end{bmatrix}$$

$$\xrightarrow[\;R_4+R_2\;]{} \begin{bmatrix} 1 & 0 & -1 & 0 & 0 \\ 0 & 1 & 1 & 2 & 2 \\ 0 & 0 & 0 & 0 & 1 \\ 0 & 0 & 0 & 0 & 0 \end{bmatrix}.$$

Since this row echelon form has 3 non-zero rows, we conclude that $\text{rank}(A) = 3$ and $\text{nullity}(A) = 5 - 3 = 2$.

The previous theorem also makes sense geometrically—if we think of $A \in \mathcal{M}_{m,n}$ as a linear transformation, then it takes n-dimensional vectors (from $\mathbb{R}^n$) as input. Of those n input dimensions, $\text{rank}(A)$ are sent to the output space and the other $\text{nullity}(A)$ dimensions are "squashed away" by A (see Figure 2.25). For example, the matrix from Example 2.4.10 can be thought of as a linear transformation from $\mathbb{R}^5$ to $\mathbb{R}^4$. Since its rank is 3, it sends a 3-dimensional subspace of $\mathbb{R}^5$ to a 3-dimensional subspace of $\mathbb{R}^4$ and it "squashes away" the other 2 dimensions of $\mathbb{R}^5$ (i.e., its nullity is 2).

It is important to keep in mind that the rank of a matrix can change drastically when its entries change just a tiny amount (see Exercise 2.4.16), and it

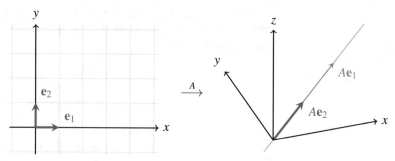

(a) A matrix $A \in \mathcal{M}_{3,2}$ with rank 1 and nullity 1 sends $\mathbb{R}^2$ to a line in $\mathbb{R}^3$.

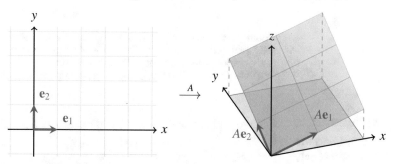

(b) A matrix $A \in \mathcal{M}_{3,2}$ with rank 2 and nullity 0 sends $\mathbb{R}^2$ to a plane in $\mathbb{R}^3$.

Figure 2.25: A non-zero matrix $A \in \mathcal{M}_{3,2}$ either has rank 1 and nullity 1 (as in (a)) or rank 2 and nullity 0 (as in (b)).

is not particularly well-behaved under matrix addition or multiplication (e.g., it is not the case that $\mathrm{rank}(A+B) = \mathrm{rank}(A) + \mathrm{rank}(B)$, nor is it the case that $\mathrm{rank}(AB) = \mathrm{rank}(A) \times \mathrm{rank}(B)$). However, there are some *in*equalities that at least loosely relate $\mathrm{rank}(A+B)$ and $\mathrm{rank}(AB)$ to the ranks of A and B:

Theorem 2.4.11 **Rank of a Sum and Product**	Let A and B be matrices (with sizes such that the operations below make sense). Then a) $\mathrm{rank}(A+B) \leq \mathrm{rank}(A) + \mathrm{rank}(B)$ b) $\mathrm{rank}(AB) \leq \min\{\mathrm{rank}(A), \mathrm{rank}(B)\}$

Proof. Denote the columns of A and B by $\mathbf{a}_1, \mathbf{a}_2, \ldots$ and $\mathbf{b}_1, \mathbf{b}_2, \ldots$, respectively. To see why part (a) holds, suppose $A, B \in \mathcal{M}_{m,n}$ and notice that the columns of $A+B$ are $\mathbf{a}_1 + \mathbf{b}_1, \mathbf{a}_2 + \mathbf{b}_2, \ldots, \mathbf{a}_n + \mathbf{b}_n$, which are each contained within the span of $\mathbf{a}_1, \mathbf{b}_1, \mathbf{a}_2, \mathbf{b}_2, \ldots, \mathbf{a}_n, \mathbf{b}_n$. It follows that

> If B is invertible then part (b) can be strengthened to $\mathrm{rank}(AB) = \mathrm{rank}(A)$ (see Exercise 2.4.17).

$$\mathrm{span}(\mathbf{a}_1 + \mathbf{b}_1, \mathbf{a}_2 + \mathbf{b}_2, \ldots, \mathbf{a}_n + \mathbf{b}_n) \subseteq \mathrm{span}(\mathbf{a}_1, \mathbf{b}_1, \mathbf{a}_2, \mathbf{b}_2, \ldots, \mathbf{a}_n, \mathbf{b}_n)$$

as well. The span on the left equals $\mathrm{range}(A+B)$, so its dimension is $\mathrm{rank}(A+B)$. The span on the right cannot have dimension larger than $\mathrm{rank}(A) + \mathrm{rank}(B)$, so the fact that $\mathrm{range}(A+B)$ is contained in the subspace on the right implies $\mathrm{rank}(A+B) \leq \mathrm{rank}(A) + \mathrm{rank}(B)$, as desired.

For part (b), suppose $A \in \mathcal{M}_{m,n}$ and $B \in \mathcal{M}_{n,p}$. We use block matrix multi-plication to express the columns of AB in terms of the columns of A:

> The notation $\sum_{j=1}^{n}$ refers to a sum: $\sum_{j=1}^{n} b_{j,1}\mathbf{a}_j = b_{1,1}\mathbf{a}_1 + \cdots + b_{n,1}\mathbf{a}_n$.

$$AB = \left[\, A\mathbf{b}_1 \mid A\mathbf{b}_2 \mid \cdots \mid A\mathbf{b}_p \,\right] = \left[\, \sum_{j=1}^{n} b_{j,1}\mathbf{a}_j \;\middle|\; \sum_{j=1}^{n} b_{j,2}\mathbf{a}_j \;\middle|\; \cdots \;\middle|\; \sum_{j=1}^{n} b_{j,p}\mathbf{a}_j \,\right].$$

This proof of part (b) shows that range(AB) $\subseteq$ range(A) and thus solves Exercise 2.3.27(a).

The important point here is that the columns of AB are linear combinations of the columns of A, so range(AB) $\subseteq$ range(A). Since the rank of a matrix is the dimension of its range, it follows that rank(AB) $\leq$ rank(A). To see that rank(AB) $\leq$ rank(B) as well (thus completing the proof), we could apply this same argument to $(AB)^T = B^T A^T$.   ■

Exercises

solutions to starred exercises on page 453

2.4.1 Determine which of the following sets B are bases of the subspace $\mathcal{S}$.

*(a) $B = \{(1,1),(1,2),(3,4)\}$, $\mathcal{S} = \mathbb{R}^2$
 (b) $B = \{(1,1),(1,-1)\}$, $\mathcal{S} = \mathbb{R}^2$
*(c) $B = \{(1,1,1),(2,1,-1)\}$, $\mathcal{S} = \mathbb{R}^3$
 (d) $B = \{(2,2,1),(0,1,1)\}$, $\mathcal{S}$ is the plane in $\mathbb{R}^3$ with equation $x+y-z = 0$.
*(e) $B = \{(1,0,1),(2,1,3)\}$, $\mathcal{S}$ is the plane in $\mathbb{R}^3$ with equation $x+y-z = 0$.
 (f) $B = \{(1,3,2),(-1,-2,1),(1,4,5)\}$, $\mathcal{S} = \text{span}(B)$.

2.4.2 Find a basis of each of the following subspaces.

*(a) The line in $\mathbb{R}^2$ with equation $y = 3x$.
 (b) The line in $\mathbb{R}^3$ going through the origin and $(2,1,4)$.
*(c) The plane in $\mathbb{R}^3$ with equation $2x+y+z = 0$.
 (d) The plane in $\mathbb{R}^3$ with equation $z = 4x-3y$.
*(e) The line in $\mathbb{R}^3$ at the intersection of the planes $x+y-z = 0$ and $2x-y+2z = 0$.
 (f) $\text{span}(\{(1,3,2,1),(-1,-2,1,0),(1,4,-1,1)\})$

💻 **2.4.3** With the help of computer software, discard 0 or more vectors from the following sets so as to turn them into bases of their span.

*(a) $\{(1,2,3,4),(3,5,7,9),(6,7,8,9),(9,8,7,6)\}$
 (b) $\{(5,5,5,1),(3,6,3,2),(4,1,4,4),(1,4,1,4)\}$
*(c) $\{(1,1,4,5,1),(2,-7,4,2,2),(5,2,4,5,3),$
       $(4,-2,6,6,3),(2,5,2,4,1)\}$
 (d) $\{(3,4,-1,1,3),(-1,1,3,0,1),(4,6,-5,3,6),$
       $(0,3,-1,2,4),(3,2,1,3,2),(-1,3,1,-2,2),$
       $(-4,5,0,1,7)\}$

💻 **2.4.4** With the help of computer software, extend the following linearly independent sets of vectors (i.e., add 0 or more vectors to them) to create bases of $\mathbb{R}^4$.

*(a) $\{(2,3,1,1),(3,3,2,3),(2,1,2,3)\}$
 (b) $\{(3,2,0,2),(0,1,3,3),(2,1,1,1),(0,2,3,2)\}$
*(c) $\{(2,4,3,4),(3,2,-1,-1)\}$
 (d) $\{(-1,3,6,3),(1,3,5,0),(3,0,-1,-3)\}$

2.4.5 For each of the following matrices A, find bases for each of range(A), null(A), range(A^T), and null(A^T).

*(a) $\begin{bmatrix} 1 & 0 \\ 0 & 1 \end{bmatrix}$
 (b) $\begin{bmatrix} 1 & 2 & 1 \\ 0 & 1 & 2 \end{bmatrix}$

*(c) $\begin{bmatrix} 0 & 0 & 0 \\ 0 & 1 & 2 \\ 0 & 0 & 0 \end{bmatrix}$
 (d) $\begin{bmatrix} 2 & 0 & 1 \\ 0 & 2 & 0 \\ 0 & 0 & 0 \end{bmatrix}$

*(e) $\begin{bmatrix} 1 & 0 \\ 0 & 2 \\ 0 & 1 \end{bmatrix}$
 (f) $\begin{bmatrix} 1 & 2 & 7 \\ 0 & 0 & 0 \\ 0 & 0 & 0 \end{bmatrix}$

*(g) $\begin{bmatrix} 1 & 2 & 0 & 3 & 0 \\ 0 & 0 & 1 & -1 & 0 \\ 0 & 0 & 0 & 0 & 1 \end{bmatrix}$
 (h) $\begin{bmatrix} 0 & 0 & 1 & 1 \\ 0 & 1 & 1 & 0 \\ 1 & 1 & 0 & 0 \end{bmatrix}$

*(i) $\begin{bmatrix} 0 & -4 & 0 & 2 & 1 \\ -1 & 2 & 1 & 2 & 1 \\ -2 & 0 & 2 & 6 & 3 \end{bmatrix}$

 (j) $\begin{bmatrix} 2 & -1 & 2 & 3 & 5 \\ -1 & 1 & -2 & -2 & -2 \\ 1 & -1 & 2 & 1 & 2 \\ 1 & -2 & 4 & 2 & 1 \end{bmatrix}$

2.4.6 Compute the rank and nullity of the following matrices.

*(a) $\begin{bmatrix} 1 & -1 \\ -1 & 1 \end{bmatrix}$
 (b) $\begin{bmatrix} 6 & 3 \\ 2 & 1 \end{bmatrix}$

*(c) $\begin{bmatrix} 1 & 2 \\ 3 & 5 \end{bmatrix}$
 (d) $\begin{bmatrix} 1 & 2 & 1 \\ 0 & 1 & 2 \end{bmatrix}$

*(e) $\begin{bmatrix} 2 & 4 & 0 \\ 1 & -2 & 0 \\ 2 & 0 & -1 \end{bmatrix}$
 (f) $\begin{bmatrix} 2 & 6 & 1 & -1 \\ 5 & 4 & 8 & 1 \\ 3 & -2 & 7 & 2 \end{bmatrix}$

*(g) $\begin{bmatrix} 1 & 0 \\ 0 & 2 \\ 3 & 1 \end{bmatrix}$
 (h) $\begin{bmatrix} 1 & 2 & 1 & 0 \\ 2 & 1 & 2 & -3 \\ 0 & 3 & 1 & 2 \\ 0 & 0 & 2 & -2 \end{bmatrix}$

🖥 **2.4.7** Use computer software to find bases for the four fundamental subspaces of the following matrices. Also compute their rank and nullity.

*(a)
$$\begin{bmatrix} 0 & 0 & 2 & 2 & -1 & -2 & 3 & -1 \\ -6 & -12 & -1 & -7 & 1 & 8 & -7 & 6 \\ 4 & 8 & -2 & 2 & 1 & -2 & 1 & -2 \\ -1 & -2 & 0 & -1 & 0 & 1 & -1 & 1 \end{bmatrix}$$

(b)
$$\begin{bmatrix} 5 & 1 & 9 & 6 & -1 & 1 & 8 \\ 2 & 2 & 2 & 4 & 1 & 0 & 0 \\ 5 & 4 & 6 & 9 & 1 & 1 & 4 \\ 6 & 3 & 9 & 9 & 0 & 1 & 7 \\ -3 & 1 & -7 & -2 & 2 & -1 & -8 \end{bmatrix}$$

2.4.8 Determine which of the following statements are true and which are false.

*(a) Every basis of $\mathbb{R}^3$ contains exactly 3 vectors.

(b) Every set of 3 vectors in $\mathbb{R}^3$ forms a basis of $\mathbb{R}^3$.

*(c) If $\mathcal{S} = \text{span}(\mathbf{v}_1, \mathbf{v}_2, \ldots, \mathbf{v}_k)$ then $\{\mathbf{v}_1, \mathbf{v}_2, \ldots, \mathbf{v}_k\}$ is a basis of $\mathcal{S}$.

(d) A basis of range(A) is given by the set of columns of A corresponding to the leading columns in a row echelon form of A.

*(e) A basis of range(A^T) is given by the set of rows of A corresponding to the non-zero rows in a row echelon form of A.

(f) The rank of the zero matrix is 0.

*(g) The $n \times n$ identity matrix has rank n.

(h) If $A \in \mathcal{M}_{3,7}$ then rank(A) ≤ 3.

*(i) If A and B are matrices of the same size then rank($A + B$) = rank(A) + rank(B).

(j) If A and B are square matrices of the same size then rank(AB) = rank(A) · rank(B).

*(k) If A and B are square matrices of the same size then rank(AB) = min{rank(A), rank(B)}.

(l) The rank of a matrix equals the number of non-zero rows that it has.

*(m) The rank of a matrix equals the number of non-zero rows in any of its row echelon forms.

(n) If $A, B \in \mathcal{M}_{m,n}$ are row equivalent then null(A) = null(B).

*(o) If $A \in \mathcal{M}_{m,n}$ then nullity(A) = nullity(A^T).

2.4.9 In this exercise, we explore a relationship between the range and null space of a matrix.

(a) Find a matrix $A \in \mathcal{M}_4$ whose range is the same as its null space.

(b) Show that there does not exist a matrix $A \in \mathcal{M}_3$ whose range is the same as its null space.

*2.4.10 Let J_n be the $n \times n$ matrix, all of whose entries are 1. Compute rank(J_n).

2.4.11 Suppose A and B are 4×4 matrices with the property that $A\mathbf{v} \neq B\mathbf{v}$ for all $\mathbf{v} \neq \mathbf{0}$. What is the rank of the matrix $A - B$?

*2.4.12 Let A_n be the $n \times n$ matrix with entries $1, 2, \ldots, n^2$ written left-to-right, top-to-bottom. For example,

$$A_2 = \begin{bmatrix} 1 & 2 \\ 3 & 4 \end{bmatrix} \quad \text{and} \quad A_3 = \begin{bmatrix} 1 & 2 & 3 \\ 4 & 5 & 6 \\ 7 & 8 & 9 \end{bmatrix}.$$

Show that rank(A_n) = 2 for all $n \geq 2$. [Hint: Write each row of A_n as a linear combination of its first two rows.]

2.4.13 Determine which values of the (real) number x lead to the following matrices having which rank.

*(a) $\begin{bmatrix} 1 & x \\ 2 & 4 \end{bmatrix}$

(b) $\begin{bmatrix} x-1 & -1 \\ 1 & 1-x \end{bmatrix}$

*(c) $\begin{bmatrix} 1 & 2 & x \\ 2 & 3-x & 3x+1 \\ -x & -2x & -1 \end{bmatrix}$

(d) $\begin{bmatrix} 1 & 1 & 1 \\ 1 & 2 & 2 \\ 1 & 2 & x \end{bmatrix}$

*2.4.14 Suppose $A \in \mathcal{M}_n$.

(a) Show that if A is diagonal then its rank is the number of non-zero diagonal entries that it has.

(b) Show that the result of part (a) is *not* necessarily true if A is just triangular. That is, find a triangular matrix whose rank does not equal its number of non-zero diagonal entries.

2.4.15 Suppose that $A, B \in \mathcal{M}_{m,n}$.

(a) Show that if A and B differ in just one entry (and all of their other entries are the same) then |rank(A) − rank(B)| ≤ 1. That is, show that rank(A) and rank(B) differ by at most 1.

(b) More generally, show that if A and B differ in exactly k entries then |rank(A) − rank(B)| $\leq k$.

2.4.16 In this exercise, we show that changing the entries of a matrix just a tiny amount can change its rank drastically. Let x be a real number, and consider the matrix

$$A = \begin{bmatrix} x & 1 & 1 & \cdots & 1 \\ 1 & x & 1 & \cdots & 1 \\ 1 & 1 & x & \cdots & 1 \\ \vdots & \vdots & \vdots & \ddots & \vdots \\ 1 & 1 & 1 & \cdots & x \end{bmatrix} \in \mathcal{M}_n.$$

(a) Show that if $x = 1$ then rank(A) = 1.

(b) Show that if $x \neq 1$ then rank(A) = n.

2.4.17 Suppose A and B are matrices with sizes such that the product AB makes sense.

(a) Show that if A is invertible (and thus square) then rank(AB) = rank(B). [Hint: Try to make use of Theorem 2.4.11 in two different ways.]

(b) Show that if $B \in \mathcal{M}_n$ is invertible then rank(AB) = rank(A).

2.4.18 Suppose $A \in \mathcal{M}_n$ is a matrix with the property that $A^2 = O$. Show that rank(A) $\leq n/2$.

2.4.19 Prove Theorem 2.4.7.

2.4.20 Suppose that S_1 and S_2 are subspaces of $\mathbb{R}^n$ with the property that $S_1 \subseteq S_2$. Show that $\dim(S_1) \leq \dim(S_2)$.

[Hint: Intuitively this is clear, but to actually prove it we have to use the definition of dimension as the number of vectors in a basis.]

2.4.21 Suppose that S_1 and S_2 are subspaces of $\mathbb{R}^n$ with the property that $S_1 \subseteq S_2$ and $\dim(S_1) = \dim(S_2)$. Show that $S_1 = S_2$.

2.4.22 Prove Theorem 2.4.3(b). That is, show that if $S = \mathrm{span}(\mathbf{v}_1, \mathbf{v}_2, \ldots, \mathbf{v}_k)$ is a non-zero subspace of $\mathbb{R}^n$ then there is a subset of $\{\mathbf{v}_1, \mathbf{v}_2, \ldots, \mathbf{v}_k\}$ that is a basis of S.

2.4.23 Consider the set

$$B = \{(1,-1,2),(-1,2,3),(2,-2,4)\}$$

and the subspace $S = \mathrm{span}(B)$.

(a) Show that B is a linearly dependent set, so a basis of S can be obtained by removing a vector from B.

(b) Why does removing $(-1,2,3)$ from B not result in a basis of S?

2.4.24 Complete the proof of Theorem 2.4.4 by showing that condition (b)(i) implies condition (b)(iii). That is, show that if S is a subspace of $\mathbb{R}^n$ and $B \subseteq S$ spans S and contains $\dim(S)$ vectors, then B is a basis of S.

2.4.25 Suppose $A, B \in \mathcal{M}_n$ are such that AB is invertible. Show that A and B must be invertible.

[Side note: This is the converse of Theorem 2.2.3, which said that if A and B are invertible, then so is AB.]

2.4.26 Suppose $A \in \mathcal{M}_{m,n}$.

(a) Show that if $\mathbf{v} \in \mathrm{range}(A)$ and $\mathbf{w} \in \mathrm{null}(A^T)$ then $\mathbf{v}$ and $\mathbf{w}$ are orthogonal.

(b) Show that if $\mathbf{v} \in \mathrm{null}(A)$ and $\mathbf{w} \in \mathrm{range}(A^T)$ then $\mathbf{v}$ and $\mathbf{w}$ are orthogonal.

2.4.27 Suppose that $A, B \in \mathcal{M}_{m,n}$. Recall from Exercise 2.1.13 that A and B are row equivalent if and only if they have the same reduced row echelon form.

(a) Show that A and B are row equivalent if and only if $\mathrm{null}(A) = \mathrm{null}(B)$.

(b) Show that A and B are row equivalent if and only if $\mathrm{range}(A^T) = \mathrm{range}(B^T)$.

2.4.28 Suppose $A \in \mathcal{M}_{m,n}$ and $B \in \mathcal{M}_{n,p}$.

(a) Show that $\mathrm{nullity}(AB) \leq \mathrm{nullity}(A) + \mathrm{nullity}(B)$. [Hint: Extend a basis of $\mathrm{null}(B)$ to a basis of $\mathrm{null}(AB)$.]

(b) Show that $\mathrm{rank}(AB) \geq \mathrm{rank}(A) + \mathrm{rank}(B) - n$.

2.4.29 Suppose $A \in \mathcal{M}_{m,n}$.

(a) Show that $\mathrm{rank}(A)$ equals the maximal number of linearly independent columns of A.

(b) Show that $\mathrm{rank}(A)$ equals the maximal number of linearly independent rows of A.

2.4.30 Show that $A \in \mathcal{M}_{m,n}$ has rank 1 if and only if there exist non-zero (column) vectors $\mathbf{v} \in \mathbb{R}^m$ and $\mathbf{w} \in \mathbb{R}^n$ such that $A = \mathbf{v}\mathbf{w}^T$.

2.4.31 Suppose $A \in \mathcal{M}_{m,n}$.

(a) Suppose $m \geq n$ and $\mathrm{rank}(A) = n$. Show that there exists a matrix $B \in \mathcal{M}_{n,m}$ such that $BA = I_n$.

(b) Suppose $n \geq m$ and $\mathrm{rank}(A) = m$. Show that there exists a matrix $C \in \mathcal{M}_{n,m}$ such that $AC = I_m$.

[Side note: The matrices B and C in parts (a) and (b) are sometimes called a **left inverse** and a **right inverse** of A, respectively.]

2.5 Summary and Review

In this chapter, we introduced Gaussian elimination and Gauss–Jordan elimination as methods for solving systems of linear equations. This led us to the concept of the reduced row echelon form $R \in \mathcal{M}_{m,n}$ of a matrix $A \in \mathcal{M}_{m,n}$, which can be thought of as the simplest coefficient matrix with the property that the linear systems $A\mathbf{x} = \mathbf{0}$ and $R\mathbf{x} = \mathbf{0}$ have the same solution sets. We saw that if A is square (i.e., $m = n$) then its RREF is equal to the identity matrix I if and only if it is invertible. That is, for such matrices we can find a matrix $A^{-1} \in \mathcal{M}_n$ with the property that $AA^{-1} = A^{-1}A = I$, and we think of A^{-1} as a linear transformation that "undoes" what A "does" to vectors.

We saw several theorems (in particular, Theorems 2.2.4, 2.3.3, 2.3.5, 2.4.7, and 2.4.9) that further characterize invertibility of matrices in terms of other linear algebraic concepts. We summarize most of these conditions here for ease of reference.

Theorem 2.5.1

The Invertible Matrix Theorem

Suppose $A \in \mathcal{M}_n$. The following are equivalent:

a) A is invertible.
b) A^T is invertible.
c) $\text{rank}(A) = n$.
d) $\text{nullity}(A) = 0$.
e) The linear system $A\mathbf{x} = \mathbf{b}$ has a solution for all $\mathbf{b} \in \mathbb{R}^n$.
f) The linear system $A\mathbf{x} = \mathbf{b}$ has a unique solution for all $\mathbf{b} \in \mathbb{R}^n$.
g) The linear system $A\mathbf{x} = \mathbf{0}$ has a unique solution ($\mathbf{x} = \mathbf{0}$).
h) The reduced row echelon form of A is I.
i) The columns of A are linearly independent.
j) The columns of A span $\mathbb{R}^n$.
k) The columns of A form a basis of $\mathbb{R}^n$.
l) The rows of A are linearly independent.
m) The rows of A span $\mathbb{R}^n$.
n) The rows of A form a basis of $\mathbb{R}^n$.

We will see yet another equivalent condition that could be added to this theorem a bit later, in Theorem 3.2.1.

We then introduced subspaces, bases, and the rank of a matrix primarily as tools for discussing matrices that are *not* invertible, and generalizing the properties of the above theorem to all matrices. For example, if a matrix $A \in \mathcal{M}_n$ is not invertible, then its range (i.e., the span of its columns) is not all of $\mathbb{R}^n$. Instead, its range is a subspace of $\mathbb{R}^n$ with dimension equal to $\text{rank}(A)$, and there is a set consisting of this many of its columns that forms a basis of $\text{range}(A)$. In a sense, invertible matrices form the realm where linear algebra is "easy": we do not have to worry about knowing about the range or null space of an invertible matrix, since its range is all of $\mathbb{R}^n$ and its null space is $\{\mathbf{0}\}$.

It is also worth reminding ourselves of some of the equivalent ways of saying that two matrices are row equivalent (i.e., there is a sequence of row operations that converts one matrix into the other) that we saw in this chapter. The following theorem summarizes these characterizations of row equivalence for easy reference.

Theorem 2.5.2

Characterization of Row Equivalence

Suppose $A, B \in \mathcal{M}_{m,n}$. The following are equivalent:

a) A and B are row equivalent.
b) A and B have the same reduced row echelon form.
c) $\text{null}(A) = \text{null}(B)$.
d) $\text{range}(A^T) = \text{range}(B^T)$.
e) There is an invertible matrix $P \in \mathcal{M}_m$ such that $A = PB$.

In particular, characterization (b) of row equivalence in the above theorem was proved in Exercise 2.1.13, while (c) and (d) were established in Exercise 2.4.27, and (e) comes from Exercise 2.5.6.

We close this section by briefly returning to some ideas that we originally presented in Section 2.2.1. Recall that we showed in Theorem 2.2.1 that row-reducing $[\, A \mid I \,]$ to $[\, R \mid E \,]$ ensures that $R = EA$. We now state this theorem in a slightly different way that takes advantage of the fact that we now understand invertible matrices.

We leave the proof of the upcoming theorem to Exercise 2.5.5.

Theorem 2.5.3

RREF Decomposition

Suppose $A \in \mathcal{M}_{m,n}$ has reduced row echelon form $R \in \mathcal{M}_{m,n}$. Then there exists an invertible matrix $P \in \mathcal{M}_m$ such that $A = PR$.

Since multiplication on the left by an invertible matrix is equivalent to

The matrix P in the statement of Theorem 2.5.3 is the inverse of the matrix E from the statement of Theorem 2.2.1.

applying a sequence of row operations to that matrix, this RREF decomposition is just another way of saying that every matrix A can be row-reduced to its reduced row echelon form. However, phrasing it in this way provides us with our first *matrix decomposition*: a way of writing a matrix as a product of two or more matrices with special properties. Being able to decompose matrices into a product of multiple simpler matrices is a very powerful technique that we will make considerable use of in the next chapter of this book. We also explore another matrix decomposition that arises from Gaussian elimination (one that lets us write a matrix as a product of a lower triangular matrix and an upper triangular matrix) in Section 2.D.

Exercises

solutions to starred exercises on page 456

2.5.1 Determine whether or not the following pairs of matrices are row equivalent.

*(a) $A = \begin{bmatrix} 1 & 1 \\ 1 & 1 \end{bmatrix}$ and $B = \begin{bmatrix} 1 & 1 \\ 1 & 2 \end{bmatrix}$

(b) $A = \begin{bmatrix} 1 & 1 & 1 \\ 1 & 2 & 3 \end{bmatrix}$ and $B = \begin{bmatrix} 0 & 1 & 2 \\ 1 & 1 & 1 \end{bmatrix}$

*(c) $A = \begin{bmatrix} 1 & 2 & 1 \\ 2 & 1 & 1 \\ 1 & -1 & 0 \end{bmatrix}$ and $B = \begin{bmatrix} 3 & 3 & 2 \\ 0 & 0 & 0 \\ 0 & 3 & 1 \end{bmatrix}$

(d) $A = \begin{bmatrix} 1 & -1 & 2 \\ 1 & 0 & 2 \\ 2 & -1 & 4 \end{bmatrix}$ and $B = \begin{bmatrix} 1 & 2 & -1 \\ 1 & 2 & 0 \\ 2 & 4 & -1 \end{bmatrix}$

*(e) $A = \begin{bmatrix} 1 & -1 & 0 \\ -1 & 1 & -3 \\ 1 & -2 & -1 \\ 1 & 1 & -3 \end{bmatrix}$ and $B = \begin{bmatrix} 2 & -1 & 3 \\ 0 & -1 & 2 \\ -1 & -1 & 1 \\ -1 & 0 & 0 \end{bmatrix}$

(f) $A = \begin{bmatrix} 1 & 1 & 4 & 3 \\ 1 & 2 & 3 & 1 \\ 3 & 3 & 3 & 0 \end{bmatrix}$ and $B = \begin{bmatrix} 1 & 2 & 3 & 4 \\ 2 & 3 & 4 & 2 \\ 4 & 2 & 0 & 1 \end{bmatrix}$

2.5.2 Determine which of the following statements are true and which are false.

*(a) The reduced row echelon form of a matrix is unique.

(b) If R is the reduced row echelon form of A then the linear systems $A\mathbf{x} = \mathbf{0}$ and $R\mathbf{x} = \mathbf{0}$ have the same solution sets.

*(c) If R is the reduced row echelon form of A then the linear systems $A\mathbf{x} = \mathbf{b}$ and $R\mathbf{x} = \mathbf{b}$ have the same solution sets.

(d) Every matrix can be written as a product of elementary matrices.

*(e) Every matrix A can be written in the form $A = PR$, where P is invertible and R is the reduced row echelon form of A.

(f) If $A = PR$, where R is the reduced row echelon form of A, then P must be invertible.

*(g) If a square matrix has range equal to $\{\mathbf{0}\}$ then it must be invertible.

(h) If $A, B \in \mathcal{M}_n$ are both invertible then they are row equivalent.

*(i) If $A, B \in \mathcal{M}_{m,n}$ are row equivalent then they have the same rank.

*$\textbf{2.5.3}$ Suppose $A, B \in \mathcal{M}_{m,n}$ and let $\{\mathbf{v}_1, \mathbf{v}_2, \ldots, \mathbf{v}_n\}$ be a basis of $\mathbb{R}^n$. Show that $A = B$ if and only if $A\mathbf{v}_j = B\mathbf{v}_j$ for all $1 \leq j \leq n$.

2.5.4 Suppose A is a 4×4 matrix, B is a 4×3 matrix, and C is a 3×4 matrix such that $A = BC$. Prove that A is not invertible.

∗∗2.5.5 Prove Theorem 2.5.3.

[Hint: Thanks to Theorem 2.2.1, all you have to do is explain why P can be chosen to be invertible.]

∗∗2.5.6 Suppose $A, B \in \mathcal{M}_{m,n}$. Show that A and B are row equivalent if and only if there exists an invertible matrix $P \in \mathcal{M}_m$ such that $A = PB$.

2.5.7 Suppose that $A \in \mathcal{M}_{m,n}$.

(a) Show that $\text{null}(A) = \text{null}(A^T A)$.
    [Hint: If $\mathbf{x} \in \text{null}(A^T A)$ then compute $\|A\mathbf{x}\|^2$.]
(b) Show that $\text{rank}(A^T A) = \text{rank}(A)$.
(b) Show that $\text{range}(A) = \text{range}(AA^T)$.
    [Hint: Use Exercise 2.4.21.]
(d) Show that $\text{rank}(AA^T) = \text{rank}(A)$.
(e) Provide an example that shows that it is *not* necessarily the case that $\text{range}(A) = \text{range}(A^T A)$ or $\text{null}(A) = \text{null}(AA^T)$.

2.A   Extra Topic: Linear Algebra Over Finite Fields

One of the most useful aspects of Gaussian elimination and Gauss–Jordan elimination is that they only rely on our ability to perform two different mathematical operations: addition and (scalar) multiplication. These algorithms do not rely on much of the specific structure of the real numbers, so they can also be used to solve systems of linear equations involving other types of numbers. Slightly more specifically, we can use numbers coming from any **field**, which roughly speaking is a set containing 0 and 1, and which has addition and multiplication operations that behave similarly to the ones on real numbers (e.g., $a(b+c) = ab + ac$ for all a, b, and c in the field).

The reader is likely at least superficially familiar with some other types of numbers, like the rational numbers (i.e., ratios of integers) and complex numbers (i.e., numbers of the form $a + ib$, where $a, b \in \mathbb{R}$ and $i^2 = -1$, which are covered in Appendix A.1). However, many other sets of numbers work just as well, and in this section we focus on the sets of non-negative integers under modular arithmetic.

2.A.1   Binary Linear Systems

For now, we work with the set $\mathbb{Z}_2$ of numbers under arithmetic mod 2. That is, $\mathbb{Z}_2$ is simply the set $\{0,1\}$, but with the understanding that addition and multiplication of these numbers work as follows:

$$0+0 = 0 \qquad 0+1 = 1 \qquad 1+0 = 1 \qquad 1+1 = 0, \quad \text{and}$$
$$0 \times 0 = 0 \qquad 0 \times 1 = 0 \qquad 1 \times 0 = 0 \qquad 1 \times 1 = 1.$$

In other words, these operations work just as they normally do, with the exception that $1 + 1 = 0$ instead of $1 + 1 = 2$ (since there is no "2" in $\mathbb{Z}_2$). Phrased differently, we can think of these operations as usual binary arithmetic, except we do not bother carrying digits when adding. As yet another interpretation, we can think of it as a simplified form of arithmetic that only keeps track of whether or not a number is even (with 0 for even and 1 for odd).

This weird form of arithmetic might seem arbitrary and silly, but it is useful when we want to use vectors to represent objects that can be toggled between 2 different states (e.g., "on" and "off"), since adding 1 to a number in $\mathbb{Z}_2$ performs this toggle—it changes 0 to 1 and 1 to 0. It is also useful in computing, since computers represent data via 0s and 1s, and the addition and multiplication operations correspond to the bitwise XOR and AND gates, respectively.

For now, we work through an example to illustrate the fact that we can solve linear systems with entries from $\mathbb{Z}_2$ in the exact same way that we solve linear systems with entries from $\mathbb{R}$.

Example 2.A.1

Solving a Binary Linear System

Solve the following linear system over $\mathbb{Z}_2$:

$$\begin{aligned} x_1 \quad\;\;\; + x_3 \quad\;\;\; + x_5 &= 1 \\ x_1 + x_2 \quad\quad\;\;\; + x_5 &= 0 \\ x_3 + x_4 \quad\;\;\; &= 1 \\ x_1 \quad\;\;\; + x_3 + x_4 + x_5 &= 0 \\ x_2 + x_3 \quad\;\;\; + x_5 &= 1 \end{aligned}$$

Solution:

We represent this linear system via an augmented matrix and then use Gauss–Jordan elimination to put it into reduced row echelon form, just like we would for any other linear system. The only thing that we have to be careful of is that $1 + 1 = 0$ in this setting:

If $1+1=0$ then subtracting 1 from both sides shows that $1 = 0-1$, which we use repeatedly when row-reducing here.

Because $-1 = 1$ in $\mathbb{Z}_2$, the row operations $R_i - R_j$ and $R_i + R_j$ are the same as each other.

$$
\begin{bmatrix}
1 & 0 & 1 & 0 & 1 & 1 \\
1 & 1 & 0 & 0 & 1 & 0 \\
0 & 0 & 1 & 1 & 0 & 1 \\
1 & 0 & 1 & 1 & 1 & 0 \\
0 & 1 & 1 & 0 & 1 & 1
\end{bmatrix}
\xrightarrow[R_4 - R_1]{R_2 - R_1}
\begin{bmatrix}
1 & 0 & 1 & 0 & 1 & 1 \\
0 & 1 & 1 & 0 & 0 & 1 \\
0 & 0 & 1 & 1 & 0 & 1 \\
0 & 0 & 0 & 1 & 0 & 1 \\
0 & 1 & 1 & 0 & 1 & 1
\end{bmatrix}
$$

$$
\xrightarrow{R_5 - R_2}
\begin{bmatrix}
1 & 0 & 1 & 0 & 1 & 1 \\
0 & 1 & 1 & 0 & 0 & 1 \\
0 & 0 & 1 & 1 & 0 & 1 \\
0 & 0 & 0 & 1 & 0 & 1 \\
0 & 0 & 0 & 0 & 1 & 0
\end{bmatrix}
\xrightarrow{R_1 - R_5}
\begin{bmatrix}
1 & 0 & 1 & 0 & 0 & 1 \\
0 & 1 & 1 & 0 & 0 & 1 \\
0 & 0 & 1 & 1 & 0 & 1 \\
0 & 0 & 0 & 1 & 0 & 1 \\
0 & 0 & 0 & 0 & 1 & 0
\end{bmatrix}
$$

$$
\xrightarrow{R_3 - R_4}
\begin{bmatrix}
1 & 0 & 1 & 0 & 0 & 1 \\
0 & 1 & 1 & 0 & 0 & 1 \\
0 & 0 & 1 & 0 & 0 & 0 \\
0 & 0 & 0 & 1 & 0 & 1 \\
0 & 0 & 0 & 0 & 1 & 0
\end{bmatrix}
\xrightarrow[R_2 - R_3]{R_1 - R_3}
\begin{bmatrix}
1 & 0 & 0 & 0 & 0 & 1 \\
0 & 1 & 0 & 0 & 0 & 1 \\
0 & 0 & 1 & 0 & 0 & 0 \\
0 & 0 & 0 & 1 & 0 & 1 \\
0 & 0 & 0 & 0 & 1 & 0
\end{bmatrix}.
$$

It follows that the unique solution of this linear system is $(x_1, x_2, x_3, x_4, x_5) = (1, 1, 0, 1, 0)$.

There is one particular case, however, where linear systems over $\mathbb{Z}_2$ are somewhat different from those over the real numbers, and that is when the linear system has multiple solutions. When working over $\mathbb{R}$, if a linear system has multiple solutions then it must have infinitely many solutions. However, no linear system over $\mathbb{Z}_2$ can possibly have infinitely many solutions, since there are only a finite number of vectors in $\mathbb{Z}_2^n$ (specifically, the only possible solutions of a linear system over $\mathbb{Z}_2$ in n variables are the 2^n vectors consisting of all possible arrangements of 0s and 1). To get a feeling for what does and does not change in this case, we work through another example.

The notation $\mathbb{Z}_2^n$ refers to the set of vectors with n entries from $\mathbb{Z}_2$, just like $\mathbb{R}^n$ refers to the set of vectors with n real entries.

Example 2.A.2

Solving a Binary Linear System with Multiple Solutions

Solve the following linear system over $\mathbb{Z}_2$:

$$
\begin{aligned}
x_1 \phantom{{}+x_2} + x_3 + x_4 \phantom{{}+x_5} &= 1 \\
x_2 + x_3 \phantom{{}+x_4+x_5} &= 1 \\
x_1 \phantom{{}+x_2} + x_3 + x_4 + x_5 &= 0 \\
x_1 + x_2 \phantom{{}+x_3+x_4} + x_5 &= 0 \\
x_2 + x_3 \phantom{{}+x_4} + x_5 &= 0
\end{aligned}
$$

Solution:

Again, we just apply Gauss–Jordan elimination to put the augmented matrix that represents this linear system into reduced row echelon form:

$$
\left[\begin{array}{ccccc|c}
1 & 0 & 1 & 1 & 0 & 1 \\
0 & 1 & 1 & 0 & 0 & 1 \\
1 & 0 & 1 & 1 & 1 & 0 \\
1 & 1 & 0 & 0 & 1 & 0 \\
0 & 1 & 1 & 0 & 1 & 0
\end{array}\right]
\xrightarrow[R_4-R_1]{R_3-R_1}
\left[\begin{array}{ccccc|c}
1 & 0 & 1 & 1 & 0 & 1 \\
0 & 1 & 1 & 0 & 0 & 1 \\
0 & 0 & 0 & 0 & 1 & 1 \\
0 & 1 & 1 & 1 & 1 & 1 \\
0 & 1 & 1 & 0 & 1 & 0
\end{array}\right]
$$

$$
\xrightarrow[R_5-R_2]{R_4-R_2}
\left[\begin{array}{ccccc|c}
1 & 0 & 1 & 1 & 0 & 1 \\
0 & 1 & 1 & 0 & 0 & 1 \\
0 & 0 & 0 & 0 & 1 & 1 \\
0 & 0 & 0 & 1 & 1 & 0 \\
0 & 0 & 0 & 0 & 1 & 1
\end{array}\right]
\xrightarrow{R_3\leftrightarrow R_4}
\left[\begin{array}{ccccc|c}
1 & 0 & 1 & 1 & 0 & 1 \\
0 & 1 & 1 & 0 & 0 & 1 \\
0 & 0 & 0 & 1 & 1 & 0 \\
0 & 0 & 0 & 0 & 1 & 1 \\
0 & 0 & 0 & 0 & 1 & 1
\end{array}\right]
$$

$$
\xrightarrow[R_5-R_4]{R_3-R_4}
\left[\begin{array}{ccccc|c}
1 & 0 & 1 & 1 & 0 & 1 \\
0 & 1 & 1 & 0 & 0 & 1 \\
0 & 0 & 0 & 1 & 0 & 1 \\
0 & 0 & 0 & 0 & 1 & 1 \\
0 & 0 & 0 & 0 & 0 & 0
\end{array}\right]
\xrightarrow{R_1-R_3}
\left[\begin{array}{ccccc|c}
1 & 0 & 1 & 0 & 0 & 0 \\
0 & 1 & 1 & 0 & 0 & 1 \\
0 & 0 & 0 & 1 & 0 & 1 \\
0 & 0 & 0 & 0 & 1 & 1 \\
0 & 0 & 0 & 0 & 0 & 0
\end{array}\right].
$$

Notice that we never need to perform a "multiplication" row operation, since $0R_j$ is not a valid elementary row operation and $1R_j$ does not change anything.

We thus see that x_3 is a free variable while the others are leading. In particular, $x_5 = 1$, $x_4 = 1$, $x_2 = 1 - x_3$, and $x_1 = x_3$ (recall that $x_3 = -x_3$ in $\mathbb{Z}_2$).

However, x_3 being free does *not* mean that there are infinitely many solutions. Since there are only two possible choices for x_3 here (0 and 1), there are only two solutions, and they can be found simply by plugging $x_3 = 0$ or $x_3 = 1$ into the equations for x_1 and x_2 that we wrote down above. In particular, the two solutions of this linear system are

$$(x_1, x_2, x_3, x_4, x_5) = (0, 1, 0, 1, 1) \quad \text{and} \quad (x_1, x_2, x_3, x_4, x_5) = (1, 0, 1, 1, 1).$$

As illustrated by the above example, linear systems with free variables do not have infinitely many solutions in $\mathbb{Z}_2$, since there are not infinitely many choices for those free variables. Instead, there are only two choices for each free variable, which gives us the following fact:

> (!)   Every linear system over $\mathbb{Z}_2$ has either no solutions or exactly 2^k solutions, where k is a non-negative integer (equal to the number of free variables in the linear system).

2.A.2   The "Lights Out" Game

As one particularly interesting application of the fact that we can solve linear systems over finite fields (and over $\mathbb{Z}_2$ in particular), we consider the "Lights Out" game. The setup of this game is that buttons are arranged in a square grid, and pressing on one of those buttons toggles it on or off, but also toggles all other buttons that touch it orthogonally (but not diagonally), as in Figure 2.26.

The game itself asks the player to press the buttons so as to put them into some pre-specified configuration, such as turning them all on. The reason that this game is challenging is that buttons interact with each other—pressing an orthogonal neighbor of an "on" button turns it back off, so we typically will not win by just pressing whichever buttons we want to turn on.

Two buttons touching each other "orthogonally" means that they share a side, not just a corner.

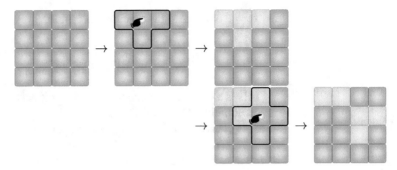

Figure 2.26: The "Lights Out" game on a 4×4 grid. Touching one of the buttons toggles (off-to-on or on-to-off) that button as well as all of its orthogonal neighbors.

In order to rephrase this game so that linear algebra can help us analyze it, we encode the states of the buttons in binary—0 for "off" and 1 for "on". We then arrange those 16 states into a vector (we choose to place them into the vector in standard left-to-right reading order: the top-left button's state is the first entry of the vector, the second button in the top row is the second entry of the vector, and so on, as in Figure 2.27).

The particular ordering of the buttons does not matter. All that matters is that we consistently use the same ordering throughout our calculations and interpretation of the final answer.

1	2	3	4
5	6	7	8
9	10	11	12
13	14	15	16

Figure 2.27: We use a vector $\mathbf{v} \in \mathbb{Z}_2^{16}$ to encode the states of the 4×4 "Lights Out" game, with its entries describing the states of the buttons in standard reading order: left-to-right and then top-to-bottom.

When we encode the game in this way, pressing one of the buttons just performs a specific vector addition in $\mathbb{Z}_2$. For example, the sequence of two button presses presented in Figure 2.26 can be represented by the vector addition

$$(0,0,0,0,\ 0,0,0,0,\ 0,0,0,0,\ 0,0,0,0) \quad \text{(all buttons start "off")}$$
$$+\ (1,1,1,0,\ 0,1,0,0,\ 0,0,0,0,\ 0,0,0,0) \quad \text{(press \#1 toggles 1, 2, 3, 6)}$$
$$+\ (0,0,1,0,\ 0,1,1,1,\ 0,0,1,0,\ 0,0,0,0) \quad \text{(press \#2 toggles 3, 6, 7, 8, 11)}$$
$$\overline{=\ (1,1,0,0,\ 0,0,1,1,\ 0,0,1,0,\ 0,0,0,0)} \quad \text{(just add, but } 1+1=0)$$

More generally, we use $\mathbf{v}_s, \mathbf{v}_e \in \mathbb{Z}_2^{16}$ to represent the vectors that encode the starting state and ending (i.e., target) state, respectively. For example, if our goal is to turn every light in the 4×4 version of the game from off to on then we have

$$\mathbf{v}_s = (0,0,0,0,\ 0,0,0,0,\ 0,0,0,0,\ 0,0,0,0) \quad \text{and}$$
$$\mathbf{v}_e = (1,1,1,1,\ 1,1,1,1,\ 1,1,1,1,\ 1,1,1,1).$$

If $\mathbf{a}_1, \mathbf{a}_2, \ldots, \mathbf{a}_{16}$ denote the 16 different vectors that are potentially added to $\mathbf{v}_s$ when we press the 16 different buttons, and we let $\mathbf{x} \in \mathbb{Z}_2^{16}$ denote the vector that encodes which buttons we press (i.e., $x_j = 1$ if we press the j-th button and $x_j = 0$ otherwise) then our goal is to find $x_1, x_2, \ldots, x_{16}$ so that

$$\mathbf{v}_s + (x_1\mathbf{a}_1 + x_2\mathbf{a}_2 + \cdots + x_{16}\mathbf{a}_{16}) = \mathbf{v}_e, \quad \text{or}$$

$$x_1\mathbf{a}_1 + x_2\mathbf{a}_2 + \cdots + x_{16}\mathbf{a}_{16} = \mathbf{v}_e - \mathbf{v}_s.$$

This is a system of linear equations, and it is convenient to write it as the matrix equation $A\mathbf{x} = \mathbf{v}_e - \mathbf{v}_s$, where $A = [\, \mathbf{a}_1 \mid \mathbf{a}_2 \mid \cdots \mid \mathbf{a}_{16} \,]$, which we can solve via Gaussian or Gauss–Jordan elimination as usual. In this particular case, the 16×16 matrix A has the following form, where (for ease of visualization) we use dots to denote entries that are equal to 0:

For example, the 5th column of A encodes the fact that pressing button 5 toggles buttons 1, 5, 6, and 9 (the non-zero entries in that column).

We just partition A as a block matrix to make it easier to visualize.

$$A = \left[\begin{array}{cccc|cccc|cccc|cccc}
1 & 1 & \cdot & \cdot & 1 & \cdot & \cdot & \cdot & \cdot & \cdot & \cdot & \cdot & \cdot & \cdot & \cdot & \cdot \\
1 & 1 & 1 & \cdot & \cdot & 1 & \cdot & \cdot & \cdot & \cdot & \cdot & \cdot & \cdot & \cdot & \cdot & \cdot \\
\cdot & 1 & 1 & 1 & \cdot & \cdot & 1 & \cdot & \cdot & \cdot & \cdot & \cdot & \cdot & \cdot & \cdot & \cdot \\
\cdot & \cdot & 1 & 1 & \cdot & \cdot & \cdot & 1 & \cdot & \cdot & \cdot & \cdot & \cdot & \cdot & \cdot & \cdot \\
\hline
1 & \cdot & \cdot & \cdot & 1 & 1 & \cdot & \cdot & 1 & \cdot & \cdot & \cdot & \cdot & \cdot & \cdot & \cdot \\
\cdot & 1 & \cdot & \cdot & 1 & 1 & 1 & \cdot & \cdot & 1 & \cdot & \cdot & \cdot & \cdot & \cdot & \cdot \\
\cdot & \cdot & 1 & \cdot & \cdot & 1 & 1 & 1 & \cdot & \cdot & 1 & \cdot & \cdot & \cdot & \cdot & \cdot \\
\cdot & \cdot & \cdot & 1 & \cdot & \cdot & 1 & 1 & \cdot & \cdot & \cdot & 1 & \cdot & \cdot & \cdot & \cdot \\
\hline
\cdot & \cdot & \cdot & \cdot & 1 & \cdot & \cdot & \cdot & 1 & 1 & \cdot & \cdot & 1 & \cdot & \cdot & \cdot \\
\cdot & \cdot & \cdot & \cdot & \cdot & 1 & \cdot & \cdot & 1 & 1 & 1 & \cdot & \cdot & 1 & \cdot & \cdot \\
\cdot & \cdot & \cdot & \cdot & \cdot & \cdot & 1 & \cdot & \cdot & 1 & 1 & 1 & \cdot & \cdot & 1 & \cdot \\
\cdot & \cdot & \cdot & \cdot & \cdot & \cdot & \cdot & 1 & \cdot & \cdot & 1 & 1 & \cdot & \cdot & \cdot & 1 \\
\hline
\cdot & \cdot & \cdot & \cdot & \cdot & \cdot & \cdot & \cdot & 1 & \cdot & \cdot & \cdot & 1 & 1 & \cdot & \cdot \\
\cdot & \cdot & \cdot & \cdot & \cdot & \cdot & \cdot & \cdot & \cdot & 1 & \cdot & \cdot & 1 & 1 & 1 & \cdot \\
\cdot & \cdot & \cdot & \cdot & \cdot & \cdot & \cdot & \cdot & \cdot & \cdot & 1 & \cdot & \cdot & 1 & 1 & 1 \\
\cdot & \cdot & \cdot & \cdot & \cdot & \cdot & \cdot & \cdot & \cdot & \cdot & \cdot & 1 & \cdot & \cdot & 1 & 1
\end{array}\right]. \quad (2.A.1)$$

Before we explicitly solve the "Lights Out" game, it is worth pointing out that a solution only depends on *which* buttons are pressed, not the order in which they are pressed, since the only things that determine the state of a button at the end of the game are its starting state and how many times it was toggled. Similarly, no solution could ever require us to press a button more than once, since pressing a button twice is the same as not pressing it at all. Mathematically, these two facts are encoded in the fact that mod-2 vector addition is commutative and associative.

Example 2.A.3

Solving the 4×4 "Lights Out" Game

Either find a set of button presses that flips all of the buttons in the 4×4 "Lights Out" game from Figure 2.26 from off to on, or show that no such set exists.

Solution:

The linear system that we wish to solve in this case is $A\mathbf{x} = \mathbf{v}_e - \mathbf{v}_s$, where A is the 16×16 coefficient matrix that we constructed in Equation (2.A.1) and $\mathbf{v}_e - \mathbf{v}_s \in \mathbb{Z}_2^{16}$ is the vector with every entry equal to 1.

The reduced row echelon form of the augmented matrix $\begin{bmatrix} A \mid \mathbf{v}_e - \mathbf{v}_s \end{bmatrix}$ is

Again, we use dots to denote entries that are equal to 0.

$$\left[\begin{array}{cccc|cccc|cccc|cccc|c}
1 & \cdot & \cdot & \cdot & \cdot & \cdot & \cdot & \cdot & \cdot & \cdot & \cdot & \cdot & \cdot & 1 & 1 & 1 & 1 \\
\cdot & 1 & \cdot & \cdot & \cdot & \cdot & \cdot & \cdot & \cdot & \cdot & \cdot & \cdot & 1 & 1 & \cdot & 1 & 1 \\
\cdot & \cdot & 1 & \cdot & \cdot & \cdot & \cdot & \cdot & \cdot & \cdot & \cdot & \cdot & 1 & \cdot & 1 & 1 & 1 \\
\cdot & \cdot & \cdot & 1 & \cdot & \cdot & \cdot & \cdot & \cdot & \cdot & \cdot & \cdot & 1 & 1 & 1 & \cdot & 1 \\
\cdot & \cdot & \cdot & \cdot & 1 & \cdot & \cdot & \cdot & \cdot & \cdot & \cdot & \cdot & 1 & \cdot & 1 & \cdot & 1 \\
\cdot & \cdot & \cdot & \cdot & \cdot & 1 & \cdot & \cdot & \cdot & \cdot & \cdot & \cdot & \cdot & \cdot & \cdot & 1 & \cdot \\
\cdot & \cdot & \cdot & \cdot & \cdot & \cdot & 1 & \cdot & \cdot & \cdot & \cdot & \cdot & 1 & \cdot & \cdot & \cdot & \cdot \\
\cdot & \cdot & \cdot & \cdot & \cdot & \cdot & \cdot & 1 & \cdot & \cdot & \cdot & \cdot & 1 & \cdot & 1 & 1 & 1 \\
\cdot & \cdot & \cdot & \cdot & \cdot & \cdot & \cdot & \cdot & 1 & \cdot & \cdot & \cdot & 1 & 1 & \cdot & \cdot & 1 \\
\cdot & \cdot & \cdot & \cdot & \cdot & \cdot & \cdot & \cdot & \cdot & 1 & \cdot & \cdot & 1 & 1 & 1 & \cdot & 1 \\
\cdot & \cdot & \cdot & \cdot & \cdot & \cdot & \cdot & \cdot & \cdot & \cdot & 1 & \cdot & \cdot & 1 & 1 & 1 & 1 \\
\cdot & \cdot & \cdot & \cdot & \cdot & \cdot & \cdot & \cdot & \cdot & \cdot & \cdot & 1 & \cdot & \cdot & 1 & 1 & 1 \\
\cdot & \cdot & \cdot & \cdot & \cdot & \cdot & \cdot & \cdot & \cdot & \cdot & \cdot & \cdot & \cdot & \cdot & \cdot & \cdot & \cdot \\
\cdot & \cdot & \cdot & \cdot & \cdot & \cdot & \cdot & \cdot & \cdot & \cdot & \cdot & \cdot & \cdot & \cdot & \cdot & \cdot & \cdot \\
\cdot & \cdot & \cdot & \cdot & \cdot & \cdot & \cdot & \cdot & \cdot & \cdot & \cdot & \cdot & \cdot & \cdot & \cdot & \cdot & \cdot \\
\cdot & \cdot & \cdot & \cdot & \cdot & \cdot & \cdot & \cdot & \cdot & \cdot & \cdot & \cdot & \cdot & \cdot & \cdot & \cdot & \cdot
\end{array}\right],$$

which tells us that this linear system has 4 free variables and thus $2^4 = 16$ solutions, which we can get our hands on by making various choices for the free variables $x_{13}, x_{14}, x_{15},$ and x_{16}. The simplest solution arises from choosing $x_{13} = 0, x_{14} = 1, x_{15} = 0, x_{16} = 0$, which gives

$$\mathbf{x} = (0,0,1,0,\ 1,0,0,0,\ 0,0,0,1,\ 0,1,0,0).$$

This solution tells us that the 4×4 "Lights Out" game can be solved by pressing the 3rd, 5th, 12th, and 14th buttons on the grid (again, in left-to-right then top-to-bottom standard reading order) as follows:

The remaining 15 solutions can be similarly constructed by making different choices for $x_{13}, x_{14}, x_{15},$ and x_{16} (i.e., we range through all possible ways of pressing buttons on the bottom row of the 4×4 grid):

Again, to describe
a solution of this
game, we just need
to specify which
buttons are
pressed—the order
in which they are
pressed is irrelevant.

Not surprisingly, many of these solutions are simply rotations and reflections of each other. If we regard solutions as "the same" if they can be mirrored or rotated to look like each other, then there are only 5 distinct solutions.

Since 16 different sets of button presses all lead to the same result on the 4×4 grid (i.e., turn all lights from "off" to "on"), most starting configurations of that grid cannot be turned into the all-"on" configuration (of the $2^{16} = 65\,536$ starting configurations, only $65\,536/16 = 4\,096$ are solvable). For example, of the 16 configurations of the 4×4 grid with the bottom 3 rows off, the only one that is solvable is the one that also has the entire top row off (see Figure 2.28).

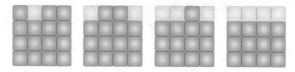

Variants of "Lights
Out" appear
frequently in video
games. Architects in
fantasy worlds have
a strange
propensity to use it
as a door-unlocking
mechanism.

Figure 2.28: Some unsolvable configurations of the "Lights Out" game. Starting from these configurations, no set of button presses can turn all of the buttons on.

These same techniques can similarly be used to solve the "Lights Out" game on larger game boards, with different starting configurations, or even when the buttons are not laid out in a square grid. We present one more example to illustrate how our methods work in these slightly more general settings work.

Example 2.A.4

Solving a Triangular "Lights Out" Game

Either find a set of button presses that flips on the remaining buttons (1, 2, 3, 4, and 9) in the "Lights Out" game displayed below, or show that no such set exists.

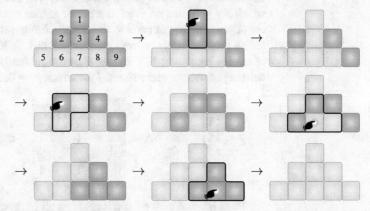

Solution:

We represent the starting state via the vector $\mathbf{v}_s = (0, 0, 0, 0, 1, 1, 1, 1, 0)$ and the target state via the vector $\mathbf{v}_e = (1, 1, 1, 1, 1, 1, 1, 1, 1)$. If we let A be the 9×9 binary matrix whose j-th column (for each $1 \leq j \leq 9$) describes which buttons are toggled when button j is pressed, then the linear system $A\mathbf{x} = \mathbf{v}_e - \mathbf{v}_s$ that we want to solve can be represented by the following augmented matrix $\left[A \mid \mathbf{v}_e - \mathbf{v}_s \right]$:

> Again, we just partition this matrix in this way for ease of visualization. Notice that each block corresponds to one row of the "Lights Out" grid.

$$\left[\begin{array}{cccc|ccccc|c} 1 & \cdot & 1 & \cdot & \cdot & \cdot & \cdot & \cdot & \cdot & 1 \\ \cdot & 1 & 1 & \cdot & \cdot & 1 & \cdot & \cdot & \cdot & 1 \\ 1 & 1 & 1 & 1 & \cdot & \cdot & 1 & \cdot & \cdot & 1 \\ \cdot & \cdot & 1 & 1 & \cdot & \cdot & \cdot & 1 & \cdot & 1 \\ \cdot & \cdot & \cdot & \cdot & 1 & 1 & \cdot & \cdot & \cdot & 0 \\ \cdot & 1 & \cdot & \cdot & 1 & 1 & 1 & \cdot & \cdot & 0 \\ \cdot & \cdot & 1 & \cdot & \cdot & 1 & 1 & 1 & \cdot & 0 \\ \cdot & \cdot & \cdot & 1 & \cdot & \cdot & 1 & 1 & 1 & 0 \\ \cdot & \cdot & \cdot & \cdot & \cdot & \cdot & \cdot & 1 & 1 & 1 \end{array} \right].$$

It is straightforward (albeit tedious) to verify that the reduced row echelon form of this augmented matrix has the form $\left[I \mid \mathbf{x} \right]$, where $\mathbf{x} = (1, 1, 0, 0, 0, 0, 1, 1, 0)$. It follows that this linear system has a unique solution, and it corresponds to pressing buttons 1, 2, 7, and 8:

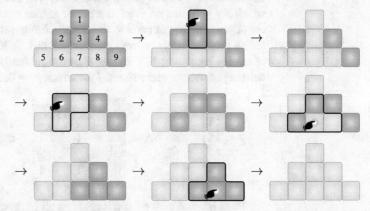

Interestingly, the RREF of A being the identity matrix tells us that the "Lights Out" game on this board can *always* be solved (and that the solution is always unique), regardless of the starting configuration of on/off buttons.

In the previous examples, we saw two very different button layouts on which the "Lights Out" game could be solved from the all-off starting configuration. One of the most remarkable facts about this game is that, even though the starting configuration of the board affects whether or not a solution exists, its size and shape do not. That is, there *always* exists a set of button presses that flips that state of every single button, even if the grid is a large and hideous mess like the one in Figure 2.29.

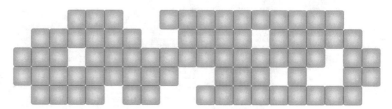

Figure 2.29: The "Lights Out" game can be solved on any game board, no matter how ugly, as long as we start from the all-off configuration.

Theorem 2.A.1

"Lights Out" Can Always Be Solved (From All-Off)

In the "Lights Out" game, there is always a set of button presses that turns the all-off configuration into the all-on configuration, regardless of the size and shape of the grid of buttons.

Proof. Recall that the linear system we want to solve has the form $A\mathbf{x} = \mathbf{1}$, where $\mathbf{1} = (1,1,\ldots,1) \in \mathbb{Z}_2^n$ is the vector with every entry equal to 1. In other words, we want to show that $\mathbf{1} \in \text{range}(A)$. Furthermore, A has the following two properties that we will need to make use of:

This proof makes use of many of the ideas from Section 2.4.2. If you have not yet read that section, it is probably best to skip this proof for now.

- Every diagonal entry of A equals 1, since each button toggles itself when pressed.
- A is symmetric (i.e., $A^T = A$), since button i is a neighbor of button j if and only if button j is a neighbor of button i (and thus they toggle each other) for all i and j.

With these two properties in mind, we now prove the theorem in three steps. The first step is to show that every $\mathbf{y} \in \text{null}(A)$ (i.e., every $\mathbf{y} \in \mathbb{Z}_2^n$ for which $A\mathbf{y} = \mathbf{0}$) has an even number of entries equal to 1 (in other words, it is not possible to get back to the all-off configuration via an odd number of button presses—only an even number will do). To his end, let k be the number of entries equal to 1 in $\mathbf{y}$ and let B be the submatrix of A that is obtained by deleting every row and column j for which $y_j = 0$. For example, if

$$
\mathbf{y} = \begin{bmatrix} 0 \\ 1 \\ 1 \\ 0 \\ 1 \end{bmatrix} \quad \text{and} \quad A = \begin{bmatrix} 1 & 0 & 1 & 1 & 0 \\ 0 & 1 & 0 & 1 & 0 \\ 1 & 0 & 1 & 1 & 1 \\ 1 & 1 & 1 & 1 & 0 \\ 0 & 0 & 1 & 0 & 1 \end{bmatrix} \quad \text{then} \quad B = \begin{bmatrix} 1 & 0 & 0 \\ 0 & 1 & 1 \\ 0 & 1 & 1 \end{bmatrix}.
$$

Then $B^T = B$ as well and it also has all of its diagonal entries equal to 1. Furthermore, the fact that $A\mathbf{y} = \mathbf{0}$ tells us that $B\mathbf{1} = \mathbf{0}$. However, the entries of $B\mathbf{1}$ just count the number of 1s in each row of B, so every row of B contains an even number of 1s. By Exercise 2.A.10, it follows that the number of rows of B (i.e., k) must be even. We thus conclude that $\mathbf{y}$ contains an even number of 1s, which completes the first step of the proof.

Be slightly careful here—this vector $\mathbf{1}$ lives in $\mathbb{Z}_2^k$, whereas the $\mathbf{1}$ that we mentioned at the very start of this proof lives in $\mathbb{Z}_2^n$.

For the second step of the proof, we show that the reduced row echelon form R of A has an odd number of 1s in each of its columns. For the leading columns of R, this fact is trivial since they (by definition) contain exactly one 1. For the non-leading columns of R, we recall from Remark 2.4.2 that a basis of $\text{null}(A)$ can be obtained by spacing out the non-zero entries in those non-leading columns appropriately and then adding one more non-zero term in their "diagonal" entry. For example, if

Since we are working in $\mathbb{Z}_2$, the extra non-zero entry that we add to these columns is 1, not -1 like in Remark 2.4.2.

$$R = \begin{bmatrix} 1 & 0 & 0 & 0 & 0 & 1 \\ 0 & 1 & 1 & 0 & 0 & 1 \\ 0 & 0 & 0 & 1 & 0 & 0 \\ 0 & 0 & 0 & 0 & 1 & 1 \\ 0 & 0 & 0 & 0 & 0 & 0 \\ 0 & 0 & 0 & 0 & 0 & 0 \end{bmatrix} \quad \text{then} \quad \text{null}(A) = \text{span}\left(\begin{bmatrix} 0 \\ 1 \\ 1 \\ 0 \\ 0 \\ 0 \end{bmatrix}, \begin{bmatrix} 1 \\ 1 \\ 0 \\ 0 \\ 1 \\ 1 \end{bmatrix} \right).$$

Since these basis vectors are contained in $\text{null}(A)$, we know from the first step of this proof that they each contain an even number of 1s, so the non-leading columns of R each contain an odd number of 1s. This completes the second step of the proof.

For the third (and final) step of the proof, we recall that the non-zero rows of R form a basis of $\text{range}(A^T)$, which equals $\text{range}(A)$ in this case since $A^T = A$. In particular, this means that any linear combination of the rows of R is in the range of A. Since each column of R contains an odd number of 1s, adding up all of the rows of R results in the all-ones vector $\mathbf{1}$. It follows that $\mathbf{1} \in \text{range}(A)$, which is exactly what we wanted to prove. ∎

In fact, the above theorem even works for variants of this game in which pressing a button toggles the state of any other given configuration of buttons—not necessarily its orthogonal neighbors. That is, if we put wires between buttons and demand that pressing a button toggles the state of every other button that is connected to it via a wire (but do not make any restriction on the physical locations of these buttons relative to each other), then there is *still* always a solution, regardless of how many buttons there are or how we connect them (as long as we start from the all-off configuration). Furthermore, we can use the exact same method that we have been using already to actually find the solutions(s)—see Exercise 2.A.9.

2.A.3 Linear Systems with More States

We can also solve linear systems that make use of modular arithmetic with a modulus larger than 2. For example, $\mathbb{Z}_3$ denotes the set $\{0, 1, 2\}$ under arithmetic mod 3. That is, we perform addition and multiplication in the usual ways, except we "cycle" them through the numbers 0, 1, and 2. For example, instead of having $2 + 2 = 4$, we have $2 + 2 = 1$ since the addition "rolls over" back to 0 when it hits 3. Explicitly, the addition and multiplication tables in $\mathbb{Z}_3$ are as follows:

+	0	1	2
0	0	1	2
1	1	2	0
2	2	0	1

×	0	1	2
0	0	0	0
1	0	1	2
2	0	2	1

Perhaps the most familiar example of modular arithmetic comes from 12-hour clocks, which split time up into 12-hour chunks and thus correspond to modular arithmetic with modulus 12. For example, if it is 9:00 now then in 7 hours it will be 4:00, since $9 + 7 = 16 = 12 + 4$, and the extra 12 is irrelevant (it only affects whether it is morning or night, or which day it is—not the hour displayed on the clock). For example, in mod-12 arithmetic, $9 + 7 = 4$ (see Figure 2.30).

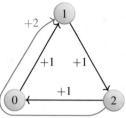
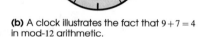

(a) In $\mathbb{Z}_3$, addition and multiplication work cyclically with the three values 0, 1, and 2. For example, $2+2=1$ (not 4).

(b) A clock illustrates the fact that $9+7=4$ in mod-12 arithmetic.

Figure 2.30: Modular arithmetic works by having addition and multiplication "wrap around" the modulus.

It turns out that we can use our standard techniques for solving linear systems over $\mathbb{Z}_p$ (i.e., using just the numbers 0, 1, 2, ..., $p-1$) when p is any prime number. We now present an example that shows how this works for $\mathbb{Z}_3$.

Example 2.A.5

Solving a Ternary Linear System

Solve the following linear system over $\mathbb{Z}_3$:

$$
\begin{aligned}
w \quad\ \ +y+2z &= 2 \\
x+y+\ z &= 1 \\
2w+x+y+\ z &= 0 \\
w \quad\ +y \quad\ &= 1
\end{aligned}
$$

Solution:

Just as we did when solving linear systems over $\mathbb{Z}_2$, we represent this linear system via an augmented matrix and then use Gauss–Jordan elimination to put it into reduced row echelon form:

Keep in mind that we are working in mod-3 arithmetic, so subtraction also works cyclically: $0-1=2$, $0-2=1$, $1-2=2$, and so on (just follow the arrows in Figure 2.30 backwards).

$$
\left[\begin{array}{cccc|c}
1 & 0 & 1 & 2 & 2 \\
0 & 1 & 1 & 1 & 1 \\
2 & 1 & 1 & 1 & 0 \\
1 & 0 & 1 & 0 & 1
\end{array}\right]
\xrightarrow[R_4-R_1]{R_3-2R_1}
\left[\begin{array}{cccc|c}
1 & 0 & 1 & 2 & 2 \\
0 & 1 & 1 & 1 & 1 \\
0 & 1 & 2 & 0 & 2 \\
0 & 0 & 0 & 1 & 2
\end{array}\right]
$$

$$
\xrightarrow{R_3-R_2}
\left[\begin{array}{cccc|c}
1 & 0 & 1 & 2 & 2 \\
0 & 1 & 1 & 1 & 1 \\
0 & 0 & 1 & 2 & 1 \\
0 & 0 & 0 & 1 & 2
\end{array}\right]
\xrightarrow[\substack{R_2-R_4 \\ R_3-2R_4}]{R_1-2R_4}
\left[\begin{array}{cccc|c}
1 & 0 & 1 & 0 & 1 \\
0 & 1 & 1 & 0 & 2 \\
0 & 0 & 1 & 0 & 0 \\
0 & 0 & 0 & 1 & 2
\end{array}\right]
$$

$$
\xrightarrow[R_2-R_3]{R_1-R_3}
\left[\begin{array}{cccc|c}
1 & 0 & 0 & 0 & 1 \\
0 & 1 & 0 & 0 & 2 \\
0 & 0 & 1 & 0 & 0 \\
0 & 0 & 0 & 1 & 2
\end{array}\right].
$$

It follows that the unique solution of this linear system is $(w,x,y,z) = (1,2,0,2)$.

Remark 2.A.1

The "Set" Card Game

One particularly interesting application of linear algebra over $\mathbb{Z}_3$ comes in the form of a card game called "Set". In this game, players use a custom deck of cards in which each card contains colored symbols that can vary in 4 different properties:

- color (red, green, or purple),
- shape (diamond, squiggle, or oval),
- shading (solid, striped, or hollow), and
- number of symbols (one, two, or three).

In particular, each possible combination of properties appears exactly once in the deck (for example, there is exactly one card in the deck with two striped red ovals), for a total of $3^4 = 81$ cards in the deck. Some examples of cards are as follows:

| green | red | purple | green | purple |

It is impossible for *all* properties to be the same among the three cards, since each card only appears once in the deck.

In this game, a "set" is any collection of three cards such that each property of the cards is either the same or different. For example, the following collections of cards form valid sets:

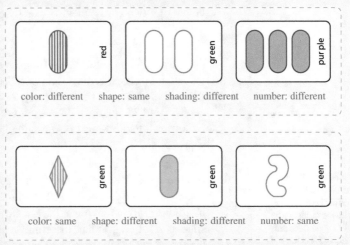

Three cards do *not* form a valid set if one of the properties is shared by two of the cards but not the other.

The game is typically played with two or more players being presented with a large collection of cards and being tasked with finding valid sets among them—whoever finds the most sets before cards or time run out is the winner.

Mathematically, this game is interesting because it can be phrased in terms of linear algebra over $\mathbb{Z}_3$. In particular, we can represent each of the 81 cards as a vector in $\mathbb{Z}_3^4$ by encoding each of the properties as one of the components of a 4-entry vector whose entries can be 0, 1, or 2. Specifically, we can write the cards as vectors of the form $\mathbf{v} = (c, p, d, n)$, where c, p, d, and n encode the **c**olor, sha**p**e, sha**d**ing, and **n**umber of symbols on the card, respectively, as follows:

Value	Color (c)	Shape (p)	Shading (d)	Number (n)
0	red	diamond	solid	one
1	green	squiggle	striped	two
2	purple	oval	hollow	three

The particular assignment of values here does not matter—all that matters is that we pick some way of assigning the properties to $\{0,1,2\}$ and we stick with it.

For example, a card with one red striped oval on it would be represented by the vector $\mathbf{v} = (c,p,d,n) = (0,2,1,0)$.

When representing "Set" cards in this way, the condition that they form a valid set (i.e., each property is either the same or different) is equivalent to the condition that their associated vectors $\mathbf{v}_1$, $\mathbf{v}_2$, and $\mathbf{v}_3$ are such that $\mathbf{v}_1 + \mathbf{v}_2 + \mathbf{v}_3 = \mathbf{0}$. To see why this is the case, just notice that the only ways to add up 3 numbers in $\mathbb{Z}_3$ to get 0 are as follows:

$$0+0+0 = 0, \quad 1+1+1 = 0, \quad 2+2+2 = 0, \quad \text{and} \quad 0+1+2 = 0.$$

That is, the three numbers must all be the same or all be different, which is exactly the rule for making sets.

We can use this representation of the "Set" game to quickly prove some interesting facts about it:

- Given any two cards, there is exactly one card that forms a valid set with them. This is because, given $\mathbf{v}_1, \mathbf{v}_2 \in \mathbb{Z}_3^4$, there is a unique $\mathbf{v}_3 \in \mathbb{Z}_3^4$ such that $\mathbf{v}_1 + \mathbf{v}_2 + \mathbf{v}_3 = \mathbf{0}$ (in particular, $\mathbf{v}_3 = -\mathbf{v}_1 - \mathbf{v}_2$).

- If 26 sets (and thus $26 \times 3 = 78$ cards) are removed from the deck, the remaining 3 cards must form a set. This is because the sum of all 81 vectors in $\mathbb{Z}_3^4$ is $\mathbf{0}$ (in each entry, 27 vectors are "0", 27 vectors are "1", and 27 vectors are "2"), so if the sum of the 78 vectors we remove is $\mathbf{0}$ then so must be the sum of the remaining 3 vectors.

Recall that an affine space is a shifted subspace. It's like a subspace, but not necessarily going through $\mathbf{0}$.

- Since $\mathbf{v}_1 + \mathbf{v}_2 + \mathbf{v}_3 = \mathbf{0}$ is equivalent to $\mathbf{v}_1 - \mathbf{v}_2 = \mathbf{v}_2 - \mathbf{v}_3$ in $\mathbb{Z}_3^4$, three cards form a set if and only if their associated vectors lie on a common line (1-dimensional affine space) in $\mathbb{Z}_3^4$. This is difficult to visualize directly since $\mathbb{Z}_3^4$ is 4-dimensional, but we can get some intuition for it by looking at the lines within a 2-dimensional affine slice of $\mathbb{Z}_3^4$. In particular, the affine space consisting of vectors of the form $(x,2,y,x)$ is displayed below:

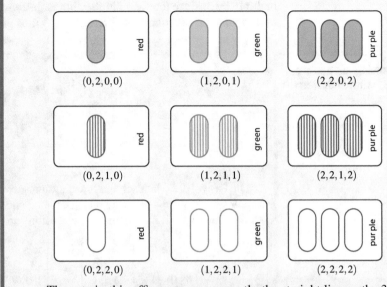

$(0,2,0,0)$		$(1,2,0,1)$		$(2,2,0,2)$	
$(0,2,1,0)$		$(1,2,1,1)$		$(2,2,1,2)$	
$(0,2,2,0)$		$(1,2,2,1)$		$(2,2,2,2)$	

The sets in this affine space are exactly the straight lines—the 3 rows, 3 columns, 3 forward diagonals (keep in mind that diagonals and lines "wrap around" in this space), and 3 backward diagonals.

We cannot directly apply these same techniques to solve linear systems via mod-4 arithmetic (or mod-n arithmetic when n is not a prime number) since in this case we run into issues like the equation $2x = 1$ not having a solution. That is, we cannot "divide by 2" in mod-4 arithmetic (see the following mod-4 addition and multiplication tables):

To be clear, modular arithmetic itself still works fine when p is not prime—it is just the process of solving linear systems that gets harder in this case.

+	0	1	2	3
0	0	1	2	3
1	1	2	3	0
2	2	3	0	1
3	3	0	1	2

×	0	1	2	3
0	0	0	0	0
1	0	1	2	3
2	0	2	0	2
3	0	3	2	1

However, this problem does not arise when working in $\mathbb{Z}_p$ when p is prime, since in this case each row of the mod-p multiplication table is a permutation of $\{0, 1, 2, \ldots, p-1\}$, so in particular we can always "divide" by any scalar—given any $y \in \mathbb{Z}_p$, we can find an $x \in \mathbb{Z}_p$ such that $xy = 1$. For example, the addition and multiplication tables in $\mathbb{Z}_5$ are as follows:

+	0	1	2	3	4
0	0	1	2	3	4
1	1	2	3	4	0
2	2	3	4	0	1
3	3	4	0	1	2
4	4	0	1	2	3

×	0	1	2	3	4
0	0	0	0	0	0
1	0	1	2	3	4
2	0	2	4	1	3
3	0	3	1	4	2
4	0	4	3	2	1

Keep in mind that fractions do not exist in $\mathbb{Z}_5$—the only scalars that we should ever see when working in $\mathbb{Z}_5$ are 0, 1, 2, 3, and 4.

With this in mind, we now present an example of how to solve a linear system over $\mathbb{Z}_5$, but we keep in mind that if we ever have an urge to divide by a scalar, we instead multiply by the scalar that makes the product equal 1. For example, to solve the equation $2x = 3$ in $\mathbb{Z}_5$, instead of dividing by 2 we multiply by 3, since the $\mathbb{Z}_5$ multiplication table above shows us that $2 \cdot 3 = 1$. Multiplying the equation $2x = 3$ through by 3 then gives us $x = 4$.

Example 2.A.6

Solving a Quinary Linear System

Find all solutions of the following linear system over $\mathbb{Z}_5$:

$$\begin{aligned} 2w \quad\quad + 4y \quad\quad &= 1 \\ 3w + x + 2y \quad\quad &= 0 \\ 4x + 4y + z &= 1 \\ 3w + 3x + 4y + 2z &= 1 \end{aligned}$$

Solution:

Just as we did when solving linear systems over $\mathbb{Z}_2$ and $\mathbb{Z}_3$, we represent this linear system via an augmented matrix and then use Gauss–Jordan elimination to put it into reduced row echelon form:

$$\left[\begin{array}{cccc|c} 2 & 0 & 4 & 0 & 1 \\ 3 & 1 & 2 & 0 & 0 \\ 0 & 4 & 4 & 1 & 1 \\ 3 & 3 & 4 & 2 & 1 \end{array}\right] \xrightarrow{3R_1} \left[\begin{array}{cccc|c} 1 & 0 & 2 & 0 & 3 \\ 3 & 1 & 2 & 0 & 0 \\ 0 & 4 & 4 & 1 & 1 \\ 3 & 3 & 4 & 2 & 1 \end{array}\right]$$

$$\xrightarrow[R_4-3R_1]{R_2-3R_1} \left[\begin{array}{cccc|c} 1 & 0 & 2 & 0 & 3 \\ 0 & 1 & 1 & 0 & 1 \\ 0 & 4 & 4 & 1 & 1 \\ 0 & 3 & 3 & 2 & 2 \end{array}\right] \xrightarrow[R_4-3R_2]{R_3-4R_2} \left[\begin{array}{cccc|c} 1 & 0 & 2 & 0 & 3 \\ 0 & 1 & 1 & 0 & 1 \\ 0 & 0 & 0 & 1 & 2 \\ 0 & 0 & 0 & 2 & 4 \end{array}\right]$$

$$\xrightarrow{R_4-2R_3} \left[\begin{array}{cccc|c} 1 & 0 & 2 & 0 & 3 \\ 0 & 1 & 1 & 0 & 1 \\ 0 & 0 & 0 & 1 & 2 \\ 0 & 0 & 0 & 0 & 0 \end{array}\right].$$

It follows that y is a free variable and w, x, and z are leading variables satisfying $z = 2$, $x = 1 - y$, and $w = 3 - 2y$. By letting y range over the five possible values 0, 1, 2, 3, and 4, we see that this linear system has exactly five solutions, which are

$$(3,1,0,2), \ (1,0,1,2), \ (4,4,2,2), \ (2,3,3,2), \ \text{and} \ (0,2,4,2).$$

Given that the linear system of the previous example has exactly 5 solutions, it seems natural to wonder how many solutions a linear system over $\mathbb{Z}_p$ can have. We already answered this question when $p = 2$: recall that, in this case, the number of solutions is always a power of 2. The following theorem establishes the natural generalization of this fact to arbitrary primes.

Theorem 2.A.2

Number of Solutions of a Linear System Over a Finite Field

Let p be a prime number. Then a linear system over $\mathbb{Z}_p$ has either no solutions or exactly p^k solutions, where k is a non-negative integer (equal to the number of free variables in the linear system).

The proof of this theorem involves nothing more than noting that each free variable can take on one of p different values, so if there are k free variables then there are p^k different combinations of values that they can take on, and each one corresponds to a different solution of the linear system.

Exercises

solutions to starred exercises on page 456

2.A.1 Find all solutions of the following systems of linear equations over $\mathbb{Z}_2$.

*(a) $x+y+z=1$
     $x+y=1$
     $x+z=0$

(b) $x+y+z=0$
     $x+y=1$

(d) $x+z=0$
     $w+y+z=1$
     $w+x+y+z=0$
     $x+y+z=1$

*(c) $w+x+z=1$
     $w+y+z=0$
     $x+y+z=1$

*(e) $v+w+x=0$
     $w+x+y=1$
     $x+y+z=0$
     $v+y+z=1$
     $v+w+z=0$

(f) $r+s+w+x=1$
     $r+v+x+y=0$
     $s+v+w+z=1$
     $r+w+y+z=1$
     $s+w+x+z=0$

2.A.2 Find all solutions of the following systems of linear equations over $\mathbb{Z}_3$.

*(a) $x+2y+2z=0$
     $2x+y+2z=1$
     $x+z=2$

(b) $2w+x+y=2$
     $w+2y+2z=1$
     $x+y+2z=0$
     $2x+2y+z=0$

2.A.3 Find all solutions of the following systems of linear equations over $\mathbb{Z}_5$.

*(a) $x+2y+3z=4$
     $2x+3y+z=1$
     $4x+y+2z=0$

(b) $4w+y+3z=3$
     $2w+x+y+2z=1$
     $w+3x+4z=3$
     $4x+y+3z=3$

2.A.4 Set up a linear system for finding a set of button presses that solves the "Lights Out" game on each of the following grids. Use computer software to solve this linear system, thus solving the game.

*(a)   (b)

*(c) 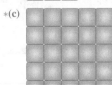   (d)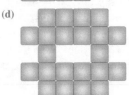

2.A.5 Set up a linear system for finding a set of button presses that solves the "Lights Out" game on each of the following grids. Use computer software to either solve this linear system, thus solving the game, or show that no solution exists.

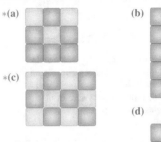

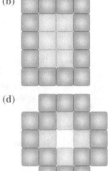

2.A.6 Determine which of the following statements are true and which are false.

*(a) Linear systems over $\mathbb{Z}_2$ must have exactly 0, 1, or infinitely many solutions.

(b) A linear system over $\mathbb{Z}_5$ with 3 variables must have exactly 0, 1, 5, 25, or 125 solutions.

*(c) Every game of "Lights Out" has a solution, regardless of the layout and on/off configuration of the buttons.

(d) Every game of "Lights Out", starting from the all-off configuration and with the goal of reaching the all-on configuration, has a solution regardless of the layout of the buttons.

*(e) If we represent a game of "Lights Out" as a linear system over $\mathbb{Z}_2$ of the form $A\mathbf{x} = \mathbf{v}_e - \mathbf{v}_s$ then A must be symmetric (i.e., $A^T = A$).

(f) There exist two cards in the "Set" game of Remark 2.A.1 that do not form a valid set with any third card.

*‌**2.A.7** Find integers v, w, x, y, z (or show that none exist) such that

$$
\begin{array}{rrrrrl}
v & 3w + & x + & y + & z & \text{is even} \\
-2v - & w + & 2x + & y & & \text{is odd} \\
v + & w & & + y + & z & \text{is odd} \\
& w - & x + & 2y - & z & \text{is even, and} \\
v + & w - & 3x + & y + & 2z & \text{is even.}
\end{array}
$$

2.A.8 Find integers v, w, x, y, z (or show that none exist) such each of the following quantities is an integer multiple of 5:

$$
\begin{array}{rrrrrrr}
v + & w + & x + & y - & z - & 1 \\
v - & 2w + & 2x + & 2y + & 2z + & 2 \\
-v + & w + & x & & + z \\
& 2w - & x + & y - & 3z + & 3 \\
2v + & w & & - 2y + & z - & 4
\end{array}
$$

∗∗2.A.9 Consider the variants of the "Lights Out" game in which the following 9 buttons toggle each other if and only if they are connected directly by a wire:

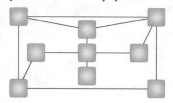

(a) Set up a linear system (over $\mathbb{Z}_2$) that can be used to solve this game (i.e., to find a set of button presses that turns every button on).

☐(b) Find a solution to the linear system from part (a) and thus the "Lights Out" game.

∗∗2.A.10 Suppose that B is a symmetric $k \times k$ matrix, all of whose entries are 0 or 1.

(a) Show that if B has every diagonal entry equal to 0 and has an odd number of 1s in each row, then k must be even. [Hint: B must have an even number of entries equal to 1 (why?).]
(b) Show that if B has every diagonal entry equal to 1 and has an even number of 1s in each row, then k must be even.

2.A.11 Recall that every solution of the "Set" card game described in Remark 2.A.1 is a solution of a certain linear system over $\mathbb{Z}_3$. It turns out that there are exactly 1080 different sets (i.e., winning 3-card combinations) possible, which is not a power of 3. Explain why this does not contradict Theorem 2.A.2.

2.A.12 Provide an example to show that Theorem 2.A.2 does not hold when p is not prime. [Hint: There is a mod-4 linear system with exactly 2 solutions.]

2.B   Extra Topic: Linear Programming

We now introduce an optimization technique that is based on the linear algebra tools that we have developed so far. In many real-world situations, we want to maximize or minimize some function under some constraints (i.e., restrictions on the variables). Standard calculus techniques can handle the cases where the function being optimized only has one or two input variables, but they quickly become cumbersome as the number of variables increases. On the other hand, linear programming is a method that lets us easily optimize functions of many variables, as long as the function and constraints are all linear.

2.B.1   The Form of a Linear Program

Loosely speaking, a linear program is a optimization problem in which the function being maximized or minimized, as well as all of the constraints, are linear in the variables. We typically use $x_1, x_2, \ldots, x_n \in \mathbb{R}$ as the variables, so an example of a linear program (in two variables x_1 and x_2) is

$$\begin{aligned} \text{maximize:} \quad & x_1 + 3x_2 \\ \text{subject to:} \quad & 2x_1 + x_2 \leq 4 \\ & x_1 + 3x_2 \leq 6 \\ & x_1, \quad x_2 \geq 0 \end{aligned}$$

The entries of A are the coefficients on the left-hand-side of the constraints, the entries of $\mathbf{b}$ are the scalars on the right-hand-side of the constraints, and the entries of $\mathbf{c}$ are the coefficients in the function being maximized.

When working with linear programs, it is usually convenient to group the various scalars into vectors and matrices (just like we write systems of linear equations as the single matrix equation $A\mathbf{x} = \mathbf{b}$ rather than writing out each scalar and variable explicitly every time). With this in mind, we would typically write the above linear program in the more compact form

$$\begin{aligned} \text{maximize:} \quad & \mathbf{c} \cdot \mathbf{x} \\ \text{subject to:} \quad & A\mathbf{x} \leq \mathbf{b} \\ & \mathbf{x} \geq \mathbf{0} \end{aligned} \qquad \text{(2.B.1)}$$

where

$$A = \begin{bmatrix} 2 & 1 \\ 1 & 3 \end{bmatrix}, \quad \mathbf{b} = (4,6), \quad \text{and} \quad \mathbf{c} = (1,3),$$

and the vector inequalities are meant entrywise (e.g., $\mathbf{x} \geq \mathbf{0}$ means that each entry of $\mathbf{x}$ is non-negative, and $A\mathbf{x} \leq \mathbf{b}$ means that each entry of $A\mathbf{x}$ is no larger than the corresponding entry of $\mathbf{b}$).

With the above example in mind, we now clarify exactly what types of optimization problems are and are not linear programs.

Definition 2.B.1 **Linear Program (Standard Form)**	A **linear program** (**LP**) is an optimization problem that can be written in the following form, where $A \in \mathcal{M}_{m,n}$, $\mathbf{b} \in \mathbb{R}^m$, $\mathbf{c} \in \mathbb{R}^n$ are fixed, and $\mathbf{x} \in \mathbb{R}^n$ is a vector of variables: $$\begin{aligned} \text{maximize:} \quad & \mathbf{c} \cdot \mathbf{x} \\ \text{subject to:} \quad & A\mathbf{x} \leq \mathbf{b} \\ & \mathbf{x} \geq \mathbf{0} \end{aligned} \quad \text{(2.B.2)}$$ Furthermore, this is called the **standard form** of the linear program.

The above definition perhaps seems somewhat restrictive. For example, what if we wanted to consider a minimization problem that contained some equality constraints, and allowed some of the variables to be negative? It turns out that this is no problem—by massaging things a little bit, we can actually write a fairly wide variety of optimization problems in the standard form (2.B.2).

Before we demonstrate these techniques though, it is useful to clarify some terminology. The **objective function** is the function that is being maximized or minimized (i.e., $\mathbf{c} \cdot \mathbf{x}$), and the **optimal value** is the maximal or minimal value that the objective function can have subject to the constraints (i.e., it is the "solution" of the linear program). A **feasible vector** is a vector $\mathbf{x} \in \mathbb{R}^n$ that satisfies all of the constraints (i.e., $A\mathbf{x} \leq \mathbf{b}$ and $\mathbf{x} \geq \mathbf{0}$), and the **feasible region** is the set of all feasible vectors.

Minimization Problems

To write a minimization problem in standard form, we just have to notice that minimizing $\mathbf{c} \cdot \mathbf{x}$ is the same as maximizing $-(\mathbf{c} \cdot \mathbf{x})$ and then multiplying the optimal value by -1. We illustrate this fact with Figure 2.31 and an example.

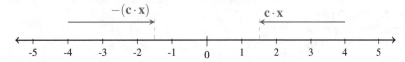

Figure 2.31: Minimizing $\mathbf{c} \cdot \mathbf{x}$ is essentially the same as maximizing $-(\mathbf{c} \cdot \mathbf{x})$; the final answers just differ by a minus sign.

Example 2.B.1

Minimization in Linear Programs

Write the following linear program in standard form:

$$\begin{array}{rl} \text{minimize:} & -x_1 - 2x_2 \\ \text{subject to:} & x_1 + x_2 \le 3 \\ & -x_1 + x_2 \le 1 \\ & x_1, \quad x_2 \ge 0 \end{array}$$

Solution:

We simply change the "minimize" to a "maximize" and flip the sign on each term in the function being maximized:

$$\begin{array}{rl} \text{maximize:} & x_1 + 2x_2 \\ \text{subject to:} & x_1 + x_2 \le 3 \\ & -x_1 + x_2 \le 1 \\ & x_1, \quad x_2 \ge 0 \end{array}$$

If we wanted to, we could then specify exactly what A, $\mathbf{b}$, and $\mathbf{c}$ are in this linear program's standard form:

$$A = \begin{bmatrix} 1 & 1 \\ -1 & 1 \end{bmatrix}, \quad \mathbf{b} = (3,1), \quad \text{and} \quad \mathbf{c} = (1,2).$$

We will talk about how to actually *find* the optimal value of a linear program shortly.

Note that we have to be slightly careful and keep in mind that the optimal value of this new linear program is the *negative* of the optimal value of the original linear program. The optimal value in both cases is attained at $\mathbf{x} = (x_1, x_2) = (1,2)$, but the original minimization problem has optimal value $-x_1 - 2x_2 = -5$, whereas its standard form has optimal value $x_1 + 2x_2 = 5$.

Equality and Flipped Constraints

Equality constraints and "$\ge$" constraints can both be converted into "$\le$" constraints, and thus pose no problem for linear programs. To convert a "$\ge$" constraint into a "$\le$" constraint, we just multiply it by -1 (recall that the sign of an inequality flips when multiplying it through by a negative number). To handle equality constraints, recall that $x = a$ is equivalent to the pair of inequalities $x \le a$ and $x \ge a$, which in turn is equivalent to the pair of inequalities $x \le a$ and $-x \le -a$, so one equality constraint can be converted into two "$\le$" constraints. Again, we illustrate these facts with an example.

Example 2.B.2

Equality and Flipped Constraints in Linear Programs

Write the following linear program in standard form:

$$\begin{array}{rl} \text{maximize:} & x_1 + 2x_2 \\ \text{subject to:} & 3x_1 - 4x_2 \ge 1 \\ & 2x_1 + x_2 = 5 \\ & x_1, \quad x_2 \ge 0 \end{array}$$

Solution:

We multiply the constraint $3x_1 - 4x_2 \ge 1$ by -1 to obtain the equivalent constraints $-3x_1 + 4x_2 \le -1$, and we split the equality constraint $2x_1 + x_2 = 5$ into the equivalent pair of inequality constraints $2x_1 + x_2 \le 5$ and $-2x_1 - x_2 \le -5$. After making these changes, the linear program has the form

$$
\begin{aligned}
\text{maximize:} \quad & x_1 + 2x_2 \\
\text{subject to:} \quad & -3x_1 + 4x_2 \le -1 \\
& 2x_1 + x_2 \le 5 \\
& -2x_1 - x_2 \le -5 \\
& x_1, \quad x_2 \ge 0
\end{aligned}
$$

In general, A has as many rows as there are constraints and as many columns as there are variables (similar to the coefficient matrix of a linear system).

If we wanted to, we could then specify exactly what A, $\mathbf{b}$, and $\mathbf{c}$ are in this linear program's standard form:

$$
A = \begin{bmatrix} -3 & 4 \\ 2 & 1 \\ -2 & -1 \end{bmatrix}, \quad \mathbf{b} = (-1, 5, -5), \quad \text{and} \quad \mathbf{c} = (1, 2).
$$

Negative Variables

If we want one of the variables in a linear program to be able to take on negative values, then we can just write it as a difference of two non-negative variables, since every real number x can be written in the form $x = x^+ - x^-$, where $x^+ \ge 0$ and $x^- \ge 0$. We illustrate how this fact can be used to write even more linear programs in standard form via another example.

Example 2.B.3

Negative Variables in Linear Programs

Write the following linear program in standard form:

$$
\begin{aligned}
\text{maximize:} \quad & 2x_1 - x_2 \\
\text{subject to:} \quad & x_1 + 3x_2 \le 4 \\
& 3x_1 - 2x_2 \le 5 \\
& x_1 \quad\quad \ge 0
\end{aligned}
$$

Solution:

The only reason this linear program is not yet in standard form is that there is no $x_2 \ge 0$ constraint. To rewrite this linear program in an equivalent form with all non-negative variables, we write $x_2 = x_2^+ - x_2^-$, where $x_2^+, x_2^- \ge 0$, and then we replace all instances of x_2 in the linear program with $x_2^+ - x_2^-$:

$$
\begin{aligned}
\text{maximize:} \quad & 2x_1 - (x_2^+ - x_2^-) \\
\text{subject to:} \quad & x_1 + 3(x_2^+ - x_2^-) \le 4 \\
& 3x_1 - 2(x_2^+ - x_2^-) \le 5 \\
& x_1, \quad x_2^+, \quad x_2^- \ge 0
\end{aligned}
$$

Be careful when computing A and $\mathbf{c}$—we have to expand out parentheses first. For example, the first inequality is $x_1 + 3(x_2^+ - x_2^-) \le 4$, which expands to $x_1 + 3x_2^+ - 3x_2^- \le 4$, so the first row of A is $(1, 3, -3)$.

This is now a linear program in 3 non-negative variables, and we can expand out all parentheses and then specify exactly what A, $\mathbf{b}$, and $\mathbf{c}$ are in this linear program's standard form:

$$
A = \begin{bmatrix} 1 & 3 & -3 \\ 3 & -2 & 2 \end{bmatrix}, \quad \mathbf{b} = (4, 5), \quad \text{and} \quad \mathbf{c} = (2, -1, 1).
$$

It is perhaps worth working through one example that makes use of all of these techniques at once to convert a linear program into standard form.

Example 2.B.4

Converting a Linear Program into Primal Standard Form

Write the following linear program in standard form:

$$\begin{aligned} \text{minimize:} \quad & 3x_1 - 2x_2 \\ \text{subject to:} \quad & x_1 + 3x_2 = 4 \\ & x_1 + 2x_2 \geq 3 \\ & x_2 \geq 0 \end{aligned}$$

Solution:

We start by converting the minimization to a maximization (by multiplying the objective function by -1) and converting the equality and "$\geq$" inequality constraints into "$\leq$" constraints:

$$\begin{aligned} \text{maximize:} \quad & -3x_1 + 2x_2 \\ \text{subject to:} \quad & x_1 + 3x_2 \leq 4 \\ & -x_1 - 3x_2 \leq -4 \\ & -x_1 - 2x_2 \leq -3 \\ & x_2 \geq 0 \end{aligned}$$

Again, we must be careful to keep in mind that the optimal value of this linear program is the negative of the optimal value of the original linear program. To complete the conversion to standard form, we replace the unconstrained variable x_1 by $x_1^+ - x_1^-$, where $x_1^+, x_1^- \geq 0$:

$$\begin{aligned} \text{maximize:} \quad & -3(x_1^+ - x_1^-) + 2x_2 \\ \text{subject to:} \quad & x_1^+ - x_1^- + 3x_2 \leq 4 \\ & -(x_1^+ - x_1^-) - 3x_2 \leq -4 \\ & -(x_1^+ - x_1^-) - 2x_2 \leq -3 \\ & x_1^+, \ x_1^-, \quad x_2 \geq 0 \end{aligned}$$

This is now a linear program in 3 non-negative variables, and we can specify exactly what A, $\mathbf{b}$, and $\mathbf{c}$ are in this linear program's standard form:

Once again, be careful and expand out parentheses before computing A and $\mathbf{c}$.

$$A = \begin{bmatrix} 1 & -1 & 3 \\ -1 & 1 & -3 \\ -1 & 1 & -2 \end{bmatrix}, \quad \mathbf{b} = (4, -4, -3), \quad \text{and} \quad \mathbf{c} = (-3, 3, 2).$$

2.B.2   Geometric Interpretation

To get a bit of intuition about linear programs, we now consider how we could interpret and solve the following one geometrically:

This is the linear program from Example 2.B.1.

$$\begin{aligned} \text{maximize:} \quad & x_1 + 2x_2 \\ \text{subject to:} \quad & x_1 + x_2 \leq 3 \\ & -x_1 + x_2 \leq 1 \\ & x_1, \quad x_2 \geq 0 \end{aligned} \qquad (2.B.3)$$

We start by graphing the constraints—there are two variables, so the feasible region $\mathbf{x} = (x_1, x_2)$ is a subset of $\mathbb{R}^2$. The constraints $x_1 \geq 0$ and $x_2 \geq 0$ are simple enough, and just force $\mathbf{x}$ to lie in the non-negative (i.e., top-right) quadrant. The other two constraints are perhaps easier to visualize if we rearrange them into the forms $x_2 \leq -x_1 + 3$ (i.e., everything on and below the line $x_2 = -x_1 + 3$)

and $x_2 \leq x_1 + 1$ (i.e., everything on and below the line $x_2 = x_1 + 1$), respectively. These lines and inequalities are plotted in Figure 2.32.

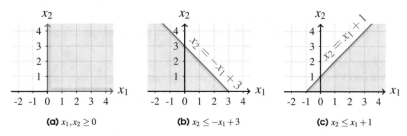

(a) $x_1, x_2 \geq 0$ **(b)** $x_2 \leq -x_1 + 3$ **(c)** $x_2 \leq x_1 + 1$

Figure 2.32: A visualization of the inequality constraints of linear program (2.B.3).

Since feasible vectors must satisfy *all* of the constraints of the linear program, the feasible region is exactly the intersection of the regions defined by these individual constraints, which is illustrated in Figure 2.33(a).

Now that we know what the feasible region looks like, we can solve the linear program by investigating how the objective function behaves on that region. To this end, we define $z = x_1 + 2x_2$, so that our goal is to maximize z. We can rearrange this equation to get $x_2 = -x_1/2 + z/2$, which is a line with slope $-1/2$ and y-intercept $z/2$. We want to find the largest possible z such that this line intersects the feasible region of the linear program, which is depicted in Figure 2.33(b). We see that the largest such value of z is $z = 5$ (corresponding to a y-intercept of $5/2$), so this is the optimal value of the linear program, and it is attained at the vector $\mathbf{x} = (x_1, x_2) = (1, 2)$.

> These lines are called the **level sets** or **level curves** of the function $f(x_1, x_2) = x_1 + 2x_2$.

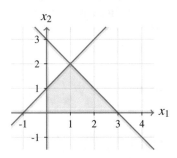

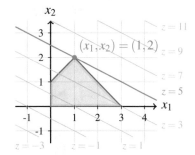

(a) The feasible region of a linear program is the intersection of the various half-planes defined by its constraints.

(b) Of all the lines of the form $x_2 = -x_1/2 + z/2$, the one with largest value of z (i.e., largest y-intercept) that intersects the feasible region is the one with $z = 5$ (i.e., y-intercept $5/2$).

Figure 2.33: A visualization of the feasible region of the linear program (2.B.3), and how we can use it to find its optimal value.

This geometric method of solving linear programs works for essentially any linear program of 2 variables, but it is much more difficult to make use of when there are 3 or more variables, since it is much more difficult to visualize a feasible region in 3 (or more!) dimensions. However, much of the intuition from 2 dimensions carries over to higher dimensions. For example, the feasible set of a linear program is always a **convex polytope** (roughly speaking, a region with flat sides, no holes, and that never bends inward), and the optimal value is always attained at a corner of the feasible set (if it exists and is finite).

Unbounded and Infeasible Problems

There are two basic ways in which a linear program can fail to have a solution. A linear program is called **unbounded** if there exist feasible vectors that make the objective function arbitrarily large (if it is a maximization problem) or arbitrarily small (if it is a minimization problem). For example, the linear program

Linear programs with unbounded feasible regions can have bounded optimal values. For example, if this linear program asked for a minimum instead of a maximum, its optimal value would be 0 (attained at $x_1 = x_2 = 0$).

$$\begin{aligned}\text{maximize:} \quad & x_1 + 2x_2 \\ \text{subject to:} \quad & -x_1 + x_2 \le 1 \\ & x_1, \quad x_2 \ge 0\end{aligned} \qquad (2.B.4)$$

is unbounded because we can choose $x_2 = x_1 + 1$ so that (x_1, x_2) is a feasible vector whenever $x_1 \ge 0$. The value of the objective function at this vector is $x_1 + 2x_2 = x_1 + 2(x_1 + 1) = 3x_1 + 2$, which can be made as large as we like by increasing x_1. The feasible region of this linear program is displayed in Figure 2.34(a).

On the other hand, a linear program is called **infeasible** if the feasible set is empty (i.e., there are no feasible vectors). For example, the linear program

$$\begin{aligned}\text{maximize:} \quad & x_1 + 2x_2 \\ \text{subject to:} \quad & x_1 - x_2 \le -1 \\ & -x_1 + x_2 \le -1 \\ & x_1, \quad x_2 \ge 0\end{aligned} \qquad (2.B.5)$$

is infeasible since there is no way to find values of x_1, x_2 that satisfy all of the constraints simultaneously, as demonstrated geometrically in Figure 2.34(b). One way of seeing that this linear program is infeasible algebraically is to add the first inequality to the second inequality, which results in the new inequality $(x_1 - x_2) + (-x_1 + x_2) = 0 \le -1 - 1 = -2$, which is not true no matter what x_1 and x_2 equal.

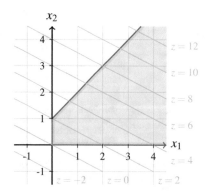

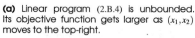

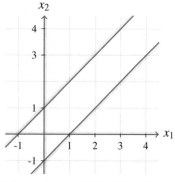

Notice the analogy between the trichotomy for linear systems (every linear system has no solution, a unique solution, or infinitely many solutions) and for linear programs (every linear program is infeasible, has a finite solution, or is unbounded).

(a) Linear program (2.B.4) is unbounded. Its objective function gets larger as (x_1, x_2) moves to the top-right.

(b) Linear program (2.B.5) is infeasible. There is no vector (x_1, x_2) that satisfies all four constraints simultaneously.

Figure 2.34: Examples of unbounded and infeasible linear programs.

If a maximization problem is unbounded then we sometimes say that its optimal value is ∞, and if it is infeasible then we might say that its optimal value is $-\infty$. Similarly, if a minimization problem is unbounded then we sometimes say that its optimal value if $-\infty$, and if it is infeasible then its optimal value is ∞. The benefit of defining things in this way is that then every linear program has an optimal value (though that optimal value might not be a real number).

2.B.3 The Simplex Method for Solving Linear Programs

We now present an algorithm, called the **simplex method**, for solving linear programs algebraically, much like Gaussian elimination can be used to solve linear systems algebraically (so that we do not have to rely on the geometric method introduced in the previous section, which breaks down when the linear program has more than 2 or 3 variables). In particular, we illustrate this algorithm with the following linear program:

$$\begin{aligned}
\text{maximize:} \quad & 2x_1 + x_2 \\
\text{subject to:} \quad & 3x_1 + 2x_2 \geq -1 \\
& x_1 - x_2 \leq 2 \\
& x_1 - 3x_2 \geq -3 \\
& x_1 \qquad\quad \geq 0
\end{aligned} \tag{2.B.6}$$

Step 1: **Put the linear program into standard form.** Fortunately, we already know how to perform this step, and if we apply it to the linear program (2.B.6) then we get

$$\begin{aligned}
\text{maximize:} \quad & 2x_1 + x_2 - x_3 \\
\text{subject to:} \quad & -3x_1 - 2x_2 + 2x_3 \leq 1 \\
& x_1 - x_2 + x_3 \leq 2 \\
& - x_1 + 3x_2 - 3x_3 \leq 3 \\
& x_1, \quad x_2, \quad x_3 \geq 0
\end{aligned} \tag{2.B.7}$$

In particular, this standard form was obtained by multiplying the first and third constraints by -1 and replacing the unconstrained variable x_2 by $x_2 - x_3$ where $x_2, x_3 \geq 0$.

Step 1.5: **Hope that $\mathbf{b} \geq \mathbf{0}$.** If $\mathbf{b}$ (the vector containing the coefficients on the right-hand side of the "$\leq$" inequalities) has a negative entry then we will run into a problem later on. We will discuss what this problem is and how to fix it later—for this particular example we just note that $\mathbf{b} = (1,2,3) \geq \mathbf{0}$, so we move on.

Step 2: **Add some new variables.** Next, we change each of the "$\leq$" constraints into an equality constraint by adding a new variable to it. For example, the inequality $x \leq 5$ is equivalent to the pair of constraints $s + x = 5, s \geq 0$. The new variable s is called a **slack variable**, and it just measures the difference between the right- and left-hand sides. Applying this technique to the linear program (2.B.7) puts it into the form

One slack variable is added for *each* "$\leq$" constraint that is being turned into an equality constraint.

$$\begin{aligned}
\text{maximize:} \quad & \qquad\qquad\qquad 2x_1 + x_2 - x_3 \\
\text{subject to:} \quad & s_1 \qquad\qquad -3x_1 - 2x_2 + 2x_3 = 1 \\
& \quad s_2 \qquad\; + x_1 - x_2 + x_3 = 2 \\
& \qquad s_3 - x_1 + 3x_2 - 3x_3 = 3 \\
& s_1, \; s_2, \; s_3, \quad x_1, \quad x_2, \quad x_3 \geq 0
\end{aligned} \tag{2.B.8}$$

We also introduce a variable z that equals the objective function (i.e., $z = \mathbf{c} \cdot \mathbf{x}$), and we rephrase the linear program as maximizing over z with the additional constraint that $z - \mathbf{c} \cdot \mathbf{x} = 0$. Making this change to the linear program (2.B.8) gives it the form

$$
\begin{array}{llrrrr}
\text{maximize:} & z \\
\text{subject to:} & z & -2x_1 - x_2 + x_3 = 0 \\
& s_1 & -3x_1 - 2x_2 + 2x_3 = 1 \\
& s_2 & + x_1 - x_2 + x_3 = 2 \\
& s_3 - x_1 + 3x_2 - 3x_3 = 3 \\
& s_1, \quad s_2, \quad s_3, \quad x_1, \quad x_2, \quad x_3 \geq 0
\end{array}
\qquad (2.B.9)
$$

This final change perhaps seems somewhat arbitrary and silly right now, but it makes the upcoming steps work out much more cleanly.

Step 3: Put the linear program into a tableau. Adding the new variables to the linear program in the previous step unfortunately made it quite large, so we now construct a matrix representation of it so that we do not have to repeatedly write down the variable names (this is the same reason that we typically represent a linear system $A\mathbf{x} = \mathbf{b}$ via the augmented matrix $[A \mid \mathbf{b}]$). In particular, the matrix form of a linear program

$$
\begin{array}{ll}
\text{maximize:} & \mathbf{c} \cdot \mathbf{x} \\
\text{subject to:} & A\mathbf{x} \leq \mathbf{b} \\
& \mathbf{x} \geq \mathbf{0}
\end{array}
$$

is called its **tableau**, and it is the block matrix

$$
\left[
\begin{array}{c|c|c|c}
1 & \mathbf{0}^T & -\mathbf{c}^T & 0 \\
\hline
\mathbf{0} & I & A & \mathbf{b}
\end{array}
\right].
$$

This matrix perhaps looks quite strange at first glance, but a little reflection shows that it is simply the augmented matrix corresponding to the equality constraints in the linear program (2.B.9) that we constructed in Step 2.

We do not actually need to explicitly carry out Step 2 at all—we can go straight from the standard form of a linear program to its tableau.

In particular, the top row simply says that $z - \mathbf{c}^T\mathbf{x} = 0$ and the bottom block row $[\mathbf{0} \mid I \mid A \mid \mathbf{b}]$ represents the equality constraints involving slack variables (the identity block comes from the slack variables themselves). For example, the tableau of the linear program (2.B.9) is

$$
z - \mathbf{c} \cdot \mathbf{x} = 0 \quad
\begin{array}{c}
\begin{array}{ccccccc}
z & s_1 & s_2 & s_3 & x_1 & x_2 & x_3
\end{array} \\
\left[
\begin{array}{cccc|ccc|c}
1 & 0 & 0 & 0 & -2 & -1 & 1 & 0 \\
\hline
0 & 1 & 0 & 0 & -3 & -2 & 2 & 1 \\
0 & 0 & 1 & 0 & 1 & -1 & 1 & 2 \\
0 & 0 & 0 & 1 & -1 & 3 & -3 & 3
\end{array}
\right],
\end{array}
$$

which is also exactly the augmented matrix corresponding to the equality constraints in that linear program.

Once nice feature of this tableau (and the main reason that we introduced the new variables in Step 2) is that it is necessarily in reduced row echelon form, so all solutions of these equality constraints can be read off from it directly. In particular, the newly-introduced variables z, s_1, s_2, and s_3 are the leading variables and the original variables x_1, x_2, and x_3 are the free variables. We can choose x_1, x_2, and x_3 to be anything we like, as long as that choice results in all variables (except for maybe z) being non-negative.

The tableau of a linear program only encodes its objective function and equality constraints. The requirement that all variables except for z must be non-negative is implicit.

Step 4: Starting from one feasible vector, find a better feasible vector. A feasible vector of the linear program corresponds to a choice of the free variables that results in each variable being non-negative. One way to find a feasible vector is to simply set all of the free variables equal to 0. In this case, that gives us

This is why we needed $\mathbf{b} \geq \mathbf{0}$ back in Step 1.5. Choosing $\mathbf{x} = \mathbf{0}$ results in the slack variables being the entries of $\mathbf{b}$ must be non-negative in order to be feasible.

$$x_1 = x_2 = x_3 = 0, \qquad s_1 = 1, s_2 = 2, s_3 = 3, \qquad z = 0.$$

While this is indeed a feasible vector, it is not a particularly good one—it gives us a value of $z = 0$ in the objective function, and we would like to do better (i.e., we would like to increase z). To do so, we increase one free variable at a time.

To increase the value of the objective function $z = 2x_1 + x_2 - x_3$, we could increase either x_1 or x_2, but since x_1 has the largest coefficient, increasing it will provide the quickest gain. We thus increase x_1 as much as possible (i.e., without violating the constraints of the linear program). Since we are leaving $x_2 = x_3 = 0$ alone for now, the constraints simply have the form

$$\begin{aligned} -3x_1 - 2x_2 + 2x_3 = -3x_1 &\leq 1 \\ x_1 - x_2 + x_3 = x_1 &\leq 2 \\ - x_1 + 3x_2 - 3x_3 = - x_1 &\leq 3. \end{aligned}$$

Since we are *increasing* x_1, the most restrictive of these constraints is the second one, so we wish to set $x_1 = 2$ and update the other variables accordingly. One way to do this is to apply row operations to the tableau so as eliminate all entries in the x_1 column other than the entry corresponding to the 2nd (i.e., the most restrictive) constraint:

This is all very analogous to how we used Gauss–Jordan elimination to solve linear systems back in Section 2.1.3. The only difference is in the choice of which entries to pivot on (i.e., which column to zero out).

$$
\begin{array}{ccccccc}
z & s_1 & s_2 & s_3 & x_1 & x_2 & x_3 \\
\end{array}
$$

$$
\left[\begin{array}{c|ccc|ccc|c}
1 & 0 & 0 & 0 & -2 & -1 & 1 & 0 \\
\hline
0 & 1 & 0 & 0 & -3 & -2 & 2 & 1 \\
0 & 0 & 1 & 0 & 1 & -1 & 1 & 2 \\
0 & 0 & 0 & 1 & -1 & 3 & -3 & 3
\end{array}\right]
$$

$$
\begin{array}{c}
R_1 + 2R_3 \\
R_2 + 3R_3 \\
R_4 + R_3 \\
\longrightarrow
\end{array}
\left[\begin{array}{c|ccc|ccc|c}
1 & 0 & 2 & 0 & 0 & -3 & 3 & 4 \\
\hline
0 & 1 & 3 & 0 & 0 & -5 & 5 & 7 \\
0 & 0 & 1 & 0 & 1 & -1 & 1 & 2 \\
0 & 0 & 1 & 1 & 0\!\uparrow & 2 & -2 & 5
\end{array}\right].
$$

new "leading" entry

Even though this new tableau is not quite in reduced row echelon form, for our purposes it is just as good—we can still read the solutions of the equality constraints directly from it. We just have to think of s_1, s_3, and x_1 as the leading variables (after all, their columns look like leading columns) and s_2, x_2, and x_3 as the free variables. Setting this new set of free variables equal to 0 gives us

$$s_2 = x_2 = x_3 = 0, \qquad s_1 = 7, x_1 = 2, s_3 = 5, \qquad z = 4.$$

In particular, we noconclude that the optimal value oftice that the value of $z = 4$ in the objective function is better than the value of $z = 0$ that we had started with.

Step 5: **Repeat until we cannot.** Now we just apply Step 4 over and over again, but more quickly now that we are a bit more comfortable with it. Succinctly, we find the most negative coefficient in the top row of the tableau, which is now the -3 in the x_2 column, and then in that column we find the positive entry that minimizes its ratio with the scalar on the right-hand side:

$$\begin{array}{c} \swarrow \text{ most negative entry in top row} \\ \left[\begin{array}{cccc|ccc|c} 1 & 0 & 2 & 0 & 0 & -3 & 3 & 4 \\ 0 & 1 & 3 & 0 & 0 & -3 & 5 & 7 \\ 0 & 0 & 1 & 0 & 1 & -1 & 1 & 2 \\ 0 & 0 & 1 & 1 & 0 & 2 & -2 & 5 \end{array} \right] \leftarrow \end{array}$$

$5/2$ is smallest (and only) positive ratio

Finally, we pivot on the positive entry that we just found (i.e., the 2 in the x_2 column), thus obtaining a tableau with a new value of z in the upper-right corner:

Keep in mind that row operations are performed sequentially—we *first* multiply row 4 by $1/2$ and then perform the other three row operations.

$$\begin{array}{cccccccc} z & s_1 & s_2 & s_3 & x_1 & x_2 & x_3 & \\ \left[\begin{array}{ccc|cccc|c} 1 & 0 & 2 & 0 & 0 & -3 & 3 & 4 \\ 0 & 1 & 3 & 0 & 0 & -5 & 5 & 7 \\ 0 & 0 & 1 & 0 & 1 & -1 & 1 & 2 \\ 0 & 0 & 1 & 1 & 0 & 2 & -2 & 5 \end{array} \right] & \text{new "z" value} \\ \downarrow \end{array}$$

$$\begin{array}{c} (1/2)R_4 \\ R_1+3R_4 \\ R_2+5R_4 \\ R_3+R_4 \\ \xrightarrow{\hspace{1cm}} \end{array} \left[\begin{array}{ccc|cccc|c} 1 & 0 & 7/2 & 3/2 & 0 & 0 & 0 & 23/2 \\ 0 & 1 & 11/2 & 5/2 & 0 & 0 & 0 & 39/2 \\ 0 & 0 & 3/2 & 1/2 & 1 & 0 & 0 & 9/2 \\ 0 & 0 & 1/2 & 1/2 & 0 & 1 & -1 & 5/2 \end{array} \right]$$

$\uparrow$ new "leading" entry

We then repeat this same procedure over and over again until we get stuck. In fact, for this particular linear program we get stuck if we try to do this even once more, since there are no longer any negative entries in the top row for us to choose. Indeed, the top row of the final tableau above tells us that $z = 23/2 - (7/2)s_2 - (3/2)s_3$. Since $s_2, s_3 \geq 0$ we conclude that $z = 23/2$ is the optimal value of this linear program, and it is attained when $x_1 = 9/2$ and $x_2 = 5/2$ (and $s_1 = 39/2$ and $s_2 = s_3 = x_3 = 0$).

Geometrically, what Steps 4 and 5 of the simplex method do is start at the origin (recall that Step 4 begins by setting each $x_j = 0$) and repeatedly look for nearby corners that produce a higher value in the objective function. For the particular linear program (2.B.7) that we just worked through, we first moved from $(x_1, x_2, x_3) = (0,0,0)$ to $(x_1, x_2, x_3) = (2,0,0)$ and then to $(x_1, x_2, x_3) = (9/2, 5/2, 0)$, as illustrated in Figure 2.35.

It is worth recalling that the 3-variable linear program that we solved was equivalent to the 2-variable linear program (2.B.6) that we were actually interested in originally. The 2D feasible region of this linear program is just the projection of the 3D feasible region from Figure 2.35 onto the plane $x_2 + x_3 = 0$ (we have not discussed how to project onto a plane, but it analogous to how we projected onto a line in Section 1.4.2; think of the 2D projected shape as the shadow of the full 3D shape). This 2D feasible region is displayed in Figure 2.36.

The feasible region
of a linear program
with n variables and
m constraints can
have up to $\binom{m}{n}$
corners. The simplex
algorithm is useful
for helping us
search through this
large number of
corners quickly.

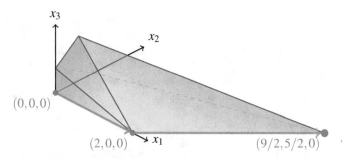

Figure 2.35: The feasible region of the linear program (2.B.7). The simplex method works by jumping from one corner of the feasible region to a neighboring one in such a way as to increase the value of the objective function by as much as possible at each step.

The 3D feasible
region in Figure 2.35
has 6 corners, but
the 2D projected
feasible region in
Figure 2.36 only has
4 corners. The 2
corners that were
lost when projecting
are the ones at
$(0,0,0)$ and $(2,0,0)$,
which were
projected down to
$(0,0)$ and $(2,0)$,
respectively.

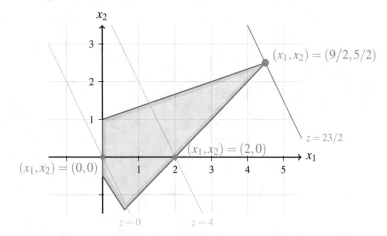

Figure 2.36: The 2D feasible region of the linear program (2.B.6), which is a projection (shadow) of the feasible region from Figure 2.35. The simplex method starts with a value of $z=0$ in the objective function at $(x_1,x_2)=(0,0)$ and then moves to $(x_1,x_2)=(2,0)$ to get a value of $z=4$, and finally to $(x_1,x_2)=(9/2,5/2)$ to get the optimal value of $z=23/2$.

Example 2.B.5

**How Best to
Sell Cookies**

Suppose a bakery bakes and sells three types of cookies: peanut butter, chocolate chip, and peanut butter chocolate chip. The profit that they make per dozen cookies of each type is as follows:

- Peanut butter: $1.00
- Chocolate chip: $0.80
- Peanut butter chocolate chip: $1.50

Due to ingredient supply limitations, they can bake no more than 100 dozen cookies containing chocolate chips and 80 dozen cookies containing peanut butter per week. Furthermore, their market research indicates that people are unwilling to buy more than 150 dozen cookies from them per week. How many dozen cookies of each type should the bakery bake each week in order to maximize profits?

Setting up the linear program:

We start by formulating this problem as a linear program. Let x_p, x_c, and x_{pc} denote the number of dozen peanut butter, chocolate chip, and peanut butter chocolate chip cookies to be baked, respectively. The bakery wants to maximize its profits, which are given by the equation

$$x_p + 0.8x_c + 1.5x_{pc},$$

so this will be the objective function of the linear program. Furthermore, the constraints described by the problem have the form

$$x_c + x_{pc} \leq 100, \quad x_p + x_{pc} \leq 80, \quad \text{and} \quad x_p + x_c + x_{pc} \leq 150,$$

and each of the variables is non-negative as well (i.e., $x_p, x_c, x_{pc} \geq 0$) since the bakery cannot bake a negative number of cookies. The linear program that we wish to solve thus has the form

$$
\begin{aligned}
\text{maximize:} \quad & x_p + 0.8x_c + 1.5x_{pc} \\
\text{subject to:} \quad & x_c + x_{pc} \leq 100 \\
& x_p + x_{pc} \leq 80 \\
& x_p + x_c + x_{pc} \leq 150 \\
& x_p, \quad x_c, \quad x_{pc} \geq 0
\end{aligned}
$$

Solving the linear program:

   Since this linear program is already in standard form, and the scalars on the right-hand side of each inequality are all non-negative, to solve it via the simplex method we can jump straight to Step 3. That is, we put this linear program into its tableau:

$$
\begin{bmatrix}
1 & 0 & 0 & 0 & -1 & -0.8 & -1.5 & 0 \\
\hline
0 & 1 & 0 & 0 & 0 & 1 & 1 & 100 \\
0 & 0 & 1 & 0 & 1 & 0 & 1 & 80 \\
0 & 0 & 0 & 1 & 1 & 1 & 1 & 150
\end{bmatrix}.
$$

Now we just repeatedly apply Step 4 of the simplex method: we select the column whose top entry is the most negative, and then we pivot on the entry in that column whose ratio with the right-hand side is the smallest positive number possible.

$$\begin{array}{cccccccc} & z & s_1 & s_2 & s_3 & x_p & x_c & x_{pc} & \\ \left[\begin{array}{c} 1 \\ 0 \\ 0 \\ 0 \end{array}\right. & \begin{array}{c} 0 \\ 1 \\ 0 \\ 0 \end{array} & \begin{array}{c} 0 \\ 0 \\ 1 \\ 0 \end{array} & \begin{array}{c|c} 0 \\ 0 \\ 0 \\ 1 \end{array} & \begin{array}{c} -1 \\ 0 \\ 1 \\ 1 \end{array} & \begin{array}{c} -0.8 \\ 1 \\ 0 \\ 1 \end{array} & \begin{array}{c|c} -1.5 & \\ 1 & \\ 1 & \\ 1 & \end{array} & \left.\begin{array}{c} 0 \\ 100 \\ 80 \\ 150 \end{array}\right] \end{array}$$

$$\xrightarrow[\substack{R_1+1.5R_3 \\ R_2-R_3 \\ R_4-R_3}]{} \left[\begin{array}{cccc|cccc} 1 & 0 & 1.5 & 0 & 0.5 & -0.8 & 0 & 120 \\ 0 & 1 & -1 & 0 & -1 & 1 & 0 & 20 \\ 0 & 0 & 1 & 0 & 1 & 0 & 1 & 80 \\ 0 & 0 & -1 & 1 & 0 & 1 & 0 & 70 \end{array}\right]$$

$$\xrightarrow[\substack{R_1+0.8R_2 \\ R_4-R_2}]{} \left[\begin{array}{cccc|cccc} 1 & 0.8 & 0.7 & 0 & -0.3 & 0 & 0 & 136 \\ 0 & 1 & -1 & 0 & -1 & 1 & 0 & 20 \\ 0 & 0 & 1 & 0 & 1 & 0 & 1 & 80 \\ 0 & -1 & 0 & 1 & 1 & 0 & 0 & 50 \end{array}\right]$$

$$\xrightarrow[\substack{R_1+0.3R_4 \\ R_2+R_4 \\ R_3-R_4}]{} \left[\begin{array}{cccc|cccc} 1 & 0.5 & 0.7 & 0.3 & 0 & 0 & 0 & 151 \\ 0 & 0 & -1 & 1 & 0 & 1 & 0 & 70 \\ 0 & 1 & 1 & -1 & 0 & 0 & 1 & 30 \\ 0 & -1 & 0 & 1 & 1 & 0 & 0 & 50 \end{array}\right]$$

Here, we highlight the entry of the tableau that we will pivot on in the upcoming set of row operations. The row operations are chosen to turn all other entries in that column into zeros.

Since we have arrived at a tableau with a non-negative top row, we are done and conclude that the optimal value of this linear program (i.e., the most money that the bakery can make per week) is \$151. Furthermore, it is attained when they bake $x_p = 50$ dozen peanut butter cookies, $x_c = 70$ dozen chocolate chip cookies, and $x_{pc} = 30$ dozen peanut butter chocolate chip cookies.

The three slanted sides of this feasible region each correspond to one of the inequality constraints in the linear program.

The feasible region of the linear program that we just solved, as well as the corners traversed by the simplex method to solve it, are displayed below:

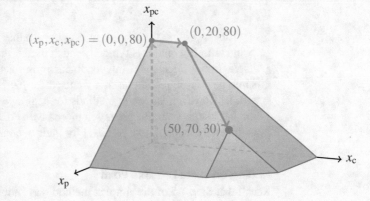

x_{pc}

$(x_p, x_c, x_{pc}) = (0, 0, 80)$ $(0, 20, 80)$

$(50, 70, 30)$ x_c

x_p

Remark 2.B.1

Integer Solutions to Linear Programs are Hard to Find

In Example 2.B.5, we were lucky that the solution to the linear program required the bakery to bake an integer number of cookies—while they might be able to *bake* a fraction of a cookie, they would certainly have a tough time *selling* it.

If the solution to a linear program is not an integer, but it really should be (e.g., due to physical constraints of the problem being modeled), we can try just rounding the various variables to the nearest integer. However, this does not always work very well. For example, the optimal value of the linear program

$$\begin{aligned}
\text{maximize:} \quad & x_1 + 3x_2 \\
\text{subject to:} \quad & x_1 + 2x_2 \leq 4 \\
& -2x_1 + x_2 \leq 0 \\
& x_1, \quad x_2 \geq 0
\end{aligned}$$

is 5.6, which is attained at the vector $(x_1, x_2) = (0.8, 1.6)$. If we try to find the optimal value when restricted to integer inputs simply by rounding these values down to $(x_1, x_2) = (0, 1)$, we run into a big problem—this vector is not even feasible (it violates the constraint $-2x_1 + x_2 \leq 0$)!

Of the four possible ways of rounding each of $x_1 = 0.8$ and $x_2 = 1.6$ up or down, only one results in a feasible vector: $(x_1, x_2) = (1, 1)$. However, this vector is not optimal—it gives a value of $x_1 + 3x_2 = 4$ in the objective function, whereas the vector $(2, 1)$ gives a value of 5. We can make sense of all of this by plotting the feasible region of this linear program:

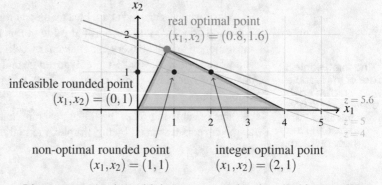

Linear programming with integer constraints is called **integer linear programming**, and it is typically much harder than linear programming itself. Low-dimensional integer linear programs can be solved via the same geometric technique that we used for low-dimensional linear programs. However, algebraic methods for dealing with higher-dimensional integer linear programs are much more involved, so the interested reader is directed to a book like [Mar99] for their treatment.

Finding an Initial Feasible Point

Recall that Step 1.5 of the simplex method was "hope that $\mathbf{b} \geq \mathbf{0}$". The reason that we required $\mathbf{b} \geq \mathbf{0}$ is that we started the simplex method at the vector $\mathbf{x} = \mathbf{0}$, which is only feasible if $\mathbf{b} \geq \mathbf{0}$. To illustrate the problem with having $\mathbf{b} \ngeq \mathbf{0}$ more clearly, and to see how we can get around it, consider the following linear program:

$$\begin{aligned}
\text{maximize:} \quad & x_1 + 2x_2 \\
\text{subject to:} \quad & x_1 + x_2 \leq 2 \\
& -x_1 - x_2 \leq -1 \\
& -2x_1 + x_2 \leq -2 \\
& x_1, \quad x_2 \geq 0
\end{aligned} \tag{2.B.10}$$

This linear program is already in standard form, but we cannot perform

---

It is also possible for a linear program with integer coefficients to have a non-empty feasible region but no feasible vectors with integer coordinates, or for it to have an integer-valued optimal vector that is arbitrarily far for the real-valued optimal vector (see Exercises 2.B.6 and 2.B.8).

Step 4 of the simplex method since we do not have an "obvious" choice of feasible vector to start from, since $x_1 = x_2 = 0$ violates the second and third constraints and is thus not feasible. To get around this problem we first solve a *different* linear program, which we call the **feasibility linear program**, that either finds a (not necessarily optimal) feasible vector for us or shows that none exist (i.e., shows that the linear program is infeasible).

*We sometimes abbreviate this term as **feasibility LP** for brevity.*

To construct this feasibility LP we introduce yet another variable $y \geq 0$, which we call the **artificial variable**, and we make two changes to the original linear program (2.B.10):

- we subtract y from each equation that has a negative right-hand side, and
- we change the objective function to "maximize $-y$".

Making these changes gives us the feasibility linear program

The idea here is that we want to make y as small as possible (hopefully 0). To minimize y, we maximize $-y$.

$$\begin{aligned} \text{maximize:} \quad & -y \\ \text{subject to:} \quad & x_1 + x_2 \quad\;\; \leq 2 \\ & -\,x_1 - x_2 - y \leq -1 \\ & -2x_1 + x_2 - y \leq -2 \\ & x_1, \;\; x_2, \;\; y \geq 0 \end{aligned} \qquad (2.B.11)$$

The reason for making these changes is that the feasibility LP (2.B.11) has an "obvious" feasible vector that we can start the simplex method from: $x_1 = x_2 = 0$, $y = 2$ (in general, we set each $x_j = 0$ and then set y large enough to satisfy all constraints). Furthermore, its optimal value equals 0 if and only if the original linear program (2.B.10) is feasible, since $(x_1, x_2, y) = (x_1, x_2, 0)$ is a feasible vector of the feasibility LP (2.B.11) if and only if (x_1, x_2) is a feasible vector of the original linear program (2.B.10).

Example 2.B.6

Finding a Feasible Point of a Linear Program

Find a feasible vector of the linear program (2.B.10), or show that it is infeasible.

Solution:

As indicated earlier, we can solve this problem by using the simplex method to solve the feasibility LP (2.B.11). Since we want to start at the "obvious" feasible vector $(x_1, x_2, y) = (0, 0, 2)$ that we noted earlier, after constructing the tableau we first pivot on the entry in the "y" column corresponding to the "$-2x_1 + x_2 - y \leq -2$" constraint (since this one places the biggest restriction on y):

*Again, keep in mind that row operations are performed sequentially. We multiply row 4 by -1 and *then* do the "addition" row operations.*

$$\begin{array}{c c} & \begin{array}{c c c c c c c c} & z & s_1 & s_2 & s_3 & x_1 & x_2 & y \end{array} \\ & \left[\begin{array}{c|c c c|c c c|c} 1 & 0 & 0 & 0 & 0 & 0 & 1 & 0 \\ \hline 0 & 1 & 0 & 0 & 1 & 1 & 0 & 2 \\ 0 & 0 & 1 & 0 & -1 & -1 & -1 & -1 \\ 0 & 0 & 0 & 1 & -2 & 1 & \boxed{-1} & -2 \end{array}\right] \begin{array}{l} \\ \\ \\ \text{pivot} \\ \text{here} \end{array} \\[4ex] \begin{array}{r} -R_4 \\ R_1 - R_4 \\ R_3 + R_4 \\ \longrightarrow \end{array} & \left[\begin{array}{c|c c c|c c c|c} 1 & 0 & 0 & 1 & -2 & 1 & 0 & -2 \\ \hline 0 & 1 & 0 & 0 & 1 & 1 & 0 & 2 \\ 0 & 0 & 1 & -1 & 1 & -2 & 0 & 1 \\ 0 & 0 & 0 & -1 & 2 & -1 & 1 & 2 \end{array}\right] \end{array}$$

We now apply the simplex method to this tableau as usual, which fortunately ends in just one step:

$$\left[\begin{array}{ccc|cccc|c} 1 & 0 & 0 & 1 & -2 & 1 & 0 & -2 \\ \hline 0 & 1 & 0 & 0 & 1 & 1 & 0 & 2 \\ 0 & 0 & 1 & -1 & 1 & -2 & 0 & 1 \\ 0 & 0 & 0 & -1 & 2 & -1 & 1 & 2 \end{array}\right]$$

$$\begin{array}{c} (1/2)R_4 \\ R_1+2R_4 \\ R_2-R_4 \\ R_3-R_4 \\ \xrightarrow{\hspace{1cm}} \end{array} \left[\begin{array}{ccc|cccc|c} 1 & 0 & 0 & 0 & 0 & 0 & 1 & 0 \\ \hline 0 & 1 & 0 & 1/2 & 0 & 3/2 & -1/2 & 1 \\ 0 & 0 & 1 & -1/2 & 0 & -3/2 & -1/2 & 0 \\ 0 & 0 & 0 & -1/2 & 1 & -1/2 & 1/2 & 1 \end{array}\right]$$

Here we could pivot in either the third or fourth row, since there is a tie for the minimal ratio with the right-hand side. We chose row 4 randomly.

The top-right entry of the final tableau being 0 tells us that the optimal value of this feasibility *LP* is 0, so the original linear program (2.B.10) is feasible. In particular, since this final tableau has $x_1 = 1$ and $x_2 = 0$, we conclude that $(x_1, x_2) = (1, 0)$ is a feasible vector of that original LP. It is perhaps worthwhile to visualize the feasible region of this feasibility LP to see what we have done here:

We have truncated the feasible region of the feasibility LP—it is actually unbounded. Its top surface (triangle) is the feasible region of the original LP (2.B.10).

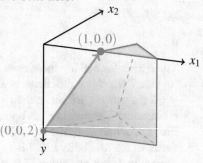

If we had instead found an optimal value of $-y < 0$ in the above example, it would have meant that the original linear program (2.B.10) is infeasible. However, that is not the case here, and we can use the feasible vector that we found to start the simplex method and solve that linear program.

Example 2.B.7

Solving a Linear Program from a Non-zero Feasible Point

Use the simplex method to solve the linear program (2.B.10).

Solution:
We proceed just like we always do, except since we want to start at the feasible vector $(x_1, x_2) = (1, 0)$ that we found in Example 2.B.6, we first pivot in the x_1 column so as to make x_1 leading:

In general, we start by pivoting in whichever columns correspond to the non-zero entries in our initial feasible vector.

$$\begin{array}{ccccccc} & z & s_1 & s_2 & s_3 & x_1 & x_2 & \end{array}$$
$$\left[\begin{array}{c|cccc|cc|c} 1 & 0 & 0 & 0 & -1 & -2 & 0 \\ \hline 0 & 1 & 0 & 0 & 1 & 1 & 2 \\ 0 & 0 & 1 & 0 & -1 & -1 & -1 \\ 0 & 0 & 0 & 1 & -2 & 1 & -2 \end{array}\right]$$

$$\begin{array}{c} R_1+R_2 \\ R_3+R_2 \\ R_4+2R_2 \\ \xrightarrow{\hspace{1cm}} \end{array} \left[\begin{array}{c|cccc|cc|c} 1 & 1 & 0 & 0 & 0 & -1 & 2 \\ \hline 0 & 1 & 0 & 0 & 1 & 1 & 2 \\ 0 & 1 & 1 & 0 & 0 & 0 & 1 \\ 0 & 2 & 0 & 1 & 0 & 3 & 2 \end{array}\right]$$

Notice that the right-hand side vector of this new tableau is non-negative—this always happens if the feasible vector that we found earlier and are trying to start from is indeed feasible.

From here, we just apply the simplex method as usual, repeatedly pivoting until the top row of the tableau is non-negative:

$$\begin{bmatrix} 1 & 1 & 0 & 0 & 0 & -1 & 2 \\ 0 & 1 & 0 & 0 & 1 & 1 & 2 \\ 0 & 1 & 1 & 0 & 0 & 0 & 1 \\ 0 & 2 & 0 & 1 & 0 & 3 & 2 \end{bmatrix}$$

$$\begin{array}{c} (1/3)R_4 \\ R_1+R_4 \\ R_2-R_4 \\ \longrightarrow \end{array} \begin{bmatrix} 1 & 5/3 & 0 & 1/3 & 0 & 0 & 8/3 \\ 0 & 1/3 & 0 & -1/3 & 1 & 0 & 4/3 \\ 0 & 1 & 1 & 0 & 0 & 0 & 1 \\ 0 & 2/3 & 0 & 1/3 & 0 & 1 & 2/3 \end{bmatrix}$$

We thus see that the optimal value of this linear program is $8/3$, and it is attained at the vector $(x_1,x_2) = (4/3,2/3)$. We can visualize the feasible region of this linear program as follows:

Notice that this feasible region is the top surface of the one from Example 2.B.6.

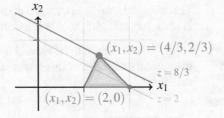

Remark 2.B.2

The Fine Details of the Simplex Method

We have glossed over countless details of the simplex method, including a proof that it always works, a proper explanation for *why* it jumps from corner to corner of the feasible region (or even what it means to be a "corner" of a set when the dimension is larger than 3!), and a discussion of some implementation details and edge cases (e.g., what to do if there is a tie between minimal positive ratios when determining which entry to pivot on in Step 4). Indeed, there are entire textbooks devoted to linear programs and how to solve them, so the interested reader is directed to any of [Chv83, Win03, HL10] for a more thorough treatment.

2.B.4   Duality

We saw geometrically back in Figure 2.33 that the linear program

$$\begin{array}{ll} \text{maximize:} & x_1 + 2x_2 \\ \text{subject to:} & x_1 + x_2 \le 3 \\ & -x_1 + x_2 \le 1 \\ & x_1, \quad x_2 \ge 0 \end{array} \qquad (2.B.12)$$

The optimal value of this linear program is attained at $(x_1,x_2) = (1,2)$.

has optimal value 5, but how could we *algebraically* convince someone that this is the optimal value of the linear program? By providing them with the vector $(x_1,x_2) = (1,2)$, we certainly convince them that the optimal value is at *least* 5, since they can check that it satisfies all constraints and gives a value of 5 when plugged into the objective function.

On the other hand, how could we convince them that it's not possible to do any better? Certainly we could solve the LP via the simplex method and show them our work, but this does not seem very efficient. After all, to convince someone that we have found a solution of a linear *system*, all we have to do is present them with that solution—we do not have to show them our work with Gaussian elimination. We would like a similarly simple method of verifying solutions of linear *programs*.

The key idea that lets us easily *verify* a solution (but not necessarily *find* it) is to add up the constraints of the linear program in a clever way so as to obtain new (more useful) constraints. For example, if we add up the first and second constraint of the linear program (2.B.12), we learn that

$$(x_1 + x_2) + (-x_1 + x_2) = 2x_2 \le 3 + 1 = 4, \quad \text{so} \quad x_2 \le 2.$$

But if we now add the first constraint $(x_1 + x_2 \le 3)$ again, we similarly see that

$$x_2 + (x_1 + x_2) = x_1 + 2x_2 \le 2 + 3 = 5.$$

In other words, we have shown that the objective function of the linear program is never larger than 5, so this is indeed the optimal value of the linear program.

A bit more generally, if we add up y_1 times the first constraint and y_2 times the second constraint of this linear program for some $y_1, y_2 \ge 0$, then it is the case that

$$y_1(x_1 + x_2) + y_2(-x_1 + x_2) \le 3y_1 + y_2.$$

We need $y_1, y_2 \ge 0$ so the signs of the inequalities do not change.

If we cleverly choose y_1 and y_2 so that

$$x_1 + 2x_2 \le y_1(x_1 + x_2) + y_2(-x_1 + x_2),$$

then $3y_1 + y_2$ must also be an upper bound on $x_1 + 2x_2$ (i.e., the objective function of the linear program). It is straightforward to check that $y_1 = 3/2$, $y_2 = 1/2$ works and establishes the bound $x_1 + 2x_2 \le 3(3/2) + (1/2) = 5$, just like before.

The following definition is the result of generalizing this line of thinking to arbitrary linear programs:

Definition 2.B.2

Dual of a Linear Program

Suppose $A \in \mathcal{M}_{m,n}(\mathbb{R})$, $\mathbf{b} \in \mathbb{R}^m$, $\mathbf{c} \in \mathbb{R}^n$ are fixed, and $\mathbf{x} \in \mathbb{R}^n, \mathbf{y} \in \mathbb{R}^m$ are vectors of variables. Then the **dual** of a linear program

$$
\begin{aligned}
\text{maximize:} \quad & \mathbf{c} \cdot \mathbf{x} \\
\text{subject to:} \quad & A\mathbf{x} \le \mathbf{b} \\
& \mathbf{x} \ge \mathbf{0}
\end{aligned}
$$

is the linear program

$$
\begin{aligned}
\text{minimize:} \quad & \mathbf{b} \cdot \mathbf{y} \\
\text{subject to:} \quad & A^T \mathbf{y} \ge \mathbf{c} \\
& \mathbf{y} \ge \mathbf{0}
\end{aligned}
$$

Four things "flip" when constructing a dual problem: "maximize" becomes "minimize", A turns into A^T, the "$\le$" constraint becomes "$\ge$", and $\mathbf{b}$ and $\mathbf{c}$ switch spots.

The original linear program in Definition 2.B.2 is called the **primal problem**, and the two of them together are called a **primal/dual pair**. As hinted at by our previous discussion, the dual problem is remarkable for the fact that it can provide us with upper bounds on the optimal value of the primal problem (and the primal problem similarly provides lower bounds on the optimal value of the dual problem):

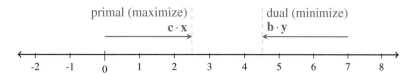

Figure 2.37: Weak duality says that the objective function of the primal (maximization) problem cannot be increased past the objective function of the dual (minimization) problem.

Theorem 2.B.1	If $\mathbf{x} \in \mathbb{R}^n$ is a feasible vector of a (primal) linear program and $\mathbf{y} \in \mathbb{R}^m$ is a feasible vector of its dual problem, then
Weak Duality	

$$\mathbf{c} \cdot \mathbf{x} \leq \mathbf{b} \cdot \mathbf{y}.$$

Proof. Since $A\mathbf{x} \leq \mathbf{b}$ and all of the entries of $\mathbf{y}$ are non-negative, it follows that $\mathbf{y} \cdot (A\mathbf{x}) \leq \mathbf{y} \cdot \mathbf{b}$. However, by using the fact that $A^T\mathbf{y} \geq \mathbf{c}$, a similar argument shows that $\mathbf{y} \cdot (A\mathbf{x}) = (A^T\mathbf{y}) \cdot \mathbf{x} \geq \mathbf{c} \cdot \mathbf{x}$. Stringing these inequalities together shows that $\mathbf{c} \cdot \mathbf{x} \leq \mathbf{y} \cdot \mathbf{b} = \mathbf{b} \cdot \mathbf{y}$, as desired. ∎

> Recall from Exercise 1.3.17 that $\mathbf{y} \cdot (A\mathbf{x}) = \mathbf{y}^T A \mathbf{x} = (A^T\mathbf{y}) \cdot \mathbf{x}$. This is a general fact about the dot product and transpose that is not specific to linear programming.

The weak duality theorem not only provides us with a way of establishing upper bounds on the optimal value of our linear program, but it also lets us easily determine when we have found its optimal value. In particular, since we are bounding a maximization problem above by a minimization problem, we know that if $\mathbf{x} \in \mathbb{R}^n$ and $\mathbf{y} \in \mathbb{R}^m$ are feasible vectors of the primal and dual problems, respectively, for which $\mathbf{c} \cdot \mathbf{x} = \mathbf{b} \cdot \mathbf{y}$, then they must be optimal. In other words, if we can find feasible vectors of each problem that give the same value when plugged into their respective objective functions, they must be optimal, since they cannot possibly be increased or decreased past each other (see Figure 2.37).

The following example illustrates how we can construct the dual of a linear program and use it to verify optimality.

Example 2.B.8	Construct the dual of the linear program
Constructing and Using a Dual Program	

$$
\begin{aligned}
\text{maximize:} \quad & x_1 + 2x_2 \\
\text{subject to:} \quad & x_1 + x_2 \leq 3 \\
& -x_1 + x_2 \leq 1 \\
& x_1, \quad x_2 \geq 0
\end{aligned}
$$

> This is the linear program (2.B.12) again.

and then verify that the optimal value of both linear programs equals 5.

Solution:

This linear program is in primal standard form with

$$A = \begin{bmatrix} 1 & 1 \\ -1 & 1 \end{bmatrix}, \quad \mathbf{b} = (3,1), \quad \text{and} \quad \mathbf{c} = (1,2),$$

and we already noted that $(x_1, x_2) = (1,2)$ is a feasible vector of this linear program that gives $\mathbf{c} \cdot \mathbf{x} = 5$, so 5 is a lower bound on the optimal value of both the primal and dual problems. The dual program is

$$
\begin{aligned}
\text{minimize:} \quad & 3y_1 + y_2 \\
\text{subject to:} \quad & y_1 - y_2 \geq 1 \\
& y_1 + y_2 \geq 2 \\
& y_1, \ \ y_2 \geq 0
\end{aligned}
$$

It is straightforward to verify that $(y_1, y_2) = (3/2, 1/2)$ is a feasible vector of the dual program, and it gives $\mathbf{b} \cdot \mathbf{y} = 5$, so 5 is an upper bound on the optimal value of both the primal and dual problems. We thus conclude that both problems have optimal value 5.

The above example perhaps does not display the true magic of duality, since we already knew the solution of that linear program. Duality becomes particularly useful when there is a seemingly "obvious" solution to a linear program that we would like to quickly and easily show *is* in fact its solution. We illustrate this technique with another example.

Example 2.B.9

Verifying an "Obvious" Solution

Solve the linear program

$$
\begin{aligned}
\text{maximize:} \quad & x_1 + x_2 + x_3 + x_4 + x_5 \\
\text{subject to:} \quad & x_1 + x_2 + x_3 \qquad\qquad\ \ \leq 3 \\
& x_2 + x_3 + x_4 \qquad\ \ \leq 3 \\
& x_3 + x_4 + x_5 \leq 3 \\
& x_1 \qquad\quad\ + x_4 + x_5 \leq 3 \\
& x_1 + x_2 \qquad\qquad + x_5 \leq 3 \\
& x_1, \ \ x_2, \ \ x_3, \ \ x_4, \ \ x_5 \geq 0
\end{aligned}
$$

by constructing its dual problem and finding feasible vectors of both problems that produce the same value in their objective functions.

Solution:

It does not take long to see that $(x_1, x_2, x_3, x_4, x_5) = (1, 1, 1, 1, 1)$ is a feasible vector of this linear program giving a value of

$$
x_1 + x_2 + x_3 + x_4 + x_5 = 1 + 1 + 1 + 1 + 1 = 5
$$

in the objective function. Also, this vector somehow "feels" optimal—to increase x_1, for example, we would have to correspondingly decrease each of $x_2 + x_3$ and $x_4 + x_5$ in order for the first and fourth constraints to still be satisfied.

To pin this intuition down and verify that this vector is in fact optimal, we could use the simplex algorithm to solve this linear program, but an easier way (especially for large linear programs) is to construct its dual. In this case, the dual program is

It can be helpful
when constructing
the dual of a linear
program to explicitly
write out the A
matrix and the **b**
and **c** vectors first.

$$
\begin{aligned}
\text{minimize:} \quad & 3y_1 + 3y_2 + 3y_3 + 3y_4 + 3y_5 \\
\text{subject to:} \quad & y_1 \qquad\qquad\;\; + y_4 + y_5 \geq 1 \\
& y_1 + y_2 \qquad\qquad + y_5 \geq 1 \\
& y_1 + y_2 + y_3 \qquad\qquad\; \geq 1 \\
& \qquad\; y_2 + y_3 + y_4 \qquad\; \geq 1 \\
& \qquad\qquad\; y_3 + y_4 + y_5 \geq 1 \\
& y_1, \quad y_2, \quad y_3, \quad y_4, \quad y_5 \geq 0
\end{aligned}
$$

Our goal now is to find a feasible vector of this dual LP that produces a value of 5 in its objective function. A bit of thought and squinting reveals that the vector $(y_1, y_2, y_3, y_4, y_5) = (1,1,1,1,1)/3$ works. Since we have found feasible vectors of both the primal and dual problems that result in their objective functions equaling the same value, we conclude that these vectors must be optimal and the optimal value of each linear program is 5.

In the previous two examples, we saw that not only did the dual problem serve as an upper bound on the primal problem, but rather we were able to find particular feasible vectors of each problem that resulted in their objective functions taking on the same value, thus proving optimality. The following theorem establishes the remarkable fact that this *always* happens—if a linear program has an optimal value (i.e., it is neither unbounded nor infeasible), then the dual program has the same optimal value. However, we present this theorem without proof, as it requires a significant amount of extra background material to prove (again, we refer the interested reader to any of the textbooks mentioned in Remark 2.B.2 for the details).

Theorem 2.B.2

Strong Duality

If $\mathbf{x}_* \in \mathbb{R}^n$ is a feasible vector of a (primal) linear program satisfying $\mathbf{c} \cdot \mathbf{x}_* \geq \mathbf{c} \cdot \mathbf{x}$ for all feasible vectors $\mathbf{x} \in \mathbb{R}^n$, then there exists a feasible vector $\mathbf{y}_* \in \mathbb{R}^m$ of its dual problem such that

$$\mathbf{c} \cdot \mathbf{x}_* = \mathbf{b} \cdot \mathbf{y}_*.$$

$\mathbf{x}_*$ and $\mathbf{y}_*$ are vectors attaining the maximum and minimum values of the primal and dual linear programs, respectively.

It is worth noting that strong duality shows that Figure 2.37 is somewhat misleading—it is not just the case that the primal and dual programs bound each other, but rather they bound each other "tightly" in the sense that their optimal values coincide (see Figure 2.38).

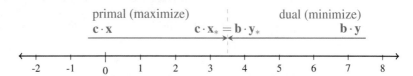

Figure 2.38: Strong duality says that the objective function of the primal (maximization) problem can be increased to the exact same value that the objective function of the dual (minimization) problem can be decreased to.

Note that the weak duality theorem says that if a primal problem is unbounded (i.e., it has feasible vectors $\mathbf{x}$ making $\mathbf{c} \cdot \mathbf{x}$ arbitrarily large), then its dual problem must be infeasible, since we could otherwise bound the primal problem by some quantity $\mathbf{b} \cdot \mathbf{y}$. A similar argument shows that if a dual problem is unbounded, then the primal problem must be infeasible.

The linear programs in a primal/dual pair are thus either both solvable, both infeasible, or one is infeasible while the other is unbounded. We saw examples where both are solvable in Examples 2.B.8 and 2.B.9. To illustrate the case where one problem is unbounded and its dual is infeasible, consider the following primal/dual pair:

Now is a good time to practice constructing the dual of a linear program. Try to construct the duals that we present here yourself.

Primal	**Dual**
maximize: $x_1 + 2x_2$	minimize: y
subject to: $-x_1 + x_2 \leq 1$	subject to: $-y \geq 1$
$x_1, \quad x_2 \geq 0$	$y \geq 2$
	$y \geq 0$

We illustrated that the above primal problem is unbounded in Figure 2.34(a), and it is straightforward to see that the dual problem is infeasible (it is not possible that $-y \geq 1$ and $y \geq 0$ simultaneously).

Finally, to illustrate the case where both problems are infeasible, consider the following primal/dual pair:

Primal	**Dual**
maximize: $x_1 + 2x_2$	minimize: $-y_1 - y_2$
subject to: $x_1 - x_2 \leq -1$	subject to: $y_1 - y_2 \geq 1$
$-x_1 + x_2 \leq -1$	$-y_1 + y_2 \geq 2$
$x_1, \quad x_2 \geq 0$	$y_1, \quad y_2 \geq 0$

We illustrated that the above primal problem is infeasible in Figure 2.34(b), and the dual problem is infeasible since adding the two constraints yields the (false) inequality $0 \geq 3$.

The possible infeasible/solvable/unbounded pairings that primal and dual problems can share are summarized in Table 2.1.

		Primal problem		
		Infeasible	Solvable	Unbounded
Dual	Infeasible	✓	·	✓
	Solvable	·	✓	·
	Unbounded	✓	·	·

Table 2.1: A summary of how the infeasibility, solvability, or unboundedness of one problem in a primal/dual pair can be used to help determine that of the other. For example, if a primal problem is unbounded then we know immediately that its dual must be infeasible.

The Dual of Problems Not in Standard Form

Strictly speaking, Definition 2.B.2 only tells us how to construct the dual of a linear program that is represented in primal standard form. If a linear program has (for example) equality constraints, then we must convert it into primal standard form before constructing its dual problem.

Example 2.B.10

Constructing the Dual of a Problem Not in Primal Standard Form

Construct the dual of the following linear program:

$$\begin{aligned} \text{maximize:} \quad & x_1 + 2x_2 \\ \text{subject to:} \quad & x_1 + x_2 = 3 \\ & -x_1 + x_2 \leq 1 \\ & x_1, \quad x_2 \geq 0 \end{aligned} \qquad (2.B.13)$$

Solution:

Before constructing the dual of this linear program, we first construct its primal standard form by turning the equality constraint into two inequality constraints:

$$\begin{aligned} \text{maximize:} \quad & x_1 + 2x_2 \\ \text{subject to:} \quad & x_1 + x_2 \leq 3 \\ & -x_1 - x_2 \leq -3 \\ & -x_1 + x_2 \leq 1 \\ & x_1, \quad x_2 \geq 0 \end{aligned}$$

When constructing the dual of a linear program, the number of constraints and variables swap. Since the primal problem here has 2 variables and 3 constraints, the dual program has 3 variables and 2 constraints (ignoring the non-negativity constraints in each problem).

Now that this linear program is in primal standard form, we can compute its dual:

$$\begin{aligned} \text{minimize:} \quad & 3y_1 - 3y_2 + y_3 \\ \text{subject to:} \quad & y_1 - y_2 - y_3 \geq 1 \\ & y_1 - y_2 + y_3 \geq 2 \\ & y_1, \quad y_2, \quad y_3 \geq 0 \end{aligned} \qquad (2.B.14)$$

However, it turns out that there are some simple rules that we can use to convert an arbitrary linear program (not necessarily in standard form) into its dual program, without having to do the extra intermediate step of converting it into its primal standard form. To get an idea of how they work, notice that in the linear program (2.B.14), the variables y_1 and y_2 never appear independently, but rather always appear together in the form $y_1 - y_2$. We can thus replace $y_1 - y_2$ with the single (potentially negative) variable y_*, which results in the following equivalent linear program:

$$\begin{aligned} \text{minimize:} \quad & 3y_* + y_3 \\ \text{subject to:} \quad & y_* - y_3 \geq 1 \\ & y_* + y_3 \geq 2 \\ & y_3 \geq 0 \end{aligned}$$

The first constraint in linear program (2.B.13) being equality (rather than "$\leq$") thus had the effect on the dual problem of making the first variable unconstrained (rather than ≥ 0). This makes a fair amount of sense intuitively—if we strengthen one of the constraints in the primal problem from "$\leq$" to equality then the optimal value will decrease, so we must correspondingly loosen one of the constraints in the dual problem. We can similarly come up with rules for other possible forms of a linear program (e.g., one having a "$\geq$" constraint or an unconstrained variable), and these rules are summarized in Table 2.2.

We close this section with one final example that illustrates how to make use of these rules to quickly construct the dual of a linear program that is not in primal standard form.

Primal problem (maximize)		Dual problem (minimize)	
i-th constraint:	$\geq$	i-th variable:	≤ 0
	$=$		unconstrained
	$\leq$		≥ 0
j-th variable:	≥ 0	j-th constraint:	$\geq$
	unconstrained		$=$
	≤ 0		$\leq$

We do not prove these rules here. Again, see Remark 2.B.2.

Table 2.2: A summary of the rules that can be used to help construct the dual of a linear program not in primal standard form. For example, if the i-th constraint of the primal problem is a "$\leq$" constraint, then the i-th variable of the dual problem must be restricted to being ≥ 0.

Example 2.B.11

Constructing and Using a Dual Program

Construct the dual of the following linear program

$$\begin{aligned} \text{maximize:} \quad & x_2 \\ \text{subject to:} \quad & x_1 + x_2 = -3 \\ & 2x_1 - x_2 \geq 1 \\ & x_1 \qquad \geq 0 \end{aligned}$$

and then find that the optimal value of both linear programs.

Solution:

By using the rules from Table 2.2, we construct the dual problem in the usual way, except keeping in mind that since the first constraint of the primal is equality, the first variable of the dual must be unconstrained, since the second constraint of the primal is "$\geq$", the second variable of the dual must be ≤ 0, and since the second variable of the primal is unconstrained, the second constraint of the dual must be equality. It follows the dual of this linear program is

Alternatively, we could convert this linear program to primal standard form and then construct its dual, just like we did in Example 2.B.10—the resulting primal/dual pair will look slightly different, but will have the same optimal value.

$$\begin{aligned} \text{minimize:} \quad & -3y_1 + y_2 \\ \text{subject to:} \quad & y_1 + 2y_2 \geq 0 \\ & y_1 - y_2 = 1 \\ & y_2 \leq 0 \end{aligned}$$

These linear programs are small and simple enough that we can find their optimal values just by fumbling and eyeballing feasible vectors for a while. It does not take long to find the vector $(x_1, x_2) = (0, -3)$, which is feasible for the primal problem and has $\mathbf{c} \cdot \mathbf{x} = -3$, and the vector $(y_1, y_2) = (1, 0)$, which is feasible for the dual problem and has $\mathbf{b} \cdot \mathbf{y} = -3$. Since these values are the same, weak duality tells us that both linear programs have optimal value -3.

Exercises

solutions to starred exercises on page 457

2.B.1 Use the geometric method of Section 2.B.2 to either solve each of the following linear programs or show that they are unbounded or infeasible.

*(a)
$$\begin{aligned} \text{maximize:} \quad & x_1 + x_2 \\ \text{subject to:} \quad & 2x_1 + x_2 \geq 1 \\ & x_1 + 2x_2 \leq 3 \\ & x_1, \quad x_2 \geq 0 \end{aligned}$$

(b) minimize: $2x_1 - x_2$
     subject to: $x_1 + x_2 \leq 4$
                   $2x_1 + 3x_2 \geq 4$
                   $x_1, \quad x_2 \geq 0$

*(c) minimize: $3x_1 + x_2$
      subject to: $x_1 + x_2 \geq 2$
                    $2x_1 - x_2 \geq 1$
                    $x_1, \ x_2 \geq 0$

(d) maximize: $2x_1 + x_2$
     subject to: $x_1 + x_2 \geq 2$
                   $x_1 + 2x_2 \geq 3$
                   $2x_1 + 2x_2 \leq 5$
                   $x_1, \quad x_2 \geq 0$

2.B.2 Use the simplex method to either solve each of the following linear programs or show that they are unbounded (they all have $\mathbf{x} = \mathbf{0}$ as a feasible point and are thus feasible).

*(a) maximize: $4x_1 + x_2$
      subject to: $2x_1 - 2x_2 \leq 5$
                    $x_1 + 3x_2 \leq 3$
                    $x_1, \quad x_2 \geq 0$

(b) maximize: $x_1 + x_2 + x_3$
     subject to: $x_1 + 2x_2 + 3x_3 \leq 2$
                   $3x_1 + 2x_2 + x_3 \leq 1$
                   $x_1, \quad x_2, \quad x_3 \geq 0$

*(c) maximize: $x_1 + x_2 + x_3$
      subject to: $2x_1 - x_2 + x_3 \leq 1$
                    $x_1 + 3x_2 \qquad \leq 2$
                    $-x_1 + x_2 + 2x_3 \leq 3$
                    $x_1, \quad x_2, \quad x_3 \geq 0$

(d) maximize: $2x_1 + x_2 + x_3 + x_4$
     subject to: $x_1 + 2x_2 - 3x_3 + 2x_4 \leq 1$
                   $2x_1 - x_2 + x_3 + x_4 \leq 0$
                   $3x_1 + x_2 - 2x_3 - x_4 \leq 1$
                   $x_1, \quad x_2, \quad x_3, \quad x_4 \geq 0$

*(e) maximize: $x_1 + x_2 + x_3 + x_4$
      subject to: $4x_1 + 3x_2 + 2x_3 \qquad \leq 1$
                    $x_1 + 4x_2 + x_3 + 2x_4 \leq 2$
                    $2x_1 + x_2 + 3x_3 + 2x_4 \leq 2$
                    $x_1, \quad x_2, \quad x_3, \quad x_4 \geq 0$

2.B.3 Use the simplex method to find a feasible point of each of the following linear programs (or show that they are infeasible) and then solve them (or show that they are unbounded).

*(a) maximize: x_2
      subject to: $2x_1 - 2x_2 \geq 2$
                    $x_1 + 3x_2 \leq 3$
                    $x_1, \quad x_2 \geq 0$

(b) minimize: $2x_1 - x_2 - 2x_3$
     subject to: $2x_1 - x_2 + 4x_3 \geq 2$
                   $x_1 + 2x_2 - 3x_3 \leq 1$
                   $x_1, \quad x_2, \quad x_3 \geq 0$

*(c) minimize: $x_1 + 2x_2 - x_3$
      subject to: $x_2 - x_3 \leq 3$
                    $2x_1 + 2x_2 - x_3 = 1$
                    $x_1 + 2x_2 + 2x_3 \leq 3$
                    $x_1, \quad x_2, \quad x_3 \geq 0$

(d) maximize: $3x_1 + 2x_2 - x_3 + 2x_4$
     subject to: $2x_1 - 2x_2 - x_3 + 2x_4 \geq 2$
                   $x_1 + 3x_2 - 2x_3 + x_4 = 2$
                   $3x_1 + x_2 - 3x_3 + 3x_4 \leq 3$
                   $x_1, \quad x_2, \qquad x_4 \geq 0$

*(e) maximize: $x_1 + 2x_2 + x_3 + 2x_4$
      subject to: $2x_1 + 4x_2 - x_3 \qquad \leq 3$
                    $3x_2 + x_3 + 2x_4 = 1$
                    $3x_1 - x_2 + 2x_3 + 2x_4 \leq 3$
                    $x_1, \qquad x_3 \qquad \geq 0$

*** 2.B.4** Construct the dual of each of the linear programs from Exercise 2.B.3.

2.B.5 Determine which of the following statements are true and which are false.

*(a) If $A\mathbf{x} \leq \mathbf{b}$ and A is invertible, then $\mathbf{x} \leq A^{-1}\mathbf{b}$.
(b) If $\mathbf{c}, \mathbf{x}, \mathbf{y} \in \mathbb{R}^n$ and $\mathbf{x} \geq \mathbf{y}$ then $\mathbf{c} \cdot \mathbf{x} \geq \mathbf{c} \cdot \mathbf{y}$.
*(c) If $\mathbf{c}, \mathbf{x}, \mathbf{y} \in \mathbb{R}^n$, $\mathbf{c} \geq \mathbf{0}$ and $\mathbf{x} \geq \mathbf{y}$, then $\mathbf{c} \cdot \mathbf{x} \geq \mathbf{c} \cdot \mathbf{y}$.
(d) If a linear program has finite optimal value, then it is attained at a unique vector (i.e., there is exactly one vector $\mathbf{x}$ such that $\mathbf{c} \cdot \mathbf{x}$ is as large or as small as possible).
*(e) If the feasible region of a linear program is unbounded (i.e., infinitely large), then the linear program is unbounded (i.e., does not have a finite optimal value).
(f) The optimal value (if it exists) of a linear program in primal standard form is always less than or equal to the optimal value of its dual problem.

****2.B.6** Find the optimal value of the linear program

$$\begin{aligned}
\text{maximize:} \quad & x_1 + x_2 \\
\text{subject to:} \quad & 2x_1 + x_2 \leq 2 \\
& 2x_1 + 3x_2 \geq 3 \\
& x_1 - 2x_2 \geq -1 \\
& x_1, \quad x_2 \geq 0
\end{aligned}$$

and then show that its feasible region contains no points with integer coordinates.

2.B.7 Two of the five constraints in the linear program (2.B.10) are redundant—which ones?

**2.B.8 Suppose $0 < c \in \mathbb{R}$ and consider the linear program

$$
\begin{array}{ll}
\text{maximize:} & x_2 \\
\text{subject to:} & cx_1 + x_2 \leq c \\
& -cx_1 + x_2 \leq 0 \\
& x_1, \ x_2 \geq 0
\end{array}
$$

(a) Show that the optimal value of this linear program is $c/2$.
(b) Show that if x_1 and x_2 are constrained to be integers then the optimal value of this integer linear program is 0.

**2.B.9 Show that the dual of the dual of a linear program is the original linear program.

[Hint: Write the dual from Definition 2.B.2 in standard form.]

2.B.10 Linear programming duality can be used to give us a deeper understanding of basic linear algebraic operations like linear combinations.

(a) Suppose $\mathbf{x}_1, \ldots, \mathbf{x}_k, \mathbf{y} \in \mathbb{R}^n$. Construct a linear program that determines whether or not $\mathbf{y} \in \text{span}(\mathbf{x}_1, \ldots, \mathbf{x}_k)$.
(b) Use the dual of the linear program from part (a) to show that $\mathbf{y} \in \text{span}(\mathbf{x}_1, \ldots, \mathbf{x}_k)$ if and only if $\mathbf{y}$ is orthogonal to every vector that $\mathbf{x}_1, \ldots, \mathbf{x}_k$ are orthogonal to (i.e., $\mathbf{x}_1 \cdot \mathbf{z} = \cdots = \mathbf{x}_k \cdot \mathbf{z} = 0$ implies $\mathbf{y} \cdot \mathbf{z} = 0$).

2.B.11 The **1-norm** and ∞ **-norm** of a vector $\mathbf{v} \in \mathbb{R}^n$ are defined by

$$
\|\mathbf{v}\|_1 \stackrel{\text{def}}{=} \sum_{j=1}^n |v_j| \quad \text{and} \quad \|\mathbf{v}\|_\infty \stackrel{\text{def}}{=} \max_{1 \leq j \leq n} \{|v_j|\},
$$

respectively, and they provide other ways (besides the usual vector length $\|\mathbf{v}\|$) of measuring how large a vector is.

(a) Construct a linear program that, given a matrix $A \in \mathcal{M}_{m,n}$ and a vector $\mathbf{b} \in \mathbb{R}^m$, finds a vector $\mathbf{x} \in \mathbb{R}^n$ that makes $\|A\mathbf{x} - \mathbf{b}\|_1$ as small as possible.
[Hint: $|a| \leq b$ if and only if $-b \leq a \leq b$.]
(b) Construct a linear program that, given a matrix $A \in \mathcal{M}_{m,n}$ and a vector $\mathbf{b} \in \mathbb{R}^m$, finds a vector $\mathbf{x} \in \mathbb{R}^n$ that makes $\|A\mathbf{x} - \mathbf{b}\|_\infty$ as small as possible.

[Side note: If the linear system $A\mathbf{x} = \mathbf{b}$ does not have a solution, these linear programs find the "closest thing" to a solution.]

2.B.12 A matrix $A \in \mathcal{M}_n$ with non-negative entries is called **column stochastic if its columns each add up to 1 (i.e., $a_{1,j} + a_{2,j} + \cdots + a_{n,j} = 1$ for each $1 \leq j \leq n$). Show that if A is column stochastic then there is a vector $\mathbf{x} \in \mathbb{R}^n$ for which $A\mathbf{x} = \mathbf{x}$ and $\mathbf{x} \geq \mathbf{0}$.

[Hint: Set up a linear program and use duality. There is an "obvious" vector $\mathbf{y} \in \mathbb{R}^n$ for which $A^T \mathbf{y} = \mathbf{y}$... what is it?]

2.C Extra Topic: More About the Rank

Recall from Section 2.4.3 that the rank of a matrix can be characterized in numerous different ways, such as the dimension of its range, the dimension of the span of its rows, or the number of non-zero rows in any of its row echelon forms. In this section, we look at the rank in a bit more depth and provide two (or three, depending on how we count) additional characterizations of it.

2.C.1 The Rank Decomposition

Our first new characterization of the rank is a kind of converse to Theorem 2.4.11(b), which said that for all matrices A and B, we have

$$
\text{rank}(AB) \leq \min \{\text{rank}(A), \text{rank}(B)\}.
$$

This inequality is particularly useful (i.e., provides a particularly strong bound) when we can write a matrix as a product of a tall and skinny matrix with a short and fat matrix. In particular, if we recall that $\text{rank}(A) \leq \min\{m,n\}$ for all $A \in \mathcal{M}_{m,n}$, then we see that if we can write $A = CR$ for some $C \in \mathcal{M}_{m,r}$ and $R \in \mathcal{M}_{r,n}$, then we must have

$$
\text{rank}(A) = \text{rank}(CR) \leq \min \{\text{rank}(C), \text{rank}(R)\} \leq \min\{m, r, n\} \leq r.
$$

In other words, it is not possible to write a matrix A as a product of a tall and skinny matrix with a short and fat matrix if they are skinnier and shorter, respectively, than $\text{rank}(A)$. The following theorem says that this skinniness/shortness threshold actually completely characterizes the rank (see Figure 2.39).

$$A = \begin{bmatrix} * & * & * & * & * & * \\ * & * & * & * & * & * \\ * & * & * & * & * & * \\ * & * & * & * & * & * \\ * & * & * & * & * & * \\ * & * & * & * & * & * \\ * & * & * & * & * & * \end{bmatrix} = \begin{bmatrix} * & * & * \\ * & * & * \\ * & * & * \\ * & * & * \\ * & * & * \\ * & * & * \\ * & * & * \end{bmatrix} \begin{bmatrix} * & * & * & * & * & * \\ * & * & * & * & * & * \\ * & * & * & * & * & * \end{bmatrix}$$

$\operatorname{rank}(A) = 3$

Figure 2.39: Theorem 2.C.1 says that every matrix can be written as a product of a tall and skinny matrix and a short and fat matrix. The rank determines exactly how skinny and short the matrices in the product can be (as shown here, A has rank 3, so it can be written as a product of a matrix C with 3 columns and a matrix R with 3 rows).

Theorem 2.C.1 Rank Decomposition

Let $A \in \mathcal{M}_{m,n}$. Then the smallest integer r for which there exist matrices $C \in \mathcal{M}_{m,r}$ and $R \in \mathcal{M}_{r,n}$ with $A = CR$ is exactly $r = \operatorname{rank}(A)$.

The names "C" and "R" stand for "column" and "row", respectively, since C has long columns and R has long rows. Alternatively, "R" could stand for "RREF", since it can be chosen to be the RREF of A with its $\mathbf{0}$ rows discarded.

Proof. We already showed that $\operatorname{rank}(A) \leq r$, so to complete the proof we just need to show that if $r = \operatorname{rank}(A)$ then we can find such a C and R. To see this, recall from Theorem 2.5.3 that we can write $A = P\hat{R}$, where P is some invertible matrix and $\hat{R}$ is the reduced row echelon form of A. If $r = \operatorname{rank}(A)$ then the first r rows of $\hat{R}$ are exactly its non-zero rows. We can thus write P and $\hat{R}$ as block matrices

$$P = \left[\, C \mid D \,\right] \quad \text{and} \quad \hat{R} = \left[\begin{array}{c} R \\ \hline O \end{array}\right],$$

where $C \in \mathcal{M}_{m,r}$ contains the leftmost r columns of P and $R \in \mathcal{M}_{r,n}$ contains the top r rows of $\hat{R}$ (i.e., its non-zero rows). Then

$$A = P\hat{R} = \left[\, C \mid D \,\right] \left[\begin{array}{c} R \\ \hline O \end{array}\right] = CR + DO = CR,$$

as claimed. ∎

Example 2.C.1 Computing a Rank Decomposition

We talk about finding *a* rank decomposition of a matrix, rather than *the* rank decomposition of a matrix, since the matrices C and R are not unique (see Exercise 2.C.6).

Compute a rank decomposition of the matrix $A = \begin{bmatrix} 1 & 1 & 1 & -1 \\ 0 & 1 & 1 & 0 \\ -1 & 1 & 1 & 1 \end{bmatrix}$.

Solution:

In order to construct a rank decomposition of a matrix, we first recall from Example 2.4.8 that row-reducing $[\, A \mid I \,]$ results in the matrix

$$\left[\begin{array}{cccc|ccc} 1 & 0 & 0 & -1 & 1 & -1 & 0 \\ 0 & 1 & 1 & 0 & 0 & 1 & 0 \\ 0 & 0 & 0 & 0 & 1 & -2 & 1 \end{array}\right].$$

If you have
forgotten this
relationship
between
row-reduction and
matrix
multiplication, refer
back to
Theorem 2.2.1.

It follows that $\hat{R} = EA$ (and equivalently, $A = P\hat{R}$, where $P = E^{-1}$), where

$$\hat{R} = \begin{bmatrix} 1 & 0 & 0 & -1 \\ 0 & 1 & 1 & 0 \\ 0 & 0 & 0 & 0 \end{bmatrix}, \quad E = \begin{bmatrix} 1 & -1 & 0 \\ 0 & 1 & 0 \\ 1 & -2 & 1 \end{bmatrix}, \quad \text{and} \quad P = \begin{bmatrix} 1 & 1 & 0 \\ 0 & 1 & 0 \\ -1 & 1 & 1 \end{bmatrix}.$$

Since $\hat{R}$ has two non-zero rows, it follows that $\operatorname{rank}(A) = 2$. The matrices C and R in the rank decomposition can thus be chosen to consist of the leftmost 2 columns of P and the topmost 2 rows of $\hat{R}$, respectively:

$$A = CR, \quad \text{where} \quad C = \begin{bmatrix} 1 & 1 \\ 0 & 1 \\ -1 & 1 \end{bmatrix} \quad \text{and} \quad R = \begin{bmatrix} 1 & 0 & 0 & -1 \\ 0 & 1 & 1 & 0 \end{bmatrix}.$$

One of the useful features of the rank decomposition of a matrix is that it lets us store low-rank matrices much more efficiently than the naïve method that involves writing down all entries in the matrix. In particular, to store a rank-r matrix $A \in \mathcal{M}_{m,n}$, we can just keep track of the mr entries of C and rn entries of R in its rank decomposition, which is much simpler than keeping track of the mn entries of A when r is small.

For example, in order to store the matrix

In a sense, low-rank
matrices contain
"less information"
than full-rank
matrices—there are
patterns in their
entries that can be
exploited to
compress them.

$$A = \begin{bmatrix} 1 & 0 & -1 & 0 & 2 & 0 & 2 & 0 & -1 & 0 & 1 \\ 0 & 2 & 0 & 1 & 0 & 3 & 0 & 1 & 0 & 2 & 0 \\ 1 & 0 & -1 & 0 & 2 & 0 & 2 & 0 & -1 & 0 & 1 \\ 0 & 4 & 0 & 2 & 0 & 6 & 0 & 2 & 0 & 4 & 0 \\ 1 & 0 & -1 & 0 & 2 & 0 & 2 & 0 & -1 & 0 & 1 \\ 0 & -2 & 0 & -1 & 0 & -3 & 0 & -1 & 0 & -2 & 0 \\ 1 & 0 & -1 & 0 & 2 & 0 & 2 & 0 & -1 & 0 & 1 \\ 0 & 4 & 0 & 2 & 0 & 6 & 0 & 2 & 0 & 4 & 0 \\ 1 & 0 & -1 & 0 & 2 & 0 & 2 & 0 & -1 & 0 & 1 \\ 0 & 2 & 0 & 1 & 0 & 3 & 0 & 1 & 0 & 2 & 0 \\ 1 & 0 & -1 & 0 & 2 & 0 & 2 & 0 & -1 & 0 & 1 \end{bmatrix}$$

we could list all $11 \times 11 = 121$ of its entries. However, a more economical way to store it is to first compute one of its rank decompositions: $A = CR$ with

We just write C^T
here (instead of C
itself) for formatting
reasons—C has 11
rows and thus
would take up a lot
of space.

$$C^T = \begin{bmatrix} 1 & 0 & 1 & 0 & 1 & 0 & 1 & 0 & 1 & 0 & 1 \\ 0 & 1 & 0 & 2 & 0 & -1 & 0 & 2 & 0 & 1 & 0 \end{bmatrix} \quad \text{and}$$

$$R = \begin{bmatrix} 1 & 0 & -1 & 0 & 2 & 0 & 2 & 0 & -1 & 0 & 1 \\ 0 & 2 & 0 & 1 & 0 & 3 & 0 & 1 & 0 & 2 & 0 \end{bmatrix},$$

and then just keep track of C and R instead. This way, we only need to keep track of $(11 \times 2) + (2 \times 11) = 44$ numbers instead of the 121 numbers required by the naïve method.

In the most extreme case when $r = 1$, Theorem 2.C.1 says that a non-zero matrix A has rank 1 if and only if it can be written as a product of a matrix with 1 column (i.e., a column vector) and a matrix with 1 row (i.e., a row vector). In other words, rank-1 matrices $A \in \mathcal{M}_{m,n}$ are exactly those that can be written in the form $A = \mathbf{v}\mathbf{w}^T$ for some non-zero column vectors $\mathbf{v} \in \mathbb{R}^m$ and $\mathbf{w} \in \mathbb{R}^n$. The following theorem provides a natural generalization of this fact to higher-rank matrices:

We originally
proved this claim
about rank-1
matrices back in
Exercise 2.4.30.

Theorem 2.C.2

Rank-One Sum Decomposition

Let $A \in \mathcal{M}_{m,n}$. Then the smallest integer r for which there exist sets of vectors $\{\mathbf{v}_j\}_{j=1}^{r} \subset \mathbb{R}^m$ and $\{\mathbf{w}_j\}_{j=1}^{r} \subset \mathbb{R}^n$ with

$$A = \sum_{j=1}^{r} \mathbf{v}_j \mathbf{w}_j^T$$

is exactly $r = \text{rank}(A)$. Furthermore, the sets $\{\mathbf{v}_i\}_{i=1}^{r}$ and $\{\mathbf{w}_i\}_{i=1}^{r}$ can be chosen to be linearly independent.

Proof. Use the rank decomposition to write $A = CR$, where $C \in \mathcal{M}_{m,r}$ and $R \in \mathcal{M}_{r,n}$, and then write C and R in terms of their columns and rows, respectively:

$$C = \begin{bmatrix} \mathbf{v}_1 \mid \mathbf{v}_2 \mid \cdots \mid \mathbf{v}_r \end{bmatrix} \quad \text{and} \quad R^T = \begin{bmatrix} \mathbf{w}_1 \mid \mathbf{w}_2 \mid \cdots \mid \mathbf{w}_r \end{bmatrix}.$$

Then performing block matrix multiplication reveals that

$$A = CR = \begin{bmatrix} \mathbf{v}_1 \mid \mathbf{v}_2 \mid \cdots \mid \mathbf{v}_r \end{bmatrix} \begin{bmatrix} \mathbf{w}_1^T \\ \hline \mathbf{w}_2^T \\ \hline \vdots \\ \hline \mathbf{w}_n^T \end{bmatrix} = \sum_{j=1}^{r} \mathbf{v}_j \mathbf{w}_j^T,$$

as claimed. The fact that $\{\mathbf{v}_i\}_{i=1}^{r}$ and $\{\mathbf{w}_i\}_{i=1}^{r}$ are linearly independent sets follows from how the matrices C and R were constructed in the proof of the rank decomposition (Theorem 2.C.1)—$\{\mathbf{v}_i\}_{i=1}^{r}$ consists of the leftmost r columns of an invertible matrix and $\{\mathbf{w}_i\}_{i=1}^{r}$ consists of the non-zero rows of a matrix in reduced row echelon form.

This theorem says that a rank-r matrix can be written as a sum of r rank-1 matrices, but not fewer.

To see that r cannot be smaller than $\text{rank}(A)$, recall that each of the matrices $\mathbf{v}_j \mathbf{w}_j^T$ have rank 1, so Theorem 2.4.11(a) tells us that

$$\text{rank}(A) = \text{rank}\left(\sum_{j=1}^{r} \mathbf{v}_j \mathbf{w}_j^T\right) \le \sum_{j=1}^{r} \text{rank}\left(\mathbf{v}_j \mathbf{w}_j^T\right) = \sum_{j=1}^{r} 1 = r,$$

which completes the proof. ∎

The above proof shows that the rank-one sum decomposition (Theorem 2.C.2) really is equivalent to the rank decomposition (Theorem 2.C.1) in a straightforward way: the vectors $\{\mathbf{v}_j\}_{j=1}^{r}$ and $\{\mathbf{w}_j\}_{j=1}^{r}$ are just the columns of the matrices C and R^T, respectively. In fact, these decompositions are so closely related to each other that they are sometimes considered the "same" decomposition, just written in a slightly different form. Nevertheless, we find it useful to have both forms of this decomposition written out explicitly.

Example 2.C.2

Computing a Rank-One Sum Decomposition

Compute a rank-one sum decomposition of the matrix

$$A = \begin{bmatrix} 1 & 1 & 1 & -1 \\ 0 & 1 & 1 & 0 \\ -1 & 1 & 1 & 1 \end{bmatrix}.$$

Solution:

We showed in Example 2.C.1 that this matrix has rank decomposition

$$A = CR, \quad \text{where} \quad C = \begin{bmatrix} 1 & 1 \\ 0 & 1 \\ -1 & 1 \end{bmatrix} \quad \text{and} \quad R = \begin{bmatrix} 1 & 0 & 0 & -1 \\ 0 & 1 & 1 & 0 \end{bmatrix}.$$

We can thus construct a rank-one sum decomposition of A by multiplying each column of C by the corresponding row of R and adding:

$$A = \begin{bmatrix} 1 \\ 0 \\ -1 \end{bmatrix} \begin{bmatrix} 1 & 0 & 0 & -1 \end{bmatrix} + \begin{bmatrix} 1 \\ 1 \\ 1 \end{bmatrix} \begin{bmatrix} 0 & 1 & 1 & 0 \end{bmatrix}.$$

2.C.2 Rank in Terms of Submatrices

We now present an alternative characterization of the rank that is particularly useful for helping us eye-ball lower bounds on the rank of large matrices. The basic idea is that if we look at just a small piece of a matrix, then the rank of that small piece cannot exceed the rank of the entire matrix.

As our first step toward making this idea precise, we say that a **submatrix** of $A \in \mathcal{M}_{m,n}$ is any matrix that can be obtained by erasing zero or more rows and/or columns of A. For example, if

> The rows and/or columns that are erased do not need to be next to each other.

$$A = \begin{bmatrix} 1 & 2 & 3 & 4 & 5 \\ 6 & 7 & 8 & 9 & 10 \\ 11 & 12 & 13 & 14 & 15 \end{bmatrix}$$

then erasing the third and fifth columns of A, as well as its second row, gives us the submatrix

$$\begin{bmatrix} 1 & 2 & 3 & 4 & 5 \\ 6 & 7 & 8 & 9 & 10 \\ 11 & 12 & 13 & 14 & 15 \end{bmatrix} \quad \longrightarrow \quad \begin{bmatrix} 1 & 2 & 4 \\ 11 & 12 & 14 \end{bmatrix}.$$

Remark 2.C.1
Geometric Interpretation of Submatrices

Geometrically, if a matrix represents a linear transformation, then its submatrices represent restrictions of that linear transformation to hyperplanes spanned by the coordinate axes. In particular, erasing the j-th column of a matrix corresponds to ignoring the j-th input variable of the linear transformation, and erasing the i-th row of a matrix corresponds to ignoring its i-th output variable.

For example, a 3×3 matrix represents a linear transformation acting on $\mathbb{R}^3$. Here, we have highlighted how the matrix

$$A = \begin{bmatrix} 1 & 2 & 2 \\ 2 & 1 & -1 \\ 1 & 1 & 0 \end{bmatrix}$$

distorts the xy-plane (since it is difficult too picture how it transforms all of $\mathbb{R}^3$):

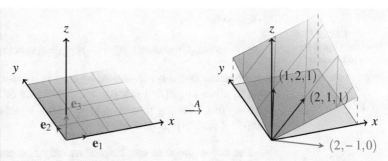

If we erase the third column of this matrix, then the resulting 3×2 sub-matrix is a linear transformation that acts on $\mathbb{R}^2$ in the same way that the original 3×3 matrix acted on the xy-plane (i.e., the input z variable is ignored):

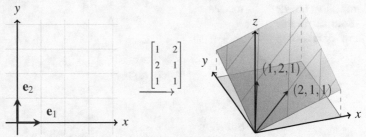

If we then erase its third *row*, the resulting 2×2 submatrix describes what the xy-plane looks like after being transformed, if we ignore the output in the direction of the z-axis. That is, it tells us what the output looks like if we look down at the xy-plane from the top of the z-axis:

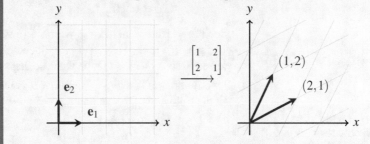

The following theorem pins down the claim that we made earlier (i.e., that a submatrix of A cannot have rank larger than that of A itself) and lets us get some easy-to-spot lower bounds on the rank of a matrix:

Theorem 2.C.3

Ranks of Submatrices

Suppose $A \in \mathcal{M}_{m,n}$. If B is a submatrix of A then $\mathrm{rank}(B) \leq \mathrm{rank}(A)$.

Here we are using Theorem 2.3.2.

Proof. Just recall that if A has columns $\mathbf{a}_1, \mathbf{a}_2, \ldots, \mathbf{a}_n$ then

$$\mathrm{rank}(A) = \dim(\mathrm{range}(A)) = \dim(\mathrm{span}(\mathbf{a}_1, \mathbf{a}_2, \ldots, \mathbf{a}_n)).$$

If a submatrix $C \in \mathcal{M}_{m,\ell}$ is created by erasing some columns of A, then $\mathrm{rank}(C) \leq \mathrm{rank}(A)$ since $\mathrm{range}(C)$ is the span of the columns of C, which are a subset of the columns of A.

On the other hand, we can also express $\mathrm{rank}(C)$ in terms of its rows

$c_1, c_2, \ldots, c_m$:

Recall from Theorem 2.4.8 that $\text{rank}(C) = \text{rank}(C^T)$.

$$\text{rank}(C) = \dim(\text{range}(C^T)) = \dim(\text{span}(c_1, c_2, \ldots, c_m)).$$

If a submatrix $B \in \mathcal{M}_{k,\ell}$ is created by erasing some rows of C, then $\text{rank}(B) \leq \text{rank}(C)$ since $\text{range}(B^T)$ is the span of the rows of B, which are a subset of the rows of C. Putting these two inequalities together shows that $\text{rank}(B) \leq \text{rank}(C) \leq \text{rank}(A)$, as desired. ∎

The above theorem can help us quickly and easily come up with crude lower bounds on the rank of a large matrix without having to do much work. We now illustrate how to make use of it with some examples.

Example 2.C.3

Computing Lower Bounds for the Rank of a Matrix

Use Theorem 2.C.3 to find lower bounds on the ranks of the following matrices:

a) $A = \begin{bmatrix} 1 & 1 & 0 & 0 & 1 \\ 1 & 0 & 1 & 0 & 1 \\ 1 & 0 & 0 & 1 & 1 \end{bmatrix}$ b) $B = \begin{bmatrix} 1 & 1 & 1 & 1 & 1 \\ 1 & 2 & 4 & 8 & 16 \\ 1 & 2 & 5 & 14 & 41 \\ 1 & 3 & 9 & 27 & 81 \end{bmatrix}$

Solutions:

a) Notice that if we ignore the first and fifth columns of A, we get the 3×3 identity matrix as a submatrix of A:

$$\begin{bmatrix} \begin{matrix} 1 & 0 & 0 \\ 0 & 1 & 0 \\ 0 & 0 & 1 \end{matrix} \end{bmatrix} \longrightarrow \begin{bmatrix} 1 & 0 & 0 \\ 0 & 1 & 0 \\ 0 & 0 & 1 \end{bmatrix}.$$

Since the identity matrix is invertible (and thus has rank 3), we conclude that $\text{rank}(A) \geq 3$ (in fact, since A is 3×5, its rank cannot exceed 3, so it must in fact be the case that $\text{rank}(A) = 3$).

b) If we ignore the last two columns of B as well as its third row, we get the following 3×3 submatrix:

Lots of other 3×3 submatrices work just as well here.

$$\begin{bmatrix} 1 & 1 & 1 & & \\ 1 & 2 & 4 & 8 & 16 \\ 1 & 2 & 5 & 14 & 41 \\ 1 & 3 & 9 & 27 & 81 \end{bmatrix} \longrightarrow \begin{bmatrix} 1 & 1 & 1 \\ 1 & 2 & 4 \\ 1 & 3 & 9 \end{bmatrix}.$$

In fact, $\text{rank}(B) = 3$ since its third row is the average of its first and fourth rows.

Since this matrix is invertible (see Example 2.2.8(d) for a direct verification of this fact, or see note that it is a Vandermonde matrix and recall from Example 2.3.8 that all such matrices are invertible), it has rank 3. It follows that $\text{rank}(B) \geq 3$.

In fact, we can rephrase the above theorem slightly in order to turn it into a new characterization of the rank of a matrix:

Corollary 2.C.4

Rank in Terms of Invertible Submatrices

Suppose $A \in \mathcal{M}_{m,n}$. Then $\text{rank}(A)$ equals the largest integer r for which there exists an invertible $r \times r$ submatrix of A.

Proof. Theorem 2.C.3 tells us that $r \leq \text{rank}(A)$, so we just need to show that we can always find an invertible submatrix of size $r \times r$ when $r = \text{rank}(A)$.

To this end, notice that if $r = \text{rank}(A)$ then we can pick a set of r columns of A that span its range. Let $C \in \mathcal{M}_{m,r}$ be the submatrix consisting of these columns (in the same order that they appear in A). Then $\text{rank}(C) = r$ as well, so $\text{rank}(C^T) = r$, so there is some set of r rows of C that span range(C^T). Let $B \in \mathcal{M}_r$ be the submatrix consisting of these rows. Then $\text{rank}(B) = r$ as well, so it is invertible, which completes the proof. ∎

Exercises

solutions to starred exercises on page 458

2.C.1 Compute a rank decomposition of each of the matrices from Exercise 2.4.6.

⌨ 2.C.2 For each of the following matrices $A \in \mathcal{M}_{m,n}$, use computer software to compute their rank r and find matrices $C \in \mathcal{M}_{m,r}$ and $R \in \mathcal{M}_{r,n}$ such that $A = CR$.

*(a)
$$\begin{bmatrix} 4 & 5 & 0 & 4 & 1 & 3 \\ 2 & 2 & 1 & 2 & 1 & 1 \\ -4 & 0 & 3 & -2 & -3 & 1 \\ 3 & 5 & 4 & 4 & 1 & 3 \end{bmatrix}$$

(b)
$$\begin{bmatrix} -1 & 0 & 3 & -1 & 2 & -1 & -2 \\ 3 & 1 & -3 & 2 & -2 & 1 & 4 \\ 3 & 2 & 3 & 1 & 2 & -1 & 2 \\ 6 & 6 & 2 & 0 & 2 & 0 & 6 \\ 6 & 4 & 6 & 2 & 4 & -2 & 4 \end{bmatrix}$$

2.C.3 For each of the following matrices, compute their rank r and then find an invertible $r \times r$ submatrix of them.

*(a) $\begin{bmatrix} 1 & 2 \\ 2 & 4 \end{bmatrix}$ (b) $\begin{bmatrix} 1 & 2 & 1 \\ 3 & 2 & 1 \end{bmatrix}$

*(c) $\begin{bmatrix} 1 & 2 \\ 2 & 4 \\ 1 & 3 \end{bmatrix}$ (d) $\begin{bmatrix} 1 & 1 & 2 \\ 1 & 1 & 2 \\ 2 & 2 & 3 \end{bmatrix}$

*(e) $\begin{bmatrix} 1 & 2 & 3 \\ 4 & 5 & 6 \\ 7 & 8 & 9 \\ 10 & 11 & 12 \end{bmatrix}$ (f) $\begin{bmatrix} 1 & 1 & 0 & 0 \\ 1 & 0 & 1 & 0 \\ 0 & 0 & 0 & 1 \\ 1 & 0 & 1 & 1 \end{bmatrix}$

⌨ 2.C.4 For each of the matrices from Exercise 2.C.2, use computer software to compute their rank r and find an invertible $r \times r$ submatrix of them.

2.C.5 Determine which of the following statements are true and which are false.

*(a) If $A \in \mathcal{M}_n$ has rank 5 then it can be written as a sum of 5 rank-1 matrices.
(b) If $A \in \mathcal{M}_n$ has rank 10 then it can be written as a sum of 5 rank-2 matrices.
*(c) If $A \in \mathcal{M}_n$ has a non-invertible 3×3 submatrix then $\text{rank}(A) \leq 3$.

(d) If $A \in \mathcal{M}_n$ has an invertible 4×4 submatrix then $\text{rank}(A) \geq 4$.
*(e) If $A \in \mathcal{M}_n$ has an $(n-1) \times (n-1)$ submatrix equal to the zero matrix then $\text{rank}(A) \leq 1$.

**2.C.6 In this exercise, we show that the rank decomposition of Theorem 2.C.1 is not unique. Suppose $A \in \mathcal{M}_{m,n}$ has rank r and rank decomposition $A = CR$ (where $C \in \mathcal{M}_{m,r}$ and $R \in \mathcal{M}_{r,n}$). Let $P \in \mathcal{M}_r$ be any invertible matrix, and define $\tilde{C} = CP^{-1}$ and $\tilde{R} = PR$. Show that $A = \tilde{C}\tilde{R}$ is also a rank decomposition of A.

2.C.7 Suppose $A \in \mathcal{M}_{m,n}$ has z non-zero entries. Show that $\text{rank}(A) \leq z$.

*2.C.8 In this exercise, we consider the problem of how a matrix can change upon multiplying it on the left and right by invertible matrices.

(a) Show that every matrix $A \in \mathcal{M}_{m,n}$ can be written in the form $A = PDQ$, where $P \in \mathcal{M}_m$ and $Q \in \mathcal{M}_n$ are invertible and
$$D = \begin{bmatrix} I_{\text{rank}(A)} & O \\ O & O \end{bmatrix}.$$

(b) Suppose $A, B \in \mathcal{M}_{m,n}$. Show that there exist invertible matrices $P \in \mathcal{M}_m$ and $Q \in \mathcal{M}_n$ such that $A = PBQ$ if and only if $\text{rank}(A) = \text{rank}(B)$.

2.C.9 Suppose $a_0, a_1, \ldots, a_m$ are distinct real numbers and let $V \in \mathcal{M}_{m+1,n+1}$ be the matrix
$$V = \begin{bmatrix} 1 & a_0 & a_0^2 & \cdots & a_0^n \\ 1 & a_1 & a_1^2 & \cdots & a_1^n \\ \vdots & \vdots & \vdots & \ddots & \vdots \\ 1 & a_m & a_m^2 & \cdots & a_m^n \end{bmatrix}.$$

Show that $\text{rank}(V) = \min\{m+1, n+1\}$ (i.e., V has full rank).

[Side note: If $m = n$ then V is a Vandermonde matrix, which we showed are invertible in Example 2.3.8.]

2.D Extra Topic: The LU Decomposition

Recall that Gauss–Jordan elimination can be interpreted entirely in terms of matrices—every row operation that we apply to a matrix is equivalent to multiplying it on the left by an elementary matrix. We saw in Theorem 2.5.3 that if we use this idea repeatedly while row-reducing a matrix $A \in \mathcal{M}_{m,n}$ to its reduced row echelon form R, then we can write A in the form $A = PR$, where P is invertible (in particular, it is the inverse of the product of all of the elementary matrices used to row-reduce A).

It is worth investigating what happens if we instead only row-reduce to *any* row echelon form, rather than going all the way to the *reduced* row echelon form. Certainly it is still the case that $A = PR$ where P is invertible and R is a row echelon form of A, but usually much more is true. To get a feeling for what special form P can be chosen to have in this case, we now find such a decomposition of the matrix

$$A = \begin{bmatrix} 1 & -2 & -1 \\ 2 & -1 & -1 \\ 3 & 6 & 2 \end{bmatrix}$$

by using Gaussian elimination to put $[\,A \mid I\,]$ in row echelon form (rather than using Gauss–Jordan elimination to put it in reduced row echelon form). In particular, we recall from Theorem 2.2.1 that row-reducing $[\,A \mid I\,]$ to $[\,R \mid E\,]$ gives $R = EA$, so $A = PR$, where $P = E^{-1}$:

$$\begin{bmatrix} 1 & -2 & -1 & | & 1 & 0 & 0 \\ 2 & -1 & -1 & | & 0 & 1 & 0 \\ 3 & 6 & 2 & | & 0 & 0 & 1 \end{bmatrix} \xrightarrow[R_3 - 3R_1]{R_2 - 2R_1} \begin{bmatrix} 1 & -2 & -1 & | & 1 & 0 & 0 \\ 0 & 3 & 1 & | & -2 & 1 & 0 \\ 0 & 12 & 5 & | & -3 & 0 & 1 \end{bmatrix}$$

$$\xrightarrow{R_3 - 4R_2} \begin{bmatrix} 1 & -2 & -1 & | & 1 & 0 & 0 \\ 0 & 3 & 1 & | & -2 & 1 & 0 \\ 0 & 0 & 1 & | & 5 & -4 & 1 \end{bmatrix}.$$

It follows that $A = PR$, where

$$R = \begin{bmatrix} 1 & -2 & -1 \\ 0 & 3 & 1 \\ 0 & 0 & 1 \end{bmatrix} \quad \text{and} \quad P = \begin{bmatrix} 1 & 0 & 0 \\ -2 & 1 & 0 \\ 5 & -4 & 1 \end{bmatrix}^{-1} = \begin{bmatrix} 1 & 0 & 0 \\ 2 & 1 & 0 \\ 3 & 4 & 1 \end{bmatrix}. \quad (2.D.1)$$

In particular, since R is a row echelon form of A it is upper triangular, so we refer to it as U from now on. Similarly, the matrix P that we constructed happened to be lower triangular, so we refer to it as L from now on. Being able to decompose a matrix into the product of lower and upper triangular matrices like this is very useful when solving systems of linear equations (as we will demonstrate later in this section), so we give this type of decomposition a name.

Definition 2.D.1

LU Decomposition

Suppose $A \in \mathcal{M}_{m,n}$. An **LU decomposition** of A is any factorization of the form $A = LU$, where $L \in \mathcal{M}_m$ is lower triangular and $U \in \mathcal{M}_{m,n}$ is upper triangular.

Despite the fact that we were already able to compute an LU decomposition of the matrix A above, we know very little about them. In particular, we spend the rest of this section exploring the following questions:

1) The matrix $U = R$ from Equation (2.D.1) was upper triangular just as a result of being chosen to be a row echelon form of A, but why was $L = P$ lower triangular?

The term "upper triangular" means the same thing for non-square matrices as it does for square ones— all non-zero entries are on and above the main diagonal, such as in

$$\begin{bmatrix} 1 & 2 & 3 & 4 \\ 0 & 5 & 6 & 7 \\ 0 & 0 & 8 & 9 \end{bmatrix}.$$

2) In order to construct L and U, we had to perform Gaussian elimination twice—once to row-reduce $[\,A \mid I\,]$ and once more to invert a matrix. Is there a more efficient method of computing L and U?

3) Can we find an LU decomposition of any matrix, or did the matrix A above have some special structure that we exploited?

Before proceeding to answer these questions, it will be useful for us to remind ourselves of some facts about triangular matrices that we proved back in Exercise 2.2.16, which we make use of repeatedly throughout this section:

- A triangular matrix is invertible if and only if its diagonal entries are non-zero.

- If a lower (upper) triangular matrix is invertible then its inverse is also lower (upper) triangular.

- If a triangular matrix has all diagonal entries equal to 1 then so does its inverse.

2.D.1   Computing an LU Decomposition

We start by investigating question (1) above—why the matrix $L = P$ from Equation (2.D.1) was lower triangular. To this end, we start by recalling that we constructed L by row-reducing $[\,A \mid I\,]$ to $[\,R \mid E\,]$, where R is a row echelon form of A, and setting $L = E^{-1}$.

This discussion (and the upcoming theorem) rely on us specifically performing Gaussian elimination, not a variant of it like Gauss–Jordan elimination.

The key fact that is responsible for the lower triangular shape of L is that we did not have to perform any "swap" row operations when row-reducing A to a row echelon form, so the only entries that we modified in the augmented right-hand-side of $[\,A \mid I\,]$ were those below the main diagonal. Specifically, it is not difficult to convince ourselves that if we start with a lower triangular matrix (like I) then "multiplication" row operations cR_j and "addition" row operations $R_i + cR_j$ with $j < i$ preserve lower triangularity, so E must be lower triangular as well.

Since $L = E^{-1}$, we then just need to recall that the inverse of a lower triangular matrix is also lower triangular, which proves the following theorem.

Theorem 2.D.1

LU Decomposition via Gaussian Elimination

If Gaussian elimination row-reduces $[\,A \mid I\,]$ to $[\,R \mid E\,]$, where R is a row echelon form of A, without using any "swap" row operations, then setting $L = E^{-1}$ and $U = R$ gives an LU decomposition $A = LU$.

While we already used the method of this theorem to construct an LU decomposition earlier, it is perhaps worth going through the procedure again now that we understand a bit better why it works.

Example 2.D.1

Computing an LU Decomposition

Compute an LU decomposition of the matrix $A = \begin{bmatrix} 2 & 1 & -2 & 1 \\ -4 & -4 & 3 & 0 \\ 2 & -5 & -2 & 8 \end{bmatrix}.$

Solution:

We start by applying Gaussian elimination to $[\,A \mid I\,]$, as suggested by

Theorem 2.D.1:

$$\left[\begin{array}{cccc|ccc} 2 & 1 & -2 & 1 & 1 & 0 & 0 \\ -4 & -4 & 3 & 0 & 0 & 1 & 0 \\ 2 & -5 & -2 & 8 & 0 & 0 & 1 \end{array}\right]$$

$$\xrightarrow[R_3 - R_1]{R_2 + 2R_1} \left[\begin{array}{cccc|ccc} 2 & 1 & -2 & 1 & 1 & 0 & 0 \\ 0 & -2 & -1 & 2 & 2 & 1 & 0 \\ 0 & -6 & 0 & 7 & -1 & 0 & 1 \end{array}\right]$$

$$\xrightarrow{R_3 - 3R_2} \left[\begin{array}{cccc|ccc} 2 & 1 & -2 & 1 & 1 & 0 & 0 \\ 0 & -2 & -1 & 3 & 2 & 1 & 0 \\ 0 & 0 & 3 & 1 & -7 & -3 & 1 \end{array}\right].$$

One LU decomposition $A = LU$ is thus given by

$$U = \begin{bmatrix} 2 & 1 & -2 & 1 \\ 0 & -2 & -1 & 3 \\ 0 & 0 & 3 & 1 \end{bmatrix} \quad \text{and} \quad L = \begin{bmatrix} 1 & 0 & 0 \\ 2 & 1 & 0 \\ -7 & -3 & 1 \end{bmatrix}^{-1} = \begin{bmatrix} 1 & 0 & 0 \\ -2 & 1 & 0 \\ 1 & 3 & 1 \end{bmatrix}.$$

In the above example, the lower triangular matrix L that we constructed was a **unit triangular matrix**—it had all of its diagonal entries equal to 1. We call an LU decomposition associated with such an L a **unit LU decomposition**, and most matrices that have an LU decomposition also have a unit LU decomposition. To see why this is the case, consider either of the following arguments:

Both of these arguments rely on L being invertible. However, there are some matrices with an LU decomposition, but no LU decomposition with an invertible L (see Exercise 2.D.13) and thus no unit LU decomposition.

- When performing Gaussian elimination, it is always possible to avoid using "multiplication" row operations—they only serve to simplify algebra in some cases. Avoiding these row operations when constructing the matrix E in Theorem 2.D.1 results in it having ones on its diagonal (since "addition" row operations of the form $R_i + cR_j$ with $j < i$ only affect the *strictly* lower triangular portion of E), so $L = E^{-1}$ has ones on its diagonal as well.

- Alternatively, suppose that $A = LU$ is any LU decomposition of A with L invertible. If we let D be the diagonal matrix with the same diagonal entries as L then LD^{-1} is a lower triangular matrix with ones on its diagonal and DU is an upper triangular matrix, so $A = (LD^{-1})(DU)$ is a unit LU decomposition of A.

Since unit LU decompositions are somewhat more elegant than general LU decompositions (for example, we will see in Exercise 2.D.6 that the unit LU decomposition of an invertible matrix is unique when it exists, whereas LU decompositions themselves are *never* unique), we restrict our attention to them from now on.

Remark 2.D.1

Redundancies in the LU Decomposition

LU decompositions themselves are quite non-unique, since if D is any invertible diagonal matrix and $A = LU$ is an LU decomposition of A then $A = (LD^{-1})(DU)$ is another one. For example, if

$$A = \begin{bmatrix} 2 & 4 \\ 2 & 7 \end{bmatrix} \quad \text{then} \quad A = \begin{bmatrix} 1 & 0 \\ 1 & 1 \end{bmatrix}\begin{bmatrix} 2 & 4 \\ 0 & 3 \end{bmatrix} = \begin{bmatrix} 2 & 0 \\ 2 & 3 \end{bmatrix}\begin{bmatrix} 1 & 2 \\ 0 & 1 \end{bmatrix}$$

are two different LU decompositions of A.

You can think of unit LU decompositions as removing redundancies in LU decompositions. An LU decomposition of a matrix $A \in \mathcal{M}_n$ is specified by the $n(n+1)/2$ potentially non-zero entries in each of L and U, for a total of $n(n+1) = n^2 + n$ numbers, so it seems reasonable that we can freely choose n of those numbers while still recovering the n^2 entries of A.

There are many other possible choices that we could make to remove this redundancy as well (for example, we could choose the diagonal entries of U to all equal 1 instead), but choosing the diagonal entries of L to equal 1 is the most convenient choice for most purposes.

Again, but Faster Now

To address our question (2) from earlier—how to construct a (unit) LU decomposition in a way that does not require us to use Gaussian elimination twice—we notice that the lower triangular entries in the matrix L in Example 2.D.1 (-2, 1, and 3) were exactly the negatives of the scalars in the "addition" row operations that we used to put $[\,A \mid I\,]$ into row echelon form $[\,U \mid E\,]$ ($R_2 + 2R_1, R_3 - R_1$, and $R_3 - 3R_2$). This is not a coincidence—roughly speaking, it is a result of the fact that if $A = LU$ then the rows of L contain the coefficients of the row operations needed to transform U into A. However, the row operations used to transform U into A are exactly the inverses of those used to convert A into U, and the row operation that undoes $R_i + cR_j$ is simply $R_i - cR_j$.

More precisely, we have the following refinement of Theorem 2.D.1 that has the advantage of only requiring us to perform Gaussian elimination once to find an LU decomposition, not twice (since we no longer need to invert a matrix to find L):

> *Similarly, the entries in the lower triangular portion of $L = P$ in Equation (2.D.1) were **2**, **3**, and **4**, and the row operations we used to find it were $R_2 - 2R_1$, $R_3 - 3R_1$, and $R_3 - 4R_2$.*

Theorem 2.D.2

Computing an LU Decomposition Quickly

Here, $\ell_{i,j}$ refers to the (i,j)-entry of L.

Suppose $A, U \in \mathcal{M}_{m,n}$. A unit lower triangular matrix $L \in \mathcal{M}_m$ is such that $A = LU$ if and only if the following sequence of row operations transforms A into U:

$$
\begin{aligned}
&R_2 - \ell_{2,1}R_1 \\
&R_3 - \ell_{3,1}R_1 \\
&\qquad\vdots \\
&R_m - \ell_{m,1}R_1
\end{aligned}
\longrightarrow
\begin{aligned}
&R_3 - \ell_{3,2}R_2 \\
&\qquad\vdots \\
&R_m - \ell_{m,2}R_2
\end{aligned}
\longrightarrow \cdots \quad R_m - \ell_{m,m-1}R_{m-1}.
$$

Proof. Recall from our block matrix multiplication techniques (and in particular, Exercise 1.3.27(b)) that the rows of $A = LU$ are linear combinations of the rows of U, and the coefficients of those linear combinations are the entries of L. In particular, if we let $\mathbf{a}_i^T$ and $\mathbf{u}_i^T$ denote the i-th rows of A and U, respectively, then L is lower triangular with ones on its diagonal if and only if

> *If $j = 1$ then the sum on the right has no terms in it. This is called the empty sum, and it equals $\mathbf{0}$ (so $\mathbf{a}_1^T = \mathbf{u}_1^T$).*

$$
\mathbf{a}_i^T = \mathbf{u}_i^T + \sum_{j=1}^{i-1} \ell_{i,j}\mathbf{u}_j^T \quad \text{for all} \quad 1 \le i \le m. \tag{2.D.2}
$$

This equation says exactly that there is a particular sequence of row operations of the form $R_i + \ell_{i,j}R_j$ (with $j < i$) that converts U into A. For example, applying the sequence of row operations

$$
R_m + \ell_{m,1}R_1, \quad R_m + \ell_{m,2}R_2, \quad \ldots, \quad R_m + \ell_{m,m-1}R_{m-1}
$$

to U turns its m-th row into

$$\mathbf{u}_m^T + \ell_{m,1}\mathbf{u}_1^T + \ell_{m,2}\mathbf{u}_2^T + \cdots + \ell_{m,m-1}\mathbf{u}_{m-1}^T,$$

which is exactly $\mathbf{a}_m^T$, via Equation (2.D.2). Similar sequences of row operations turn each of the rows of U into the rows of A, but we have to be careful to apply them in an order that does not modify a row before adding it to another row. To this end, we work from the bottom to the top, first applying any row operations that involves adding a multiple of R_{m-1}, then all row operations that involve adding a multiple of R_{m-2}, and so on. In particular, U is transformed into A via the following sequence of row operations:

Here the row operations are just read left-to-right and then top-to-bottom, in usual reading order.

$$R_m + \ell_{m,m-1}R_{m-1},$$
$$R_m + \ell_{m,m-2}R_{m-2}, \qquad R_{m-1} + \ell_{m-1,m-2}R_{m-2}$$
$$\vdots \qquad\qquad \vdots \qquad\qquad \ddots$$
$$R_m + \ell_{m,1}R_1, \qquad R_{m-1} + \ell_{m-1,1}R_1, \qquad \cdots \qquad R_2 + \ell_{2,1}R_1.$$

Recall that the inverse of the row operation $R_i + \ell_{i,j}R_j$ is $R_i - \ell_{i,j}R_j$.

To transform A into U, we thus apply the inverses of these row operations in the opposite order, which is exactly the sequence of row operations described by the theorem. ∎

If we apply the above theorem in the case when U is a row echelon form of A, then we get exactly a unit LU decomposition of A. We illustrate this quicker method of constructing an LU decomposition with an example.

Example 2.D.2

Computing an LU Decomposition Quickly

Construct a unit LU decomposition of the matrix

$$A = \begin{bmatrix} 2 & 4 & -1 & -1 \\ 4 & 9 & 0 & -1 \\ -6 & -9 & 7 & 6 \\ -2 & -2 & 9 & 0 \end{bmatrix}.$$

Solution:

As always, our goal is to row-reduce A into an upper triangular (row echelon) form without using any "swap" row operations (though this time we want to avoid "multiplication" row operations as well). Thanks to Theorem 2.D.2, we no longer need to keep track of the augmented right-hand-side (i.e., we just row-reduce A, not $[A \mid I]$):

$$\begin{bmatrix} 2 & 4 & -1 & -1 \\ 4 & 9 & 0 & -1 \\ -6 & -9 & 7 & 6 \\ -2 & -2 & 9 & 0 \end{bmatrix} \xrightarrow[\substack{R_3+3R_1 \\ R_4+R_1}]{R_2-2R_1} \begin{bmatrix} 2 & 4 & -1 & -1 \\ 0 & 1 & 2 & 1 \\ 0 & 3 & 4 & 3 \\ 0 & 2 & 8 & -1 \end{bmatrix}$$

For example, the first set of row operations that we performed were $R_2 - 2R_1$, $R_3 + 3R_1$, and $R_4 + R_1$, so the entries in the first column of L are 2, −3, and −1.

$$\xrightarrow[R_4-2R_2]{R_3-3R_2} \begin{bmatrix} 2 & 4 & -1 & -1 \\ 0 & 1 & 2 & 1 \\ 0 & 0 & -2 & 0 \\ 0 & 0 & 4 & -3 \end{bmatrix} \xrightarrow{R_4+2R_3} \begin{bmatrix} 2 & 4 & -1 & -1 \\ 0 & 1 & 2 & 1 \\ 0 & 0 & -2 & 0 \\ 0 & 0 & 0 & -3 \end{bmatrix}.$$

A unit LU decomposition of A is thus obtained by choosing U to be this row echelon form and L to be the unit lower triangular matrix whose entries are the negatives of the coefficients that we used in the "addition" row operations above:

$$L = \begin{bmatrix} 1 & 0 & 0 & 0 \\ 2 & 1 & 0 & 0 \\ -3 & 3 & 1 & 0 \\ -1 & 2 & -2 & 1 \end{bmatrix} \quad \text{and} \quad U = \begin{bmatrix} 2 & 4 & -1 & -1 \\ 0 & 1 & 2 & 1 \\ 0 & 0 & -2 & 0 \\ 0 & 0 & 0 & -3 \end{bmatrix}.$$

One of the other nice features of this faster method of computing a unit LU decomposition is that it provides us with a converse of Theorem 2.D.1. That is, it shows that not only does a matrix have a unit LU decomposition if it can be made upper triangular via Gaussian elimination without any swap row operations, but in fact these are the *only* matrices with a unit LU decomposition.

Corollary 2.D.3

Characterization of Unit LU Decompositions

A matrix $A \in \mathcal{M}_{m,n}$ has a unit LU decomposition if and only if it can be row-reduced to an upper triangular matrix entirely via "addition" row operations of the form $R_i + cR_j$ with $i > j$.

Proof. Theorem 2.D.1 already established the "if" direction of this theorem, so we just need to prove the "only if" implication. To this end, suppose that $A = LU$ is a unit LU decomposition of A. Theorem 2.D.2 then provides a list of elementary row operations (which are all of the type described by the statement of this theorem) that row-reduces A to U. Since U is upper triangular, we are done. ∎

Notice that Corollary 2.D.3 specifies that A can be row-reduced to an upper triangular matrix, but not necessarily a row echelon form. This is because of matrices like

$$A = \begin{bmatrix} 0 & 0 \\ 0 & 1 \end{bmatrix},$$

Every matrix in row echelon form is upper triangular, but not every upper triangular matrix is in row echelon form.

which has a unit LU decomposition (since A itself is upper triangular, we can just choose $L = I$ and $U = A$) but requires a swap row operation (or an operation of the form $R_i + cR_j$ for some $i < j$ instead of $i > j$) to put it in row echelon form.

2.D.2 Solving Linear Systems

One of the primary uses of the LU decomposition is as a method for solving multiple systems of linear equations more quickly than we could by repeatedly applying Gaussian elimination or Gauss–Jordan elimination. To see how this works, suppose we have already computed an LU decomposition $A = LU$ of the coefficient matrix of the linear system $A\mathbf{x} = \mathbf{b}$. Then $LU\mathbf{x} = \mathbf{b}$, which is a linear system that we can solve via the following two-step procedure:

- First, solve the linear system $L\mathbf{y} = \mathbf{b}$ for the vector $\mathbf{y}$. This linear system is straightforward to solve via "forward" elimination (i.e., first solving for y_1 then substituting that so as to solve for y_2, and so on) due to the lower triangular shape of L.
- Next, solve the linear system $U\mathbf{x} = \mathbf{y}$ for the vector $\mathbf{x}$. This linear system is similarly straightforward to solve via backward elimination due to the upper triangular shape of U.

Once we have obtained the vector $\mathbf{x}$ via this procedure, it is the case that

$$A\mathbf{x} = LU\mathbf{x} = L(U\mathbf{x}) = L\mathbf{y} = \mathbf{b},$$

so $\mathbf{x}$ is indeed a solution of the original linear system, as desired.

Although solving a linear system in this way might seem like more work than just using Gaussian elimination to solve it directly (after all, we have to solve *two* linear systems via this method rather than just one), the triangular shape of the linear systems saves us a lot of work in the end. In particular, if $A \in \mathcal{M}_n$ and n is large then solving the linear system $A\mathbf{x} = \mathbf{b}$ directly via Gaussian elimination requires roughly $2n^3/3$ operations (i.e., additions, subtractions, multiplications, and/or divisions), whereas each of the triangular systems $L\mathbf{y} = \mathbf{b}$ and $U\mathbf{x} = \mathbf{y}$ can be solved in roughly n^2 operations. When n is large, solving these two triangular systems via roughly $2n^2$ operations is considerably quicker than solving the system directly via roughly $2n^3/3$ operations.

Example 2.D.3

Solving Linear Systems via an LU Decomposition

Use the LU decomposition to solve the linear system

$$\begin{bmatrix} 2 & 4 & -1 & -1 \\ 4 & 9 & 0 & -1 \\ -6 & -9 & 7 & 6 \\ -2 & -2 & 9 & 0 \end{bmatrix} \begin{bmatrix} w \\ x \\ y \\ z \end{bmatrix} = \begin{bmatrix} 0 \\ 2 \\ 0 \\ 1 \end{bmatrix}.$$

Solution:

We constructed an LU decomposition $A = LU$ of this coefficient matrix in Example 2.D.2, so we start by solving the lower triangular system $L\mathbf{y} = \mathbf{b}$:

Even though we still have to perform a lot of row operations to find the solution, they are very simple due to the triangular shapes of L and U. Each "addition" row operation only affects one entry of the coefficient matrix (plus the augmented right-hand side) rather than the entire row.

$$\left[\begin{array}{cccc|c} 1 & 0 & 0 & 0 & 0 \\ 2 & 1 & 0 & 0 & 2 \\ -3 & 3 & 1 & 0 & 0 \\ -1 & 2 & -2 & 1 & 1 \end{array}\right] \begin{array}{c} R_2-2R_1 \\ R_3+3R_1 \\ R_4+R_1 \\ \xrightarrow{} \end{array} \left[\begin{array}{cccc|c} 1 & 0 & 0 & 0 & 0 \\ 0 & 1 & 0 & 0 & 2 \\ 0 & 3 & 1 & 0 & 0 \\ 0 & 2 & -2 & 1 & 1 \end{array}\right]$$

$$\begin{array}{c} R_3-3R_2 \\ R_4-2R_2 \\ \xrightarrow{} \end{array} \left[\begin{array}{cccc|c} 1 & 0 & 0 & 0 & 0 \\ 0 & 1 & 0 & 0 & 2 \\ 0 & 0 & 1 & 0 & -6 \\ 0 & 0 & -2 & 1 & -3 \end{array}\right] \begin{array}{c} R_4+2R_3 \\ \xrightarrow{} \end{array} \left[\begin{array}{cccc|c} 1 & 0 & 0 & 0 & 0 \\ 0 & 1 & 0 & 0 & 2 \\ 0 & 0 & 1 & 0 & -6 \\ 0 & 0 & 0 & 1 & -15 \end{array}\right],$$

so $\mathbf{y} = (0, 2, -6, -15)$. Next, we solve the upper triangular system $U\mathbf{x} = \mathbf{y}$:

$$\left[\begin{array}{cccc|c} 2 & 4 & -1 & -1 & 0 \\ 0 & 1 & 2 & 1 & 2 \\ 0 & 0 & -2 & 0 & -6 \\ 0 & 0 & 0 & -3 & -15 \end{array}\right] \begin{array}{c} R_1/2 \\ -R_3/2 \\ -R_4/3 \\ \xrightarrow{} \end{array} \left[\begin{array}{cccc|c} 1 & 2 & -1/2 & -1/2 & 0 \\ 0 & 1 & 2 & 1 & 2 \\ 0 & 0 & 1 & 0 & 3 \\ 0 & 0 & 0 & 1 & 5 \end{array}\right]$$

$$\begin{array}{c} R_1+\frac{1}{2}R_4 \\ R_2-R_4 \\ \xrightarrow{} \end{array} \left[\begin{array}{cccc|c} 1 & 2 & -1/2 & 0 & 5/2 \\ 0 & 1 & 2 & 0 & -3 \\ 0 & 0 & 1 & 0 & 3 \\ 0 & 0 & 0 & 1 & 5 \end{array}\right] \begin{array}{c} R_1+\frac{1}{2}R_3 \\ R_2-2R_3 \\ \xrightarrow{} \end{array} \left[\begin{array}{cccc|c} 1 & 2 & 0 & 0 & 4 \\ 0 & 1 & 0 & 0 & -9 \\ 0 & 0 & 1 & 0 & 3 \\ 0 & 0 & 0 & 1 & 5 \end{array}\right]$$

$$\begin{array}{c} R_1-2R_2 \\ \xrightarrow{} \end{array} \left[\begin{array}{cccc|c} 1 & 0 & 0 & 0 & 22 \\ 0 & 1 & 0 & 0 & -9 \\ 0 & 0 & 1 & 0 & 3 \\ 0 & 0 & 0 & 1 & 5 \end{array}\right].$$

> It follows that the (unique) solution to the original linear system $A\mathbf{x} = \mathbf{b}$ is $\mathbf{x} = (w,x,y,z) = (22,-9,3,5)$.

Even though we can solve these triangular linear systems more quickly than other linear systems, using the LU decomposition like we did in the previous example is still somewhat silly, since constructing the LU decomposition in the first place takes just as long as solving the linear system directly. After all, both of these tasks are carried out via Gaussian elimination.

For this reason, the LU decomposition is typically only used in the case when we wish to solve *multiple* linear systems, each of which have the same coefficient matrix but different right-hand-side vectors. In this case, we can pre-compute the LU decomposition of the coefficient matrix (which is time-consuming and requires a full application of Gaussian elimination) and then use that LU decomposition to (much more quickly) solve each of the linear systems. As an example to illustrate this method, we show how it can be used to fit polynomials to data sets.

Example 2.D.4

Repeated Polynomial Interpolation

Find degree-3 polynomials whose graphs go through the following sets of points:

a) $\{(0,3),(1,-1),(2,-3),(3,9)\}$
b) $\{(0,-1),(1,-1),(2,-1),(3,5)\}$

Solutions:

a) If the polynomial that we seek is $p(x) = c_3 x^3 + c_2 x^2 + c_1 x + c_0$, then the equations $p(0) = 3$, $p(1) = 0$, $p(2) = 1$, and $p(3) = 18$ can be written explicitly as the following linear system:

We originally explored how to use linear systems to fit polynomials to sets of points in Remark 2.3.1.

$$\begin{bmatrix} 1 & 0 & 0 & 0 \\ 1 & 1 & 1 & 1 \\ 1 & 2 & 4 & 8 \\ 1 & 3 & 9 & 27 \end{bmatrix} \begin{bmatrix} c_0 \\ c_1 \\ c_2 \\ c_3 \end{bmatrix} = \begin{bmatrix} 3 \\ -1 \\ -3 \\ 9 \end{bmatrix}.$$

In particular, notice that the y-values in the given data set only appear on the right-hand-side of the linear system—the coefficient matrix is determined only by the x-values and thus will be the same in part (b). For this reason, we start by constructing the LU decomposition of this coefficient matrix, rather than solving the linear system directly:

$$\begin{bmatrix} 1 & 0 & 0 & 0 \\ 1 & 1 & 1 & 1 \\ 1 & 2 & 4 & 8 \\ 1 & 3 & 9 & 27 \end{bmatrix} \xrightarrow[\substack{R_2-R_1 \\ R_3-R_1 \\ R_4-R_1}]{} \begin{bmatrix} 1 & 0 & 0 & 0 \\ 0 & 1 & 1 & 1 \\ 0 & 2 & 4 & 8 \\ 0 & 3 & 9 & 27 \end{bmatrix}$$

$$\xrightarrow[\substack{R_3-2R_2 \\ R_4-3R_2}]{} \begin{bmatrix} 1 & 0 & 0 & 0 \\ 0 & 1 & 1 & 1 \\ 0 & 0 & 2 & 6 \\ 0 & 0 & 6 & 24 \end{bmatrix} \xrightarrow[R_4-3R_3]{} \begin{bmatrix} 1 & 0 & 0 & 0 \\ 0 & 1 & 1 & 1 \\ 0 & 0 & 2 & 6 \\ 0 & 0 & 0 & 6 \end{bmatrix}.$$

It thus follows from Theorem 2.D.2 that one unit LU decomposition is given by choosing

$$L = \begin{bmatrix} 1 & 0 & 0 & 0 \\ 1 & 1 & 0 & 0 \\ 1 & 2 & 1 & 0 \\ 1 & 3 & 3 & 1 \end{bmatrix} \quad \text{and} \quad U = \begin{bmatrix} 1 & 0 & 0 & 0 \\ 0 & 1 & 1 & 1 \\ 0 & 0 & 2 & 6 \\ 0 & 0 & 0 & 6 \end{bmatrix}.$$

Then solving the lower triangular linear system $L\mathbf{y} = \mathbf{b} = (3, -1, 3, 9)$ gives $\mathbf{y} = (3, -4, 2, 12)$, and solving the upper triangular linear system $U\mathbf{x} = \mathbf{y} = (3, -4, 2, 12)$ gives $\mathbf{x} = (c_0, c_1, c_2, c_3) = (3, -1, -5, 2)$. It follows that the (unique) degree-3 polynomial going through the 4 given points is

$$p(x) = 2x^3 - 5x^2 - x + 3.$$

The fact that this polynomial goes through the four given points is straightforward to verify by plugging $x = 0, 1, 2$ and 3 into it, and it is displayed below:

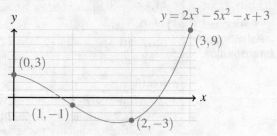

b) In this case, the linear system that we wish to solve is

$$\begin{bmatrix} 1 & 0 & 0 & 0 \\ 1 & 1 & 1 & 1 \\ 1 & 2 & 4 & 8 \\ 1 & 3 & 9 & 27 \end{bmatrix} \begin{bmatrix} c_0 \\ c_1 \\ c_2 \\ c_3 \end{bmatrix} = \begin{bmatrix} -1 \\ -1 \\ -1 \\ 5 \end{bmatrix}.$$

Since the coefficient matrix in this linear system is that same as the one from part (a), we can re-use the LU decomposition that we already computed there.

In particular, we jump straight to solving the lower triangular linear system $L\mathbf{y} = \mathbf{b} = (-1, -1, -1, 5)$, which gives $\mathbf{y} = (-1, 0, 0, 6)$. Solving the upper triangular linear system $U\mathbf{x} = \mathbf{y} = (-1, 0, 0, 6)$ then gives $\mathbf{x} = (c_0, c_1, c_2, c_3) = (-1, 2, -3, 1)$, so the (unique) degree-3 polynomial going through the 4 given points is

$$p(x) = x^3 - 3x^2 + 2x - 1.$$

Again, it is straightforward to verify that the graph of this polynomial goes through the four points $(0, -1)$, $(1, -1)$, $(2, -1)$, and $(3, 5)$:

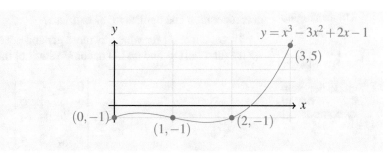

The idea of using a matrix decomposition to offload the difficult part of a calculation (like we did with the LU decomposition in the previous example) is a very widely applicable one, and there are many different matrix decompositions that are each suited to different tasks. The LU decomposition is well-suited to solving linear systems, and we will investigate another decomposition in Section 3.4 that lets us quickly compute matrix powers.

Remark 2.D.2

The Inverse Works, But LU is Better

There is a seemingly simpler way to solve multiple linear systems that have the same coefficient matrix but different right-hand sides—compute the inverse of the coefficient matrix and multiply it by the right-hand-side vectors. For example, if we want to solve the linear systems

$$A\mathbf{x} = \mathbf{b}, \quad A\mathbf{y} = \mathbf{c}, \quad \text{and} \quad A\mathbf{z} = \mathbf{d},$$

we could pre-compute A^{-1} and then multiply by it three times to get $\mathbf{x} = A^{-1}\mathbf{b}$, $\mathbf{y} = A^{-1}\mathbf{c}$, and $\mathbf{z} = A^{-1}\mathbf{d}$. Both this approach and the one based on the LU decomposition work, but we favor the LU decomposition for a few reasons:

We explore another reason for preferring the LU decomposition in Exercise 2.D.9.

- The LU decomposition works for non-square linear systems, whereas inverses do not exist in this case.

- Even some square matrices are not invertible—the LU decomposition can be used in these cases to determine whether or not the associated linear system has solutions. While some matrices also do not have an LU decomposition, we will see in the next subsection that *every* matrix has a slightly more general "PLU decomposition" that works just as well.

- If $A \in \mathcal{M}_n$ then computing A^{-1} via Gauss–Jordan elimination requires roughly $2n^3$ operations (additions, subtractions, multiplications, and/or divisions), making it about three times as computationally expensive as the LU decomposition, which can be computed in roughly $2n^3/3$ operations via Gaussian elimination.

- The LU decomposition is more numerically stable than the inverse—problems can arise when computing the inverse of a matrix that is close to a non-invertible one (though the details of this claim are outside of the scope of this book).

2.D.3 The PLU Decomposition

Since there are matrices that do not have a unit LU decomposition (in particular, these are the matrices that requires a swap row operation to be put in upper triangular form), it seems natural to ask what the next best thing is. That is, can we find a matrix decomposition that is slightly more general than the unit LU

There are also
matrices like

$$\begin{bmatrix} 0 & 1 \\ 1 & 0 \end{bmatrix}$$

that do not have
any LU
decomposition.

decomposition and applies to *all* matrices?

To get a feeling for what this more general decomposition might look like, we try (and fail) to find an LU decomposition of the matrix

$$A = \begin{bmatrix} 0 & 2 & 4 & 2 \\ 2 & 1 & 1 & 2 \\ 1 & 1 & 2 & 3 \\ 1 & 2 & 3 & 4 \end{bmatrix}$$

by simply doing what we did at the start of this section—row-reducing $[\,A \mid I\,]$ until it is in row echelon form $[\,R \mid E\,]$ and noting that $A = E^{-1}R$:

We swap with row 3
instead of row 2 just
to make the
arithmetic work out
more cleanly—we
like having a 1
instead of a 2 in the
leading entry (i.e.,
pivot position).

$$\left[\begin{array}{cccc|cccc} 0 & 2 & 4 & 2 & 1 & 0 & 0 & 0 \\ 2 & 1 & 1 & 2 & 0 & 1 & 0 & 0 \\ 1 & 1 & 2 & 3 & 0 & 0 & 1 & 0 \\ 1 & 2 & 3 & 4 & 0 & 0 & 0 & 1 \end{array}\right] \xrightarrow{R_1 \leftrightarrow R_3} \left[\begin{array}{cccc|cccc} 1 & 1 & 2 & 3 & 0 & 0 & 1 & 0 \\ 2 & 1 & 1 & 2 & 0 & 1 & 0 & 0 \\ 0 & 2 & 4 & 2 & 1 & 0 & 0 & 0 \\ 1 & 2 & 3 & 4 & 0 & 0 & 0 & 1 \end{array}\right]$$

$$\xrightarrow[R_4 - R_1]{R_2 - 2R_1} \left[\begin{array}{cccc|cccc} 1 & 1 & 4 & 3 & 0 & 0 & 1 & 0 \\ 0 & -1 & -3 & -4 & 0 & 1 & -2 & 0 \\ 0 & 2 & 4 & 2 & 1 & 0 & 0 & 0 \\ 0 & 1 & 1 & 1 & 0 & 0 & -1 & 1 \end{array}\right]$$

$$\xrightarrow[R_4 + R_2]{R_3 + 2R_2} \left[\begin{array}{cccc|cccc} 1 & 1 & 4 & 3 & 0 & 0 & 1 & 0 \\ 0 & -1 & -3 & -4 & 0 & 1 & -2 & 0 \\ 0 & 0 & -2 & -6 & 1 & 2 & -4 & 0 \\ 0 & 0 & -2 & -3 & 0 & 1 & -3 & 1 \end{array}\right]$$

$$\xrightarrow{R_4 - R_3} \left[\begin{array}{cccc|cccc} 1 & 1 & 4 & 3 & 0 & 0 & 1 & 0 \\ 0 & -1 & -3 & -4 & 0 & 1 & -2 & 0 \\ 0 & 0 & -2 & -6 & 1 & 2 & -4 & 0 \\ 0 & 0 & 0 & 3 & -1 & -1 & 1 & 1 \end{array}\right].$$

It follows that $A = E^{-1}R$, where

$$R = \begin{bmatrix} 1 & 1 & 4 & 3 \\ 0 & -1 & -3 & -4 \\ 0 & 0 & -2 & -6 \\ 0 & 0 & 0 & 3 \end{bmatrix} \quad \text{and}$$

$$E^{-1} = \begin{bmatrix} 0 & 0 & 1 & 0 \\ 0 & 1 & -2 & 0 \\ 1 & 2 & -4 & 0 \\ -1 & -1 & 1 & 1 \end{bmatrix}^{-1} = \begin{bmatrix} 0 & -2 & 1 & 0 \\ 2 & 1 & 0 & 0 \\ 1 & 0 & 0 & 0 \\ 1 & -1 & 1 & 1 \end{bmatrix}.$$

Notice in particular that R is upper triangular (after all, it is a row echelon form), so we refer to it as U from now on (just like we did with the LU decomposition). On the other hand, E^{-1} is not lower triangular, but its rows can be permuted to make it lower triangular. In particular, if we swap its first and third rows, then we get

$$\begin{bmatrix} 0 & -2 & 1 & 0 \\ 2 & 1 & 0 & 0 \\ 1 & 0 & 0 & 0 \\ 1 & -1 & 1 & 1 \end{bmatrix} \xrightarrow{R_1 \leftrightarrow R_3} \begin{bmatrix} 1 & 0 & 0 & 0 \\ 2 & 1 & 0 & 0 \\ 0 & -2 & 1 & 0 \\ 1 & -1 & 1 & 1 \end{bmatrix},$$

We investigate permutation matrices in much more depth in Section 3.A.3.

which is lower triangular. We can encode this row swapping operation as multiplication on the left by a **permutation matrix**, which is a matrix that contains exactly one 1 in each row and column (and all of its other entries are 0). In this case, we can write

$$E^{-1} = PL, \quad \text{where} \quad P = \begin{bmatrix} 0 & 0 & 1 & 0 \\ 0 & 1 & 0 & 0 \\ 1 & 0 & 0 & 0 \\ 0 & 0 & 0 & 1 \end{bmatrix} \quad \text{and} \quad L = \begin{bmatrix} 1 & 0 & 0 & 0 \\ 2 & 1 & 0 & 0 \\ 0 & -2 & 1 & 0 \\ 1 & -1 & 1 & 1 \end{bmatrix}.$$

It follows that $A = E^{-1}U = PLU$, where P is a permutation matrix, L is lower triangular, and U is upper triangular. We now give this decomposition a name:

Definition 2.D.2

PLU Decomposition

Suppose $A \in \mathcal{M}_{m,n}$. A **PLU decomposition** of A is any factorization of the form $A = PLU$, where $P \in \mathcal{M}_m$ is a permutation matrix, $L \in \mathcal{M}_m$ is lower triangular, and $U \in \mathcal{M}_{m,n}$ is upper triangular.

The identity matrix is a permutation matrix, and if we set $P = I$ then we get the LU decomposition as a special case of the PLU decomposition.

The biggest advantage of the PLU decomposition over the LU decomposition is that it applies to every matrix, not just those that can be row-reduced without using swap row operations (as we show in the next theorem). In fact, the permutation matrix P simply encodes which swap row operations were performed along the way to row echelon form.

Before proceeding with the main theorem of this subsection, which tells us how to construct a PLU decomposition of any matrix, we note that permutation matrices $P \in \mathcal{M}_n$ are special for the fact that are not only invertible, but in fact their inverse is simply their transpose: $P^{-1} = P^T$. We prove this fact in Exercise 2.D.15, and also a bit later in Theorem 3.A.3.

Theorem 2.D.4

Computing a PLU Decomposition

If Gaussian elimination row-reduces $[A \mid I]$ to $[U \mid E]$, where U is a row echelon form of A, then there exists a permutation matrix P and a lower triangular matrix L such that $E^{-1} = PL$. In particular,

$$A = PLU$$

is a PLU decomposition of A.

Proof. Most parts of this theorem have already been proved—all that we have yet to show is that there exists a permutation matrix P and a lower triangular matrix L such that $E^{-1} = PL$. To this end, notice that using Gaussian elimination to turn $[A \mid I]$ into $[U \mid E]$ results in E having at most 1 non-zero entry in its first row (from whichever row of I is swapped into the first row of E).

Similarly, the second row of E contains at most 2 non-zero entries (one of which is below the non-zero entry in the first row, and one of which comes from whichever row of I is swapped into the second row of E). Similarly, there are at most 3 non-zero entries in the third row of E (two of which must below the non-zero entries of the second row, and one of which can be elsewhere),

If we avoid "multiplication" row operations during Gaussian elimination then the matrix L described by this theorem will in fact be *unit* lower triangular.

and so on. Visually, this means that E looks something like

$$E = \begin{bmatrix} 0 & 0 & \star & 0 & 0 \\ 0 & 0 & * & \star & 0 \\ \star & 0 & * & * & 0 \\ * & 0 & * & * & \star \\ * & \star & * & * & * \end{bmatrix}.$$

In particular, it follows that we can permute the *columns* of E so as to put it into lower triangular form. That is, we can find a permutation matrix $P_* \in \mathcal{M}_m$ and a lower triangular matrix $L_* \in \mathcal{M}_m$ such that $E = L_* P_*$. Taking the inverse of both sides of this equation shows that

$$E^{-1} = P_*^{-1} L_*^{-1}.$$

Since $P_*^{-1} = P_*^T$ is also a permutation matrix, and L_*^{-1} is also lower triangular, it follows that we can set $P = P_*^{-1}$ and $L = L_*^{-1}$ to complete the proof. ■

Before we work through an example, we note that after row-reducing $[A \mid I]$ into $[U \mid E]$, there there is actually a very simple and straightforward way to find the permutation matrix P in the PLU decomposition of A: place a "1" in the entry corresponding to the first non-zero entry in each column of E, and then take the transpose of the resulting matrix. For example, if

We say "the" PLU decomposition of A, but PLU decompositions are actually very non-unique (see Exercise 2.D.14). Here we just mean the PLU decomposition that is generated by Gaussian elimination as in Theorem 2.D.4.

$$E = \begin{bmatrix} 0 & 0 & \star & 0 & 0 \\ 0 & 0 & * & \star & 0 \\ \star & 0 & * & * & 0 \\ * & 0 & * & * & \star \\ * & \star & * & * & * \end{bmatrix} \quad \text{then} \quad P^T = \begin{bmatrix} 0 & 0 & 1 & 0 & 0 \\ 0 & 0 & 0 & 1 & 0 \\ 1 & 0 & 0 & 0 & 0 \\ 0 & 0 & 0 & 0 & 1 \\ 0 & 1 & 0 & 0 & 0 \end{bmatrix}.$$

The reason this that works is simply that EP is lower triangular (which is straightforward but annoying to check from the definition of matrix multiplication), so its inverse $P^{-1}E^{-1}$ is lower triangular too. If we define $L = P^{-1}E^{-1}$ then multiplying on the left by P shows that $E^{-1} = PL$, as desired.

Example 2.D.5

Constructing a PLU Decomposition

Construct a PLU decomposition of the matrix $A = \begin{bmatrix} 1 & 1 & 2 & 1 \\ 1 & 1 & 1 & 2 \\ 1 & 2 & 1 & 1 \\ 0 & 1 & 1 & 1 \end{bmatrix}$.

Solution:

Applying Gaussian elimination to $[A \mid I]$ results in the following matrix $[U \mid E]$ in row echelon form:

Depending on the exact sequence of row operations we perform to get to row echelon form, we could get a different $[U \mid E]$. That's OK.

$$[U \mid E] = \left[\begin{array}{cccc|cccc} 1 & 1 & 2 & 1 & 1 & 0 & 0 & 0 \\ 0 & 1 & -1 & 0 & -1 & 0 & 1 & 0 \\ 0 & 0 & 2 & 1 & 1 & 0 & -1 & 1 \\ 0 & 0 & 0 & 3/2 & -1/2 & 1 & -1/2 & 1/2 \end{array} \right].$$

We then choose P^T to have ones in its entries corresponding to the topmost non-zero entries of each column of E, and we also compute E^{-1}:

$$P^T = \begin{bmatrix} 1 & 0 & 0 & 0 \\ 0 & 0 & 1 & 0 \\ 0 & 0 & 0 & 1 \\ 0 & 1 & 0 & 0 \end{bmatrix} \quad \text{and} \quad E^{-1} = \begin{bmatrix} 1 & 0 & 0 & 0 \\ 1 & 0 & -1/2 & 1 \\ 1 & 1 & 0 & 0 \\ 0 & 1 & 1 & 0 \end{bmatrix}.$$

Since we have now found the matrices P (it is the transpose of P^T) and U (it is the left half of $[\, U \mid E \,]$) in the PLU decomposition $A = PLU$, all that remains is to compute L, which we know from Theorem 2.D.4 equals

If we construct P correctly, L will necessarily be lower triangular when we compute it in this way.

$$L = P^{-1}E^{-1} = P^T E^{-1} = \begin{bmatrix} 1 & 0 & 0 & 0 \\ 0 & 0 & 1 & 0 \\ 0 & 0 & 0 & 1 \\ 0 & 1 & 0 & 0 \end{bmatrix} \begin{bmatrix} 1 & 0 & 0 & 0 \\ 1 & 0 & -1/2 & 1 \\ 1 & 1 & 0 & 0 \\ 0 & 1 & 1 & 0 \end{bmatrix}$$

$$= \begin{bmatrix} 1 & 0 & 0 & 0 \\ 1 & 1 & 0 & 0 \\ 0 & 1 & 1 & 0 \\ 1 & 0 & -1/2 & 1 \end{bmatrix}.$$

We can think of the PLU decomposition as the matrix form of Gaussian elimination just like we think of the RREF decomposition of Theorem 2.5.3 as the matrix form of Gauss–Jordan elimination. The P matrix encodes the swap row operations used when row-reducing, the L matrix encodes the addition row operations (and the multiplication row operations, if it is not chosen to be unit triangular), and U is the row echelon form reached via these row operations. The analogy between the RREF decomposition and the PLU decomposition is summarized in Table 2.3.

Algorithm	Resulting form	Matrix decomposition
Gaussian elimination	row echelon form	PLU decomposition
Gauss–Jordan elim.	reduced REF	RREF decomposition

Table 2.3: The PLU decomposition is created from using Gaussian elimination to put a matrix into row echelon form in a manner completely analogous to how the RREF decomposition is created from using Gauss–Jordan elimination to put a matrix into *reduced* row echelon form.

Just like the LU decomposition can be used to (repeatedly) solve linear systems, so can the PLU decomposition. The basic idea is the same—if we want to solve the linear system $A\mathbf{x} = \mathbf{b}$ and we have already computed a PLU decomposition $A = PLU$, then $PLU\mathbf{x} = \mathbf{b}$ is a linear system that we can solve as follows:

The second and third steps here are the same as they were for solving linear systems via the LU decomposition.

- Set $\mathbf{c} = P^T \mathbf{b}$.
- Solve the linear system $L\mathbf{y} = \mathbf{c}$ for $\mathbf{y}$.
- Solve the linear system $U\mathbf{x} = \mathbf{y}$ for $\mathbf{x}$.

It is then straightforward to check that

Recall that $P^{-1} = P^T$, so $\mathbf{c} = P^T \mathbf{b}$ means $P\mathbf{c} = \mathbf{b}$.

$$A\mathbf{x} = PLU\mathbf{x} = PL(U\mathbf{x}) = PL\mathbf{y} = P(L\mathbf{y}) = P\mathbf{c} = \mathbf{b},$$

so $\mathbf{x}$ is indeed a solution of the original linear system, as desired.

Example 2.D.6

Using the PLU Decomposition to Solve a Linear System

Let A be the matrix from Example 2.D.5. Use a PLU decomposition of A to solve the linear system $A\mathbf{x} = \mathbf{b}$, where $\mathbf{b} = (0,1,2,3)$.

Solution:

We let P, L, and U be the matrices that we computed in Example 2.D.5 so that $A = PLU$ is a PLU decomposition of A. We then start by computing

Multiplying $\mathbf{b}$ by P^T just permutes its entries around, hence the term "permutation matrix".

$$\mathbf{c} = P^T \mathbf{b} = \begin{bmatrix} 1 & 0 & 0 & 0 \\ 0 & 0 & 1 & 0 \\ 0 & 0 & 0 & 1 \\ 0 & 1 & 0 & 0 \end{bmatrix} \begin{bmatrix} 0 \\ 1 \\ 2 \\ 3 \end{bmatrix} = \begin{bmatrix} 0 \\ 2 \\ 3 \\ 1 \end{bmatrix}.$$

Next, we solve the lower triangular linear system $L\mathbf{y} = \mathbf{c}$ via forward substitution:

$$\left[\begin{array}{cccc|c} 1 & 0 & 0 & 0 & 0 \\ 1 & 1 & 0 & 0 & 2 \\ 0 & 1 & 1 & 0 & 3 \\ 1 & 0 & -1/2 & 1 & 1 \end{array} \right] \xrightarrow{\text{row-reduce}} \left[\begin{array}{cccc|c} 1 & 0 & 0 & 0 & 0 \\ 0 & 1 & 0 & 0 & 2 \\ 0 & 0 & 1 & 0 & 1 \\ 0 & 0 & 0 & 1 & 3/2 \end{array} \right],$$

so $\mathbf{y} = (0,2,1,3/2)$.

While this perhaps seems like a lot of work, each of these three steps is significantly faster than applying Gaussian elimination to the coefficient matrix directly (refer back to Remark 2.D.2).

Finally, we solve the upper triangular linear system $U\mathbf{x} = \mathbf{y}$ via backward substitution:

$$\left[\begin{array}{cccc|c} 1 & 1 & 2 & 1 & 0 \\ 0 & 1 & -1 & 0 & 2 \\ 0 & 0 & 2 & 1 & 1 \\ 0 & 0 & 0 & 3/2 & 3/2 \end{array} \right] \xrightarrow{\text{row-reduce}} \left[\begin{array}{cccc|c} 1 & 0 & 0 & 0 & -3 \\ 0 & 1 & 0 & 0 & 2 \\ 0 & 0 & 1 & 0 & 0 \\ 0 & 0 & 0 & 1 & 1 \end{array} \right],$$

so the unique solution of this linear system is $\mathbf{x} = (-3,2,0,1)$.

2.D.4   Another Characterization of the LU Decomposition

We saw in Theorem 2.D.2 that a matrix $A \in \mathcal{M}_{m,n}$ has a unit LU decomposition if and only if Gaussian elimination makes it upper triangular without using any "swap" row operations. This criterion is quite useful in practice if we have an explicit matrix to work with, but it is a somewhat clunky condition to work with theoretically—what connection (if any) is there between matrix properties like its rank and the fact that it can be row-reduced without any "swap" row operations? We now present another method of showing that a matrix has a unit LU decomposition that bridges this gap, as it is based only on rank and invertibility, both of which are deeply intertwined with all other aspects of linear algebra.

Refer back to Theorems 2.4.8 and 2.5.1 for the connections between rank, invertibility, and other linear algebraic concepts.

Theorem 2.D.5

Existence of an LU Decomposition

Suppose $A \in \mathcal{M}_n$ has rank r. If the top-left $k \times k$ submatrices of A are invertible for all $1 \leq k \leq r$ then A has a unit LU decomposition.

Proof. We prove this result by induction on the dimension n. If $n = 1$ then the result is trivial, since any 1×1 matrix $A = [a_{1,1}]$ can be written in the form $A = LU$ with $L = [1]$ and $U = [a_{1,1}]$.

For the inductive step of the proof, we assume that every $(n-1) \times (n-1)$

matrix satisfying the hypotheses of the theorem has a unit LU decomposition, and our goal is to use that fact to show that every matrix $n \times n$ matrix has a unit LU decomposition as well. We start by partitioning $A \in \mathcal{M}_n$ as a block matrix whose top-left block B is $(n-1) \times (n-1)$ and whose bottom-right block c is 1×1:

Here, $\mathbf{v}$ and $\mathbf{w}$ are column vectors and c is a scalar.

$$A = \left[\begin{array}{c|c} B & \mathbf{v} \\ \hline \mathbf{w}^T & c \end{array} \right].$$

Since $B \in \mathcal{M}_{n-1}$, we know by the inductive hypothesis that it has a unit LU decomposition $B = LU$. Our goal is to extend this to a unit LU decomposition of A. To this end, we want to find column vectors $\mathbf{x}, \mathbf{y} \in \mathbb{R}^{n-1}$ and a scalar $d \in \mathbb{R}$ such that

Given an LU decomposition in which the diagonal entries of L are *not* all ones, it can be "fixed" by letting D be the diagonal matrix with the same diagonal entries as L and noticing that $LU = (LD^{-1})(DU)$ are both LU decompositions, but LD^{-1} has ones on its diagonal.

$$A = \left[\begin{array}{c|c} B & \mathbf{v} \\ \hline \mathbf{w}^T & c \end{array} \right] = \underbrace{\left[\begin{array}{c|c} L & \mathbf{0} \\ \hline \mathbf{x}^T & 1 \end{array} \right]}_{\text{"new" } L} \underbrace{\left[\begin{array}{c|c} U & \mathbf{y} \\ \hline \mathbf{0}^T & d \end{array} \right]}_{\text{"new" } U} = \left[\begin{array}{c|c} LU & L\mathbf{y} \\ \hline \mathbf{x}^T U & d + \mathbf{x}^T \mathbf{y} \end{array} \right].$$

The top-left block of this equation requires $B = LU$, which holds by how we chose L and U in the first place. The bottom-right block requires $c = d + \mathbf{x}^T \mathbf{y}$, which can be satisfied just by choosing $d = c - \mathbf{x}^T \mathbf{y}$. The top-right block requires us to find a $\mathbf{y}$ such that $L\mathbf{y} = \mathbf{v}$, which can be be done straightforwardly since L has ones on its diagonal and is thus invertible (by Exercise 2.2.16): $\mathbf{y} = L^{-1}\mathbf{v}$.

All that remains is to see that there exists a vector $\mathbf{x}$ such that the bottom-left block is correct: $\mathbf{w}^T = \mathbf{x}^T U$. To this end, we split into two separate cases:

Case 1: B is invertible. In this case, $B = LU$, so $U = BL^{-1}$ is also invertible, so we can simply choose $\mathbf{x} = (U^{-1})^T \mathbf{w}$.

Case 2: B is not invertible. Since the top-left $r \times r$ submatrix of B is invertible (by one of the hypotheses of the theorem), we conclude that $\text{rank}(A) = \text{rank}(B) = r$. It follows that $\mathbf{w}^T$ must be a linear combination of the rows of B, since otherwise Exercise 2.4.29(b) would imply that

$$\text{rank}\left(\left[\begin{array}{c} B \\ \hline \mathbf{w}^T \end{array} \right] \right) > r,$$

which in turn would imply $\text{rank}(A) > r$.

Furthermore, the equation $B = LU$ tells us that the rows of B are linear combinations of the rows of U (via Exercise 1.3.27(b)), so in fact $\mathbf{w}^T$ is a linear combination of the rows of U. But this is exactly what the equation $\mathbf{w}^T = \mathbf{x}^T U$ says—in particular, we choose the entries of $\mathbf{x}^T$ to be exactly the coefficients of that linear combination.

This theorem also works even if A is not square—see Exercise 2.D.5.

Now that we know how to construct each of d, $\mathbf{x}$, and $\mathbf{y}$, the inductive step and the proof are complete. ∎

Example 2.D.7

Showing a Matrix Has an LU Decomposition

Show that the matrix $A = \begin{bmatrix} 3 & -2 & -1 \\ 3 & -1 & 4 \\ 1 & 5 & 2 \end{bmatrix}$ has a unit LU decomposition.

Solution:

It is straightforward to show that $\text{rank}(A) = 3$, so we must check invertibility of each of the top-left 1×1, 2×2 and 3×3 submatrices of A:

Forgotten how to
compute the rank
of a matrix or check
whether or not it's
invertible? Have a
look back at
Sections 2.2.3
and 2.4.3.

$$[3], \quad \begin{bmatrix} 3 & -2 \\ 3 & -1 \end{bmatrix}, \quad \text{and} \quad \begin{bmatrix} 3 & -2 & -1 \\ 3 & -1 & 4 \\ 1 & 5 & 2 \end{bmatrix}.$$

It is straightforward to show that these matrices are all invertible, so Theorem 2.D.7 tells us that A has a unit LU decomposition.

Although we could mimic the proof of Theorem 2.D.5 in order to *construct* an LU decomposition of a matrix as well, it is typically quicker and easier to just use the method based on Gaussian elimination from Theorem 2.D.2. In other words, it is best to think of Theorem 2.D.5 just as a tool for checking existence of an LU decomposition, not as a computational tool.

It is also perhaps worth noting that the converse of Theorem 2.D.5 holds in the special case when $A \in \mathcal{M}_n$ is invertible. That is, an invertible matrix has a unit LU decomposition if *and only if* all of its top-left square submatrices are invertible (see Exercise 2.D.7), and furthermore it is necessarily unique (see Exercise 2.D.6). However, the converse of Theorem 2.D.5 does not hold for non-invertible matrices. For example, the matrices

$$A = \begin{bmatrix} 0 & 1 & 0 \\ 0 & 1 & 0 \\ 0 & 0 & 1 \end{bmatrix} \quad \text{and} \quad B = \begin{bmatrix} 0 & 1 & 0 \\ 0 & 1 & 0 \\ 1 & 0 & 0 \end{bmatrix}$$

have the same rank and the same top-left 1×1 and 2×2 submatrices, but A has a unit LU decomposition while B does not.

Exercises

solutions to starred exercises on page 459

2.D.1 Compute a unit LU decomposition of each of the following matrices.

*(a) $\begin{bmatrix} 1 & 2 \\ 2 & 5 \end{bmatrix}$

(b) $\begin{bmatrix} 2 & -1 \\ 4 & 1 \end{bmatrix}$

*(c) $\begin{bmatrix} 3 & 1 & 2 \\ -3 & -3 & -1 \end{bmatrix}$

(d) $\begin{bmatrix} -1 & 2 & 0 \\ -3 & 5 & 2 \end{bmatrix}$

*(e) $\begin{bmatrix} 1 & 2 \\ 2 & 3 \\ 1 & -1 \end{bmatrix}$

(f) $\begin{bmatrix} 1 & 2 & -1 \\ -1 & -3 & -2 \\ 3 & 5 & -8 \end{bmatrix}$

*(g) $\begin{bmatrix} 1 & -4 & 5 \\ 3 & -9 & 8 \\ -2 & 5 & -2 \end{bmatrix}$

(h) $\begin{bmatrix} 2 & -1 & 4 & 3 \\ -4 & 4 & -7 & -6 \\ 6 & -7 & 12 & 10 \end{bmatrix}$

2.D.2 Compute a PLU decomposition of each of the following matrices.

*(a) $\begin{bmatrix} 0 & 2 \\ 1 & 3 \end{bmatrix}$

(b) $\begin{bmatrix} 1 & 2 & 3 \\ 4 & 5 & 6 \end{bmatrix}$

*(c) $\begin{bmatrix} 1 & 1 & 1 \\ 1 & 1 & 2 \\ 1 & 2 & 3 \end{bmatrix}$

(d) $\begin{bmatrix} 0 & 2 & 3 & 3 \\ 1 & 2 & 1 & 2 \\ 1 & 2 & 1 & 4 \end{bmatrix}$

2.D.3 Use the given LU or PLU decomposition to find the solution of the following linear systems.

*(a) $\begin{bmatrix} 1 & 0 \\ 2 & 1 \end{bmatrix} \begin{bmatrix} 2 & 3 \\ 0 & 1 \end{bmatrix} \begin{bmatrix} x \\ y \end{bmatrix} = \begin{bmatrix} 1 \\ -1 \end{bmatrix}$

(b) $\begin{bmatrix} 1 & 0 & 0 \\ 2 & 1 & 0 \\ -1 & 1 & 1 \end{bmatrix} \begin{bmatrix} 2 & 1 & 0 \\ 0 & 2 & 1 \\ 0 & 0 & 2 \end{bmatrix} \begin{bmatrix} x \\ y \\ z \end{bmatrix} = \begin{bmatrix} 3 \\ 1 \\ 2 \end{bmatrix}$

*(c) $\begin{bmatrix} 0 & 0 & 1 \\ 0 & 1 & 0 \\ 1 & 0 & 0 \end{bmatrix} \begin{bmatrix} 1 & 0 & 0 \\ 0 & 1 & 0 \\ 1 & 0 & 1 \end{bmatrix} \begin{bmatrix} 1 & 2 & 3 \\ 0 & 1 & 2 \\ 0 & 0 & 1 \end{bmatrix} \begin{bmatrix} x \\ y \\ z \end{bmatrix} = \begin{bmatrix} 1 \\ 2 \\ 3 \end{bmatrix}$

(d) $\begin{bmatrix} 1 & 0 & 0 & 0 \\ 1 & 1 & 0 & 0 \\ 2 & 1 & 1 & 0 \\ 1 & 1 & 3 & 1 \end{bmatrix} \begin{bmatrix} 2 & 1 & 0 & 1 \\ 0 & 1 & 2 & 0 \\ 0 & 0 & 1 & 1 \\ 0 & 0 & 0 & 3 \end{bmatrix} \begin{bmatrix} w \\ x \\ y \\ z \end{bmatrix} = \begin{bmatrix} 1 \\ 1 \\ 1 \\ 1 \end{bmatrix}$

2.D.4 Determine which of the following statements are true and which are false.

*(a) Every matrix has a unit LU decomposition.
(b) Every matrix has a PLU decomposition.
*(c) A permutation matrix is one that has a single 1 in each row and column, and every other entry equal to 0.
(d) Every permutation matrix is its own inverse.
*(e) Every matrix that is in row echelon form is upper triangular.
(f) Every matrix that is upper triangular is in row echelon form.

2.D.5 Show that Theorem 2.D.5 holds even if A is not square.

[Hint: Start with an LU decomposition of a maximal square submatrix of A.]

2.D.6 In this exercise, we investigate the uniqueness of unit LU decompositions.

(a) Show that if $A \in \mathcal{M}_n$ is invertible then its unit LU decomposition is unique (if it exists).
(b) Find two different unit LU decompositions of the (non-invertible) matrix
$$A = \begin{bmatrix} 0 & 0 \\ 0 & 1 \end{bmatrix}.$$

2.D.7 Show that if $A \in \mathcal{M}_n$ is invertible then the converse of Theorem 2.D.5 holds (i.e., A has a unit LU decomposition if *and only if* all of its top-left square submatrices are invertible).

2.D.8 Use Exercise 2.D.7 to determine whether or not the following matrices have a unit LU decomposition (you do not need to *compute* this decomposition).

*(a) $\begin{bmatrix} 1 & 3 \\ 3 & 2 \end{bmatrix}$ (b) $\begin{bmatrix} 0 & 1 \\ 1 & -2 \end{bmatrix}$

*(c) $\begin{bmatrix} 3 & 1 & -1 \\ -1 & 1 & 3 \\ 2 & -2 & 1 \end{bmatrix}$ (d) $\begin{bmatrix} 4 & 5 & 5 & 0 \\ 3 & 4 & 3 & 3 \\ 1 & 1 & 2 & 3 \\ 3 & 3 & 4 & 2 \end{bmatrix}$

2.D.9 Let
$$A = \begin{bmatrix} 2 & -1 & 0 & 0 & 0 & 0 \\ -1 & 3 & -1 & 0 & 1 & 0 \\ 0 & -1 & 4 & -1 & 0 & 0 \\ 0 & 0 & -1 & 4 & -1 & 0 \\ 0 & 0 & 0 & -1 & 3 & -1 \\ 0 & 1 & 0 & 0 & -1 & 2 \end{bmatrix}.$$

(a) Use computer software to compute a unit LU decomposition of A.
(b) Use computer software to compute A^{-1}.
[Side note: Since A is **sparse** (i.e., has many zeros in it), so is its LU decomposition. However, A^{-1} is not sparse, which is another reason that we typically prefer using the LU decomposition for solving linear systems instead of the inverse.]

*2.D.10 An **LDU decomposition** of a matrix $A \in \mathcal{M}_{m,n}$ is any factorization of the form $A = LDU$, where L and U are *unit* lower and upper triangular matrices, respectively, and D is diagonal.

Suppose $A \in \mathcal{M}_n$ is invertible. Show that A has an LDU decomposition if and only if it has a unit LU decomposition.

2.D.11 Compute an LDU decomposition (see Exercise 2.D.10) of the matrices from the following exercises.

*(a) Exercise 2.D.1(a).
(b) Exercise 2.D.1(b).
*(c) Exercise 2.D.1(f).
(d) Exercise 2.D.1(g).

2.D.12 Show that if a symmetric and invertible matrix $A \in \mathcal{M}_n$ has an LDU decomposition (see Exercise 2.D.10) then it is unique and has the form $A = LDL^T$.

[Hint: Use Exercise 2.D.6(a).]

2.D.13 Consider the matrix $A = \begin{bmatrix} 0 & 0 \\ 1 & 1 \end{bmatrix}$.

(a) Find an LU decomposition of A.
(b) Show that A does not have a unit LU decomposition (and thus does not have an LU decomposition in which the lower triangular matrix L is invertible).
(c) Find a PLU decomposition of A.

2.D.14 PLU decompositions are quite non-unique, even if the lower triangular matrix is restricted to having 1s on its diagonal (in which case we call it a *unit* PLU decomposition).

(a) Find two different unit PLU decompositions of the matrix
$$\begin{bmatrix} 2 & -1 \\ 4 & -1 \end{bmatrix}.$$
(b) Based on your answer to part (a) and Exercise 2.D.6, how many different unit PLU decompositions do you think a typical invertible 3×3 matrix has? A 4×4 invertible matrix?

2.D.15 Suppose that $P \in \mathcal{M}_n$ is a permutation matrix (i.e., exactly one entry in each of its rows and column equals 1, and the rest equal 0). Show that P is invertible and $P^{-1} = P^T$.

2.D.16 Recall from Example 2.3.8 that a **Vandermonde matrix** is a square matrix of the form
$$V = \begin{bmatrix} 1 & a_0 & a_0^2 & \cdots & a_0^n \\ 1 & a_1 & a_1^2 & \cdots & a_1^n \\ \vdots & \vdots & \vdots & \ddots & \vdots \\ 1 & a_n & a_n^2 & \cdots & a_n^n \end{bmatrix},$$
where $a_0, a_1, \ldots, a_n$ are real numbers. Show that if $a_0, a_1, \ldots, a_n$ are distinct then V has a unit LU decomposition.

3. Unraveling Matrices

The shortest path between two truths in the real domain passes through the complex domain.

Jacques Hadamard

We now start investigating the properties of matrices that can be used to simplify our understanding of the linear transformations that they represent. Most of the matrix properties that we will look at in this chapter come from thinking about how they act on certain subspaces, and trying to find particular subspaces of $\mathbb{R}^n$ where they behave "nicely". For example, consider the matrix

$$A = \begin{bmatrix} 2 & 1 \\ 1 & 2 \end{bmatrix},$$

which acts as a linear transformation on $\mathbb{R}^2$ in the manner shown in Figure 3.1.

Recall that this A sends $\mathbf{e}_1 = (1,0)$ to $A\mathbf{e}_1 = (2,1)$, which is its first column, and it sends $\mathbf{e}_2 = (0,1)$ to $A\mathbf{e}_2 = (1,2)$, which is its second column.

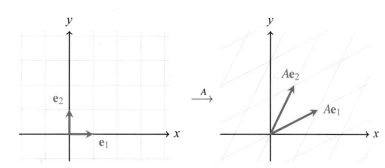

Figure 3.1: The matrix $A = \begin{bmatrix} 2 & 1 \\ 1 & 2 \end{bmatrix}$ acting as a linear transformation on $\mathbb{R}^2$.

When represented in this way, this matrix does not have a particularly nice geometric interpretation—it distorts the square grid defined by $\mathbf{e}_1$ and $\mathbf{e}_2$ into a rotated parallelogram grid. However, if we change our perspective a bit and instead consider how it acts on the input grid defined by $\mathbf{v}_1 = (1,1)$ and $\mathbf{v}_2 = (-1,1)$, then this matrix transforms $\mathbb{R}^2$ in the manner shown in Figure 3.2, which is much easier to understand visually.

In particular, since A just scales $\mathbf{v}_1$ by a factor of 3 (it does not change its direction) and it does not change $\mathbf{v}_2$ at all, the square grid determined by $\mathbf{v}_1$ and $\mathbf{v}_2$ is just stretched into a rectangular grid with the same orientation. This grid is not rotated or skewed at all, but instead it is just stretched along one of its sides.

© Springer Nature Switzerland AG 2021
N. Johnston, *Introduction to Linear and Matrix Algebra*,
https://doi.org/10.1007/978-3-030-52811-9_3

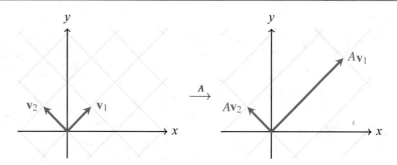

Figure 3.2: When viewed in this way, the matrix $A = \begin{bmatrix} 2 & 1 \\ 1 & 2 \end{bmatrix}$ is a bit easier to visualize.

In this chapter, we flesh out the mathematics behind this example. We introduce exactly how we can represent a matrix in a different basis (i.e., a different input grid), we investigate how to find a particular basis ($\{\mathbf{v}_1, \mathbf{v}_2\}$ in this example) that makes a matrix easier to work with and visualize, and we look at what types of computations can be made simpler by representing them in this way.

3.1   Coordinate Systems

Our starting point is the concept of a coordinate system, which is something that lets us write a vector or matrix so that it still represents the same geometric object (either an arrow in space or a linear transformation), but viewed through a different lens that makes it easier to perform calculations with. For now we just focus on the mechanics of what coordinate systems are and how to use them—we return to the problem of finding *useful* coordinate systems in Sections 3.3 and 3.4.

3.1.1   Representations of Vectors

One of the most useful features of bases is that they remove ambiguity in linear combinations. While we can always write a vector in a subspace as a linear combination of vectors in any set that spans that subspace, that linear combination will typically not be unique.

For example, it is the case that $\mathbb{R}^2 = \mathrm{span}\big((1,0),(0,1),(1,1)\big)$, so every vector $(x,y) \in \mathbb{R}^2$ can be written as a linear combination of $(1,0),(0,1)$, and $(1,1)$. In particular, some ways to write $(2,1)$ as a linear combination of these vectors include

$$
\begin{aligned}
(2,1) &= 2(1,0) + (0,1) \\
&= -(0,1) + 2(1,1) \\
&= (1,0) + (1,1),
\end{aligned}
$$

as well as infinitely many other possibilities (see Figure 3.3).

The reason for the non-uniqueness of the above linear combinations is that the set $\{(1,0),(0,1),(1,1)\}$ is not linearly independent (and thus not a basis). If we throw away the vector $(1,1)$ then we arrive at the standard basis $\{(1,0),(0,1)\}$, and linear combinations of the standard basis vectors *are* unique

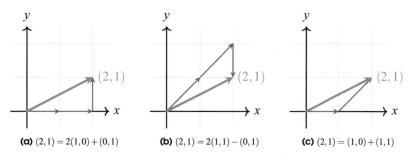

(a) $(2,1) = 2(1,0) + (0,1)$ **(b)** $(2,1) = 2(1,1) - (0,1)$ **(c)** $(2,1) = (1,0) + (1,1)$

Figure 3.3: The vector $(2,1)$ can be written as a linear combination of $(1,0), (0,1)$, and $(1,1)$ in many different ways. In a sense, the vector $(1,1)$ is "not needed" in the linear combinations and only serves to make them non-unique.

(and in this case, the unique way to write $(2,1)$ as a linear combination of the standard basis vectors is simply $(2,1) = 2(1,0) + (0,1)$).

The main result of this subsection shows that this uniqueness claim is true of bases in general.

Theorem 3.1.1 **Uniqueness of Linear Combinations**	Let S be a subspace of $\mathbb{R}^n$ with basis B. For every vector $\mathbf{v} \in S$, there is exactly one way to write $\mathbf{v}$ as a linear combination of the vectors from B.

Proof. Since $B = \{\mathbf{v}_1, \mathbf{v}_2, \ldots, \mathbf{v}_k\}$ is a basis, it spans S, so $\mathbf{v}$ can be written as a linear combination of the vectors from B in *at least* one way. To show uniqueness, suppose that

$$\mathbf{v} = c_1\mathbf{v}_1 + c_2\mathbf{v}_2 + \cdots + c_k\mathbf{v}_k \quad \text{and}$$
$$\mathbf{v} = d_1\mathbf{v}_1 + d_2\mathbf{v}_2 + \cdots + d_k\mathbf{v}_k$$

are two ways of writing $\mathbf{v}$ as a linear combination of $\mathbf{v}_1, \mathbf{v}_2, \ldots, \mathbf{v}_k$. Subtracting those equations gives

$$\mathbf{0} = \mathbf{v} - \mathbf{v} = (c_1 - d_1)\mathbf{v}_1 + (c_2 - d_2)\mathbf{v}_2 + \cdots + (c_k - d_k)\mathbf{v}_k.$$

Since B is a basis, and is thus linearly independent, it follows that

> The converse of this theorem also holds. That is, if B is a set with the property that every vector $\mathbf{v} \in S$ can be written as a linear combination of the members of B in exactly one way, then B must be a basis of S (see Exercise 3.1.10).

$$c_1 - d_1 = 0, \qquad c_2 - d_2 = 0, \qquad \ldots, \qquad c_k - d_k = 0, \quad \text{so}$$
$$c_1 = d_1, \qquad c_2 = d_2, \qquad \ldots, \qquad c_k = d_k.$$

The two linear combinations for $\mathbf{v}$ are thus actually the *same* linear combination, which proves uniqueness. ∎

It is worth emphasizing that the above proof shows that the two defining properties of bases—spanning S and being linearly independent—each correspond to half of the above theorem. The fact that a basis B spans S tells us that every vector can be written as a linear combination of the members of B, and the fact that B is linearly independent tells us that those linear combinations are unique.

The fact that linear combinations of basis vectors are unique means that we can treat them as distinct directions that uniquely identify each point in space. Much like we can specify points on the surface of the Earth uniquely by their latitude and longitude (i.e., where that point is in the east–west direction and in the north–south direction), we can also specify vectors in a subspace uniquely by how far they extend in the direction of each basis vector. In fact, we even use terminology that is familiar to us from when we specify a point on the surface of the Earth: the (unique!) coefficients c_1, c_2, ..., c_k described by the previous theorem are called the **coordinates** of the vector $\mathbf{v}$:

Definition 3.1.1

Coordinates with Respect to a Basis

Suppose S is a subspace of $\mathbb{R}^n$, $B = \{\mathbf{v}_1, \mathbf{v}_2, \ldots, \mathbf{v}_k\}$ is a basis of S, and $\mathbf{v} \in S$. Then the unique scalars c_1, c_2, ..., c_k for which

$$\mathbf{v} = c_1 \mathbf{v}_1 + c_2 \mathbf{v}_2 + \cdots + c_k \mathbf{v}_k$$

are called the **coordinates** of $\mathbf{v}$ with respect to B, and the vector

$$[\mathbf{v}]_B \overset{\text{def}}{=} (c_1, c_2, \ldots, c_k)$$

is called the **coordinate vector** of $\mathbf{v}$ with respect to B.

Remark 3.1.1

Ordered Bases

There is actually a problem with Definition 3.1.1, and that is the fact that order does not matter for bases (e.g., $\{\mathbf{e}_1, \mathbf{e}_2\}$ and $\{\mathbf{e}_2, \mathbf{e}_1\}$ are the exact same basis of $\mathbb{R}^2$). However, order *does* matter for vectors (e.g., (c_1, c_2) and (c_2, c_1) are different vectors). In other words, we want to use coordinates to talk about how far $\mathbf{v}$ extended in the direction of the *first* basis vector, how far it extends in the direction of the *second* basis vector, and so on, but bases (and sets in general) have no "first" or "second" vectors.

To get around this problem, whenever we work with coordinates or coordinate vectors, we simply use the basis vectors in the order written. That is, if we write $B = \{\mathbf{v}_1, \mathbf{v}_2, \ldots, \mathbf{v}_k\}$ then we understand that the "first" basis vector is $\mathbf{v}_1$, the "second" basis vector is $\mathbf{v}_2$, and so on. This could be made precise by defining an **ordered basis** of a subspace $S \subseteq \mathbb{R}^n$ to be a linearly independent *tuple* (rather than a set) of vectors that span S, but we do not explicitly do so.

When working with the standard basis of $\mathbb{R}^n$ (i.e., when $S = \mathbb{R}^n$ and $B = \{\mathbf{e}_1, \mathbf{e}_2, \ldots, \mathbf{e}_n\}$), coordinate vectors coincide exactly with how we think of vectors already—every vector simply equals its coordinate vector with respect to the standard basis. This is because the unique scalars c_1, c_2, ..., c_n for which

$$\mathbf{v} = c_1 \mathbf{e}_1 + c_2 \mathbf{e}_2 + \cdots + c_n \mathbf{e}_n$$

are just the entries of $\mathbf{v}$, so when working with the standard basis we just have $\mathbf{v} = (c_1, c_2, \ldots, c_n) = [\mathbf{v}]_B$. The true utility of coordinates and coordinate vectors becomes more apparent when we work with proper subspaces of $\mathbb{R}^n$ rather than with $\mathbb{R}^n$ itself.

Example 3.1.1

Finding Coordinate Vectors

Find the coordinate vector of $\mathbf{v} \in S$ with respect to the basis B of S.

a) $\mathbf{v} = (5, 1)$, $B = \{(2, 1), (1, 2)\}$, $S = \mathbb{R}^2$
b) $\mathbf{v} = (5, 4, 3)$, $B = \{(1, 2, 1), (2, 1, 1)\}$, $S = \text{span}(B)$

Solutions:
a) We want to find the scalars c_1 and c_2 for which

$$(5, 1) = c_1(2, 1) + c_2(1, 2).$$

Since we are told
that B is a basis, we
know in advance
that this linear
system will have a
unique solution.

We solved this type of problem repeatedly in the previous chapter—it is a linear system that can be solved by placing $(2,1)$ and $(1,2)$ as columns in a matrix with $(5,1)$ as the augmented right-hand side:

$$\left[\begin{array}{cc|c} 2 & 1 & 5 \\ 1 & 2 & 1 \end{array}\right] \xrightarrow{R_2-\frac{1}{2}R_1} \left[\begin{array}{cc|c} 2 & 1 & 5 \\ 0 & 3/2 & -3/2 \end{array}\right].$$

From here we can solve the linear system via back substitution to see that the unique solution is $c_1 = 3, c_2 = -1$, so $[\mathbf{v}]_B = (3,-1)$:

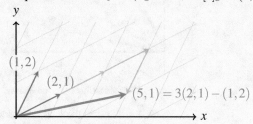

b) We want to find the scalars c_1 and c_2 for which

$$(5,4,3) = c_1(1,2,1) + c_2(2,1,1).$$

Again, we solve this problem by viewing it as a linear system:

$$\left[\begin{array}{cc|c} 1 & 2 & 5 \\ 2 & 1 & 4 \\ 1 & 1 & 3 \end{array}\right] \xrightarrow[R_3-R_1]{R_2-2R_1} \left[\begin{array}{cc|c} 1 & 2 & 5 \\ 0 & -3 & -6 \\ 0 & -1 & -2 \end{array}\right] \xrightarrow[R_3+R_2]{\frac{-1}{3}R_2} \left[\begin{array}{cc|c} 1 & 2 & 5 \\ 0 & 1 & 2 \\ 0 & 0 & 0 \end{array}\right].$$

From here we can solve the linear system via back substitution to see that the unique solution is $c_1 = 1, c_2 = 2$, so $[\mathbf{v}]_B = (1,2)$:

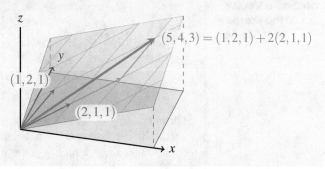

It is worth noting that if $B = \{\mathbf{v}_1, \mathbf{v}_2, \ldots, \mathbf{v}_k\}$ is a basis then $[\mathbf{v}_j]_B = \mathbf{e}_j$ (the j-th standard basis vector) for all $1 \leq j \leq k$, since the unique linear combination of $\{\mathbf{v}_1, \mathbf{v}_2, \ldots, \mathbf{v}_k\}$ resulting in $\mathbf{v}_j$ is quite trivial:

$$\mathbf{v}_j = 0\mathbf{v}_1 + 0\mathbf{v}_2 + \cdots + 0\mathbf{v}_{j-1} + 1\mathbf{v}_j + 0\mathbf{v}_{j+1} + \cdots + 0\mathbf{v}_k.$$

We thus think of coordinate vectors as being a lens through which we look at $\mathcal{S}$ so that $\mathbf{v}_1$ looks like $\mathbf{e}_1$, $\mathbf{v}_2$ looks like $\mathbf{e}_2$, and so on (see Figure 3.4). It is worth stating this observation more prominently, for easy reference:

> (!) If $B = \{\mathbf{v}_1, \mathbf{v}_2, \ldots, \mathbf{v}_k\}$ then $[\mathbf{v}_j]_B = \mathbf{e}_j$ for all $1 \leq j \leq k$.

We saw in Example 3.1.1(b) that, even though the vector $(5,4,3)$ lives in

$\mathbb{R}^3$, it also lives in the 2-dimensional plane $S = \text{span}\big((1,2,1),(2,1,1)\big)$ and thus can be described *in that plane* via just 2 coordinates. In a sense, we think of that plane as a copy of $\mathbb{R}^2$ that is embedded in $\mathbb{R}^3$, and we give the vectors in that plane coordinates as if they just lived in $\mathbb{R}^2$.

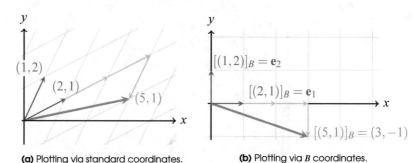

(a) Plotting via standard coordinates.   **(b)** Plotting via B coordinates.

Figure 3.4: We can think of coordinate vectors as distorting space in such a way as to make the given basis look like the standard basis. In this case, the space is $S = \mathbb{R}^2$ and the basis is $B = \{(2,1),(1,2)\}$, as in Example 3.1.1(a).

While reducing a 3-dimensional vector down to a 2-dimensional one like this might not seem particularly important, many real-world problems are much higher-dimensional. If we only care about a particular 3-dimensional subspace of $\mathbb{R}^{85}$ (for example), then we can use coordinate vectors to describe vectors in that subspace via just 3 coordinates rather than 85, which makes computations significantly easier.

Example 3.1.2

Computing a Coordinate Vector in the Range of a Matrix

Find a basis B of the range of the following matrix A and then compute the coordinate vector $[\mathbf{v}]_B$ of the vector $\mathbf{v} = (2,1,-3,1,2) \in \text{range}(A)$:

$$A = \begin{bmatrix} 1 & 0 & 1 \\ 2 & 1 & 1 \\ 0 & 1 & -1 \\ 2 & 1 & 1 \\ 1 & 0 & 1 \end{bmatrix}.$$

Solution:

Recall from Example 2.4.6 that one way to find the basis of the range of A is to take the columns of A that are leading in one of its row echelon forms:

$$\begin{bmatrix} 1 & 0 & 1 \\ 2 & 1 & 1 \\ 0 & 1 & -1 \\ 2 & 1 & 1 \\ 1 & 0 & 1 \end{bmatrix} \xrightarrow[\substack{R_4-2R_1 \\ R_5-R_1}]{R_2-2R_1} \begin{bmatrix} 1 & 0 & 1 \\ 0 & 1 & -1 \\ 0 & 1 & -1 \\ 0 & 1 & -1 \\ 0 & 0 & 0 \end{bmatrix} \xrightarrow[R_4-R_2]{R_3-R_2} \begin{bmatrix} 1 & 0 & 1 \\ 0 & 1 & -1 \\ 0 & 0 & 0 \\ 0 & 0 & 0 \\ 0 & 0 & 0 \end{bmatrix}.$$

Be careful here—remember that we take the leading columns from A itself, not from its row echelon form.

Since the first two columns of this row echelon form are leading, we choose B to consist of the first two columns of A:

$$B = \{(1,2,0,2,1),(0,1,1,1,0)\}.$$

To compute $[\mathbf{v}]_B$, we then solve the linear system

$$(2, 1, -3, 1, 2) = c_1(1, 2, 0, 2, 1) + c_2(0, 1, 1, 1, 0)$$

for c_1 and c_2:

$$\begin{bmatrix} 1 & 0 & 2 \\ 2 & 1 & 1 \\ 0 & 1 & -3 \\ 2 & 1 & 1 \\ 1 & 0 & 2 \end{bmatrix} \xrightarrow[\substack{R_2-2R_1 \\ R_4-2R_1 \\ R_5-R_1}]{} \begin{bmatrix} 1 & 0 & 2 \\ 0 & 1 & -3 \\ 0 & 1 & -3 \\ 0 & 1 & -3 \\ 0 & 0 & 0 \end{bmatrix} \xrightarrow[\substack{R_3-R_2 \\ R_4-R_2}]{} \begin{bmatrix} 1 & 0 & 2 \\ 0 & 1 & -3 \\ 0 & 0 & 0 \\ 0 & 0 & 0 \\ 0 & 0 & 0 \end{bmatrix}.$$

It follows that $c_1 = 2$ and $c_2 = -3$, so $[\mathbf{v}]_B = (2, -3)$.

One way to think about the previous example is that the range of A is just 2-dimensional (i.e., its rank is 2), so representing the vector $\mathbf{v} = (2, 1, -3, 1, 2)$ in its range via 5 coordinates is somewhat wasteful—we can fix a basis B of the range and then represent $\mathbf{v}$ via just two coordinates, as in $[\mathbf{v}]_B = (2, -3)$.

The following theorem essentially says that the function that sends $\mathbf{v}$ to $[\mathbf{v}]_B$ is a linear transformation.

Furthermore, once we have represented vectors more compactly via coordinate vectors, we can work with them naïvely and still get correct answers. That is, a coordinate vector $[\mathbf{v}]_B$ can be manipulated (i.e., added and scalar multiplied) in the exact same way as the underlying vector $\mathbf{v}$ that it represents:

Theorem 3.1.2

Vector Operations on Coordinate Vectors

Suppose $\mathcal{S}$ is a subspace of $\mathbb{R}^n$ with basis B, and let $\mathbf{v}, \mathbf{w} \in \mathcal{S}$ and $c \in \mathbb{R}$. Then

a) $[\mathbf{v} + \mathbf{w}]_B = [\mathbf{v}]_B + [\mathbf{w}]_B$, and
b) $[c\mathbf{v}]_B = c[\mathbf{v}]_B$.

As the proof of this theorem is not terribly enlightening, we leave it as Exercise 3.1.11 and instead jump straight to an example.

Example 3.1.3

Linear Combinations of Coordinate Vectors

Let $B = \{(1, 2, 0, 2, 1), (0, 1, 1, 1, 0)\}$ be a basis of the subspace

$$\mathcal{S} = \text{span}(B) \subset \mathbb{R}^5,$$

and suppose $\mathbf{v} = (2, 1, -3, 1, 2)$ and $\mathbf{w} = (1, 0, -2, 0, 1)$. Compute the following quantities directly from their definitions:

a) $[2\mathbf{v} - 5\mathbf{w}]_B$
b) $2[\mathbf{v}]_B - 5[\mathbf{w}]_B$

We know from Theorem 3.1.2 that we must get the same answer to parts (a) and (b) of this example.

Solutions:

a) It is straightforward to compute

$$2\mathbf{v} - 5\mathbf{w} = 2(2, 1, -3, 1, 2) - 5(1, 0, -2, 0, 1) = (-1, 2, 4, 2, -1).$$

Next, we can compute the coordinate vector of $2\mathbf{v} - 5\mathbf{w}$ in the same way that we did in Example 3.1.2:

$$\left[\begin{array}{cc|c} 1 & 0 & -1 \\ 2 & 1 & 2 \\ 0 & 1 & 4 \\ 2 & 1 & 2 \\ 1 & 0 & -1 \end{array}\right] \xrightarrow[\substack{R_4-2R_1 \\ R_5-R_1}]{R_2-2R_1} \left[\begin{array}{cc|c} 1 & 0 & -1 \\ 0 & 1 & 4 \\ 0 & 1 & 4 \\ 0 & 1 & 4 \\ 0 & 0 & 0 \end{array}\right] \xrightarrow[R_4-R_2]{R_3-R_2} \left[\begin{array}{cc|c} 1 & 0 & -1 \\ 0 & 1 & 4 \\ 0 & 0 & 0 \\ 0 & 0 & 0 \\ 0 & 0 & 0 \end{array}\right].$$

It follows that

$$2\mathbf{v} - 5\mathbf{w} = (-1,2,4,2,-1) = -(1,2,0,2,1) + 4(0,1,1,1,0),$$

so $[2\mathbf{v} - 5\mathbf{w}]_B = (-1,4)$.

b) We already know from Example 3.1.2 that $[\mathbf{v}]_B = (2,-3)$, so we just need to compute $[\mathbf{w}]_B$:

$$\left[\begin{array}{cc|c} 1 & 0 & 1 \\ 2 & 1 & 0 \\ 0 & 1 & -2 \\ 2 & 1 & 0 \\ 1 & 0 & 1 \end{array}\right] \xrightarrow[\substack{R_4-2R_1 \\ R_5-R_1}]{R_2-2R_1} \left[\begin{array}{cc|c} 1 & 0 & 1 \\ 0 & 1 & -2 \\ 0 & 1 & -2 \\ 0 & 1 & -2 \\ 0 & 0 & 0 \end{array}\right] \xrightarrow[R_4-R_2]{R_3-R_2} \left[\begin{array}{cc|c} 1 & 0 & 1 \\ 0 & 1 & -2 \\ 0 & 0 & 0 \\ 0 & 0 & 0 \\ 0 & 0 & 0 \end{array}\right].$$

It follows that $[\mathbf{w}]_B = (1,-2)$, so we see that

$$2[\mathbf{v}]_B - 5[\mathbf{w}]_B = 2(2,-3) - 5(1,-2) = (-1,4),$$

which agrees with the answer that we found in part (a).

Remark 3.1.2

Orthonormal Bases and the Size of Coordinate Vectors

When performing computations, bases that are "almost" linearly dependent can sometimes be numerically undesirable. To illustrate what we mean by this, suppose $\mathbf{v}_1 = (1,0.2)$ and $\mathbf{v}_2 = (1,0.1)$, and consider the basis $B = \{\mathbf{v}_1, \mathbf{v}_2\}$ of $\mathbb{R}^2$. When we represent vectors in this basis, their coordinate vectors can have surprisingly large entries that make them difficult to work with.

For example, it is straightforward to verify that

$$(-1,1) = 11\mathbf{v}_1 - 12\mathbf{v}_2, \quad \text{so} \quad [(-1,1)]_B = (11,-12).$$

Geometrically, the reason for these large entries is that the parallelogram grid defined by $\mathbf{v}_1$ and $\mathbf{v}_2$ is extremely thin, so we need large coefficients in order to represent vectors that do not point in the same direction as the parallelograms:

If $\mathbf{v}_1$ and $\mathbf{v}_2$ are even closer, then coordinate vectors can become even larger. For example, if $\mathbf{v}_1 = (1,0.002)$ and $\mathbf{v}_2 = (1,0.001)$ then $[(-1,1)]_B = (1001,-1002)$.

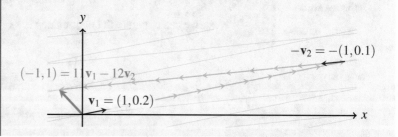

In order to avoid problems like this one, we typically prefer to work with **orthonormal bases** whenever possible, which are bases B with the following additional properties:

a) $\mathbf{v} \cdot \mathbf{w} = 0$ for all $\mathbf{v} \neq \mathbf{w} \in B$, and
b) $\|\mathbf{v}\| = 1$ for all $\mathbf{v} \in B$.

The standard basis is an example of an orthonormal basis, and in general the parallelogram grid that they define is in fact a (potentially rotated) unit square grid:

More generally, the grid is made up of unit squares, unit cubes, or unit hypercubes, depending on the dimension.

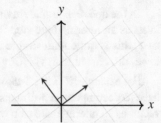

If B is an orthonormal basis then $\|\mathbf{v}\| = \|[\mathbf{v}]_B\|$, so $[\mathbf{v}]_B$ cannot be large if $\mathbf{v}$ is small (in contrast with the non-orthonormal example $[(-1,1)]_B = (11,-12)$ above). We briefly explore other nice properties of orthonormal bases in Exercises 3.1.21–3.1.24, but we defer most of their discussion to [Joh20].

3.1.2   Change of Basis

In order to make matrix multiplication work with coordinate vectors, we have to also know how to represent matrices in different bases. Before we can get there though, we need to develop some additional machinery to help us convert between different coordinate systems.

Given a particular basis of a subspace, Theorem 3.1.1 tells us that coordinate vectors with respect to that basis are unique. However, if we change the basis, then the coordinate vectors change too. We have seen this (at least implicitly) a few times already, but it's worth working through an example to illustrate this fact.

Example 3.1.4

Computing a Coordinate Vector from Another Coordinate Vector

Let $B = \{(1,1),(1,-1)\}$ and $C = \{(3,1),(0,1)\}$ be bases of $\mathbb{R}^2$, and suppose that $[\mathbf{v}]_B = (5,1)$. Compute $[\mathbf{v}]_C$.

Solution:

Since $[\mathbf{v}]_B = (5,1)$ we know that $\mathbf{v} = 5(1,1) + (1,-1) = (6,4)$. Next, we simply represent $\mathbf{v} = (6,4)$ in the basis C by finding scalars c_1 and c_2 so that $(6,4) = c_1(3,1) + c_2(0,1)$:

$$\left[\begin{array}{cc|c} 3 & 0 & 6 \\ 1 & 1 & 4 \end{array}\right] \xrightarrow{\frac{1}{3}R_1} \left[\begin{array}{cc|c} 1 & 0 & 2 \\ 1 & 1 & 4 \end{array}\right] \xrightarrow{R_2 - R_1} \left[\begin{array}{cc|c} 1 & 0 & 2 \\ 0 & 1 & 2 \end{array}\right].$$

It follows that $c_1 = 2$ and $c_2 = 2$, so we conclude that $[\mathbf{v}]_C = (2,2)$, as illustrated below. Notice that even though $[\mathbf{v}]_B$ and $[\mathbf{v}]_C$ look quite different, the underlying vector $\mathbf{v}$ is the same in both cases:

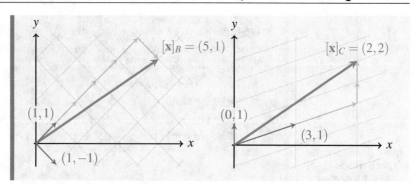

In the above example, in order to compute $[\mathbf{v}]_C$ from $[\mathbf{v}]_B$, we computed $\mathbf{v}$ itself as an intermediate step. While this was not a problem in this particular case, this is quite undesirable in general since $[\mathbf{v}]_B$ and $[\mathbf{v}]_C$ can have a much lower dimension than $\mathbf{v}$.

For example, if S is a 3-dimensional subspace of $\mathbb{R}^{85}$ then we do not want to have to decompress a 3-dimensional coordinate vector into an 85-dimensional vector as an intermediate step when changing basis—we would prefer a method of jumping directly from one 3-dimensional coordinate vector to another. The following definition introduces the object (a matrix) that lets us do exactly this.

Definition 3.1.2 **Change-of-Basis** **Matrix**	Let S be a subspace of $\mathbb{R}^n$ with bases $B = \{\mathbf{v}_1, \mathbf{v}_2, \ldots, \mathbf{v}_k\}$ and C. The **change-of-basis matrix** from B to C, denoted by $P_{C \leftarrow B}$, is the $k \times k$ matrix whose columns are the coordinate vectors $[\mathbf{v}_1]_C, [\mathbf{v}_2]_C, \ldots, [\mathbf{v}_k]_C$: $$P_{C \leftarrow B} \stackrel{\text{def}}{=} \big[\ [\mathbf{v}_1]_C \mid [\mathbf{v}_2]_C \mid \cdots \mid [\mathbf{v}_k]_C\ \big].$$

It might seem strange to use the notation $P_{C \leftarrow B}$ instead of $P_{B \to C}$, but we will soon see that it makes our life a lot easier this way.

As its name suggests, a change-of-basis matrix converts coordinate vectors with respect to one basis B into coordinate vectors with respect to another basis C. For example, if we return to the vectors and bases from Example 3.1.4, we can write the vectors in B in terms of the vectors in C as follows:

$$\mathbf{v}_1 = (1,1) = \frac{1}{3}(3,1) + \frac{2}{3}(0,1) \quad \text{and} \quad \mathbf{v}_2 = (1,-1) = \frac{1}{3}(3,1) - \frac{4}{3}(0,1).$$

It follows that

$$[\mathbf{v}_1]_C = (1,2)/3 \quad \text{and} \quad [\mathbf{v}_2]_C = (1,-4)/3, \quad \text{so} \quad P_{C \leftarrow B} = \frac{1}{3}\begin{bmatrix} 1 & 1 \\ 2 & -4 \end{bmatrix}.$$

With this change-of-basis matrix in hand, it is now trivial to use the fact that $[\mathbf{v}]_B = (5,1)$ to compute $[\mathbf{v}]_C$—we just multiply $[\mathbf{v}]_B$ by $P_{C \leftarrow B}$ to get

$$[\mathbf{v}]_C = P_{C \leftarrow B}[\mathbf{v}]_B = \frac{1}{3}\begin{bmatrix} 1 & 1 \\ 2 & -4 \end{bmatrix}\begin{bmatrix} 5 \\ 1 \end{bmatrix} = \begin{bmatrix} 2 \\ 2 \end{bmatrix}, \tag{3.1.1}$$

which agrees with the answer that we found in Example 3.1.4. The following theorem shows that this method of changing the basis of a coordinate vector always works.

Theorem 3.1.3

Change-of-Basis Matrices

Suppose B and C are bases of a subspace S of $\mathbb{R}^n$, and let $P_{C \leftarrow B}$ be the change-of-basis matrix from B to C. Then

   a) $P_{C \leftarrow B}[\mathbf{v}]_B = [\mathbf{v}]_C$ for all $\mathbf{v} \in S$, and
   b) $P_{C \leftarrow B}$ is invertible and $(P_{C \leftarrow B})^{-1} = P_{B \leftarrow C}$.

Furthermore, $P_{C \leftarrow B}$ is the unique matrix with property (a).

This theorem is why we use the notation $P_{C \leftarrow B}$ instead of $P_{B \to C}$: in part (a) of the theorem, the middle Bs "cancel out" and leave just the Cs behind.

Proof. Let $B = \{\mathbf{v}_1, \mathbf{v}_2, \ldots, \mathbf{v}_k\}$ so that we have names for the vectors in B. To see that property (a) holds, suppose $\mathbf{v} \in S$ and write $\mathbf{v} = c_1 \mathbf{v}_1 + c_2 \mathbf{v}_2 + \cdots + c_k \mathbf{v}_k$, so that $[\mathbf{v}]_B = (c_1, c_2, \ldots, c_k)$. We can then directly compute

$$P_{C \leftarrow B}[\mathbf{v}]_B = \left[\, [\mathbf{v}_1]_C \mid [\mathbf{v}_2]_C \mid \cdots \mid [\mathbf{v}_k]_C \,\right] \begin{bmatrix} c_1 \\ c_2 \\ \vdots \\ c_k \end{bmatrix} \qquad \text{(definition of } P_{C \leftarrow B})$$

$$= c_1 [\mathbf{v}_1]_C + c_2 [\mathbf{v}_2]_C + \cdots + c_k [\mathbf{v}_k]_C \qquad \text{(block matrix mult.)}$$

$$= [c_1 \mathbf{v}_1 + c_2 \mathbf{v}_2 + \cdots + c_k \mathbf{v}_k]_C \qquad \text{(by Theorem 3.1.2)}$$

$$= [\mathbf{v}]_C. \qquad \text{(since } [\mathbf{v}]_B = (c_1, c_2, \ldots, c_k))$$

This theorem can be thought of as another characterization of invertible matrices. That is, a matrix is invertible if and only if it is a change-of-basis matrix (see Exercise 3.1.13).

On the other hand, to see why $P_{C \leftarrow B}$ is the *unique* matrix with property (a), suppose $P \in \mathcal{M}_k$ is any matrix for which $P[\mathbf{v}]_B = [\mathbf{v}]_C$ for all $\mathbf{v} \in S$. For every $1 \leq j \leq k$, if $\mathbf{v} = \mathbf{v}_j$ then we see that $[\mathbf{v}]_B = [\mathbf{v}_j]_B = \mathbf{e}_j$ (the j-th standard basis vector), so $P[\mathbf{v}]_B = P\mathbf{e}_j$ is the j-th column of P. On the other hand, it is also the case that $P[\mathbf{v}]_B = [\mathbf{v}]_C = [\mathbf{v}_j]_C$. The j-th column of P thus equals $[\mathbf{v}_j]_C$ for each $1 \leq j \leq k$, so $P = P_{C \leftarrow B}$.

To see why property (b) holds, we now use property (a) twice to see that, for each $1 \leq j \leq k$, we have

$$(P_{B \leftarrow C} P_{C \leftarrow B})\mathbf{e}_j = P_{B \leftarrow C}(P_{C \leftarrow B}[\mathbf{v}_j]_B) = P_{B \leftarrow C}[\mathbf{v}_j]_C = [\mathbf{v}_j]_B = \mathbf{e}_j.$$

It follows that the j-th column of $P_{B \leftarrow C} P_{C \leftarrow B}$ equals $\mathbf{e}_j$ for each $1 \leq j \leq k$, so $P_{B \leftarrow C} P_{C \leftarrow B} = I$. Theorem 2.2.7 then tells us that $P_{C \leftarrow B}$ is invertible with $(P_{C \leftarrow B})^{-1} = P_{B \leftarrow C}$. ∎

One of the useful features of change-of-basis matrices is that they can be re-used to change the basis of multiple different vectors. For example, in Equation (3.1.1) we used the change-of-basis matrix $P_{C \leftarrow B}$ to change $[\mathbf{v}]_B = (5, 1)$ from the basis $B = \{(1, 1), (1, -1)\}$ to $C = \{(3, 1), (0, 1)\}$. If we also have $[\mathbf{w}]_B = (-1, 4)$ then we can re-use that same change-of-basis matrix to see that

$$[\mathbf{w}]_C = P_{C \leftarrow B}[\mathbf{w}]_B = \frac{1}{3}\begin{bmatrix} 1 & 1 \\ 2 & -4 \end{bmatrix}\begin{bmatrix} -1 \\ 4 \end{bmatrix} = \begin{bmatrix} 1 \\ -6 \end{bmatrix}.$$

Example 3.1.5

Constructing and Using a Change of Basis Matrix

Suppose

$$\mathbf{v}_1 = (2, 1, 0, 1, 2), \qquad \mathbf{v}_2 = (1, 2, -1, 1, 0), \qquad \mathbf{v}_3 = (0, 1, -3, 1, 1),$$
$$\mathbf{w}_1 = (1, -1, 1, 0, 2), \qquad \mathbf{w}_2 = (1, 1, 2, 0, -1), \qquad \mathbf{w}_3 = (2, 3, 1, 1, -1).$$

Then $B = \{\mathbf{v}_1, \mathbf{v}_2, \mathbf{v}_3\}$ and $C = \{\mathbf{w}_1, \mathbf{w}_2, \mathbf{w}_3\}$ are bases of $S = \text{span}(B)$.

It is also the case that $S = \text{span}(C)$.

   a) Compute $P_{C \leftarrow B}$.

   b) Compute $[\mathbf{v}]_C$ if $[\mathbf{v}]_B = (1, 2, 3)$.

c) Compute $[\mathbf{w}]_C$ if $[\mathbf{w}]_B = (2,0,-1)$.

Solutions:

a) We must first compute $[\mathbf{v}_1]_C$, $[\mathbf{v}_2]_C$, and $[\mathbf{v}_3]_C$ (i.e., we must write $\mathbf{v}_1$, $\mathbf{v}_2$, and $\mathbf{v}_3$ each as a linear combination of $\mathbf{w}_1$, $\mathbf{w}_2$, and $\mathbf{w}_3$). For $\mathbf{v}_1$, we can solve the linear system

> This linear system is $\mathbf{v}_1 = c_1\mathbf{w}_1 + c_2\mathbf{w}_2 + c_3\mathbf{w}_3$.

$$(2,1,0,1,2)$$
$$= c_1(1,-1,1,0,2) + c_2(1,1,2,0,-1) + c_3(2,3,1,1,-1)$$

as follows:

$$\left[\begin{array}{ccc|c} 1 & 1 & 2 & 2 \\ -1 & 1 & 3 & 1 \\ 1 & 2 & 1 & 0 \\ 0 & 0 & 1 & 1 \\ 2 & -1 & -1 & 2 \end{array}\right] \xrightarrow[\substack{R_2+R_1 \\ R_3-R_1 \\ R_5-2R_1}]{} \left[\begin{array}{ccc|c} 1 & 1 & 2 & 2 \\ 0 & 2 & 5 & 3 \\ 0 & 1 & -1 & -2 \\ 0 & 0 & 1 & 1 \\ 0 & -3 & -5 & -2 \end{array}\right]$$

> We could swap some rows around here to put this matrix into row echelon form, but it's easy enough to read off the solution as-is.

$$\xrightarrow[\substack{R_2-2R_3 \\ R_5+3R_3}]{} \left[\begin{array}{ccc|c} 1 & 1 & 2 & 2 \\ 0 & 0 & 7 & 7 \\ 0 & 1 & -1 & -2 \\ 0 & 0 & 1 & 1 \\ 0 & 0 & -8 & -8 \end{array}\right] \xrightarrow[\substack{R_2-7R_4 \\ R_5+8R_4}]{} \left[\begin{array}{ccc|c} 1 & 1 & 2 & 2 \\ 0 & 0 & 0 & 0 \\ 0 & 1 & -1 & -2 \\ 0 & 0 & 1 & 1 \\ 0 & 0 & 0 & 0 \end{array}\right].$$

We can now solve this linear system via back substitution and see that $c_1 = 1$, $c_2 = -1$, $c_3 = 1$, so $[\mathbf{v}_1]_C = (1,-1,1)$. Similar computations show that $[\mathbf{v}_2]_C = (0,-1,1)$ and $[\mathbf{v}_3]_C = (0,-2,1)$, so

$$P_{C\leftarrow B} = \begin{bmatrix} [\mathbf{v}_1]_C \mid [\mathbf{v}_2]_C \mid [\mathbf{v}_3]_C \end{bmatrix} = \begin{bmatrix} 1 & 0 & 0 \\ -1 & -1 & -2 \\ 1 & 1 & 1 \end{bmatrix}.$$

b) We just multiply:

$$[\mathbf{v}]_C = P_{C\leftarrow B}[\mathbf{v}]_B = \begin{bmatrix} 1 & 0 & 0 \\ -1 & -1 & -2 \\ 1 & 1 & 1 \end{bmatrix}\begin{bmatrix} 1 \\ 2 \\ 3 \end{bmatrix} = \begin{bmatrix} 1 \\ -9 \\ 6 \end{bmatrix}.$$

> Constructing $P_{C\leftarrow B}$ took some work, but converting between bases is trivial now that we have it.

c) Again, we just multiply:

$$[\mathbf{w}]_C = P_{C\leftarrow B}[\mathbf{w}]_B = \begin{bmatrix} 1 & 0 & 0 \\ -1 & -1 & -2 \\ 1 & 1 & 1 \end{bmatrix}\begin{bmatrix} 2 \\ 0 \\ -1 \end{bmatrix} = \begin{bmatrix} 2 \\ 0 \\ 1 \end{bmatrix}.$$

The schematic displayed in Figure 3.5 illustrates the relationship between coordinate vectors and change-of-basis matrices: the bases B and C provide two different ways of making the subspace $\mathcal{S}$ look like $\mathbb{R}^k$, and the change-of-basis matrices $P_{C\leftarrow B}$ and $P_{B\leftarrow C}$ are linear transformations that convert these two different representations into each other.

> **Remark 3.1.3**
>
> **Change-of-Basis Matrices and Standard Matrices**

It is worthwhile to compare the definition of a change-of-basis matrix $P_{C\leftarrow B}$ (Definition 3.1.2) to that of the standard matrix $[T]$ of a linear

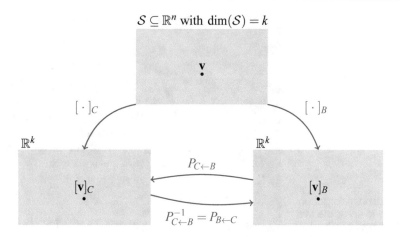

Figure 3.5: A visualization of the relationship between vectors, their coordinate vectors, and change-of-basis matrices. There are many different bases that let us think of $\mathcal{S}$ as $\mathbb{R}^k$, and change-of-basis matrices let us convert between them.

transformation $T : \mathbb{R}^k \to \mathbb{R}^k$ (Theorem 1.4.1):

$$P_{C \leftarrow B} = \left[\ [\mathbf{v}_1]_C \mid [\mathbf{v}_2]_C \mid \cdots \mid [\mathbf{v}_k]_C\ \right] \quad \text{and}$$
$$[T] = \left[\ T(\mathbf{e}_1) \mid T(\mathbf{e}_2) \mid \cdots \mid T(\mathbf{e}_k)\ \right].$$

The fact that these definitions look so similar (in both cases, we modify basis vectors in a linear way and stick them as columns into a matrix) is not a coincidence—$P_{C \leftarrow B}$ is exactly the standard matrix of the linear transformation that sends $[\mathbf{v}_j]_B$ to $[\mathbf{v}_j]_C$ for all $1 \leq j \leq k$.

This interpretation of $P_{C \leftarrow B}$ can be useful when trying to remember how to construct it (i.e., to keep track of which vectors, represented in which basis, make up its columns). Since $P_{C \leftarrow B}$ sends coordinate vectors represented in the B basis to the C basis, it should send $[\mathbf{v}_j]_B = \mathbf{e}_j$ to $[\mathbf{v}_j]_C$ (i.e., its j-th column should be $[\mathbf{v}_j]_C$).

We already saw how to construct change-of-basis matrices in Example 3.1.5, but there is one special case where the computation simplifies considerably—when the subspace $\mathcal{S}$ is all of $\mathbb{R}^n$ and one of the bases is simply the standard basis. Since this case will be particularly important for us in upcoming sections, it is worth working through an example.

Example 3.1.6

Change-of-Basis Matrices to and from the Standard Basis

Let $B = \{\mathbf{v}_1, \mathbf{v}_2, \mathbf{v}_3\}$ and $E = \{\mathbf{e}_1, \mathbf{e}_2, \mathbf{e}_3\}$ be bases of $\mathbb{R}^3$, where

$$\mathbf{v}_1 = (1, 1, -1), \quad \mathbf{v}_2 = (2, 2, -1), \quad \text{and} \quad \mathbf{v}_3 = (-1, -2, 1).$$

Compute $P_{E \leftarrow B}$ and $P_{B \leftarrow E}$.

Solution:

To compute $P_{E \leftarrow B}$, we must first write the vectors from the "old" basis B in the "new" basis E. However, since E is the standard basis, this has already been done for us (i.e., recall that $[\mathbf{v}]_E = \mathbf{v}$ for all vectors $\mathbf{v}$):

$$[\mathbf{v}_1]_E = (1, 1, -1), \quad [\mathbf{v}_2]_E = (2, 2, -1), \quad \text{and} \quad [\mathbf{v}_3]_E = (-1, -2, 1).$$

Then $P_{E \leftarrow B}$ is simply the matrix that has these vectors as its columns:

$$P_{E \leftarrow B} = \big[\, [\mathbf{v}_1]_E \mid [\mathbf{v}_2]_E \mid [\mathbf{v}_3]_E \,\big] = \begin{bmatrix} 1 & 2 & -1 \\ 1 & 2 & -2 \\ -1 & -1 & 1 \end{bmatrix}.$$

To construct $P_{B \leftarrow E}$, we could compute each of $[\mathbf{e}_1]_B$, $[\mathbf{e}_2]_B$, and $[\mathbf{e}_3]_B$ and then place them as columns in a matrix, but this would require us to solve three linear systems (one for each of those vectors). An easier way to get our hands on $P_{B \leftarrow E}$ is to recall from Theorem 3.1.3(b) that $P_{B \leftarrow E} = P_{E \leftarrow B}^{-1}$, so we can just invert the change of basis matrix that we computed above:

If you have forgotten
how to compute the
inverse of a matrix,
refer back to
Section 2.2.3.

$$P_{B \leftarrow E} = P_{E \leftarrow B}^{-1} = \begin{bmatrix} 1 & 2 & -1 \\ 1 & 2 & -2 \\ -1 & -1 & 1 \end{bmatrix}^{-1} = \begin{bmatrix} 0 & -1 & -2 \\ 1 & 0 & 1 \\ 1 & -1 & 0 \end{bmatrix}.$$

There are some similar tricks that can be used to simplify the computation of change-of-basis matrices when neither of the bases are the standard basis, but they are somewhat less important so we defer them to Exercise 3.1.14.

3.1.3 Similarity and Representations of Linear Transformations

We now do for matrices and linear transformations what we did for vectors in Section 3.1.1: we give them coordinates so that we can describe how they act on vectors that are represented in different bases. Our starting point is the following definition/theorem that tells us how to represent a linear transformation in this way:

Theorem 3.1.4

Standard Matrix of a Linear Transformation with Respect to a Basis

Suppose $B = \{\mathbf{v}_1, \mathbf{v}_2, \ldots, \mathbf{v}_n\}$ is a basis of $\mathbb{R}^n$ and $T : \mathbb{R}^n \to \mathbb{R}^n$ is a linear transformation. Then there exists a unique matrix $[T]_B \in \mathcal{M}_n$ for which

$$[T(\mathbf{v})]_B = [T]_B [\mathbf{v}]_B \quad \text{for all} \quad \mathbf{v} \in \mathbb{R}^n.$$

This matrix is called the **standard matrix** of T with respect to B, and it is

$$[T]_B \overset{\text{def}}{=} \big[\, [T(\mathbf{v}_1)]_B \mid [T(\mathbf{v}_2)]_B \mid \cdots \mid [T(\mathbf{v}_n)]_B \,\big].$$

This theorem can be
extended to the
case when
$T : \mathbb{R}^n \to \mathbb{R}^m$ (with
$m \neq n$), even with
different bases on
the input and output
spaces, but the
version given here is
enough for us.

In other words, this theorem tells us that instead of applying T to $\mathbf{v}$ and then computing the coordinate vector of the output, we can equivalently compute the coordinate vector of the input and then multiply by the matrix $[T]_B$ (see Figure 3.6). This fact is a direct generalization of Theorem 1.4.1, which is what we get if we use the standard basis $B = \{\mathbf{e}_1, \mathbf{e}_2, \ldots, \mathbf{e}_n\}$.

Since the proof of the above theorem is almost identical to that of Theorem 1.4.1 (just replace the standard basis vectors $\mathbf{e}_1, \mathbf{e}_2, \ldots, \mathbf{e}_n$ by $\mathbf{v}_1, \mathbf{v}_2, \ldots, \mathbf{v}_n$ throughout the proof), we leave it to Exercise 3.1.16. In fact, almost all of the basic properties of standard matrices that we explored in Section 1.4 carry over straightforwardly to this slightly more general setting. For example, if $S, T : \mathbb{R}^n \to \mathbb{R}^n$ are linear transformations then $[S \circ T]_B = [S]_B [T]_B$, which is the natural generalization of Theorem 1.4.2. We leave proofs of facts like these to the exercises as well, and we instead jump right into an example.

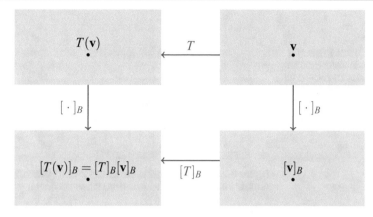

Read this figure as starting at the top-right corner and moving to the bottom-left.

Figure 3.6: A visualization of the relationship between linear transformations and their standard matrices. Just like T sends $\mathbf{v}$ to $T(\mathbf{v})$, the standard matrix $[T]_B$ sends $[\mathbf{v}]_B$ to $[T(\mathbf{v})]_B$. In other words, Theorem 3.1.4 says that we can first compute $T(\mathbf{v})$ and then find its coordinate vector (the top-left path in this figure), or we can compute the coordinate vector $[\mathbf{v}]_B$ and then multiply by $[T]_B$ (the bottom-right path in this figure)—the answer we get will be the same either way.

Example 3.1.7

Representing a Linear Transformation in a Strange Basis

Let $T : \mathbb{R}^2 \to \mathbb{R}^2$ be defined by $T(x,y) = (2x+y, x+2y)$. Compute $[T]_B$ for each of the following bases:

a) $B = \{\mathbf{e}_1, \mathbf{e}_2\}$
b) $B = \{(1,1),(1,-1)\}$

Solutions:

a) To construct $[T]_B$ when B is the standard basis, we just do what we have been doing ever since Section 1.4: we compute $T(\mathbf{e}_1)$ and $T(\mathbf{e}_2)$ and place them as columns in a matrix:

$$[T]_B = \big[\, [T(\mathbf{e}_1)]_B \mid [T(\mathbf{e}_2)]_B \,\big] = \begin{bmatrix} 2 & 1 \\ 1 & 2 \end{bmatrix}.$$

b) First, we need to compute $T(1,1)$ and $T(1,-1)$:

$$T(1,1) = (3,3) \quad \text{and} \quad T(1,-1) = (1,-1).$$

Next, we must represent these vectors in the basis B:

$$(3,3) = 3(1,1)+0(1,-1) \quad \text{so} \quad [(3,3)]_B = (3,0), \quad \text{and}$$
$$(1,-1) = 0(1,1)+1(1,-1) \quad \text{so} \quad [(1,-1)]_B = (0,1).$$

Finally, we place these coordinate vectors as columns into a matrix:

As we see here, changing the basis B can drastically change what $[T]_B$ looks like.

$$[T]_B = \big[\, [T(1,1)]_B \mid [T(1,-1)]_B \,\big] = \begin{bmatrix} 3 & 0 \\ 0 & 1 \end{bmatrix}.$$

In part (a) of the previous example, the standard matrix of the linear transformation T (with respect to the standard basis) did not seem to have any particularly nice structure. The reason for this is that T just does not do anything particularly interesting to the standard basis vectors, as illustrated in Figure 3.7.

On the other hand, when we changed to a different basis in part (b) we

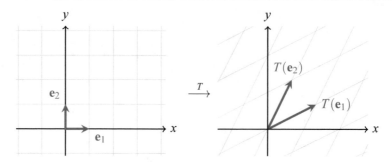

This is the same
linear transformation
that was depicted in
Figure 3.1.

Figure 3.7: The linear transformation T from Example 3.1.7 does not do anything terribly interesting to the standard basis vectors.

saw that the standard matrix of this linear transformation became diagonal. Geometrically, this corresponds to the fact that T just stretches, but does not rotate, the vectors in this new basis (see Figure 3.8). In other words, if we skew the input and output spaces so that the members of this basis point along the x- and y-axes, then T just stretches along those axes (like a diagonal matrix).

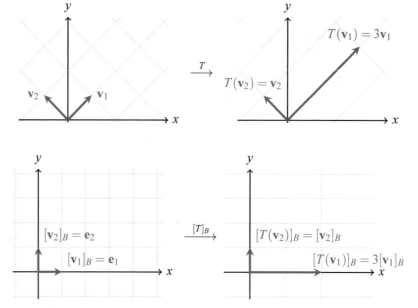

Compare this image
with Figure 3.4. Just
like in that figure, we
are distorting space
so that the basis
vectors point along
the x- and y-axes.

Figure 3.8: The linear transformation T from Example 3.1.7 looks diagonal if we view it in the basis $B = \{(1,1),(1,-1)\}$.

The above example illustrates why it is sometimes useful to work with bases other than the standard basis—if we represent a linear transformation with respect to a certain basis, it might be much easier to work with than if we use the standard basis (since, for example, diagonal matrices are much easier to work with than general matrices).

Just like we sometimes want to convert a vector from being represented in one basis to another, we often want to do the same with linear transformations (for example, if we have already represented a linear transformation as a matrix in the standard basis, we would like to have a direct way of changing it into another basis). Fortunately, we already did most of the hard that goes into solving this problem when we introduced change-of-basis matrices, so we can just stitch things together to make them work in this setting.

Theorem 3.1.5

Change-of-Basis for Linear Transformations

Let B and C be bases of $\mathbb{R}^n$ and let $T : \mathbb{R}^n \to \mathbb{R}^n$ be a linear transformation. Then

$$[T]_C = P_{C \leftarrow B}[T]_B P_{B \leftarrow C},$$

where $P_{C \leftarrow B}$ and $P_{B \leftarrow C}$ are change-of-basis matrices.

As before, notice that adjacent subscripts in this theorem match (e.g., the three Bs that are next to each other).

Proof. We simply multiply the matrix $P_{C \leftarrow B}[T]_B P_{B \leftarrow C}$ on the right by an arbitrary coordinate vector $[\mathbf{v}]_C$:

$$\begin{aligned}
P_{C \leftarrow B}[T]_B P_{B \leftarrow C}[\mathbf{v}]_C &= P_{C \leftarrow B}[T]_B[\mathbf{v}]_B && (P_{B \leftarrow C}[\mathbf{v}]_C = [\mathbf{v}]_B \text{ by Theorem 3.1.3(a))} \\
&= P_{C \leftarrow B}[T(\mathbf{v})]_B && ([T]_B[\mathbf{v}]_B = [T(\mathbf{v})]_B \text{ by Theorem 3.1.4)} \\
&= [T(\mathbf{v})]_C. && (\text{by Theorem 3.1.3(a) again})
\end{aligned}$$

However, we know from Theorem 3.1.4 that $[T]_C$ is the *unique* matrix for which $[T]_C[\mathbf{v}]_C = [T(\mathbf{v})]_C$ for all $\mathbf{v} \in \mathbb{R}^n$, so it follows that $P_{C \leftarrow B}[T]_B P_{B \leftarrow C} = [T]_C$, as claimed. ∎

A schematic that illustrates the statement of the above theorem, as well as all of the other relationships between standard matrices and change-of-basis matrices that we have seen, is provided by Figure 3.9. All it says is that there are numerous different ways of computing $[T(\mathbf{v})]_C$ from $\mathbf{v}$, such as:

- We could apply T and then represent $T(\mathbf{v})$ in the basis C.
- We could represent $\mathbf{v}$ in the basis C and then multiply by $[T]_C$.
- We could represent $\mathbf{v}$ in the basis B, multiply by $[T]_B$, and then multiply by $P_{C \leftarrow B}$.

Again, read this figure as starting at the top-right corner and moving to the bottom-left.

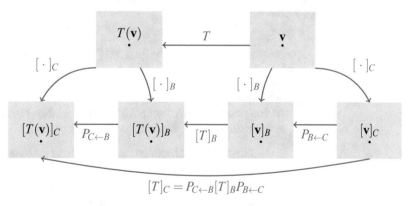

Figure 3.9: A visualization of the relationship between linear transformations, standard matrices, change-of-basis matrices, and coordinate vectors. In particular, the bottom row illustrates Theorem 3.1.5, which says that we can construct $[T]_C$ from $[T]_B$ by multiplying on the right and left by appropriate change-of-basis matrices. This image is basically just a combination of Figures 3.5 and 3.6.

Example 3.1.8

Changing the Basis of a Linear Transformation

Let $B = \{(1,2,3), (2,-1,0), (-1,1,2)\}$ be a basis of $\mathbb{R}^3$ and let $T : \mathbb{R}^3 \to \mathbb{R}^3$ be the linear transformation with standard matrix (with respect to the standard basis) equal to

$$[T] = \begin{bmatrix} 2 & 1 & -3 \\ 1 & 0 & 1 \\ 0 & 1 & 2 \end{bmatrix}.$$

Compute $[T]_B$.

Solution:

If we let $E = \{\mathbf{e}_1, \mathbf{e}_2, \mathbf{e}_3\}$ denote the standard basis of $\mathbb{R}^3$ then Theorem 3.1.5 tells us that $[T]_B = P_{B \leftarrow E}[T]P_{E \leftarrow B}$. Well, we recall from Example 3.1.6 that $P_{E \leftarrow B}$ is the matrix whose columns are the vectors from B:

> We could have also written $[T]$ as $[T]_E$.

$$P_{E \leftarrow B} = \begin{bmatrix} 1 & 2 & -1 \\ 2 & -1 & 1 \\ 3 & 0 & 2 \end{bmatrix}, \quad \text{so} \quad P_{B \leftarrow E} = P_{E \leftarrow B}^{-1} = \frac{1}{7} \begin{bmatrix} 2 & 4 & -1 \\ 1 & -5 & 3 \\ -3 & -6 & 5 \end{bmatrix}.$$

It follows that

$$[T]_B = P_{B \leftarrow E}[T]P_{E \leftarrow B}$$

$$= \frac{1}{7} \begin{bmatrix} 2 & 4 & -1 \\ 1 & -5 & 3 \\ -3 & -6 & 5 \end{bmatrix} \begin{bmatrix} 2 & 1 & -3 \\ 1 & 0 & 1 \\ 0 & 1 & 2 \end{bmatrix} \begin{bmatrix} 1 & 2 & -1 \\ 2 & -1 & 1 \\ 3 & 0 & 2 \end{bmatrix}$$

> Most of the examples that we work through are carefully constructed to avoid ugly fractions, but it's good to be reminded that they exist every now and then.

$$= \frac{1}{7} \begin{bmatrix} 2 & 4 & -1 \\ 1 & -5 & 3 \\ -3 & -6 & 5 \end{bmatrix} \begin{bmatrix} -5 & 3 & -7 \\ 4 & 2 & 1 \\ 8 & -1 & 5 \end{bmatrix}$$

$$= \frac{1}{7} \begin{bmatrix} -2 & 15 & -15 \\ -1 & -10 & 3 \\ 31 & -26 & 40 \end{bmatrix}.$$

Similarity

At this point, we have two fairly natural questions in front of us, which serve as the motivation for the remainder of this chapter:

1) Given a linear transformation, how can we find a basis C such that $[T]_C$ is as simple as possible (e.g., diagonal)?

2) If we are given two square matrices A and B, how can we determine whether or not they represent the same linear transformation? That is, how can we determine whether or not there exist bases C and D and a *common* linear transformation T such that $A = [T]_C$ and $B = [T]_D$?

We need some more machinery before we can tackle question (1), so we return to it in Section 3.4. For now, we consider question (2). By Theorem 3.1.5, we know that $A = [T]_C$ and $B = [T]_D$ if and only if $B = P_{D \leftarrow C}AP_{C \leftarrow D}$. However, we also know from Theorem 3.1.3(b) that $P_{C \leftarrow D} = P_{D \leftarrow C}^{-1}$, so we have

$$B = P_{D \leftarrow C}AP_{D \leftarrow C}^{-1}.$$

Finally, Exercise 3.1.13 tells us that every invertible matrix is a change-of-basis matrix, which means that the answer to question (2) is that matrices $A, B \in \mathcal{M}_n$

is an invertible matrix $P \in \mathcal{M}_n$ such that $B = PAP^{-1}$. This property is important enough that we give it a name:

Definition 3.1.3
Similarity

We say that two matrices $A, B \in \mathcal{M}_n$ are **similar** if there exists an invertible $P \in \mathcal{M}_n$ such that $A = PBP^{-1}$.

We sometimes say that A is **similar to** B if $A = PBP^{-1}$. However, if A is similar to B then B is also similar to A, since $B = P^{-1}AP$.

For example, the matrices

$$A = \begin{bmatrix} 1 & 2 \\ 3 & 4 \end{bmatrix} \quad \text{and} \quad B = \begin{bmatrix} 5 & -1 \\ -2 & 0 \end{bmatrix} \tag{3.1.2}$$

are similar, since straightforward calculation shows that if $P = \begin{bmatrix} 1 & 1 \\ 1 & -1 \end{bmatrix}$ then

$$PBP^{-1} = \begin{bmatrix} 1 & 1 \\ 1 & -1 \end{bmatrix} \begin{bmatrix} 5 & -1 \\ -2 & 0 \end{bmatrix} \begin{bmatrix} 1/2 & 1/2 \\ 1/2 & -1/2 \end{bmatrix}$$

$$= \begin{bmatrix} 1 & 1 \\ 1 & -1 \end{bmatrix} \begin{bmatrix} 2 & 3 \\ -1 & -1 \end{bmatrix} = \begin{bmatrix} 1 & 2 \\ 3 & 4 \end{bmatrix} = A.$$

We do not yet have the tools required to actually *find* the matrix P when it exists and thus show that two matrices are similar, but there are some simple things that we can do to (sometimes) show that they are *not* similar. For example, we showed in Exercise 2.4.17 that multiplying A on the left or right by an invertible matrix does not change its rank. It follows that if A and B are similar (i.e., there exists an invertible matrix P such that $A = PBP^{-1}$), then

$$\text{rank}(A) = \text{rank}(PBP^{-1}) = \text{rank}(PB) = \text{rank}(B).$$

For example, the two matrices from Equation (3.1.2) are similar so they must have the same rank, and indeed they both have rank 2. We state this general observation a bit more prominently for emphasis:

Since the rank is shared by similar matrices, we say that it is **similarity invariant**.

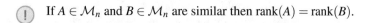 If $A \in \mathcal{M}_n$ and $B \in \mathcal{M}_n$ are similar then $\text{rank}(A) = \text{rank}(B)$.

Intuitively, the reason for this fact is that the rank is actually a property of linear transformations, not of matrices. Indeed, the rank of a matrix is the dimension of its range, which does not depend on the particular basis used to represent the underlying linear transformation (e.g., a plane is still a plane no matter which basis we view it in—no basis turns it into a line).

Example 3.1.9

Using the Rank to Show Matrices are Not Similar

Show that the matrices $A = \begin{bmatrix} 1 & 2 \\ 2 & 4 \end{bmatrix}$ and $B = \begin{bmatrix} 1 & 2 \\ 3 & 4 \end{bmatrix}$ are not similar.

Solution:
We start by using row operations to find row echelon forms of these matrices:

$$\begin{bmatrix} 1 & 2 \\ 2 & 4 \end{bmatrix} \xrightarrow{R_2 - 2R_1} \begin{bmatrix} 1 & 2 \\ 0 & 0 \end{bmatrix} \quad \text{and} \quad \begin{bmatrix} 1 & 2 \\ 3 & 4 \end{bmatrix} \xrightarrow{R_2 - 3R_1} \begin{bmatrix} 1 & 2 \\ 0 & -2 \end{bmatrix}.$$

It follows that $\text{rank}(A) = 1$ and $\text{rank}(B) = 2$, so they are not similar.

The Trace

There are several other properties of matrices that, like the rank, are shared by similar matrices. We now introduce another such property that has the nice feature of being much easier to compute than the rank.

| Definition 3.1.4 | Suppose $A \in \mathcal{M}_n$ is a square matrix. Then the **trace** of A, denoted by $\text{tr}(A)$, is the sum of its diagonal entries: |
| The Trace | |

$$\text{tr}(A) \overset{\text{def}}{=} a_{1,1} + a_{2,2} + \cdots + a_{n,n}.$$

For example, the two matrices from Equation (3.1.2) both have trace 5. We now work through another example to make sure that we understand this definition properly.

Example 3.1.10	Compute the trace of each of the following matrices:
Computing	
the Trace	

a) $\begin{bmatrix} 2 & 2 \\ 4 & 5 \end{bmatrix}$
b) $\begin{bmatrix} 3 & 6 & -3 \\ 0 & -2 & 3 \\ 2 & 4 & -1 \end{bmatrix}$

Solutions:

Both of these traces are straightforward to compute from the definition: we just add up the diagonal entries of the matrices.

a) $\text{tr}\left(\begin{bmatrix} 2 & 2 \\ 4 & 5 \end{bmatrix} \right) = 2 + 5 = 7.$

b) $\text{tr}\left(\begin{bmatrix} 3 & 6 & -3 \\ 0 & -2 & 3 \\ 2 & 4 & -1 \end{bmatrix} \right) = 3 - 2 - 1 = 0.$

At first glance, the trace perhaps seems like a rather arbitrary function—why should we care about the sum of the diagonal entries of a matrix, and why would this quantity be shared among similar matrices? The answer is that, roughly speaking, the trace is the "nicest" linear function on matrices. In particular, the trace satisfies the following desirable properties:

Theorem 3.1.6	Suppose A and B are matrices whose sizes are such that the following operations make sense, and let c be a scalar. Then
Properties of	
the Trace	a) $\text{tr}(A + B) = \text{tr}(A) + \text{tr}(B)$
	b) $\text{tr}(cA) = c\,\text{tr}(A)$
	c) $\text{tr}(AB) = \text{tr}(BA)$

In fact, the trace and its scalar multiples are the *only* functions that satisfy these properties.

Proof. Properties (a) and (b) are hopefully clear enough, so we leave them to Exercise 3.1.19.

The surprising property in this theorem is property (c). To verify that it is true, suppose $A \in \mathcal{M}_{m,n}$ and $B \in \mathcal{M}_{n,m}$, and use the definition of matrix multiplication to compute the diagonal entries of AB and BA:

Recall that the notation $[AB]_{i,i}$ means the (i,i)-entry of AB.

$$[AB]_{i,i} = \sum_{j=1}^{n} a_{i,j} b_{j,i} \quad \text{for all} \quad 1 \leq i \leq m, \quad \text{and}$$

$$[BA]_{j,j} = \sum_{i=1}^{m} b_{j,i} a_{i,j} \quad \text{for all} \quad 1 \leq j \leq n.$$

While these diagonal entries individually may be different, when we add them up we get the same quantity:

$$\operatorname{tr}(AB) = \sum_{i=1}^{m} [AB]_{i,i} = \sum_{i=1}^{m} \sum_{j=1}^{n} a_{i,j} b_{j,i}$$

$$= \sum_{j=1}^{n} \sum_{i=1}^{m} b_{j,i} a_{i,j} = \sum_{j=1}^{n} [BA]_{j,j} = \operatorname{tr}(BA),$$

where the central step is accomplished by noting that $a_{i,j} b_{j,i} = b_{j,i} a_{i,j}$ and the fact that we can swap the order of summation. ■

Example 3.1.11
Trace of a Product

Suppose $A = \begin{bmatrix} 1 & 2 \\ 3 & 4 \end{bmatrix}$ and $B = \begin{bmatrix} 1 & -1 \\ -2 & 1 \end{bmatrix}$. Compute $\operatorname{tr}(AB)$ and $\operatorname{tr}(BA)$.

Solution:

We compute both of these quantities directly by performing the indicated matrix multiplications and then adding up the diagonal entries of the resulting matrices:

$$\operatorname{tr}(AB) = \operatorname{tr}\left(\begin{bmatrix} 1 & 2 \\ 3 & 4 \end{bmatrix} \begin{bmatrix} 1 & -1 \\ -2 & 1 \end{bmatrix}\right) = \operatorname{tr}\left(\begin{bmatrix} -3 & 1 \\ -5 & 1 \end{bmatrix}\right) = -2 \quad \text{and}$$

$$\operatorname{tr}(BA) = \operatorname{tr}\left(\begin{bmatrix} 1 & -1 \\ -2 & 1 \end{bmatrix} \begin{bmatrix} 1 & 2 \\ 3 & 4 \end{bmatrix}\right) = \operatorname{tr}\left(\begin{bmatrix} -2 & -2 \\ 1 & 0 \end{bmatrix}\right) = -2,$$

which agrees with the fact (Theorem 3.1.6(c)) that $\operatorname{tr}(AB) = \operatorname{tr}(BA)$.

The fact that $\operatorname{tr}(AB) = \operatorname{tr}(BA)$ is what makes the trace so useful for us—even though matrix multiplication is not commutative, the trace lets us treat it is if it were commutative in some situations. For example, it tells us that if A and B are similar (i.e., there exists an invertible matrix P such that $A = PBP^{-1}$) then

$$\operatorname{tr}(A) = \operatorname{tr}(PBP^{-1}) = \operatorname{tr}\left(P(BP^{-1})\right) = \operatorname{tr}\left((BP^{-1})P\right) = \operatorname{tr}\left(B(P^{-1}P)\right) = \operatorname{tr}(B).$$

$\uparrow$

by Theorem 3.1.6(c)

As with the analogous fact about the rank, we state this observation a bit more prominently for emphasis:

> (!) If $A \in \mathcal{M}_n$ and $B \in \mathcal{M}_n$ are similar then $\operatorname{tr}(A) = \operatorname{tr}(B)$.

Example 3.1.12

Using the Trace to Show Matrices are Not Similar

Show that the matrices $A = \begin{bmatrix} 1 & 2 \\ 3 & 4 \end{bmatrix}$ and $B = \begin{bmatrix} 1 & 2 \\ 3 & 5 \end{bmatrix}$ are not similar.

Solution:
The rank does not help us here, since $\text{rank}(A) = \text{rank}(B) = 2$. On the other hand, $\text{tr}(A) = 5$ and $\text{tr}(B) = 6$, so we know that A and B are not similar.

Although it is true that $\text{tr}(AB) = \text{tr}(BA)$, we need to be careful when taking the trace of the product of three or more matrices. By grouping matrix products together, we can see that

$$\text{tr}(ABC) = \text{tr}\big(A(BC)\big) = \text{tr}\big((BC)A\big) = \text{tr}(BCA),$$

and similarly that $\text{tr}(ABC) = \text{tr}(CAB)$. However, the other similar-looking equalities like $\text{tr}(ABC) = \text{tr}(ACB)$ do *not* hold in general. For example, if

$$A = \begin{bmatrix} 1 & 0 \\ 0 & 0 \end{bmatrix}, \quad B = \begin{bmatrix} 0 & 1 \\ 0 & 0 \end{bmatrix}, \quad \text{and} \quad C = \begin{bmatrix} 0 & 0 \\ 1 & 0 \end{bmatrix},$$

then direct calculation reveals that $\text{tr}(ABC) = 1$, but $\text{tr}(ACB) = 0$.

In general, all that we can say about the trace when it acts on the product of many matrices is that it is "cyclically commutative": it remains unchanged when the first matrix in the product wraps around to become the last matrix in the product, as long as the other matrices stay in the same order. For example, if $A, B, C, D \in \mathcal{M}_n$ then

$$\text{tr}(ABCD) = \text{tr}(BCDA) = \text{tr}(CDAB) = \text{tr}(DABC).$$

However, $\text{tr}(ABCD)$ does not necessarily equal the trace of any of the other 20 possible products of A, B, C, and D.

Exercises

solutions to starred exercises on page 461

3.1.1 Compute the coordinate vector $[\mathbf{v}]_B$ of the indicated vector $\mathbf{v}$ in the subspace $\mathcal{S}$ with respect to the basis B. You do not need to prove that B is indeed a basis.

* *(a) $\mathbf{v} = (6,1)$, $B = \{(3,0),(0,2)\}$, $\mathcal{S} = \mathbb{R}^2$
  (b) $\mathbf{v} = (4,2)$, $B = \{(1,1),(1,-1)\}$, $\mathcal{S} = \mathbb{R}^2$
* *(c) $\mathbf{v} = (2,3,1)$, $B = \{(1,0,-1),(1,2,1)\}$, $\mathcal{S}$ is the plane in $\mathbb{R}^3$ with equation $x - y + z = 0$.
  (d) $\mathbf{v} = (1,1,1)$, $B = \{(1,2,3),(-1,0,1),(2,2,1)\}$, $\mathcal{S} = \mathbb{R}^3$.

3.1.2 Suppose $\mathbf{v} = (2,3)$ and $\mathbf{w} = (2,-1)$.

* *(a) Find a basis B of $\mathbb{R}^2$ such that $[\mathbf{v}]_B = (1,0)$ and $[\mathbf{w}]_B = (0,1)$.
  (b) Find a basis B of $\mathbb{R}^2$ such that $[\mathbf{v}]_B = (3,-1)$ and $[\mathbf{w}]_B = (-1,2)$.

💻 **3.1.3** Use computer software to show that if

$$B = \big\{(3,1,4,-2,1),(2,3,-1,1,2),(-1,2,-2,4,-2)\big\}$$

and $\mathbf{v} = (1,5,6,8,-9)$, then $\mathbf{v} \in \text{span}(B)$, and compute $[\mathbf{v}]_B$.

3.1.4 Compute the change-of-basis matrix $P_{C \leftarrow B}$ for each of the following pairs of bases B and C.

* *(a) $B = \{(1,2),(3,4)\}$, $C = \{(1,0),(0,1)\}$
  (b) $B = \{(1,0),(0,1)\}$, $C = \{(2,1),(-4,3)\}$
* *(c) $B = \{(1,2),(3,4)\}$, $C = \{(2,1),(-4,3)\}$
  (d) $B = \{(4,0,0),(0,4,0),(1,4,1)\}$, $C = \{(1,2,3),(1,0,1),(-1,1,2)\}$

3.1.5 Find the standard matrix $[T]_B$ of the following linear transformations with respect to the basis $B = \{(2,3),(1,-1)\}$:

* *(a) $T(v_1,v_2) = (v_1 - v_2, v_1 + 3v_2)$.
  (b) $T(v_1,v_2) = (v_1 + 2v_2, v_1 + 2v_2)$.
* *(c) $T(v_1,v_2) = (-v_1 + 2v_2, 3v_1)$.
  (d) $T(v_1,v_2) = (4v_1 + 3v_2, 5v_1 + v_2)$.

☐ **3.1.6** Use computer software to find the standard matrix $[T]_B$ of the following linear transformations with respect to the basis $B = \{(3,1,2),(0,-1,-1),(2,-1,0)\}$:

*(a) $T(v_1,v_2,v_3) =$
$(2v_2 + 3v_3, 3v_1 + v_2 + 3v_3, 3v_1 + 2v_2 + 2v_3)$.

(b) $T(v_1,v_2,v_3) =$
$(4v_1 + v_2 - v_3, 2v_1 - v_2 - v_3, 2v_1 - v_2 + v_3)$.

3.1.7 Show that the following pairs of matrices A and B are not similar.

*(a) $A = \begin{bmatrix} 1 & -1 \\ -1 & 1 \end{bmatrix}$ and $B = \begin{bmatrix} 2 & -1 \\ -1 & 0 \end{bmatrix}$

(b) $A = \begin{bmatrix} 1 & -1 \\ -1 & 1 \end{bmatrix}$ and $B = \begin{bmatrix} 1 & -2 \\ -2 & 4 \end{bmatrix}$

*(c) $A = \begin{bmatrix} 0 & 1 & 2 \\ 1 & 2 & 3 \\ 2 & 3 & 4 \end{bmatrix}$ and $B = \begin{bmatrix} 1 & -1 & 0 \\ 3 & 1 & 2 \\ 1 & 1 & 1 \end{bmatrix}$

(d) $A = \begin{bmatrix} 2 & 0 & 1 \\ 1 & 1 & 3 \\ -1 & 2 & 1 \end{bmatrix}$ and $B = \begin{bmatrix} 1 & -1 & 1 \\ -1 & 1 & 1 \\ 0 & 0 & 2 \end{bmatrix}$

3.1.8 Determine whether or not the following pairs of matrices A and B are similar.

[Hint: Neither the rank nor the trace will help you here. Work directly from the definition of similarity.]

*(a) $A = \begin{bmatrix} 1 & 0 \\ 0 & 2 \end{bmatrix}$ and $B = \begin{bmatrix} 3 & -2 \\ 1 & 0 \end{bmatrix}$

(b) $A = \begin{bmatrix} 1 & 2 \\ 0 & 3 \end{bmatrix}$ and $B = \begin{bmatrix} 2 & 0 \\ 1 & 2 \end{bmatrix}$

3.1.9 Determine which of the following statements are true and which are false.

*(a) If B is a basis of a subspace S of $\mathbb{R}^n$ then $P_{B\leftarrow B} = I$.

(b) Every square matrix is similar to itself.

*(c) The identity matrix I is only similar to itself.

(d) If A and B are similar matrices then $\text{rank}(A) = \text{rank}(B)$.

*(e) If $\text{rank}(A) = \text{rank}(B)$ and $\text{tr}(A) = \text{tr}(B)$ then A and B are similar.

(f) If A and B are square matrices of the same size then $\text{tr}(A+B) = \text{tr}(A) + \text{tr}(B)$.

*(g) If A and B are square matrices of the same size then $\text{tr}(AB) = \text{tr}(A)\text{tr}(B)$.

(h) If A is a square matrix then $\text{tr}(A^T) = \text{tr}(A)$.

∗∗3.1.10 Show that the converse of Theorem 3.1.1 holds. That is, show that if S is a subspace of $\mathbb{R}^n$ and $B \subseteq S$ is a finite set with the property that every vector $\mathbf{v} \in S$ can be written as a linear combination of the members of B in exactly one way, then B must be a basis of S.

∗∗3.1.11 In this exercise, we prove Theorem 3.1.2. Suppose that S is a subspace of $\mathbb{R}^n$ with basis B, and $\mathbf{v}, \mathbf{w} \in S$ and $c \in \mathbb{R}$.

(a) Show that $[\mathbf{v} + \mathbf{w}]_B = [\mathbf{v}]_B + [\mathbf{w}]_B$.

(b) Show that $[c\mathbf{v}]_B = c[\mathbf{v}]_B$.

∗∗3.1.12 Let S be a k-dimensional subspace of $\mathbb{R}^n$ with basis B. Let $C = \{\mathbf{w}_1, \ldots, \mathbf{w}_m\} \subset S$ be a set of vectors and let $D = \{[\mathbf{w}_1]_B, \ldots, [\mathbf{w}_m]_B\}$ be the set of corresponding coordinate vectors.

(a) Show that C is linearly independent if and only if D is linearly independent.

(b) Show that C spans S if and only if D spans $\mathbb{R}^k$.

(c) Show that C is a basis of S if and only if D is a basis of $\mathbb{R}^k$.

∗∗3.1.13 Show that if $S \subseteq \mathbb{R}^n$ is a k-dimensional subspace, $P \in \mathcal{M}_k$ is invertible, and B is a basis of S, then there exists a basis C of S such that $P = P_{B\leftarrow C}$.

[Side note: This shows that the converse of Theorem 3.1.3(b) holds: not only is every change-of-basis matrix invertible, but also every invertible matrix is a change-of-basis matrix too.]

∗∗3.1.14 Let S be a subspace of $\mathbb{R}^n$ with bases B, C, and E. We now present two methods that simplify the computation of $P_{C\leftarrow B}$ when E is chosen so as to be easy to convert into (e.g., if $S = \mathbb{R}^n$ and E is the standard basis).

(a) Show that $P_{C\leftarrow B} = P_{C\leftarrow E}P_{E\leftarrow B}$.

(b) Show that the reduced row echelon form of $[\, P_{E\leftarrow C} \mid P_{E\leftarrow B} \,]$ is $[\, I \mid P_{C\leftarrow B} \,]$.

☐ **3.1.15** Use computer software to compute the change-of-basis matrix $P_{C\leftarrow B}$ for each of the following pairs of bases B and C, both directly from the definition and also via the method of Exercise 3.1.14(b). Which method do you prefer?

*(a) $B = \{(3,0,0,3),(3,0,1,1),$
$(2,-1,0,1),(2,-1,3,-1)\}$,
$C = \{(0,1,3,2),(-1,1,2,-1),$
$(1,-1,-1,-1),(0,2,2,0)\}$

(b) $B = \{(-1,-1,1,0,2),(-1,-1,0,1,0),$
$(2,2,2,-1,-1),(1,1,1,2,0),(1,2,0,1,2)\}$,
$C = \{(2,-1,3,0,1),(0,1,0,1,0),$
$(2,-1,2,-1,0),(1,1,0,1,0),(3,-1,2,3,3)\}$

∗∗3.1.16 Prove Theorem 3.1.4.

∗∗3.1.17 Show that if A and B are square matrices that are similar, then $\text{nullity}(A) = \text{nullity}(B)$.

3.1.18 Suppose $A, B \in \mathcal{M}_n$.

(a) Show that if at least one of A or B is invertible then AB and BA are similar.

(b) Provide an example to show that if A and B are not invertible then AB and BA may not be similar.

∗∗3.1.19 Recall Theorem 3.1.6, which established some of the basic properties of the trace.

(a) Prove part (a) of the theorem. That is, prove that $\text{tr}(A+B) = \text{tr}(A) + \text{tr}(B)$ for all $A, B \in \mathcal{M}_n$.

(b) Prove part (b) of the theorem. That is, prove that $\text{tr}(cA) = c\,\text{tr}(A)$ for all $A \in \mathcal{M}_n$ and all $c \in \mathbb{R}$.

3.1.20 Show that the only linear function $f : \mathcal{M}_n \to \mathbb{R}$ with the property that $f(ABC) = f(ACB)$ for all $A, B, C \in \mathcal{M}_n$ is the zero function (i.e., the function defined by $f(A) = 0$ for all $A \in \mathcal{M}_n$).

[Hint: Compute $f(\mathbf{e}_i \mathbf{e}_j^T)$ in multiple different ways.]

[Side note: This exercise shows that the cyclic commutativity of the trace is "as good as it gets": no non-zero linear function is unchanged under arbitrary permutations of the matrices in a product on which the function acts.]

3.1.21 Suppose that $\mathcal{S}$ is a subspace of $\mathbb{R}^n$ and $\mathbf{v}, \mathbf{w} \in \mathcal{S}$.

(a) Show that if B is an orthonormal basis of $\mathcal{S}$ (see Remark 3.1.2) then $\mathbf{v} \cdot \mathbf{w} = [\mathbf{v}]_B \cdot [\mathbf{w}]_B$.
(b) Show that if B is an orthonormal basis of $\mathcal{S}$ then $\|\mathbf{v}\| = \|[\mathbf{v}]_B\|$.
(c) Provide an example to show that parts (a) and (b) are not necessarily true if B is a non-orthonormal basis.

3.1.22 Suppose $B \subset \mathbb{R}^n$ is a set consisting of non-zero vectors that are mutually orthogonal (i.e., $\mathbf{v} \cdot \mathbf{w} = 0$ for all $\mathbf{v} \neq \mathbf{w} \in B$). Show that B is linearly independent.

3.1.23 Suppose $B = \{\mathbf{v}_1, \mathbf{v}_2, \ldots, \mathbf{v}_n\}$ is an orthonormal basis of $\mathbb{R}^n$, and let A be the matrix with the basis vectors as its columns:
$$A = \left[\, \mathbf{v}_1 \mid \mathbf{v}_2 \mid \cdots \mid \mathbf{v}_n \,\right].$$
Show that $A^T A = I$.

3.1.24 Let $\{\mathbf{e}_1, \mathbf{e}_2\}$ be the standard basis of $\mathbb{R}^2$ and let R^θ be the linear transformation that rotates $\mathbb{R}^2$ counterclockwise by θ radians.

(a) Show that $\{R^\theta(\mathbf{e}_1), R^\theta(\mathbf{e}_2)\}$ is an orthonormal basis of $\mathbb{R}^2$.
(b) Show that by choosing θ appropriately, *every* orthonormal basis of $\mathbb{R}^2$ can be written in the form $\{R^\theta(\mathbf{e}_1), R^\theta(\mathbf{e}_2)\}$.

3.2 Determinants

We now introduce one of the most important properties of a matrix: its **determinant**, which roughly is a measure of how "large" the matrix is. More specifically, recall that every square matrix $A \in \mathcal{M}_n$ can be thought of as a linear transformation that sends $\mathbf{x} \in \mathbb{R}^n$ to $A\mathbf{x} \in \mathbb{R}^n$. That linear transformation sends the unit square (or cube, or hypercube, ...) with sides $\mathbf{e}_1, \mathbf{e}_2, \ldots, \mathbf{e}_n$ to the parallelogram (or parallelepiped, or hyperparallelepiped, ...) with side vectors equal to the columns of A: $A\mathbf{e}_1, A\mathbf{e}_2, \ldots, A\mathbf{e}_n$ (see Figure 3.10).

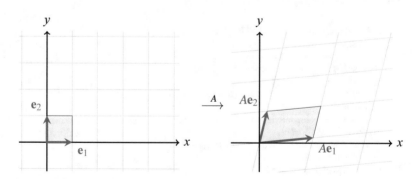

Figure 3.10: A 2×2 matrix A stretches the unit square (with sides $\mathbf{e}_1$ and $\mathbf{e}_2$) into a parallelogram with sides $A\mathbf{e}_1$ and $A\mathbf{e}_2$ (the columns of A). The determinant of A is the area of this parallelogram.

Some books denote the determinant of a matrix $A \in \mathcal{M}_n$ by $|A|$ instead of $\det(A)$.

The determinant of A, which we denote by $\det(A)$, is the area (or volume, or hypervolume, depending on the dimension n) of this distortion of the unit square. In other words, it measures how much A expands space when acting as a linear transformation—it is the ratio

$$\frac{\text{volume of output region}}{\text{volume of input region}}.$$

Since linear transformations behave so uniformly on $\mathbb{R}^n$, this definition does not rely specifically on using the unit square as the input to A—if A doubles the area of the unit square (i.e., $\det(A) = 2$), then it also doubles the area of any square, and in fact it doubles the area of any region for which it makes sense to talk about "area" (see Figure 3.11).

The determinant generalizes many of the ways of computing areas and volumes that we introduced in Section 1.A.

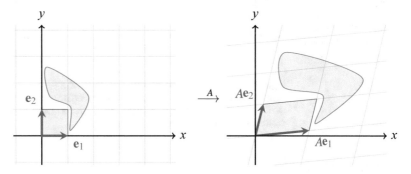

Figure 3.11: Both of the regions on the left have their areas stretched by the same amount by the matrix A ($\det(A) = 2.46$ in this case).

It is not yet at all clear how to actually *compute* the determinant of a matrix. It turns out that there are several different formulas for computing it, and they are all somewhat ugly and involved, so we start by looking at some of the basic properties of the determinant instead.

3.2.1 Definition and Basic Properties

Before we even define the determinant algebraically, we think about some properties that it should have. The first important property is that, since the identity matrix does not stretch or shrink $\mathbb{R}^n$ at all, it must be the case that $\det(I) = 1$. Secondly, since every $A \in \mathcal{M}_n$ expands space by a factor of $\det(A)$, and similarly each $B \in \mathcal{M}_n$ expands space by a factor of $\det(B)$, it must be the case that AB stretches space by the product of these two factors. That is,

$$\det(AB) = \det(A)\det(B) \quad \text{for all} \quad A, B \in \mathcal{M}_n.$$

To illustrate the third and final property of the determinant that we need, consider what happens to $\det(A)$ if we multiply one of the columns of A by a scalar $c \in \mathbb{R}$. Geometrically, this corresponds to multiplying one of the sides of the output parallelogram by c, which in turn scales the area of that parallelogram by c (see Figure 3.12). That is, we have

$$\det\left(\left[\, \mathbf{a}_1 \mid \cdots \mid c\mathbf{a}_j \mid \cdots \mid \mathbf{a}_n \,\right]\right) = c \cdot \det\left(\left[\, \mathbf{a}_1 \mid \cdots \mid \mathbf{a}_j \mid \cdots \mid \mathbf{a}_n \,\right]\right).$$

Similarly, if we add a vector to one of the columns of a matrix, then that corresponds geometrically to adding that vector to one of the side vectors of the output parallelogram, while leaving all of the other sides alone. We should thus expect that the area of this parallelogram is the sum of the areas of the two individual parallelograms:

$$\det\left(\left[\, \mathbf{a}_1 \mid \cdots \mid \mathbf{v} + \mathbf{w} \mid \cdots \mid \mathbf{a}_n \,\right]\right)$$
$$= \det\left(\left[\, \mathbf{a}_1 \mid \cdots \mid \mathbf{v} \mid \cdots \mid \mathbf{a}_n \,\right]\right) + \det\left(\left[\, \mathbf{a}_1 \mid \cdots \mid \mathbf{w} \mid \cdots \mid \mathbf{a}_n \,\right]\right).$$

This property is perhaps a bit more difficult to visualize, but it is illustrated when $n = 2$ in Figure 3.13.

Throughout this section, when we talk about area or volume in $\mathbb{R}^n$, we mean whichever of area, volume, or hypervolume is actually relevant depending on n (and similarly when we say words like "square" or "cube").

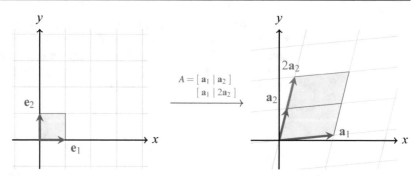

Figure 3.12: Scaling one of the columns of $A = [\ \mathbf{a}_1\ |\ \mathbf{a}_2\ ]$ scales the area of the output of the corresponding linear transformation (and thus $\det(A)$) by that same amount. Here, the parallelogram with sides $\mathbf{a}_1$ and $2\mathbf{a}_2$ has twice the area of the parallelogram with sides $\mathbf{a}_1$ and $\mathbf{a}_2$.

The equality of these two areas is sometimes called **Cavalieri's principle**. It perhaps helps to think of each area as made up of many small slices that are parallel to the side $\mathbf{a}_1$.

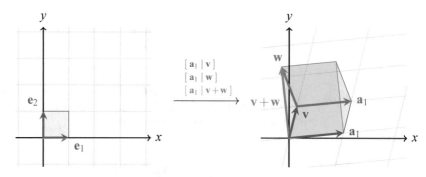

Figure 3.13: The blue and green parallelograms share the side $\mathbf{a}_1$ but not their other sides ($\mathbf{v}$ and $\mathbf{w}$, respectively). The sum of their areas equals the area of the orange parallelogram with sides $\mathbf{a}_1$ (the shared side) and $\mathbf{v} + \mathbf{w}$ (the sum of the non-shared sides).

In other words, the determinant is linear in the columns of a matrix. In order to be able to work with the determinant algebraically, we *define* it to be the function that satisfies this linearity property, as well as the other two properties that we discussed earlier:

Definition 3.2.1

Determinant

The **determinant** is the (unique!) function $\det : \mathcal{M}_n \to \mathbb{R}$ that satisfies the following three properties:

a) $\det(I) = 1$,

b) $\det(AB) = \det(A)\det(B)$ for all $A, B \in \mathcal{M}_n$, and

c) for all $c \in \mathbb{R}$ and all $\mathbf{v}, \mathbf{w}, \mathbf{a}_1, \mathbf{a}_2, \ldots, \mathbf{a}_n \in \mathbb{R}^n$, it is the case that

Property (c) is sometimes called **multilinearity** of the determinant, which in this context just means "linear in each of the columns".

$$\det\left([\ \mathbf{a}_1\ |\ \cdots\ |\ \mathbf{v} + c\mathbf{w}\ |\ \cdots\ |\ \mathbf{a}_n\ ]\right)$$
$$= \det\left([\ \mathbf{a}_1\ |\ \cdots\ |\ \mathbf{v}\ |\ \cdots\ |\ \mathbf{a}_n\ ]\right) + c \cdot \det\left([\ \mathbf{a}_1\ |\ \cdots\ |\ \mathbf{w}\ |\ \cdots\ |\ \mathbf{a}_n\ ]\right).$$

We note that it is not yet at all obvious that these three properties uniquely define the determinant—what prevents there from being multiple different functions with these properties? However, as we proceed throughout this section, we will use these properties to develop an explicit formula for the determinant, from which uniqueness follows.

To begin our investigation of the consequences of the three properties in the above definition, we look at what they tell us about invertible matrices. If

$A \in \mathcal{M}_n$ is invertible then properties (a) and (b) tell us that

$$1 = \det(I) = \det(AA^{-1}) = \det(A)\det(A^{-1}).$$

It follows that $\det(A) \neq 0$ and $\det(A^{-1}) = 1/\det(A)$. This makes sense geometrically, since if A expands space by a factor of $\det(A)$ then A^{-1} must shrink it back down by that same amount.

We looked at this geometric interpretation of invertibility at the end of Section 2.3.2.

On the other hand, if A is not invertible, then we recall that it squashes all of $\mathbb{R}^n$ down into some $(n-1)$-dimensional (or smaller) subspace. That is, A squashes the unit square/cube/hypercube into something that is "flat" and thus has 0 volume (for example, a 1-dimensional line in $\mathbb{R}^2$ has 0 area, a 2-dimensional plane in $\mathbb{R}^3$ has 0 volume, and so on). We thus conclude that if A is not invertible then $\det(A) = 0$ (see Figure 3.14).

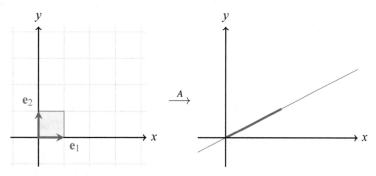

Figure 3.14: Non-invertible matrices send $\mathbb{R}^n$ to a subspace of dimension less than n, and thus have determinant 0. In this figure, $\mathbb{R}^2$ is sent to a (1-dimensional) line, and thus the unit square is sent to a line segment (which has 0 area).

We summarize our observations about the determinant of invertible and non-invertible matrices in the following theorem:

Theorem 3.2.1

Determinants and Invertibility

Suppose $A \in \mathcal{M}_n$. Then A is invertible if and only if $\det(A) \neq 0$, and if it is invertible then $\det(A^{-1}) = 1/\det(A)$.

Proof. We have already demonstrated most pieces of this theorem. To be rigorous, we still need to give a proof of the fact that non-invertible matrices have 0 determinant that only uses the three properties from Definition 3.2.1, rather than using our original geometric interpretation of the determinant.

To this end, we recall from Theorem 2.2.4 that if A is not invertible then there exists a non-zero vector $\mathbf{x} \in \mathbb{R}^n$ such that $A\mathbf{x} = \mathbf{0}$. If we let P be any invertible matrix that has $\mathbf{x}$ as its first column then

To construct an invertible matrix P with $\mathbf{x}$ as its first column, recall from Theorem 2.4.3(a) that we can extend $\mathbf{x}$ to a basis of $\mathbb{R}^n$, and then Theorem 2.5.1 tells us that placing those basis vectors as columns into a matrix P will ensure that it is invertible.

$$\det(P^{-1}AP) = \det(P^{-1})\det(A)\det(P)$$
$$= \frac{1}{\det(P)}\det(A)\det(P) = \det(A), \qquad (3.2.1)$$

so our goal now is to show that $\det(P^{-1}AP) = 0$. By recalling that the first column of P is $\mathbf{x}$ (i.e., $P\mathbf{e}_1 = \mathbf{x}$), we see that

$$(P^{-1}AP)\mathbf{e}_1 = (P^{-1}A)(P\mathbf{e}_1) = (P^{-1}A)\mathbf{x} = P^{-1}(A\mathbf{x}) = P^{-1}\mathbf{0} = \mathbf{0}.$$

In other words, the first column of $P^{-1}AP$ is the zero vector $\mathbf{0}$, so our goal now is to show that every such matrix has 0 determinant. To see this, we use multilinearity (i.e., defining property (c)) of the determinant:

$$\det\left(\left[\,\mathbf{0}\mid\mathbf{a}_2\mid\cdots\mid\mathbf{a}_n\,\right]\right) = \det\left(\left[\,\mathbf{a}_1-\mathbf{a}_1\mid\mathbf{a}_2\mid\cdots\mid\mathbf{a}_n\,\right]\right)$$
$$= \det\left(\left[\,\mathbf{a}_1\mid\mathbf{a}_2\mid\cdots\mid\mathbf{a}_n\,\right]\right) - \det\left(\left[\,\mathbf{a}_1\mid\mathbf{a}_2\mid\cdots\mid\mathbf{a}_n\,\right]\right)$$
$$= 0,$$

which completes the proof. ∎

It is worth briefly focusing on the fact that we showed in Equation (3.2.1) that $\det(P^{-1}AP) = \det(A)$ for every invertible matrix P. In other words, the determinant (just like the rank and the trace) is shared between similar matrices:

In other words, the determinant is similarity invariant.

(!)   If $A \in \mathcal{M}_n$ and $B \in \mathcal{M}_n$ are similar then $\det(A) = \det(B)$.

Morally, the reason for this similarity invariance of the determinant is the fact that, just like the rank, the determinant is really a property of linear transformations, not of matrices. That is, the particular basis that we are using to view $\mathbb{R}^n$ does not affect how much the linear transformation stretches space.

Once we actually know how to compute the determinant, we will be able to use it to show that certain matrices are not similar, just like we did with the rank and trace in Examples 3.1.9 and 3.1.12. For now though, we close this section with a couple of other useful properties of the determinant that are a bit easier to demonstrate than the invertibility properties that we investigated above. We do not prove either of the following properties, but instead leave them to Exercise 3.2.13.

Theorem 3.2.2

Other Properties of the Determinant

Suppose $A \in \mathcal{M}_n$ and $c \in \mathbb{R}$. Then
a) $\det(cA) = c^n \det(A)$, and
b) $\det(A^T) = \det(A)$.

Property (b) of this theorem is actually best proved with techniques from the next subsection. It is just more convenient to state it here.

It is perhaps worth providing a geometric interpretation for property (a) of the above theorem: since the determinant of an $n \times n$ matrix is an n-dimensional volume, it makes sense that scaling the matrix by a factor of c scales the resulting volume by c^n (rather than just by c, which is what we might mistakenly guess at first). See Figure 3.15 for an illustration in the $n = 2$ case.

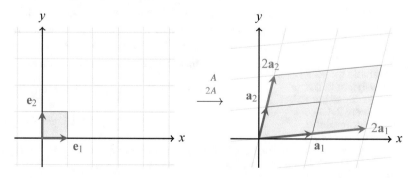

Figure 3.15: Scaling $A \in \mathcal{M}_2$ by a factor of 2 increases the resulting area (i.e., the determinant) by a factor of $2^2 = 4$.

Example 3.2.1
Determinant Properties

Suppose that $A, B, C \in \mathcal{M}_3$ are matrices with $\det(A) = 2$, $\det(B) = 3$, and $\det(C) = 5$. Compute each of the following determinants:

a) $\det(AB)$
b) $\det(A^2 C^T B^{-1})$
c) $\det(3AB^{-2}C^2)$
d) $\det\left(2A^{-3}B^{-2}(C^T B)^4 (C/5)^2\right)$

Solutions:

a) $\det(AB) = \det(A)\det(B) = 2 \cdot 3 = 6$.

b) $\det(A^2 C^T B^{-1}) = \det(A)^2 \det(C^T) \det(B^{-1}) = 2^2 \cdot 5 \cdot (1/3) = 20/3$.

The exponent 3 that appears in parts (c) and (d) is because the matrices have size 3×3.

c) $\det(3AB^{-2}C^2) = 3^3 \det(A)\det(B)^{-2}\det(C)^2 = 27 \cdot 2 \cdot (1/9) \cdot 25 = 150$.

d) By combining all of the determinant rules that we have seen, we get

$$\det\left(2A^{-3}B^{-2}(C^T B)^4 (C/5)^2\right)$$
$$= 2^3 \det(A)^{-3} \det(B)^{-2} \det(C^T)^4 \det(B)^4 \det(C/5)^2$$
$$= 8 \cdot (1/8) \cdot (1/9) \cdot 5^4 \cdot 3^4 \cdot (1/25)^2 = 9.$$

3.2.2 Computation

We already know that the determinant of every non-invertible matrix is 0, so our goal now is to develop a method of computing the determinant of invertible matrices. As a preliminary step toward this goal, we first consider the simpler problem of computing the determinant of elementary matrices. We consider the three different types of elementary matrices one at a time.

Multiplication

The "multiplication" row operation cR_i corresponds to an elementary matrix of the form

$$cR_i \quad \begin{bmatrix} 1 & & & & & & \\ & \ddots & & & & & \\ & & 1 & & & & \\ & & & c & & & \\ & & & & 1 & & \\ & & & & & \ddots & \\ & & & & & & 1 \end{bmatrix},$$

Even though we originally constructed the elementary matrices via *row* operations, we are investigating what their *columns* look like here, since property (c) of the determinant depends on columns, not rows.

where the scalar c appears in the i-th column. It is thus obtained by multiplying the i-th column of the identity matrix (which has determinant 1) by c, so multilinearity of the determinant (i.e., defining property (c)) tells us that the determinant of this elementary matrix is c. Geometrically, this makes sense because this elementary matrix just stretches the unit square in the direction of one of the standard basis vectors by a factor of c (see Figure 3.16).

Addition

The "addition" row operation $R_i + cR_j$ corresponds to an elementary matrix of the form

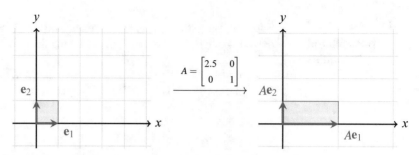

Figure 3.16: The elementary matrix with diagonal entries 2.5 and 1 stretches $\mathbf{e}_1$ by a factor of 2.5 and thus stretches the area of the unit square by 2.5. It follows that its determinant is 2.5.

$$R_i + cR_j \begin{bmatrix} 1 & & & & & & \\ & \ddots & & & & & \\ & & 1 & & & & \\ & & \vdots & \ddots & & & \\ & & c & \cdots & 1 & & \\ & & & & & \ddots & \\ & & & & & & 1 \end{bmatrix},$$

where the scalar c appears in the i-th row and j-th column. It is thus obtained by adding $c\mathbf{e}_i$ to the j-th column of I, so we can use multilinearity (defining property (c) of determinants) to see that the determinant of this elementary matrix is

$$\det \left(\begin{bmatrix} \mathbf{e}_1 & | & \cdots & | & \mathbf{e}_j + c\mathbf{e}_i & | & \cdots & | & \mathbf{e}_n \end{bmatrix} \right)$$
$$= \det \left(\begin{bmatrix} \mathbf{e}_1 & | & \cdots & | & \mathbf{e}_j & | & \cdots & | & \mathbf{e}_n \end{bmatrix} \right) + c \cdot \det \left(\begin{bmatrix} \mathbf{e}_1 & | & \cdots & | & \mathbf{e}_i & | & \cdots & | & \mathbf{e}_n \end{bmatrix} \right).$$

In the above sum, the matrix on the left is simply the identity matrix, which has determinant 1. The matrix on the right contains $\mathbf{e}_i$ as two of its columns (both the i-th column and the j-th column) and is thus not invertible, so its determinant is 0. This elementary matrix thus has determinant $1 + c \cdot 0 = 1$.

This determinant equaling 1 perhaps seems somewhat counter-intuitive at first, but it makes sense geometrically: this elementary matrix is a shear matrix that slants the unit square but does not change its area (see Figure 3.17).

Shear matrices were investigated in Exercise 1.4.23.

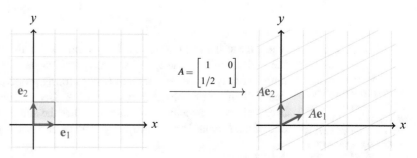

Figure 3.17: The elementary matrix corresponding to the row operation $R_2 + \frac{1}{2}R_1$ shears the unit square but does not change its area, so its determinant equals 1.

Swap

The "swap" row operation $R_i \leftrightarrow R_j$, which corresponds to an elementary matrix of the form

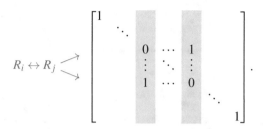

is perhaps the most surprising. Geometrically, it seems as though it should have determinant 1, since it does not actually change the unit square at all, but rather just re-orders its side vectors.

To actually compute its determinant algebraically, we consider the matrix that is the same as the identity matrix, except its i-th and j-th columns both equal $\mathbf{e}_i + \mathbf{e}_j$. Since this matrix has two equal columns, it is not invertible, so its determinant equals 0. Then using multilinearity (i.e., defining property (c) of determinants) repeatedly shows that

$$
\begin{aligned}
0 &= \det\left(\left[\, \mathbf{e}_1 \mid \cdots \mid \mathbf{e}_i + \mathbf{e}_j \mid \cdots \mid \mathbf{e}_i + \mathbf{e}_j \mid \cdots \mid \mathbf{e}_n \,\right]\right) \\
&= \det\left(\left[\, \mathbf{e}_1 \mid \cdots \mid \mathbf{e}_i \mid \cdots \mid \mathbf{e}_i + \mathbf{e}_j \mid \cdots \mid \mathbf{e}_n \,\right]\right) \\
&\quad + \det\left(\left[\, \mathbf{e}_1 \mid \cdots \mid \mathbf{e}_j \mid \cdots \mid \mathbf{e}_i + \mathbf{e}_j \mid \cdots \mid \mathbf{e}_n \,\right]\right) \\
&= \det\left(\left[\, \mathbf{e}_1 \mid \cdots \mid \mathbf{e}_i \mid \cdots \mid \mathbf{e}_i \mid \cdots \mid \mathbf{e}_n \,\right]\right) \qquad \text{(not invertible, } \det = 0\text{)} \\
&\quad + \det\left(\left[\, \mathbf{e}_1 \mid \cdots \mid \mathbf{e}_i \mid \cdots \mid \mathbf{e}_j \mid \cdots \mid \mathbf{e}_n \,\right]\right) \qquad \text{(identity matrix, } \det = 1\text{)} \\
&\quad + \det\left(\left[\, \mathbf{e}_1 \mid \cdots \mid \mathbf{e}_j \mid \cdots \mid \mathbf{e}_i \mid \cdots \mid \mathbf{e}_n \,\right]\right) \qquad \text{(elementary swap matrix)} \\
&\quad + \det\left(\left[\, \mathbf{e}_1 \mid \cdots \mid \mathbf{e}_j \mid \cdots \mid \mathbf{e}_j \mid \cdots \mid \mathbf{e}_n \,\right]\right). \qquad \text{(not invertible, } \det = 0\text{)}
\end{aligned}
$$

After simplifying three of the four determinants above (one of the terms is just the determinant of the identity matrix and thus equals 1, and two of the determinants equal 0 since the matrix has a repeated column and is thus not invertible), we see that

$$
0 = 0 + 1 + \det\left(\left[\, \mathbf{e}_1 \mid \cdots \mid \mathbf{e}_j \mid \cdots \mid \mathbf{e}_i \mid \cdots \mid \mathbf{e}_n \,\right]\right) + 0,
$$

where the remaining term is exactly the determinant of the elementary swap matrix that we want. We thus conclude that this determinant must be -1.

While this result feels "wrong" at first, it is actually correct and completely necessary if we want the determinant to have nice mathematical properties—instead of interpreting the determinant just as the area of the parallelogram that the unit square is mapped to, we must interpret it as a *signed* area that is negative if the linear transformation reverses (i.e., reflects) the orientation of space (see Figure 3.18).

In fact, the elementary matrix corresponding to the multiplication row operation can also have a negative determinant, if the scalar c on the diagonal is negative. This also agrees with the interpretation of a negative determinant meaning that the orientation of space has been flipped, since multiplying a vector by a negative scalar reverses the direction of that vector.

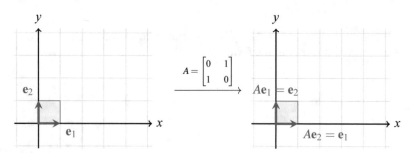

Figure 3.18: The elementary matrix corresponding to the row operation $R_1 \leftrightarrow R_2$ swaps the standard basis vectors. The unit square is mapped to the exact same square, which thus has the same area, but the determinant of this linear transformation is -1 (instead of 1) since the orientation of space is flipped.

Remark 3.2.1

Negative Determinants

While it might seem uncomfortable at first that determinants can be negative, this idea of an area or volume being signed is nothing new if you have taken an integral calculus course. In particular, recall that the definite integral

$$\int_a^b f(x)\,dx$$

is defined as the area under the graph of $f(x)$ from $x = a$ to $x = b$:

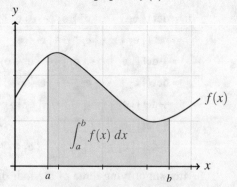

However, this geometric definition is only accurate if $f(x) \geq 0$ for all x between a and b. If $f(x)$ is potentially negative, then in order for the definite integral to satisfy nice properties like

$$\int_a^b f(x)\,dx + \int_b^c f(x)\,dx = \int_a^c f(x)\,dx,$$

we must instead define it to be the *signed* area under the graph of $f(x)$. That is, areas above the x-axis are considered positive and areas under the x-axis are considered negative:

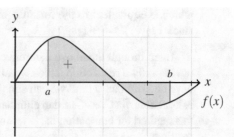

The definite integral depicted here would be computed as the area of the "$+$" region minus the area of the "$-$" region.

Similarly, we initially motivated the determinant geometrically in terms of areas and volumes, but for it to have nice mathematical properties (i.e., the three properties given in Definition 3.2.1), we must be willing to accept that it can be negative.

Arbitrary Matrices

For clarity, we now summarize the determinant calculations that we just completed for the three types of elementary matrices.

 The elementary matrix corresponding to the row operation cR_i has determinant c, the row operation $R_i + cR_j$ has determinant 1, and the row operation $R_i \leftrightarrow R_j$ has determinant -1.

Remarkably, now that we are able to compute the determinant of elementary matrices, we know everything that we need to know in order to develop a general method for computing the determinant of arbitrary matrices. The following theorem says that we can compute the determinant of a matrix just by performing Gauss–Jordan elimination on it and keeping track of which "multiplication" and "swap" row operations we used in the process.

Theorem 3.2.3

Computing Determinants via Gauss–Jordan Elimination

Let $A \in \mathcal{M}_n$ be invertible, and suppose that A can be row-reduced to I via k "multiplication" row operations $c_1 R_{i_1}, c_2 R_{i_2}, \ldots, c_k R_{i_k}$ as well as s "swap" row operations (and some "addition" row operations that we do not care about). Then

$$\det(A) = \frac{(-1)^s}{c_1 c_2 \cdots c_k}.$$

Proof. Recall that if A is invertible then we can write $E_m \cdots E_2 E_1 A = I$, where $E_1, E_2, \ldots, E_m$ are the elementary matrices corresponding to the row operations used to row-reduce A to I, so

$$\det(E_m \cdots E_2 E_1 A) = \det(I) = 1.$$

Using multiplicativity of the determinant and re-arranging shows that

$$\det(A) = \frac{1}{\det(E_1)\det(E_2)\cdots\det(E_m)}.$$

Now we just use the fact that we know what the determinants of elementary matrices are: the "multiplication" row operations $c_1 R_{i_1}, c_2 R_{i_2}, \ldots, c_k R_{i_k}$ correspond to elementary matrices with determinants $c_1, c_2, \ldots, c_k$, the "swap" row operations correspond to elementary matrices with determinant -1, and the "addition" row operations correspond to elementary matrices with determinant 1 (which we thus ignore). We thus conclude that

$$\det(A) = \frac{1}{(-1)^s c_1 c_2 \cdots c_k},$$

which is equivalent to the formula presented in the statement of the theorem since $1/(-1)^s = (-1)^s$. ∎

Even though the above theorem only applies to invertible matrices, this is not actually a restriction, since we already know that non-invertible matrices have determinant 0 (and we can simultaneously check whether or not a matrix is invertible via Gauss–Jordan elimination as well). The next example illustrates this method for computing the determinant of both invertible and non-invertible matrices.

Example 3.2.2

Computing Determinants via Gauss–Jordan Elimination

Compute the determinant of each of the following matrices:

a) $\begin{bmatrix} 2 & 2 \\ 4 & 5 \end{bmatrix}$

b) $\begin{bmatrix} 0 & 2 \\ 3 & 4 \end{bmatrix}$

c) $\begin{bmatrix} 1 & 2 & 3 \\ 4 & 5 & 6 \\ 7 & 8 & 9 \end{bmatrix}$

d) $\begin{bmatrix} 1 & 1 & 1 \\ 1 & 2 & 4 \\ 1 & 3 & 9 \end{bmatrix}$

Solutions:

a) To compute the determinant of this matrix, we try to row-reduce it to the identity matrix, and keep track of the row operations that we perform along the way:

The two "addition" row operations that we perform here do not matter when computing the determinant.

$$\begin{bmatrix} 2 & 2 \\ 4 & 5 \end{bmatrix} \xrightarrow{R_2-2R_1} \begin{bmatrix} 2 & 2 \\ 0 & 1 \end{bmatrix} \xrightarrow{R_1-2R_2} \begin{bmatrix} 2 & 0 \\ 0 & 1 \end{bmatrix} \xrightarrow{\frac{1}{2}R_1} \begin{bmatrix} 1 & 0 \\ 0 & 1 \end{bmatrix}.$$

Since we did not need to perform any "swap" row operations, and the only "multiplication" row operation that we performed was $\frac{1}{2}R_1$, we conclude that the determinant of this matrix is $1/(1/2) = 2$.

b) Again, we row-reduce and keep track of the row operations as we apply them:

$$\begin{bmatrix} 0 & 2 \\ 3 & 4 \end{bmatrix} \xrightarrow{R_1 \leftrightarrow R_2} \begin{bmatrix} 3 & 4 \\ 0 & 2 \end{bmatrix} \xrightarrow{R_1-2R_2} \begin{bmatrix} 3 & 0 \\ 0 & 2 \end{bmatrix} \xrightarrow[\substack{(1/2)R_2}]{(1/3)R_1} \begin{bmatrix} 1 & 0 \\ 0 & 1 \end{bmatrix}.$$

Be careful with fractions within fractions when computing determinants in this way.

Since we performed $s = 1$ "swap" row operation, as well as the two "multiplication" row operations $\frac{1}{3}R_1$ and $\frac{1}{2}R_2$, it follows that the determinant of this matrix is $(-1)/((1/2)(1/3)) = -6$.

c) Yet again, we perform Gauss–Jordan elimination and keep track of the row operations that we perform:

If we ever get a zero row at any point while row-reducing, we can immediately stop and conclude the matrix has determinant 0.

$$\begin{bmatrix} 1 & 2 & 3 \\ 4 & 5 & 6 \\ 7 & 8 & 9 \end{bmatrix} \xrightarrow[\substack{R_3-7R_1}]{R_2-4R_1} \begin{bmatrix} 1 & 2 & 3 \\ 0 & -3 & -6 \\ 0 & -6 & -12 \end{bmatrix} \xrightarrow{R_3-2R_2} \begin{bmatrix} 1 & 2 & 3 \\ 0 & -3 & -6 \\ 0 & 0 & 0 \end{bmatrix}.$$

Since we have row-reduced this matrix to one with a zero row, we conclude that it is not invertible and thus has determinant 0.

d) As always, we row-reduce to the identity matrix and keep track of the row operations that we perform along the way:

$$\begin{bmatrix} 1 & 1 & 1 \\ 1 & 2 & 4 \\ 1 & 3 & 9 \end{bmatrix} \xrightarrow[R_3-R_1]{R_2-R_1} \begin{bmatrix} 1 & 1 & 1 \\ 0 & 1 & 3 \\ 0 & 2 & 8 \end{bmatrix} \xrightarrow[R_3-2R_2]{R_1-R_2} \begin{bmatrix} 1 & 0 & -2 \\ 0 & 1 & 3 \\ 0 & 0 & 2 \end{bmatrix}$$

$$\xrightarrow{\frac{1}{2}R_3} \begin{bmatrix} 1 & 0 & -2 \\ 0 & 1 & 3 \\ 0 & 0 & 1 \end{bmatrix} \xrightarrow[R_2-3R_3]{R_1+2R_3} \begin{bmatrix} 1 & 0 & 0 \\ 0 & 1 & 0 \\ 0 & 0 & 1 \end{bmatrix}.$$

At this point it is maybe a good idea to look back at Example 2.2.8, where we checked invertibility of many of these same matrices. Almost no part of the calculations changed.

Since we did not perform any "swap" row operations, and the only "multiplication" row operation that we performed was $\frac{1}{2}R_3$, we conclude that the determinant of this matrix is $1/(1/2) = 2$.

We can make this method of computing the determinant slightly faster by modifying it so that we just need to compute a row echelon form of the matrix, rather than row-reducing it all the way to its *reduced* row echelon form I. The key insight that makes this possible is provided by the following theorem, which shows that it is straightforward to compute the determinant of triangular matrices (and thus of matrices that are in row echelon form):

Theorem 3.2.4

Determinant of a Triangular Matrix

Suppose $A \in \mathcal{M}_n$ is a triangular matrix. Then $\det(A)$ is the product of its diagonal entries:

$$\det(A) = a_{1,1}a_{2,2}\cdots a_{n,n}.$$

Proof. Throughout this proof, we assume that A is upper triangular, but the same method works if A instead is lower triangular.

First, we consider the case when all of the diagonal entries of A are non-zero: $a_{1,1}, a_{2,2}, \ldots, a_{n,n} \neq 0$. Then we can construct a sequence of row operations that row-reduces A to I. First, divide the n-th row of A by $a_{n,n}$ and then, for each $1 \leq i \leq n-1$, subtract $a_{i,n}R_n$ from R_i:

All we are doing here is applying Gauss–Jordan elimination to A, while being careful to list exactly which row operations we are applying in the process.

$$\begin{bmatrix} a_{1,1} & \cdots & a_{1,n-1} & a_{1,n} \\ \vdots & \ddots & \vdots & \vdots \\ 0 & \cdots & a_{n-1,n-1} & a_{n-1,n} \\ 0 & \cdots & 0 & a_{n,n} \end{bmatrix} \xrightarrow{\frac{1}{a_{n,n}}R_n} \begin{bmatrix} a_{1,1} & \cdots & a_{1,n-1} & a_{1,n} \\ \vdots & \ddots & \vdots & \vdots \\ 0 & \cdots & a_{n-1,n-1} & a_{n-1,n} \\ 0 & \cdots & 0 & 1 \end{bmatrix}$$

$$\xrightarrow[\substack{R_2-a_{2,n}R_n \\ \vdots \\ R_{n-1}-a_{n-1,n}R_n}]{R_1-a_{1,n}R_n} \begin{bmatrix} a_{1,1} & \cdots & a_{1,n-1} & 0 \\ \vdots & \ddots & \vdots & \vdots \\ 0 & \cdots & a_{n-1,n-1} & 0 \\ 0 & \cdots & 0 & 1 \end{bmatrix}$$

We now repeat this procedure on the matrix from right to left: we divide the $(n-1)$-th row by $a_{n-1,n-1}$ and then, for each $1 \leq i \leq n-2$, subtract $a_{i,n-1}R_{n-1}$ from R_i. By repeating in this way, we eventually row-reduce all the way to the identity matrix without performing any "swap" row operations, and the only "multiplication" row operations performed are

$$\frac{1}{a_{n,n}}R_n, \quad \frac{1}{a_{n-1,n-1}}R_{n-1}, \quad \ldots, \quad \frac{1}{a_{1,1}}R_1.$$

It then follows from Theorem 3.2.3 that

$$\det(A) = \frac{1}{\frac{1}{a_{1,1}} \cdot \frac{1}{a_{2,2}} \cdots \frac{1}{a_{n,n}}} = a_{1,1}a_{2,2} \cdots a_{n,n},$$

as desired.

On the other hand, if A has a diagonal entry equal to 0, then $a_{1,1}a_{2,2} \cdots a_{n,n} = 0$, so our goal is to show that $\det(A) = 0$ as well (i.e., we want to show that A is not invertible). To see this, suppose that the $a_{i,i} = 0$ (and furthermore, i is the largest subscript for which $a_{i,i} = 0$). Then applying the same row-reduction procedure as in the first half of this proof would turn the i-th row into a zero row, thus showing that A is not invertible and completing the proof. ∎

We also showed that a triangular matrix is invertible if and only if its diagonal entries are all non-zero in Exercise 2.2.16.

The above result leads immediately to the following slight improvement of our method for computing determinants:

Theorem 3.2.5

Computing Determinants via Gaussian Elimination

We apologize for using "R" here to refer both to a row echelon form of A and to elementary row operations.

Let $A \in \mathcal{M}_n$, and suppose that A can be row-reduced to a row echelon form R via k "multiplication" row operations $c_1 R_{i_1}, c_2 R_{i_2}, \ldots, c_k R_{i_k}$ as well as s "swap" row operations (and some "addition" row operations that we do not care about). Then

$$\det(A) = \frac{(-1)^s r_{1,1} r_{2,2} \cdots r_{n,n}}{c_1 c_2 \cdots c_k},$$

where $r_{1,1}, r_{2,2}, \ldots, r_{n,n}$ are the diagonal entries of the row echelon form R.

Proof. The proof is almost identical to that of Theorem 3.2.3, but instead of applying Gauss–Jordan elimination to row-reduce A to I, we apply Gaussian elimination to row-reduce A to *any* row echelon form R and then apply Theorem 3.2.4 to R. ∎

Example 3.2.3

Computing Determinants via Gaussian Elimination

Compute the determinant of each of the following matrices:

a) $\begin{bmatrix} 2 & 2 \\ 4 & 5 \end{bmatrix}$

b) $\begin{bmatrix} 1 & -1 \\ -1 & 1 \end{bmatrix}$

c) $\begin{bmatrix} 3 & 6 & -3 \\ 0 & -2 & 3 \\ 2 & 4 & -1 \end{bmatrix}$

d) $\begin{bmatrix} 0 & 2 & 1 & 2 \\ 1 & -1 & 1 & 0 \\ 2 & 1 & 0 & 1 \\ -2 & 0 & 1 & 1 \end{bmatrix}$

Solutions:

a) To compute the determinant of this matrix, we apply Gaussian elimination and keep track of the row operations that we perform along the way:

$$\begin{bmatrix} 2 & 2 \\ 4 & 5 \end{bmatrix} \xrightarrow{R_2 - 2R_1} \begin{bmatrix} 2 & 2 \\ 0 & 1 \end{bmatrix}.$$

This first example is the same as Example 3.2.2(a). Notice that this method is much quicker and only requires one row operation instead of

Since we did not need to perform any row operations that contribute to the determinant, we conclude that the determinant of the original matrix is the product of the diagonal entries of its row echelon form: $2 \cdot 1 = 2$.

b) Again, we apply Gaussian elimination and keep track of the row operations that we perform along the way:

$$\begin{bmatrix} 1 & -1 \\ -1 & 1 \end{bmatrix} \xrightarrow{R_2+R_1} \begin{bmatrix} 1 & -1 \\ 0 & 0 \end{bmatrix}.$$

Since this row echelon form has a 0 diagonal entry, we conclude that its determinant equals 0 (equivalently, as soon as we see a zero row when row-reducing, we know it must not be invertible and thus has determinant equal to 0).

c) Yet again, we row-reduce the matrix to row echelon form and keep track of the row operations that we perform:

$$\begin{bmatrix} 3 & 6 & -3 \\ 0 & -2 & 3 \\ 2 & 4 & -1 \end{bmatrix} \xrightarrow{\frac{1}{3}R_1} \begin{bmatrix} 1 & 2 & -1 \\ 0 & -2 & 3 \\ 2 & 4 & -1 \end{bmatrix} \xrightarrow{R_3-2R_1} \begin{bmatrix} 1 & 2 & -1 \\ 0 & -2 & 3 \\ 0 & 0 & 1 \end{bmatrix}.$$

This matrix is now in row echelon form. Since we did not apply any "swap" row operations, and the only "multiplication" row operation that we applied was $(1/3)R_1$, we conclude that the determinant of the original matrix is $(1 \cdot (-2) \cdot 1)/(1/3) = -6$.

d) As always, we row-reduce and keep track of the row operations as we perform them:

Many other sequences of row operations can be used to put this matrix into row echelon form. They all give the same value for the determinant.

$$\begin{bmatrix} 0 & 2 & 1 & 2 \\ 1 & -1 & 1 & 0 \\ 2 & 1 & 0 & 1 \\ -2 & 0 & 1 & 1 \end{bmatrix} \xrightarrow{R_1 \leftrightarrow R_2} \begin{bmatrix} 1 & -1 & 1 & 0 \\ 0 & 2 & 1 & 2 \\ 2 & 1 & 0 & 1 \\ -2 & 0 & 1 & 1 \end{bmatrix}$$

$$\xrightarrow[R_4+2R_1]{R_3-2R_1} \begin{bmatrix} 1 & -1 & 1 & 0 \\ 0 & 2 & 1 & 2 \\ 0 & 3 & -2 & 1 \\ 0 & -2 & 3 & 1 \end{bmatrix} \xrightarrow[R_4+R_2]{R_3-\frac{3}{2}R_2} \begin{bmatrix} 1 & -1 & 1 & 0 \\ 0 & 2 & 1 & 2 \\ 0 & 0 & -7/2 & -2 \\ 0 & 0 & 4 & 3 \end{bmatrix}$$

$$\xrightarrow{R_4+\frac{8}{7}R_3} \begin{bmatrix} 1 & -1 & 1 & 0 \\ 0 & 2 & 1 & 2 \\ 0 & 0 & -7/2 & -2 \\ 0 & 0 & 0 & 5/7 \end{bmatrix}.$$

Since we performed one "swap" row operation, but no "multiplication" row operations, we conclude that the determinant of the original matrix is $(-1) \cdot 1 \cdot 2 \cdot (-7/2) \cdot (5/7) = 5$.

3.2.3 Explicit Formulas and Cofactor Expansions

One of the most remarkable properties of the determinant is that there exists an explicit formula for computing it just in terms of multiplication and addition of the entries of the matrix. Before presenting the main theorem of this section, which presents the formula for general $n \times n$ matrices, we discuss the 2×2 and 3×3 special cases.

The determinant of a 1×1 matrix $[a]$ is just a itself.

Theorem 3.2.6

Determinant of 2 × 2 Matrices

The determinant of a 2×2 matrix is given by

$$\det\left(\begin{bmatrix} a & b \\ c & d \end{bmatrix}\right) = ad - bc.$$

Proof. We prove this theorem by making use of multilinearity (i.e., defining property (c) of the determinant). We can write

In particular, notice that this theorem says that a 2×2 matrix is invertible if and only if $ad - bc \neq 0$. Compare this statement with Theorem 2.2.6.

$$\det\left(\begin{bmatrix} a & b \\ c & d \end{bmatrix}\right) = \det\left(\begin{bmatrix} a & b \\ 0 & d \end{bmatrix}\right) + \det\left(\begin{bmatrix} 0 & b \\ c & d \end{bmatrix}\right),$$

since the two matrices on the right differ only in their leftmost column. Well,

$$\det\left(\begin{bmatrix} a & b \\ 0 & d \end{bmatrix}\right) = ad$$

since that matrix is upper triangular, so its determinant is the product of its diagonal entries (by Theorem 3.2.4). Similarly,

$$\det\left(\begin{bmatrix} 0 & b \\ c & d \end{bmatrix}\right) = -\det\left(\begin{bmatrix} c & d \\ 0 & b \end{bmatrix}\right) = -bc,$$

since the "swap" row operation multiplies the determinant of a matrix by -1. By adding these two quantities together, the theorem follows. ∎

The above theorem is perhaps best remembered in terms of diagonals of the matrix—the determinant of a 2×2 matrix is the product of its forward diagonal minus the product of its backward diagonal:

$$\det\left(\begin{bmatrix} a & b \\ c & d \end{bmatrix}\right) = ad - bc.$$

Example 3.2.4

Using the Determinant to Show Matrices are Not Similar

We justified the fact that similar matrices have the same determinant on page 262.

Show that the matrices $A = \begin{bmatrix} 1 & 2 \\ 3 & 4 \end{bmatrix}$ and $B = \begin{bmatrix} 0 & 2 \\ 3 & 5 \end{bmatrix}$ are not similar.

Solution:

The rank and trace cannot help us show that these matrices are not similar, since $\text{rank}(A) = \text{rank}(B) = 2$ and $\text{tr}(A) = \text{tr}(B) = 5$. However, their determinants are different:

$$\det(A) = 1 \cdot 4 - 2 \cdot 3 = -2 \quad \text{and} \quad \det(B) = 0 \cdot 5 - 2 \cdot 3 = -6.$$

Since $\det(A) \neq \det(B)$, we conclude that A and B are not similar.

The formula for the determinant of a 3×3 matrix is similar in flavor, but somewhat more complicated:

Theorem 3.2.7

Determinant of 3 × 3 Matrices

The determinant of a 3×3 matrix is given by

$$\det\left(\begin{bmatrix} a & b & c \\ d & e & f \\ g & h & i \end{bmatrix}\right) = aei + bfg + cdh - afh - bdi - ceg.$$

Proof. The proof is very similar to the proof of the 2×2 determinant formula given earlier, just with more steps. We make use of multilinearity (i.e., defining property (c) of the determinant) to write

$$\det\left(\begin{bmatrix} a & b & c \\ d & e & f \\ g & h & i \end{bmatrix}\right)$$

$$= \det\left(\begin{bmatrix} a & b & c \\ 0 & e & f \\ 0 & h & i \end{bmatrix}\right) + \det\left(\begin{bmatrix} 0 & b & c \\ d & e & f \\ 0 & h & i \end{bmatrix}\right) + \det\left(\begin{bmatrix} 0 & b & c \\ 0 & e & f \\ g & h & i \end{bmatrix}\right),$$

since the three matrices on the right differ only in their leftmost column. We now compute the first of the three determinants on the right by using a similar trick on its second column:

$$\det\left(\begin{bmatrix} a & b & c \\ 0 & e & f \\ 0 & h & i \end{bmatrix}\right) = \det\left(\begin{bmatrix} a & b & c \\ 0 & e & f \\ 0 & 0 & i \end{bmatrix}\right) + \det\left(\begin{bmatrix} a & 0 & c \\ 0 & 0 & f \\ 0 & h & i \end{bmatrix}\right),$$

since the two matrices on the right differ only in their middle column. Well, the first matrix on the right is upper triangular so its determinant is the product of its diagonal entries (by Theorem 3.2.4), and the second matrix on the right can be made upper triangular by swapping two of its rows (which multiplies the determinant by -1), so

$$\det\left(\begin{bmatrix} a & b & c \\ 0 & e & f \\ 0 & h & i \end{bmatrix}\right) = aei - \det\left(\begin{bmatrix} a & 0 & c \\ 0 & h & i \\ 0 & 0 & f \end{bmatrix}\right) = aei - afh.$$

This same approach works for any matrix of any size by working left-to-right through its columns and swapping rows until all matrices being considered are upper triangular.

Similar arguments show that

$$\det\left(\begin{bmatrix} 0 & b & c \\ d & e & f \\ 0 & h & i \end{bmatrix}\right) = cdh - bdi, \quad \text{and} \quad \det\left(\begin{bmatrix} 0 & b & c \\ 0 & e & f \\ g & h & i \end{bmatrix}\right) = bfg - ceg,$$

and adding up these three answers gives the formula stated by the theorem. ∎

Compare this to the mnemonic used for computing the cross product in Figure 1.24.

Much like we did for the 2×2 determinants, we can think of the formula for determinants of 3×3 matrices in terms of diagonals of the matrix – it is the sum of the products of its forward diagonals minus the sum of the products of its backward diagonals, with the understanding that the diagonals "loop around" the matrix (see Figure 3.19).

Determinant: $aei + bfg + cdh - afh - bdi - ceg$

Figure 3.19: A mnemonic for computing the determinant of a 3×3 matrix: we add along the forward (green) diagonals and subtract along the backward (purple) diagonals.

Example 3.2.5

Computing Determinants via Explicit Formulas

The matrices in parts (a)–(c) here are the same as the those that we computed determinants of (via Gaussian elimination) in Example 3.2.3. We of course get the same answer using either method.

Compute the determinant of each of the following matrices via Theorems 3.2.6 and 3.2.7:

a) $\begin{bmatrix} 2 & 2 \\ 4 & 5 \end{bmatrix}$

b) $\begin{bmatrix} 1 & -1 \\ -1 & 1 \end{bmatrix}$

c) $\begin{bmatrix} 3 & 6 & -3 \\ 0 & -2 & 3 \\ 2 & 4 & -1 \end{bmatrix}$

d) $\begin{bmatrix} 0 & 2 & 1 \\ 1 & -1 & 1 \\ 2 & 1 & 0 \end{bmatrix}$

Solutions:

a) The determinant of a 2×2 matrix is the product of its forward diagonal minus the product of its backward diagonal:

$$2 \cdot 5 - 2 \cdot 4 = 10 - 8 = 2.$$

b) Again, we just apply Theorem 3.2.6 to see that the determinant of this 2×2 matrix is

$$1 \cdot 1 - (-1) \cdot (-1) = 1 - 1 = 0.$$

c) The determinant of a 3×3 matrix can be computed as the sum of the products of its 3 forward diagonals, minus the sum of the products of its 3 backward diagonals:

$$(3 \cdot (-2) \cdot (-1)) + (6 \cdot 3 \cdot 2) + ((-3) \cdot 0 \cdot 4)$$
$$- (3 \cdot 3 \cdot 4) - (6 \cdot 0 \cdot (-1)) - ((-3) \cdot (-2) \cdot 2)$$
$$= 6 + 36 + 0 - 36 - 0 - 12 = -6.$$

d) Once again, we just apply Theorem 3.2.7 to compute the determinant of this 3×3 matrix:

$$(0 \cdot (-1) \cdot 0) + (2 \cdot 1 \cdot 2) + (1 \cdot 1 \cdot 1)$$
$$- (0 \cdot 1 \cdot 1) - (2 \cdot 1 \cdot 0) - (1 \cdot (-1) \cdot 2)$$
$$= 0 + 4 + 1 - 0 - 0 - (-2) = 7.$$

The determinant formulas provided by Theorems 3.2.6 and 3.2.7 in fact can be generalized to matrices of any (square) size, but doing so requires a bit of setup first, since the resulting formulas become much too ugly to just write out explicitly (for example, the determinant of a 4×4 matrix is the sum and difference of 24 terms, each of which is the product of 4 of the entries of the matrix).

The idea behind the general determinant formula will be to compute determinants of large matrices in terms of determinants of smaller matrices. The following definition provides some terminology that we will need before we can introduce the general formula.

Definition 3.2.2

Minors and Cofactors

Suppose $A \in \mathcal{M}_n$ and let $1 \leq i, j \leq n$. Then

a) the $(\mathbf{i}, \mathbf{j})$-**minor** of A is the determinant of the $(n-1) \times (n-1)$ matrix that is obtained by deleting the i-th row and j-th column of A, and

b) the $(\mathbf{i}, \mathbf{j})$-**cofactor** of A is the quantity $(-1)^{i+j}m_{i,j}$, where $m_{i,j}$ is the (i,j)-minor of A.

We typically use $m_{i,j}$ to denote the (i, j)-minor of A and $c_{i,j}$ to denote the (i, j)-cofactor of A, but there is not really a standard notation for these quantities.

For example, to find the $(1,2)$-minor of the matrix

$$A = \begin{bmatrix} 1 & 2 & 3 \\ 4 & 5 & 6 \\ 7 & 8 & 9 \end{bmatrix},$$

we erase all of the entries in its first row and second column and then compute the determinant of the resulting 2×2 matrix:

For the (i, j)-minor, we erase the row and column containing the (i, j)-entry.

$$m_{1,2} = \det \left(\begin{bmatrix} 1 & 2 & 3 \\ 4 & 5 & 6 \\ 7 & 8 & 9 \end{bmatrix} \right) = \det \left(\begin{bmatrix} 4 & 6 \\ 7 & 9 \end{bmatrix} \right) = 4 \times 9 - 6 \times 7 = 36 - 42 = -6.$$

To similarly find the (i, j)-cofactor $c_{i,j}$ of A, we just multiply the corresponding minor $m_{i,j}$ by $(-1)^{i+j}$. In this case, the $(1,2)$-cofactor of A is

$$c_{1,2} = (-1)^{1+2}m_{1,2} = (-1)(-6) = 6.$$

To help us quickly and easily determine the sign of the (i, j)-cofactor, we notice that $(-1)^{i+j}$ alternates in a checkerboard pattern if we interpret i as the row and j as the column (as we usually do when working with matrices), as illustrated in Figure 3.20. The (i, j)-cofactor is thus the same as the (i, j)-minor, but with its sign multiplied by the sign in the (i, j)-entry of this checkerboard pattern. For example, the $(1,2)$-entry of this checkerboard is "$-$", so the $(1,2)$-cofactor is the *negative* of the $(1,2)$-minor (which agrees with the calculation $c_{1,2} = -m_{1,2} = -(-6) = 6$ that we did earlier).

Figure 3.20: The expression $(-1)^{i+j}$ that determines the sign of the (i, j)-cofactor of a matrix alternates back and forth between $+1$ and -1 in a checkerboard pattern.

Example 3.2.6

Computing Minors and Cofactors

Compute all minors and cofactors of the following matrices:

a) $\begin{bmatrix} 2 & 2 \\ 4 & 5 \end{bmatrix}$

b) $\begin{bmatrix} 0 & 2 & 1 \\ 1 & -1 & 1 \\ 2 & 1 & 0 \end{bmatrix}$

Solutions:

a) This matrix has 4 minors (and 4 cofactors). Its $(1,1)$-minor is

$$m_{1,1} = \det\left(\begin{bmatrix} 2 & 2 \\ 4 & 5 \end{bmatrix}\right) = \det([5]) = 5,$$

and its 3 other minors are computed in a similar manner:

$$m_{1,2} = 4, \quad m_{2,1} = 2, \quad \text{and} \quad m_{2,2} = 2.$$

Its cofactors are the same, except the $(1,2)$- and $(2,1)$-cofactors have their signs switched:

$$c_{1,1} = 5, \quad c_{1,2} = -4, \quad c_{2,1} = -2, \quad \text{and} \quad c_{2,2} = 2.$$

b) This matrix has 9 minors (and 9 cofactors). Its $(1,1)$-minor is

$$m_{1,1} = \det\left(\begin{bmatrix} 0 & 2 & 1 \\ 1 & -1 & 1 \\ 2 & 1 & 0 \end{bmatrix}\right) = \det\left(\begin{bmatrix} -1 & 1 \\ 1 & 0 \end{bmatrix}\right) = -1,$$

and similar computations show that its 9 minors are as follows:

$$m_{1,1} = -1 \qquad m_{1,2} = -2 \qquad m_{1,3} = 2$$
$$m_{2,1} = -1 \qquad m_{2,2} = -2 \qquad m_{2,3} = 5$$
$$m_{3,1} = 3 \qquad m_{3,2} = -1 \qquad m_{3,3} = -2.$$

Its cofactors are the same, except the $(1,2)$-, $(2,1)$-, $(2,3)$-, and $(3,2)$-cofactors have their signs switched (we have highlighted these four cofactors below):

$$c_{1,1} = -1 \qquad c_{1,2} = 2 \qquad c_{1,3} = 2$$
$$c_{2,1} = 1 \qquad c_{2,2} = -2 \qquad c_{2,3} = -5$$
$$c_{3,1} = 3 \qquad c_{3,2} = 1 \qquad c_{3,3} = -2.$$

> The cofactors are exactly the same as the minors, but the signs switch according to the checkerboard pattern described earlier.

We are now able to state the main result of this section, which provides us with a recursive formula for computing the determinant of a matrix of any size in terms of determinants of smaller sizes. By repeatedly applying this result until we get down to 2×2 or 3×3 determinants (at which point we can use the formulas provided by Theorems 3.2.6 and 3.2.7), we can compute the determinant of a matrix of any size "directly" from its entries (without having to row-reduce as in the previous section).

Theorem 3.2.8

Cofactor Expansion

Suppose $A \in \mathcal{M}_n$ has cofactors $\{c_{i,j}\}_{i,j=1}^{n}$. Then

$$\det(A) = a_{i,1}c_{i,1} + a_{i,2}c_{i,2} + \cdots + a_{i,n}c_{i,n} \quad \text{for all} \quad 1 \leq i \leq n \quad \text{and}$$
$$\det(A) = a_{1,j}c_{1,j} + a_{2,j}c_{2,j} + \cdots + a_{n,j}c_{n,j} \quad \text{for all} \quad 1 \leq j \leq n.$$

Before proving this theorem, we clarify that it provides us with numerous different ways of computing the determinant of a matrix. The first formula is called a **cofactor expansion along the i-th row**, and it tells us how to compute the determinant of A from the entries in its i-th row as well as the cofactors coming from that row. Similarly, the second formula is called a **cofactor expansion along the j-th column**, and it tells us how to compute the determinant of A from the entries in its j-th column as well as the cofactors coming from that column. Remarkably, the same answer is obtained no matter which row or column is chosen.

Some other books call cofactor expansions **Laplace expansions** instead.

Example 3.2.7

Determinant via Multiple Cofactor Expansions

Compute the determinant of the following matrix via the 3 indicated cofactor expansions:

$$A = \begin{bmatrix} 2 & 1 & -1 & 0 \\ 0 & -2 & 1 & 3 \\ 0 & 0 & 1 & 0 \\ -1 & 3 & 0 & 2 \end{bmatrix}.$$

a) Along its first row.
b) Along its third row.
c) Along its second column.

Solutions:

There's no direct benefit to computing the determinant via multiple different cofactor expansions— just one suffices. This example is just meant to illustrate the fact that we really do get the same answer no matter which cofactor expansion we choose.

a) We add and subtract four 3×3 determinants (cofactors), being careful to adjust their signs according to the checkerboard pattern discussed earlier:

$$\begin{bmatrix} \overset{+}{2} & \overset{-}{1} & \overset{+}{-1} & 0 \\ 0 & -2 & 1 & 3 \\ 0 & 0 & 1 & 0 \\ -1 & 3 & 0 & 2 \end{bmatrix}$$

Then

$$\det(A) = 2\det\left(\begin{bmatrix} -2 & 1 & 3 \\ 0 & 1 & 0 \\ 3 & 0 & 2 \end{bmatrix}\right) - \det\left(\begin{bmatrix} 0 & 1 & 3 \\ 0 & 1 & 0 \\ -1 & 0 & 2 \end{bmatrix}\right)$$
$$+ (-1)\det\left(\begin{bmatrix} 0 & -2 & 3 \\ 0 & 0 & 0 \\ -1 & 3 & 2 \end{bmatrix}\right) - 0\det\left(\begin{bmatrix} 0 & -2 & 1 \\ 0 & 0 & 1 \\ -1 & 3 & 0 \end{bmatrix}\right)$$
$$= 2 \cdot (-13) - 3 + 0 - 0$$
$$= -29.$$

b) Again, we add and subtract four 3×3 determinants (cofactors), but this time along the third row of A instead of its first row:

$$\begin{bmatrix} 2 & 1 & -1 & 0 \\ 0 & -2 & 1 & 3 \\ 0 & 0 & 1 & 0 \\ -1 & 3 & 0 & 2 \end{bmatrix}$$

Then

$$\det(A) = 0\det\left(\begin{bmatrix} 1 & -1 & 0 \\ -2 & 1 & 3 \\ 3 & 0 & 2 \end{bmatrix}\right) - 0\det\left(\begin{bmatrix} 2 & -1 & 0 \\ 0 & 1 & 3 \\ -1 & 0 & 2 \end{bmatrix}\right)$$

$$+ \det\left(\begin{bmatrix} 2 & 1 & 0 \\ 0 & -2 & 3 \\ -1 & 3 & 2 \end{bmatrix}\right) - 0\det\left(\begin{bmatrix} 2 & 1 & -1 \\ 0 & -2 & 1 \\ -1 & 3 & 0 \end{bmatrix}\right)$$

$$= 0 - 0 + (-29) - 0$$

$$= -29.$$

We will get an answer of −29 no matter which row or column we expand along. Determinants are basically magic.

Notice that we did not actually need to compute 3 of the 4 determinants above, since they were multiplied by zeros. This illustrates an important technique to keep in mind when computing determinants via cofactor expansions: choose a cofactor expansion along a row or column with lots of zeros in it.

c) This time, we add and subtract multiples of the four cofactors coming from the second column of A:

$$\begin{bmatrix} 2 & 1 & -1 & 0 \\ 0 & -2 & 1 & 3 \\ 0 & 0 & 1 & 0 \\ -1 & 3 & 0 & 2 \end{bmatrix}$$

Be careful! When expanding along the 2nd column, the cofactors signs start with a negative instead of a positive. Again, refer to the checkerboard sign pattern of Figure 3.20.

Then

$$\det(A) = -\det\left(\begin{bmatrix} 0 & 1 & 3 \\ 0 & 1 & 0 \\ -1 & 0 & 2 \end{bmatrix}\right) + (-2)\det\left(\begin{bmatrix} 2 & -1 & 0 \\ 0 & 1 & 0 \\ -1 & 0 & 2 \end{bmatrix}\right)$$

$$- 0\det\left(\begin{bmatrix} 2 & -1 & 0 \\ 0 & 1 & 3 \\ -1 & 0 & 2 \end{bmatrix}\right) + 3\det\left(\begin{bmatrix} 2 & -1 & 0 \\ 0 & 1 & 3 \\ 0 & 1 & 0 \end{bmatrix}\right)$$

$$= -3 + (-2)\cdot 4 - 0 + 3\cdot(-6)$$

$$= -29,$$

which agrees with the answer that we computed in parts (a) and (b).

Proof of Theorem 3.2.8. We start by proving the theorem in the $j=1$ case (i.e., we show that $\det(A)$ equals its cofactor expansion along the first column of A). To this end, we use multilinearity of the determinant on the first column of A to write $\det(A)$ as the sum of n determinants:

$$\det(A) = \det \left(\begin{bmatrix} a_{1,1} & a_{1,2} & \cdots & a_{1,n} \\ 0 & a_{2,2} & \cdots & a_{2,n} \\ \vdots & \vdots & \ddots & \vdots \\ 0 & a_{n,2} & \cdots & a_{n,n} \end{bmatrix} \right)$$

$$+ \det \left(\begin{bmatrix} 0 & a_{1,2} & \cdots & a_{1,n} \\ a_{2,1} & a_{2,2} & \cdots & a_{2,n} \\ \vdots & \vdots & \ddots & \vdots \\ 0 & a_{n,2} & \cdots & a_{n,n} \end{bmatrix} \right) + \cdots$$

(3.2.2)

The goal now is to show that the n terms in the above sum equal the n terms in the cofactor expansion. The first determinant in the sum (3.2.2) can be computed as

$$\det \left(\begin{bmatrix} a_{1,1} & a_{1,2} & \cdots & a_{1,n} \\ 0 & a_{2,2} & \cdots & a_{2,n} \\ \vdots & \vdots & \ddots & \vdots \\ 0 & a_{n,2} & \cdots & a_{n,n} \end{bmatrix} \right) = a_{1,1} \det \left(\begin{bmatrix} a_{2,2} & \cdots & a_{2,n} \\ \vdots & \ddots & \vdots \\ a_{n,2} & \cdots & a_{n,n} \end{bmatrix} \right) = a_{1,1} c_{1,1},$$

with the second equality following exactly from the definition of the cofactor $c_{1,1}$. To see why the first equality holds, notice that if the matrix

$$\begin{bmatrix} a_{2,2} & \cdots & a_{2,n} \\ \vdots & \ddots & \vdots \\ a_{n,2} & \cdots & a_{n,n} \end{bmatrix} \quad \text{has row echelon form} \quad R,$$

then

$$\begin{bmatrix} a_{1,1} & a_{1,2} & \cdots & a_{1,n} \\ 0 & a_{2,2} & \cdots & a_{2,n} \\ \vdots & \vdots & \ddots & \vdots \\ 0 & a_{n,2} & \cdots & a_{n,n} \end{bmatrix} \quad \text{has row echelon form} \quad \left[\begin{array}{c|ccc} a_{1,1} & a_{1,2} & \cdots & a_{1,n} \\ 0 & & & \\ \vdots & & R & \\ 0 & & & \end{array} \right],$$

and furthermore the same set of row operations (shifted down by one row for the latter matrix) can be used to put these two matrices into these row echelon forms. Since the latter row echelon form has the same diagonal entries as R, with the addition of $a_{1,1}$, it follows from Theorem 3.2.5 that

$$\det \left(\begin{bmatrix} a_{1,1} & a_{1,2} & \cdots & a_{1,n} \\ 0 & a_{2,2} & \cdots & a_{2,n} \\ \vdots & \vdots & \ddots & \vdots \\ 0 & a_{n,2} & \cdots & a_{n,n} \end{bmatrix} \right) = a_{1,1} \det \left(\begin{bmatrix} a_{2,2} & \cdots & a_{2,n} \\ \vdots & \ddots & \vdots \\ a_{n,2} & \cdots & a_{n,n} \end{bmatrix} \right),$$

as claimed.

The other terms in the sum (3.2.2) can be computed in a similar manner. For example, the second term can be computed by first swapping the first two rows in the matrix:

$$\det\left(\begin{bmatrix} 0 & a_{1,2} & \cdots & a_{1,n} \\ a_{2,1} & a_{2,2} & \cdots & a_{2,n} \\ \vdots & \vdots & \ddots & \vdots \\ 0 & a_{n,2} & \cdots & a_{n,n} \end{bmatrix}\right) = -\det\left(\begin{bmatrix} a_{2,1} & a_{2,2} & a_{2,3} & \cdots & a_{2,n} \\ 0 & a_{1,2} & a_{1,3} & \cdots & a_{1,n} \\ 0 & a_{3,2} & a_{3,3} & \cdots & a_{3,n} \\ \vdots & \vdots & \vdots & \ddots & \vdots \\ 0 & a_{n,2} & a_{n,3} & \cdots & a_{n,n} \end{bmatrix}\right)$$

$$= -a_{2,1}\det\left(\begin{bmatrix} a_{1,2} & a_{1,3} & \cdots & a_{1,n} \\ a_{3,2} & a_{3,3} & \cdots & a_{3,n} \\ \vdots & \vdots & \ddots & \vdots \\ a_{n,2} & a_{n,3} & \cdots & a_{n,n} \end{bmatrix}\right)$$

$$= a_{2,1}c_{2,1}.$$

The remaining terms in the sum are similar, with the only change being that, for the i-th term in the sum, we perform $i-1$ swap operations: first we swap rows i and $i-1$, then $i-1$ and $i-2$, then $i-2$ and $i-3$, and so on until rows 2 and 1. These swaps introduce a sign of $(-1)^{i-1}$, which explains why the $c_{i,1}$ cofactor is signed the way it is.

We have thus showed that the determinant equals its cofactor expansion along its first column:

$$\det(A) = a_{1,1}c_{1,1} + a_{2,1}c_{2,1} + \cdots + a_{n,1}c_{n,1}.$$

To see that it also equals its cofactor expansion along its second column, just swap the first and second column of A before performing a cofactor expansion along its first column (which switches the signs of the cofactors in the second column, as expected). A similar argument applies for the other columns. To see that it even equals its cofactor expansions along any of its rows (instead of columns), recall from Theorem 3.2.2(b) that $\det(A) = \det(A^T)$, and a cofactor expansion along a row of A is equivalent to an expansion along a column of A^T. ∎

To make use of cofactor expansions for large matrices, we apply it recursively. For example, to compute the determinant of a 5×5 matrix, a cofactor expansion requires us to compute the determinant of five 4×4 matrices, each of which could be computed via cofactor expansions involving four 3×3 determinants, each of which can be computed by the explicit formula of Theorem 3.2.7. However, cofactor expansions quickly become very computationally expensive, so the method of computing the determinant based on Gaussian elimination (Theorem 3.2.5) is typically used in practice.

However, cofactor expansions are nevertheless very useful in certain situations. For example, if the matrix has a row or column consisting almost entirely of zeros, then performing a cofactor expansion along it is fairly quick. We illustrated this fact in Example 3.2.7(b), but it is worthwhile to do one more example to drive this point home.

Example 3.2.8

Determinant of a Matrix with Many Zeros

Compute the determinant of the following matrix:

$$\begin{bmatrix} 0 & -1 & 2 & 1 & 3 \\ 0 & 0 & 0 & 2 & 0 \\ -2 & 1 & 1 & -1 & 0 \\ 1 & 0 & -3 & 1 & 0 \\ 2 & 1 & -1 & 0 & 0 \end{bmatrix}$$

Solution:

Since the rightmost column of this matrix contains many zeros, we perform a cofactor expansion along that column, and then we perform a cofactor expansion along the top row of the resulting 4×4 matrix:

$$\det\left(\begin{bmatrix} 0 & -1 & 2 & 1 & 3 \\ 0 & 0 & 0 & 2 & 0 \\ -2 & 1 & 1 & -1 & 0 \\ 1 & 0 & -3 & 1 & 0 \\ 2 & 1 & -1 & 0 & 0 \end{bmatrix}\right) = 3\det\left(\begin{bmatrix} 0 & 0 & 0 & 2 \\ -2 & 1 & 1 & -1 \\ 1 & 0 & -3 & 1 \\ 2 & 1 & -1 & 0 \end{bmatrix}\right)$$

The determinant of this 3×3 matrix is -10.

$$= -6\det\left(\begin{bmatrix} -2 & 1 & 1 \\ 1 & 0 & -3 \\ 2 & 1 & -1 \end{bmatrix}\right) = 60.$$

Cofactor expansions are also a very useful theoretical tool, and can be used to construct an explicit formula for the determinant of a matrix of any size. For example, we can use cofactor expansions, along with the formula for the determinant of a 2×2 matrix, to re-derive the formula for the determinant of a 3×3 matrix:

Here we are performing a cofactor expansion along the first row of the matrix.

$$\det\left(\begin{bmatrix} a & b & c \\ d & e & f \\ g & h & i \end{bmatrix}\right) = a\det\left(\begin{bmatrix} e & f \\ h & i \end{bmatrix}\right) - b\det\left(\begin{bmatrix} d & f \\ g & i \end{bmatrix}\right) + c\det\left(\begin{bmatrix} d & e \\ g & h \end{bmatrix}\right)$$

$$= a(ei - fh) - b(di - fg) + c(dh - eg)$$
$$= aei + bfg + cdh - afh - bdi - ceg,$$

which matches the formula of Theorem 3.2.7. Similarly, if we then use this same method on a 4×4 matrix, we can derive the formula

It only gets worse as matrices get larger—the formula for the determinant of a 5×5 matrix has 120 terms in it.

$$\det\left(\begin{bmatrix} a & b & c & d \\ e & f & g & h \\ i & j & k & \ell \\ m & n & o & p \end{bmatrix}\right) = afkp - af\ell o - agjp + ag\ell n + ahjo - ahkn$$
$$- bekp + be\ell o + bgip - bg\ell m - bhio + bhkm$$
$$+ cejp - ce\ell n - cfip + cf\ell m + chin - chjm$$
$$- dejo + dekn + dfio - dfkm - dgin + dgjm,$$

which illustrates why we do not typically use explicit formulas for matrices of size 4×4 and larger.

Our primary purpose for introducing cofactor expansions is that they are extremely useful for computing and working with eigenvalues, which we learn about in the next section. Other uses of cofactors (and determinants in general) are also explored in Section 3.A.

Remark 3.2.2

Existence and Uniqueness of the Determinant

Throughout this section, we used the three defining properties of the determinant from Definition 3.2.1 to develop several different formulas that can be used to compute it. Those formulas can be thought of as showing that the determinant is unique—there can only be one function satisfying those three defining properties since we used those properties to come up with formulas (like the cofactor expansion along the first row of the matrix) for computing it.

However, we never actually showed that the determinant *exists* (i.e., that it is *well-defined*). When defining something abstractly via properties that it has (rather than an explicit formula), this is a subtle issue that must be dealt with, as it is possible to "define" things that do not actually make any sense or do not exist. To illustrate what we mean by this, consider the "function" $f : \mathcal{M}_n \to \mathbb{R}$ that is defined just like the determinant, except we require that $f(I) = 2$ instead of $f(I) = 1$:

  a) $f(I) = 2$, and
  b) $f(AB) = f(A)f(B)$ for all $A, B \in \mathcal{M}_n$.

We can come up with formulas for this function f just like we did for the determinant in this section, but there is one big problem—f does not actually exist. That is, no function f actually satisfies both of the properties (a) and (b) described above. To see why, simply let $A = B = I$ in property (b) above to see that

$$2 = f(I) = f(I)f(I) = 2 \cdot 2 = 4,$$

For other examples of how a function defined by the properties it satisfies might fail to exist, see Exercises 3.1.20 and 3.2.22.

which makes no sense.

Showing that the determinant does not suffer from this same problem is somewhat fiddly and not terribly enlightening, so we leave it to Appendix B.4. The idea behind the proof is to show that another formula that we have not yet seen (but we will see in Section 3.A) satisfies the three defining properties of the determinant, so the determinant could instead be defined via that explicit formula rather than in terms of the properties that it has.

Exercises

solutions to starred exercises on page 464

3.2.1 Compute the determinant of the following matrices.

*(a) $\begin{bmatrix} 1 & -1 \\ -1 & 1 \end{bmatrix}$

(b) $\begin{bmatrix} 1 & 2 \\ 3 & 5 \end{bmatrix}$

*(c) $\begin{bmatrix} 2 & 4 & 0 \\ 1 & -2 & 0 \\ 2 & 0 & -1 \end{bmatrix}$

(d) $\begin{bmatrix} 1 & 3 & 0 \\ 0 & -2 & 2 \\ -1 & 0 & 1 \end{bmatrix}$

*(e) $\begin{bmatrix} 2 & 0 & 1 \\ 5 & 2 & 8 \\ 3 & -2 & 7 \end{bmatrix}$

(f) $\begin{bmatrix} 1 & 1 & 3 \\ -4 & 2 & 1 \\ 3 & 1 & 2 \end{bmatrix}$

*(g) $\begin{bmatrix} 3 & 2 & -6 \\ 0 & 2 & 2 \\ 0 & 0 & 3 \end{bmatrix}$

(h) $\begin{bmatrix} 1 & 2 & 1 & 0 \\ 2 & 1 & 2 & -3 \\ 4 & 3 & 1 & 2 \\ 0 & 0 & 2 & -2 \end{bmatrix}$

□ 3.2.2 Use computer software to compute the determinant of the following matrices.

*(a) $\begin{bmatrix} 3 & 1 & 6 & 4 \\ -2 & 5 & 3 & 3 \\ 6 & -2 & 4 & 1 \\ 3 & 4 & -2 & 6 \end{bmatrix}$

(b) $\begin{bmatrix} -2 & 6 & -1 & 6 & -1 \\ 5 & 4 & -1 & 3 & -1 \\ 4 & 3 & 4 & -2 & 1 \\ 6 & 6 & 5 & 4 & 2 \\ -1 & 2 & -2 & -1 & -2 \end{bmatrix}$

*(c) $\begin{bmatrix} 2 & 2 & 5 & -1 & 1 & -1 \\ 1 & 3 & 0 & 0 & 1 & 1 \\ 1 & 2 & 5 & -1 & 0 & 3 \\ 4 & 0 & 5 & 5 & -1 & 3 \\ 5 & -1 & 5 & 2 & 0 & -1 \\ 3 & 0 & 3 & 3 & 4 & 3 \end{bmatrix}$

(d) $\begin{bmatrix} -2 & 3 & 2 & 5 & 2 & 4 \\ -1 & 2 & -1 & -3 & 2 & 0 \\ 6 & -2 & 3 & 5 & 4 & -1 \\ -2 & 2 & -2 & -2 & -3 & 3 \\ 3 & 3 & 2 & 2 & -3 & 1 \\ 1 & 6 & 0 & -1 & -3 & -2 \end{bmatrix}$

3.2.3 Suppose that $A, B, C \in \mathcal{M}_5$ are such that $\det(A) = 2$, $\det(B) = 3$, and $\det(C) = 0$. Compute the determinant of the following matrices.

*(a) AB^T (b) ABC
*(c) A^2B (d) $2A^2$
*(e) $A^{-1}B^3$ (f) $A^3B^{-2}A^{-4}B^2A$
*(g) $3A^2(B^TB/2)^{-1}$ (h) $A^{64}B^{-14}C^9A^{-6}B^{13}$

3.2.4 Suppose that $A \in \mathcal{M}_4$ is a matrix with $\det(A) = 6$. Compute the determinant of the matrices that are obtained from A by applying the following sequences of row operations:

*(a) $R_1 - 2R_2$
(b) $3R_3$
*(c) $R_1 \leftrightarrow R_4$
(d) $R_2 - 2R_1, R_3 - 4R_1$
*(e) $2R_1, R_2 \leftrightarrow R_4$
(f) $R_1 \leftrightarrow R_2, R_2 \leftrightarrow R_3$
*(g) $2R_2, 3R_3, 4R_4$
(h) $R_1 \leftrightarrow R_3, 2R_2, R_3 - 3R_1, R_4 - 7R_2, 3R_4$

3.2.5 Suppose that

$$\det\left(\begin{bmatrix} a & b & c \\ d & e & f \\ g & h & i \end{bmatrix}\right) = 4.$$

Compute the determinant of the following matrices.

*(a) $\begin{bmatrix} g & h & i \\ d & e & f \\ a & b & c \end{bmatrix}$ (b) $\begin{bmatrix} a & b & c \\ 2d & 2e & 2f \\ 3g & 3h & 3i \end{bmatrix}$

*(a) $\begin{bmatrix} a & b+a & 2c \\ d & e+d & 2f \\ g & h+g & 2i \end{bmatrix}$ (d) $\begin{bmatrix} a & 2b & 3c \\ 2d & 4e & 6f \\ 3g & 6h & 9i \end{bmatrix}$

3.2.6 Determine which of the following statements are true and which are false.

*(a) For all square matrices A, it is true that $\det(-A) = -\det(A)$.
(b) If A and B are square matrices of the same size, then $\det(AB) = \det(BA)$.
*(c) If A and B are square matrices of the same size, then $\det(A+B) = \det(A) + \det(B)$.
(d) If A is a 3×3 matrix with $\det(A) = 3$, then $\text{rank}(A) = 3$.
*(e) If A is a square matrix whose reduced row echelon form is I, then $\det(A) = 1$.
(f) If A is a square matrix whose reduced row echelon form is I, then $\det(A) \neq 0$.
*(g) If the columns of a square matrix A are linearly dependent, then $\det(A) = 0$.
(h) If a matrix B is obtained from $A \in \mathcal{M}_4$ by swapping rows 1 and 2 and also swapping rows 3 and 4, then $\det(A) = \det(B)$.

3.2.7 Let B_n be the $n \times n$ "backwards identity" matrix whose entries on the diagonal from its top-right to bottom-left equal 1, and whose other entries equal 0. For example,

$$B_2 = \begin{bmatrix} 0 & 1 \\ 1 & 0 \end{bmatrix} \quad \text{and} \quad B_3 = \begin{bmatrix} 0 & 0 & 1 \\ 0 & 1 & 0 \\ 1 & 0 & 0 \end{bmatrix}.$$

Find a formula for $\det(B_n)$ in terms of n.

*3.2.8 Provide an example to show that it is not necessarily the case that $\det(AB) = \det(BA)$ when A and B are not square. Both of the products AB and BA should exist in your example.

3.2.9 Compute the determinant of the matrix

$$\begin{bmatrix} \cos(\theta) & -\sin(\theta) \\ \sin(\theta) & \cos(\theta) \end{bmatrix}$$

and provide a geometric interpretation of your answer.

*3.2.10 Let $[P_{\mathbf{u}}]$ be the standard matrix of a projection onto a line (see Section 1.4.2). Compute $\det([P_{\mathbf{u}}])$.

3.2.11 Suppose $A \in \mathcal{M}_n$ and r is any integer. Show that $\det(A^r) = (\det(A))^r$.

3.2.12 A matrix $A \in \mathcal{M}_n$ is called **skew-symmetric** if $A^T = -A$. Show that skew-symmetric matrices are not invertible if n is odd.

[Hint: Determinants.]

∗∗3.2.13 Recall Theorem 3.2.2, which established some of the basic properties of the determinant.

(a) Prove part (a) of the theorem.
(b) Prove part (b) of the theorem. [Hint: Write A as a product of elementary matrices.]

3.2.14 Given a matrix $A \in \mathcal{M}_n$, let A^R denote the matrix that is obtained by rotating the entries of A in the clockwise direction by 90 degrees. For example, for 2×2 and 3×3 matrices we have

$$\begin{bmatrix} a & b \\ c & d \end{bmatrix}^R = \begin{bmatrix} c & a \\ d & b \end{bmatrix}, \quad \begin{bmatrix} a & b & c \\ d & e & f \\ g & h & i \end{bmatrix}^R = \begin{bmatrix} g & d & a \\ h & e & b \\ i & f & c \end{bmatrix}.$$

Find a formula for $\det(A^R)$ in terms of $\det(A)$ and n. [Hint: Try to make use of Exercise 3.2.7.]

□ **3.2.15** Let A_n be the $n \times n$ matrix

$$A_n = \begin{bmatrix} 1 & 1 & 1 & \cdots & 1 \\ 1 & 2 & 1 & \cdots & 1 \\ 1 & 1 & 3 & \cdots & 1 \\ \vdots & \vdots & \vdots & \ddots & \vdots \\ 1 & 1 & 1 & \cdots & n \end{bmatrix}.$$

(a) Use computer software to compute $\det(A_n)$ when $n = 2, 3, 4, 5$.
(b) Based on the computations from part (a), guess a formula for $\det(A_n)$ that works for all n.
(c) Prove that your formula from part (b) is correct.

∗∗3.2.16 In this exercise, we show that Theorem 3.2.4 can be generalized to *block* triangular matrices.

(a) Show that

$$\det\left(\begin{bmatrix} A & B \\ O & C \end{bmatrix}\right) = \det(A)\det(C).$$

(b) Show that

$$\det\left(\begin{bmatrix} A_1 & * & \cdots & * \\ O & A_2 & \cdots & * \\ \vdots & \vdots & \ddots & \vdots \\ O & O & \cdots & A_n \end{bmatrix}\right) = \prod_{j=1}^{n} \det(A_j),$$

where asterisks ($*$) denote blocks whose values are irrelevant but potentially non-zero.
[Side note: The notation $\prod_{j=1}^{n}$ refers to a product (multiplication) in the same way that the notation $\sum_{j=1}^{n}$ refers to a sum (addition).]

3.2.17 Suppose $A \in \mathcal{M}_m$, $B \in \mathcal{M}_{m,n}$, $C \in \mathcal{M}_{n,m}$, and $D \in \mathcal{M}_n$. Provide an example to show that it is *not* necessarily the case that

$$\det\left(\begin{bmatrix} A & B \\ C & D \end{bmatrix}\right) = \det(A)\det(D) - \det(B)\det(C).$$

∗3.2.18 Suppose $A \in \mathcal{M}_{m,n}$ and $B \in \mathcal{M}_{n,m}$. Show that

$$\det(I_m + AB) = \det(I_n + BA).$$

[Note: This is called **Sylvester's determinant identity**.]
[Hint: Use Exercise 3.2.16(a) to compute the determinant of the product

$$\begin{bmatrix} I_m & -A \\ B & I_n \end{bmatrix}\begin{bmatrix} I_m & A \\ O & I_n \end{bmatrix}$$

in two different ways.]

3.2.19 Recall from Example 2.3.8 that a **Vandermonde matrix** is a square matrix of the form

$$V = \begin{bmatrix} 1 & a_0 & a_0^2 & \cdots & a_0^n \\ 1 & a_1 & a_1^2 & \cdots & a_1^n \\ \vdots & \vdots & \vdots & \ddots & \vdots \\ 1 & a_n & a_n^2 & \cdots & a_n^n \end{bmatrix},$$

where $a_0, a_1, \ldots, a_n$ are real numbers. Show that

$$\det(V) = \prod_{0 \le i < j \le n} (a_j - a_i).$$

[Side note: This exercise generalizes Example 2.3.8, which showed that Vandermonde matrices are invertible whenever $a_0, a_1, \ldots, a_n$ are distinct.]

[Hint: Try a proof by induction, and use row operations to turn an $(n+1) \times (n+1)$ Vandermonde matrix into something that resembles an $n \times n$ Vandermonde matrix.]

∗3.2.20 The **matrix determinant lemma** is a result that says that if $A \in \mathcal{M}_n$ is invertible and $\mathbf{v}, \mathbf{w} \in \mathbb{R}^n$, then

$$\det(A + \mathbf{v}\mathbf{w}^T) = (1 + \mathbf{w}^T A^{-1} \mathbf{v})\det(A).$$

This exercise guides you through a proof of this lemma.

(a) Use block matrix multiplication to compute the product

$$\begin{bmatrix} I_n & \mathbf{0} \\ \mathbf{w}^T & 1 \end{bmatrix}\begin{bmatrix} I_n + \mathbf{v}\mathbf{w}^T & \mathbf{v} \\ \mathbf{0}^T & 1 \end{bmatrix}\begin{bmatrix} I_n & \mathbf{0} \\ -\mathbf{w}^T & 1 \end{bmatrix}.$$

(b) Use part (a) to show that $\det(I_n + \mathbf{v}\mathbf{w}^T) = 1 + \mathbf{w}^T \mathbf{v}$.
[Hint: Use Exercise 3.2.16 to compute the determinant of block triangular matrices.]
(c) Show that $\det(A + \mathbf{v}\mathbf{w}^T) = (1 + \mathbf{w}^T A^{-1} \mathbf{v})\det(A)$.
[Hint: Write $A + \mathbf{v}\mathbf{w}^T = A(I_n + A^{-1}\mathbf{v}\mathbf{w}^T)$.]

[Side note: This lemma is useful in situations where you know lots of information about A (its inverse and determinant) and want to know the determinant of a slight perturbation of it. See also Exercise 2.2.23.]

3.2.21 Use the matrix determinant lemma (Exercise 3.2.20) to show that the standard matrix of a reflection through a line in $\mathbb{R}^n$ (see Section 1.4.2) has determinant equal to $(-1)^{n+1}$.

3.2.22 Show that there does not exist a function $f : \mathcal{M}_n \to \mathbb{R}$ that has the same defining properties as the determinant, except with multilinearity replaced by linearity:

a) $f(I) = 1$,
b) $f(AB) = f(A)f(B)$ for all $A, B \in \mathcal{M}_n$, and
c) for all $c \in \mathbb{R}$ and $A, B \in \mathcal{M}_n$, we have $f(A + cB) = f(A) + cf(B)$.

[Hint: Compute $f(\mathbf{e}_i \mathbf{e}_j^T)$ in multiple different ways.]

3.3 Eigenvalues and Eigenvectors

Diagonal matrices and linear transformations were introduced in Section 1.4.2.

Some linear transformations behave very well when they act on certain specific vectors. For example, diagonal linear transformations simply stretch the standard basis vectors, but do not change their direction, as illustrated in Figure 3.21. Algebraically, this property corresponds to the fact that if $A \in \mathcal{M}_n$ is diagonal then $A\mathbf{e}_j = a_{j,j}\mathbf{e}_j$ for all $1 \leq j \leq n$.

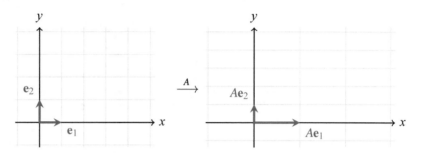

Figure 3.21: A diagonal matrix can stretch the standard basis vectors, but cannot change their direction.

This situation, where matrix multiplication behaves just like scalar multiplication, is extremely desirable since scalar multiplication is so much simpler to work with than matrix multiplication in general. This idea leads naturally to the following definition.

Definition 3.3.1

Eigenvalues and Eigenvectors

We sometimes say that $\mathbf{v}$ corresponds to λ. It does not matter which corresponds to which—we think of them as a pair.

Suppose A is a square matrix. A non-zero vector $\mathbf{v}$ is called an **eigenvector** of A if there is a scalar λ such that

$$A\mathbf{v} = \lambda\mathbf{v}.$$

Such a scalar λ is called the **eigenvalue** of A corresponding to $\mathbf{v}$.

In other words, an eigenvector of a matrix A is a vector that is just stretched by A, but not rotated by it, and the corresponding eigenvalue describes how much it is stretched by. In the case of diagonal matrices, the standard basis vectors $\mathbf{e}_1, \mathbf{e}_2, \ldots, \mathbf{e}_n$ are eigenvectors, and their corresponding eigenvalues are A's diagonal entries $a_{1,1}, a_{2,2}, \ldots, a_{n,n}$.

Remark 3.3.1

Eigenvalues, Yeah!

It is impossible to overstate the importance of eigenvalues and eigenvectors. Everything that we have learned so far in this book has been in preparation for this section—we will need to know how to solve linear systems, find a basis of the null space of a matrix, compute determinants and use them to check invertibility of matrices, construct coordinate vectors and use linear transformations to manipulate them, and so on. This section together with Section 3.4 are the punchline of this book.

3.3.1 Computation of Eigenvalues and Eigenvectors

Computing eigenvalues and eigenvectors of non-diagonal matrices is a somewhat involved process, so we first present some examples to illustrate how we can find an eigenvalue if we are given an eigenvector, or vice-versa.

Example 3.3.1

Computing an Eigenvalue, Given an Eigenvector

Suppose that

$$A = \begin{bmatrix} 2 & 1 \\ 1 & 2 \end{bmatrix} \quad \text{and} \quad \mathbf{v} = \begin{bmatrix} 1 \\ 1 \end{bmatrix}.$$

Show that $\mathbf{v}$ is an eigenvector of A, and find its corresponding eigenvalue.

Solution:

We just compute $A\mathbf{v}$ and see how much $\mathbf{v}$ is stretched:

$$A\mathbf{v} = \begin{bmatrix} 2 & 1 \\ 1 & 2 \end{bmatrix} \begin{bmatrix} 1 \\ 1 \end{bmatrix} = \begin{bmatrix} 3 \\ 3 \end{bmatrix} = 3\mathbf{v}.$$

Since $A\mathbf{v} = 3\mathbf{v}$, the eigenvalue corresponding to $\mathbf{v}$ is $\lambda = 3$. The effect of this matrix as a linear transformation is illustrated below. Notice that $\mathbf{v}$ is just stretched by A, but its direction remains unchanged:

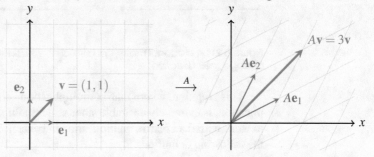

Example 3.3.2

Computing an Eigenvector, Given an Eigenvalue

After moving all variables to the left, this linear system has the form

$v_1 + v_2 = 0$

$v_1 + v_2 = 0$

Recall that eigenvectors are, by definition, non-zero. The reason for this is that $A\mathbf{0} = \lambda\mathbf{0}$ for all matrices A and all scalars λ, so if $\mathbf{0}$ were allowed as an eigenvector then every number λ would be an eigenvalue of every matrix.

Suppose that A is as in the previous example (Example 3.3.1). Show that $\lambda = 1$ is an eigenvalue of A, and find an eigenvector that it corresponds to.

Solution:

We want to find a non-zero vector $\mathbf{v} \in \mathbb{R}^2$ such that $A\mathbf{v} = 1\mathbf{v}$. This is a system of linear equations in the entries v_1 and v_2 of $\mathbf{v}$:

$$2v_1 + v_2 = v_1$$
$$v_1 + 2v_2 = v_2$$

If we move all of the variables in this linear system to the left-hand side, we can solve it via Gaussian elimination:

$$\begin{bmatrix} 1 & 1 & 0 \\ 1 & 1 & 0 \end{bmatrix} \xrightarrow{R_2 - R_1} \begin{bmatrix} 1 & 1 & 0 \\ 0 & 0 & 0 \end{bmatrix}$$

We thus see that v_2 is a free variable and v_1 is a leading variable with $v_1 = -v_2$. In other words, the eigenvectors are the (non-zero) vectors of the form $\mathbf{v} = (-v_2, v_2)$. We (arbitrarily) choose $v_2 = 1$ so that we get $\mathbf{v} = (-1, 1)$ as an eigenvector of A with corresponding eigenvalue $\lambda = 1$.

The effect of this matrix as a linear transformation is once again illustrated below. Notice that $\mathbf{v}$ is unaffected by A:

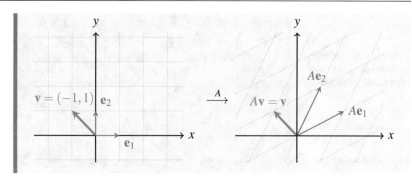

The previous examples illustrate how to find *either* an eigenvalue *or* an eigenvector, if we are given the other one. However, if we are given neither the eigenvalues nor the eigenvectors (as is generally the case), it is not quite so straightforward to find them. We thus typically use the following two-step procedure, which starts by computing the eigenvalues, and then computes the eigenvectors corresponding to those eigenvalues.

Step 1: Compute the eigenvalues. Recall that λ is an eigenvalue of A if and only if there is a non-zero vector $\mathbf{v}$ such that $A\mathbf{v} = \lambda\mathbf{v}$. By moving both terms to the same side of this equation and then factoring, we see that

$$A\mathbf{v} = \lambda\mathbf{v} \iff A\mathbf{v} - \lambda\mathbf{v} = \mathbf{0} \iff (A - \lambda I)\mathbf{v} = \mathbf{0}.$$

> Be careful when factoring $A\mathbf{v} - \lambda\mathbf{v}$. It is tempting to write $(A - \lambda)\mathbf{v} = \mathbf{0}$, but this does not make sense (what is a matrix minus a scalar?).

It follows that λ is an eigenvalue of A if and only if the linear system $(A - \lambda I)\mathbf{v} = \mathbf{0}$ has a non-zero solution. By Theorem 2.2.4, this is equivalent to $A - \lambda I$ not being invertible. We have thus arrived at the following important observation:

> (!) A scalar λ is an eigenvalue of a square matrix A if and only if $A - \lambda I$ is not invertible.

While there are multiple different ways to determine which values of λ make $A - \lambda I$ not invertible, one of the conceptually simplest methods is to compute $\det(A - \lambda I)$ via one of the explicit formulas for the determinant from Section 3.2.3, and then set it equal to 0 (since $A - \lambda I$ is not invertible if and only if $\det(A - \lambda I) = 0$, by Theorem 3.2.1). We illustrate this method via an example.

Example 3.3.3

Computing the Eigenvalues of a Matrix

> Recall that the determinant of a 2×2 matrix is
> $$\det\left(\begin{bmatrix} a & b \\ c & d \end{bmatrix}\right) = ad - bc.$$

Compute all of the eigenvalues of the matrix $A = \begin{bmatrix} 1 & 2 \\ 5 & 4 \end{bmatrix}$.

Solution:
To find the eigenvalues of A, we first compute $\det(A - \lambda I)$ via Theorem 3.2.6:

$$\det(A - \lambda I) = \det\left(\begin{bmatrix} 1-\lambda & 2 \\ 5 & 4-\lambda \end{bmatrix}\right)$$
$$= (1-\lambda)(4-\lambda) - 10 = \lambda^2 - 5\lambda - 6.$$

Setting this determinant equal to 0 then gives

$$\lambda^2 - 5\lambda - 6 = 0 \iff (\lambda+1)(\lambda-6) = 0$$
$$\iff \lambda = -1 \quad \text{or} \quad \lambda = 6,$$

so the eigenvalues of A are $\lambda = -1$ and $\lambda = 6$.

As an alternative to factoring the polynomial $\lambda^2 - 5\lambda - 6$, we could have instead found its roots via the quadratic formula, which says that the solutions of $a\lambda^2 + b\lambda + c = 0$ are

A **root** of a polynomial is a value that makes it equal 0.

$$\lambda = \frac{-b \pm \sqrt{b^2 - 4ac}}{2a}.$$

In this case, the quadratic formula gives $\lambda = (5 \pm 7)/2$, which of course agrees with the $\lambda = -1$ and $\lambda = 6$ answer that we found via factoring.

In the above example, $\det(A - \lambda I)$ was a quadratic in λ, and finding the roots of that quadratic gave us the eigenvalues of A. We will return to this idea of eigenvalues being the roots of a polynomial shortly, so keep it in the back of your mind.

Step 2: **Compute the eigenvectors.** Once we know the eigenvalues of a matrix (from Step 1 above), the eigenvectors corresponding to them can be found via the method of Example 3.3.2. That is, they can be found by solving the linear system $A\mathbf{v} = \lambda\mathbf{v}$ for $\mathbf{v}$. By moving all terms over to the left-hand side of this equation, we see that this is equivalent to solving the linear system $(A - \lambda I)\mathbf{v} = \mathbf{0}$ (i.e., computing the null space of the matrix $A - \lambda I$).

> ! A non-zero vector $\mathbf{v}$ is an eigenvector of A with corresponding eigenvalue λ if and only if $\mathbf{v} \in \text{null}(A - \lambda I)$.

Example 3.3.4

Computing the Eigenvectors of a Matrix

Compute all of the eigenvectors of the matrix from the previous example (Example 3.3.3), and state which eigenvalues correspond to them.

Solution:
We already saw that the eigenvalues of A are $\lambda = -1$ and $\lambda = 6$. To find the eigenvectors corresponding to these eigenvalues, we solve the linear systems $(A + I)\mathbf{v} = \mathbf{0}$ and $(A - 6I)\mathbf{v} = \mathbf{0}$, respectively:

$\lambda = -1$: In this case, we want to solve the linear system $(A - \lambda I)\mathbf{v} = (A + I)\mathbf{v} = \mathbf{0}$, which we can write explicitly as follows:

$$2v_1 + 2v_2 = 0$$
$$5v_1 + 5v_2 = 0$$

To solve this linear system, we use Gaussian elimination as usual:

If we computed the eigenvalues correctly, we can always find corresponding eigenvectors, so the linear system $(A - \lambda I)\mathbf{v} = \mathbf{0}$ always has a free variable.

$$\begin{bmatrix} 2 & 2 & | & 0 \\ 5 & 5 & | & 0 \end{bmatrix} \xrightarrow{R_2 - \frac{5}{2}R_1} \begin{bmatrix} 2 & 2 & | & 0 \\ 0 & 0 & | & 0 \end{bmatrix},$$

It follows that v_2 is a free variable and v_1 is a leading variable with $v_1 = -v_2$. The eigenvectors with corresponding eigenvalue $\lambda = -1$ are thus exactly the non-zero vectors of the form $\mathbf{v} = (-v_2, v_2) = v_2(-1, 1)$.

$\lambda = 6$: Similarly, we now want to solve the linear system $(A - \lambda I)\mathbf{v} = (A - 6I)\mathbf{v} = \mathbf{0}$, which we can do as follows:

$$\left[\begin{array}{cc|c} -5 & 2 & 0 \\ 5 & -2 & 0 \end{array}\right] \xrightarrow{R_2+R_1} \left[\begin{array}{cc|c} -5 & 2 & 0 \\ 0 & 0 & 0 \end{array}\right],$$

By multiplying $(2/5,1)$ by 5, we could also say that the eigenvectors here are the multiples of $(2,5)$, which is a slightly cleaner answer.

We thus conclude that v_2 is a free variable and v_1 is a leading variable with $v_1 = 2v_2/5$, so the eigenvectors with corresponding eigenvalue $\lambda = 6$ are the non-zero vectors of the form $\mathbf{v} = (2v_2/5, v_2) = v_2(2/5, 1)$.

Eigenvalues and eigenvectors can help us better understand what a matrix looks like as a linear transformation. For example, to understand the geometric effect of the matrix

$$A = \begin{bmatrix} 1 & 2 \\ 5 & 4 \end{bmatrix}$$

from Examples 3.3.3 and 3.3.4, we could show how it transforms a unit square grid into a parallelogram grid as usual (by recalling that its columns $A\mathbf{e}_1$ and $A\mathbf{e}_2$ are the sides of one of the parallelograms in this grid, as in Figure 3.22).

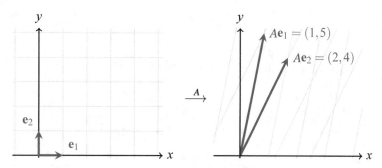

Figure 3.22: The effect of the matrix A from Example 3.3.3 on the standard basis vectors $\mathbf{e}_1$ and $\mathbf{e}_2$ is not terribly enlightening.

However, it is perhaps more illuminating to see how the parallelogram grid defined by the eigenvectors of A is transformed. In particular, because each vector in this grid is just stretched or shrunk, the grid is not rotated or skewed at all, as illustrated in Figure 3.23.

In order to make sure that we are comfortable computing eigenvalues and eigenvectors, we now work through a 3×3 example. We will see that the general procedure is the same as it was for a 2×2 matrix, but the details are a bit uglier.

Example 3.3.5

Eigenvalues and Eigenvectors of a 3×3 Matrix

Compute the eigenvalues and eigenvectors of the matrix

$$A = \begin{bmatrix} 1 & 3 & 3 \\ 3 & 1 & -1 \\ 0 & 0 & 2 \end{bmatrix}.$$

Solution:

We start by computing $\det(A - \lambda I)$ via the explicit formula of

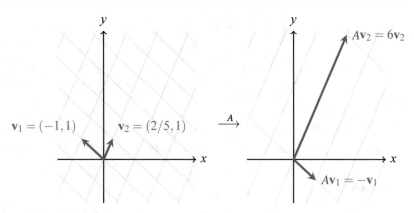

Figure 3.23: The effect of the matrix A from Example 3.3.3 on its eigenvectors $\mathbf{v}_1 = (-1,1)$ and $\mathbf{v}_2 = (2/5,1)$, which have corresponding eigenvalues -1 and 6, respectively.

Theorem 3.2.7:

$$\det(A - \lambda I) = \det\left(\begin{bmatrix} 1-\lambda & 3 & 3 \\ 3 & 1-\lambda & -1 \\ 0 & 0 & 2-\lambda \end{bmatrix}\right)$$

$$= (1-\lambda)(1-\lambda)(2-\lambda) + 0 + 0 - 0 - 0 - 9(2-\lambda)$$

$$= (2-\lambda)\big((1-\lambda)^2 - 9\big)$$

$$= (2-\lambda)(\lambda^2 - 2\lambda - 8)$$

$$= (2-\lambda)(\lambda+2)(\lambda-4).$$

Instead of multiplying out this polynomial, we factored out the common term $(2-\lambda)$, since this saved us from having to factor a cubic polynomial later (which is a pain).

To find the eigenvalues of A, we now set this polynomial equal to 0, which gives us $\lambda = -2$, $\lambda = 2$, and $\lambda = 4$. To find eigenvectors corresponding to these three eigenvalues, we solve the linear systems $(A - \lambda I)\mathbf{v} = \mathbf{0}$ for each of $\lambda = -2$, $\lambda = 2$, and $\lambda = 4$:

$\lambda = -2$: In this case, the linear system is $(A + 2I)\mathbf{v} = \mathbf{0}$, which we can solve as follows:

$$\begin{bmatrix} 3 & 3 & 3 & | & 0 \\ 3 & 3 & -1 & | & 0 \\ 0 & 0 & 4 & | & 0 \end{bmatrix} \xrightarrow{R_2-R_1} \begin{bmatrix} 3 & 3 & 3 & | & 0 \\ 0 & 0 & -4 & | & 0 \\ 0 & 0 & 4 & | & 0 \end{bmatrix}$$

$$\xrightarrow{R_3+R_2} \begin{bmatrix} 3 & 3 & 3 & | & 0 \\ 0 & 0 & -4 & | & 0 \\ 0 & 0 & 0 & | & 0 \end{bmatrix}.$$

We thus see that v_2 is a free variable and v_1 and v_3 are leading variables with $v_3 = 0$ and $v_1 = -v_2$. The eigenvectors corresponding to the eigenvalue $\lambda = -2$ are thus the non-zero multiples of $(-1,1,0)$.

$\lambda = 2$: We work through this eigenvector calculation a bit more quickly. In this case, the linear system that we want to solve is

$(A - 2I)\mathbf{v} = \mathbf{0}$. Since the reduced row echelon form of $A - 2I$ is

Now is a good time
to remind ourselves
of how to solve linear
systems. Try
computing this RREF
on your own.

$$\begin{bmatrix} 1 & 0 & 0 \\ 0 & 1 & 1 \\ 0 & 0 & 0 \end{bmatrix},$$

we see that v_3 is a free variable and v_1 and v_2 are leading variables with $v_1 = 0$ and $v_2 = -v_3$. The eigenvectors corresponding to the eigenvalue $\lambda = 2$ are thus the non-zero multiples of $(0, -1, 1)$.

$\lambda = 4$: In this case, the linear system is $(A - 4I)\mathbf{v} = \mathbf{0}$. Since the reduced row echelon form of $A - 4I$ is

Remember that, if
we computed the
eigenvalues
correctly, the RREFs
of these linear
systems *must* have a
zero row, or else we
wouldn't be able to
find an eigenvector
at all.

$$\begin{bmatrix} 1 & -1 & 0 \\ 0 & 0 & 1 \\ 0 & 0 & 0 \end{bmatrix},$$

we see that v_2 is a free variable and v_1 and v_3 are leading variables. Furthermore, $v_1 = v_2$ and $v_3 = 0$, so the eigenvectors corresponding to the eigenvalue $\lambda = 4$ are the non-zero multiples of $(1, 1, 0)$.

The eigenvalue computation in the previous example simplified a fair bit thanks to the zeros at the matrix's bottom-left corner. We now work through one last 3×3 example to illustrate how we can find the eigenvalues of a matrix whose entries are all non-zero, and for which the determinant and eigenvalue computations are thus not so straightforward.

Example 3.3.6

**Eigenvalues of an
Ugly 3×3 Matrix**

Compute the eigenvalues of the matrix $A = \begin{bmatrix} 1 & 2 & 3 \\ 1 & -2 & 1 \\ 3 & 2 & 1 \end{bmatrix}$.

Solution:

We start by computing $\det(A - \lambda I)$:

No, simplifying this
polynomial is not fun,
but it is "just" routine
algebra: we multiply
out the terms in
parentheses and
then group powers
of λ.

$$\det(A - \lambda I) = \det\left(\begin{bmatrix} 1 - \lambda & 2 & 3 \\ 1 & -2 - \lambda & 1 \\ 3 & 2 & 1 - \lambda \end{bmatrix} \right)$$

$$= (1 - \lambda)(-2 - \lambda)(1 - \lambda) + 6 + 6$$
$$- 2(1 - \lambda) - 2(1 - \lambda) - 9(-2 - \lambda)$$
$$= -\lambda^3 + 16\lambda + 24.$$

To find the eigenvalues of A, we now set this polynomial equal to 0 and solve for λ. However, since this is a cubic equation, factoring it and finding its roots requires a bit more work than solving a quadratic equation. The techniques that we use to solve cubic equations (and other higher-degree polynomial equations) like $-\lambda^3 + 16\lambda + 24 = 0$ are covered in Appendix A.2.

First, to find one solution of this equation, we recall that the rational root theorem (Theorem A.2.1) says that if a rational number solves this equation, it must be one of the divisors of 24 (the constant term in the polynomial): $\pm 1, \pm 2, \pm 3, \pm 4, \pm 6, \pm 8, \pm 12$, or ± 24. By plugging these values into the equation $-\lambda^3 + 16\lambda + 24 = 0$, we see that $\lambda = -2$ is the only one of them that is actually a solution (indeed, $-(-2)^3 + 16(-2) + 24 = 8 - 32 + 24 = 0$).

To find the other (necessarily irrational) solutions of this equation, we use polynomial long division to obtain the result of dividing $-\lambda^3 + 16\lambda + 24$ by $\lambda + 2$ (which we know is a factor since $\lambda = -2$ is a root of the cubic polynomial):

If you are not familiar with polynomial long division and this looks cryptic to you, have a look at Appendix A.2.

$$
\begin{array}{r}
-\lambda^2 + 2\lambda + 12 \\
\lambda + 2 \overline{)\,-\lambda^3 + 0\lambda^2 + 16\lambda + 24} \\
\underline{-\lambda^3 - 2\lambda^2} \\
2\lambda^2 + 16\lambda + 24 \\
\underline{2\lambda^2 + 4\lambda} \\
12\lambda + 24 \\
\underline{12\lambda + 24} \\
0
\end{array}
$$

It follows that $-\lambda^3 + 16\lambda + 24 = (\lambda + 2)(-\lambda^2 + 2\lambda + 12)$, so the roots of $-\lambda^3 + 16\lambda + 24$ (i.e., the eigenvalues of A) are $\lambda = -2$ as well as the two roots of $-\lambda^2 + 2\lambda + 12$, which we can compute via the quadratic formula:

$$
\lambda = \frac{-2 \pm \sqrt{4 + 48}}{-2} = 1 \pm \sqrt{13}.
$$

The eigenvalues of A are thus $\lambda = -2, \lambda = 1 + \sqrt{13}$, and $\lambda = 1 - \sqrt{13}$. If we were feeling adventurous, we could also find the eigenvectors corresponding to these eigenvalues by solving the linear system $(A - \lambda I)\mathbf{v} = \mathbf{0}$ for each of these three choices of λ. However, it is perhaps a better use of our time to move on to new things from here.

The previous example illustrates a key problem that arises when trying to analytically compute eigenvalues—if the matrix is 3×3 or larger, it can be extremely difficult to find the roots of the polynomial $\det(A - \lambda I)$. In fact, it can be *impossible* to do so if the matrix is 5×5 or larger (see Remark A.2.1). For this reason, methods of numerically approximating eigenvalues are typically used in practice instead, and one such method is explored in Section 3.B.

3.3.2 The Characteristic Polynomial and Algebraic Multiplicity

When we computed the eigenvalues of a matrix in the previous section, we always ended up having to find the roots of a polynomial. More specifically, we computed $\det(A - \lambda I)$, which was a polynomial in λ, and we set that polynomial equal to 0 and solved for λ.

This happens in general, since the eigenvalues of a matrix A are the solutions λ to the equation $\det(A - \lambda I) = 0$, and this determinant can be written as a sum of products of entries of $A - \lambda I$ (via the cofactor expansions of Theorem 3.2.8). It follows that $\det(A - \lambda I)$ is indeed always a polynomial in λ, and we now give it a name:

Definition 3.3.2

Characteristic Polynomial

Suppose A is a square matrix. The function $p_A : \mathbb{R} \to \mathbb{R}$ defined by

$$p_A(\lambda) = \det(A - \lambda I)$$

is called the **characteristic polynomial** of A.

Some other books instead define the characteristic polynomial to be $\det(\lambda I - A)$. These two polynomials only differ by a minus sign in odd dimensions, so they have the same roots.

For example, we showed in Examples 3.3.3 and 3.3.6 that the characteristic polynomials of

$$A = \begin{bmatrix} 1 & 2 \\ 5 & 4 \end{bmatrix} \quad \text{and} \quad B = \begin{bmatrix} 1 & 2 & 3 \\ 1 & -2 & 1 \\ 3 & 2 & 1 \end{bmatrix}$$

are

$$p_A(\lambda) = \lambda^2 - 5\lambda - 6 \quad \text{and} \quad p_B(\lambda) = -\lambda^3 + 16\lambda + 24,$$

respectively.

Examples like these ones seem to suggest that the characteristic polynomial of an $n \times n$ matrix always has degree n. This fact can be verified by noting that in a cofactor expansion of $A - \lambda I$, we add up several terms, each of which is the product of n entries of $A - \lambda I$. One of the terms being added up is the product of the diagonal entries, $(a_{1,1} - \lambda)(a_{2,2} - \lambda) \cdots (a_{n,n} - \lambda)$, which has degree n, and no other term in the sum has higher degree. We now re-state this observation a bit more prominently, since it is so important:

> (!) The characteristic polynomial of an $n \times n$ matrix has degree n.

Thanks to the factor theorem (Theorem A.2.2), we know that every degree-n polynomial has at most n distinct roots. Since the eigenvalues of a matrix are the roots of its characteristic polynomial, this immediately tells us the following fact:

> (!) Every $n \times n$ matrix has at most n distinct eigenvalues.

Just like it is useful to talk about the multiplicity of a root of a polynomial (which we roughly think of as "how many times" that root occurs), it is similarly useful to talk about the multiplicity of an eigenvalue.

Definition 3.3.3

Algebraic Multiplicity

Suppose A is a square matrix with eigenvalue λ. The **algebraic multiplicity** of λ is its multiplicity as a root of A's characteristic polynomial.

A bit more explicitly, if the characteristic polynomial of A is p_A and λ_0 is a particular eigenvalue of A, then the algebraic multiplicity of λ_0 is the exponent of the term $(\lambda_0 - \lambda)$ in the factored form of $p_A(\lambda)$. We now work through an example to clarify exactly what we mean by this.

Example 3.3.7

Algebraic Multiplicity

Compute the eigenvalues, and their algebraic multiplicities, of the matrix

$$A = \begin{bmatrix} 2 & 0 & -3 \\ 1 & -1 & -1 \\ 0 & 0 & -1 \end{bmatrix}.$$

Solution:
The characteristic polynomial of this matrix is

Since our goal is to eventually find the roots of the characteristic polynomial, we do *not* multiply it out here, since it is already factored for us.

$$p_A(\lambda) = \det(A - \lambda I) = \det\left(\begin{bmatrix} 2-\lambda & 0 & -3 \\ 1 & -1-\lambda & -1 \\ 0 & 0 & -1-\lambda \end{bmatrix}\right)$$

$$= (2-\lambda)(-1-\lambda)(-1-\lambda)+0+0-0-0-0$$

$$= (2-\lambda)(-1-\lambda)^2.$$

Now that we have a factored form of the characteristic polynomial of A, we can read off its algebraic multiplicities as the exponents of its factors. In particular, the eigenvalues of A are 2 and -1, with algebraic multiplicities of 1 and 2, respectively.

If we allow complex eigenvalues, then the fundamental theorem of algebra (Theorem A.2.3) tells us the following even stronger fact about the number of eigenvalues that matrices have:

Just a quick note on terminology: real numbers are complex. The real line is a subset of the complex plane. However, there are many non-real complex numbers too.

> (!) Every $n \times n$ matrix has *exactly* n complex eigenvalues, counting algebraic multiplicity.

For example, the 3×3 matrix from Example 3.3.7 had eigenvalues 2 and -1, with algebraic multiplicities of 1 and 2, respectively. We thus say that matrix has 3 eigenvalues "counting algebraic multiplicity" (since the algebraic multiplicities of its eigenvalues add up to 3), and we similarly list the eigenvalues "according to algebraic multiplicity" as 2, -1, and -1 (i.e., the number of times that we list an eigenvalue is its algebraic multiplicity).

We now work through an example to demonstrate why working with complex numbers is so natural when dealing with eigenvalues and eigenvectors—a real matrix may not have any real eigenvalues.

Example 3.3.8

A Real Matrix with no Real Eigenvalues

Compute the eigenvalues, and their algebraic multiplicities, of the matrix

$$A = \begin{bmatrix} -3 & -2 \\ 4 & 1 \end{bmatrix}.$$

Solution:
The characteristic polynomial of this matrix is

For the most part, you can work with complex numbers how you would expect if you accept $\sqrt{-1} = i$. See Appendix A.1 for details.

$$p_A(\lambda) = \det(A - \lambda I) = \det\left(\begin{bmatrix} -3-\lambda & -2 \\ 4 & 1-\lambda \end{bmatrix}\right)$$

$$= (-3-\lambda)(1-\lambda)+8$$

$$= \lambda^2 + 2\lambda + 5.$$

This characteristic polynomial does not factor "nicely", so we use the quadratic formula to find its roots (i.e., the eigenvalues of A):

$$\lambda = \frac{-2 \pm \sqrt{4-20}}{2} = -1 \pm \sqrt{-4} = -1 \pm 2i.$$

These eigenvalues are distinct so they each have algebraic multiplicity 1 (i.e., the two $\pm$ branches of the quadratic formula do not give us the same eigenvalue).

The above example highlights why we really need to specify that a matrix has exactly n *complex* eigenvalues counting algebraic multiplicity—even if its entries are real, it might have no real eigenvalues at all (or it might have some number of real eigenvalues strictly between 0 and n).

The complex conjugate of $a + bi$ (where $a, b \in \mathbb{R}$) is $\overline{a + bi} = a - bi$.

It is also worth observing that the eigenvalues in the previous example are complex conjugates of each other: $\overline{-1 + 2i} = -1 - 2i$. This always happens for real matrices, since the characteristic polynomial p_A of a real matrix A has real coefficients, so if $p_A(\lambda) = 0$ then $p_A(\overline{\lambda}) = \overline{p_A(\lambda)} = 0$ as well.

Remark 3.3.2

Complex Eigenvalues

A matrix having eigenvalues is an extremely desirable property and is a starting point for many advanced linear algebraic techniques. For this reason, from now on our default viewpoint will be to think in terms of complex numbers rather than real numbers. For example, instead of thinking of

$$\begin{bmatrix} 1 & 2 \\ -4 & -3 \end{bmatrix}$$

as a real matrix, we think of it as a complex matrix whose entries just happen to be real. If we wish to clarify exactly what types of entries a matrix can have, we use the notation $\mathcal{M}_{m,n}(\mathbb{R})$ to denote the set of $m \times n$ matrices with real entries and $\mathcal{M}_{m,n}(\mathbb{C})$ to denote the set of $m \times n$ matrices with complex entries.

More generally, $\mathcal{M}_{m,n}(X)$ is the set of $m \times n$ matrices whose entries come from the set X. Also, we use $\mathcal{M}_n(X)$ to denote the set of $n \times n$ (i.e., square) matrices with entries from X.

Typically, we do not shy away from using complex numbers in calculations in order to deepen our understanding of the topic being investigated. However, we do occasionally discuss the (typically weaker) statements we can prove if we restrict our attention to real numbers. For example, in Section 3.C we discuss how to interpret complex eigenvalues of real matrices geometrically.

Much like most of the other properties of matrices that we have introduced in this chapter, characteristic polynomials are similarity invariant. That is, if A and B are similar matrices (i.e., there exists an invertible matrix P such that $B = PAP^{-1}$) then

$$p_B(\lambda) = \det(B - \lambda I) = \det(PAP^{-1} - \lambda I)$$
$$= \det\left(P(A - \lambda I)P^{-1}\right) = \det(A - \lambda I) = p_A(\lambda),$$

where the second-to-last equality makes use of the fact that the determinant is similarity invariant. Again, we re-state this important observation more prominently:

> ! If $A \in \mathcal{M}_n$ and $B \in \mathcal{M}_n$ are similar then their characteristic polynomials are the same: $p_A(\lambda) = p_B(\lambda)$.

Since the eigenvalues of a matrix are the roots of its characteristic polynomial, an immediate corollary of the above observation is the fact that eigenvalues (and their algebraic multiplicities) are also similarity invariant.

Example 3.3.9

Using Characteristic Polynomials to Show Matrices are Not Similar

Show that $A = \begin{bmatrix} 4 & 0 & 2 \\ -2 & 6 & 2 \\ 2 & 0 & 4 \end{bmatrix}$ and $B = \begin{bmatrix} 8 & -5 & -5 \\ 5 & -2 & -5 \\ -5 & 5 & 8 \end{bmatrix}$ are not similar.

Solution:

The rank, trace, and determinant cannot help us show that these matrices are not similar, since

$$\text{rank}(A) = \text{rank}(B) = 3, \quad \text{tr}(A) = \text{tr}(B) = 14, \quad \text{and} \quad \det(A) = \det(B) = 72.$$

However, their characteristic polynomials are different:

$$p_A(\lambda) = \det(A - \lambda I) = \det\left(\begin{bmatrix} 4-\lambda & 0 & 2 \\ -2 & 6-\lambda & 2 \\ 2 & 0 & 4-\lambda \end{bmatrix}\right)$$

$$= (4-\lambda)(6-\lambda)(4-\lambda) + 0 + 0$$
$$- 0 - 0 - 4(6-\lambda)$$

The eigenvalues of A (listed according to algebraic multiplicity) are 2, 6, and 6, while the eigenvalues of B are 3, 3, and 8.

$$= -\lambda^3 + 14\lambda^2 - 60\lambda + 72, \quad \text{but}$$

$$p_B(\lambda) = \det(B - \lambda I) = \det\left(\begin{bmatrix} 8-\lambda & -5 & -5 \\ 5 & -2-\lambda & -5 \\ -5 & 5 & 8-\lambda \end{bmatrix}\right)$$

$$= (8-\lambda)(-2-\lambda)(8-\lambda) - 125 - 125$$
$$+ 25(8-\lambda) - 25(-2-\lambda) + 25(8-\lambda)$$

$$= -\lambda^3 + 14\lambda^2 - 57\lambda + 72.$$

Since $p_A(\lambda) \neq p_B(\lambda)$, we conclude that A and B are not similar.

Notice in the above example that the coefficient of λ^2 in the characteristic polynomials was 14, which equaled the trace of the matrices, and the constant term in the characteristic polynomials was 72, which equaled their determinant. This is not a coincidence—the following theorem shows that these coefficients *always* equal the trace and determinant, and they can also be expressed in terms of the eigenvalues of the matrix.

Theorem 3.3.1

Trace and Determinant in Terms of Eigenvalues

The coefficient of λ^n in the characteristic polynomial of an $n \times n$ matrix is always $(-1)^n$.

Denote the n eigenvalues of $A \in \mathcal{M}_n(\mathbb{C})$ (listed according to algebraic multiplicity) by $\lambda_1, \lambda_2, \ldots, \lambda_n$, and its characteristic polynomial by

$$p_A(\lambda) = (-1)^n \lambda^n + c_{n-1}\lambda^{n-1} + \cdots + c_1\lambda + c_0.$$

Then

$$c_0 = \det(A) = \lambda_1 \lambda_2 \cdots \lambda_n \quad \text{and}$$
$$(-1)^{n-1} c_{n-1} = \text{tr}(A) = \lambda_1 + \lambda_2 + \cdots + \lambda_n.$$

Proof. We start by proving the claim that $c_0 = \det(A) = \lambda_1 \lambda_2 \cdots \lambda_n$. To see why this holds, we recall the various formulas for $p_A(\lambda)$ that we know:

$$p_A(\lambda) = \det(A - \lambda I) = (-1)^n \lambda^n + c_{n-1}\lambda^{n-1} + \cdots + c_1 \lambda + c_0$$
$$= (\lambda_1 - \lambda)(\lambda_2 - \lambda)\cdots(\lambda_n - \lambda),$$

with the final formula coming from the fact that the factor theorem (Theorem A.2.2) lets us completely factor a polynomial via its n roots. Plugging $\lambda = 0$ into the above formulas shows that

$$p_A(0) = \det(A) = c_0 = \lambda_1 \lambda_2 \cdots \lambda_n,$$

as desired.

Can we take a moment to admire the beauty of this theorem? It relates the trace, determinant, and eigenvalues (the topics of the last 3 sections of this book) so well.

To prove the claim that $(-1)^{n-1}c_{n-1} = \lambda_1 + \lambda_2 + \cdots + \lambda_n$, we simply multiply out the expression

$$p_A(\lambda) = (\lambda_1 - \lambda)(\lambda_2 - \lambda)\cdots(\lambda_n - \lambda)$$

and observe that the coefficient of λ^{n-1} on the right-hand side (i.e., c_{n-1}) is $(-1)^{n-1}(\lambda_1 + \lambda_2 + \cdots + \lambda_n)$.

On the other hand, we can show that $(-1)^{n-1}c_{n-1} = \text{tr}(A)$ by considering how we would compute the coefficient of λ^{n-1} in $p_A(\lambda)$ via the determinant:

$$p_A(\lambda) = \det(A - \lambda I) = \det\left(\begin{bmatrix} a_{1,1} - \lambda & a_{1,2} & \cdots & a_{1,n} \\ a_{2,1} & a_{2,2} - \lambda & \cdots & a_{2,n} \\ \vdots & \vdots & \ddots & \vdots \\ a_{n,1} & a_{n,2} & \cdots & a_{n,n} - \lambda \end{bmatrix}\right).$$

If we use a cofactor expansion along the first column of this matrix, we can express the above determinant as a sum of n determinants of $(n-1) \times (n-1)$ matrices:

$$\det(A - \lambda I) = (a_{1,1} - \lambda)\det\left(\begin{bmatrix} a_{2,2} - \lambda & a_{2,3} & \cdots & a_{2,n} \\ a_{3,2} & a_{3,3} - \lambda & \cdots & a_{3,n} \\ \vdots & \vdots & \ddots & \vdots \\ a_{n,2} & a_{n,3} & \cdots & a_{n,n} - \lambda \end{bmatrix}\right)$$

$$- a_{2,1}\det\left(\begin{bmatrix} a_{1,2} & a_{1,3} & \cdots & a_{1,n} \\ a_{3,2} & a_{3,3} - \lambda & \cdots & a_{3,n} \\ \vdots & \vdots & \ddots & \vdots \\ a_{n,2} & a_{n,3} & \cdots & a_{n,n} - \lambda \end{bmatrix}\right)$$

$$\vdots$$

$$+ (-1)^{n+1} a_{n,1}\det\left(\begin{bmatrix} a_{1,2} & a_{1,3} & \cdots & a_{1,n} \\ a_{2,2} - \lambda & a_{2,3} & \cdots & a_{2,n} \\ \vdots & \vdots & \ddots & \vdots \\ a_{n-1,2} & a_{n-1,3} & \cdots & a_{n-1,n} \end{bmatrix}\right).$$

Notice that the only one of these n smaller determinants that can produce a term of the form λ^{n-1} is the first one, since it is the only one containing $n-1$

or more λs. The coefficient of λ^{n-1} in $p_A(\lambda)$ (i.e., c_{n-1}) is thus the same as the coefficient of λ^{n-1} in

We are not saying that the entire characteristic polynomial $p_A(\lambda)$ equals this smaller determinant—just that they have the same coefficient of λ^{n-1}.

$$(a_{1,1} - \lambda)\det\left(\begin{bmatrix} a_{2,2}-\lambda & a_{2,3} & \cdots & a_{2,n} \\ a_{3,2} & a_{3,3}-\lambda & \cdots & a_{3,n} \\ \vdots & \vdots & \ddots & \vdots \\ a_{n,2} & a_{n,3} & \cdots & a_{n,n}-\lambda \end{bmatrix}\right).$$

By repeating this argument a total of $n-1$ times (i.e., using a cofactor expansion along the first column repeatedly until we are left with a 1×1 matrix), we similarly see that c_{n-1} equals the coefficient of λ^{n-1} in

$$(a_{1,1} - \lambda)(a_{2,2} - \lambda)\cdots(a_{2,2} - \lambda).$$

Multiplying out this expression then shows that the coefficient of λ^{n-1} in it is

$$c_{n-1} = (-1)^{n-1}(a_{1,1} + a_{2,2} + \cdots + a_{n,n}) = (-1)^{n-1}\mathrm{tr}(A),$$

which completes the proof.   ∎

The fact that the determinant of a matrix equals the product of its eigenvalues has a very natural geometric interpretation. If $A \in \mathcal{M}_n$ has n distinct eigenvalues then they specify how much A stretches space in each of n different directions. Multiplying these eigenvalues together thus gives the amount by which A stretches space as a whole (i.e., its determinant—see Figure 3.24).

This geometric interpretation is only "mostly" correct—it is much more subtle if a matrix has repeated eigenvalues.

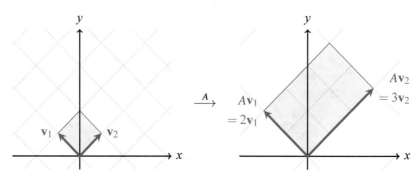

Figure 3.24: A matrix $A \in \mathcal{M}_2$ with eigenvalues 2 and 3, and thus determinant $2 \times 3 = 6$.

Example 3.3.10

Computing the Trace and Determinant in Two Ways

Compute the trace and determinant of the matrix

$$A = \begin{bmatrix} 3 & 1 & 1 \\ 0 & 1 & 0 \\ -2 & -1 & 0 \end{bmatrix}$$

directly, and also via Theorem 3.3.1.

Solution:

Computing the trace of A directly from the definition and its determinant from the explicit formula of Theorem 3.2.7 is straightforward:

$$\mathrm{tr}(A) = 3 + 1 + 0 = 4 \quad \text{and} \quad \det(A) = 0 + 0 + 0 - 0 - (-2) - 0 = 2.$$

To instead compute the trace and determinant of A via Theorem 3.3.1, we must first calculate its eigenvalues:

> When computing the trace and determinant in this way, each eigenvalue must be listed according to its algebraic multiplicity.

$$\det(A - \lambda I) = \det\left(\begin{bmatrix} 3-\lambda & 1 & 1 \\ 0 & 1-\lambda & 0 \\ -2 & -1 & -\lambda \end{bmatrix}\right)$$
$$= (3-\lambda)(1-\lambda)(-\lambda) + 2(1-\lambda)$$
$$= (1-\lambda)(\lambda^2 - 3\lambda + 2)$$
$$= (1-\lambda)^2(2-\lambda).$$

It follows that A has eigenvalues $\lambda = 1, 1$, and 2, so its trace is $\mathrm{tr}(A) = 1 + 1 + 2 = 4$ and its determinant is $\det(A) = 1 \times 1 \times 2 = 2$, which agree with the values that we computed earlier.

It is also worth noting that Theorem 3.3.1 tells us that the characteristic polynomial of a 2×2 matrix A is completely determined by its determinant and trace, since

$$p_A(\lambda) = \lambda^2 - \mathrm{tr}(A)\lambda + \det(A).$$

In particular, this means that we can explicitly compute the eigenvalues of a 2×2 matrix from just its trace and determinant (see Exercise 3.3.10). However, this is not true for larger matrices (refer back to Example 3.3.9).

The Conjugate Transpose and Hermitian Matrices

It is often desirable to work with matrices whose eigenvalues are real. There is one particularly important family of such matrices, but before we can talk about them, we need to briefly discuss how to "properly" generalize the transpose of a matrix to matrices with complex entries.

> Recall that $A \in \mathcal{M}_{m,n}(\mathbb{C})$ means that A is an $m \times n$ matrix with complex entries.

When working with a complex matrix $A \in \mathcal{M}_{m,n}(\mathbb{C})$, it is often useful to construct its **conjugate transpose** A^* (instead of its standard transpose A^T), which is computed by taking the conjugate transpose of the entries of A and then transposing it:

$$A^* \overset{\text{def}}{=} \overline{A}^T.$$

Example 3.3.11

Computing Conjugate Transposes

Compute the conjugate transpose of each of the following matrices:

a) $\begin{bmatrix} 2i & 1-3i \\ -3 & 5+2i \end{bmatrix}$

b) $\begin{bmatrix} 0 & 2+3i \\ 2-3i & 4 \end{bmatrix}$

c) $\begin{bmatrix} 1 & 2 & 3 \\ 4 & 5 & 6 \\ 7 & 8 & 9 \end{bmatrix}$

d) $\begin{bmatrix} i & 1-i \\ -2 & -2-i \\ 1+i & 2+3i \end{bmatrix}$

Solutions:

a) First, we take the complex conjugate of each entry in the matrix,

The reason *why* we typically use the conjugate transpose (instead of the standard transpose) when working with complex matrices is so that the property $\mathbf{v} \cdot (A\mathbf{w}) = (A^*\mathbf{v}) \cdot \mathbf{w}$ holds (see Exercise 3.3.15).

and then we transpose the result:

$$\overline{\begin{bmatrix} 2i & 1-3i \\ -3 & 5+2i \end{bmatrix}} = \begin{bmatrix} -2i & 1+3i \\ -3 & 5-2i \end{bmatrix}, \quad \text{so}$$

$$\begin{bmatrix} 2i & 1-3i \\ -3 & 5+2i \end{bmatrix}^* = \begin{bmatrix} -2i & -3 \\ 1+3i & 5-2i \end{bmatrix}.$$

b) This matrix equals its own conjugate transpose:

$$\begin{bmatrix} 0 & 2+3i \\ 2-3i & 4 \end{bmatrix}^* = \begin{bmatrix} 0 & 2+3i \\ 2-3i & 4 \end{bmatrix}.$$

c) Since the entries of this matrix are all real, taking the complex conjugate has no effect, so its conjugate transpose equals its standard transpose:

$$\begin{bmatrix} 1 & 2 & 3 \\ 4 & 5 & 6 \\ 7 & 8 & 9 \end{bmatrix}^* = \begin{bmatrix} 1 & 4 & 7 \\ 2 & 5 & 8 \\ 3 & 6 & 9 \end{bmatrix}.$$

d) The conjugate transpose changes this 3×2 matrix into a 2×3 matrix:

$$\begin{bmatrix} i & 1-i \\ -2 & -2-i \\ 1+i & 2+3i \end{bmatrix}^* = \begin{bmatrix} -i & -2 & 1-i \\ 1+i & -2+i & 2-3i \end{bmatrix}.$$

In the special case when a matrix $A \in \mathcal{M}_n(\mathbb{C})$ is such that $A^* = A$, like in part (b) of the above example, it is called **Hermitian**. If A is real and Hermitian (i.e., it is real and $A^T = A$) then it is called **symmetric**. Hermitian matrices (and thus symmetric matrices) are special since all of their eigenvalues are real, which we state as the final theorem of this subsection.

Theorem 3.3.2

Eigenvalues of Hermitian Matrices

If $A \in \mathcal{M}_n(\mathbb{C})$ is Hermitian (i.e., $A^* = A$) then all of its eigenvalues are real.

Proof. Suppose that $\mathbf{v} \in \mathbb{C}^n$ is an eigenvector of A with corresponding eigenvalue $\lambda \in \mathbb{C}$. Then computing $\mathbf{v}^*A\mathbf{v}$ shows that

$$\mathbf{v}^*A\mathbf{v} = \mathbf{v}^*(A\mathbf{v}) = \mathbf{v}^*(\lambda \mathbf{v}) = \lambda \mathbf{v}^*\mathbf{v}.$$

Also, taking the conjugate transpose of $\mathbf{v}^*A\mathbf{v}$ shows that

Since $\mathbf{v}^*A\mathbf{v}$ is a scalar, its conjugate transpose is just its complex conjugate.

$$(\mathbf{v}^*A\mathbf{v})^* = \mathbf{v}^*A^*\mathbf{v} = \mathbf{v}^*A\mathbf{v}.$$

By putting the two above facts together, we see that

$$\lambda \mathbf{v}^*\mathbf{v} = \mathbf{v}^*A\mathbf{v} = (\mathbf{v}^*A\mathbf{v})^* = (\lambda \mathbf{v}^*\mathbf{v})^* = \overline{\lambda}\mathbf{v}^*\mathbf{v}.$$

Since $\mathbf{v} \neq \mathbf{0}$ (recall that eigenvectors are non-zero by definition), it follows that

Recall from Appendix A.1 that $\overline{z}z = |z|^2$ for all $z \in \mathbb{C}$.

$$\mathbf{v}^*\mathbf{v} = \begin{bmatrix} \overline{v_1} & \overline{v_2} & \cdots & \overline{v_n} \end{bmatrix} \begin{bmatrix} v_1 \\ v_2 \\ \vdots \end{bmatrix} = |v_1|^2 + |v_2|^2 + \cdots + |v_n|^2$$

is a non-zero real number. We can thus divide both sides of the equation $\lambda \mathbf{v}^* \mathbf{v} = \overline{\lambda} \mathbf{v}^* \mathbf{v}$ by $\mathbf{v}^* \mathbf{v}$ to see that $\lambda = \overline{\lambda}$, which shows that $\lambda \in \mathbb{R}$. ∎

Example 3.3.12

Computing Eigenvalues of Hermitian Matrices

Compute all of the eigenvalues of each of the following matrices:

a) $\begin{bmatrix} 2 & 1+i \\ 1-i & 3 \end{bmatrix}$

b) $\begin{bmatrix} 0 & 1 & -1 \\ 1 & -3 & 2 \\ -1 & 2 & 1 \end{bmatrix}$

Solutions:

a) This matrix is Hermitian, so we know that its eigenvalues must be real. We can verify this by explicitly computing them via our usual method:

$$\det\left(\begin{bmatrix} 2-\lambda & 1+i \\ 1-i & 3-\lambda \end{bmatrix}\right) = (2-\lambda)(3-\lambda) - (1+i)(1-i)$$
$$= \lambda^2 - 5\lambda + 6 - 2$$
$$= \lambda^2 - 5\lambda + 4$$
$$= (\lambda - 1)(\lambda - 4).$$

Setting this characteristic polynomial equal to 0 then shows that the eigenvalues of the matrix are $\lambda = 1$ and $\lambda = 4$.

b) Again, since this matrix is Hermitian (in fact, it is real and thus symmetric), its eigenvalues must be real. To find its eigenvalues explicitly, we compute

$$\det\left(\begin{bmatrix} -\lambda & 1 & -1 \\ 1 & -3-\lambda & 2 \\ -1 & 2 & 1-\lambda \end{bmatrix}\right) = -\lambda(-3-\lambda)(1-\lambda) - 2 - 2$$
$$- 4(-\lambda) - (1-\lambda) - (-3-\lambda)$$
$$= -\lambda^3 - 2\lambda^2 + 9\lambda - 2.$$

Finding the roots of this cubic is a bit of a pain, but it's doable. We start by using the rational root theorem (Theorem A.2.2) to see that $\lambda = 2$ is its only rational root. To find its two other roots, we use polynomial long division to divide $-\lambda^3 - 2\lambda^2 + 9\lambda - 2$ by $\lambda - 2$:

To use the rational root theorem, we plug each of ± 1 and ± 2 into the cubic and see which of those numbers make it equal 0. This technique is covered in Appendix A.2.

$$\begin{array}{r} -\lambda^2 - 4\lambda + 1 \\ \lambda - 2 \overline{)\, -\lambda^3 - 2\lambda^2 + 9\lambda - 2} \\ \underline{-\lambda^3 + 2\lambda^2} \\ -4\lambda^2 + 9\lambda - 2 \\ \underline{-4\lambda^2 + 8\lambda} \\ \lambda - 2 \\ \underline{\lambda - 2} \\ 0 \end{array}$$

It follows that $-\lambda^3 - 2\lambda^2 + 9\lambda - 2 = (\lambda - 2)(-\lambda^2 - 4\lambda + 1)$, so the roots of $-\lambda^3 - 2\lambda^2 + 9\lambda - 2$ (i.e., the eigenvalues of this matrix) are $\lambda = 2$ as well as the two roots of $-\lambda^2 - 4\lambda + 1$, which can be computed via the quadratic formula:

$$\lambda = \frac{4 \pm \sqrt{16+4}}{-2} = -2 \pm \sqrt{5}.$$

In particular, all three of these eigenvalues are indeed real.

3.3.3   Eigenspaces and Geometric Multiplicity

Recall that the set of all eigenvectors of a matrix $A \in \mathcal{M}_n$ corresponding to a particular eigenvalue λ (together with the zero vector) is the null space of $A - \lambda I$. Since the null space of any matrix is a subspace of $\mathbb{R}^n$ (or $\mathbb{C}^n$, if it has complex entries), this set of eigenvectors corresponding to λ forms a subspace (as long as we add the zero vector to it). We will be working with these subspaces of eigenvectors extensively, so we give them a name:

Definition 3.3.4 **Eigenspace**	Suppose A is a square matrix with eigenvalue λ. The set of all eigenvectors of A corresponding to λ, together with the zero vector, is called the **eigenspace** of A corresponding to λ.

In other words, the eigenspace of A corresponding to its eigenvalue λ is $\text{null}(A - \lambda I)$.

Since eigenspaces are subspaces, we can use all of the subspace-related machinery that we developed in the previous chapter to help us better understand eigenspaces. In particular, instead of finding *all* eigenvectors corresponding to a particular eigenvalue, we typically just find a basis of that eigenspace, since that provides a more compact description of it.

For example, we showed that the eigenvalues of the matrix

$$A = \begin{bmatrix} 1 & 3 & 3 \\ 3 & 1 & -1 \\ 0 & 0 & 2 \end{bmatrix}$$

from Example 3.3.5 are -2, 2, and 4, with corresponding eigenvectors that are the non-zero multiples of $(-1,1,0)$, $(0,1,-1)$, and $(1,1,0)$, respectively. It follows that $\{(-1,1,0)\}$, $\{(0,1,-1)\}$, and $\{(1,1,0)\}$ are bases of the eigenspaces of A. We now work through an example to illustrate the fact that eigenspaces can be larger (i.e., not just 1-dimensional).

Example 3.3.13

Computing Bases of Eigenspaces

Compute bases of the eigenspaces of the matrix $A = \begin{bmatrix} 2 & 0 & -3 \\ 1 & -1 & -1 \\ 0 & 0 & -1 \end{bmatrix}$.

Solution:

We computed the eigenvalues of this matrix to be 2 (with algebraic multiplicity 1) and -1 (with algebraic multiplicity 2) in Example 3.3.7. We now compute bases of the eigenspaces corresponding to these eigenvalues one at a time:

$\lambda = 2$: We compute bases of the eigenspaces in the same way that we computed eigenvectors in the previous subsections—we start by solving the linear system $(A - 2I)\mathbf{v} = \mathbf{0}$ as follows:

$$\begin{bmatrix} 0 & 0 & -3 & | & 0 \\ 1 & -3 & -1 & | & 0 \\ 0 & 0 & -3 & | & 0 \end{bmatrix} \xrightarrow{R_3 - R_1} \begin{bmatrix} 0 & 0 & -3 & | & 0 \\ 1 & -3 & -1 & | & 0 \\ 0 & 0 & 0 & | & 0 \end{bmatrix}.$$

We could do some more row operations to get this matrix in row echelon form, but we can also perform back substitution to solve the linear system from here.

We thus see that v_2 is a free variable and v_1 and v_3 are leading variables with $v_3 = 0$ and $v_1 = 3v_2$. The eigenvectors corresponding to the eigenvalue $\lambda = 2$ thus have the form $v_2(3, 1, 0)$, so $\{(3, 1, 0)\}$ is a basis of this eigenspace.

$\lambda = -1$: This time, we solve the linear system $(A + I)\mathbf{v} = \mathbf{0}$:

$$\begin{bmatrix} 3 & 0 & -3 & | & 0 \\ 1 & 0 & -1 & | & 0 \\ 0 & 0 & 0 & | & 0 \end{bmatrix} \xrightarrow{R_2 - \frac{1}{3}R_1} \begin{bmatrix} 3 & 0 & -3 & | & 0 \\ 0 & 0 & 0 & | & 0 \\ 0 & 0 & 0 & | & 0 \end{bmatrix}.$$

We thus see that v_2 and v_3 are free variables and v_1 is a leading variable with $v_1 = v_3$. The eigenvectors corresponding to the eigenvalue $\lambda = -1$ thus have the form $v_2(0, 1, 0) + v_3(1, 0, 1)$, so $\{(0, 1, 0), (1, 0, 1)\}$ is a basis of this eigenspace.

Geometrically, this means that the eigenspace corresponding to $\lambda = 2$ is a line, while the eigenspace corresponding to $\lambda = -1$ is a plane. Every vector on that line is stretched by a factor of $\lambda = 2$, and every vector in that plane is reflected through the origin (i.e., multiplied by $\lambda = -1$):

Any (non-zero) linear combination of eigenvectors corresponding to a particular eigenvalue is also an eigenvector corresponding to that eigenvalue.

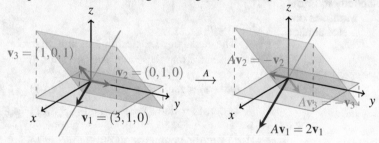

The dimension of an eigenspace is an important quantity that gives us another way of capturing the idea of "how many times" an eigenvalue occurs (much like the algebraic multiplicity did):

Definition 3.3.5

Geometric Multiplicity

Suppose A is a square matrix with eigenvalue λ. The **geometric multiplicity** of λ is the dimension of its corresponding eigenspace.

In other words, the geometric multiplicity of the eigenvalue λ of A is $\text{nullity}(A - \lambda I)$.

For example, we showed in Example 3.3.7 that the matrix

$$A = \begin{bmatrix} 2 & 0 & -3 \\ 1 & -1 & -1 \\ 0 & 0 & -1 \end{bmatrix}$$

has eigenvalues 2 and -1, with algebraic multiplicities 1 and 2, respectively. We then showed in Example 3.3.13 that their geometric multiplicities are also 1 and 2, respectively, since the eigenspace corresponding to $\lambda = 2$ is a line (which is 1-dimensional), while the eigenspace corresponding to $\lambda = -1$ is a plane (which is 2-dimensional).

One important observation about the geometric multiplicity of an eigenvalue is that it is similarity invariant (just like algebraic multiplicity). That is, if A and B are square matrices that are similar (so there exists an invertible matrix P such that $B = PAP^{-1}$), then

We already know from the previous section that A and B have the same characteristic polynomial and thus the same eigenvalues and the same *algebraic* multiplicities.

$$\text{nullity}(A - \lambda I) = \text{nullity}\big(P(A - \lambda I)P^{-1}\big)$$
$$= \text{nullity}(PAP^{-1} - \lambda PP^{-1}) = \text{nullity}(B - \lambda I),$$

where the first equality follows from the fact that the nullity of a matrix is similarity invariant (see Exercise 3.1.17). Since $\text{nullity}(A - \lambda I)$ is the geometric multiplicity of λ as an eigenvalue of A (and $\text{nullity}(B - \lambda I)$ is its geometric multiplicity as an eigenvalue of B), we have demonstrated the following fact:

> ⚠ If $A \in \mathcal{M}_n$ and $B \in \mathcal{M}_n$ are similar then they have the same eigenvalues, with the same geometric multiplicities.

Again, this observation makes geometric sense, since changing the basis through which we look at A or B only changes the orientation of its eigenspaces—it does not change their dimensionality (e.g., skewing and stretching a line results in a line, not a plane).

Example 3.3.14

Using Geometric Multiplicity to Show Matrices are Not Similar

Show that $A = \begin{bmatrix} 2 & 0 & -3 \\ 1 & -1 & -1 \\ 0 & 0 & -1 \end{bmatrix}$ and $B = \begin{bmatrix} 2 & 2 & 3 \\ 0 & -1 & 0 \\ 0 & 1 & -1 \end{bmatrix}$ are not similar.

Solution:

The rank, trace, determinant, and characteristic polynomial cannot help us show that these matrices are not similar, since

$$\text{rank}(A) = \text{rank}(B) = 3, \qquad \text{tr}(A) = \text{tr}(B) = 0,$$
$$\det(A) = \det(B) = 2, \quad \text{and} \quad p_A(\lambda) = p_B(\lambda) = -\lambda^3 + 3\lambda + 2.$$

However, the geometric multiplicities of their eigenvalue $\lambda = -1$ differ between A and B, so they are not similar.

More explicitly, we showed in Example 3.3.13 that $\lambda = -1$ has geometric multiplicity 2 as an eigenvalue of A. To compute its geometric multiplicity as an eigenvalue of B, we solve the linear system $(B + I)\mathbf{v} = \mathbf{0}$ to find a basis of its corresponding eigenspace as usual:

In general, the geometric multiplicity of λ as an eigenvalue of A is the number of zero rows in a row echelon form of $A - \lambda I$.

$$\begin{bmatrix} 3 & 2 & 3 & | & 0 \\ 0 & 0 & 0 & | & 0 \\ 0 & 1 & 0 & | & 0 \end{bmatrix} \xrightarrow{R_1 - 2R_3} \begin{bmatrix} 3 & 0 & 3 & | & 0 \\ 0 & 0 & 0 & | & 0 \\ 0 & 1 & 0 & | & 0 \end{bmatrix}.$$

It follows that v_3 is a free variable and v_1 and v_2 are leading variables with $v_1 = -v_3$ and $v_2 = 0$. The eigenvectors corresponding to $\lambda = -1$ as an eigenvalue of B are thus of the form $v_3(-1, 0, 1)$, so $\{(-1, 0, 1)\}$ is a basis of this (1-dimensional) eigenspace. It follows that the geometric multiplicity of $\lambda = -1$ as an eigenvalue of B is 1.

The previous example demonstrates that if an eigenvalue has algebraic multiplicity 2, it may have geometric multiplicity 1 or 2. In particular, this means that the algebraic and geometric multiplicities of an eigenvalue might be equal to each other, or the geometric multiplicity might be smaller than the algebraic multiplicity. The following theorem says that these are the *only* possibilities—the geometric multiplicity of an eigenvalue can never be larger than its algebraic multiplicity.

Theorem 3.3.3

Geometric Multiplicity Cannot Exceed Algebraic Multiplicity

For each eigenvalue of a square matrix, the geometric multiplicity is less than or equal to the algebraic multiplicity.

We call the eigenvalue λ_1 (instead of just λ) since we will use λ as a variable later in this proof.

Proof. Suppose that λ_1 is an eigenvalue of $A \in \mathcal{M}_n$ with geometric multiplicity k, so that there exists a basis $\{\mathbf{v}_1, \ldots, \mathbf{v}_k\}$ of the eigenspace nullity$(A - \lambda_1 I)$. Since this set is linearly independent, we can construct an invertible matrix

$$P = \left[\, \mathbf{v}_1 \mid \cdots \mid \mathbf{v}_k \mid V \,\right] \in \mathcal{M}_n$$

that has these vectors as its first k columns, by extending them to a basis of $\mathbb{R}^n$ via Theorem 2.4.3(a) (we do not care about the particular entries of $V \in \mathcal{M}_{n,n-k}$; they are just chosen to make P invertible).

We now show that $P^{-1}AP$ has a block structure that makes it much easier to work with than A itself:

$$
\begin{aligned}
P^{-1}AP &= P^{-1}A\left[\, \mathbf{v}_1 \mid \cdots \mid \mathbf{v}_k \mid V \,\right] && \text{(definition of } P) \\
&= P^{-1}\left[\, A\mathbf{v}_1 \mid \cdots \mid A\mathbf{v}_k \mid AV \,\right] && \text{(block matrix mult.)} \\
&= P^{-1}\left[\, \lambda_1\mathbf{v}_1 \mid \cdots \mid \lambda_1\mathbf{v}_k \mid AV \,\right] && (\mathbf{v}_1, \ldots, \mathbf{v}_k \text{ are eigenvecs.)} \\
&= \left[\, \lambda_1 P^{-1}\mathbf{v}_1 \mid \cdots \mid \lambda_1 P^{-1}\mathbf{v}_k \mid P^{-1}AV \,\right] && \text{(block matrix mult.)} \\
&= \left[\, \lambda_1\mathbf{e}_1 \mid \cdots \mid \lambda_1\mathbf{e}_k \mid P^{-1}AV \,\right]. && (P\mathbf{e}_j = \mathbf{v}_j, \text{ so } P^{-1}\mathbf{v}_j = \mathbf{e}_j)
\end{aligned}
$$

In particular, since the first k columns of this matrix are just λ_1 times the first k standard basis vectors, we can write $P^{-1}AP$ in the block matrix form

$$P^{-1}AP = \begin{bmatrix} \lambda_1 I_k & B \\ O & C \end{bmatrix},$$

where $B = \mathcal{M}_{k,n-k}$ and $C \in \mathcal{M}_{n-k,n-k}$ are ugly matrices whose entries we do not care about.

Then, since the characteristic polynomial of a matrix is similarity invariant, we have

$$
\begin{aligned}
p_A(\lambda) &= \det\left(P^{-1}AP - \lambda I\right) && \text{(since } p_A(\lambda) = p_{P^{-1}AP}(\lambda)) \\
&= \det\left(\begin{bmatrix} \lambda_1 I_k - \lambda I_k & B \\ O & C - \lambda I_{n-k} \end{bmatrix}\right) && \text{(block form of } P^{-1}AP) \\
&= \det\left((\lambda_1 - \lambda)I_k\right)\det(C - \lambda I_{n-k}) && \text{(by Exercise 3.2.16(a))} \\
&= (\lambda_1 - \lambda)^k p_C(\lambda) && \text{(since } (\lambda_1 - \lambda)I_k \text{ is diagonal)}
\end{aligned}
$$

The characteristic polynomial of A thus has $(\lambda_1 - \lambda)^k$ as a factor, so the algebraic multiplicity of λ_1 is at least k. ∎

Example 3.3.15

**Computing
Algebraic and
Geometric
Multiplicities**

Find the eigenvalues, as well as their algebraic and geometric multiplicities, of the matrix

$$A = \begin{bmatrix} -6 & 0 & -2 & 1 \\ 0 & -3 & -2 & 1 \\ 3 & 0 & 1 & 1 \\ -3 & 0 & 2 & 2 \end{bmatrix}.$$

Solution:

We start by finding the characteristic polynomial of this matrix by performing a cofactor expansion along the second column of $A - \lambda I$:

Unfortunately, having to factor a cubic is unavoidable here. But at least we do not have to factor a quartic!

$$p_A(\lambda) = \det(A - \lambda I) = \det\left(\begin{bmatrix} -6-\lambda & 0 & -2 & 1 \\ 0 & -3-\lambda & -2 & 1 \\ 3 & 0 & 1-\lambda & 1 \\ -3 & 0 & 2 & 2-\lambda \end{bmatrix}\right)$$

$$= (-3-\lambda)\big((-6-\lambda)(1-\lambda)(2-\lambda)+6+6$$
$$- 2(-6-\lambda)+3(1-\lambda)+6(2-\lambda)\big)$$

$$= (3+\lambda)(\lambda^3 + 3\lambda^2 - 9\lambda - 27)$$

$$= (3+\lambda)(3-\lambda)(\lambda^2 + 6\lambda + 9)$$

$$= (3+\lambda)^3(3-\lambda).$$

The eigenvalues of this matrix are thus -3 and 3, with algebraic multiplicities 3 and 1, respectively.

All that remains is to find the geometric multiplicity of these eigenvalues. For $\lambda = 3$ we know from Theorem 3.3.3 that its geometric multiplicity is at most 1 (its algebraic multiplicity). Since every eigenvalue must have geometric multiplicity at *least* 1 (after all, every eigenvalue has at least one corresponding eigenvector), we can conclude that its geometric multiplicity is exactly 1. If we wanted to, we could explicitly show that $\lambda = 3$ has a 1-dimensional eigenspace with $\{(0,0,1,2)\}$ as a basis, but this is more work than is necessary.

For $\lambda = -3$, all we know is that its geometric multiplicity is between 1 and 3 inclusive, so we have to do more work to find its exact value. To this end, we solve the linear system $(A + 3I)\mathbf{v} = \mathbf{0}$ to find a basis of this eigenspace:

$$\begin{bmatrix} -3 & 0 & -2 & 1 & | & 0 \\ 0 & 0 & -2 & 1 & | & 0 \\ 3 & 0 & 4 & 1 & | & 0 \\ -3 & 0 & 2 & 5 & | & 0 \end{bmatrix} \xrightarrow[R_4-R_1]{R_3+R_1} \begin{bmatrix} -3 & 0 & -2 & 1 & | & 0 \\ 0 & 0 & -2 & 1 & | & 0 \\ 0 & 0 & 2 & 2 & | & 0 \\ 0 & 0 & 4 & 4 & | & 0 \end{bmatrix}$$

$$\xrightarrow[\substack{R_4-2R_3}]{\substack{R_1+R_3 \\ R_2+R_3}} \begin{bmatrix} -3 & 0 & 0 & 3 & | & 0 \\ 0 & 0 & 0 & 3 & | & 0 \\ 0 & 0 & 2 & 2 & | & 0 \\ 0 & 0 & 0 & 0 & | & 0 \end{bmatrix} \xrightarrow{R_1-R_2} \begin{bmatrix} -3 & 0 & 0 & 0 & | & 0 \\ 0 & 0 & 0 & 3 & | & 0 \\ 0 & 0 & 2 & 2 & | & 0 \\ 0 & 0 & 0 & 0 & | & 0 \end{bmatrix}.$$

From here it is straightforward to use back substitution to solve the linear system and see that $\{(0,1,0,0)\}$ is a basis of the eigenspace, so the geometric multiplicity of $\lambda = -3$ is 1.

Eigenvalues of Triangular Matrices

We close this section by noting that, just as was the case with determinants, eigenvalues become much easier to deal with if we restrict our attention to triangular matrices. To illustrate why this is the case, suppose we wanted to find the eigenvalues of the matrix

$$A = \begin{bmatrix} 1 & 2 & 3 \\ 0 & 4 & 5 \\ 0 & 0 & 6 \end{bmatrix}.$$

Since the determinant of a triangular matrix is the product of its diagonal entries (refer back to Theorem 3.2.4), we can compute the characteristic polynomial of A as follows:

$$p_A(\lambda) = \det(A - \lambda I) = \det\left(\begin{bmatrix} 1-\lambda & 2 & 3 \\ 0 & 4-\lambda & 5 \\ 0 & 0 & 6-\lambda \end{bmatrix}\right)$$
$$= (1-\lambda)(4-\lambda)(6-\lambda).$$

The roots of this characteristic polynomial (and thus the eigenvalues of A) are 1, 4, and 6, which are the diagonal entries of A. The following theorem tells us that this is not a coincidence:

Theorem 3.3.4 **Eigenvalues of Triangular Matrices**	Suppose $A \in \mathcal{M}_n$ is a triangular matrix. Then its eigenvalues, listed according to algebraic multiplicity, are exactly its diagonal entries (i.e., $a_{1,1}$, $a_{2,2}, \ldots, a_{n,n}$).

We leave the proof of this theorem to Exercise 3.3.18, but we note that it follows almost immediately from Theorem 3.2.4 (i.e., the fact that the determinant of a triangular matrix is the product of its diagonal entries). We note, however, that there is no general shortcut for computing all of the eigen*vectors* of a triangular matrix.

Example 3.3.16 **Eigenvalues and Eigenvectors of Triangular Matrices**	Compute the eigenvalues of the following matrices, as well as bases for their corresponding eigenspaces:

a) $\begin{bmatrix} 1 & 2 & 3 \\ 0 & 4 & 5 \\ 0 & 0 & 6 \end{bmatrix}$ b) $\begin{bmatrix} 7 & 0 & 0 \\ 0 & 2 & 0 \\ 0 & 0 & -3 \end{bmatrix}$

c) $\begin{bmatrix} 2 & 1 & 0 \\ 0 & 2 & 1 \\ 0 & 0 & 2 \end{bmatrix}$

Solutions:
  a) As we already discussed, this matrix (which we called A above) has eigenvalues 1, 4, and 6. We find the corresponding eigenvectors by considering these eigenvalues one at a time, just like in previous examples:

   $\lambda = 1$: We want to find the null space of the matrix $A - \lambda I =$

$A - I$:

$$\begin{bmatrix} 0 & 2 & 3 & | & 0 \\ 0 & 3 & 5 & | & 0 \\ 0 & 0 & 5 & | & 0 \end{bmatrix} \xrightarrow{R_2 - \frac{3}{2}R_1} \begin{bmatrix} 0 & 2 & 3 & | & 0 \\ 0 & 0 & 1/2 & | & 0 \\ 0 & 0 & 5 & | & 0 \end{bmatrix}$$

$$\xrightarrow{R_3 - 10R_2} \begin{bmatrix} 0 & 2 & 3 & | & 0 \\ 0 & 0 & 1/2 & | & 0 \\ 0 & 0 & 0 & | & 0 \end{bmatrix}.$$

Here, v_1 is a free variable and v_2 and v_3 are leading.

One basis of this null space is thus $\{(1,0,0)\}$.

$\lambda = 4$: We want to find the null space of the matrix $A - \lambda I = A - 4I$:

Here, v_2 is a free variable and v_1 and v_3 are leading.

$$\begin{bmatrix} -3 & 2 & 3 & | & 0 \\ 0 & 0 & 5 & | & 0 \\ 0 & 0 & 2 & | & 0 \end{bmatrix} \xrightarrow{R_3 - \frac{2}{5}R_2} \begin{bmatrix} -3 & 2 & 3 & | & 0 \\ 0 & 0 & 5 & | & 0 \\ 0 & 0 & 0 & | & 0 \end{bmatrix}.$$

One basis of this null space is thus $\{(2,3,0)\}$.

$\lambda = 6$: We want to find the null space of the matrix $A - \lambda I = A - 6I$:

$$\begin{bmatrix} -5 & 2 & 3 & | & 0 \\ 0 & -2 & 5 & | & 0 \\ 0 & 0 & 0 & | & 0 \end{bmatrix} \xrightarrow{R_1 + R_2} \begin{bmatrix} -5 & 0 & 8 & | & 0 \\ 0 & -2 & 5 & | & 0 \\ 0 & 0 & 0 & | & 0 \end{bmatrix}.$$

One basis of this null space is thus $\{(16,25,10)\}$.

b) This matrix is triangular (in fact, it is diagonal), so its eigenvalues are its diagonal entries: 7, 2, and -3. It is straightforward to check that bases of the corresponding eigenspaces are $[v,]\{e_1\}$, $\{e_2\}$, and $\{e_3\}$, respectively.

This happens for all diagonal matrices—the standard basis vectors make up bases of their eigenspaces.

c) This matrix is triangular, so its eigenvalues all equal 2 (i.e., its only eigenvalue is 2, with algebraic multiplicity 3). To find a basis of the corresponding eigenspace, we find the null space of

For this matrix, the eigenvalue 2 has algebraic multiplicity 3 and geometric multiplicity 1.

$$\begin{bmatrix} 2 & 1 & 0 \\ 0 & 2 & 1 \\ 0 & 0 & 2 \end{bmatrix} - 2 \begin{bmatrix} 1 & 0 & 0 \\ 0 & 1 & 0 \\ 0 & 0 & 1 \end{bmatrix} = \begin{bmatrix} 0 & 1 & 0 \\ 0 & 0 & 1 \\ 0 & 0 & 0 \end{bmatrix}.$$

It is straightforward to check that $\{(1,0,0)\}$ is a basis of this null space.

Exercises

solutions to starred exercises on page 466

3.3.1 For each of the following matrices A and vectors $\mathbf{v}$, show that $\mathbf{v}$ is an eigenvector of A and find the corresponding eigenvalue.

*(a) $A = \begin{bmatrix} 1 & 5 \\ 4 & 2 \end{bmatrix}$, $\mathbf{v} = \begin{bmatrix} 1 \\ 1 \end{bmatrix}$.

(b) $A = \begin{bmatrix} 1 & -1 \\ 3 & 6 \end{bmatrix}$, $\mathbf{v} = \begin{bmatrix} -2 \\ 5 + \sqrt{13} \end{bmatrix}$.

*(c) $A = \begin{bmatrix} 4 & 1 & 1 \\ 3 & 3 & -1 \\ 4 & 1 & 1 \end{bmatrix}$, $\mathbf{v} = \begin{bmatrix} -4 \\ 7 \\ 9 \end{bmatrix}$.

(d) $A = \begin{bmatrix} 2 & -1 & -1 \\ -1 & 2 & 0 \\ 3 & 2 & 4 \end{bmatrix}$, $\mathbf{v} = \begin{bmatrix} 2 \\ -1 + i \\ -1 - 3i \end{bmatrix}$.

*(e) $A = \begin{bmatrix} 3 & 1 & 5 & 5 \\ 3 & 2 & 2 & 5 \\ 3 & 3 & 5 & 5 \\ 2 & 5 & 2 & 0 \end{bmatrix}, \mathbf{v} = \begin{bmatrix} 13 \\ 14 \\ 6 \\ -27 \end{bmatrix}$.

3.3.2 For each of the following matrices A and scalars λ, show that λ is an eigenvalue of A and find a corresponding eigenvector.

*(a) $\lambda = -3, A = \begin{bmatrix} 1 & 5 \\ 4 & 2 \end{bmatrix}$.

(b) $\lambda = \sqrt{2}, A = \begin{bmatrix} 0 & 1 \\ 2 & 0 \end{bmatrix}$.

*(c) $\lambda = 2, A = \begin{bmatrix} 0 & 3 & -1 \\ 2 & -1 & -1 \\ -2 & 3 & 0 \end{bmatrix}$.

(d) $\lambda = 2+i, A = \begin{bmatrix} -1 & 0 & -1 \\ 2 & 2 & 1 \\ 2 & -2 & 3 \end{bmatrix}$.

*(e) $\lambda = -1, A = \begin{bmatrix} 3 & 2 & 3 & 2 \\ 0 & 0 & 3 & -1 \\ 4 & 1 & 3 & 3 \\ 2 & 1 & 1 & 0 \end{bmatrix}$.

3.3.3 For each of the following matrices, compute all (potentially complex) eigenvalues, state their algebraic and geometric multiplicities, and find bases for their corresponding eigenspaces.

*(a) $\begin{bmatrix} 1 & 2 \\ -1 & -2 \end{bmatrix}$

(b) $\begin{bmatrix} 6 & 3 \\ 2 & 1 \end{bmatrix}$

*(c) $\begin{bmatrix} 0 & 1 \\ 0 & 0 \end{bmatrix}$

(d) $\begin{bmatrix} 2 & 1 \\ 3 & -1 \end{bmatrix}$

*(e) $\begin{bmatrix} 3 & 0 & 0 \\ 0 & -2 & 0 \\ 0 & 0 & 7 \end{bmatrix}$

(f) $\begin{bmatrix} 2 & 3 & 0 \\ 3 & 0 & 1 \\ 0 & 1 & 2 \end{bmatrix}$

*(g) $\begin{bmatrix} 2 & 1 & 0 & 0 \\ 0 & -3 & 2 & 0 \\ 0 & 0 & 1 & -1 \\ 0 & 0 & 0 & 2 \end{bmatrix}$

(h) $\begin{bmatrix} 2 & 1 & -1 & -1 \\ 0 & 3 & 1 & 1 \\ 0 & 1 & 3 & 1 \\ 0 & 0 & 0 & 2 \end{bmatrix}$

🖥 **3.3.4** Use computer software to compute all eigenvalues of the given matrix. Also state their algebraic and geometric multiplicities, and find bases for their corresponding eigenspaces.

*(a) $\begin{bmatrix} 19 & 10 & 5 & 22 \\ 9 & 17 & 5 & 19 \\ -8 & -10 & 2 & -18 \\ -11 & -10 & -5 & -14 \end{bmatrix}$

(b) $\begin{bmatrix} 12 & -3 & -6 & -3 \\ 1 & 16 & 2 & 1 \\ -3 & -3 & 9 & -3 \\ -1 & -1 & -2 & 14 \end{bmatrix}$

*(c) $\begin{bmatrix} 6 & -4 & 4 & 0 & 4 \\ 0 & 10 & -6 & 0 & -6 \\ 16 & 20 & -20 & -8 & -26 \\ -8 & -8 & 11 & 10 & 11 \\ -16 & -20 & 24 & 8 & 30 \end{bmatrix}$

(d) $\begin{bmatrix} -4 & -12 & 28 & 44 & -4 \\ 1 & 8 & -16 & -30 & 2 \\ -10 & -12 & 30 & 36 & -4 \\ 5 & 6 & -14 & -16 & 2 \\ -2 & -2 & 0 & -4 & 4 \end{bmatrix}$

3.3.5 Determine which of the following statements are true and which are false.

*(a) Every matrix has at least one real eigenvalue.

(b) A set of two eigenvectors corresponding to the same eigenvalue of a matrix must be linearly dependent.

*(c) If two matrices are row equivalent then they must have the same eigenvalues.

(d) If $\mathbf{v}$ and $\mathbf{w}$ are eigenvectors of a matrix $A \in \mathcal{M}_n$ corresponding to an eigenvalue λ then so is any non-zero linear combination of $\mathbf{v}$ and $\mathbf{w}$.

*(e) If $A \in \mathcal{M}_n$ then $\det(A - \lambda I) = \det(\lambda I - A)$.

(f) If $A \in \mathcal{M}_n(\mathbb{C})$ is Hermitian then its diagonal entries must be real.

*(g) Every diagonal matrix $A \in \mathcal{M}_n(\mathbb{R})$ is symmetric.

(h) The eigenvalues of every matrix $A \in \mathcal{M}_n(\mathbb{C})$ come in complex conjugate pairs.

*(i) The geometric multiplicity of an eigenvalue λ of $A \in \mathcal{M}_n$ equals nullity$(A - \lambda I)$.

(j) It is possible for an eigenvalue to have geometric multiplicity equal to 0.

*(k) If an eigenvalue has algebraic multiplicity 1 then its geometric multiplicity must also equal 1.

(l) If an eigenvalue has geometric multiplicity 1 then its algebraic multiplicity must also equal 1.

*(m) If two matrices $A, B \in \mathcal{M}_n$ have the same characteristic polynomial then they have the same trace.

(n) If two matrices $A, B \in \mathcal{M}_n$ have the same characteristic polynomial then they have the same determinant.

*3.3.6 Let $k \in \mathbb{R}$ and consider the following matrix:

$$A = \begin{bmatrix} k & 1 \\ -1 & 1 \end{bmatrix}.$$

For which values of k does A have (i) two distinct real eigenvalues, (ii) only one distinct real eigenvalue, and (iii) no real eigenvalues?

3.3.7 Let $A = \begin{bmatrix} 1 & 0 & 0 \\ -8 & 4 & -5 \\ 8 & 0 & 9 \end{bmatrix}$.

(a) Find all eigenvalues of A.

(b) Find a basis for each of the eigenspaces of A.

(c) Find all eigenvalues of $A^2 - 3A + I$. [Hint: You can explicitly calculate the matrix $A^2 - 3A + I$ and find its eigenvalues, but there is a much faster and easier way.]

(d) Find a basis for each eigenspace of $A^2 - 3A + I$.

*3.3.8 Suppose that all entries of $A \in \mathcal{M}_n$ are real. Show that if $\mathbf{v}$ is an eigenvector of A corresponding to the eigenvalue λ then $\overline{\mathbf{v}}$ is an eigenvector of A corresponding to the eigenvalue $\overline{\lambda}$.

3.3.9 Suppose $A \in \mathcal{M}_n$ and n is odd. Show that A has at least one real eigenvalue.

∗∗3.3.10 Suppose $A \in \mathcal{M}_2(\mathbb{C})$. Find a formula for the eigenvalues of A in terms of $\text{tr}(A)$ and $\det(A)$.

3.3.11 Suppose that $A \in \mathcal{M}_n$ is a matrix such that $A^2 = O$. Find all possible eigenvalues of A.

∗3.3.12 Suppose that $A \in \mathcal{M}_n(\mathbb{C})$ is a matrix such that $A^4 = I$. Find all possible eigenvalues of A.

∗∗3.3.13 Recall that if R^θ is the linear transformation that rotates vectors in $\mathbb{R}^2$ by an angle of θ counter-clockwise around the origin, then its standard matrix is

$$\left[R^\theta\right] = \begin{bmatrix} \cos(\theta) & -\sin(\theta) \\ \sin(\theta) & \cos(\theta) \end{bmatrix}.$$

(a) Compute the eigenvalues of $\left[R^\theta\right]$ (keep in mind that the eigenvalues might be complex, and that's OK).
(b) For which values of θ are the eigenvalues of $\left[R^\theta\right]$ real? Provide a geometric interpretation of your answer.

3.3.14 A matrix $A \in \mathcal{M}_n(\mathbb{C})$ is called **skew-Hermitian** if $A^* = -A$. Show that the eigenvalues of skew-Hermitian matrices are imaginary (i.e., of the form bi for some $b \in \mathbb{R}$).

∗∗3.3.15 Suppose that $A \in \mathcal{M}_{m,n}(\mathbb{C})$ and recall from Exercise 1.2.15 that the dot product on $\mathbb{C}^n$ is given by

$$\mathbf{v} \cdot \mathbf{w} \stackrel{\text{def}}{=} \overline{v_1}w_1 + \overline{v_2}w_2 + \cdots + \overline{v_n}w_n,$$

which equals $\mathbf{v}^*\mathbf{w}$ if $\mathbf{v}$ and $\mathbf{w}$ are column vectors.

(a) Show that $\mathbf{v} \cdot (A\mathbf{w}) = (A^*\mathbf{v}) \cdot \mathbf{w}$ for all $\mathbf{v} \in \mathbb{C}^m$ and $\mathbf{w} \in \mathbb{C}^n$.
(b) Show that if $B \in \mathcal{M}_{n,m}(\mathbb{C})$ is such that $\mathbf{v} \cdot (A\mathbf{w}) = (B\mathbf{v}) \cdot \mathbf{w}$ for all $\mathbf{v} \in \mathbb{C}^m$ and $\mathbf{w} \in \mathbb{C}^n$, then $B = A^*$.

∗∗3.3.16 Show that a square matrix is invertible if and only if 0 is not an eigenvalue of it.

3.3.17 Show that if square matrices A and B are similar (i.e., $A = PBP^{-1}$ for some invertible matrix P), then $\mathbf{v}$ is an eigenvector of B if and only if $P\mathbf{v}$ is an eigenvector of A.

∗∗3.3.18 Prove Theorem 3.3.4. That is, show that the eigenvalues of a triangular matrix, listed according to algebraic multiplicity, are its diagonal entries.

3.3.19 Suppose $A \in \mathcal{M}_{m,n}$ and $B \in \mathcal{M}_{n,m}$. In this exercise, we show that AB and BA have (essentially) the same eigenvalues.

(a) Show that if $\lambda \neq 0$ is an eigenvalue of AB then it is also an eigenvalue of BA. [Hint: Let $\mathbf{v}$ be an eigenvector of AB corresponding to λ and simplify $BAB\mathbf{v}$.]
(b) Provide an example for which 0 is an eigenvalue of AB, but not of BA. [Hint: Choose A and B so that AB is larger than BA.]

3.3.20 Suppose $A \in \mathcal{M}_n$.

(a) Show that A and A^T have the same characteristic polynomials and thus the same eigenvalues.

(b) Provide an example to show that A and A^T might have different eigenvectors.

∗∗3.3.21 Suppose $A \in \mathcal{M}_n(\mathbb{C})$.

(a) Show that λ is an eigenvalue of A if and only if $\overline{\lambda}$ is an eigenvalue of A^*.
(b) Let $\mathbf{v}$ be an eigenvector corresponding to an eigenvalue λ of A and $\mathbf{w}$ be an eigenvector corresponding to an eigenvalue $\mu \neq \overline{\lambda}$ of A^*. Show that $\mathbf{v} \cdot \mathbf{w} = 0$.

∗∗3.3.22 Suppose $A \in \mathcal{M}_n(\mathbb{C})$ has eigenvalues $\lambda_1, \lambda_2, \ldots, \lambda_n$ with corresponding unit (i.e., length 1) eigenvectors $\mathbf{v}_1, \mathbf{v}_2, \ldots, \mathbf{v}_n$. Let $B = A - \lambda_1 \mathbf{v}_1 \mathbf{v}_1^*$.

(a) Show that $\mathbf{v}_1$ is an eigenvector of B corresponding to the eigenvalue 0.
(b) Show that B has eigenvalues $0, \lambda_2, \lambda_3, \ldots, \lambda_n$. [Side note: If it helps, you can assume that $\lambda_1 \neq \lambda_2$, $\lambda_1 \neq \lambda_3$, …, $\lambda_1 \neq \lambda_n$.] [Hint: Use Exercise 3.3.21.]
(c) Give an example to show that none of $\mathbf{v}_2, \ldots, \mathbf{v}_n$ are necessarily eigenvectors of B.

∗∗3.3.23 This exercise demonstrates how to construct a matrix whose characteristic polynomial is any given polynomial (with leading term $(-1)^n \lambda^n$) of our choosing.

(a) Show that the characteristic polynomial of the matrix

$$C = \begin{bmatrix} 0 & 1 & 0 \\ 0 & 0 & 1 \\ -a_0 & -a_1 & -a_2 \end{bmatrix}$$

is $p_C(\lambda) = -(\lambda^3 + a_2\lambda^2 + a_1\lambda + a_0)$.
(b) More generally, show that the characteristic polynomial of the matrix

$$C = \begin{bmatrix} 0 & 1 & 0 & \cdots & 0 \\ 0 & 0 & 1 & \cdots & 0 \\ \vdots & \vdots & \vdots & \ddots & \vdots \\ 0 & 0 & 0 & \cdots & 1 \\ -a_0 & -a_1 & -a_2 & \cdots & -a_{n-1} \end{bmatrix}$$

is $p_C(\lambda) = (-1)^n(\lambda^n + a_{n-1}\lambda^{n-1} + \cdots + a_1\lambda + a_0)$. [Hint: Try a proof by induction.]

[Side note: This matrix C is called the **companion matrix** of the polynomial p_C. Although we have used polynomials to compute eigenvalues, in practice the opposite is often done. That is, to find the roots of a high-degree polynomial, people construct its companion matrix and then compute its eigenvalues using numerical methods like the one introduced in Section 3.B.]

∗∗3.3.24 A matrix $A \in \mathcal{M}_n(\mathbb{R})$ with non-negative entries is called **column stochastic** if its columns each add up to 1, and it is called **row stochastic** if its rows each add up to 1. For example, the matrices

$$\begin{bmatrix} 1/2 & 0 \\ 1/2 & 1 \end{bmatrix} \quad \text{and} \quad \begin{bmatrix} 1/3 & 2/3 \\ 1/2 & 1/2 \end{bmatrix}$$

are column stochastic and row stochastic, respectively.

(a) Show that every row stochastic matrix has an eigenvalue equal to 1. [Hint: There is an "obvious" eigenvector corresponding to this eigenvalue.]

(b) Show that every eigenvalue of a row stochastic matrix has absolute value no larger than 1.
[Hint: Suppose **v** is an eigenvector that is scaled so that its largest entry v_i has $v_i = 1$.]

(c) Show that the claims from parts (a) and (b) also apply to column stochastic matrices.
[Side note: We showed in Exercise 2.B.12 that the eigenvalue 1 of a column stochastic matrix has a corresponding eigenvector with all of its entries non-negative.]

3.4 Diagonalization

Recall that our motivation for this chapter was that we wanted to find bases that make linear transformations easier to work with than the standard basis. Equivalently, we wanted to transform a matrix via similarity to make it as simple as possible. In this section, we finally answer the question of when we can make linear transformations and matrices diagonal in this way, and how to choose the basis so as to make it happen.

Definition 3.4.1
Diagonalizable Matrices

A matrix $A \in \mathcal{M}_n$ is called **diagonalizable** if there is a diagonal matrix $D \in \mathcal{M}_n$ and an invertible matrix $P \in \mathcal{M}_n$ such that $A = PDP^{-1}$.

The idea behind diagonalization is that a diagonalized matrix is almost as easy to work with as a diagonal matrix. To illustrate what we mean by this, consider the problem of computing A^k, where $A \in \mathcal{M}_n$ and k is a very large integer.

- If A is a general matrix (i.e., it does not have any special structure that we can exploit), computing A^k requires a whole bunch of matrix multiplications. Since matrix multiplication is ugly and time-consuming, this seems undesirable.

- If A is a diagonal matrix, computing A^k is fairly straightforward, since diagonal matrices multiply entry-wise, so we can just compute the k-th power of each diagonal entry of A:

$$A^k = \begin{bmatrix} a_{1,1} & 0 & \cdots & 0 \\ 0 & a_{2,2} & \cdots & 0 \\ \vdots & \vdots & \ddots & \vdots \\ 0 & 0 & \cdots & a_{n,n} \end{bmatrix}^k = \begin{bmatrix} a_{1,1}^k & 0 & \cdots & 0 \\ 0 & a_{2,2}^k & \cdots & 0 \\ \vdots & \vdots & \ddots & \vdots \\ 0 & 0 & \cdots & a_{n,n}^k \end{bmatrix}.$$

- If A is diagonalizable, so $A = PDP^{-1}$ for some invertible P and diagonal D, then

$$A^k = \overbrace{(PDP^{-1})}^{} \overbrace{(PDP^{-1})}^{P^{-1}P=I} \overbrace{(PDP^{-1})}^{P^{-1}P=I} \cdots {(PDP^{-1})} = PD^kP^{-1}.$$

$$\underbrace{\phantom{(PDP^{-1})(PDP^{-1})(PDP^{-1})\cdots(PDP^{-1})}}_{k \text{ times}}$$

Since D is diagonal, D^k can be computed entry-wise, so we can compute A^k via just 2 matrix multiplications: one on the left by P and one on the right by P^{-1}.

However, this is just the first taste of how diagonalization is useful—we will also see that we can use diagonalization to do things like take the square root of a matrix, find an explicit formula for terms in the Fibonacci sequence, or even apply exponential or trigonometric functions (among others) to matrices. In a sense, once we have a diagonalization of a matrix, that matrix has been "unlocked" for us, making it significantly easier to solve problems involving it.

3.4.1   How to Diagonalize

We now stitch together the tools that we developed throughout this chapter in order to answer the questions of which matrices can be diagonalized (or equivalently, which linear transformations are diagonal in some basis), and how to do the diagonalization.

We start with the latter question—it turns out that eigenvalues and eigenvectors can be used to diagonalize matrices, and in fact they are the *only* way to do so:

Theorem 3.4.1

Diagonalizability

Let $A \in \mathcal{M}_n$ and suppose $P, D \in \mathcal{M}_n$ are such that P is invertible and D is diagonal. Then $A = PDP^{-1}$ if and only if the columns of P are eigenvectors of A whose corresponding eigenvalues are the diagonal entries of D in the same order.

Proof. If we multiply the equation $A = PDP^{-1}$ on the right by P, we see that it is equivalent to $AP = PD$. If we write P in terms of its columns $\mathbf{v}_1, \mathbf{v}_2, \ldots, \mathbf{v}_n$ and use block matrix multiplication, we see that

$$AP = A\begin{bmatrix} \mathbf{v}_1 \mid \mathbf{v}_2 \mid \cdots \mid \mathbf{v}_n \end{bmatrix} = \begin{bmatrix} A\mathbf{v}_1 \mid A\mathbf{v}_2 \mid \cdots \mid A\mathbf{v}_n \end{bmatrix}$$

and

This way of writing *PD* in terms of scaled columns of *P* only works because *D* is diagonal.

$$PD = \begin{bmatrix} \mathbf{v}_1 \mid \mathbf{v}_2 \mid \cdots \mid \mathbf{v}_n \end{bmatrix} D = \begin{bmatrix} d_{1,1}\mathbf{v}_1 \mid d_{2,2}\mathbf{v}_2 \mid \cdots \mid d_{n,n}\mathbf{v}_n \end{bmatrix}.$$

Since $AP = PD$ if and only if the columns of AP equal the columns of PD in the same order, we conclude that $AP = PD$ if and only if $A\mathbf{v}_j = d_{j,j}\mathbf{v}_j$ for all $1 \leq j \leq n$. In other words, $A = PDP^{-1}$ if and only if $\mathbf{v}_1, \mathbf{v}_2, \ldots, \mathbf{v}_n$ (the columns of P) are eigenvectors of A with corresponding eigenvalues $d_{1,1}, d_{2,2}, \ldots, d_{n,n}$, respectively. ∎

Example 3.4.1

Our First Diagonalization

Diagonalize the matrix $A = \begin{bmatrix} 1 & 2 \\ 5 & 4 \end{bmatrix}$.

Solution:

We showed in Examples 3.3.3 and 3.3.4 that this matrix has eigenvalues $\lambda_1 = -1$ and $\lambda_2 = 6$ corresponding to the eigenvectors $\mathbf{v}_1 = (-1, 1)$ and $\mathbf{v}_2 = (2, 5)$, , respectively. We thus stick these eigenvalues along the diagonal of a diagonal matrix D, and the corresponding eigenvectors as columns into a matrix P in the same order, as suggested by Theorem 3.4.1:

We could have also chosen $\mathbf{v}_2 = (2/5, 1)$, but our choice here is prettier. Which multiple of each eigenvector we choose does not matter.

$$D = \begin{bmatrix} \lambda_1 & 0 \\ 0 & \lambda_2 \end{bmatrix} = \begin{bmatrix} -1 & 0 \\ 0 & 6 \end{bmatrix} \quad \text{and} \quad P = \begin{bmatrix} \mathbf{v}_1 \mid \mathbf{v}_2 \end{bmatrix} = \begin{bmatrix} -1 & 2 \\ 1 & 5 \end{bmatrix}.$$

It is straightforward to check that P is invertible, and

$$P^{-1} = \frac{1}{7} \begin{bmatrix} -5 & 2 \\ 1 & 1 \end{bmatrix},$$

so Theorem 3.4.1 tells us that A is diagonalized by this D and P (i.e., $A = PDP^{-1}$).

Since this is our first diagonalization, it is worth doing a sanity check—let's multiply out PDP^{-1} and see that it does indeed equal A:

$$PDP^{-1} = \begin{bmatrix} -1 & 2 \\ 1 & 5 \end{bmatrix} \begin{bmatrix} -1 & 0 \\ 0 & 6 \end{bmatrix} \left(\frac{1}{7} \begin{bmatrix} -5 & 2 \\ 1 & 1 \end{bmatrix} \right)$$

$$= \frac{1}{7} \begin{bmatrix} -1 & 2 \\ 1 & 5 \end{bmatrix} \begin{bmatrix} 5 & -2 \\ 6 & 6 \end{bmatrix} = \frac{1}{7} \begin{bmatrix} 7 & 14 \\ 35 & 28 \end{bmatrix} = \begin{bmatrix} 1 & 2 \\ 5 & 4 \end{bmatrix} = A.$$

It is perhaps also worth illustrating what this diagonalization means geometrically. The matrix A is the standard matrix (in the standard basis) of some linear transformation T that acts on $\mathbb{R}^2$. If we look at this linear transformation in the basis $B = \{\mathbf{v}_1, \mathbf{v}_2\} = \{(-1,1), (2,5)\}$ consisting of eigenvectors of A then $[T]_B = D$ is diagonal:

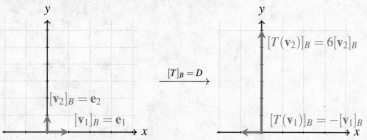

Compare this image with Figures 3.22 and 3.23, where we depicted the action of $A = [T]$ in the standard basis.

<table>
<tr><td>

Example 3.4.2

Our Second Diagonalization

</td><td>

Diagonalize the matrix $A = \begin{bmatrix} 2 & 0 & -3 \\ 1 & -1 & -1 \\ 0 & 0 & -1 \end{bmatrix}$.

Solution:

We showed in Examples 3.3.7 that this matrix has eigenvalues (listed according to algebraic multiplicity) $\lambda_1 = 2$ and $\lambda_2 = \lambda_3 = -1$, and we showed in Example 3.3.13 that bases of its eigenspaces can be made from the corresponding eigenvectors $\mathbf{v}_1 = (3,1,0)$, $\mathbf{v}_2 = (0,1,0)$, and $\mathbf{v}_3 = (1,0,1)$. We thus stick these eigenvalues along the diagonal of a diagonal matrix D, and the eigenspaces basis vectors into a matrix P as columns:

</td></tr>
</table>

It does not matter whether we choose $\mathbf{v}_2 = (0,1,0)$ and $\mathbf{v}_3 = (1,0,1)$, or $\mathbf{v}_2 = (1,0,1)$ and $\mathbf{v}_3 = (0,1,0)$. Either choice works.

$$D = \begin{bmatrix} \lambda_1 & 0 & 0 \\ 0 & \lambda_2 & 0 \\ 0 & 0 & \lambda_3 \end{bmatrix} = \begin{bmatrix} 2 & 0 & 0 \\ 0 & -1 & 0 \\ 0 & 0 & -1 \end{bmatrix}, \quad P = \begin{bmatrix} \mathbf{v}_1 \mid \mathbf{v}_2 \mid \mathbf{v}_3 \end{bmatrix} = \begin{bmatrix} 3 & 0 & 1 \\ 1 & 1 & 0 \\ 0 & 0 & 1 \end{bmatrix}.$$

It is straightforward to check that P is invertible with inverse

$$P^{-1} = \frac{1}{3}\begin{bmatrix} 1 & 0 & -1 \\ -1 & 3 & 1 \\ 0 & 0 & 3 \end{bmatrix},$$

so Theorem 3.4.1 tells us that A is diagonalized by this D and P.

Again, we multiply out PDP^{-1} to double-check that this is indeed a diagonalization of A:

The scalar $1/3$ here comes from P^{-1}; we just pulled it in front of all three matrices.

$$PDP^{-1} = \frac{1}{3}\begin{bmatrix} 3 & 0 & 1 \\ 1 & 1 & 0 \\ 0 & 0 & 1 \end{bmatrix}\begin{bmatrix} 2 & 0 & 0 \\ 0 & -1 & 0 \\ 0 & 0 & -1 \end{bmatrix}\begin{bmatrix} 1 & 0 & -1 \\ -1 & 3 & 1 \\ 0 & 0 & 3 \end{bmatrix}$$

$$= \frac{1}{3}\begin{bmatrix} 3 & 0 & 1 \\ 1 & 1 & 0 \\ 0 & 0 & 1 \end{bmatrix}\begin{bmatrix} 2 & 0 & -2 \\ 1 & -3 & -1 \\ 0 & 0 & -3 \end{bmatrix} = \begin{bmatrix} 2 & 0 & -3 \\ 1 & -1 & -1 \\ 0 & 0 & -1 \end{bmatrix} = A.$$

Geometrically, A is the standard matrix (in the standard basis) of some linear transformation T that acts on $\mathbb{R}^3$. If we instead look at this linear transformation in the basis $B = \{\mathbf{v}_1, \mathbf{v}_2, \mathbf{v}_3\}$ consisting of eigenvectors of A, then $[T]_B = D$ is diagonal:

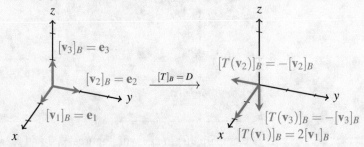

Compare this image with the one from Example 3.3.13, where we depicted the action of $A = [T]$ in the standard basis.

The method of diagonalizing a matrix that was presented in the previous examples works in general: after finding the eigenvalues of the matrix and bases of the corresponding eigenspaces, we place those eigenvalues as the diagonal entries in the diagonal matrix D, and we place the eigenspace basis vectors in the matrix P as columns in the same order. The following theorem shows that this procedure always leads to P being invertible, as long as there are enough vectors in the bases of the eigenspaces to fill out all n columns of P:

Theorem 3.4.2

Bases of Eigenspaces

Suppose A is a square matrix. If $B_1, B_2, \ldots, B_m$ are bases of eigenspaces of A corresponding to distinct eigenvalues, then

$$B_1 \cup B_2 \cup \cdots \cup B_m$$

is linearly independent.

Proof. First, let $\mathbf{v}_1 \in \text{span}(B_1), \mathbf{v}_2 \in \text{span}(B_2), \ldots, \mathbf{v}_m \in \text{span}(B_m)$ be arbitrary vectors from those eigenspaces (i.e., they either equal $\mathbf{0}$ or they are eigenvectors of A corresponding to eigenvalues $\lambda_1, \lambda_2, \ldots, \lambda_m$, respectively). We first show that the following two facts concerning these vectors hold:

Showing fact (a) proves the theorem in the case when the eigenspaces of A are all 1-dimensional.

a) If $1 \leq \ell \leq m$ is such that $\{v_1, v_2, \ldots, v_\ell\}$ is linearly dependent, then at least one of $v_1, v_2, \ldots, v_\ell$ is the zero vector.

b) If $v_1 + v_2 + \cdots + v_m = 0$ then $v_1 = v_2 = \cdots = v_m = 0$.

To prove fact (a), suppose that $\{v_1, v_2, \ldots, v_\ell\}$ is linearly dependent and assume that $v_1, v_2, \ldots, v_\ell$ are all non-zero (otherwise we are done). If we let $k \leq \ell$ be the smallest integer such that $\{v_1, v_2, \ldots, v_k\}$ is linearly dependent then there exist scalars $c_1, c_2, \ldots, c_{k-1}$ such that

$$v_k = c_1 v_1 + c_2 v_2 + \cdots + c_{k-1} v_{k-1}. \tag{3.4.1}$$

Oof, the proof of this theorem is a beast. I promise it's the worst one in this section.

Multiplying on the left by A shows that

$$\begin{aligned} \lambda_k v_k = A v_k &= c_1 A v_1 + c_2 A v_2 + \cdots + c_{k-1} A v_{k-1} \\ &= c_1 \lambda_1 v_1 + c_2 \lambda_2 v_2 + \cdots + c_{k-1} \lambda_{k-1} v_{k-1}. \end{aligned} \tag{3.4.2}$$

Also, multiplying both sides of Equation (3.4.1) by λ_k gives

$$\lambda_k v_k = c_1 \lambda_k v_1 + c_2 \lambda_k v_2 + \cdots + c_{k-1} \lambda_k v_{k-1}. \tag{3.4.3}$$

Subtracting Equation (3.4.3) from Equation (3.4.2) then shows that

$$0 = c_1(\lambda_1 - \lambda_k)v_1 + c_2(\lambda_2 - \lambda_k)v_2 + \cdots + c_{k-1}(\lambda_{k-1} - \lambda_k)v_{k-1}.$$

This proof still works even if $B_1, B_2, \ldots, B_m$ are just linearly independent (but not necessarily spanning).

Since k was chosen to be the smallest integer such that $\{v_1, v_2, \ldots, v_k\}$ is linearly dependent, $\{v_1, v_2, \ldots, v_{k-1}\}$ is linearly independent, which implies $c_j(\lambda_j - \lambda_k) = 0$ for all $1 \leq j \leq k-1$. Since $\lambda_1, \lambda_2, \ldots, \lambda_m$ are distinct, this is only possible if $c_1 = c_2 = \cdots = c_{k-1} = 0$, so $v_k = 0$, which proves fact (a).

Fact (b) follows by using fact (a), which tells us that if $v_1 + v_2 + \cdots + v_m = 0$ then at least one of $v_1, v_2, \ldots, v_m$ is the zero vector. Suppose that $v_m = 0$ (but a similar argument works if it's one of $v_1, v_2, \ldots, v_{m-1}$ that equals 0). Then

$$v_1 + v_2 + \cdots + v_{m-1} + v_m = v_1 + v_2 + \cdots + v_{m-1} + 0 = 0,$$

so using fact (a) again tells us that one of $v_1, v_2, \ldots, v_{m-1}$ is the zero vector. Repeating in this way shows that, in fact, $v_1 = v_2 = \cdots = v_m = 0$.

Here, γ_j is the geometric multiplicity of the eigenvalue corresponding to the eigenspace with basis B_j.

To now show that $B_1 \cup B_2 \cup \cdots \cup B_m$ is linearly independent, we give names to the vectors in each of these bases via $B_j = \{v_{j,1}, v_{j,2}, \ldots, v_{j,\gamma_j}\}$ for each $1 \leq j \leq m$. Now suppose that some linear combination of these vectors equals the zero vector:

$$\underbrace{\sum_{\ell=1}^{\gamma_1} c_{1,\ell} v_{1,\ell}}_{\text{call this } v_1} + \underbrace{\sum_{\ell=1}^{\gamma_2} c_{2,\ell} v_{2,\ell}}_{\text{call this } v_2} + \cdots + \underbrace{\sum_{\ell=1}^{\gamma_m} c_{m,\ell} v_{m,\ell}}_{\text{call this } v_m} = 0. \tag{3.4.4}$$

If we define $v_1, v_2, \ldots, v_m$ to be the m sums indicated above, then fact (b) tells us that $v_1 = v_2 = \cdots = v_m = 0$. But since each B_j is a basis (and thus linearly independent), the fact that

$$v_j = \sum_{\ell=1}^{\gamma_j} c_{j,\ell} v_{j,\ell} = 0 \quad \text{for all} \quad 1 \leq j \leq m$$

tells us that $c_{j,1} = c_{j,2} = \cdots = c_{j,\gamma_j} = 0$ for all $1 \leq j \leq m$. It follows that every coefficient in the linear combination (3.4.4) must equal 0, so $B_1 \cup B_2 \cup \cdots \cup B_m$ is linearly independent. ∎

In particular, since matrices with linearly independent columns are invertible (refer back to Theorem 2.3.5), the above theorem tells us that we can diagonalize any matrix $A \in \mathcal{M}_n$ for which we can find bases of its eigenspaces consisting of a total of n vectors. However, there are two problems that may make it impossible to find such bases, and thus two ways that a matrix can fail to be diagonalizable:

If you do not like complex diagonalizations of real matrices, look at Section 3.C, where we investigate how "close" to diagonal we can make D if we restrict to real entries.

- If $A \in \mathcal{M}_n(\mathbb{R})$, it may not be possible to construct D and P so that they have real entries, since A might have complex (non-real) eigenvalues and eigenvectors. For the purposes of this section (and most of this book), we ignore this problem, as we do not mind placing complex entries in D and P.

- Even if we allow for complex eigenvalues and eigenvectors, there may not be a way to choose eigenvectors so that the matrix P (whose columns are eigenvectors) is invertible. We illustrate how this problem can arise with an example.

Example 3.4.3

Not All Matrices Can Be Diagonalized

Show that the matrix $A = \begin{bmatrix} 3 & -1 \\ 1 & 1 \end{bmatrix}$ cannot be diagonalized.

Solution:

We first must compute the eigenvalues and eigenvectors of A. For its eigenvalues, we compute

$$p_A(\lambda) = \det\left(\begin{bmatrix} 3-\lambda & -1 \\ 1 & 1-\lambda \end{bmatrix}\right) = (3-\lambda)(1-\lambda)+1$$
$$= \lambda^2 - 4\lambda + 4$$
$$= (\lambda-2)^2.$$

It follows that $\lambda = 2$ is the only eigenvalue of A, and it has algebraic multiplicity 2.

To find the eigenvectors corresponding to this eigenvalue, we solve the linear system $(A - 2I)\mathbf{v} = \mathbf{0}$ as follows:

$$\begin{bmatrix} 1 & -1 & | & 0 \\ 1 & -1 & | & 0 \end{bmatrix} \xrightarrow{R_2-R_1} \begin{bmatrix} 1 & -1 & | & 0 \\ 0 & 0 & | & 0 \end{bmatrix}.$$

We thus see that v_2 is a free variable and v_1 is a leading variable with $v_1 = v_2$, so all eigenvectors corresponding to $\lambda = 2$ are of the form $\mathbf{v} = v_2(1,1)$.

If A were diagonalizable there would be two lines unchanged by A, and they would become the coordinate axes when viewed in the basis of eigenvectors.

We can now see why A cannot possibly be diagonalized—we would have to put eigenvectors of A as columns into a matrix P in a way that makes P invertible. However, this is not possible, since every eigenvector of A is a multiple of $(1,1)$.

Geometrically, the fact that A cannot be diagonalized is a result of the fact that the line in the direction of the eigenvector $\mathbf{v} = (1,1)$ is the only line whose direction is unchanged by A—every other line through the origin is rotated counter-clockwise toward that one:

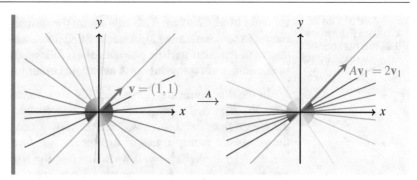

The previous theorem and example suggest that, for a matrix to be diagonalizable, it must have "enough" eigenvectors so that they span all of $\mathbb{R}^n$ (or $\mathbb{C}^n$, if we allow complex numbers). The following theorem makes this observation precise in several different ways.

Before stating the theorem, we note that if a matrix $A \in \mathcal{M}_n$ can be diagonalized as $A = PDP^{-1}$ via *real* matrices $D, P \in \mathcal{M}_n(\mathbb{R})$ then we say that A is **diagonalizable over** $\mathbb{R}$, and if it can be diagonalized via *complex* matrices $D, P \in \mathcal{M}_n(\mathbb{C})$ then we say that A is **diagonalizable over** $\mathbb{C}$. If we just say "diagonalizable" without specifying either $\mathbb{R}$ or $\mathbb{C}$ then we do not particularly care which type of diagonalization we mean (i.e., whatever statement we are making applies to both diagonalization over $\mathbb{R}$ and over $\mathbb{C}$). Furthermore, diagonalizability over $\mathbb{R}$ implies diagonalizability over $\mathbb{C}$ since real matrices are complex.

Theorem 3.4.3

Characterization of Diagonalizability

Suppose $A \in \mathcal{M}_n$. The following are equivalent:

    a) A is diagonalizable over $\mathbb{R}$ (or $\mathbb{C}$).
    b) There exists a basis of $\mathbb{R}^n$ (or $\mathbb{C}^n$) consisting of eigenvectors of A.
    c) The set of eigenvectors of A spans all of $\mathbb{R}^n$ (or $\mathbb{C}^n$).
    d) The sum of the geometric multiplicities of the real (or complex) eigenvalues of A is n.

To prove this theorem, we show that the 4 properties imply each other as follows:

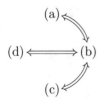

Proof. We prove this result by showing that condition (b) is equivalent to each of the others. To see why (b) $\Longleftrightarrow$ (a), just recall that Theorem 3.4.1 tells us that $A = PDP^{-1}$ is diagonalizable if and only if the columns of P are eigenvectors of A, and Theorem 2.5.1 tells us that P^{-1} actually exists (so this is a valid diagonalization of A) if and only if those columns (i.e., the eigenvectors of A) form a basis of $\mathbb{R}^n$ (or $\mathbb{C}^n$, as appropriate).

The fact that (b) $\Longleftrightarrow$ (c) follows immediately from the facts that a basis of $\mathbb{R}^n$ (or $\mathbb{C}^n$) spans it, and conversely any spanning set can be reduced down to a basis via Theorem 2.4.3(b).

For the remainder of the proof, we denote the distinct eigenvalues of A by $\lambda_1, \lambda_2, \ldots, \lambda_m$. To see why (b) $\Longrightarrow$ (d), suppose B is a basis of $\mathbb{R}^n$ (or $\mathbb{C}^n$) consisting of eigenvectors of A. Partition these eigenvectors into sets according to their corresponding eigenvalues. Specifically, let $B_1, B_2, \ldots, B_m \subseteq B$ be the sets of eigenvectors from B corresponding to $\lambda_1, \lambda_2, \ldots, \lambda_m$, respectively, and note that the fact that $\lambda_1, \lambda_2, \ldots, \lambda_m$ are distinct ensures that $B_1, B_2, \ldots, B_m$ are disjoint (after all, no non-zero vector can be scaled by two different eigenvalues).

Since B is linearly independent, so is each B_j, so the geometric multiplicity of λ_i is at least $|B_i|$. Then the fact that $n = |B| = |B_1| + \cdots + |B_m|$ implies that

Recall that $|B|$ denotes the size of (i.e., the number of vectors in) B.

the other hand, Theorem 3.3.3 tells us that the sum of geometric multiplicities cannot exceed the sum of algebraic multiplicities, which cannot exceed n (the degree of the characteristic polynomial). It follows that the sum of geometric multiplicities of eigenvalues of A must equal exactly n.

In the other direction, to see that (d) $\implies$ (b) and finish the proof, suppose B_1, B_2, ..., B_m are bases of the eigenspaces of A corresponding to the eigenvalues λ_1, λ_2, ..., λ_m, respectively. Since $|B_1|$, $|B_2|$, ..., $|B_m|$ are the geometric multiplicities of λ_1, λ_2, ..., λ_m, respectively, we know that $|B_1| + |B_2| + \cdots + |B_m| = n$. If we define $B = B_1 \cup B_2 \cup \cdots \cup B_m$, then $|B| = n$ too, since B_1, B_2, ..., B_m are disjoint. Since Theorem 3.4.2 tells us that B is linearly independent, Theorem 2.4.4 then implies that it must be a basis of $\mathbb{R}^n$ (or $\mathbb{C}^n$), which completes the proof. ∎

For example, this theorem tells us that the 2×2 matrix from Example 3.4.3 cannot be diagonalized since its only eigenvalue is $\lambda = 2$ with geometric multiplicity 1 (which is less than $n = 2$). On the other hand, we were able to diagonalize the 3×3 matrix from Example 3.4.2, since the geometric multiplicities of its eigenvalues were 1 and 2, and $1 + 2 = 3 = n$.

Example 3.4.4

Checking Diagonalizability

Determine whether or not $A = \begin{bmatrix} 5 & 1 & -1 \\ 1 & 3 & -1 \\ 2 & 0 & 2 \end{bmatrix}$ can be diagonalized.

Solution:
We start by computing and factoring A's characteristic polynomial:

The algebra here maybe looks a bit strange—we simplify things in this way because we really do not want to have to factor a cubic equation, so we never fully multiply it out.

$$\det(A - \lambda I) = \det\left(\begin{bmatrix} 5-\lambda & 1 & -1 \\ 1 & 3-\lambda & -1 \\ 2 & 0 & 2-\lambda \end{bmatrix} \right)$$

$$= (5-\lambda)(3-\lambda)(2-\lambda) - 2 + 0 - 0 + 2(3-\lambda) - (2-\lambda)$$
$$= (5-\lambda)(3-\lambda)(2-\lambda) + (2-\lambda)$$
$$= (2-\lambda)\big((5-\lambda)(3-\lambda) + 1\big)$$
$$= (2-\lambda)(\lambda^2 - 8\lambda + 16)$$
$$= (2-\lambda)(4-\lambda)^2.$$

The eigenvalues of A are thus $\lambda = 2$ and $\lambda = 4$, with algebraic multiplicities 1 and 2, respectively.

Since the eigenvalue $\lambda = 2$ has algebraic multiplicity 1, its geometric multiplicity must also equal 1 (thanks to Theorem 3.3.3). However, we have to do more work to compute the geometric multiplicity of $\lambda = 4$. In particular, we solve the linear system $(A - 4I)\mathbf{v} = \mathbf{0}$ to find a basis of the corresponding eigenspace, as usual:

$$\begin{bmatrix} 1 & 1 & -1 & | & 0 \\ 1 & -1 & -1 & | & 0 \\ 2 & 0 & -2 & | & 0 \end{bmatrix} \xrightarrow[R_3 - 2R_1]{R_2 - R_1} \begin{bmatrix} 1 & 1 & -1 & | & 0 \\ 0 & -2 & 0 & | & 0 \\ 0 & -2 & 0 & | & 0 \end{bmatrix}$$

$$\xrightarrow{\frac{-1}{2}R_2} \begin{bmatrix} 1 & 1 & -1 & | & 0 \\ 0 & 1 & 0 & | & 0 \\ 0 & -2 & 0 & | & 0 \end{bmatrix} \xrightarrow[R_3 + 2R_2]{R_1 - R_2} \begin{bmatrix} 1 & 0 & -1 & | & 0 \\ 0 & 1 & 0 & | & 0 \\ 0 & 0 & 0 & | & 0 \end{bmatrix}.$$

Since this linear system has just 1 free variable, the eigenspace is 1-dimensional, so the geometric multiplicity of $\lambda = 4$ is 1.

Since the sum of the geometric multiplicities of the eigenvalues of this matrix is $1 + 1 = 2$, but the matrix is 3×3, we conclude from Theorem 3.4.3 that A is not diagonalizable.

Theorem 3.4.3 is useful since it completely characterizes which matrices are diagonalizable. However, there is one special case that is worth pointing out where it is actually much easier to show that a matrix is diagonalizable.

Corollary 3.4.4

Matrices with Distinct Eigenvalues

If $A \in \mathcal{M}_n$ has n distinct eigenvalues then it is diagonalizable.

Proof. The eigenvalues of A being distinct means that they each have algebraic multiplicity 1, and thus also have geometric multiplicity 1 by Theorem 3.3.3. Since there are n distinct eigenvalues, their geometric multiplies thus sum to $1 + 1 + \cdots + 1 = n$. It then follows from Theorem 3.4.3 that A is diagonalizable. ∎

Keep in mind that the above corollary only works in one direction. If A has n distinct eigenvalues then it is diagonalizable, but if it has fewer distinct eigenvalues then A may or may not be diagonalizable—we must use Theorem 3.4.3 to distinguish the two possibilities in this case. For example, the matrix from Example 3.4.2 was diagonalizable despite having $\lambda = -1$ as a repeated (i.e., non-distinct) eigenvalue, since the geometric multiplicity of that eigenvalue was large enough that we could still find a basis of $\mathbb{R}^n$ consisting of eigenvectors.

Example 3.4.5

Checking Diagonalizability (Again)

Determine whether or not $A = \begin{bmatrix} 1 & 1 \\ 1 & 0 \end{bmatrix}$ is diagonalizable.

Solution:

We start by computing the characteristic polynomial of A, as usual:

$$p_A(\lambda) = \det(A - \lambda I) = \det\left(\begin{bmatrix} 1-\lambda & 1 \\ 1 & -\lambda \end{bmatrix}\right) = (1-\lambda)(-\lambda) - 1 = \lambda^2 - \lambda - 1.$$

This polynomial does not factor nicely, so we use the quadratic formula to find its roots (i.e., the eigenvalues of A):

Recall that the quadratic formula says that the roots of $a\lambda^2 + b\lambda + c$ are

$$\lambda = \frac{-b \pm \sqrt{b^2 - 4ac}}{2a}.$$

$$\lambda = \frac{1 \pm \sqrt{1+4}}{2} = \frac{1 \pm \sqrt{5}}{2}.$$

Since these 2 eigenvalues are distinct (i.e., they occur with algebraic multiplicity 1), Corollary 3.4.4 tells us that A is diagonalizable.

One of the useful features of the above corollary is that we can use it to show that many matrices are diagonalizable based only on their eigenvalues (without having to know anything about their corresponding eigenvectors). We only need to compute the corresponding eigenvectors if we actually want to compute the matrix P in the diagonalization.

Remark 3.4.1

Most Matrices are Diagonalizable

The rest of this section is devoted to exploring what we can do with diagonalizable matrices. Before we explore these applications of diagonalization though, we should emphasize that "most" matrices are diagonalizable, so assuming diagonalizability is typically not much of a restriction.

Making this idea precise is outside of the scope of this book, but we can think about it intuitively as follows: if we were to randomly generate the entries of a matrix, its eigenvalues would be a random mess, so it would be an astonishing coincidence for one of them to occur twice or more (just like if we were to randomly generate 10 real numbers, it would be exceedingly unlikely for 2 or more of them to be the same). Corollary 3.4.4 then implies that most randomly-generated matrices are diagonalizable.

Similarly, we can wiggle the entries of a non-diagonalizable matrix by an arbitrarily small amount to turn it into a matrix with distinct eigenvalues (which is thus diagonalizable). For example, we showed in Example 3.4.3 that the matrix

$$A = \begin{bmatrix} 3 & -1 \\ 1 & 1 \end{bmatrix}$$

is not diagonalizable. However, changing the bottom-right entry of A to $1 + \varepsilon$ results in its eigenvalues being

$$\lambda = \frac{4 + \varepsilon \pm \sqrt{\varepsilon^2 - 4\varepsilon}}{2},$$

which are distinct as long as $\varepsilon \notin \{0, 4\}$. It follows that nearby matrices like

$$B = \begin{bmatrix} 3 & -1 \\ 1 & 1.0001 \end{bmatrix}$$

are diagonalizable.

> Another way of dealing with non-diagonalizable matrices is via something called the "Jordan decomposition", which is covered in advanced linear algebra books like (Joh20).

3.4.2 Matrix Powers

Recall from the start of this section that if $A \in \mathcal{M}_n$ is diagonalizable (i.e., $A = PDP^{-1}$ for some diagonal D and invertible P) then

$$A^k = \overbrace{(PD\overbrace{P^{-1})(P}^{P^{-1}P=I}D\overbrace{P^{-1})(P}^{P^{-1}P=I}DP^{-1})\cdots(PDP^{-1})}^{k \text{ times}} = PD^k P^{-1} \quad \text{for all} \quad k \geq 0,$$

and D^k is typically much easier to compute directly than A^k is. We can use this technique to come up with explicit formulas for powers of diagonalizable matrices.

> **Example 3.4.6**
>
> **Using Diagonalization to Find a Formula for Powers of a Matrix**

Find an explicit formula for A^k if $A = \begin{bmatrix} 1 & 2 \\ 5 & 4 \end{bmatrix}$.

Solution:

Recall from Example 3.4.1 that A can be diagonalized as $A = PDP^{-1}$ via

$$D = \begin{bmatrix} -1 & 0 \\ 0 & 6 \end{bmatrix}, \quad P = \begin{bmatrix} -1 & 2 \\ 1 & 5 \end{bmatrix}, \quad \text{and} \quad P^{-1} = \frac{1}{7}\begin{bmatrix} -5 & 2 \\ 1 & 1 \end{bmatrix}.$$

It follows that

$$A^k = PD^kP^{-1}$$

The 1/7 comes from P^{-1}. We just pulled it in front of all 3 matrices.

$$= \frac{1}{7}\begin{bmatrix} -1 & 2 \\ 1 & 5 \end{bmatrix}\begin{bmatrix} (-1)^k & 0 \\ 0 & 6^k \end{bmatrix}\begin{bmatrix} -5 & 2 \\ 1 & 1 \end{bmatrix}$$

$$= \frac{1}{7}\begin{bmatrix} -1 & 2 \\ 1 & 5 \end{bmatrix}\begin{bmatrix} 5(-1)^{k+1} & 2(-1)^k \\ 6^k & 6^k \end{bmatrix}$$

$$= \frac{1}{7}\begin{bmatrix} 5(-1)^k + 2 \times 6^k & 2(-1)^{k+1}+2\times 6^k \\ 5(-1)^{k+1}+5\times 6^k & 2(-1)^k + 5\times 6^k \end{bmatrix} \quad \text{for all } k \geq 0.$$

Although the formula that we derived in the above example looks somewhat messy, it is much easier to use than multiplying A by itself repeatedly. Indeed, it is much simpler to compute A^{500} by performing this diagonalization and then plugging $k = 500$ into that formula, rather than multiplying A by itself 499 times.

In fact, we can make the formula from the previous example look a bit cleaner by factoring out the powers of -1 and 6 as follows:

The matrices in Equation (3.4.5) have rank 1 since their rows are multiples of each other.

$$A^k = \frac{1}{7}\left((-1)^k \begin{bmatrix} 5 & -2 \\ -5 & 2 \end{bmatrix} + 6^k \begin{bmatrix} 2 & 2 \\ 5 & 5 \end{bmatrix} \right). \tag{3.4.5}$$

This formula seems to have a lot of structure: the terms being raised to the power of k are exactly the eigenvalues of A, and they are multiplying some fixed rank-1 matrices. The following theorem shows that something analogous happens for every diagonalizable matrix, and also tells us exactly where those rank-1 matrices come from.

Theorem 3.4.5

Diagonalization as a Rank-One Sum

Suppose $P, Q, D \in \mathcal{M}_n$ are such that D is diagonal with diagonal entries $d_1, d_2, \ldots, d_n$, in that order. Write P and Q in terms of their columns:

$$P = \begin{bmatrix} \mathbf{p}_1 & | & \mathbf{p}_2 & | & \cdots & | & \mathbf{p}_n \end{bmatrix} \quad \text{and} \quad Q = \begin{bmatrix} \mathbf{q}_1 & | & \mathbf{q}_2 & | & \cdots & | & \mathbf{q}_n \end{bmatrix}.$$

Then

$$PDQ^* = \sum_{j=1}^n d_j\mathbf{p}_j\mathbf{q}_j^*.$$

Recall that Q^* is the conjugate transpose of Q, which was introduced in Section 3.3.2. This theorem is still true if we replace both conjugate transposes by the standard transpose.

Proof. We just use block matrix multiplication, with each of the matrices partitioned as indicated by the statement of the theorem:

$$PDQ^* = \begin{bmatrix} \mathbf{p}_1 & | & \mathbf{p}_2 & | & \cdots & | & \mathbf{p}_n \end{bmatrix}\begin{bmatrix} d_1 & 0 & \cdots & 0 \\ 0 & d_2 & \cdots & 0 \\ \vdots & \vdots & \ddots & \vdots \\ 0 & 0 & \cdots & d_n \end{bmatrix}\begin{bmatrix} \mathbf{q}_1^* \\ \mathbf{q}_2^* \\ \vdots \\ \mathbf{q}_n^* \end{bmatrix}$$

$$= \begin{bmatrix} \mathbf{p}_1 & | & \mathbf{p}_2 & | & \cdots & | & \mathbf{p}_n \end{bmatrix}\begin{bmatrix} d_1\mathbf{q}_1^* \\ d_2\mathbf{q}_2^* \\ \vdots \\ d_n\mathbf{q}_n^* \end{bmatrix} = \sum_{j=1}^n d_j\mathbf{p}_j\mathbf{q}_j^*,$$

which is what we wanted to prove. ∎

In particular, the above theorem tells us that A is diagonalizable via $A = PDP^{-1}$ if and only if

$$A = \sum_{j=1}^{n} \lambda_j \mathbf{p}_j \mathbf{q}_j^*,$$

where $\mathbf{p}_j$ is the j-th column of P and $\mathbf{q}_j^*$ is the j-th row of P^{-1}. Furthermore, in this case we have $A^k = PD^kP^{-1}$, from which the above theorem implies

$$A^k = \sum_{j=1}^{n} \lambda_j^k \mathbf{p}_j \mathbf{q}_j^* \quad \text{for all integers} \quad k \geq 0. \tag{3.4.6}$$

This is exactly the form of Equation (3.4.5) that we found earlier—since each $\mathbf{p}_j$ is a column vector and each $\mathbf{q}_j^*$ is a row vector, the products $\mathbf{p}_j \mathbf{q}_j^*$ are rank-1 matrices (we proved that a matrix has rank 1 if and only if it is of this form in Exercise 2.4.30 as well as Theorem 2.C.2).

Example 3.4.7

Finding a Rank-One Sum Formula for Powers of a Matrix

Find a formula of the form (3.4.6) for A^k if $A = \begin{bmatrix} 2 & 0 & -3 \\ 1 & -1 & -1 \\ 0 & 0 & -1 \end{bmatrix}$.

Solution:

Recall from Example 3.4.2 that A can be diagonalized as $A = PDP^{-1}$ via

$$D = \begin{bmatrix} 2 & 0 & 0 \\ 0 & -1 & 0 \\ 0 & 0 & -1 \end{bmatrix}, \quad P = \begin{bmatrix} 3 & 0 & 1 \\ 1 & 1 & 0 \\ 0 & 0 & 1 \end{bmatrix}, \quad \text{and } P^{-1} = \frac{1}{3}\begin{bmatrix} 1 & 0 & -1 \\ -1 & 3 & 1 \\ 0 & 0 & 3 \end{bmatrix}.$$

If we write P and P^{-1} in terms of their columns and rows, respectively,

$$P = \begin{bmatrix} \mathbf{p}_1 & | & \mathbf{p}_2 & | & \mathbf{p}_3 \end{bmatrix} = \begin{bmatrix} 3 & 0 & 1 \\ 1 & 1 & 0 \\ 0 & 0 & 1 \end{bmatrix}, \quad P^{-1} = \begin{bmatrix} \mathbf{q}_1^* \\ \mathbf{q}_2^* \\ \mathbf{q}_3^* \end{bmatrix} = \frac{1}{3}\begin{bmatrix} 1 & 0 & -1 \\ -1 & 3 & 1 \\ 0 & 0 & 3 \end{bmatrix},$$

then

Be careful when computing products like $\mathbf{p}_1 \mathbf{q}_1^*$. A column vector times a row vector results in a *matrix* (not a scalar).

$$\mathbf{p}_1 \mathbf{q}_1^* = \frac{1}{3}\begin{bmatrix} 3 \\ 1 \\ 0 \end{bmatrix}\begin{bmatrix} 1 & 0 & -1 \end{bmatrix} = \frac{1}{3}\begin{bmatrix} 3 & 0 & -3 \\ 1 & 0 & -1 \\ 0 & 0 & 0 \end{bmatrix}$$

$$\mathbf{p}_2 \mathbf{q}_2^* = \frac{1}{3}\begin{bmatrix} 0 \\ 1 \\ 0 \end{bmatrix}\begin{bmatrix} -1 & 3 & 1 \end{bmatrix} = \frac{1}{3}\begin{bmatrix} 0 & 0 & 0 \\ -1 & 3 & 1 \\ 0 & 0 & 0 \end{bmatrix}, \quad \text{and}$$

$$\mathbf{p}_3 \mathbf{q}_3^* = \frac{1}{3}\begin{bmatrix} 1 \\ 0 \\ 1 \end{bmatrix}\begin{bmatrix} 0 & 0 & 3 \end{bmatrix} = \frac{1}{3}\begin{bmatrix} 0 & 0 & 3 \\ 0 & 0 & 0 \\ 0 & 0 & 3 \end{bmatrix}.$$

Theorem 3.4.5 (more specifically, Equation (3.4.6)) then tells us that

We could add up the two matrices that are multiplied by $(-1)^k$ here if we wanted to.

$$A^k = \frac{1}{3}\left(2^k\begin{bmatrix} 3 & 0 & -3 \\ 1 & 0 & -1 \\ 0 & 0 & 0 \end{bmatrix} + (-1)^k\begin{bmatrix} 0 & 0 & 0 \\ -1 & 3 & 1 \\ 0 & 0 & 0 \end{bmatrix} + (-1)^k\begin{bmatrix} 0 & 0 & 3 \\ 0 & 0 & 0 \\ 0 & 0 & 3 \end{bmatrix}\right)$$

for all integers $k \geq 0$.

We can think of Equation (3.4.6) as saying that every power of a diagonalizable matrix can be written as a linear combination of the n fixed matrices $\mathbf{p}_1\mathbf{q}_1^*$, $\mathbf{p}_2\mathbf{q}_2^*$, $\ldots$, $\mathbf{p}_n\mathbf{q}_n^*$ (we have not yet discussed linear combinations of matrices, but it is completely analogous to linear combinations of vectors).

We now look at some examples of the types of problems that we can solve with our new-found method of constructing formulas for large powers of matrices.

Example 3.4.8

Fibonacci Sequence

The **Fibonacci sequence** is the sequence of positive integers defined recursively via

$$F_0 = 0, \quad F_1 = 1, \quad F_{n+1} = F_n + F_{n-1} \quad \text{for all} \quad n \geq 1.$$

For example,
$F_3 = F_2 + F_1 = 1 + 1 = 2,$
$F_4 = F_3 + F_2 = 2 + 1 = 3,$
and so on.

The first few terms of this sequence are $0, 1, 1, 2, 3, 5, 8, 13, 21, 34, \ldots$. Find an explicit formula for F_n that does not depend on the previous terms in the sequence.

Solution:

The key observation is to notice that this sequence can be represented in terms of matrix multiplication—the fact that $F_{n+1} = F_n + F_{n-1}$ is equivalent to the matrix-vector equation

$$\begin{bmatrix} F_{n+1} \\ F_n \end{bmatrix} = \begin{bmatrix} 1 & 1 \\ 1 & 0 \end{bmatrix} \begin{bmatrix} F_n \\ F_{n-1} \end{bmatrix} \quad \text{for all} \quad n \geq 1.$$

We generalize this example to many other integer sequences in Section 3.D.

If we iterate this matrix equation, we find that

$$\begin{bmatrix} F_{n+1} \\ F_n \end{bmatrix} = \begin{bmatrix} 1 & 1 \\ 1 & 0 \end{bmatrix} \begin{bmatrix} F_n \\ F_{n-1} \end{bmatrix}$$

$$= \begin{bmatrix} 1 & 1 \\ 1 & 0 \end{bmatrix}^2 \begin{bmatrix} F_{n-1} \\ F_{n-2} \end{bmatrix} = \cdots = \begin{bmatrix} 1 & 1 \\ 1 & 0 \end{bmatrix}^n \begin{bmatrix} F_1 \\ F_0 \end{bmatrix} = \begin{bmatrix} 1 & 1 \\ 1 & 0 \end{bmatrix}^n \begin{bmatrix} 1 \\ 0 \end{bmatrix}.$$

It follows that if we can find a closed-form formula for powers of the matrix

$$A = \begin{bmatrix} 1 & 1 \\ 1 & 0 \end{bmatrix},$$

then a formula for F_n will follow immediately.

To find a formula for the powers of A, we diagonalize it. To this end, we recall from Example 3.4.5 that its characteristic polynomial is $\lambda^2 - \lambda - 1$, and its eigenvalues are

$$\lambda = \frac{1}{2}\left(1 \pm \sqrt{5}\right).$$

ϕ is sometimes called the "golden ratio," and its decimal expansion starts $\phi = 1.61803\ldots$

For convenience, we define $\phi = (1 + \sqrt{5})/2$, so that the eigenvalues of A are ϕ and $1 - \phi$.

Next, we find the corresponding eigenvectors. For $\lambda = \phi$, we find a basis of $\text{null}(A - \phi I)$ as follows:

$$\begin{bmatrix} 1-\phi & 1 & 0 \\ 1 & -\phi & 0 \end{bmatrix} \xrightarrow{R_1 \leftrightarrow R_2} \begin{bmatrix} 1 & -\phi & 0 \\ 1-\phi & 1 & 0 \end{bmatrix}$$

$$\xrightarrow{R_2 + (\phi-1)R_1} \begin{bmatrix} 1 & -\phi & 0 \\ 0 & 0 & 0 \end{bmatrix}.$$

When computing the bottom-right entry after row-reducing, we used the fact that $1-(1-\phi)(-\phi) = -\phi^2 + \phi + 1 = 0$, since ϕ was chosen specifically to satisfy this equation.

One basis of this null space is thus $\{(\phi, 1)\}$. On the other hand, for $\lambda = 1-\phi$, we find a basis of $\text{null}(A - (1-\phi)I)$:

$$\begin{bmatrix} \phi & 1 & 0 \\ 1 & \phi-1 & 0 \end{bmatrix} \xrightarrow{R_1 \leftrightarrow R_2} \begin{bmatrix} 1 & \phi-1 & 0 \\ \phi & 1 & 0 \end{bmatrix} \xrightarrow{R_2 - \phi R_1} \begin{bmatrix} 1 & \phi-1 & 0 \\ 0 & 0 & 0 \end{bmatrix}.$$

It follows that one basis of this null space is $\{(1-\phi, 1)\}$.

By putting this all together, we see that we can diagonalize A via $A = PDP^{-1}$, where

$$D = \begin{bmatrix} \phi & 0 \\ 0 & 1-\phi \end{bmatrix}, \quad P = \begin{bmatrix} \phi & 1-\phi \\ 1 & 1 \end{bmatrix}, \quad \text{and} \quad P^{-1} = \frac{1}{\sqrt{5}}\begin{bmatrix} 1 & \phi-1 \\ -1 & \phi \end{bmatrix}.$$

Finally, we can use this diagonalization to compute arbitrary powers of A and thus find our desired formula for F_n:

Here we use the fact that if $A = PDP^{-1}$ then $A^n = PD^n P^{-1}$.

$$\begin{bmatrix} F_{n+1} \\ F_n \end{bmatrix} = \begin{bmatrix} 1 & 1 \\ 1 & 0 \end{bmatrix}^n \begin{bmatrix} 1 \\ 0 \end{bmatrix}$$

$$= \frac{1}{\sqrt{5}}\begin{bmatrix} \phi & 1-\phi \\ 1 & 1 \end{bmatrix}\begin{bmatrix} \phi^n & 0 \\ 0 & (1-\phi)^n \end{bmatrix}\begin{bmatrix} 1 & \phi-1 \\ -1 & \phi \end{bmatrix}\begin{bmatrix} 1 \\ 0 \end{bmatrix}$$

$$= \frac{1}{\sqrt{5}}\begin{bmatrix} \phi & 1-\phi \\ 1 & 1 \end{bmatrix}\begin{bmatrix} \phi^n & 0 \\ 0 & (1-\phi)^n \end{bmatrix}\begin{bmatrix} 1 \\ -1 \end{bmatrix}$$

$$= \frac{1}{\sqrt{5}}\begin{bmatrix} \phi & 1-\phi \\ 1 & 1 \end{bmatrix}\begin{bmatrix} \phi^n \\ -(1-\phi)^n \end{bmatrix}$$

$$= \frac{1}{\sqrt{5}}\begin{bmatrix} \phi^{n+1} - (1-\phi)^{n+1} \\ \phi^n - (1-\phi)^n \end{bmatrix}.$$

This formula is remarkable for the fact that it combines powers of two irrational numbers (ϕ and $1-\phi$) in such a way that the result is an integer (F_n) for all n. It is called **Binet's formula**.

We thus obtain the following simple formula for the n-th Fibonacci number:

$$F_n = \frac{1}{\sqrt{5}}\left(\phi^n - (1-\phi)^n\right).$$

The idea used in the previous example works in a lot of generality—if we can represent a problem via powers of a matrix, then there's a good chance that diagonalization can shed light on it. The following example builds upon this idea and shows how we can make use of diagonalization to solve a graph theory problem that we introduced back in Section 1.B.

Example 3.4.9

Counting Paths in Graphs via Diagonalization

Find a formula for the number of paths of length k from vertex A to vertex D in the following graph:

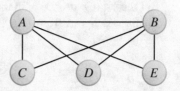

If you did not read
Section 1.B, you
should either do so
now or skip this
example.

Solution:

Recall from Theorem 1.B.1 that we can count the number of paths between vertices in this graph by first constructing the adjacency matrix $A \in \mathcal{M}_5$ of this graph—the matrix whose (i,j)-entry equals 1 if there is an edge from the i-th vertex to the j-th vertex, and equals 0 otherwise:

$$A = \begin{bmatrix} 0 & 1 & 1 & 1 & 1 \\ 1 & 0 & 1 & 1 & 1 \\ 1 & 1 & 0 & 0 & 0 \\ 1 & 1 & 0 & 0 & 0 \\ 1 & 1 & 0 & 0 & 0 \end{bmatrix}.$$

In particular, the number of paths of length k from vertex A to vertex D is equal to $[A^k]_{1,4}$.

To find an explicit formula for this quantity, we diagonalize A. Its eigenvalues, along with corresponding eigenvectors, are as follows:

The repeated
eigenvalue
$\lambda_3 = \lambda_4 = 0$ might
worry us a bit, but A
is still diagonalizable
since its
corresponding
eigenspace is
2-dimensional.

$$\begin{aligned} \lambda_1 &= -2 & \mathbf{v}_1 &= (1,1,-1,-1,-1) \\ \lambda_2 &= -1 & \mathbf{v}_2 &= (1,-1,0,0,0) \\ \lambda_3 &= 0 & \mathbf{v}_3 &= (0,0,1,-1,0) \\ \lambda_4 &= 0 & \mathbf{v}_4 &= (0,0,0,1,-1) \\ \lambda_5 &= 3 & \mathbf{v}_5 &= (3,3,2,2,2). \end{aligned}$$

The matrices P and D in the diagonalization $A = PDP^{-1}$ are thus

$$D = \begin{bmatrix} -2 & 0 & 0 & 0 & 0 \\ 0 & -1 & 0 & 0 & 0 \\ 0 & 0 & 0 & 0 & 0 \\ 0 & 0 & 0 & 0 & 0 \\ 0 & 0 & 0 & 0 & 3 \end{bmatrix} \quad \text{and} \quad P = \begin{bmatrix} 1 & 1 & 0 & 0 & 3 \\ 1 & -1 & 0 & 0 & 3 \\ -1 & 0 & 1 & 0 & 2 \\ -1 & 0 & -1 & 1 & 2 \\ -1 & 0 & 0 & -1 & 2 \end{bmatrix}.$$

It is then straightforward (albeit rather tedious) to compute

$$P^{-1} = \frac{1}{30} \begin{bmatrix} 6 & 6 & -6 & -6 & -6 \\ 15 & -15 & 0 & 0 & 0 \\ 0 & 0 & 20 & -10 & -10 \\ 0 & 0 & 10 & 10 & -20 \\ 3 & 3 & 2 & 2 & 2 \end{bmatrix},$$

so we now have a complete diagonalization of A. It follows that $A^k = PD^kP^{-1}$ for all integers $k \geq 0$, so the formula that we desire is simply the $(1,4)$-entry of PD^kP^{-1}. Since we do not require this entire matrix, we can save some effort by recalling that $[A^k]_{1,4} = [PD^kP^{-1}]_{1,4}$ is the dot product

Alternatively,
$[PD^kP^{-1}]_{1,4}$ is the dot
product of the 1st
row of P with the 4th
row of D^kP^{-1}.

of the 1st row of PD^k with the 4th column of P^{-1}. That is, the number of paths of length k from vertex A to vertex D is

$$[A^k]_{1,4} = [PD^kP^{-1}]_{1,4} = \frac{1}{30} \overbrace{((-2)^k,(-1)^k,0,0,3^{k+1})}^{\text{1st row of } PD^k} \cdot \overbrace{(-6,0,-10,10,2)}^{30\times(\text{4th column of } P^{-1})}$$

$$= \frac{1}{5}\big(3^k - (-2)^k\big) \quad \text{for all integers} \quad k \geq 0.$$

Non-Integer Powers of Matrices

Recall that when we originally defined matrix powers back in Section 1.3.2, we only did so for positive integer exponents. After all, what would it mean to multiply a matrix by itself, for example, 2.3 times? We are now in a position to fill this gap and define the r-th power of a diagonalizable matrix for *any* real number r (i.e., r does not need to be an integer):

Definition 3.4.2 **Powers of Diagonalizable Matrices** Actually, this definition works fine if $r \in \mathbb{C}$ too.	Suppose $A \in \mathcal{M}_n$ is diagonalizable (i.e., $A = PDP^{-1}$ for some invertible $P \in \mathcal{M}_n$ and diagonal $D \in \mathcal{M}_n$). We then define $$A^r \overset{\text{def}}{=} PD^rP^{-1} \quad \text{for all} \quad r \in \mathbb{R},$$ where D^r is obtained by raising each of its diagonal entries to the r-th power.

In other words, we define non-integer powers of diagonalizable matrices so as to extend the pattern $A^k = PD^kP^{-1}$ that we observed for positive integer exponents. Before proceeding to examples, we note that there are two technicalities that arise from this definition that we have to be slightly careful of:

- First, how do we know that Definition 3.4.2 is well-defined (i.e., how do we know that A^r does not change if we choose a different diagonalization of A)? Fortunately, it *is* well-defined, which we prove in Exercise 3.4.16.

- Second, if the eigenvalues of A are negative or complex (non-real) then we may have to be somewhat careful about how we compute powers of the diagonal entries of D (i.e., the eigenvalues of A). The result of such an exponentiation will be complex in general, since we have $(-1)^{1/2} = i$, for example. See Appendix A.1 for a discussion of how to exponentiate negative and complex numbers.

Example 3.4.10 **Computing Weird Matrix Powers** Try computing these eigenvalues and eigenvectors (and thus this diagonalization) on your own.	Use Definition 3.4.2 to compute the following powers of $A = \begin{bmatrix} 0 & 4 \\ -1 & 5 \end{bmatrix}$. a) $A^{1/2}$ b) A^{-1} c) A^{π} **Solutions:** a) The eigenvalues of A are 1 and 4, which correspond to the eigenvectors $(4,1)$ and $(1,1)$, respectively. We can thus diagonalize A as

follows:

$$A = \begin{bmatrix} 4 & 1 \\ 1 & 1 \end{bmatrix} \underbrace{\begin{bmatrix} 1 & 0 \\ 0 & 4 \end{bmatrix}}_{} \underbrace{\left(\frac{1}{3} \begin{bmatrix} 1 & -1 \\ -1 & 4 \end{bmatrix} \right)}_{}.$$

$$\underbrace{\phantom{\begin{bmatrix} 4 & 1 \\ 1 & 1 \end{bmatrix}}}_{P} \quad \underbrace{\phantom{\begin{bmatrix} 1 & 0 \\ 0 & 4 \end{bmatrix}}}_{D} \quad \underbrace{\phantom{\frac{1}{3} \begin{bmatrix} 1 & -1 \\ -1 & 4 \end{bmatrix}}}_{P^{-1}}$$

Using Definition 3.4.2 then tells us that

Here we moved the $1/3$ from in front of P^{-1} to the front of the product.

$$A^{1/2} = \frac{1}{3} \begin{bmatrix} 4 & 1 \\ 1 & 1 \end{bmatrix} \begin{bmatrix} 1^{1/2} & 0 \\ 0 & 4^{1/2} \end{bmatrix} \begin{bmatrix} 1 & -1 \\ -1 & 4 \end{bmatrix}$$

$$= \frac{1}{3} \begin{bmatrix} 4 & 1 \\ 1 & 1 \end{bmatrix} \begin{bmatrix} 1 & 0 \\ 0 & 2 \end{bmatrix} \begin{bmatrix} 1 & -1 \\ -1 & 4 \end{bmatrix}$$

$$= \frac{1}{3} \begin{bmatrix} 4 & 1 \\ 1 & 1 \end{bmatrix} \begin{bmatrix} 1 & -1 \\ -2 & 8 \end{bmatrix} = \frac{1}{3} \begin{bmatrix} 2 & 4 \\ -1 & 7 \end{bmatrix}.$$

b) We re-use the diagonalization from part (a). We could just raise the diagonal entries of D to the exponent -1 rather than $1/2$, but we instead find an explicit formula for A^r and then plug $r = -1$ into that formula afterward:

$$A^r = \frac{1}{3} \begin{bmatrix} 4 & 1 \\ 1 & 1 \end{bmatrix} \begin{bmatrix} 1^r & 0 \\ 0 & 4^r \end{bmatrix} \begin{bmatrix} 1 & -1 \\ -1 & 4 \end{bmatrix}$$

If we plug $r = 1/2$ into this formula, we get the above answer to part (a).

$$= \frac{1}{3} \begin{bmatrix} 4 & 1 \\ 1 & 1 \end{bmatrix} \begin{bmatrix} 1 & -1 \\ -4^r & 4^{r+1} \end{bmatrix} = \frac{1}{3} \begin{bmatrix} 4 - 4^r & 4^{r+1} - 4 \\ 1 - 4^r & 4^{r+1} - 1 \end{bmatrix}.$$

Plugging $r = -1$ into this formula gives

$$A^{-1} = \frac{1}{3} \begin{bmatrix} 4 - 4^{-1} & 4^0 - 4 \\ 1 - 4^{-1} & 4^0 - 1 \end{bmatrix} = \frac{1}{3} \begin{bmatrix} 15/4 & -3 \\ 3/4 & 0 \end{bmatrix} = \frac{1}{4} \begin{bmatrix} 5 & -4 \\ 1 & 0 \end{bmatrix}.$$

As a side note, this matrix really is the inverse of A (as we would hope based on the notation), since

$$A^{-1}A = \frac{1}{4} \begin{bmatrix} 5 & -4 \\ 1 & 0 \end{bmatrix} \begin{bmatrix} 0 & 4 \\ -1 & 5 \end{bmatrix} = \begin{bmatrix} 1 & 0 \\ 0 & 1 \end{bmatrix} = I.$$

c) Plugging $r = \pi$ into the formula that we found in part (b) gives

$$A^\pi = \frac{1}{3} \begin{bmatrix} 4 - 4^\pi & 4^{\pi+1} - 4 \\ 1 - 4^\pi & 4^{\pi+1} - 1 \end{bmatrix},$$

which cannot be simplified much further than that.

When defined in this way, matrix powers "just work" and behave much like powers of numbers in many ways. After all, multiplication of diagonal matrices works entry-wise (i.e., just like multiplication of scalars), and multiplication of diagonalizable matrices works just like multiplication of diagonal matrices.

For example, if $r = -1$ then the matrix A^{-1}, that we computed via Definition 3.4.2 in Example 3.4.10(b) was indeed the inverse of A. To see that this happens in general, note that if $A = PDP^{-1}$ is a diagonalization of A (and thus

Recall from
Exercise 3.3.16 that A
is invertible if and
only if it does not
have 0 as an
eigenvalue. This
makes sense here
since we can
compute A^{-1} using
this method if and
only if each λ_j^{-1}
exists.

D is diagonal with the eigenvalues $\lambda_1, \lambda_2, \ldots, \lambda_n$ of A along its diagonal) then

$$A^{-1}A = \left(P \begin{bmatrix} \lambda_1^{-1} & \cdots & 0 \\ \vdots & \ddots & \vdots \\ 0 & \cdots & \lambda_n^{-1} \end{bmatrix} P^{-1} \right) \left(P \begin{bmatrix} \lambda_1 & \cdots & 0 \\ \vdots & \ddots & \vdots \\ 0 & \cdots & \lambda_n \end{bmatrix} P^{-1} \right)$$

$$= P \begin{bmatrix} \lambda_1^{-1}\lambda_1 & \cdots & 0 \\ \vdots & \ddots & \vdots \\ 0 & \cdots & \lambda_n^{-1}\lambda_n \end{bmatrix} P^{-1} = PIP^{-1} = I.$$

We can also use Definition 3.4.2 to compute roots of matrices just like we compute roots of numbers. In particular, a **square root** of a matrix $A \in \mathcal{M}_n$ is a matrix $B \in \mathcal{M}_n$ such that $B^2 = A$. Well, if $A = PDP^{-1}$ is a diagonalization of A then $A^{1/2}$ is one such matrix, since

$$(A^{1/2})^2 = \left(P \begin{bmatrix} \lambda_1^{1/2} & \cdots & 0 \\ \vdots & \ddots & \vdots \\ 0 & \cdots & \lambda_n^{1/2} \end{bmatrix} P^{-1} \right) \left(P \begin{bmatrix} \lambda_1^{1/2} & \cdots & 0 \\ \vdots & \ddots & \vdots \\ 0 & \cdots & \lambda_n^{1/2} \end{bmatrix} P^{-1} \right)$$

$$= P \begin{bmatrix} \lambda_1^{1/2}\lambda_1^{1/2} & \cdots & 0 \\ \vdots & \ddots & \vdots \\ 0 & \cdots & \lambda_n^{1/2}\lambda_n^{1/2} \end{bmatrix} P^{-1} = P \begin{bmatrix} \lambda_1 & \cdots & 0 \\ \vdots & \ddots & \vdots \\ 0 & \cdots & \lambda_n \end{bmatrix} P^{-1} = A.$$

For example, if we multiply the matrix $A^{1/2}$ that we computed in Example 3.4.10(a) by itself, we get

$$(A^{1/2})^2 = \left(\frac{1}{3} \begin{bmatrix} 2 & 4 \\ -1 & 7 \end{bmatrix} \right) \left(\frac{1}{3} \begin{bmatrix} 2 & 4 \\ -1 & 7 \end{bmatrix} \right) = \frac{1}{9} \begin{bmatrix} 0 & 36 \\ -9 & 45 \end{bmatrix} = \begin{bmatrix} 0 & 4 \\ -1 & 5 \end{bmatrix} = A,$$

as expected. Slightly more generally, we say that $B \in \mathcal{M}_n$ is a **k-th root** of $A \in \mathcal{M}_n$ if $B^k = A$, and the same argument as above shows that $B = A^{1/k}$ is one such k-th root.

Example 3.4.11

**Computing a
Matrix Cube Root**

Find a cube root of the matrix $A = \begin{bmatrix} -2 & 6 \\ 3 & -5 \end{bmatrix}$.

Solution:

The eigenvalues of A are 1 and -8, with corresponding eigenvectors $(2,1)$ and $(1,-1)$, respectively. We can thus diagonalize A as follows:

$$A = \underbrace{\begin{bmatrix} 2 & 1 \\ 1 & -1 \end{bmatrix}}_{P} \underbrace{\begin{bmatrix} 1 & 0 \\ 0 & -8 \end{bmatrix}}_{D} \underbrace{\left(\frac{1}{3} \begin{bmatrix} 1 & 1 \\ 1 & -2 \end{bmatrix} \right)}_{P^{-1}}.$$

We can then find a cube root B of A simply by taking the cube root of the diagonal entries of D (1 and -8), leaving the rest of the diagonalization alone:

$$B = \frac{1}{3} \begin{bmatrix} 2 & 1 \\ 1 & -1 \end{bmatrix} \begin{bmatrix} 1 & 0 \\ 0 & -2 \end{bmatrix} \begin{bmatrix} 1 & 1 \\ 1 & -2 \end{bmatrix} = \begin{bmatrix} 0 & 2 \\ 1 & -1 \end{bmatrix}.$$

To double-check our work, we can verify that it is indeed the case that $B^3 = A$:

There are also 8
other cube roots of
this matrix A, but they
all have complex
(non-real) entries.

$$B^3 = \begin{bmatrix} 0 & 2 \\ 1 & -1 \end{bmatrix} \begin{bmatrix} 0 & 2 \\ 1 & -1 \end{bmatrix} \begin{bmatrix} 0 & 2 \\ 1 & -1 \end{bmatrix}$$

$$= \begin{bmatrix} 0 & 2 \\ 1 & -1 \end{bmatrix} \begin{bmatrix} 2 & -2 \\ -1 & 3 \end{bmatrix} = \begin{bmatrix} -2 & 6 \\ 3 & -5 \end{bmatrix} = A.$$

It is important to keep in mind that, just like roots of numbers are not unique (for example, 2 and −2 are both square roots of 4), roots of matrices are also not unique. However, this non-uniqueness is much more pronounced for matrices, as we now demonstrate.

Example 3.4.12

Multiple Square Roots of a Matrix

Find four different square roots of $A = \begin{bmatrix} 0 & 4 \\ -1 & 5 \end{bmatrix}$.

Solution:

We already found one square root of A back in Example 3.4.10(a):

$$A^{1/2} = \frac{1}{3} \begin{bmatrix} 2 & 4 \\ -1 & 7 \end{bmatrix}$$

is one, and $-A^{1/2}$ is another one. To find two more square roots of A, we can use the same procedure that we used to compute $A^{1/2}$, but take different square roots of its eigenvalues (when computing $A^{1/2}$, we took the positive square root of both of its eigenvalues). In particular, since one diagonalization of A is

$$A = \frac{1}{3} \begin{bmatrix} 4 & 1 \\ 1 & 1 \end{bmatrix} \begin{bmatrix} 1 & 0 \\ 0 & 4 \end{bmatrix} \begin{bmatrix} 1 & -1 \\ -1 & 4 \end{bmatrix},$$

two other square roots of A are

Despite looking
absolutely nothing
like $A^{1/2}$, B really is a
square root of A.
Compute B^2 to
check.

$$B = \frac{1}{3} \begin{bmatrix} 4 & 1 \\ 1 & 1 \end{bmatrix} \begin{bmatrix} 1 & 0 \\ 0 & -2 \end{bmatrix} \begin{bmatrix} 1 & -1 \\ -1 & 4 \end{bmatrix} = \frac{1}{3} \begin{bmatrix} 4 & 1 \\ 1 & 1 \end{bmatrix} \begin{bmatrix} 1 & -1 \\ 2 & -8 \end{bmatrix} = \begin{bmatrix} 2 & -4 \\ 1 & -3 \end{bmatrix},$$

as well as $-B$.

Example 3.4.13

Infinitely Many Square Roots of a Matrix

We learned about
reflections and their
standard matrices
back in Section 1.4.2.

Show that the identity matrix has infinitely many different square roots.

Solution:

Perhaps the simplest way to demonstrate this fact is by thinking geometrically. We want to find infinitely many linear transformations with the property that, if we apply them to a vector twice, they do nothing. Reflections are one such family of linear transformations: reflecting a vector across a line once changes that vector, but reflecting it again moves it back to where it started.

It follows that if $F_{\mathbf{u}}$ is the reflection across the line in the direction of a unit vector $\mathbf{u}$, then $F_{\mathbf{u}}^2 = I$:

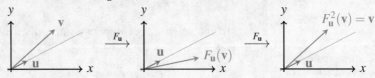

We made this exact same observation about the square of a reflection way back in Exercise 1.4.17.

Since the standard matrix of $F_{\mathbf{u}}$ is $[F_{\mathbf{u}}] = 2\mathbf{u}\mathbf{u}^T - I$, every matrix of the form $2\mathbf{u}\mathbf{u}^T - I$ is a square root of I (i.e., $(2\mathbf{u}\mathbf{u}^T - I)^2 = I$ for all unit vectors $\mathbf{u}$).

The following theorem explains the key difference that led to the identity matrix having infinitely many square roots, versus the matrix from Example 3.4.12 only having finitely many square roots (indeed, the four square roots that we found in that example are all of them). It also shows that the method we used in Example 3.4.12 (i.e., taking all possible roots of the diagonal entries of D in a diagonalization of A) gives us all roots of A—there are no other roots that require different techniques to be found.

> **Theorem 3.4.6**
>
> **Counting Roots of Matrices**
>
> Suppose $A \in \mathcal{M}_n(\mathbb{C})$ is diagonalizable and let $k > 1$ be an integer.
>
> a) If A's eigenvalues are distinct, it has exactly k^n different k-th roots.
> b) If A's eigenvalues are not distinct, it has infinitely many k-th roots.

Our discussion of matrix powers has perhaps gone on long enough, so we leave the proof of the above theorem as Exercise 3.4.15. Instead, we close this subsection by noting that the "usual" exponent laws that we would hope for matrix powers to satisfy do still hold, as long as we restrict our attention to matrices with non-negative real eigenvalues. That is, $A^{r+s} = A^r A^s$ and $(A^r)^s = A^{rs}$ for all real numbers r and s, as long as A has non-negative real eigenvalues (see Exercise 3.4.19).

Seriously, we have to be careful when taking non-integer powers of negative or complex (non-real) numbers— see Appendix A.1.

However, if we do not restrict to matrices with non-negative real eigenvalues, these exponent rules are not even true for 1×1 matrices (i.e., scalars). For example,

$$\left((-1)^{1/2}\right)^2 = i^2 = -1, \quad \text{but} \quad \left((-1)^2\right)^{1/2} = 1^{1/2} = 1.$$

Furthermore, keep in mind that the theorem only applies to diagonalizable matrices. We will see in Exercise 3.5.4 that some non-diagonalizable matrices do not have any roots at all.

3.4.3 Matrix Functions

Now that we understand how powers of matrices work, it is not a huge leap to imagine that we can apply polynomials to matrices. In particular, we say that if p is a polynomial defined by

$$p(x) = c_k x^k + \cdots + c_2 x^2 + c_1 x + c_0,$$

Recall that $A^0 = I$ for all $A \in \mathcal{M}_n$.

then we can apply p to a square matrix A via

$$p(A) = c_k A^k + \cdots + c_2 A^2 + c_1 A + c_0 I.$$

That is, we just raise A to the same powers that x was raised to in the scalar-valued version of p.

Just like we can use diagonalization to easily compute large powers of matrices, we can also use it to quickly compute a polynomial applied to a matrix. In particular, we just apply the polynomial to the diagonal part of the diagonalization:

Theorem 3.4.7

Polynomial of a Diagonalizable Matrix

Suppose $A \in \mathcal{M}_n$ is diagonalizable (i.e., $A = PDP^{-1}$ for some invertible $P \in \mathcal{M}_n$ and diagonal $D \in \mathcal{M}_n$) and p is a polynomial. Then

$$p(A) = Pp(D)P^{-1},$$

where $p(D)$ is obtained by applying p to each diagonal entry of D.

Proof. We just write $p(x) = c_k x^k + \cdots + c_2 x^2 + c_1 x + c_0$, and then make use of the fact that $\left(PDP^{-1}\right)^r = PD^r P^{-1}$ for all r:

$$
\begin{aligned}
p(A) = p\left(PDP^{-1}\right) &= c_k\left(PDP^{-1}\right)^k + \cdots + c_2\left(PDP^{-1}\right)^2 + c_1\left(PDP^{-1}\right) + c_0 I \\
&= c_k\left(PD^k P^{-1}\right) + \cdots + c_2\left(PD^2 P^{-1}\right) + c_1\left(PDP^{-1}\right) + c_0 I \\
&= P\left(c_k D^k + \cdots + c_2 D^2 + c_1 D + c_0 I\right)P^{-1} \\
&= Pp(D)P^{-1},
\end{aligned}
$$

as desired. ∎

One immediate consequence of Theorem 3.4.7 is that if a matrix $A \in \mathcal{M}_n$ has an eigenvalue λ with corresponding eigenvector $\mathbf{v}$, then $p(\lambda)$ is an eigenvalue of $p(A)$ corresponding to the same eigenvector $\mathbf{v}$. In fact, this is true even for non-diagonalizable matrices (see Exercise 3.4.18).

Example 3.4.14

Polynomial of a Matrix

Compute $p(A)$ directly from the definition and also via Theorem 3.4.7, if $p(x) = x^3 - 3x^2 + 2x - 4$ and $A = \begin{bmatrix} 0 & 4 \\ -1 & 5 \end{bmatrix}$.

Solution:

To compute $p(A)$ directly, we first compute A^2 and A^3:

$$A^2 = \begin{bmatrix} 0 & 4 \\ -1 & 5 \end{bmatrix}\begin{bmatrix} 0 & 4 \\ -1 & 5 \end{bmatrix} = \begin{bmatrix} -4 & 20 \\ -5 & 21 \end{bmatrix} \quad \text{and}$$

$$A^3 = A^2 A = \begin{bmatrix} -4 & 20 \\ -5 & 21 \end{bmatrix}\begin{bmatrix} 0 & 4 \\ -1 & 5 \end{bmatrix} = \begin{bmatrix} -20 & 84 \\ -21 & 85 \end{bmatrix}.$$

It follows that

$$p(A) = A^3 - 3A^2 + 2A - 4I$$

$$= \begin{bmatrix} -20 & 84 \\ -21 & 85 \end{bmatrix} - 3 \begin{bmatrix} -4 & 20 \\ -5 & 21 \end{bmatrix} + 2 \begin{bmatrix} 0 & 4 \\ -1 & 5 \end{bmatrix} - 4 \begin{bmatrix} 1 & 0 \\ 0 & 1 \end{bmatrix}$$

$$= \begin{bmatrix} -12 & 32 \\ -8 & 28 \end{bmatrix}.$$

To instead use Theorem 3.4.7 to compute $p(A)$, recall from Example 3.4.10 that A can be diagonalized as

$$A = \frac{1}{3} \begin{bmatrix} 4 & 1 \\ 1 & 1 \end{bmatrix} \underbrace{\begin{bmatrix} 1 & 0 \\ 0 & 4 \end{bmatrix}}_{D} \begin{bmatrix} 1 & -1 \\ -1 & 4 \end{bmatrix}.$$

We thus have

$$p(A) = \frac{1}{3} \begin{bmatrix} 4 & 1 \\ 1 & 1 \end{bmatrix} \begin{bmatrix} p(1) & 0 \\ 0 & p(4) \end{bmatrix} \begin{bmatrix} 1 & -1 \\ -1 & 4 \end{bmatrix}$$

$$= \frac{1}{3} \begin{bmatrix} 4 & 1 \\ 1 & 1 \end{bmatrix} \begin{bmatrix} -4 & 0 \\ 0 & 20 \end{bmatrix} \begin{bmatrix} 1 & -1 \\ -1 & 4 \end{bmatrix}$$

$$= \frac{1}{3} \begin{bmatrix} 4 & 1 \\ 1 & 1 \end{bmatrix} \begin{bmatrix} -4 & 4 \\ -20 & 80 \end{bmatrix} = \begin{bmatrix} -12 & 32 \\ -8 & 28 \end{bmatrix},$$

which agrees with the answer that we found earlier via direct computation.

Well, if we can apply a polynomial to a matrix just by applying it to each of the diagonal entries in its diagonalization, why not just do the same thing for functions in general? Since diagonal matrices (and thus diagonalizable matrices) multiply so much like scalars, doing this ensures that matrix functions retain most of the useful properties of their scalar-valued counterparts.

Definition 3.4.3	Suppose $A \in \mathcal{M}_n$ is diagonalizable (i.e., $A = PDP^{-1}$ for some invertible $P \in \mathcal{M}_n$ and diagonal $D \in \mathcal{M}_n$) and f is a scalar-valued function. Then we define
Functions of Diagonalizable Matrices	$$f(A) \stackrel{\text{def}}{=} Pf(D)P^{-1},$$
Don't worry—$f(A)$ is well-defined (see Exercise 3.4.17).	where $f(D)$ is obtained by applying f to each of the diagonal entries of D.

In other words, to compute a function of a matrix, we just do the exact same thing that we did for powers and for polynomials: diagonalize, apply the function to the diagonal part, and then un-diagonalize.

Example 3.4.15	Compute the following functions of $A = \begin{bmatrix} 0 & 4 \\ -1 & 5 \end{bmatrix}$.
Computing Weird Matrix Functions	a) e^A b) $\sin(A)$

Solutions:

a) Recall from Example 3.4.10 that A can be diagonalized as

$$A = \frac{1}{3}\begin{bmatrix} 4 & 1 \\ 1 & 1 \end{bmatrix}\underbrace{\begin{bmatrix} 1 & 0 \\ 0 & 4 \end{bmatrix}}_{D}\begin{bmatrix} 1 & -1 \\ -1 & 4 \end{bmatrix}.$$

Definition 3.4.3 thus tells us that

$$e^A = \frac{1}{3}\begin{bmatrix} 4 & 1 \\ 1 & 1 \end{bmatrix}\begin{bmatrix} e & 0 \\ 0 & e^4 \end{bmatrix}\begin{bmatrix} 1 & -1 \\ -1 & 4 \end{bmatrix}$$

$$= \frac{1}{3}\begin{bmatrix} 4 & 1 \\ 1 & 1 \end{bmatrix}\begin{bmatrix} e & -e \\ -e^4 & 4e^4 \end{bmatrix} = \frac{1}{3}\begin{bmatrix} 4e - e^4 & 4e^4 - 4e \\ e - e^4 & 4e^4 - e \end{bmatrix}.$$

> Keep in mind that we only apply the function to the diagonal entries of D, not *all* entries of D.

b) Repeating the exact same calculation from part (a), but applying the function $\sin(x)$ instead of e^x to the diagonal entries of D, gives

$$\sin(A) = \frac{1}{3}\begin{bmatrix} 4 & 1 \\ 1 & 1 \end{bmatrix}\begin{bmatrix} \sin(1) & 0 \\ 0 & \sin(4) \end{bmatrix}\begin{bmatrix} 1 & -1 \\ -1 & 4 \end{bmatrix}$$

$$= \frac{1}{3}\begin{bmatrix} 4 & 1 \\ 1 & 1 \end{bmatrix}\begin{bmatrix} \sin(1) & -\sin(1) \\ -\sin(4) & 4\sin(4) \end{bmatrix}$$

$$= \frac{1}{3}\begin{bmatrix} 4\sin(1) - \sin(4) & 4\sin(4) - 4\sin(1) \\ \sin(1) - \sin(4) & 4\sin(4) - \sin(1) \end{bmatrix}.$$

> Do not try to simplify $\sin(A)$ any further—sometimes an ugly mess is just an ugly mess.

Since matrix functions are defined via diagonalizations, which in turn work via eigenvalues and eigenvectors, we know that if a square matrix A is diagonalizable with λ as an eigenvalue corresponding to an eigenvector $\mathbf{v}$, then $f(\lambda)$ must be an eigenvalue of $f(A)$ corresponding to the same eigenvector $\mathbf{v}$.

Remark 3.4.2

Matrix Functions via Power Series

Another way to define matrix functions, which has the advantage of also applying to non-diagonalizable matrices, is to recall that many functions can be written as a power series—an infinite sum of powers of the input variable—on some open interval (a, b):

> A function that can be written as a power series is called **analytic**.

$$f(x) = c_0 + c_1 x + c_2 x^2 + c_3 x^3 + \cdots = \sum_{k=0}^{\infty} c_k x^k \quad \text{for all} \quad x \in (a, b).$$

Most named functions from calculus courses are of this type. For example,

> This all works for complex-valued analytic functions too.

$$e^x = \sum_{k=0}^{\infty} \frac{1}{k!} x^k \qquad \text{for all} \quad x \in \mathbb{R},$$

$$\sin(x) = \sum_{k=0}^{\infty} \frac{(-1)^k}{(2k+1)!} x^{2k+1} \qquad \text{for all} \quad x \in \mathbb{R},$$

$$\frac{1}{1-x} = \sum_{k=0}^{\infty} x^k \qquad \text{for all} \quad x \in (-1, 1), \quad \text{and}$$

$$-\ln(1-x) = \sum_{k=0}^{\infty} \frac{1}{k} x^k \qquad \text{for all} \quad x \in (-1, 1).$$

By simply replacing powers of x by powers of A in these power series, we get analogous matrix functions. For example, we could define e^A by

$$e^A = \sum_{k=0}^{\infty} \frac{1}{k!} A^k \quad \text{for all} \quad A \in \mathcal{M}_n.$$

The tricky thing about defining matrix functions in this way is dealing with (a) convergence concerns (how do we know that the entries of A^k decrease fast enough that every entry of this infinite sum approaches a fixed limit?), and (b) how to actually compute these sums (we certainly need a better method than adding up infinitely many large powers of A).

However, if we restrict our attention to diagonalizable matrices and ignore convergence concerns, this method is equivalent to Definition 3.4.3 since

$$f(A) = f(PDP^{-1}) = \sum_{k=0}^{\infty} c_k (PDP^{-1})^k = P\left(\sum_{k=0}^{\infty} c_k D^k \right) P^{-1} = Pf(D)P^{-1}$$

and each diagonal entry of $f(D)$ can be computed by applying f to the diagonal entries of D.

The most well-known and frequently-used matrix function (besides powers and polynomials) is the matrix exponential $f(A) = e^A$. The following theorem establishes some of the basic properties of the matrix exponential, which are analogous to the facts that, for real numbers, $e^0 = 1$ and $1/e^x = e^{-x}$.

Theorem 3.4.8	Suppose $A \in \mathcal{M}_n$ is diagonalizable. Then
Properties of the Matrix Exponential	a) $e^O = I$, and b) e^A is invertible and $\left(e^A\right)^{-1} = e^{-A}$.

Proof. For part (a), just notice that the zero matrix is diagonal, so

$$e^O = \begin{bmatrix} e^0 & 0 & \cdots & 0 \\ 0 & e^0 & \cdots & 0 \\ \vdots & \vdots & \ddots & \vdots \\ 0 & 0 & \cdots & e^0 \end{bmatrix} = \begin{bmatrix} 1 & 0 & \cdots & 0 \\ 0 & 1 & \cdots & 0 \\ \vdots & \vdots & \ddots & \vdots \\ 0 & 0 & \cdots & 1 \end{bmatrix} = I.$$

Similarly, for part (b) we just multiply e^A by e^{-A} and work with their diagonal parts entry-by-entry:

The main practical use of the matrix exponential is for solving systems of linear differential equations—a topic that is outside the scope of this book.

$$e^A e^{-A} = \left(Pe^D P^{-1}\right)\left(Pe^{-D}P^{-1}\right)$$

$$= P \begin{bmatrix} e^{d_{1,1}} & 0 & \cdots & 0 \\ 0 & e^{d_{2,2}} & \cdots & 0 \\ \vdots & \vdots & \ddots & \vdots \\ 0 & 0 & \cdots & e^{d_{n,n}} \end{bmatrix} \begin{bmatrix} e^{-d_{1,1}} & 0 & \cdots & 0 \\ 0 & e^{-d_{2,2}} & \cdots & 0 \\ \vdots & \vdots & \ddots & \vdots \\ 0 & 0 & \cdots & e^{-d_{n,n}} \end{bmatrix} P^{-1}$$

$$= P \begin{bmatrix} 1 & 0 & \cdots & 0 \\ 0 & 1 & \cdots & 0 \\ \vdots & \vdots & \ddots & \vdots \\ 0 & 0 & \cdots & 1 \end{bmatrix} P^{-1}$$

$$= I.$$

We thus conclude that e^A and e^{-A} are inverses of each other. ∎

Note, however, that it is typically not the case that $e^{A+B} = e^A e^B$ (even though $e^{x+y} = e^x e^y$ for all real numbers x and y). The reason that this property does not extend to matrices is the exact same reason that matrix powers do not satisfy properties like $(A+B)^2 = A^2 + 2AB + B^2$ or $(AB)^2 = A^2 B^2$: non-commutativity of matrix multiplication gets in the way. For example, if

$$A = \begin{bmatrix} 1 & 1 \\ 0 & 0 \end{bmatrix} \quad \text{and} \quad B = \begin{bmatrix} 0 & 0 \\ 1 & 1 \end{bmatrix},$$

then

$$e^{A+B} = \frac{1}{2} \begin{bmatrix} e^2 + 1 & e^2 - 1 \\ e^2 - 1 & e^2 + 1 \end{bmatrix}, \quad \text{but}$$

$$e^A e^B = \begin{bmatrix} e & e-1 \\ 0 & 1 \end{bmatrix} \begin{bmatrix} 1 & 0 \\ e-1 & e \end{bmatrix} = \begin{bmatrix} e^2 - e + 1 & e^2 - e \\ e-1 & e \end{bmatrix}.$$

Roughly speaking, properties of functions carry over to matrices as long as they only depend on one input (a single number x or matrix A), but not if they depend on two or more inputs (numbers x and y, or matrices A and B).

The matrix exponential also provides an interesting connection between the trace and determinant of a matrix:

Theorem 3.4.9
Determinant of the Matrix Exponential

If $A \in \mathcal{M}_n$ is diagonalizable then $\det\left(e^A\right) = e^{\operatorname{tr}(A)}$.

Proof. Recall from Theorem 3.3.1 that if the eigenvalues of A, listed according to algebraic multiplicity, are $\lambda_1, \lambda_2, \ldots, \lambda_n$, then

$$\det(A) = \lambda_1 \lambda_2 \cdots \lambda_n \quad \text{and} \quad \operatorname{tr}(A) = \lambda_1 + \lambda_2 + \cdots + \lambda_n.$$

The equation that we want comes from combining that theorem with the fact that the eigenvalues of e^A are $e^{\lambda_1}, e^{\lambda_2}, \ldots, e^{\lambda_n}$:

$$\det\left(e^A\right) = e^{\lambda_1} e^{\lambda_2} \cdots e^{\lambda_n} = e^{\lambda_1 + \lambda_2 + \cdots + \lambda_n} = e^{\operatorname{tr}(A)}. \quad ∎$$

The previous two theorems are actually true for non-diagonalizable matrices as well, but even defining functions of non-diagonalizable matrices requires some technicalities that are beyond the scope of this book. They are typically handled via something called the "Jordan decomposition" of a matrix, which is covered in [Joh20].

Exercises

solutions to starred exercises on page 468

3.4.1 Diagonalize the following matrices over $\mathbb{R}$, or give a reason why that is not possible.

*(a) $\begin{bmatrix} 1 & 1 \\ 1 & 1 \end{bmatrix}$

(b) $\begin{bmatrix} 2 & 1 \\ -1 & 2 \end{bmatrix}$

*(c) $\begin{bmatrix} 1 & 1 \\ 0 & 1 \end{bmatrix}$

(d) $\begin{bmatrix} -2 & 6 \\ -3 & 7 \end{bmatrix}$

*(e) $\begin{bmatrix} 2 & 0 & 0 \\ 0 & 0 & 1 \\ 0 & 1 & 0 \end{bmatrix}$

(f) $\begin{bmatrix} 5 & 0 & -3 \\ -3 & 2 & 3 \\ 6 & 0 & -4 \end{bmatrix}$

*(g) $\begin{bmatrix} 3 & 0 & 1 \\ 0 & -1 & 2 \\ 0 & 0 & 2 \end{bmatrix}$

(h) $\begin{bmatrix} -3 & 4 & 5 \\ -3 & 5 & 3 \\ 1 & -2 & 1 \end{bmatrix}$

3.4.2 Diagonalize the following matrices over $\mathbb{C}$, or give a reason why that is not possible.

*(a) $\begin{bmatrix} 1 & 1 \\ 1 & 1 \end{bmatrix}$

(b) $\begin{bmatrix} 2+2i & 1-i \\ 2 & 1-i \end{bmatrix}$

*(c) $\begin{bmatrix} 2 & 1 \\ -1 & 2 \end{bmatrix}$

(d) $\begin{bmatrix} 1 & 1 \\ -9 & -5 \end{bmatrix}$

*(e) $\begin{bmatrix} 2 & 0 & 0 \\ 0 & 0 & 1 \\ 0 & -1 & 0 \end{bmatrix}$

(f) $\begin{bmatrix} -2 & -3 & -2 \\ 4 & 6 & 4 \\ -4 & -5 & -4 \end{bmatrix}$

*(g) $\begin{bmatrix} 1+i & -1 & 0 \\ 1 & -1+i & 0 \\ 2 & -2 & i \end{bmatrix}$

(h) $\begin{bmatrix} 2 & 4 & -2 \\ 1 & 3 & -1 \\ 3 & -1 & 1 \end{bmatrix}$

▢ 3.4.3 Use computer software to diagonalize the following matrices over $\mathbb{R}$ (if possible) or $\mathbb{C}$ (if possible), or explain why this is not possible.

*(a) $\begin{bmatrix} 4 & -2 & 2 & -3 & 2 \\ 0 & 4 & -1 & 1 & 0 \\ 2 & 0 & 3 & -2 & 2 \\ 2 & 0 & 0 & 1 & 2 \\ -1 & 2 & -2 & 3 & 1 \end{bmatrix}$

(b) $\begin{bmatrix} -2 & 7 & 9 & 3 & 10 \\ -1 & 4 & 3 & 2 & 5 \\ 2 & -4 & -5 & -3 & -7 \\ 1 & -1 & -1 & 0 & -2 \\ -2 & 3 & 5 & 2 & 6 \end{bmatrix}$

*(c) $\begin{bmatrix} 11 & 4 & 7 & -18 & 5 \\ 7 & 3 & 8 & -15 & 5 \\ 20 & 5 & 23 & -43 & 15 \\ 15 & 5 & 14 & -29 & 10 \\ -1 & 1 & -3 & 4 & 0 \end{bmatrix}$

(d) $\begin{bmatrix} -4 & 4 & 2 & 2 & 1 \\ -5 & 5 & 2 & 2 & 1 \\ -5 & 4 & 3 & 2 & 1 \\ 5 & -4 & -2 & -1 & -1 \\ -5 & 4 & 2 & 2 & 2 \end{bmatrix}$

3.4.4 Compute the indicated function of the matrix A that was diagonalized in the specified earlier exercise.

*(a) e^A, where A is the matrix from Exercise 3.4.1(a).

*(b) $\sin(A)$, Exercise 3.4.1(a).

(c) $\sqrt{A}$, Exercise 3.4.1(d).

(d) e^A, Exercise 3.4.1(e).

▢ 3.4.5 Use computer software to diagonalize the following matrices and then find a real square root of them.

*(a) $\begin{bmatrix} 18 & 5 & 4 & 9 & -8 \\ -17 & -4 & -4 & -9 & 8 \\ 20 & 8 & 5 & 12 & -8 \\ 3 & 3 & 0 & 4 & 0 \\ 20 & 8 & 4 & 12 & -7 \end{bmatrix}$

(b) $\begin{bmatrix} -2 & 3 & 11 & -1 & 3 \\ -14 & -6 & -4 & -7 & 3 \\ 6 & -9 & 3 & -15 & -9 \\ 1 & 12 & -1 & 5 & 12 \\ 3 & 0 & 15 & -12 & -9 \end{bmatrix}$

3.4.6 Determine which of the following statements are true and which are false.

*(a) If there exists $P \in \mathcal{M}_n$ and a diagonal $D \in \mathcal{M}_n$ such that $AP = PD$ then A is diagonalizable.

(b) If $A \in \mathcal{M}_n(\mathbb{C})$ has n eigenvectors then A is diagonalizable.

*(c) A matrix $A \in \mathcal{M}_n(\mathbb{C})$ is diagonalizable if and only if it has n eigenvalues, counting algebraic multiplicity.

(d) A matrix $A \in \mathcal{M}_n(\mathbb{C})$ is diagonalizable if and only if it has n eigenvalues, counting geometric multiplicity.

*(f) If $A \in \mathcal{M}_n(\mathbb{C})$ is diagonalizable then it has n distinct eigenvalues.

(g) If $A \in \mathcal{M}_n(\mathbb{C})$ has n distinct eigenvalues then it is diagonalizable.

*(h) Every diagonalizable matrix is invertible.

(i) If $A \in \mathcal{M}_n$ is diagonalizable then A^2 is diagonalizable too.

*(j) If $A \in \mathcal{M}_n$ is diagonalizable with all eigenvalues equal to each other (i.e., a single eigenvalue with multiplicity n) then it must be a scalar multiple of the identity matrix.

3.4.7 Let $k \in \mathbb{R}$ and consider the following matrix:

$$A = \begin{bmatrix} 5 & 3 \\ k & -1 \end{bmatrix}.$$

(a) For which values of k is A diagonalizable over $\mathbb{R}$?

(b) For which values of k is A diagonalizable over $\mathbb{C}$?

***3.4.8** Suppose

$$A = \begin{bmatrix} 5 & -6 \\ -3 & 2 \end{bmatrix}.$$

Find a cube root of A (i.e., find a matrix B such that $B^3 = A$).

*3.4.9 The **Lucas sequence** is the sequence of positive integers defined recursively via

$$L_0 = 2, \quad L_1 = 1, \quad L_{n+1} = L_n + L_{n-1} \quad \text{for all} \quad n \geq 1.$$

Find an explicit formula for L_n. [Hint: This sequence satisfies the same recurrence as the Fibonacci numbers, but just has different initial conditions, so the diagonalization from Example 3.4.8 can be used here as well.]

3.4.10 The **Pell sequence** is the sequence of positive integers defined recursively via

$$P_0 = 0, \quad P_1 = 1, \quad P_{n+1} = 2P_n + P_{n-1} \quad \text{for all} \quad n \geq 1.$$

Find an explicit formula for P_n.

[Hint: Mimic the method used in Example 3.4.8.]

3.4.11 Let F_n be the n-th term of the Fibonacci sequence from Example 3.4.8.

(a) Show that

$$\begin{bmatrix} F_{n+1} & F_n \\ F_n & F_{n-1} \end{bmatrix} = \begin{bmatrix} 1 & 1 \\ 1 & 0 \end{bmatrix}^n \quad \text{for all} \quad n \geq 1.$$

(b) Show that $F_{n+1}F_{n-1} - F_n^2 = (-1)^n$ for all $n \geq 1$. [Hint: Determinants.]

(c) Show that $F_{m+n} = F_{m+1}F_n + F_m F_{n-1}$ for all $m, n \geq 1$. [Hint: $A^m A^n = A^{m+n}$ for all square matrices A.]

**3.4.12 Suppose $A \in \mathcal{M}_n(\mathbb{C})$.

(a) Show that if A is diagonalizable then $\text{rank}(A)$ equals the number of non-zero eigenvalues of A.

(b) Give an example to show that part (a) is not necessarily true if A is not diagonalizable.

3.4.13 Suppose $A = \mathbf{x}\mathbf{y}^T$ for some non-zero column vectors $\mathbf{x}, \mathbf{y} \in \mathbb{R}^n$.

[Side note: Recall that a matrix has rank 1 if and only if it can be written in this form.]

(a) Show that A has at most one non-zero eigenvalue, equal to $\mathbf{x} \cdot \mathbf{y}$, with corresponding eigenvector $\mathbf{x}$.

(b) Show that A is diagonalizable if and only if $\mathbf{x} \cdot \mathbf{y} \neq 0$.

*3.4.14 Suppose $A, B \in \mathcal{M}_n$ are matrices with the same eigenvalues and corresponding eigenvectors.

(a) Show that if A is diagonalizable then $A = B$.

(b) Give an example to show that if A is not diagonalizable then it might be the case that $A \neq B$.

**3.4.15 Recall Theorem 3.4.6, which counted the number of roots of a diagonalizable matrix.

(a) Prove part (a) of the theorem.

(b) Prove part (b) of the theorem.

**3.4.16 Show that matrix powers are well-defined by Definition 3.4.2. That is, show that if $P_1 D_1 P_1^{-1} = P_2 D_2 P_2^{-1}$ are two diagonalizations of a matrix then $P_1 D_1^r P_1^{-1} = P_2 D_2^r P_2^{-1}$ for all real r.

**3.4.17 Show that matrix functions are well-defined by Definition 3.4.3. That is, show that if $P_1 D_1 P_1^{-1} = P_2 D_2 P_2^{-1}$ are two diagonalizations of a matrix then $P_1 f(D_1) P_1^{-1} = P_2 f(D_2) P_2^{-1}$ for all functions f.

**3.4.18 Suppose $A \in \mathcal{M}_n$ has eigenvector $\mathbf{v}$ with corresponding eigenvalue λ.

(a) Show that $\mathbf{v}$ is an eigenvector of A^2 with eigenvalue λ^2.

(b) Show that if $k \geq 0$ is an integer then $\mathbf{v}$ is an eigenvector of A^k with eigenvalue λ^k.

(c) Show that if p is a polynomial then $\mathbf{v}$ is an eigenvector of $p(A)$ with eigenvalue $p(\lambda)$.

**3.4.19 Suppose $A \in \mathcal{M}_n$ is diagonalizable and has non-negative real eigenvalues. Show that if r and s are real numbers then

(a) $A^{r+s} = A^r A^s$, and

(b) $(A^r)^s = A^{rs}$.

3.4.20 Suppose $A \in \mathcal{M}_n$ is diagonalizable. Show that $e^{(A^T)} = (e^A)^T$.

*3.4.21 Suppose $A, B \in \mathcal{M}_n$ are diagonalizable and so is $A + B$. Show that

$$\det(e^{A+B}) = \det(e^A e^B).$$

[Be careful: remember that it is *not* the case that $e^{A+B} = e^A e^B$ in general.]

3.4.22 Suppose $A \in \mathcal{M}_n$ is diagonalizable. Show that $\sin^2(A) + \cos^2(A) = I$.

3.4.23 Recall from Exercise 3.3.23 that the **companion matrix of the polynomial $p(\lambda) = \lambda^n + a_{n-1}\lambda^{n-1} + \cdots + a_1\lambda + a_0$ is the matrix

$$C = \begin{bmatrix} 0 & 1 & 0 & \cdots & 0 \\ 0 & 0 & 1 & \cdots & 0 \\ \vdots & \vdots & \vdots & \ddots & \vdots \\ 0 & 0 & 0 & \cdots & 1 \\ -a_0 & -a_1 & -a_2 & \cdots & -a_{n-1} \end{bmatrix}.$$

(a) Show that if the eigenvalues $\lambda_1, \lambda_2, \ldots, \lambda_n$ of C (i.e., the roots of p) are distinct, then C can be diagonalized by the matrix

$$V = \begin{bmatrix} 1 & 1 & \cdots & 1 \\ \lambda_1 & \lambda_2 & \cdots & \lambda_n \\ \lambda_1^2 & \lambda_2^2 & \cdots & \lambda_n^2 \\ \vdots & \vdots & \ddots & \vdots \\ \lambda_1^{n-1} & \lambda_2^{n-1} & \cdots & \lambda_n^{n-1} \end{bmatrix}.$$

That is, show that $C = VDV^{-1}$, where D is the diagonal matrix with diagonal entries $\lambda_1, \lambda_2, \ldots, \lambda_n$, in that order.

[Side note: Recall from Example 2.3.8 that V is the transpose of a Vandermonde matrix, and it is invertible if $\lambda_1, \lambda_2, \ldots, \lambda_n$ are distinct.]

[Hint: What should the eigenvectors of C be if this diagonalization is correct?]

(b) Show that if the eigenvalues of C are not distinct then it is not diagonalizable.

[Hint: Show that the geometric multiplicity of an eigenvalue of C is always 1, regardless of its algebraic multiplicity.]

3.5 Summary and Review

In this chapter, we introduced coordinate systems as a method of viewing vectors and linear transformations in different ways that can make them easier to work with. This led to the question of how we can determine whether or not two matrices $A, B \in \mathcal{M}_n$ are similar (i.e., whether or not there exists an invertible matrix $P \in \mathcal{M}_n$ such that $A = PBP^{-1}$), since they are similar if and only if they represent the same linear transformation in different bases.

We noted that several important properties of matrices and linear transformations are similarity invariant, which means that if two matrices $A, B \in \mathcal{M}_n$ are similar then they share that property. Some examples of similarity invariant

The fact that the characteristic polynomial is similarity invariant implies that the trace, determinant, eigenvalues, and algebraic multiplicities are as well.

properties of matrices include their:

- rank,
- trace,
- determinant,
- eigenvalues,
- algebraic and geometric multiplicities of their eigenvalues, and
- characteristic polynomial.

One particularly important case is when $A \in \mathcal{M}_n$ is similar to a diagonal matrix D (i.e., $A = PDP^{-1}$ for some invertible P and diagonal D), in which case we call A diagonalizable. Geometrically, we can think of such matrices as un-skewing space, then stretching along the standard basis vectors, and then re-skewing space back to how it started, as in Figure 3.25.

Compare this image to Figure 3.8.

A basis of $\mathbb{R}^n$ (or $\mathbb{C}^n$) consisting eigenvectors of A is sometimes called an **eigenbasis** *of A.*

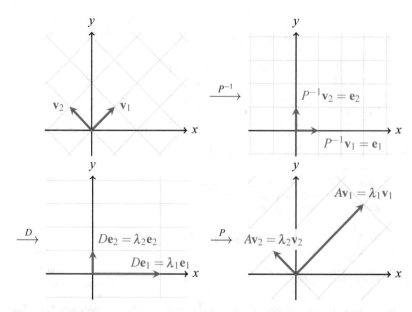

Figure 3.25: If $A = PDP^{-1} \in \mathcal{M}_2$ has eigenvectors $\mathbf{v}_1$ and $\mathbf{v}_2$ with corresponding eigenvalues λ_1 and λ_2, respectively, then its diagonalization skews the basis $\{\mathbf{v}_1, \mathbf{v}_2\}$ into the standard basis, then stretches along the coordinate axes by the eigenvalues, and then skews the standard basis back into $\{\mathbf{v}_1, \mathbf{v}_2\}$.

If A is diagonalizable then it can be manipulated almost as easily as diagonal matrices, so we can do things like compute large powers of A very quickly, or even define things like square roots of A or e^A. The idea is to simply do all of the hard work to the diagonal piece D and then un-skew and skew space after the fact (i.e., multiply on the left by P and multiply on the right by P^{-1} after manipulating D appropriately).

In addition to making computations significantly easier, diagonalization can also make it much simpler to prove theorems (as long as we are willing to take the trade-off that we are only proving things about diagonalizable matrices, not all matrices).

For example, we showed in Theorem 3.3.1 that if $A \in \mathcal{M}_n(\mathbb{C})$ has eigenvalues (listed according to algebraic multiplicity) $\lambda_1, \lambda_2, \ldots, \lambda_n$ then

$$\det(A) = \lambda_1 \lambda_2 \cdots \lambda_n \quad \text{and} \quad \operatorname{tr}(A) = \lambda_1 + \lambda_2 + \cdots + \lambda_n,$$

but the proof of these facts (especially the fact about the trace) was quite long and involved. However, these facts can be proved in just one line via similarity invariance of the determinant and trace if $A = PDP^{-1}$ is diagonalizable:

$$\det(A) = \det(PDP^{-1}) = \det(D) = \lambda_1 \lambda_2 \cdots \lambda_n \quad \text{and}$$
$$\operatorname{tr}(A) = \operatorname{tr}(PDP^{-1}) = \operatorname{tr}(D) = \lambda_1 + \lambda_2 + \cdots + \lambda_n,$$

where the final equalities both follow from the fact that the diagonal entries of D are $\lambda_1, \lambda_2, \ldots, \lambda_n$.

Remark 3.5.1

Does a Property Hold for Diagonalizable Matrices or All Matrices?

When trying to solve a problem in linear algebra, it is often tempting to assume that the matrix we are working with is diagonalizable. However, this can be slightly dangerous too, as we might mislead ourselves into thinking that some nice property of diagonalizable matrices holds for all matrices (even though it does not).

In situations like this, when we know that a property holds for diagonalizable matrices and we want to get an idea of whether or not it holds in general, a good trick is to check whether or not the (non-diagonalizable) matrix

$$N = \begin{bmatrix} 0 & 1 \\ 0 & 0 \end{bmatrix}$$

has that property.

For example, we showed in Exercise 3.4.12 that the rank of a diagonalizable matrix $A \in \mathcal{M}_n(\mathbb{C})$ equals the number of non-zero eigenvalues that it has (counting algebraic multiplicity). But we can see that this property does not hold for non-diagonalizable matrices, since N has no non-zero eigenvalues yet has rank 1. Similarly, since the method of Section 3.4.2 can be used to find a square root of every diagonalizable matrix $A \in \mathcal{M}_n(\mathbb{C})$, it might be tempting to think that all matrices have square roots. However, N does not—see Exercise 3.5.4.

Much like invertible matrices, diagonalizable matrices provide a place where linear algebra is "easy". While it might be difficult to prove something or to perform a particular matrix calculation in general, doing it just for diagonalizable matrices is much simpler. For example, if we restrict our attention to diagonalizable matrices then we can provide a complete answer to the question of when two matrices are similar:

Theorem 3.5.1

Similarity of Diagonalizable Matrices

Suppose $A, B \in \mathcal{M}_n(\mathbb{C})$ are diagonalizable. Then A and B are similar if and only if they have the same characteristic polynomial.

Proof. We already showed that similar matrices have the same characteristic polynomial back on page 295, so we just need to show that the converse holds for diagonalizable matrices.

To this end, suppose that A and B have the same characteristic polynomial. Then they have the same eigenvalues with the same algebraic multiplicities, so they are similar to a common diagonal matrix D (whose diagonal entries are those shared eigenvalues):

$$A = PDP^{-1} \quad \text{and} \quad B = QDQ^{-1} \quad \text{for some invertible} \quad P, Q \in \mathcal{M}_n(\mathbb{C}).$$

Multiplying the equation $B = QDQ^{-1}$ on the left by Q^{-1} and on the right by Q shows that $Q^{-1}BQ = D$. Substituting this into the equation $A = PDP^{-1}$ shows that

Recall that the inverse of a product is the product of the inverses in the opposite order, so $(PQ^{-1})^{-1} = QP^{-1}$.

$$A = PDP^{-1} = P\left(Q^{-1}BQ\right)P^{-1} = \left(PQ^{-1}\right)B\left(PQ^{-1}\right)^{-1},$$

which demonstrates that A and B are similar. ∎

We saw back in Example 3.3.14 that this theorem is not true if A or B is not diagonalizable, since the geometric multiplicities of eigenvalues are also similarity invariant, but not determined by the characteristic polynomial. It turns out that it is even possible for two non-diagonalizable matrices to have the same characteristic polynomial *and* the same geometric multiplicities of all of their eigenvalues, yet still not be similar (see Exercise 3.5.3). A complete method of determining whether or not two matrices are similar is provided by something called the "Jordan decomposition" of a matrix, which is covered in advanced linear algebra books like [Joh20].

Exercises

solutions to starred exercises on page 471

3.5.1 Determine whether or not the following pairs of matrices are similar.

*(a) $A = \begin{bmatrix} 1 & 1 \\ 1 & 1 \end{bmatrix}$ and $B = \begin{bmatrix} 1 & 2 \\ 2 & 1 \end{bmatrix}$

(b) $A = \begin{bmatrix} 1 & 1 \\ 1 & 1 \end{bmatrix}$ and $B = \begin{bmatrix} 2 & 0 \\ 3 & 0 \end{bmatrix}$

*(c) $A = \begin{bmatrix} 3 & 1 \\ 2 & 3 \end{bmatrix}$ and $B = \begin{bmatrix} 2 & -1 \\ 3 & 5 \end{bmatrix}$

(d) $A = \begin{bmatrix} 2 & 1 \\ -3 & 3 \end{bmatrix}$ and $B = \begin{bmatrix} 4 & 2 \\ -1 & 1 \end{bmatrix}$

*(e) $A = \begin{bmatrix} 3 & 3 \\ 0 & 2 \end{bmatrix}$ and $B = \begin{bmatrix} 4 & 2 \\ -1 & 1 \end{bmatrix}$

(f) $A = \begin{bmatrix} 1 & 2 & 3 \\ 4 & 5 & 6 \\ 7 & 8 & 9 \end{bmatrix}$ and $B = \begin{bmatrix} 1 & 2 & 3 \\ 8 & 9 & 4 \\ 7 & 6 & 5 \end{bmatrix}$

*(g) $A = \begin{bmatrix} 3 & 1 & -1 \\ 0 & 2 & 1 \\ -1 & 0 & 2 \end{bmatrix}$ and $B = \begin{bmatrix} 3 & 1 & 2 \\ -2 & 3 & 0 \\ 1 & -1 & 1 \end{bmatrix}$

3.5.2 Determine which of the following statements are true and which are false.

*(a) If A is a square matrix with 3 as one of its eigenvalues, then 9 must be an eigenvalue of A^2.

(b) If $A, B \in \mathcal{M}_n$ are similar then they have the same range.

*(c) If $A, B \in \mathcal{M}_n$ are similar then their eigenvalues have the same geometric multiplicities.

(d) If $A, B \in \mathcal{M}_n$ are similar then they have the same eigenvectors.

*(e) If $A \in \mathcal{M}_2$ and $B \in \mathcal{M}_2$ each have eigenvalues 4 and 5 then they must be similar.

(f) If $A \in \mathcal{M}_3$ and $B \in \mathcal{M}_3$ each have eigenvalues 1, 2, and 2, listed according to algebraic multiplicity, then they must be similar.

3.5.3 Consider the matrices

$$A = \begin{bmatrix} 0 & 0 & 1 & 0 \\ 0 & 0 & 0 & 1 \\ 0 & 0 & 0 & 0 \\ 0 & 0 & 0 & 0 \end{bmatrix} \quad \text{and} \quad B = \begin{bmatrix} 0 & 0 & 0 & 0 \\ 0 & 0 & 1 & 0 \\ 0 & 0 & 0 & 1 \\ 0 & 0 & 0 & 0 \end{bmatrix}.$$

(a) Show that A and B have the same rank.
(b) Show that A and B have the same characteristic polynomial.
(c) Show that A and B have the same eigenvectors (and thus in particular their eigenvalues have the same geometric multiplicities).
(d) Show that A and B are not similar.
 [Hint: Rewrite $A = PBP^{-1}$ as $AP = PB$ and solve a linear system to find P.]

3.5.4 Show that the matrix

$$N = \begin{bmatrix} 0 & 1 \\ 0 & 0 \end{bmatrix}$$

does not have a square root. That is, show that there does not exist a matrix $A \in \mathcal{M}_2(\mathbb{C})$ such that $A^2 = N$.

3.A Extra Topic: More About Determinants

We now explore some more properties of determinants that go a bit beyond the "core" results of Section 3.2 that should be known by all students of linear algebra, but are nonetheless interesting and/or useful in their own right.

3.A.1 The Cofactor Matrix and Inverses

Recall from Section 3.2.3 that the determinant of a matrix can be computed explicitly in terms of its cofactors. This section is devoted to showing that cofactors can also be used to construct an explicit formula for the inverse of a matrix. To this end, recall from Definition 3.2.2 that the (i, j)-cofactor of a matrix $A \in \mathcal{M}_n$ is the quantity $c_{i,j} = (-1)^{i+j} m_{i,j}$, where $m_{i,j}$ is the determinant of the $(n-1) \times (n-1)$ matrix obtained by removing the i-th row and j-th column of A. Placing these cofactors into a matrix in the natural way creates a matrix that it is worth giving a name to.

The quantity $m_{i,j}$ was called the (i,j)-minor of A.

| Definition 3.A.1 | The **cofactor matrix** of $A \in \mathcal{M}_n$, denoted by $\mathrm{cof}(A)$, is the matrix whose |
| **Cofactor Matrix** | entries are the cofactors $\{c_{i,j}\}_{i,j=1}^n$ of A: |

$$\mathrm{cof}(A) \stackrel{\text{def}}{=} \begin{bmatrix} c_{1,1} & c_{1,2} & \cdots & c_{1,n} \\ c_{2,1} & c_{2,2} & \cdots & c_{2,n} \\ \vdots & \vdots & \ddots & \vdots \\ c_{n,1} & c_{n,2} & \cdots & c_{n,n} \end{bmatrix} \in \mathcal{M}_n.$$

Example 3.A.1
Computing Cofactor Matrices

Compute the cofactor matrix of the following matrices:

a) $\begin{bmatrix} 2 & 3 \\ -1 & 4 \end{bmatrix}$

b) $\begin{bmatrix} 1 & 2 & 3 \\ 4 & 5 & 6 \\ 7 & 8 & 9 \end{bmatrix}$

Solutions:

a) The cofactors of this matrix are

$$c_{1,1} = \det([4]) = 4, \qquad\qquad c_{1,2} = -\det([-1]) = 1,$$
$$c_{2,1} = -\det([3]) = -3, \quad \text{and} \quad c_{2,2} = \det([2]) = 2.$$

It follows that

$$\text{cof}\left(\begin{bmatrix} 2 & 3 \\ -1 & 4 \end{bmatrix}\right) = \begin{bmatrix} c_{1,1} & c_{1,2} \\ c_{2,1} & c_{2,2} \end{bmatrix} = \begin{bmatrix} 4 & 1 \\ -3 & 2 \end{bmatrix}.$$

b) The $(1,1)$-cofactor of this matrix is

$$c_{1,1} = \det\left(\begin{bmatrix} 5 & 6 \\ 8 & 9 \end{bmatrix}\right) = 5 \cdot 9 - 6 \cdot 8 = 45 - 48 = -3.$$

Cofactor matrices
quickly become
cumbersome to
compute as we work
with larger matrices.
For example,
constructing the
cofactor matrix of a
4×4 involves
computing 16
determinants of 3×3
matrices.

Similarly, the other cofactors are

$$\begin{array}{lll} c_{1,1} = -3 & c_{1,2} = 6 & c_{1,3} = -3 \\ c_{2,1} = 6 & c_{2,2} = -12 & c_{2,3} = 6 \\ c_{3,1} = -3 & c_{3,2} = 6 & c_{3,3} = -3, \end{array}$$

so

$$\text{cof}\left(\begin{bmatrix} 1 & 2 & 3 \\ 4 & 5 & 6 \\ 7 & 8 & 9 \end{bmatrix}\right) = \begin{bmatrix} -3 & 6 & -3 \\ 6 & -12 & 6 \\ -3 & 6 & -3 \end{bmatrix}.$$

Our primary interest in cofactor matrices comes from the following theorem, which establishes a surprising connection between cofactor matrices and matrix inverses.

Theorem 3.A.1	If $A \in \mathcal{M}_n$ is an invertible matrix then
Matrix Inverse via Cofactor Matrix	$$A^{-1} = \frac{1}{\det(A)} \text{cof}(A)^T.$$

Recall from
Theorem 3.2.1 that if
A is invertible then
$\det(A) \neq 0$, so the
division by $\det(A)$ in
this theorem is not a
problem.

Proof. This theorem is essentially just a restatement of the cofactor method of computing the determinant (Theorem 3.2.8). If we compute the (i,i)-entry of the product $A\text{cof}(A)^T$ straight from the definition of matrix multiplication, we get

$$[A\text{cof}(A)^T]_{i,i} = a_{i,1}[\text{cof}(A)^T]_{1,i} + a_{i,2}[\text{cof}(A)^T]_{2,i} + \cdots + a_{i,n}[\text{cof}(A)^T]_{n,i}$$
$$= a_{i,1}c_{i,1} + a_{i,2}c_{i,2} + \cdots + a_{i,n}c_{i,n}$$
$$= \det(A).$$

We have thus shown that all of the diagonal entries of $A\text{cof}(A)^T$ are equal to $\det(A)$.

On the other hand, if $i \neq j$ then the (i,j)-entry of the product $A\text{cof}(A)^T$ is

$$[A\text{cof}(A)^T]_{i,j} = a_{i,1}c_{j,1} + a_{i,2}c_{j,2} + \cdots + a_{i,n}c_{j,n}. \qquad (3.A.1)$$

The matrix $\text{cof}(A)^T$ is
sometimes called
the **adjugate matrix**
of A and denoted by
$\text{adj}(A)$.

We claim that this sum equals 0. To see why this is the case, define $B \in \mathcal{M}_n$ to be the matrix that is identical to A except its i th row equals the i th row of A

(in particular, this means that the i-th and j-th rows of B are identical to each other):

$$A = \begin{bmatrix} a_{1,1} & a_{1,2} & \cdots & a_{1,n} \\ a_{2,1} & a_{2,2} & \cdots & a_{2,n} \\ \vdots & \vdots & \ddots & \vdots \\ a_{i,1} & a_{i,2} & \cdots & a_{i,n} \\ \vdots & \vdots & \ddots & \vdots \\ a_{j,1} & a_{j,2} & \cdots & a_{j,n} \\ \vdots & \vdots & \ddots & \vdots \\ a_{n,1} & a_{n,2} & \cdots & a_{n,n} \end{bmatrix}, \quad B = \begin{bmatrix} a_{1,1} & a_{1,2} & \cdots & a_{1,n} \\ a_{2,1} & a_{2,2} & \cdots & a_{2,n} \\ \vdots & \vdots & \ddots & \vdots \\ a_{i,1} & a_{i,2} & \cdots & a_{i,n} \\ \vdots & \vdots & \ddots & \vdots \\ a_{i,1} & a_{i,2} & \cdots & a_{i,n} \\ \vdots & \vdots & \ddots & \vdots \\ a_{n,1} & a_{n,2} & \cdots & a_{n,n} \end{bmatrix}.$$

$\leftarrow$ cofactor expansion

Since A and B are identical except in their j-th rows, their cofactors along that row are the same, so computing $\det(B)$ via a cofactor expansion along that row gives

$$\det(B) = a_{i,1}c_{j,1} + a_{i,2}c_{j,2} + \cdots + a_{i,n}c_{j,n},$$

The fact that a matrix with two identical rows cannot be invertible follows from Theorem 2.3.5.

which is exactly the quantity $[A\mathrm{cof}(A)^T]_{i,j}$ from Equation (3.A.1). Since B has two identical rows, it is not invertible, so we conclude that $0 = \det(B) = [A\mathrm{cof}(A)^T]_{i,j}$

.

We have thus shown that all off-diagonal entries of $A\mathrm{cof}(A)^T$ equal 0, so $A\mathrm{cof}(A)^T = \det(A)I$. By dividing both sides by $\det(A)$, we see that

$$A\left(\frac{1}{\det(A)}\mathrm{cof}(A)^T\right) = I,$$

so the inverse of A is as claimed. ∎

The beauty of the above theorem is that, since we have explicit formulas for the determinant of a matrix, we can now construct explicit formulas for the inverse of a matrix as well. For example, the cofactor matrix of a 2×2 matrix

$$A = \begin{bmatrix} a & b \\ c & d \end{bmatrix} \quad \text{is} \quad \mathrm{cof}(A) = \begin{bmatrix} d & -c \\ -b & a \end{bmatrix}.$$

Since $\det(A) = ad - bc$, it follows that

$$A^{-1} = \frac{1}{\det(A)}\mathrm{cof}(A)^T = \frac{1}{ad-bc}\begin{bmatrix} d & -b \\ -c & a \end{bmatrix},$$

which is exactly the formula for the inverse of a 2×2 matrix that we found back in Theorem 2.2.6.

We can also use this theorem to compute the inverse of larger matrices, which we illustrate with an example.

Example 3.A.2

Using Cofactors to Compute an Inverse

Use Theorem 3.A.1 to compute the inverse of the matrix

$$A = \begin{bmatrix} 1 & 1 & 1 \\ 1 & 2 & 4 \\ 1 & 3 & 9 \end{bmatrix}.$$

Solution:
In order to use Theorem 3.A.1, we must compute $\det(A)$ as well as all of its cofactors. We summarize these calculations here:

$$\det(A) = 2 \qquad c_{1,1} = 6 \qquad c_{1,2} = -5 \qquad c_{1,3} = 1$$
$$c_{2,1} = -6 \qquad c_{2,2} = 8 \qquad c_{2,3} = -2$$
$$c_{3,1} = 2 \qquad c_{3,2} = -3 \qquad c_{3,3} = 1.$$

We also computed the inverse of this matrix via Gauss–Jordan elimination way back in Example 2.2.8(d).

It follows that

$$\operatorname{cof}(A) = \begin{bmatrix} 6 & -5 & 1 \\ -6 & 8 & -2 \\ 2 & -3 & 1 \end{bmatrix}, \quad \text{so} \quad A^{-1} = \frac{1}{2}\begin{bmatrix} 6 & -6 & 2 \\ -5 & 8 & -3 \\ 1 & -2 & 1 \end{bmatrix}.$$

By generalizing the previous example, we can also use Theorem 3.A.1 to come up with explicit formulas for the inverse of larger matrices, but they quickly become so ugly as to not be practical (and thus the method of computing inverses based on Gauss–Jordan elimination, from Theorem 2.2.5, is preferred). For example, the cofactor matrix of a 3×3 matrix

$$A = \begin{bmatrix} a & b & c \\ d & e & f \\ g & h & i \end{bmatrix} \quad \text{is} \quad \operatorname{cof}(A) = \begin{bmatrix} ei-fh & fg-di & dh-eg \\ ch-bi & ai-cg & bg-ah \\ bf-ce & cd-af & ae-bd \end{bmatrix}.$$

Theorem 3.A.1 then gives us the following explicit formula for the inverse of a 3×3 matrix:

Here we used the explicit formula for the determinant of a 3×3 matrix from Theorem 3.2.7.

$$A^{-1} = \frac{1}{\det(A)}\operatorname{cof}(A)^{T}$$

$$= \frac{1}{aei+bfg+cdh-afh-bdi-ceg}\begin{bmatrix} ei-fh & ch-bi & bf-ce \\ fg-di & ai-cg & cd-af \\ dh-eg & bg-ah & ae-bd \end{bmatrix}.$$

3.A.2   Cramer's Rule

Now that we know how to write the inverse of a matrix in terms of determinants (thanks to Theorem 3.A.1), and we already know how to solve linear systems in terms of matrix inverses (recall that if $A\mathbf{x} = \mathbf{b}$ and A is invertible, then $\mathbf{x} = A^{-1}\mathbf{b}$ is the unique solution), we can very quickly derive a method for solving linear systems in terms of determinants.

Theorem 3.A.2

Cramer's Rule

Suppose $A \in \mathcal{M}_n$ is invertible and $\mathbf{b} \in \mathbb{R}^n$. Define A_j to be the matrix that equals A, except its j-th column is replaced by $\mathbf{b}$. Then the linear system $A\mathbf{x} = \mathbf{b}$ has a unique solution $\mathbf{x}$, whose entries are

$$x_j = \frac{\det(A_j)}{\det(A)} \quad \text{for all} \quad 1 \le j \le n.$$

Proof. We know from Theorem 2.2.4 that, since A is invertible, the linear system $A\mathbf{x} = \mathbf{b}$ has a unique solution $\mathbf{x} = A^{-1}\mathbf{b}$. We thus just need to compute the entries of $A^{-1}\mathbf{b}$, which we do via Theorem 3.A.1:

$$x_j = [A^{-1}\mathbf{b}]_j = \left[\frac{1}{\det(A)}\text{cof}(A)^T\mathbf{b}\right]_j = \frac{1}{\det(A)}(c_{1,j}b_1 + c_{2,j}b_2 + \cdots + c_{n,j}b_n),$$

where $c_{1,j}, c_{2,j}, \ldots, c_{n,j}$ are cofactors of A and the final equality above follows from the definition of matrix multiplication. Well, the quantity $c_{1,j}b_1 + c_{2,j}b_2 + \cdots + c_{n,j}b_n$ above on the right is exactly the cofactor expansion of A_j along its j-th column:

The key idea here is that A and A_j have the same cofactors corresponding to their j-th column, since they are identical everywhere except for their j-th column.

$$A = \begin{bmatrix} a_{1,1} & \cdots & a_{1,j} & \cdots & a_{1,n} \\ a_{2,1} & \cdots & a_{2,j} & \cdots & a_{2,n} \\ \vdots & \ddots & \vdots & \ddots & \vdots \\ a_{n,1} & \cdots & a_{n,j} & \cdots & a_{n,n} \end{bmatrix}, \quad A_j = \begin{bmatrix} a_{1,1} & \cdots & b_1 & \cdots & a_{1,n} \\ a_{2,1} & \cdots & b_2 & \cdots & a_{2,n} \\ \vdots & \ddots & \vdots & \ddots & \vdots \\ a_{n,1} & \cdots & b_n & \cdots & a_{n,n} \end{bmatrix}$$

cofactor expansion ———↑

It follows from the cofactor expansion theorem (Theorem 3.2.8) that $c_{1,j}b_1 + c_{2,j}b_2 + \cdots + c_{n,j}b_n = \det(A_j)$, so $x_j = \det(A_j)/\det(A)$, as desired. ∎

As with our other results that involve determinants, we can substitute the explicit formulas for the determinant of small matrices into Cramer's rule to obtain explicit formulas for the solution of small linear systems. For example, if $A \in \mathcal{M}_2$ is invertible then Cramer's rule tells us that the unique solution of the linear system $A\mathbf{x} = \mathbf{b}$ is given by the formulas

$$x_1 = \frac{\det\left(\begin{bmatrix} b_1 & a_{1,2} \\ b_2 & a_{2,2} \end{bmatrix}\right)}{\det\left(\begin{bmatrix} a_{1,1} & a_{1,2} \\ a_{2,1} & a_{2,2} \end{bmatrix}\right)} = \frac{b_1 a_{2,2} - a_{1,2} b_2}{a_{1,1}a_{2,2} - a_{1,2}a_{2,1}} \quad \text{and}$$

$$x_2 = \frac{\det\left(\begin{bmatrix} a_{1,1} & b_1 \\ a_{2,1} & b_2 \end{bmatrix}\right)}{\det\left(\begin{bmatrix} a_{1,1} & a_{1,2} \\ a_{2,1} & a_{2,2} \end{bmatrix}\right)} = \frac{a_{1,1} b_2 - b_1 a_{2,1}}{a_{1,1}a_{2,2} - a_{1,2}a_{2,1}}.$$

Example 3.A.3

Solving Linear Systems via Cramer's Rule

Use Cramer's rule to solve the following linear systems:

a) $\begin{aligned} x + 2y &= 4 \\ 3x + 4y &= 6 \end{aligned}$

b) $\begin{aligned} 3x - 2y + z &= -3 \\ 2x + 3y - 2z &= 5 \\ y + z &= 4 \end{aligned}$

Solutions:

a) We use the explicit formulas for x and y that were introduced immediately above this example:

$$x = \frac{4 \cdot 4 - 2 \cdot 6}{1 \cdot 4 - 2 \cdot 3} = \frac{4}{-2} = -2 \quad y = \frac{1 \cdot 6 - 4 \cdot 3}{1 \cdot 4 - 2 \cdot 3} = \frac{-6}{-2} = 3.$$

b) For this 3×3 system, we are a bit more explicit with our calculations. We start by writing out the four matrices $A, A_1, A_2,$ and A_3 whose determinants are used in the computation of x, y, and z:

$$A = \begin{bmatrix} 3 & -2 & 1 \\ 2 & 3 & -2 \\ 0 & 1 & 1 \end{bmatrix}, \quad A_1 = \begin{bmatrix} -3 & -2 & 1 \\ 5 & 3 & -2 \\ 4 & 1 & 1 \end{bmatrix},$$

$$A_2 = \begin{bmatrix} 3 & -3 & 1 \\ 2 & 5 & -2 \\ 0 & 4 & 1 \end{bmatrix}, \quad A_3 = \begin{bmatrix} 3 & -2 & -3 \\ 2 & 3 & 5 \\ 0 & 1 & 4 \end{bmatrix}.$$

> Recall that the determinant of a 3×3 matrix can be computed via the formula of Theorem 3.2.7.

The determinants of these four matrices are

$$\det(A) = 21, \quad \det(A_1) = 4, \quad \det(A_2) = 53, \quad \text{and} \quad \det(A_3) = 31.$$

Cramer's rule thus tells us that the unique solution of this linear system is

$$x = \frac{\det(A_1)}{\det(A)} = \frac{4}{21}, \quad y = \frac{\det(A_2)}{\det(A)} = \frac{53}{21}, \quad z = \frac{\det(A_3)}{\det(A)} = \frac{31}{21}.$$

Similar to how we used Cramer's rule to construct an explicit formula for the solution of a linear system with two equations and two variables, we could use it to construct such a formula for larger linear systems as well. For example, if $A \in \mathcal{M}_3$ is invertible then the linear system $A\mathbf{x} = \mathbf{b}$ has a unique solution, whose first entry is given by

$$x_1 = \frac{\det\left(\begin{bmatrix} b_1 & a_{1,2} & a_{1,3} \\ b_2 & a_{2,2} & a_{2,3} \\ b_3 & a_{3,2} & a_{3,3} \end{bmatrix} \right)}{\det\left(\begin{bmatrix} a_{1,1} & a_{1,2} & a_{1,3} \\ a_{2,1} & a_{2,2} & a_{2,3} \\ a_{3,1} & a_{3,2} & a_{3,3} \end{bmatrix} \right)} =$$

$$\frac{b_1\, a_{2,2}a_{3,3} + a_{1,2}a_{2,3}\, b_3 + a_{1,3}\, b_2\, a_{3,2} - b_1\, a_{2,3}a_{3,2} - a_{1,2}\, b_2\, a_{3,3} - a_{1,3}a_{2,2}\, b_3}{a_{1,1}a_{2,2}a_{3,3} + a_{1,2}a_{2,3}a_{3,1} + a_{1,3}a_{2,1}a_{3,2} - a_{1,1}a_{2,3}a_{3,2} - a_{1,2}a_{2,1}a_{3,3} - a_{1,3}a_{2,2}a_{3,1}},$$

and whose other two entries x_2 and x_3 are given by similar formulas. However, these formulas are quite unwieldy, so it's typically easier to solve linear systems of size 3×3 or larger just by using Gaussian elimination as usual.

> **Remark 3.A.1**
>
> **Computational (In)Efficiency of Cramer's Rule**
>
> This same disclaimer applies to matrix inverses and Theorem 3.A.1.

Keep in mind that Cramer's rule is computationally inefficient and thus not often used to solve linear systems in practice. In particular, to use Cramer's rule we must compute the determinant of $n + 1$ different $n \times n$ matrices, and the quickest method that we have seen for computing the determinant of a matrix is the method based on Gaussian elimination (Theorem 3.2.5). However, we could just use Gaussian elimination (once, instead of $n + 1$ times) to directly solve the linear system in the first place.

Nevertheless, Cramer's rule is still a useful theoretical tool that lets us use the many nice properties of determinants to prove things about linear systems.

3.A.3 Permutations

There is one final widely-used formula for the determinant that provides it with an alternate algebraic interpretation. It is based on permutations, which are functions that shuffle objects around without duplicating or erasing any of those objects:

Definition 3.A.2
Permutation

A function $\sigma : \{1,2,\ldots,n\} \to \{1,2,\ldots,n\}$ is called a **permutation** if $\sigma(i) \neq \sigma(j)$ whenever $i \neq j$.

Recall that the notation $\sigma : \{1,2,\ldots,n\} \to \{1,2,\ldots,n\}$ means that σ is a function that takes $1,2,\ldots,n$ as inputs and also produces them as possible outputs.

For example, the function $\sigma : \{1,2,3\} \to \{1,2,3\}$ defined by $\sigma(1) = 2$, $\sigma(2) = 3$, and $\sigma(3) = 1$ is a permutation (it just shuffles 1, 2, and 3 around), but the function $f : \{1,2,3\} \to \{1,2,3\}$ defined by $f(1) = 2$, $f(2) = 2$, and $f(3) = 1$ is not (since $f(1) = f(2)$).

In order to describe permutations a bit more efficiently, we use **one-line notation**, where we simply list the outputs of the permutation in order within parentheses: $(\sigma(1)\ \sigma(2)\ \cdots\ \sigma(n))$. For example, the permutation discussed earlier with $\sigma(1) = 2$, $\sigma(2) = 3$, and $\sigma(3) = 1$ is described succinctly by its one-line notation: $(2\ 3\ 1)$. In the other direction, if a permutation is described via its one-line notation $\sigma = (4\ 1\ 3\ 2)$ then we know that it acts on $\{1,2,3,4\}$ via $\sigma(1) = 4$, $\sigma(2) = 1$, $\sigma(3) = 3$, and $\sigma(4) = 2$.

There are exactly $n! = n \cdot (n-1) \cdot (n-2) \cdots 2 \cdot 1$ permutations σ that act on the set $\{1,2,\ldots,n\}$. To see this, notice that $\sigma(1)$ can be any of n different values, $\sigma(2)$ can be any of the $n-1$ remaining values (since it cannot be the same as $\sigma(1)$), $\sigma(3)$ can be any of the $n-2$ remaining values not equal to $\sigma(1)$ or $\sigma(2)$, and so on. The set of all of these permutations acting on $\{1,2,\ldots,n\}$ is denoted by S_n.

Example 3.A.4
Constructing All Permutations

List all permutations in the following sets:

a) S_2
b) S_3

Solutions:
a) There are only $2! = 2$ permutations that act on the set $\{1,2\}$, and they are $(1\ 2)$ and $(2\ 1)$.

b) There are $3! = 6$ permutations that act on the set $\{1,2,3\}$, and they are

$$
\begin{array}{ccc}
(1\ 2\ 3) & (2\ 1\ 3) & (3\ 1\ 2) \\
(1\ 3\ 2) & (2\ 3\ 1) & (3\ 2\ 1).
\end{array}
$$

Proving that the composition of two permutations is a permutation should be believable enough—we skip over an explicit proof, since the details are not too enlightening.

Since permutations are functions, we can compose them with each other, just like we do with linear transformations. Specifically, if $\sigma, \tau \in S_n$ then we define their **composition** $\sigma \circ \tau$ by

$$(\sigma \circ \tau)(j) = \sigma(\tau(j)) \quad \text{for all} \quad 1 \leq j \leq n.$$

Similarly, the **identity permutation**, which we denote by ι, is the one that does nothing at all: $\iota = (1\ 2\ \cdots\ n)$, so $\iota(j) = j$ for all $1 \leq j \leq n$, and the **inverse** σ^{-1} of a permutation σ is the one that "undoes" it:

$$\sigma \circ \sigma^{-1} = \iota \quad \text{and} \quad \sigma^{-1} \circ \sigma = \iota.$$

Example 3.A.5

Composing and Inverting Permutations

Suppose $\sigma = (3\ 2\ 5\ 1\ 4)$ and $\tau = (5\ 1\ 4\ 2\ 3)$. Compute the one-line notations for the following permutations:

a) $\sigma \circ \tau$

c) σ^{-1}

b) $\tau \circ \sigma$

d) τ^{-1}

Solutions:

a) We just directly compute $\sigma(\tau(j))$ for each $1 \le j \le 5$:

$$\sigma(\tau(1)) = \sigma(5) = 4 \quad \sigma(\tau(2)) = \sigma(1) = 3 \quad \sigma(\tau(3)) = \sigma(4) = 1$$
$$\sigma(\tau(4)) = \sigma(2) = 2 \quad \sigma(\tau(5)) = \sigma(3) = 5.$$

The one-line notation for $\sigma \circ \tau$ is constructed simply by writing $\sigma(\tau(1))$, $\sigma(\tau(2)), \ldots, \sigma(\tau(5))$ in order.

We thus conclude that $\sigma \circ \tau = (4\ 3\ 1\ 2\ 5)$.

b) Again, we just directly compute $\tau(\sigma(j))$ for each $1 \le j \le 5$:

$$\tau(\sigma(1)) = \tau(3) = 4 \quad \tau(\sigma(2)) = \tau(2) = 1 \quad \tau(\sigma(3)) = \tau(5) = 3$$
$$\tau(\sigma(4)) = \tau(1) = 5 \quad \tau(\sigma(5)) = \tau(4) = 2.$$

It follows that $\tau \circ \sigma = (4\ 1\ 3\ 5\ 2)$.

If $\sigma(j) = k$ then $\sigma^{-1}(k) = j$.

c) Since σ^{-1} takes the output of σ back to its input, one way to construct σ^{-1} is to write down the location of the numbers $1, 2, \ldots,$ n in the one-line notation of σ, in order. For this particular σ, the numbers $1, 2, 3, 4, 5$ appear in positions $4, 2, 1, 5, 3$ in σ's one-line notation, so $\sigma^{-1} = (4\ 2\ 1\ 5\ 3)$.

d) The numbers $1, 2, 3, 4, 5$ appear in positions $2, 4, 5, 3, 1$ in τ's one-line notation, so $\tau^{-1} = (2\ 4\ 5\ 3\ 1)$.

Every permutation can be associated with a particular matrix by permuting the columns of the identity matrix in the manner described by the permutation. The following definition makes this idea precise.

Definition 3.A.3

Permutation Matrix

Given a permutation $\sigma : \{1, 2, \ldots, n\} \to \{1, 2, \ldots, n\}$, the associated **permutation matrix** is the matrix

$$P_\sigma = \big[\ \mathbf{e}_{\sigma(1)} \mid \mathbf{e}_{\sigma(2)} \mid \cdots \mid \mathbf{e}_{\sigma(n)}\ \big].$$

We first introduced permutation matrices when discussing the PLU decomposition in Section 2.D.3.

In other words, the permutation matrix P_σ is the standard matrix of the linear transformation that sends $\mathbf{e}_j$ to $\mathbf{e}_{\sigma(j)}$ for each $1 \le j \le n$. Geometrically, this linear transformation permutes the sides of the unit square (or cube, or hypercube...) and thus changes its orientation, but does not stretch or skew it at all (see Figure 3.26).

We now work through a couple of examples to get more comfortable with permutation matrices.

Example 3.A.6

Constructing Permutation Matrices

Construct the permutation matrices corresponding to the following permutations, which have been written in one-line notation:

a) $\sigma = (3\ 4\ 2\ 1)$

b) $\sigma = (2\ 5\ 1\ 3\ 4)$

c) $\iota \in S_6$

This cube is rotating, pivoting on its corner at the origin, so that each standard basis vector lands on a different axis.

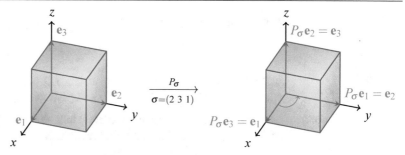

Figure 3.26: If $\sigma = (2\ 3\ 1)$ then the permutation matrix P_σ sends $\mathbf{e}_1$ to $\mathbf{e}_2$, $\mathbf{e}_2$ to $\mathbf{e}_3$, and $\mathbf{e}_3$ to $\mathbf{e}_1$.

Solutions:

Permutation matrices are exactly the matrices with a single 1 in each row and column, and every other entry equal to 0.

a) As described by Definition 3.A.3, we just place $\mathbf{e}_3$, $\mathbf{e}_4$, $\mathbf{e}_2$, and $\mathbf{e}_1$ as columns into a matrix, in that order:

$$P_\sigma = \begin{bmatrix} \mathbf{e}_3 \mid \mathbf{e}_4 \mid \mathbf{e}_2 \mid \mathbf{e}_1 \end{bmatrix} = \begin{bmatrix} 0 & 0 & 0 & 1 \\ 0 & 0 & 1 & 0 \\ 1 & 0 & 0 & 0 \\ 0 & 1 & 0 & 0 \end{bmatrix}.$$

b) Similarly, for this permutation we place $\mathbf{e}_2$, $\mathbf{e}_5$, $\mathbf{e}_1$, $\mathbf{e}_3$, and $\mathbf{e}_4$ as columns into a matrix:

$$P_\sigma = \begin{bmatrix} \mathbf{e}_2 \mid \mathbf{e}_5 \mid \mathbf{e}_1 \mid \mathbf{e}_3 \mid \mathbf{e}_4 \end{bmatrix} = \begin{bmatrix} 0 & 0 & 1 & 0 & 0 \\ 1 & 0 & 0 & 0 & 0 \\ 0 & 0 & 0 & 1 & 0 \\ 0 & 0 & 0 & 0 & 1 \\ 0 & 1 & 0 & 0 & 0 \end{bmatrix}.$$

c) For the identity permutation on $\{1,2,3,4,5,6\}$, we place $\mathbf{e}_1$, $\mathbf{e}_2$, ..., $\mathbf{e}_6$ as columns into a matrix. Doing so gives us exactly the 6×6 identity matrix:

$$P_\sigma = \begin{bmatrix} \mathbf{e}_1 \mid \mathbf{e}_2 \mid \cdots \mid \mathbf{e}_6 \end{bmatrix} = I_6.$$

In part (c) of the above example, we saw that the permutation matrix associated with the identity permutation in S_6 was the identity matrix in $\mathcal{M}_6$. It's not difficult to show that this is true in general, regardless of n. There are also some other pleasant properties of permutation matrices, all of which basically say that permutation matrices interact with each other in the same way that the permutations themselves interact with each other. The following theorem summarizes these properties.

Theorem 3.A.3

Properties of Permutation Matrices

Let $\sigma, \tau \in S_n$ be permutations with corresponding permutation matrices $P_\sigma, P_\tau \in \mathcal{M}_n$. Then

a) $P_\sigma P_\tau = P_{\sigma \circ \tau}$
b) P_σ is invertible, and $P_\sigma^{-1} = P_\sigma^T = P_{\sigma^{-1}}$.

Proof. For part (a), we just recall that P_σ acts on the standard basis vectors via

$P_\sigma \mathbf{e}_j = \mathbf{e}_{\sigma(j)}$ for all j, so

$$(P_\sigma P_\tau)\mathbf{e}_j = P_\sigma(P_\tau \mathbf{e}_j) = P_\sigma \mathbf{e}_{\tau(j)} = \mathbf{e}_{\sigma(\tau(j))} = \mathbf{e}_{(\sigma\circ\tau)(j)}.$$

Recall that $P_{\sigma\circ\tau}$ is *defined* to have its j-th column equal to $\mathbf{e}_{(\sigma\circ\tau)(j)}$.

In particular, this means that, for all $1 \le j \le n$, the j-th column of $P_\sigma P_\tau$ is $\mathbf{e}_{(\sigma\circ\tau)(j)}$. We thus conclude that $P_\sigma P_\tau = P_{\sigma\circ\tau}$, as desired.

For part (b), we first note that part (a) tells us that $P_\sigma P_{\sigma^{-1}} = P_{\sigma\circ\sigma^{-1}} = P_\iota = I$, so $P_\sigma^{-1} = P_{\sigma^{-1}}$, as claimed. All that remains is to show that $P_\sigma^T = P_\sigma^{-1}$. To this end, we compute the entries of $P_\sigma^T P_\sigma$:

$$[P_\sigma^T P_\sigma]_{i,j} = \mathbf{e}_{\sigma(i)}^T \mathbf{e}_{\sigma(j)} = \mathbf{e}_{\sigma(i)} \cdot \mathbf{e}_{\sigma(j)}.$$

A matrix whose inverse is its transpose (like P_σ) is called **unitary**.

If $i = j$ then $\sigma(i) = \sigma(j)$, so the above dot product equals 1 (i.e., the diagonal entries of $P_\sigma^T P_\sigma$ all equal 1). On the other hand, if $i \ne j$ then $\sigma(i) \ne \sigma(j)$, so $\mathbf{e}_{\sigma(i)} \cdot \mathbf{e}_{\sigma(j)} = 0$ (i.e., the off-diagonal entries of $P_\sigma^T P_\sigma$ all equal 0). It follows that $P_\sigma^T P_\sigma = I$, so $P_\sigma^T = P_\sigma^{-1}$, which completes the proof. ■

We just need one final property of permutations before we can establish their connection with the determinant.

Definition 3.A.4 **Sign of a** **Permutation**	The **sign of a permutation** σ, denoted by $\text{sgn}(\sigma)$, is the quantity $$\text{sgn}(\sigma) \overset{\text{def}}{=} \det(P_\sigma).$$

Example 3.A.7 **Computing the** **Sign of a** **Permutation**	Compute the signs of the following permutations, which have been written in one-line notation: a) $\sigma = (3\ 4\ 2\ 1)$ b) $\sigma = (2\ 5\ 1\ 3\ 4)$

Solutions:
   a) We computed the permutation matrix P_σ of this permutation back in Example 3.A.6(a), so we now just need to compute its determinant:

These determinants are straightforward to compute via cofactor expansions (Theorem 3.2.8) since they have so many zeros.

$$\text{sgn}(\sigma) = \det(P_\sigma) = \det\left(\begin{bmatrix} 0 & 0 & 0 & 1 \\ 0 & 0 & 1 & 0 \\ 1 & 0 & 0 & 0 \\ 0 & 1 & 0 & 0 \end{bmatrix}\right) = -1.$$

   b) Again, we just compute the determinant of the permutation matrix P_σ that we already constructed for this permutation:

$$\text{sgn}(\sigma) = \det(P_\sigma) = \det\begin{bmatrix} 0 & 0 & 1 & 0 & 0 \\ 1 & 0 & 0 & 0 & 0 \\ 0 & 0 & 0 & 1 & 0 \\ 0 & 0 & 0 & 0 & 1 \\ 0 & 1 & 0 & 0 & 0 \end{bmatrix} = 1.$$

Permutations are sometimes called **even** or **odd**, depending on whether their sign is 1 or −1, respectively.

In the previous example, the signs of the determinants were ±1. This is the case for every permutation, since permutation matrices can be obtained from the identity matrix (which has determinant 1) by repeatedly swapping columns, and swapping two columns of a matrix multiplies its determinant by

-1. In fact, this argument tells us the following fact that makes the sign of a permutation easier to compute:

> ⚠️ The sign of a permutation equals $(-1)^s$, where s is the number of times that two of its outputs must be swapped to turn it into the identity permutation.

Example 3.A.8

Computing the Sign of a Permutation More Quickly

It is possible to use a different number of swaps (e.g., 5 or 7 instead of 3 in part (a) here) to get the identity permutation, but the parity of the number of swaps will always be the same (i.e., always odd or always even).

Compute the signs of the following permutations (the same ones from Example 3.A.7), without making use of determinants or permutation matrices:

a) $\sigma = (3\,4\,2\,1)$
b) $\sigma = (2\,5\,1\,3\,4)$

Solutions:

a) To compute the sign of σ, we swap numbers in its one-line notation until they are in the order $1, 2, 3, 4$ (i.e., until we have created the identity permutation):

$$(3\,4\,2\,1) \to (1\,4\,2\,3) \to (1\,2\,4\,3) \to (1\,2\,3\,4).$$

Since we swapped 3 times, we conclude that $\text{sgn}(\sigma) = (-1)^3 = -1$.

b) Again, we swap numbers in σ's one-line notation until they are in the order $1, 2, 3, 4, 5$ (i.e., until we have created the identity permutation):

$$(2\,5\,1\,3\,4) \to (1\,5\,2\,3\,4) \to (1\,2\,5\,3\,4) \to (1\,2\,3\,5\,4) \to (1\,2\,3\,4\,5).$$

Since we swapped 4 times, we conclude that $\text{sgn}(\sigma) = (-1)^4 = 1$.

Now that we understand permutations and permutation matrices a bit better, we are (finally!) in a position to present the main result of this subsection, which provides a formula for the determinant of a matrix in terms of permutations.

Theorem 3.A.4

Determinants via Permutations

The formula provided by this theorem is sometimes called the **Leibniz formula** for the determinant.

Suppose $A \in \mathcal{M}_n$. Then $\det(A) = \sum_{\sigma \in S_n} \text{sgn}(\sigma) a_{\sigma(1),1} a_{\sigma(2),2} \cdots a_{\sigma(n),n}$.

While this theorem might seem like a mouthful, it really is just a non-recursive way of generalizing the formulas for the determinant that we developed in Section 3.2.3 (similar to how cofactor expansions provided us with a recursive way of generalizing those formulas). For example, when $n = 2$, there are two permutations in S_2: $(1\,2)$, which has sign $+1$, and $(2\,1)$, which has sign -1. The above theorem thus tells us that $\det(A) = a_{1,1}a_{2,2} - a_{2,1}a_{1,2}$, which is exactly the formula that we originally derived in Theorem 3.2.6.

Similarly, when $n = 3$, there are six permutations in S_3, which we now list along with their signs and the term that they contribute to the determinant calculation in the above theorem:

σ	$\text{sgn}(\sigma)$	contribution to $\det(A)$
(1 2 3)	1	$a_{1,1}a_{2,2}a_{3,3}$
(1 3 2)	-1	$-a_{1,1}a_{3,2}a_{2,3}$
(2 1 3)	-1	$-a_{2,1}a_{1,2}a_{3,3}$
(2 3 1)	1	$a_{2,1}a_{3,2}a_{1,3}$
(3 1 2)	1	$a_{3,1}a_{1,2}a_{2,3}$
(3 2 1)	-1	$-a_{3,1}a_{2,2}a_{1,3}$

In the third column of this table, the second subscripts are always 1, 2, 3, in that order, and the first subscripts come from σ.

Adding up the six terms in the rightmost column of the table above gives exactly the formula for the determinant that we originally derived in Theorem 3.2.7.

Proof of Theorem 3.A.4. Write $A = [\, \mathbf{a}_1 \mid \mathbf{a}_2 \mid \cdots \mid \mathbf{a}_n \,]$, so that $\mathbf{a}_j$ denotes the j-th column of A. We can write each $\mathbf{a}_j$ as a linear combination of the standard basis vectors as follows:

$$\mathbf{a}_j = \sum_{i=1}^{n} a_{i,j}\mathbf{e}_i.$$

By repeatedly using this fact, along with multilinearity of the determinant (i.e., property (c) of Definition 3.2.1), it follows that

$$\det(A) = \det\left([\, \mathbf{a}_1 \mid \mathbf{a}_2 \mid \cdots \mid \mathbf{a}_n \,]\right)$$

$$= \det\left(\left[\, \sum_{i_1=1}^{n} a_{i_1,1}\mathbf{e}_{i_1} \mid \mathbf{a}_2 \mid \cdots \mid \mathbf{a}_n \,\right]\right)$$

$$= \sum_{i_1=1}^{n} a_{i_1,1} \det\left([\, \mathbf{e}_{i_1} \mid \mathbf{a}_2 \mid \cdots \mid \mathbf{a}_n \,]\right)$$

Yes, this is hideous. Try not to think about how hideous it is—just work one column at a time and breathe.

$$= \sum_{i_1=1}^{n} a_{i_1,1} \det\left(\left[\, \mathbf{e}_{i_1} \mid \sum_{i_2=1}^{n} a_{i_2,2}\mathbf{e}_{i_2} \mid \cdots \mid \mathbf{a}_n \,\right]\right)$$

$$= \sum_{i_1,i_2=1}^{n} (a_{i_1,1}a_{i_2,2}) \det\left([\, \mathbf{e}_{i_1} \mid \mathbf{e}_{i_2} \mid \cdots \mid \mathbf{a}_n \,]\right)$$

$$\vdots$$

$$= \sum_{i_1,i_2,\ldots,i_n=1}^{n} (a_{i_1,1}a_{i_2,2}\cdots a_{i_n,n}) \det\left([\, \mathbf{e}_{i_1} \mid \mathbf{e}_{i_2} \mid \cdots \mid \mathbf{e}_{i_n} \,]\right).$$

The fact that matrices with repeated columns are not invertible follows from Theorem 2.3.5.

If any of the indices $i_1, i_2, \ldots, i_n$ in the above sum are equal to each other then the matrix $[\, \mathbf{e}_{i_1} \mid \mathbf{e}_{i_2} \mid \cdots \mid \mathbf{e}_{i_n} \,]$ has a repeated column and is thus not invertible and has determinant 0. It follows that the sum over all of those indices can be replaced by a sum over *non-repeating* indices (i.e., permutations σ with $\sigma(1) = i_1$, $\sigma(2) = i_2$, and so on), so that

$$\det(A) = \sum_{\sigma \in S_n} \left(a_{\sigma(1),1}a_{\sigma(2),2}\cdots a_{\sigma(n),n}\right) \det\left([\, \mathbf{e}_{\sigma(1)} \mid \mathbf{e}_{\sigma(2)} \mid \cdots \mid \mathbf{e}_{\sigma(n)} \,]\right)$$

$$= \sum_{\sigma \in S_n} \left(a_{\sigma(1),1}a_{\sigma(2),2}\cdots a_{\sigma(n),n}\right) \det(P_\sigma)$$

$$= \sum_{\sigma \in S_n} \text{sgn}(\sigma)a_{\sigma(1),1}a_{\sigma(2),2}\cdots a_{\sigma(n),n},$$

which completes the proof. ∎

Theorem 3.A.4 is typically not used numerically for the same reason that
cofactor expansions are not used numerically, the method of computing the

determinant based on Gaussian elimination or Gauss–Jordan elimination is much faster for matrices of size 4×4 or larger. However, it is often useful as a theoretical tool that can be used to establish connections between the determinant and other areas of mathematics (especially abstract algebra, where permutations are a common object of study).

To illustrate the type of result that we can prove with this new approach to determinants based on permutations, recall that if $A, B \in \mathcal{M}_n$ are square then $\det(AB) = \det(A)\det(B)$. The following theorem provides a natural generalization of this fact to non-square matrices $A \in \mathcal{M}_{m,n}$ and $B \in \mathcal{M}_{n,m}$ that involves the determinants of their square submatrices.

> Keep in mind that if A and B are not square then $\det(A)$ and $\det(B)$ are not even defined.

Theorem 3.A.5

Cauchy–Binet Formula

Suppose $A \in \mathcal{M}_{m,n}$ and $B \in \mathcal{M}_{n,m}$ with $m \leq n$, and let $A_{j_1,\ldots,j_m}$ and $B_{j_1,\ldots,j_m}$ denote the $m \times m$ submatrices of A and B consisting of the $j_1,\ldots,j_m$-th columns of A and $j_1,\ldots,j_m$-th rows of B, respectively. Then

$$\det(AB) = \sum_{1 \leq j_1 < \cdots < j_m \leq n} \det\left(A_{j_1,\ldots,j_m}\right) \det\left(B_{j_1,\ldots,j_m}\right).$$

Proof. We use Theorem 3.A.4 to write

> Recall that $[AB]_{i,j}$ denotes the (i,j)-entry of AB.

$$\det(AB) = \sum_{\sigma \in S_m} \operatorname{sgn}(\sigma) [AB]_{\sigma(1),1} \cdots [AB]_{\sigma(m),m}$$

$$= \sum_{\sigma \in S_m} \operatorname{sgn}(\sigma) \left(\sum_{j=1}^n a_{\sigma(1),j} b_{j,1}\right) \cdots \left(\sum_{j=1}^n a_{\sigma(m),j} b_{j,m}\right)$$

$$= \sum_{1 \leq j_1,\ldots,j_m \leq n} b_{j_1,1} \cdots b_{j_m,m} \left(\sum_{\sigma \in S_m} \operatorname{sgn}(\sigma) a_{\sigma(1),j} \cdots a_{\sigma(m),j_m}\right)$$

$$= \sum_{1 \leq j_1,\ldots,j_m \leq n} b_{j_1,1} \cdots b_{j_m,m} \det\left(A_{j_1,\ldots,j_m}\right).$$

From here, we note that this sum is over *unordered* tuples $(j_1, j_2, \ldots, j_m)$, so we can split it into two nested sums: an outer sum over *ordered* tuples with $j_1 \leq j_2 \leq \cdots \leq j_m$ and an inner sum over permutations of the members of that tuple. That is, the above sum can be written in the form

> The $\operatorname{sgn}(\sigma)$ term after the first equality here comes from pulling the permutation σ out of $\det\left(A_{j_1,\ldots,j_m}\right)$ and thus permuting the columns of $A_{j_1,\ldots,j_m}$.

$$\sum_{1 \leq j_1 \leq \cdots \leq j_m \leq n} \left(\sum_{\sigma \in S_m} b_{j_{\sigma(1)},1} \cdots b_{j_{\sigma(m)},m} \det\left(A_{j_{\sigma(1)},\ldots,j_{\sigma(m)}}\right)\right)$$

$$= \sum_{1 \leq j_1 \leq \cdots \leq j_m \leq n} \left(\sum_{\sigma \in S_m} \operatorname{sgn}(\sigma) b_{j_{\sigma(1)},1} \cdots b_{j_{\sigma(m)},m}\right) \det\left(A_{j_1,\ldots,j_m}\right)$$

$$= \sum_{1 \leq j_1 \leq \cdots \leq j_m \leq n} \det\left(B_{j_1,\ldots,j_m}\right) \det\left(A_{j_1,\ldots,j_m}\right).$$

By recalling that $\det\left(A_{j_1,\ldots,j_m}\right) = 0$ whenever two of the "j" indices are the same (since that causes $A_{j_1,\ldots,j_m}$ to have a repeated column), we see that we can replace the sum over $1 \leq j_1 \leq \cdots \leq j_m \leq n$ above with a sum over $1 \leq j_1 < \cdots < j_m \leq n$, which completes the proof. ∎

We already noted that in the $m = n$ case, the Cauchy–Binet formula simply says that $\det(AB) = \det(A)\det(B)$, since the only tuple $(j_1, j_2, \ldots, j_n)$ with $1 \leq j_1 < j_2 < \cdots < j_n \leq n$ is $(j_1, j_2, \ldots, j_n) = (1, 2, \ldots, n)$. At the other extreme, if $m = 1$ then $A \in \mathcal{M}_{1,n}$ is a row vector and $B \in \mathcal{M}_{n,1}$ is a column vector, and the Cauchy–Binet formula simply says that

$$\det(AB) = a_{1,1}b_{1,1} + a_{1,2}b_{2,1} + \cdots + a_{1,n}b_{n,1},$$

which makes sense since AB in this case is just the dot product of A with B and the determinant of a scalar (1×1 matrix) just equals that scalar. We close this section with an example to illustrate how the intermediate cases of the Cauchy–Binet formula work.

Example 3.A.9

Using the Cauchy–Binet Formula

Use the Cauchy–Binet formula to compute $\det(AB)$ if

$$A = \begin{bmatrix} 1 & 0 & 2 \\ 2 & 1 & -1 \end{bmatrix} \quad \text{and} \quad B = \begin{bmatrix} 3 & 1 \\ -2 & 3 \\ -1 & 2 \end{bmatrix}.$$

Solution:

We first construct all 2×2 submatrices of A and B:

$$A_{1,2} = \begin{bmatrix} 1 & 0 \\ 2 & 1 \end{bmatrix}, \quad A_{1,3} = \begin{bmatrix} 1 & 2 \\ 2 & -1 \end{bmatrix}, \quad A_{2,3} = \begin{bmatrix} 0 & 2 \\ 1 & -1 \end{bmatrix},$$

$$B_{1,2} = \begin{bmatrix} 3 & 1 \\ -2 & 3 \end{bmatrix}, \quad B_{1,3} = \begin{bmatrix} 3 & 1 \\ -1 & 2 \end{bmatrix}, \quad \text{and} \quad B_{2,3} = \begin{bmatrix} -2 & 3 \\ -1 & 2 \end{bmatrix}.$$

The Cauchy–Binet formula is a useful as a theoretical result, but not as a computational tool. It is much quicker and easier to compute $\det(AB)$ directly than it is to compute the determinants of all of these submatrices.

The determinants of these 2×2 submatrices are

$$\det(A_{1,2}) = 1, \qquad \det(A_{1,3}) = -5, \qquad \det(A_{2,3}) = -2,$$
$$\det(B_{1,2}) = 11, \qquad \det(B_{1,3}) = 7, \quad \text{and} \quad \det(B_{2,3}) = -1.$$

The Cauchy–Binet formula then says that

$$\det(AB) = 1 \cdot 11 + (-5) \cdot 7 + (-2) \cdot (-1) = 11 - 35 + 2 = -22.$$

Of course, this fact can be verified directly by multiplying A and B and then computing the determinant of AB directly:

$$\det(AB) = \det\left(\begin{bmatrix} 1 & 5 \\ 5 & 3 \end{bmatrix}\right) = -22.$$

Exercises

solutions to starred exercises on page 472

3.A.1 Compute the cofactor matrix of each of the following matrices.

*(a) $\begin{bmatrix} 1 & 0 \\ 0 & 1 \end{bmatrix}$

(b) $\begin{bmatrix} 6 & 3 \\ 2 & 1 \end{bmatrix}$

*(c) $\begin{bmatrix} 2 & 3 \\ 3 & 2 \end{bmatrix}$

(d) $\begin{bmatrix} 1 & 2 \\ 0 & 1 \end{bmatrix}$

*(e) $\begin{bmatrix} 2 & 4 & 0 \\ 1 & -2 & 0 \\ 2 & 0 & -1 \end{bmatrix}$

(f) $\begin{bmatrix} 2 & 6 & 1 \\ 0 & 0 & 0 \\ 3 & -2 & 7 \end{bmatrix}$

*(g) $\begin{bmatrix} 0 & 0 & 1 \\ 0 & 2 & 3 \\ 4 & 5 & 6 \end{bmatrix}$

(h) $\begin{bmatrix} 1 & -2 & -2 \\ -2 & 1 & -2 \end{bmatrix}$

💻 **3.A.2** Use computer software to compute the cofactor matrix of each of the following matrices.

*(a) $\begin{bmatrix} 3 & 2 & 4 & 3 \\ 1 & 1 & 3 & 2 \\ 4 & 0 & 1 & 1 \\ 3 & 1 & 2 & 4 \end{bmatrix}$ (b) $\begin{bmatrix} 6 & 2 & 1 & 0 & 4 \\ 4 & 2 & 0 & 4 & 6 \\ 0 & 6 & 3 & 4 & 4 \\ 0 & 3 & 6 & 0 & 3 \\ 2 & 6 & 1 & 6 & 6 \end{bmatrix}$

3.A.3 Use Cramer's rule to find the unique solution of each of the following linear systems.

*(a) $\begin{aligned} x + 2y &= 3 \\ 2x + y &= 3 \end{aligned}$ (b) $\begin{aligned} x - y &= 2 \\ 2x - y &= 5 \end{aligned}$

*(c) $\begin{aligned} x + y + z &= 4 \\ x - y + z &= 0 \\ x + y - z &= 1 \end{aligned}$ (d) $\begin{aligned} 2x + y - z &= 1 \\ x - 3y + z &= -2 \\ 2x - 2y - z &= 6 \end{aligned}$

*(e) $\begin{aligned} x + y + z &= 1 \\ -x + y + z &= 2 \\ 2x + y &= 0 \end{aligned}$

(f) $\begin{aligned} 2w - x - 2y - z &= 1 \\ w - 2x - y - 3z &= 2 \\ 4w + y - 2z &= 3 \\ 2x - y + 2z &= 2 \end{aligned}$

3.A.4 Compute the inverse of each of the following permutations.

*(a) $(1\,2\,3)$ (b) $(2\,3\,1)$
*(c) $(1\,3\,2\,4)$ (d) $(2\,4\,1\,3)$
*(e) $(5\,4\,3\,2\,1)$ (f) $(4\,3\,1\,5\,2)$
*(g) $(3\,4\,1\,5\,2\,6)$ (h) $(5\,6\,3\,2\,1\,4)$

3.A.5 Compute the indicated compositions of permutations.

*(a) $(1\,2\,3) \circ (3\,2\,1)$
(b) $(3\,1\,2) \circ (2\,3\,1)$
*(c) $(1\,3\,2\,4) \circ (4\,2\,3\,1)$
(d) $(2\,4\,1\,3) \circ (4\,1\,3\,2)$
*(e) $(3\,4\,5\,1\,2) \circ (2\,4\,3\,1\,5)$
(f) $(2\,3\,1\,5\,4) \circ (1\,2\,5\,3\,4)$
*(g) $(6\,4\,2\,1\,3\,5) \circ (2\,4\,6\,1\,3\,5)$
(h) $(2\,4\,1\,5\,3\,6) \circ (3\,6\,1\,5\,2\,4)$

3.A.6 Determine which of the following statements are true and which are false.

*(a) If A is invertible then so is $\text{cof}(A)$.
(b) If A is not invertible then $\text{cof}(A)$ does not exist.
*(c) $\text{cof}(I) = I$
(d) If $A, B \in \mathcal{M}_n$ are such that $\text{cof}(A) = \text{cof}(B)$ then $A = B$.
*(e) For all $A \in \mathcal{M}_n$, it is the case that $\text{rank}(A) = \text{rank}(\text{cof}(A))$.
(f) Cramer's rule applies to every linear system with a unique solution.

*(g) The inverse of the identity permutation is itself: $\iota^{-1} = \iota$.
(h) Composition of permutations is commutative: $\sigma \circ \tau = \tau \circ \sigma$ for all $\sigma, \tau \in S_n$.
*(i) There are exactly 120 permutations in S_5.

3.A.7 Suppose $A, B \in \mathcal{M}_n$ are invertible. Show that $\text{cof}(AB) = \text{cof}(A)\text{cof}(B)$.

[Side note: This result is also true even if A and B are not invertible, but this more general fact is hard to prove.]

*3.A.8** Suppose $A \in \mathcal{M}_n$ is invertible and k is an integer. Show that $\text{cof}(A^k) = (\text{cof}(A))^k$. [Side note: This result is also true when A is not invertible as long as $k \geq 0$.]

3.A.9 Suppose $A \in \mathcal{M}_n$ and $c \in \mathbb{R}$. Show that $\text{cof}(cA) = c^{n-1}\text{cof}(A)$.

* **3.A.10** Suppose $A \in \mathcal{M}_n$ in invertible. Show that $\det(\text{cof}(A)) = (\det(A))^{n-1}$.

[Side note: Again, this is true even if A is not invertible.]

3.A.11 Suppose $A \in \mathcal{M}_n$ is invertible. Show that $\text{cof}(\text{cof}(A)) = (\det(A))^{n-2}A$.

[Side note: Yes, this one is *also* true even if A is not invertible, and it shows in particular that $\text{cof}(\text{cof}(A)) = O$ whenever A is not invertible and $n \geq 3$.]

3.A.12 This exercise guides you through an alternate proof of Cramer's rule (Theorem 3.A.2). Let $A \in \mathcal{M}_n$, $A\mathbf{x} = \mathbf{b}$, and A_j is the matrix equal to A but with its j-th column replaced by $\mathbf{b}$.

(a) Compute the columns of the matrix $A^{-1}A_j$.
(b) Compute $\det(A^{-1}A_j)$ in two different ways: by using what you learned in part (a) and by using multiplicativity of the determinant.

*3.A.13** Suppose $\sigma, \tau \in S_n$ are permutations.

(a) Show that $\text{sgn}(\sigma \circ \tau) = \text{sgn}(\sigma) \cdot \text{sgn}(\tau)$.
(b) Show that exactly half of the permutations in S_n have sign 1, and the other half have sign -1.

*3.A.14** Suppose $A \in \mathcal{M}_{m,n}$ and $B \in \mathcal{M}_{n,m}$. Explain why the Cauchy–Binet formula (Theorem 3.A.5) only tells us how to compute $\det(AB)$ in the case when $m \leq n$. What is $\det(AB)$ if $m > n$?

3.A.15 Suppose $A \in \mathcal{M}_{m,n}$ with $m \leq n$, and let $A_{j_1, \ldots, j_m}$ denote the $m \times m$ submatrix of A consisting of the $j_1, \ldots, j_m$-th columns of A. Show that

$$\det(AA^T) = \sum_{1 \leq j_1 < \cdots < j_m \leq n} \det\left(A_{j_1, \ldots, j_m}\right)^2.$$

3.A.16 The **permanent** of a matrix $A \in \mathcal{M}_n$, denoted by per(A), is the quantity defined by

$$\text{per}(A) = \sum_{\sigma \in S_n} a_{\sigma(1),1} a_{\sigma(2),2} \cdots a_{\sigma(n),n}.$$

That is, it has the same formula as the one that Theorem 3.A.4 provides for the determinant, except the signs of the permutations are ignored.

Compute the permanent of each of the following matrices.

*(a) $\begin{bmatrix} 1 & 0 \\ 0 & 1 \end{bmatrix}$

(b) $\begin{bmatrix} 1 & 3 \\ 0 & -2 \end{bmatrix}$

*(c) $\begin{bmatrix} 1 & -1 \\ -1 & 1 \end{bmatrix}$

(d) $\begin{bmatrix} 1 & 2 \\ 3 & 5 \end{bmatrix}$

*(e) $\begin{bmatrix} 2 & 0 & 1 \\ 5 & 2 & 8 \\ 3 & -2 & 7 \end{bmatrix}$

(f) $\begin{bmatrix} 1 & 1 & 3 \\ -4 & 2 & 1 \\ 3 & 1 & 2 \end{bmatrix}$

*(g) $\begin{bmatrix} 3 & 2 & -6 \\ 0 & 2 & 2 \\ 0 & 0 & 3 \end{bmatrix}$

(h) $\begin{bmatrix} 1 & 2 & 1 & 0 \\ 2 & 1 & 2 & -3 \\ 4 & 3 & 1 & 2 \\ 0 & 0 & 2 & -2 \end{bmatrix}$

*3.A.17** Show that the permanent (see Exercise 3.A.16) satisfies per(A^T) = per(A) for all $A \in \mathcal{M}_n$.

3.A.18 Show that the permanent (see Exercise 3.A.16) of a triangular matrix equals the product of its diagonal entries: per(A) = $a_{1,1} a_{2,2} \cdots a_{n,n}$.

*3.A.19** Give an example to show that the permanent (see Exercise 3.A.16) does *not* satisfy the property per(AB) = per(A)per(B).

3.B   Extra Topic: Power Iteration

Eigenvalues and eigenvectors help clarify our intuition that linear transformations stretch space in a somewhat "uniform" way, but possibly by different amounts in different directions/dimensions. The different directions are determined by its eigenvectors, and the amounts by which it stretches are the corresponding eigenvalues.

If a linear transformation is applied multiple times in succession, then its features get exaggerated—directions in which it stretches get stretched even more and directions that it squishes get squished even more. In particular this means that, most of the time, most of space gets pushed toward whichever direction gets stretched the most—that is, the direction of the eigenspace corresponding to the maximal eigenvalue (see Figure 3.27).

In this section, we use this idea to come up with another method of computing the eigenvalues and corresponding eigenvectors of a matrix. The advantage of the method that we introduce here is that it does not require us to find a root of the characteristic polynomial (which is quite a nasty task when the matrix, and thus the degree of the characteristic polynomial, is large).

3.B.1   The Method

Since repeatedly applying a matrix to $\mathbb{R}^n$ skews it in the direction of the eigenspace corresponding to its largest eigenvalue, we can get a good idea of what this eigenvalue/eigenvector pair is just by applying A to a randomly-chosen starting vector. That is, if we start with a randomly-chosen vector $\mathbf{v}_0 \in \mathbb{R}^n$, we then define

$$\mathbf{v}_k = \frac{A\mathbf{v}_{k-1}}{\|A\mathbf{v}_{k-1}\|} \quad \text{for all} \quad k \geq 1.$$

In this figure, we zoom out by a factor of 6 every time we apply A in order to prevent the vectors from becoming so large that we cannot see what is going on.

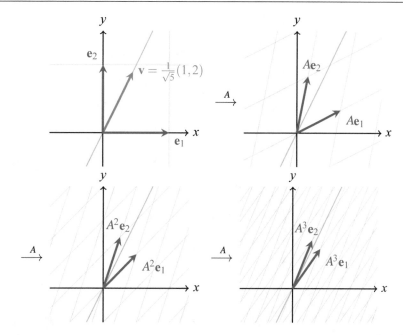

Figure 3.27: Repeatedly applying the matrix (linear transformation) $A = \begin{bmatrix} 4 & 1 \\ 2 & 5 \end{bmatrix}$ to $\mathbb{R}^2$ results in space getting more and more skewed. In particular, space is squished closer and closer to the line in the direction of $\mathbf{v} = (1,2)/\sqrt{5}$, which is a unit eigenvector corresponding to the maximal eigenvalue $\lambda = 6$ of A.

We zoomed out after each application of A in Figure 3.27 for the same reason that we renormalize $\mathbf{v}_k$ after each iteration—it helps us better see the directions in which interesting things happen, while keeping the scale consistent.

The idea here is that multiplying $\mathbf{v}_{k-1}$ by A skews it a bit closer toward the eigenspace corresponding to the maximal eigenvalue, and rescaling it to have length 1 just keeps its scaling consistent (i.e., it prevents these vectors from blowing up if the maximal eigenvalue is larger than 1 or decreasing toward 0 if the maximal eigenvalue is smaller than 1).

We claim that, as k gets large, $\mathbf{v}_k^T A \mathbf{v}_k$ typically becomes a better and better approximation of the maximal (in absolute value) eigenvalue of A, which we call the **dominant eigenvalue** of A. Furthermore, $\mathbf{v}_k$ gets closer and closer to the eigenspace corresponding to that dominant eigenvalue, which we call the **dominant eigenspace** of A. This procedure is called **power iteration**, and before analyzing how well it works (and sometimes does not work) in detail, we present some examples to illustrate how to use it.

Example 3.B.1

Power Iteration for a 2×2 Matrix

Use power iteration to approximate the dominant eigenvalue λ and corresponding eigenvector $\mathbf{v}$ of the matrix $A = \begin{bmatrix} 1 & 2 \\ 5 & 4 \end{bmatrix}$.

Solution:

We start by picking $\mathbf{v}_0$ arbitrarily. If we select $\mathbf{v}_0 = \mathbf{e}_1$ then power iteration tells us that

k	$A\mathbf{v}_{k-1}$	$\|A\mathbf{v}_{k-1}\|$	$\mathbf{v}_k = A\mathbf{v}_{k-1}/\|A\mathbf{v}_{k-1}\|$
0	–	–	$\mathbf{v}_0 = \mathbf{e}_1 = (1,0)$
1	$(1.00, 5.00)$	5.10	$(0.20, 0.98)$
2	$(2.16, 4.90)$	5.36	$(0.40, 0.92)$
3	$(2.23, 5.68)$	6.10	$(0.37, 0.93)$
4	$(2.23, 5.55)$	5.98	$(0.37, 0.93)$
5	$(2.23, 5.57)$	6.00	$(0.37, 0.93)$

We could of course keep going and compute $\mathbf{v}_6$, $\mathbf{v}_7$, $\mathbf{v}_8$, and so on, but very little is changing from one iteration to the next at this point. In particular, if power iteration works the way that we claim it does, this means that the dominant eigenvalue of A should be approximately

We only display two
decimal places of
each of these
numbers, but when
we perform the
calculations we
keep as many digits
of precision as our
computer lets us
(typically 16 or so by
default).

$$\lambda \approx \mathbf{v}_5^T A \mathbf{v}_5 \approx \begin{bmatrix} 0.37 & 0.93 \end{bmatrix} \begin{bmatrix} 2.23 \\ 5.57 \end{bmatrix} \approx 6.00,$$

with corresponding unit eigenvector $\mathbf{v} \approx \mathbf{v}_5 \approx (0.37, 0.93)$.

As a sanity check to verify our work, we can have a look back at Examples 3.3.3 and 3.3.4, where we found that the eigenvalues of A are $\lambda = 6$ and $\lambda = -1$, and a unit eigenvector corresponding to $\lambda = 6$ is $\mathbf{v} = (2,5)/\sqrt{29} \approx (0.37, 0.93)$.

Also, we can visualize the vectors $\mathbf{v}_0$, $\mathbf{v}_1$, $\mathbf{v}_2$, ... that we computed via power iteration as shown below. Note that these vectors bounce back and forth around the dominant eigenspace (which is a line), and $\mathbf{v}_3$ already lies so close to that eigenspace as to be almost indistinguishable from an actual eigenvector.

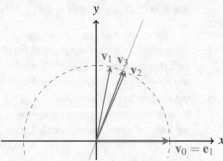

In the previous example, using power iteration perhaps seemed somewhat silly, since we already saw that we can just use the characteristic polynomial and other techniques that we have already developed to construct the eigenvalues and eigenvectors exactly. The real advantage of power iteration becomes apparent when working with larger matrices—especially ones with lots of zero entries (called **sparse** matrices).

Example 3.B.2

Power Iteration for a Large Matrix

Use power iteration to approximate the dominant eigenvalue λ and a corresponding eigenvector $\mathbf{v}$ of the matrix

$$A = \begin{bmatrix} 0 & -1 & 1 & 0 & 1 & 0 & 1 \\ -1 & 0 & 1 & 1 & 1 & -1 & 0 \\ 1 & 1 & 0 & -1 & -2 & 1 & 0 \\ 0 & 1 & -1 & 0 & -1 & 1 & 1 \\ 1 & 1 & -2 & -1 & 0 & 1 & 0 \\ 0 & -1 & 1 & 1 & 1 & 0 & -1 \\ 1 & 0 & 0 & 1 & 0 & -1 & 0 \end{bmatrix}.$$

Solution:

As before, we start by picking $\mathbf{v}_0$ arbitrarily, so we just select $\mathbf{v}_0 = \mathbf{e}_1$ again. Then applying power iteration (on a computer, not by hand!) gives us

k	$\mathbf{v}_k = A\mathbf{v}_{k-1}/\|A\mathbf{v}_{k-1}\|$	$\mathbf{v}_k^T A \mathbf{v}_k$
0	(1.00, 0.00, 0.00, 0.00, 0.00, 0.00, 0.00)	0.00
1	(0.00, −0.50, 0.50, 0.00, 0.50, 0.00, 0.50)	−2.00
2	(0.59, 0.29, −0.44, −0.29, −0.44, 0.29, 0.00)	−4.26
3	(−0.26, −0.45, 0.51, 0.32, 0.51, −0.32, 0.00)	−4.83
4	(0.30, 0.39, −0.49, −0.37, −0.49, 0.37, 0.08)	−4.91
5	(−0.26, −0.41, 0.49, 0.37, 0.49, −0.37, −0.09)	−4.91
6	(0.27, 0.40, −0.49, −0.38, −0.49, 0.38, 0.10)	−4.92
7	(−0.26, −0.41, 0.49, 0.38, 0.49, −0.38, −0.10)	−4.92

Because the maximal eigenvalue is negative, the vectors $\mathbf{v}_k$ do not approach any fixed eigenvector (though they do still approach the eigenspace). Rather, their sign changes after each iteration.

We thus conclude that the maximal eigenvalue of A (in absolute value) is approximately -4.92, and there is a corresponding unit eigenvector close to $\mathbf{v}_7 \approx (-0.26, -0.41, 0.49, 0.38, 0.49, -0.38, -0.10)$.

In this case, we do not have an exact closed-form expression of the maximal eigenvalue of A that we can compare our decimal approximation -4.92 to. However, we can at least note that the characteristic polynomial of A is

$$p_A(\lambda) = -\lambda^7 + 19\lambda^5 - 36\lambda^4 - 24\lambda^3 + 116\lambda^2 - 102\lambda + 28.$$

Factoring this polynomial is out of the question, but numerical software can be used to verify that its largest (in absolute value) root is indeed located near $\lambda = -4.92$.

The reason that power iteration is so useful for sparse matrices in particular is that multiplying a vector by a sparse matrix can be done much more efficiently than general matrix multiplication can. In particular, if each row of $A \in \mathcal{M}_n$ has s non-zero entries, then each entry of $A\mathbf{v}$ can be computed via just s multiplications and $s-1$ additions, rather than n multiplications and $n-1$ additions. Because of this, power iteration can be used to estimate eigenvalues and eigenvectors of absolutely humongous matrices, as long as they do not have too many non-zero entries.

Remark 3.B.1

Eigenvalues are Used to Find Roots of Polynomials, Not Vice-Versa

Our primary method for finding the eigenvalues of a matrix has been to construct its characteristic polynomial and then find the roots of that polynomial. While this method works fine for small (e.g., 2×2, 3×3, and maybe even 4×4) matrices, it is not really how things are done in practice. In fact, roots of high-degree polynomials are typically computed by computing the eigenvalues of a matrix, not vice-versa.

Specifically, to find the roots of $p(x) = \lambda^n + a_{n-1}\lambda^{n-1} + \cdots + a_1\lambda + a_0$, we first construct its **companion matrix** (introduced in Exercise 3.3.23)

$$C = \begin{bmatrix} 0 & 1 & 0 & \cdots & 0 \\ 0 & 0 & 1 & \cdots & 0 \\ \vdots & \vdots & \vdots & \ddots & \vdots \\ 0 & 0 & 0 & \cdots & 1 \\ -a_0 & -a_1 & -a_2 & \cdots & -a_{n-1} \end{bmatrix},$$

> Actually, the characteristic polynomial of C is $p_C(\lambda) = (-1)^n p(\lambda)$, but this has the same roots as p itself, so we do not care much about the $(-1)^n$ coefficient in front.

which has p as its characteristic polynomial. Then we apply some numerical method like power iteration to C to find one or more of its eigenvalues, which are the roots of p. In fact, power iteration is particularly well-suited to this task since C is so sparse (it only has $2n-1$ non-zero entries, despite being an $n \times n$ matrix).

For example, to find the largest root (in absolute value) of the polynomial $p(x) = x^6 - 2x^5 + 2x^4 - 3x^3 + x^2 + x - 2$, we first construct its companion matrix

> In fact, computing $C\mathbf{v}_k$ only requires n multiplications (all in the last row)— the first $n-1$ entries can be computed simply by shifting up the bottom $n-1$ entries of $\mathbf{v}_k$.

$$C = \begin{bmatrix} 0 & 1 & 0 & 0 & 0 & 0 \\ 0 & 0 & 1 & 0 & 0 & 0 \\ 0 & 0 & 0 & 1 & 0 & 0 \\ 0 & 0 & 0 & 0 & 1 & 0 \\ 0 & 0 & 0 & 0 & 0 & 1 \\ 2 & -1 & -1 & 3 & -2 & 2 \end{bmatrix},$$

and then we apply power iteration to it:

k	$\mathbf{v}_k = C\mathbf{v}_{k-1}/\|C\mathbf{v}_{k-1}\|$	$\mathbf{v}_k^T C\mathbf{v}_k$
0	$(1.00, 0.00, 0.00, 0.00, 0.00, 0.00)$	0.00
1	$(0.00, 0.00, 0.00, 0.00, 0.00, 1.00)$	2.00
2	$(0.00, 0.00, 0.00, 0.00, 0.45, 0.89)$	1.20
3	$(0.00, 0.00, 0.00, 0.33, 0.67, 0.67)$	1.33
$\vdots$		
26	$(0.06, 0.10, 0.17, 0.28, 0.48, 0.81)$	1.69
27	$(0.06, 0.10, 0.17, 0.28, 0.48, 0.81)$	1.68

We thus conclude that the largest root of p is approximately 1.68 (and indeed, it is–the roots of p are approximately 1.68, -0.72, $-0.17 \pm 1.32i$, and $0.69 \pm 0.67i$).

3.B.2 When it Does (and Does Not) Work

Now that we have seen how to perform power iteration, it is time to clarify exactly what conditions need to be satisfied in order for it to work. There are essentially two ways in which power iteration can fail, and we start by illustrating them with examples.

The first way in which power iteration can fail is if the starting vector $\mathbf{v}_0$ is contained in the span of the non-maximal eigenvectors, since then it must necessarily stay in that span no matter how many times we apply A to it, so it cannot possibly be stretched in the direction of the maximal eigenspace. Fortunately, if we choose $\mathbf{v}_0$ randomly then this only happens extremely rarely.

Example 3.B.3

Failed Power Iteration due to Poor Choice of Starting Vector

Apply power iteration to the matrix $A = \begin{bmatrix} 1 & 2 \\ 0 & 3 \end{bmatrix}$.

Solution:
   If we start by choosing $\mathbf{v}_0 = \mathbf{e}_1$ like usual and then iteratively setting $\mathbf{v}_k = A\mathbf{v}_{k-1}/\|A\mathbf{v}_{k-1}\|$ for $k = 1, 2, 3, \ldots$, then absolutely nothing interesting happens:

k	$\mathbf{v}_k = A\mathbf{v}_{k-1}/\|A\mathbf{v}_{k-1}\|$	$\mathbf{v}_k^T A \mathbf{v}_k$
0	$(1.00, 0.00)$	1.00
1	$(1.00, 0.00)$	1.00
2	$(1.00, 0.00)$	1.00

Since A is triangular, we can read its eigenvalues off of its diagonal: 1 and 3.

The reason for this behavior is that $\mathbf{v}_0$ is an eigenvector of A (corresponding to the non-maximal eigenvalue 1). If we instead choose $\mathbf{v}_0$ randomly to be $\mathbf{v}_0 = (0.80, 0.60)$ then power iteration does indeed converge to the maximum eigenvalue 3:

Also, $\mathbf{v}_k$ converges to the unit eigenvector $(1,1)/\sqrt{2}$ corresponding to the eigenvalue 3.

k	$\mathbf{v}_k = A\mathbf{v}_{k-1}/\|A\mathbf{v}_{k-1}\|$	$\mathbf{v}_k^T A \mathbf{v}_k$
0	$(0.80, 0.60)$	2.68
1	$(0.74, 0.67)$	2.89
2	$(0.72, 0.69)$	2.96
3	$(0.71, 0.70)$	2.99
4	$(0.71, 0.71)$	3.00

   The other way in which power iteration can fail is if the matrix has multiple distinct eigenvalues with the same maximal absolute value. In this case, power iteration pulls the $\mathbf{v}_k$ vectors toward *each* of the maximal eigenspaces, so they often end up just bouncing around aimlessly rather than being stretched in one particular direction. Unfortunately, there is not much that we can do to fix this problem, as it is intrinsic to the matrix itself rather than the starting vector $\mathbf{v}_0$.

Example 3.B.4

Failed Power Iteration due to No Maximal Eigenvalue

Apply power iteration to the matrix $A = \begin{bmatrix} 1 & 1 \\ 3 & -1 \end{bmatrix}$.

Solution:
   We start by choosing $\mathbf{v}_0 = \mathbf{e}_1$, just like usual, and then we iteratively set $\mathbf{v}_k = A\mathbf{v}_{k-1}/\|A\mathbf{v}_{k-1}\|$ for $k = 1, 2, 3, \ldots$:

The exact values of the vectors in this table are $(1,0)$ and $(1,3)/\sqrt{10}$. Similarly, the values of $\mathbf{v}_k^T A \mathbf{v}_k$ are *exactly* 1 and 0.4.

k	$\mathbf{v}_k = A\mathbf{v}_{k-1}/\|A\mathbf{v}_{k-1}\|$	$\mathbf{v}_k^T A\mathbf{v}_k$
0	$(1.00, 0.00)$	1.00
1	$(0.32, 0.95)$	0.40
2	$(1.00, 0.00)$	1.00
3	$(0.32, 0.95)$	0.40

It thus appears that $\mathbf{v}_k$ just alternates back and forth between $(1.00, 0.00)$ and $(0.32, 0.95)$, and $\mathbf{v}_k^T A\mathbf{v}_k$ similarly alternates back and forth between 1.00 and 0.40 (and in particular, does not approach a single value).

It is straightforward to check that the eigenvalues of A are actually 2 and -2, which the power iteration above did not even come close to finding. We might at first guess that the problem is a result of a poorly-chosen starting vector $\mathbf{v}_0$ (as was the case in Example 3.B.3), but something very similar happens if we start with a different vector like $\mathbf{v}_0 = (0.80, 0.60)$:

k	$\mathbf{v}_k = A\mathbf{v}_{k-1}/\|A\mathbf{v}_{k-1}\|$	$\mathbf{v}_k^T A\mathbf{v}_k$
0	$(0.80, 0.60)$	2.20
1	$(0.61, 0.79)$	1.69
2	$(0.80, 0.60)$	2.20
3	$(0.61, 0.79)$	1.69

The problem in this case is that the maximal eigenvalues 2 and -2 have the same absolute value, so vectors are pulled back and forth between these two different eigenspaces with equal "force" and neither one ends up "winning" as A is applied more times. In particular, $A^2 = 4I$, so applying A twice (and rescaling appropriately) simply moves each vector $A\mathbf{v}$ back to $\mathbf{v}$, where it started, much like a reflection:

The lines $y = x$ and $y = -3x$ are the eigenspaces corresponding to the eigenvalues 2 and -2, respectively.

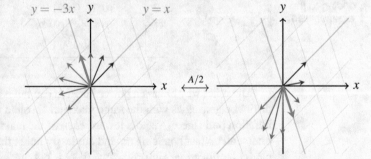

It is also worth noting that power iteration can never help us find complex (non-real) eigenvalues or eigenvectors of real matrices, since if we start with a real vector $\mathbf{v}_0$ then $\mathbf{v}_1 = A\mathbf{v}_0/\|A\mathbf{v}_0\|$ will also be real, as will $\mathbf{v}_2$, $\mathbf{v}_3$, and so on. This is actually a special case of the problem described above—recall that complex eigenvalues of real matrices come in complex conjugate pairs, and complex conjugates have the same absolute value of each other: $|\lambda| = |\overline{\lambda}|$ for all $\lambda \in \mathbb{C}$. For example, the matrix

$$\begin{bmatrix} 1 & 1 \\ -1 & 1 \end{bmatrix}$$

has eigenvalues $1 \pm i$, which have the same absolute value as each other:

However, we might get lucky and have power iteration converge if we start with a random $\mathbf{v}_0 \in \mathbb{C}^n$ instead.

$|1+i| = |1-i| = \sqrt{2}$. Power iteration thus does not necessarily converge to these maximal eigenvalues when applied to this matrix (and of course cannot if we start with a real vector $\mathbf{v}_0$).

Fortunately, these are the *only* cases in which power iteration fails. That is, as long as there is only one eigenvalue of A with maximal absolute value and we are not unlucky in our choice of $\mathbf{v}_0$, power iteration converges to that maximal eigenvalue as we expect:

Theorem 3.B.1 **Power Iteration**	Suppose $A \in \mathcal{M}_n$ has a linearly independent set of eigenvectors $\{\mathbf{w}_1, \mathbf{w}_2, \ldots, \mathbf{w}_n\}$ with corresponding eigenvalues $\lambda_1, \lambda_2, \ldots, \lambda_n$, respectively. Also let $\mathbf{v}_0 \in \mathbb{R}^n$ be a vector and suppose that the following two conditions hold: • $\|\lambda_1\| > \|\lambda_j\|$ for all $2 \leq j \leq n$, and • $\mathbf{v}_0 \notin \text{span}\{\mathbf{w}_2, \mathbf{w}_3, \ldots, \mathbf{w}_n\}$. If we define $\mathbf{v}_k = A\mathbf{v}_{k-1}/\|A\mathbf{v}_{k-1}\|$ for all $k \geq 1$ then $\lim_{k \to \infty} \mathbf{v}_k^* A \mathbf{v}_k = \lambda_1$.

In other words, power iteration always converges to the maximal eigenvalue of A, except possibly in the two problematic cases we discussed earlier.

Proof. We begin by noting that $\mathbf{v}_k$ is obtained via multiplying $\mathbf{v}_0$ by A a total of k times and normalizing after each multiplication. Since the normalization can instead be deferred to the end of the calculation, we can write the formula for $\mathbf{v}_k$ a bit more explicitly as $\mathbf{v}_k = A^k\mathbf{v}_0/\|A^k\mathbf{v}_0\|$. With this in mind, we begin by investigating what happens to $A^k\mathbf{v}_0$ as k gets large.

Without loss of generality, we can assume that each eigenvector $\mathbf{w}_j$ is a unit vector, since rescaling vectors does not affect linear independence (see Exercise 2.3.28). Furthermore, since $\{\mathbf{w}_1, \mathbf{w}_2, \ldots, \mathbf{w}_n\}$ is linearly independent, it is in fact a basis of $\mathbb{R}^n$, so we can write $\mathbf{v}_0$ as a linear combination of these basis vectors:

$$\mathbf{v}_0 = c_1\mathbf{w}_1 + c_2\mathbf{w}_2 + \cdots + c_n\mathbf{w}_n.$$

As stated, this theorem only applies to diagonalizable matrices, as those are the matrices with a linearly independent set of n eigenvectors (by Theorem 3.4.3). It does apply to non-diagonalizable matrices too, but the proof of that fact is beyond the scope of this book.

Then repeatedly using the fact that $A\mathbf{w}_j = \lambda_j\mathbf{w}_j$ for each $1 \leq j \leq n$ gives

$$\begin{aligned} A^k\mathbf{v}_0 &= A^k(c_1\mathbf{w}_1 + c_2\mathbf{w}_2 + \cdots + c_n\mathbf{w}_n) \\ &= c_1\lambda_1^k\mathbf{w}_1 + c_2\lambda_2^k\mathbf{w}_2 + \cdots + c_n\lambda_n^k\mathbf{w}_n \\ &= \lambda_1^k\left(c_1\mathbf{w}_1 + c_2\left(\frac{\lambda_2}{\lambda_1}\right)^k\mathbf{w}_2 + \cdots + c_n\left(\frac{\lambda_n}{\lambda_1}\right)^k\mathbf{w}_n\right). \end{aligned}$$

We then notice that, since $|\lambda_1| > |\lambda_j|$ for all $2 \leq j \leq n$, we have $|\lambda_j/\lambda_1| < 1$ and so $(\lambda_j/\lambda_1)^k \to 0$ as $k \to \infty$ for all $2 \leq j \leq n$ as well. For simplicity, we then define $\mathbf{r}_k$ to be the "remainder" vector that consists of terms that become small as k gets large:

$$\mathbf{r}_k = c_2\left(\frac{\lambda_2}{\lambda_1}\right)^k\mathbf{w}_2 + \cdots + c_n\left(\frac{\lambda_n}{\lambda_1}\right)^k\mathbf{w}_n.$$

Putting all of this together shows that

$$\mathbf{v}_k = \frac{A^k\mathbf{v}_0}{\|A^k\mathbf{v}_0\|} = \frac{\lambda_1^k(c_1\mathbf{w}_1 + \mathbf{r}_k)}{\|\lambda_1^k(c_1\mathbf{w}_1 + \mathbf{r}_k)\|} \quad \text{for all} \quad k \geq 0.$$

In particular, by using the fact that $\lim_{k\to\infty} \mathbf{r}_k = \mathbf{0}$ we can now compute the desired limit:

The $|\lambda_1|^{2k}$ term here comes from combining $\overline{\lambda_1^k}$ from the $\mathbf{v}_k^*$ on the left with λ_1^k from the $\mathbf{v}_k$ on the right.

$$\lim_{k\to\infty} \mathbf{v}_k^* A \mathbf{v}_k = \lim_{k\to\infty} \frac{|\lambda_1|^{2k}\left(c_1\mathbf{w}_1 + \mathbf{r}_k\right)^* A\left(c_1\mathbf{w}_1 + \mathbf{r}_k\right)}{\left\|\lambda_1^k\left(c_1\mathbf{w}_1 + \mathbf{r}_k\right)\right\|^2}$$

$$= \frac{\left(c_1\mathbf{w}_1 + \lim_{k\to\infty}\mathbf{r}_k\right)^* A\left(c_1\mathbf{w}_1 + \lim_{k\to\infty}\mathbf{r}_k\right)}{\left\|c_1\mathbf{w}_1 + \lim_{k\to\infty}\mathbf{r}_k\right\|^2}$$

$$= \frac{|c_1|^2 \mathbf{w}_1^* A \mathbf{w}_1}{\left\|c_1\mathbf{w}_1\right\|^2} = \mathbf{w}_1^* A \mathbf{w}_1 = \mathbf{w}_1^*(\lambda_1\mathbf{w}_1) = \lambda_1,$$

where at the end we use the facts that $\mathbf{w}_1$ is a unit vector (twice) and that it is an eigenvector of A with corresponding eigenvalue λ_1.   ∎

Theorem 3.B.1 works fine if the largest eigenvalue is repeated (i.e., has algebraic multiplicity larger than 1). There is only a problem if there are two *different* maximal eigenvalues with the same absolute value.

It is worth noting that the above proof shows that

$$\mathbf{v}_k = \frac{\lambda_1^k\left(c_1\mathbf{w}_1 + \mathbf{r}_k\right)}{\left\|\lambda_1^k\left(c_1\mathbf{w}_1 + \mathbf{r}_k\right)\right\|} = \left(\frac{\lambda_1}{|\lambda_1|}\right)^k \frac{c_1\mathbf{w}_1 + \mathbf{r}_k}{\left\|c_1\mathbf{w}_1 + \mathbf{r}_k\right\|} \approx \frac{c_1}{|c_1|}\left(\frac{\lambda_1}{|\lambda_1|}\right)^k \mathbf{w}_1$$

when k is large. It follows that $\mathbf{v}_k$ approaches the maximal eigen*space* of A, but it might not approach any of its particular fixed maximal eigen*vectors*, since the scalar $(\lambda_1/|\lambda_1|)^k$ can change sign (or in the case of complex matrices, bounce around the unit circle in the complex plane) as k varies. We already saw an example of this behavior in Example 3.B.2, where $\mathbf{v}_k$ was multiplied by roughly -1 from one iteration to the next (as a result of the maximal eigenvalue λ_1 being negative, and thus $\lambda_1/|\lambda_1| = -1$).

It is also worth noting that, even if power iteration converges to the maximal eigenvalue of a matrix, it might converge quite slowly. The proof of Theorem 3.B.1 above shows that the speed of convergence is determined by the ratio $|\lambda_2/\lambda_1|$ of the two largest (in absolute value) eigenvalues of A—the closer this ratio is to 1, the slower convergence is, and the closer it is to 0, the quicker convergence is.

Example 3.B.5

Power Iteration Might Converge Very Slowly

Apply power iteration to the matrix $A = \begin{bmatrix} 8 & -1 & 0 \\ -1 & 2 & 5 \\ 0 & 5 & -6 \end{bmatrix}$.

Solution:

If we apply power iteration starting from $\mathbf{v}_0 = \mathbf{e}_1$ as usual, we notice that things do not change very quickly:

k	$\mathbf{v}_k = A\mathbf{v}_{k-1}/\|A\mathbf{v}_{k-1}\|$	$\mathbf{v}_k^T A \mathbf{v}_k$
0	$(1.00,\ 0.00,\ 0.00)$	8.00
1	$(0.99, -0.12,\ 0.00)$	8.15
2	$(0.99, -0.15, -0.07)$	8.19
3	$(0.98, -0.20, -0.04)$	8.21
4	$(0.98, -0.19, -0.10)$	8.21
5	$(0.97, -0.22, -0.05)$	8.21

Based on the above calculations, it looks like the maximal eigenvalue of A

is approximately 8.21, but running power iteration for longer reveals this to be very, very wrong. After $k = 63$ iterations, the eigenvalue estimate dips back down below 8.00, after $k = 156$ iterations it dips down below 0.00, and it is finally correct to two decimal places after we reach iteration $k = 320$:

The vectors $\mathbf{v}_{318}$ and $\mathbf{v}_{320}$ are actually slightly different at later decimal places not displayed here (and similarly for $\mathbf{v}_{319}$ and $\mathbf{v}_{321}$).

k	$\mathbf{v}_k = A\mathbf{v}_{k-1}/\|A\mathbf{v}_{k-1}\|$	$\mathbf{v}_k^T A\mathbf{v}_k$
318	$(0.05, \quad 0.43, -0.90)$	-8.40
319	$(0.00, -0.44, \quad 0.90)$	-8.40
320	$(0.05, \quad 0.43, -0.90)$	-8.41
321	$(0.00, -0.44, \quad 0.90)$	-8.41

The other eigenvalue of A is $\lambda_3 \approx 4.19$.

The reason for this strange behavior is that the two largest eigenvalues $\lambda_1 \approx -8.41$ and $\lambda_2 \approx 8.22$ of A are so close to each other in absolute value (so the ratio $|\lambda_2/\lambda_1| \approx 0.98$ is close to 1). Furthermore, the starting vector $\mathbf{v}_0 = \mathbf{e}_1$ is very close to the eigenspace corresponding to λ_2 (which is $\mathrm{span}\{(0.97, -0.22, -0.08)\}$), so it starts off by getting pulled toward that eigenspace before λ_1 dominates and pulls it back toward its own. A plot of the eigenvalue estimate $\mathbf{v}_k^* A\mathbf{v}_k$, as a function of the iteration k, is provided below:

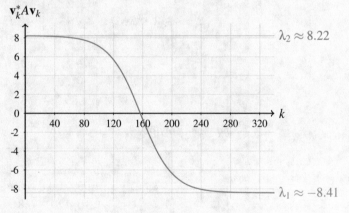

Finding Non-Maximal Eigenvalues

Power iteration can actually be adapted to find *all* eigenvalues of a matrix, not just the dominant one. To see how this works, recall from Exercise 3.3.22 that if $A \in \mathcal{M}_n$ has eigenvalues $\lambda_1, \lambda_2, \ldots, \lambda_n$ and $\mathbf{w}_1$ is a unit eigenvector corresponding to λ_1, then the matrix

$$A - \lambda_1 \mathbf{w}_1 \mathbf{w}_1^T$$

However, the eigen*vectors* of $A - \lambda_1 \mathbf{w}_1 \mathbf{w}_1^T$ may be different than those of A. See Exercise 3.3.22.

has eigenvalues $0, \lambda_2, \lambda_3, \ldots, \lambda_n$. In particular, if λ_1 is the dominant eigenvalue of A then the dominant eigenvalue of $A - \lambda_1 \mathbf{w}_1 \mathbf{w}_1^T$ is the second-largest eigenvalue (in absolute value) of A, which can be found via power iteration. Then we can just repeat this process to find the third-largest eigenvalue (in absolute value) and so on.

Example 3.B.6

Finding All Eigenvalues via Power Iteration

Use power iteration to find all eigenvalues of $A = \begin{bmatrix} 1 & 2 & 3 \\ 4 & 5 & 4 \\ 3 & 2 & 1 \end{bmatrix}$.

Solution:

We start by applying power iteration to A to find its dominant eigenvalue. As usual, we just start the iteration with $\mathbf{v}_0 = \mathbf{e}_1$ for convenience:

k	$\mathbf{v}_k = A\mathbf{v}_{k-1}/\|A\mathbf{v}_{k-1}\|$	$\mathbf{v}_k^T A \mathbf{v}_k$
0	$(1.00, 0.00, 0.00)$	1.00
1	$(0.20, 0.78, 0.59)$	7.85
2	$(0.42, 0.84, 0.33)$	8.49
3	$(0.36, 0.85, 0.39)$	8.53
4	$(0.38, 0.85, 0.37)$	8.53

Indeed, the dominant eigenvalue of A is $\lambda_1 \approx 8.53$ with corresponding unit eigenvector $\mathbf{w}_1 \approx (0.38, 0.85, 0.37)$.

To find the second-largest eigenvalue of A, we set

$$B = A - \lambda_1 \mathbf{w}_1 \mathbf{w}_1^T \approx \begin{bmatrix} 1 & 2 & 3 \\ 4 & 5 & 4 \\ 3 & 2 & 1 \end{bmatrix} - 8.53 \begin{bmatrix} 0.14 & 0.32 & 0.14 \\ 0.32 & 0.72 & 0.32 \\ 0.14 & 0.32 & 0.14 \end{bmatrix}$$

$$= \begin{bmatrix} -0.21 & -0.73 & 1.80 \\ 1.27 & -1.14 & 1.31 \\ 1.80 & -0.69 & -0.18 \end{bmatrix}$$

and then apply power iteration to B:

k	$\mathbf{v}_k = B\mathbf{v}_{k-1}/\|B\mathbf{v}_{k-1}\|$	$\mathbf{v}_k^T B \mathbf{v}_k$
0	$(\ \ 1.00, 0.00, \ \ 0.00)$	-0.21
1	$(-0.10, 0.57, \ \ 0.81)$	-0.52
2	$(\ \ 0.81, 0.22, -0.54)$	-1.81
3	$(-0.68, 0.04, \ \ 0.73)$	-1.99
4	$(\ \ 0.72, 0.03, -0.69)$	-2.00

Keep in mind that $\mathbf{w}_2$ is an eigenvector of B, but not necessarily of A.

The dominant eigenvalue of B (and the second-to-largest eigenvalue of A in absolute value) is thus $\lambda_2 \approx -2.00$, with corresponding unit eigenvector $\mathbf{w}_2 \approx (0.72, 0.03, -0.69)$.

To finally find the smallest eigenvalue of A, we set

$$C = B - \lambda_2 \mathbf{w}_2 \mathbf{w}_2^T \approx \begin{bmatrix} 0.82 & -0.69 & 0.81 \\ 1.31 & -1.14 & 1.27 \\ 0.81 & -0.73 & 0.78 \end{bmatrix}$$

and then apply power iteration to C:

k	$\mathbf{v}_k = C\mathbf{v}_{k-1}/\|C\mathbf{v}_{k-1}\|$	$\mathbf{v}_k^T C\mathbf{v}_k$
0	$(1.00, 0.00, 0.00)$	0.82
1	$(0.47, 0.75, 0.46)$	0.47
2	$(0.51, 0.75, 0.42)$	0.47
3	$(0.51, 0.75, 0.42)$	0.47

The dominant eigenvalue of C (and the smallest eigenvalue of A in absolute value) is thus $\lambda_3 \approx 0.47$.

We have thus found that the eigenvalues of A are approximately 8.53, -2.00, and 0.47. Since this matrix is fairly small, we can verify our answers explicitly by computing the characteristic polynomial of A to be

$$p_A(\lambda) = -\lambda^3 + 7\lambda^2 + 14\lambda - 8,$$

which indeed has roots -2, $(9 + \sqrt{65})/2 \approx 8.53$, and $(9 - \sqrt{65})/2 \approx 0.47$.

Keep in mind that, when finding all eigenvalues of a matrix in this way, the conditions of Theorem 3.B.1 must be satisfied each and every time we apply power iteration. For example, the matrix

$$A = \begin{bmatrix} 1 & 2 & 2 & 2 \\ 3 & -1 & 2 & -1 \\ 2 & 3 & 2 & 2 \\ 3 & 3 & 0 & -1 \end{bmatrix}$$

has eigenvalues equal to approximately 6.25, $-2.55 \pm 1.88i$, and -0.16. However, if we try to use this method based on power iteration to find these eigenvalues, we will only be able to find the dominant eigenvalue 6.25. The problem is that the next two eigenvalues $-2.55 \pm 1.88i$ are equal to each other in absolute value, so power iteration gets stuck and is unable to find them (in fact, power iteration can never find *any* complex eigenvalue of a real matrix since it only involves multiplication, addition, and division of real numbers).

3.B.3 Positive Matrices and Ranking Algorithms

Since Theorem 3.B.1 tells us that power iteration only applies to matrices that have an eigenvalue that is strictly larger than the rest, it is important to find families of matrices that have this property. With this in mind, we say that a matrix is **positive** if all of its entries are positive (i.e., strictly bigger than 0). For example, here are some matrices that are and are not positive:

The rightmost matrix here is **non-negative** though, since all of its entries are bigger than *or equal to* 0.

Positive

$$\begin{bmatrix} 3 & 1 \\ 2 & 7 \end{bmatrix} \quad \begin{bmatrix} 2 & 4 & 1 \\ 8 & 1 & 3 \end{bmatrix}$$

Not positive

$$\begin{bmatrix} 1 & -1 \\ -2 & 3 \end{bmatrix} \quad \begin{bmatrix} 4 & 3 & 2 \\ 1 & 0 & 1 \end{bmatrix}.$$

Our main result about positive matrices is that they have exactly the property that we want—they have one eigenvalue (which happens to always be positive and real) that is strictly larger than the absolute value of the rest. The following theorem establishes this fact, as well as some simple properties of the eigenvectors to which that largest eigenvalue corresponds.

Theorem 3.B.2

Perron–Frobenius

Suppose $A \in \mathcal{M}_n$ is positive with eigenvalues $\lambda_1, \lambda_2, \ldots, \lambda_n$ (listed according to algebraic multiplicity). Then:

a) One of the eigenvalues, say λ_1, is positive and larger than the rest: $\lambda_1 > |\lambda_j|$ for all $2 \leq j \leq n$.

Properties (b)–(d) tell us that $\mathbf{v}_1$ is (up to scaling) the unique eigenvector of A (corresponding to *any* eigenvalue) with all entries positive.

b) λ_1 has algebraic multiplicity 1.

c) There is an eigenvector $\mathbf{v}_1$ corresponding to the eigenvalue λ_1 with all of its entries positive.

d) Every eigenvector corresponding to an eigenvalue λ_j ($2 \leq j \leq n$) has at least one negative or non-real entry.

Proof. Before being able to prove any of the statements of the theorem, we need to establish some notation and machinery that we will use throughout the proof. For brevity, we write $\mathbf{v} \geq \mathbf{w}$ and $\mathbf{v} > \mathbf{w}$ to indicate that $v_j \geq w_j$ or $v_j > w_j$ for all j, respectively (just like we did in Section 2.B). In particular, this means that non-negativity of $\mathbf{v}$ can be denoted by $\mathbf{v} \geq \mathbf{0}$, and positivity of $\mathbf{v}$ can be denoted by $\mathbf{v} > \mathbf{0}$. We also let $S^n \subset \mathbb{R}^n$ be the set of unit vectors with positive entries, which we can think of geometrically as the intersection of the unit (hyper)sphere with the non-negative orthant (see Figure 3.28).

Keep in mind that, whenever we say "positive" in this section, we mean *strictly* positive—no zeros allowed.

This theorem actually applies to some matrices that are not positive as well—see Exercise 3.B.8.

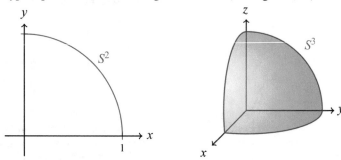

Figure 3.28: The set S^n is the portion of the unit circle/sphere/hypersphere that is contained in the non-negative quadrant/orthant.

Note that if $\mathbf{v} \in S^n$ then $A\mathbf{v} > \mathbf{0}$, since it is computed by multiplying together and adding up non-negative numbers, and at least one entry of $\mathbf{v}$ is strictly positive. We can thus define a function $L : S^n \to \mathbb{R}$ by

$$L(\mathbf{v}) = \max\{c \in \mathbb{R} : A\mathbf{v} \geq c\mathbf{v}\}. \tag{3.B.1}$$

In particular, there exists a $c > 0$ such that $A\mathbf{v} \geq c\mathbf{v}$ since $A\mathbf{v} > \mathbf{0}$ (so $L(\mathbf{v}) > 0$ for all $\mathbf{v} \in S^n$). Furthermore, once we have found *some* c that works, the maximal such c (i.e., the value of $L(\mathbf{v})$) can be found by increasing c until one of the entries of $c\mathbf{v}$ equals the corresponding entry of $A\mathbf{v}$. That is, a somewhat more explicit formula for L is given by

The notation $[A\mathbf{v}]_j$ means the j-th entry of the vector $A\mathbf{v}$.

$$L(\mathbf{v}) = \min_{1 \leq j \leq n}\left\{\frac{[A\mathbf{v}]_j}{v_j} : v_j \neq 0\right\}$$

In words, $L(\mathbf{v})$ tells us the smallest amount by which any entry of $\mathbf{v}$ is stretched by A.

Notice that L is a continuous function—slightly changing $\mathbf{v}$ can only slightly change $L(\mathbf{v})$—and the set S^n is both closed (i.e., it contains its boundary/edges) and bounded (i.e., it does not contain vectors of arbitrarily large length). It follows from the Extreme Value Theorem (a theorem from analysis that we do not explore here) that L attains a maximum value somewhere on S^n. That is, there is a particular number $\mu \in \mathbb{R}$ and a particular vector $\mathbf{v}_* \in S^n$ with the property that $L(\mathbf{v}_*) = \mu \geq L(\mathbf{v})$ for all $\mathbf{v} \in S^n$.

Before proceeding, we must prove four claims about the relationships between A, $\mathbf{v}_*$, and μ that we will use repeatedly.

Claim 1: If $\mathbf{v} \in S^n$ and $L(\mathbf{v}) = \mu$ then $A\mathbf{v} = \mu\mathbf{v}$. To see why this claim holds, notice that $L(\mathbf{v}) = \mu$ implies $A\mathbf{v} \geq \mu\mathbf{v}$ and thus $A\mathbf{v} - \mu\mathbf{v} \geq \mathbf{0}$. Suppose (for the purpose of establishing a contradiction) that $A\mathbf{v} - \mu\mathbf{v} \neq \mathbf{0}$. Since A is positive, this implies that $A(A\mathbf{v} - \mu\mathbf{v}) > \mathbf{0}$, and re-arranging then gives $A(A\mathbf{v}) > \mu A\mathbf{v}$. It follows that

$$\frac{A(A\mathbf{v})}{\|A\mathbf{v}\|} > \frac{\mu A\mathbf{v}}{\|A\mathbf{v}\|}, \quad \text{so} \quad L\left(\frac{A\mathbf{v}}{\|A\mathbf{v}\|}\right) > \mu.$$

However, this contradicts the fact that μ is the maximal value of L, so $\mu \geq L(A\mathbf{v}/\|A\mathbf{v}\|)$. It follows that our assumption that $A\mathbf{v} - \mu\mathbf{v} \neq \mathbf{0}$ must be mistaken, so $A\mathbf{v} = \mu\mathbf{v}$, as desired.

The above claim shows that μ is in fact an eigenvalue of A with corresponding eigenvector $\mathbf{v}^*$. The remaining claims will help us understand this particular eigenvalue/eigenvector pair a bit better.

Claim 2: If $\mathbf{v} \geq \mathbf{0}$ then $A\mathbf{v} > \mathbf{0}$. This follows simply from the fact that A is positive, so for each $1 \leq i \leq n$, the sum $[A\mathbf{v}]_i = a_{i,1}v_2 + a_{i,2}v_2 + \cdots + a_{i,n}v_n$ has at least one strictly positive term.

Claim 3: If $\mathbf{v} \in S^n$ and $L(\mathbf{v}) = \mu$ then $\mathbf{v} > \mathbf{0}$. The fact that $\mathbf{v} \geq \mathbf{0}$ is a direct consequence of the definition of S^n, so we just need to show that *all* of its entries are non-zero (and it is thus positive). To this end, notice that since A is positive it is the case that $A\mathbf{v} > \mathbf{0}$ (by Claim 2). However, Claim 1 tells us that $A\mathbf{v} = \mu\mathbf{v}$, so dividing both sides by μ shows us that $\mathbf{v} = (A\mathbf{v})/\mu > \mathbf{0}$ as well.

Claim 3 above shows that the eigenvector $\mathbf{v}_*$ of A is positive. Since we want the only eigenvalue of A with a positive eigenvector to be λ_1, we now know that we must prove that $\mu = \lambda_1$. The fourth and final claim demonstrates this fact:

Claim 4: If $B \in \mathcal{M}_n(\mathbb{R})$ is such that $O \leq B \leq A$ and λ is an eigenvalue of B then $|\lambda| \leq \mu$. To see why this property holds, notice that if $\mathbf{w}$ is a unit eigenvector of B corresponding to the eigenvalue λ and we fix an integer $1 \leq i \leq n$ then

Ugh, we apologize for the unfortunate notation here. To be clear, $|[B\mathbf{w}]_i|$ is the absolute value of the i-th entry of $B\mathbf{w}$, while $[B|\mathbf{w}|]_i$ is the i-th entry of $B|\mathbf{w}|$.

$$|\lambda||w_i| = \left|[B\mathbf{w}]_i\right| = \left|\sum_{j=1}^{n} b_{i,j}w_j\right| \leq \sum_{j=1}^{n} b_{i,j}|w_j| = \left[B|\mathbf{w}|\right]_i, \quad (3.B.2)$$

where the inequality is just the triangle inequality, together with the fact that A is positive. If we let $|\mathbf{w}|$ denote the vector whose entries are the absolute values of the entries of $\mathbf{w}$ (i.e., $|\mathbf{w}| = (|w_1|, |w_2|, \ldots, |w_n|)$) then the above inequalities implies $B|\mathbf{w}| \geq |\lambda||\mathbf{w}|$. However, since $A - B$ is entry-wise non-negative, it is also the case that $(A - B)|\mathbf{w}| \geq \mathbf{0}$, so $A|\mathbf{w}| \geq B|\mathbf{w}| \geq |\lambda||\mathbf{w}|$. It follows that $L(|\mathbf{w}|) \geq |\lambda|$, but since $\mu \geq L(|\mathbf{w}|)$ we then have $\mu \geq |\lambda|$.

In particular, if we choose $B = A$ in Claim 4 then we see that μ is indeed the dominant eigenvalue of A (i.e., $\mu = \lambda_1$). Putting this all together reveals that we have successfully proved part (c) of the theorem and *almost* proved part (a) of it (we still have to show that μ is *strictly* larger than the absolute value of the other eigenvalues of A, but we leave that to later). We thus now turn our attention to parts (b) and (d) of the theorem.

Yes, this proof really is this long. Buckle up...

The proof that μ has algebraic multiplicity 1 (i.e., part (b) of the theorem) is somewhat technical and involves some clever techniques from multivariable calculus, so we leave it to Appendix B.5.

To see that eigenvectors corresponding to eigenvalues of A other than μ must contain at least one non-positive entry (i.e., part (d) of the theorem), suppose that λ is an eigenvalue of A with corresponding eigenvector $\mathbf{v} \geq \mathbf{0}$ (our goal is to show that this implies $\lambda = \mu$). Then $A\mathbf{v} = \lambda\mathbf{v}$, and since $A\mathbf{v} > \mathbf{0}$ (by Claim 2), we conclude that $\lambda > 0$. Then there exists some real scalar $c \neq 0$ such that (i) $c\mu \geq c\lambda$, and (ii) $\mathbf{v}_1 > c\mathbf{v}$, so

If $\mu \leq \lambda$ then we can choose $c < 0$ arbitrarily. If $\mu \geq \lambda$ then we can choose $c > 0$ to be sufficiently small so that $\mathbf{v}_1 > c\mathbf{v}$.

$$A(\mathbf{v}_1 - c\mathbf{v}) = A\mathbf{v}_1 - cA\mathbf{v} = \mu\mathbf{v}_1 - c\lambda\mathbf{v} \geq \mu\mathbf{v}_1 - c\mu\mathbf{v} = \mu(\mathbf{v}_1 - c\mathbf{v}).$$

It follows that

$$\frac{A(\mathbf{v}_1 - c\mathbf{v})}{\|\mathbf{v}_1 - c\mathbf{v}\|} \geq \frac{\mu(\mathbf{v}_1 - c\mathbf{v})}{\|\mathbf{v}_1 - c\mathbf{v}\|}, \quad \text{so} \quad L\left(\frac{\mathbf{v}_1 - c\mathbf{v}}{\|\mathbf{v}_1 - c\mathbf{v}\|}\right) \geq \mu. \tag{3.B.3}$$

However, since μ is the maximal value of L on S^n, it follows that Inequality (3.B.3) is actually an equality. Claim 1 then tells us that $A(\mathbf{v}_1 - c\mathbf{v}) = \mu(\mathbf{v}_1 - c\mathbf{v})$, so $\mathbf{v}_1 - c\mathbf{v}$ is an eigenvector of A corresponding to the eigenvalue μ. Since μ has algebraic multiplicity 1, it also has geometric multiplicity 1, so $\mathbf{v}_1 - c\mathbf{v}$ must be a multiple of $\mathbf{v}_1$. However, this is only possible if $\mathbf{v}$ itself is a multiple of $\mathbf{v}_1$, so these eigenvectors correspond to the same eigenvalue: $\lambda = \mu$.

We have thus shown that μ satisfies properties (b), (c), and (d) of the theorem that we claimed λ_1 satisfies. All that remains is to show that μ is *strictly* larger than the absolute value of all other eigenvalues of A. To this end, we return to Claim 4 and set $B = A$. We already showed that $|\lambda| \leq \mu$, and our goal is to show that if $|\lambda| = \mu$ then $\lambda = \mu$.

The equality condition of the triangle inequality was explored in Exercise 1.2.17.

Well, if $|\lambda| = \mu$ then Inequality (3.B.2) would have to be equality for each i, but equality holds in the triangle inequality if and only if, for each $1 \leq j \leq n$, the terms $b_{i,j}w_j$ are non-negative real multiples of each other. Since $b_{i,j} > 0$ for all $1 \leq i, j \leq n$, it follows that $\mathbf{w}$ is a scalar multiple of $|\mathbf{w}|$, so $|\mathbf{w}|$ is also an eigenvector corresponding to the eigenvalue λ. However, we already showed that the only eigenvalue of $A = B$ with a non-negative corresponding eigenvector is μ, so this shows that $\lambda = \mu$ and (finally!) completes the proof. ∎

Theorem 3.B.2 guarantees that the dominant eigenvalue λ_1 of a positive matrix is itself positive (and real) and *strictly* larger than the absolute value of the other eigenvalues, and that its dominant eigenspace contains an entry-wise positive vector $\mathbf{v}_1$. Power iteration is thus particularly useful when applied to positive matrices, since combining Theorems 3.B.1 and 3.B.2 shows that power iteration always finds their dominant eigenvalue (as long as the starting vector $\mathbf{v}_0$ is not chosen to start in the span of the non-dominant eigenspaces).

Example 3.B.7

Applying Power Iteration to a Positive Matrix

Estimate the dominant eigenvalue and eigenvectors of the positive matrix

$$A = \begin{bmatrix} 2 & 1 & 4 \\ 4 & 1 & 2 \\ 1 & 2 & 4 \end{bmatrix}$$

via power iteration and then confirm analytically that they are indeed an eigenvalue/eigenvector pair.

Solution:

If we apply power iteration starting from $\mathbf{v}_0 = \mathbf{e}_1$ as usual, we get

k	$\mathbf{v}_k = A\mathbf{v}_{k-1}/\|A\mathbf{v}_{k-1}\|$	$\mathbf{v}_k^T A \mathbf{v}_k$
0	$(1.00, 0.00, 0.00)$	2.00
1	$(0.44, 0.87, 0.22)$	4.48
2	$(0.52, 0.60, 0.60)$	6.96
3	$(0.58, 0.56, 0.59)$	7.06
4	$(0.58, 0.58, 0.58)$	7.00
5	$(0.58, 0.58, 0.58)$	7.00

The vector $\mathbf{v} = (1,1,1)$ comes from rescaling $\mathbf{v}_5 = (0.58, 0.58, 0.58)$.

The calculation above suggests that $\lambda = 7$ is the dominant eigenvalue of A and that the dominant eigenspace consists of the multiples of $\mathbf{v} = (1,1,1)$. To verify that these are indeed an eigenvalue and eigenvector of A, respectively, we could factor the characteristic polynomial of A and then solve a linear system as we did repeatedly in Section 3.3. However, it is much easier to just compute $A\mathbf{v}$ and see that it equals $\lambda\mathbf{v}$:

$$A\mathbf{v} = \begin{bmatrix} 2 & 1 & 4 \\ 4 & 1 & 2 \\ 1 & 2 & 4 \end{bmatrix} \begin{bmatrix} 1 \\ 1 \\ 1 \end{bmatrix} = \begin{bmatrix} 7 \\ 7 \\ 7 \end{bmatrix} = 7\mathbf{v} = \lambda\mathbf{v},$$

as claimed.

Google's PageRank Algorithm

Back in the early days of the World Wide Web, there were numerous search engines that could be used to find web pages of interest, and there was not a clear "winner" until Google came onto the scene in 1998. The primary reason for Google's success was its clever algorithm for ranking the "importance" of web pages, called **PageRank**, which is based on power iteration.

Prior to Google, search engines sorted web pages by things like how often the search words appeared on them.

The idea behind PageRank is that the rank (i.e., "importance") of a web page should be determined by the ranks of the pages that link to it—after all, a link to our web page from a Wikipedia is probably a better indicator of its importance than a link from my personal homepage. In particular, if there are n web pages to be ranked ($n \approx 5$ billion at the time of this writing) and we let r_i denote the rank of page i, p_j denote the number of pages that page j links to, and

This seems completely circular (we are using the importance of web pages to determine the importance of web pages), but we just run with it for now.

$$a_{i,j} = \begin{cases} \frac{1}{p_j} & \text{if page } j \text{ links to page } i, \text{ and} \\ 0 & \text{otherwise,} \end{cases}$$

then we set

$$r_i = a_{i,1}r_1 + a_{i,2}r_2 + \cdots + a_{i,n}r_n \quad \text{for all} \quad 1 < i < n. \tag{3.B.4}$$

That is, we are thinking of each web page as sharing its own rank evenly among the other pages that it links to. We illustrate how to construct this linear system with an example.

Example 3.B.8

A Linear System for Determining PageRank

Suppose that 5 web pages, which we call A, B, C, D, and E, link to each other as follows:

For example, page A links to pages B and C. Construct and solve the linear system (3.B.4) for determining the ranks r_A, r_B, r_C, r_D, and r_E of these pages.

Solution:

For example, page A shares half of its rank with each of pages B and C. On the other hand, page A receives $1/3$ of page B's rank, as well as all of page C's, $1/2$ of page D's, and $1/4$ of page E's. We can express this relationship via the linear equation

$$r_A = \tfrac{1}{3}r_B + r_C + \tfrac{1}{2}r_D + \tfrac{1}{4}r_E.$$

If we similarly construct linear equations based on which pages link into pages B, C, D, and E, we arrive at the linear system

$$\tfrac{1}{3}r_B + r_C + \tfrac{1}{2}r_D + \tfrac{1}{4}r_E = r_A,$$
$$\tfrac{1}{2}r_A + \tfrac{1}{4}r_E = r_B,$$
$$\tfrac{1}{2}r_A + \tfrac{1}{4}r_E = r_C,$$
$$\tfrac{1}{3}r_B + \tfrac{1}{4}r_E = r_D,$$
$$\tfrac{1}{3}r_B + \tfrac{1}{2}r_D = r_E.$$

It is straightforward to use Gaussian elimination to see that the solutions of this linear system are exactly the multiples of $(r_A, r_B, r_C, r_D, r_E) = (36, 21, 21, 10, 12)$. The exact values of the entries of this vector are not particularly important—what matters is their relative ordering. In particular, we have $r_A > r_B = r_C > r_E > r_D$ as the resulting ranking of how important each page is.

Remark 3.B.2

The World Wide Web as a Graph

Refer back to Section 1.B for an introduction to graphs and adjacency matrices.

One convenient way to represent web pages and links between them is as a (weighted, directed) graph, where the vertices of the graph represent the pages and an edge from vertex i to vertex j means that page i links to page j. Furthermore, we give each edge coming out of vertex j the "weight" $1/p_j$, where p_j is the total number of edges coming out of vertex j. For example, the graph that represents the 5 pages from Example 3.B.8 looks like

For each vertex, the sum of the weights of edges coming out of that vertex is 1.

The linear system (3.B.4) that is used to compute the ranks of web pages then simply has the form $A\mathbf{r} = \mathbf{r}$, where A is the adjacency matrix of this graph. However, since this graph is weighted, instead of placing 1 in the (i, j)-entry of A if there is an edge from vertex i to vertex j, we place the weight of that edge. For example, the 5-vertex graph above has adjacency matrix

$$A = \begin{bmatrix} 0 & 1/3 & 1 & 1/2 & 1/4 \\ 1/2 & 0 & 0 & 0 & 1/4 \\ 1/2 & 0 & 0 & 0 & 1/4 \\ 0 & 1/3 & 0 & 0 & 1/4 \\ 0 & 1/3 & 0 & 1/2 & 0 \end{bmatrix}.$$

It is straightforward to verify that if $\mathbf{r} = (r_A, r_B, r_C, r_D, r_E)$ then the linear system $A\mathbf{r} = \mathbf{r}$ is exactly the same as the one that we constructed in Example 3.B.8.

We can of course solve the linear system (3.B.4) directly via Gaussian elimination, thus finding the ranks of each web page. However, in practice it is much quicker to use power iteration to do so (after all, the linear system has the form $A\mathbf{r} = \mathbf{r}$, so we are actually finding an eigenvector of the matrix A corresponding to the eigenvalue 1).

The reason that power iteration is faster than Gaussian elimination here is that the World Wide Web is absolutely huge—it has about 5 billion web pages at the time of this writing—and no computer in the world is fast enough to use Gaussian elimination to solve a linear system with that many equations and variables. On the other hand, power iteration only requires us to repeatedly multiply a vector by the coefficient matrix A, which is much faster in this case thanks to A being sparse (i.e., since each web page only links to a few others).

Recall that an entry-wise non-negative matrix whose columns add up to 1 is called **column stochastic**.

The reason that power iteration *works* to find an eigenvector corresponding to the eigenvalue $\lambda = 1$ is that the columns of the matrix A all add up to 1, and we showed in Exercise 3.3.24 that (a) $\lambda = 1$ is an eigenvalue of every such matrix, and (b) all eigenvalues of these matrices have absolute value at most 1. It follows that $\lambda = 1$ is the dominant eigenvalue of A, so it (and its corresponding eigenvector $\mathbf{r}$) is what power iteration converges to.

Example 3.B.9

Finding PageRank via Power Iteration

Use power iteration to solve the linear system from Example 3.B.8.

Solution:

If we apply power iteration to this linear system, starting with $\mathbf{v}_0 = \mathbf{e}_1$, we get

k	$\mathbf{v}_k = A\mathbf{v}_{k-1}/\|A\mathbf{v}_{k-1}\|$	$\mathbf{v}_k^T A\mathbf{v}_k$
0	$(1.00, 0.00, 0.00, 0.00, 0.00)$	0.00
1	$(0.00, 0.71, 0.71, 0.00, 0.00)$	0.00
2	$(0.94, 0.00, 0.00, 0.24, 0.24)$	0.21
3	$(0.23, 0.68, 0.68, 0.08, 0.15)$	0.49
$\vdots$		
16	$(0.74, 0.41, 0.41, 0.21, 0.25)$	0.99
17	$(0.72, 0.44, 0.44, 0.20, 0.24)$	1.00
18	$(0.74, 0.42, 0.42, 0.21, 0.24)$	1.00

We thus conclude that the vector

$$\mathbf{r} = (r_A, r_B, r_C, r_D, r_E) \approx (0.74, 0.42, 0.42, 0.21, 0.24)$$

This particular power iteration does not converge very quickly, due to having a negative eigenvalue $\lambda_2 = -0.77$ that is somewhat close to the dominant eigenvalue $\lambda_1 = 1$ in absolute value.

is close to the eigenspace corresponding to eigenvalue 1, which is what we wanted. Use as when we found $\mathbf{r}$ analytically in Example 3.B.8, the exact values of its entries are not particularly important (so it is okay that they are just approximations here)—what matters is their relative ordering $r_A > r_B = r_C > r_E > r_D$.

It is worth looking back at the pages themselves from Example 3.B.8 and trying to convince ourselves that this ranking makes sense. For example, page A is linked to by every other page, so it should have the highest rank, and pages B and C are linked to by the exact same sets of pages (A and E), so they should have the same rank as each other. We look at why it makes sense for pages D and E to have the lowest ranks in Exercise 3.B.14.

There is actually one important issue that we glossed over when applying power iteration in the above example—how did we know that power iteration would converge in the first place? That is, how did we know that the matrix A did not have another eigenvalue with absolute value equal to 1? The answer is simply that we were lucky—there are many column stochastic matrices with multiple distinct eigenvalues of absolute value 1, such as

Matrices like these correspond to the case when the web pages all link to each other in a cycle, so power iteration has no idea how to assign their ranks (i.e., the assignment of ranks is too circular for it to ever make progress).

$$\begin{bmatrix} 0 & 1 \\ 1 & 0 \end{bmatrix} \quad \text{and} \quad \begin{bmatrix} 0 & 1 & 0 \\ 0 & 0 & 1 \\ 1 & 0 & 0 \end{bmatrix}.$$

In particular, the first matrix has eigenvalues 1 and -1, and the second matrix has eigenvalues 1 and $(-1 \pm i\sqrt{3})/2$, each of which have absolute value 1.

To get around this problem, instead of applying power iteration to A itself, it is typically applied to $A + \varepsilon J$ instead, where J is the matrix of appropriate size with every entry equal to 1 and ε is some very small (but positive) real number. If ε is small enough then this change does not change the rankings of the web pages at all, and the matrix $A + \varepsilon J$ has the advantage of being positive, rather than just non-negative, so Theorem 3.B.2 applies to it. In particular, this guarantees that power iteration converges, and the eigenvector that it finds has positive entries (and thus the web page ranks are positive).

Exercises

solutions to starred exercises on page 473

▢ 3.B.1 With the help of computer software, use power iteration to approximate the dominant eigenvalue and a corresponding unit eigenvector of each of the following matrices.

*(a) $\begin{bmatrix} 1 & 2 \\ 3 & 4 \end{bmatrix}$ 　　(b) $\begin{bmatrix} 3 & -1 \\ -4 & 5 \end{bmatrix}$

*(c) $\begin{bmatrix} -2 & 1 & 1 \\ 4 & 1 & 2 \\ 1 & 0 & -2 \end{bmatrix}$ 　(d) $\begin{bmatrix} 3 & -1 & 2 \\ 0 & 4 & 1 \\ -4 & 1 & 2 \end{bmatrix}$

*(e) $\begin{bmatrix} 3 & 2 & 4 & 3 \\ 1 & 1 & 1 & 3 \\ 3 & 4 & 2 & 0 \\ 0 & 3 & 1 & 2 \end{bmatrix}$ 　(f) $\begin{bmatrix} 0 & 5 & 5 & 2 \\ 5 & 4 & 7 & 2 \\ 5 & 4 & 0 & 6 \\ 7 & 5 & 5 & 4 \end{bmatrix}$

▢ 3.B.2 With the help of computer software, use power iteration to approximate *all* eigenvalues, as well as a corresponding unit eigenvector for each, of the following matrices.

*(a) $\begin{bmatrix} 1 & 2 \\ 3 & 4 \end{bmatrix}$ 　　(b) $\begin{bmatrix} -2 & 4 \\ 5 & 5 \end{bmatrix}$

*(c) $\begin{bmatrix} 1 & 2 & 3 \\ 4 & 5 & 6 \\ 7 & 8 & 9 \end{bmatrix}$ 　(d) $\begin{bmatrix} -2 & 3 & -1 \\ 5 & 2 & 3 \\ 3 & 3 & 0 \end{bmatrix}$

▢ 3.B.3 With the help of computer software, use power iteration to approximate the dominant eigenvalue and a corresponding unit eigenvector of each of the following matrices.

*(a) $\begin{bmatrix} -3 & -4 & 3 & 7 & 1 & -2 & 1 \\ 4 & 6 & 4 & -1 & -4 & 3 & -2 \\ -2 & 1 & -3 & -1 & -3 & 1 & -1 \\ 0 & 4 & 1 & 5 & -3 & 5 & -1 \\ 0 & 3 & 4 & 0 & 2 & -1 & -5 \\ 3 & -1 & 3 & 5 & 6 & 0 & 7 \\ 4 & 8 & 3 & 7 & 7 & 6 & -1 \end{bmatrix}$

(b) $\begin{bmatrix} 0 & 1 & 2 & 3 & 4 & 5 & 6 & 7 \\ 1 & 2 & 3 & 4 & 5 & 6 & 7 & 6 \\ 2 & 3 & 4 & 5 & 6 & 7 & 6 & 5 \\ 3 & 4 & 5 & 6 & 7 & 6 & 5 & 4 \\ 4 & 5 & 6 & 7 & 6 & 5 & 4 & 3 \\ 5 & 6 & 7 & 6 & 5 & 4 & 3 & 2 \\ 6 & 7 & 6 & 5 & 4 & 3 & 2 & 1 \\ 7 & 6 & 5 & 4 & 3 & 2 & 1 & 0 \end{bmatrix}$

3.B.4 Determine which of the following statements are true and which are false.

*(a) If $A \in \mathcal{M}_n(\mathbb{R})$ is diagonalizable then power iteration, applied to A, always converges to its dominant eigenvalue.

(b) If $A \in \mathcal{M}_3(\mathbb{R})$ has eigenvalues 3, 2, and 1, then power iteration, applied to A, always converges to the eigenvalue 3.

*(c) If $A \in \mathcal{M}_3(\mathbb{R})$ has eigenvalues 5, 3, and -3, then power iteration, applied to A, converges to the eigenvalue 5 as long as the initial vector $\mathbf{v}_0$ is not in the span of the eigenspaces corresponding to the eigenvalues 3 and -3.

(d) power iteration produces a sequence of values

$$\mathbf{v}_0^T A \mathbf{v}_0, \mathbf{v}_1^T A \mathbf{v}_1, \mathbf{v}_2^T A \mathbf{v}_2, \ldots$$

that converges to the dominant eigenvalue of a matrix $A \in \mathcal{M}_n(\mathbb{R})$ then the sequence of vectors $\mathbf{v}_0, \mathbf{v}_1, \mathbf{v}_2,$ ... converges to a corresponding eigenvector.

*(e) If $A \in \mathcal{M}_n(\mathbb{R})$ is positive then it has an eigenvector with all of its entries positive.

(f) If $A \in \mathcal{M}_n(\mathbb{R})$ is positive then all of its eigenvalues are positive.

3.B.5 Recall the matrix A from Example 3.B.1, which we applied power iteration to with the starting vector $\mathbf{v}_0 = \mathbf{e}_1$ so as to find its maximal eigenvalue and corresponding eigenvector.

(a) Repeat the computations from that example starting with $\mathbf{v}_0 = \mathbf{e}_2$ instead of $\mathbf{v}_0 = \mathbf{e}_1$.

(b) Find a starting vector $\mathbf{v}_0$ for which power iteration, starting with $\mathbf{v}_0$, does not converge.

∗∗3.B.6 Suppose $A \in \mathcal{M}_n$ satisfies the hypotheses of Theorem 3.B.1 and has dominant eigenvalue λ_1 that is positive and real. Explain why

$$\lim_{k \to \infty} \|A \mathbf{v}_k\| = \lambda_1.$$

[Side note: A rough explanation is enough—do not feel the need to rigorously prove this limit like we did in Theorem 3.B.1.]

3.B.7 Show that if $A \in \mathcal{M}_2(\mathbb{R})$ is positive then all of its eigenvalues are real.

∗∗3.B.8 A matrix $A \in \mathcal{M}_n$ is called **primitive** if A^k is positive for some integer $k \geq 1$. Show that the Perron–Frobenius theorem (Theorem 3.B.2) applies to primitive matrices, as long as we change part (a) of the theorem to say that $|\lambda_1| > |\lambda_j|$ instead of $\lambda_1 > |\lambda_j|$ for all $2 \leq j \leq n$ (i.e., λ_1 is no longer necessarily positive and real).

3.B.9 Verify that each of the following matrices are primitive (see Exercise 3.B.8). That is, find a positive integer k such their k-th powers are positive.

*(a) $\begin{bmatrix} 0 & 1 \\ 1 & 1 \end{bmatrix}$

(b) $\begin{bmatrix} 2 & 1 \\ 1 & -1 \end{bmatrix}$

*(c) $\begin{bmatrix} i & i \\ i & i \end{bmatrix}$

(d) $\begin{bmatrix} \sqrt{5}+i & \sqrt{5}-i \\ \sqrt{5}-i & \sqrt{5}+i \end{bmatrix}$

*(e) $\begin{bmatrix} 0 & 1 & 1 \\ 1 & 1 & 0 \\ 1 & 0 & 0 \end{bmatrix}$

(f) $\begin{bmatrix} 0 & 1 & 0 \\ 1 & 1 & 1 \\ 0 & 1 & -1 \end{bmatrix}$

*□ **3.B.10** For each of the matrices in Exercise 3.B.9, use power iteration (with the help of computer software) to approximate its dominant eigenvalue λ_1 and an eigenvector v_1 with positive entries to which it corresponds (which are guaranteed to exist by Exercise 3.B.8).

3.B.11 An entry-wise non-negative matrix $A \in \mathcal{M}_n(\mathbb{R})$ is called **irreducible** if, for each (i, j), there exists an integer k (that may depend on i and j) such that $[A^k]_{i,j} > 0$.

(a) Show that the following matrix is irreducible:

$$\begin{bmatrix} 0 & 1 \\ 1 & 0 \end{bmatrix}.$$

(b) Show that the matrix from part (a) is not primitive (see Exercise 3.B.8).

(c) Show that if a matrix A is irreducible then $I + A$ is primitive.

3.B.12 Determine whether the following matrices are primitive (see Exercise 3.B.8), irreducible (see Exercise 3.B.11), both, or neither.

*(a) $\begin{bmatrix} -1 & 1 \\ 1 & 1 \end{bmatrix}$

(b) $\begin{bmatrix} 1 & 1 \\ 0 & 1 \end{bmatrix}$

*(c) $\begin{bmatrix} 0 & 1 \\ 2 & 3 \end{bmatrix}$

(d) $\begin{bmatrix} -2 & 1 \\ 1 & 3 \end{bmatrix}$

*(e) $\begin{bmatrix} 1 & -1 & 1 \\ -1 & 1 & 1 \\ 1 & 1 & 1 \end{bmatrix}$

(f) $\begin{bmatrix} 0 & 1 & 0 \\ 1 & 0 & 1 \\ 0 & 1 & 0 \end{bmatrix}$

□ **3.B.13** Suppose 5 web pages link to each other as indicated below:

Use computer software and power iteration to find the ranks of these 5 pages (i.e., mimic Example 3.B.9).

3.B.14 The web pages B, C, D, and E in Example 3.B.8 are all linked to by exactly 2 other web pages, yet we showed in Example 3.B.9 that pages D and E have lower ranks than pages B and C. Provide an intuitive explanation for why this makes sense (i.e., explain why pages D and E "should" have lower ranks than pages B and C).

3.C   Extra Topic: Complex Eigenvalues of Real Matrices

Recall that real matrices may have complex (non-real) eigenvalues and eigenvectors. For example, we showed in Example 3.3.8 that the matrix

$$A = \begin{bmatrix} -3 & -2 \\ 4 & 1 \end{bmatrix}$$

If you are not comfortable with complex numbers, have a look at Appendix A.1 before reading this section.

has eigenvalues $-1 \pm 2i$. While this is a fine algebraic result, there are some situations where we really want to restrict our attention to real numbers (e.g., if we are thinking of A as a linear transformation acting on $\mathbb{R}^2$), in which case these complex eigenvalues do not seem to be of much use to us. For example, our geometric interpretation of eigenvalues is as an amount that A stretches certain vectors (the corresponding eigenvectors), but A clearly does not stretch any vector in $\mathbb{R}^2$ by a factor of $-1 + 2i$.

To illustrate how we will get around this apparent problem and come up with a geometric interpretation of complex eigenvalues of real matrices, we first find the eigenvectors of the above matrix A.

Example 3.C.1

Computing Complex Eigenvectors

Compute bases of the eigenspaces of the matrix $A = \begin{bmatrix} -3 & -2 \\ 4 & 1 \end{bmatrix}$.

Solution:

We showed in Example 3.3.8 that A has eigenvalues $\lambda_\pm = -1 \pm 2i$. We thus want to find the non-zero vectors $\mathbf{v}_\pm$ satisfying the linear systems $(A - (-1 \pm 2i)I)\mathbf{v}_\pm = \mathbf{0}$. We start by solving the "+" system:

The bottom-left entry after the row operation was computed via
$4 + (1-i)(-2-2i) =$
$4 + (-2+2i-2i-2) =$
$4 - 4 = 0.$

$$\left[\begin{array}{cc|c} -2-2i & -2 & 0 \\ 4 & 2-2i & 0 \end{array}\right] \xrightarrow{R_2 + (1-i)R_1} \left[\begin{array}{cc|c} -2-2i & -2 & 0 \\ 0 & 0 & 0 \end{array}\right].$$

It follows that one basis of the eigenspace corresponding to $\lambda_+ = -1 + 2i$ consists of the single vector $\mathbf{v}_+ = (1, -1-i)$. A similar calculation shows that one basis of the eigenspace corresponding to $\lambda_- = -1 - 2i$ consists of the vector $\mathbf{v}_- = (1, -1+i)$.

In particular, the eigenvectors in the previous example are also not real, so we cannot plot them in $\mathbb{R}^2$ and talk about the geometric effect that A has on them. The key observation that we make in order to get around this problem is that, just like the eigenvalues of a real matrix always come in complex conjugate pairs (as we noted back in Section 3.3.2), so do its eigenvectors. In this case, its eigenvalues are complex conjugates of each other, since $\overline{-1+2i} = -1 - 2i$, and its eigenvectors are also complex conjugates of each other, since $\overline{(1, -1-i)} = (1, -1+i)$.

The complex conjugate of a complex number is $\overline{a+ib} = a - ib.$

To see that this phenomenon occurs for *every* real matrix B, suppose that $\mathbf{v}$ is an eigenvector of B with corresponding eigenvalue λ. Then $B\mathbf{v} = \lambda\mathbf{v}$, which we can use to compute $B\overline{\mathbf{v}}$ as follows:

The leftmost equality holds because B has real entries, so $\overline{B} = B$.

$$B\overline{\mathbf{v}} = \overline{B}\overline{\mathbf{v}} = \overline{B\mathbf{v}} = \overline{\lambda\mathbf{v}} = \overline{\lambda}\,\overline{\mathbf{v}}.$$

Since $B\overline{\mathbf{v}} = \overline{\lambda}\,\overline{\mathbf{v}}$, we conclude that $\overline{\mathbf{v}}$ is an eigenvector of B with corresponding eigenvalue $\overline{\lambda}$. This observation is important, so we re-state it a bit more prominently:

> ⓘ The non-real eigenvalues and eigenvectors of real matrices come in complex conjugate pairs.

3.C.1 Geometric Interpretation for 2×2 Matrices

Since the main result of this section combines several pieces of machinery that we have either not explicitly covered yet (mostly facts concerning complex numbers) or have not used in quite some time, we briefly remind ourselves of some useful tidbits of math that we may have forgotten.

The polar form (and other facts about complex numbers) are covered in Appendix A.1.

First, recall that every complex number λ can be written in polar form $\lambda = re^{i\theta}$, which means that λ is a distance of r from the origin in the complex plane and is rotated up from the real axis by an angle of θ (see Figure 3.29). In particular, r and θ can be computed via $r = \sqrt{a^2 + b^2}$ and $\theta = \arctan(b/a)$ if $a > 0$ (if $a \leq 0$ then we have to be more careful when computing θ to make sure that we choose it to be in the correct quadrant).

It is also worth knowing that $e^{i\theta} = \cos(\theta) + i\sin(\theta)$ for all $\theta \in \mathbb{R}$. This is called **Euler's formula**.

Second, we use $\mathrm{Re}(\mathbf{v})$ and $\mathrm{Im}(\mathbf{v})$ to denote the real and imaginary parts of a vector, respectively. That is, if $\mathbf{v} = \mathbf{x} + i\mathbf{y}$ for some $\mathbf{x}, \mathbf{y} \in \mathbb{R}^n$ then $\mathrm{Re}(\mathbf{v}) = \mathbf{x}$ and $\mathrm{Im}(\mathbf{v}) = \mathbf{y}$. Also, by recalling that the complex conjugate of $\mathbf{v}$ is $\overline{\mathbf{v}} = \mathrm{Re}(\mathbf{v}) - i\mathrm{Im}(\mathbf{v})$, we can solve for $\mathrm{Re}(\mathbf{v})$ and $\mathrm{Im}(\mathbf{v})$ in terms of $\mathbf{v}$ and $\overline{\mathbf{v}}$:

$$\mathrm{Re}(\mathbf{v}) = \frac{1}{2}(\mathbf{v} + \overline{\mathbf{v}}) \quad \text{and} \quad \mathrm{Im}(\mathbf{v}) = \frac{1}{2i}(\mathbf{v} - \overline{\mathbf{v}}).$$

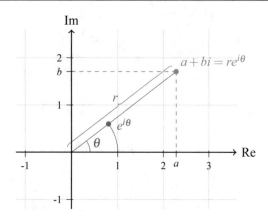

Figure 3.29: Every complex number $a+ib$ can be written in the form $a+ib=re^{i\theta}$, where $r=\sqrt{a^2+b^2}$ is the distance from $a+ib$ to the origin and θ is the angle that $a+ib$ makes with the real axis.

Third, recall from Section 3.1 that if B is a basis of a subspace then $[\mathbf{x}]_B$ refers to the coordinate vector of $\mathbf{x}$ with respect to the basis B. In particular, if $B=\{\mathbf{v},\mathbf{w}\}$ is a basis of a 2-dimensional subspace (i.e., a plane) and $\mathbf{x}=c\mathbf{v}+d\mathbf{w}$, then $[\mathbf{x}]_B=(c,d)$.

Finally, recall that $[R^\theta]$ is the 2×2 matrix that rotates vectors in $\mathbb{R}^2$ counter-clockwise by an angle of θ. Specifically, this matrix has the form

$$[R^\theta]=\begin{bmatrix}\cos(\theta) & -\sin(\theta) \\ \sin(\theta) & \cos(\theta)\end{bmatrix}.$$

With all of these observations and technicalities taken care of, we are now able to present a geometric interpretation of the non-real eigenvalues of real matrices, at least in the 2×2 case. In particular, if we recall from Exercise 3.3.13 that every rotation matrix $[R^\theta]$ has complex (non-real) eigenvalues, unless θ is an integer multiple of π, then the upcoming theorem can be thought of as the converse statement that every real matrix with non-real eigenvalues is, up to scaling and similarity, a rotation.

> **Theorem 3.C.1**
>
> **2×2 Matrices with Complex Eigenvalues Look Like Rotations**
>
> Suppose $A\in\mathcal{M}_2(\mathbb{R})$ has a complex (non-real) eigenvalue $re^{i\theta}\in\mathbb{C}$ with corresponding eigenvector $\mathbf{v}\in\mathbb{C}^2$. If we let $Q=\big[\,\mathrm{Re}(\mathbf{v})\mid -\mathrm{Im}(\mathbf{v})\,\big]$ then Q is invertible and
> $$A=Q\big(r[R^\theta]\big)Q^{-1}.$$

Just like being able to diagonalize a matrix via $A=PDP^{-1}$ means that A "looks diagonal" when viewed in the basis consisting of the columns of P, the above theorem means that A "looks like" a rotation counter-clockwise by an angle of θ composed with a stretch by a factor of r when viewed in the basis $B=\{\mathrm{Re}(\mathbf{v}),-\mathrm{Im}(\mathbf{v})\}$.

For instance, if we return to the matrix

$$A=\begin{bmatrix}-3 & -2 \\ 4 & 1\end{bmatrix}.$$

from Example 3.3.8, we recall that one of its eigenvalues is $\lambda=-1+2i=\sqrt{5}e^{i\theta}$, where $\theta\approx\pm(0.6476)\pi$. In Example 3.C.1, we found that one corresponding eigenvector is $\mathbf{v}=(1,-1-i)$. It follows that this matrix can be

Rotation matrices (and other similar linear transformations) were introduced in Section 1.4.2.

viewed as rotating $\mathbb{R}^2$ in the basis $B = \{\text{Re}(\mathbf{v}), -\text{Im}(\mathbf{v})\} = \{(1,-1),(0,1)\}$ counter-clockwise by approximately 0.6476 half-rotations (i.e., an angle of $\theta \approx (0.6476)\pi$) and stretching it by a factor of $r = \sqrt{5} \approx 2.2361$, as in Figure 3.30.

In Figure 3.30(a), the ellipse itself is not rotated, but rather vectors are rotated around its edge. The ellipse is stretched by a factor of $r = \sqrt{5}$, but its orientation remains the same.

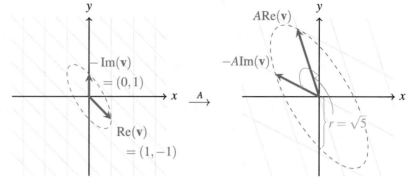

(a) The matrix A rotates vectors around the edge of an ellipse and then stretches them by a factor of $r = \sqrt{5}$.

Figure 3.30(b) can be thought of as an "un-skewed" version of Figure 3.30(a)—they are qualitatively the same, but (b) has had the specifics smoothed out.

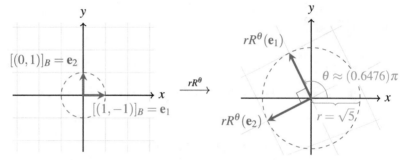

(b) In the basis $B = \{\text{Re}(\mathbf{v}), -\text{Im}(\mathbf{v})\} = \{(1,-1),(0,1)\}$, the matrix A can be seen as rotating counter-clockwise by an angle of $\theta \approx (0.6476)\pi$ and stretching by a factor of $r = \sqrt{5}$.

Figure 3.30: A visualization of the geometric effect of the matrix $A = \begin{bmatrix} -3 & -2 \\ 4 & 1 \end{bmatrix}$ from Example 3.C.1.

Proof Theorem 3.C.1. First, we note that since the eigenvalue $re^{i\theta}$ is not real, it does not equal its complex conjugate $re^{-i\theta}$, so A is diagonalizable via $A = PDP^{-1}$, where

Recall from Corollary 3.4.4 that every $n \times n$ matrix with n distinct eigenvalues is diagonalizable.

$$D = \begin{bmatrix} re^{i\theta} & 0 \\ 0 & re^{-i\theta} \end{bmatrix} \quad \text{and} \quad P = \begin{bmatrix} \mathbf{v} \mid \overline{\mathbf{v}} \end{bmatrix}$$

have complex entries.

Our goal now is to transform this diagonalization into the form described by the statement of the theorem (which in particular consists entirely of matrices with real entries). To this end, let $H \in \mathcal{M}_2(\mathbb{C})$ be the matrix

$$H = \begin{bmatrix} 1 & 1 \\ -i & i \end{bmatrix} \quad \text{with inverse} \quad H^{-1} = \frac{1}{2}\begin{bmatrix} 1 & i \\ 1 & -i \end{bmatrix}. \tag{3.C.1}$$

It is straightforward to verify that

$$QH = \begin{bmatrix} \text{Re}(\mathbf{v}) \mid -\text{Im}(\mathbf{v}) \end{bmatrix} \begin{bmatrix} 1 & 1 \\ -i & i \end{bmatrix}$$
$$= \begin{bmatrix} \text{Re}(\mathbf{v}) + i\text{Im}(\mathbf{v}) \mid \text{Re}(\mathbf{v}) - i\text{Im}(\mathbf{v}) \end{bmatrix} = \begin{bmatrix} \mathbf{v} \mid \overline{\mathbf{v}} \end{bmatrix} = P, \tag{3.C.2}$$

which tells us that Q is invertible since it implies $Q = PH^{-1}$, which is the product of two invertible matrices.

The remainder of the proof is devoted to showing that we can use this relationship between P and Q to transform $D = P^{-1}AP$ into $Q^{-1}AQ$ and see that it has the stretched rotation form that we claimed in the statement of the theorem. In particular, we now compute HDH^{-1} in two different ways. First,

> In the final equality
> we use the fact that
> $QH = P$, so
> $HP^{-1} = Q^{-1}$.

$$HDH^{-1} = H(P^{-1}AP)H^{-1} = (HP^{-1})A(HP^{-1})^{-1} = Q^{-1}AQ,$$

so if we can also show that $HDH^{-1} = r[R^\theta]$ then we will be done. Equivalently, we want to show that $HD = r[R^\theta]H$, which can be done simply by multiplying the indicated matrices together:

$$HD = \begin{bmatrix} 1 & 1 \\ -i & i \end{bmatrix} \begin{bmatrix} re^{i\theta} & 0 \\ 0 & re^{-i\theta} \end{bmatrix} = r \begin{bmatrix} e^{i\theta} & e^{-i\theta} \\ -ie^{i\theta} & ie^{-i\theta} \end{bmatrix}, \quad \text{and}$$

> In the final equality
> here, we use the
> fact that
> $e^{i\theta} = \cos(\theta) + i\sin(\theta)$.
> If we recall that
> $\sin(-\theta) = -\sin(\theta)$
> then we similarly see
> that
> $e^{-i\theta} = \cos(\theta) - i\sin(\theta)$.

$$r[R^\theta]H = r \begin{bmatrix} \cos(\theta) & -\sin(\theta) \\ \sin(\theta) & \cos(\theta) \end{bmatrix} \begin{bmatrix} 1 & 1 \\ -i & i \end{bmatrix}$$

$$= r \begin{bmatrix} \cos(\theta) + i\sin(\theta) & \cos(\theta) - i\sin(\theta) \\ \sin(\theta) - i\cos(\theta) & \sin(\theta) + i\cos(\theta) \end{bmatrix} = r \begin{bmatrix} e^{i\theta} & e^{-i\theta} \\ -ie^{i\theta} & ie^{-i\theta} \end{bmatrix}.$$

Since these two matrices do equal each other, we conclude that

$$Q^{-1}AQ = HDH^{-1} = r[R^\theta],$$

which completes the proof. ∎

The matrix H from Equation 3.C.1 is quite remarkable for the fact that it does not depend on A. That is, no matter what matrix $A \in \mathcal{M}_2(\mathbb{R})$ is chosen, H can be used to convert its complex diagonalization into the stretched rotation form of Theorem 3.C.1, and vice-versa.

Example 3.C.2

Converting a Complex Diagonalization

Use the matrix H from Equation 3.C.1 to convert a complex diagonalization of the matrix

$$A = \begin{bmatrix} -3 & -2 \\ 4 & 1 \end{bmatrix}$$

from Example 3.C.1 to its stretched rotation form of Theorem 3.C.1.

Solution:

We start by recalling that the eigenvalues of this matrix are $\lambda_\pm = -1 \pm 2i$ and a pair of corresponding eigenvectors is given by $\mathbf{v}_\pm = (1, -1 \mp i)$. It follows that A can be diagonalized as $A = PDP^{-1}$, where

$$D = \begin{bmatrix} -1+2i & 0 \\ 0 & -1-2i \end{bmatrix} \quad \text{and} \quad P = \begin{bmatrix} 1 & 1 \\ -1-i & -1+i \end{bmatrix}.$$

To convert this diagonalization into the stretched rotation form described by Theorem 3.C.1, we compute

$$Q = PH^{-1} = \frac{1}{2} \begin{bmatrix} 1 & 1 \\ -1-i & -1+i \end{bmatrix} \begin{bmatrix} 1 & i \\ 1 & -i \end{bmatrix} = \begin{bmatrix} 1 & 0 \\ -1 & 1 \end{bmatrix},$$

We could also compute this stretched rotation form of A directly from the statement of Theorem 3.C.1 (without using the matrix H at all).

which is exactly $\big[\, \mathrm{Re}(\mathbf{v}_+) \mid -\mathrm{Im}(\mathbf{v}_+) \,\big]$, as we would hope (we want our answer to match up with the statement of Theorem 3.C.1). Similarly, we notice that

$$HDH^{-1} = \frac{1}{2}\begin{bmatrix} 1 & 1 \\ -i & i \end{bmatrix}\begin{bmatrix} -1+2i & 0 \\ 0 & -1-2i \end{bmatrix}\begin{bmatrix} 1 & i \\ 1 & -i \end{bmatrix} = \begin{bmatrix} -1 & -2 \\ 2 & -1 \end{bmatrix},$$

which is exactly the stretched rotation matrix depicted in Figure 3.30(b) (i.e., it equals $\sqrt{5}\big[R^\theta\big]$, where $\theta = \arccos(-1/\sqrt{5}) \approx (0.6476)\pi$).

3.C.2 Block Diagonalization of Real Matrices

We now ramp up the ideas of the previous subsection to real matrices of arbitrary size. Fortunately, not much changes when we do this—the idea is that instead of the matrix $A \in \mathcal{M}_n(\mathbb{R})$ looking like a rotation on all of $\mathbb{R}^n$, it looks like a rotation if we only focus our attention on the two-dimension subspaces defined by its eigenvectors that come in complex conjugate pairs.

Before stating the theorem, we note that we use the notation $\mathrm{diag}(a,b,c,\ldots)$ to denote the diagonal matrix with diagonal entries a, b, c, …. In fact, we even allow a, b, c, … to be matrices themselves, in which case $\mathrm{diag}(a,b,c,\ldots)$ is a *block* diagonal matrix with diagonal blocks equal to a, b, c, …. For example,

$$\mathrm{diag}\Big(I_2, 5, \big[R^{\pi/7}\big]\Big) = \left[\begin{array}{cc|c|cc} 1 & 0 & 0 & 0 & 0 \\ 0 & 1 & 0 & 0 & 0 \\ \hline 0 & 0 & 5 & 0 & 0 \\ \hline 0 & 0 & 0 & \cos(\pi/7) & -\sin(\pi/7) \\ 0 & 0 & 0 & \sin(\pi/7) & \cos(\pi/7) \end{array}\right].$$

With this notation out of the way, we can now state the main result of this section.

Theorem 3.C.2

Block Diagonalization of Real Matrices

There are m real eigenvalues and 2ℓ complex (non-real) eigenvalues, so $2\ell + m = n$.

In particular, notice that B and Q are both real and the diagonal blocks of B all have size 2×2 or 1×1.

Suppose $A = \mathcal{M}_n(\mathbb{R})$ has ℓ complex conjugate eigenvalue pairs and m real eigenvalues (counting algebraic multiplicity) and is diagonalizable via $A = PDP^{-1}$, where $D, P \in \mathcal{M}_n(\mathbb{C})$ have the forms

$$D = \mathrm{diag}\big(r_1 e^{i\theta_1}, r_1 e^{-i\theta_1}, \ldots, r_\ell e^{i\theta_\ell}, r_\ell e^{-i\theta_\ell},\ \lambda_1, \ldots, \lambda_m\big) \quad \text{and}$$
$$P = \big[\, \mathbf{v}_1 \mid \overline{\mathbf{v}_1} \mid \cdots \mid \mathbf{v}_\ell \mid \overline{\mathbf{v}_\ell} \mid \mathbf{w}_1 \mid \cdots \mid \mathbf{w}_m \,\big],$$

with $r_j, \lambda_j, \theta_j \in \mathbb{R}$, $\mathbf{w}_j \in \mathbb{R}^n$, and $\mathbf{v}_j \in \mathbb{C}^n$ for all j.

Then $A = QBQ^{-1}$, where $B, Q \in \mathcal{M}_n(\mathbb{R})$ are defined by

$$B = \mathrm{diag}\Big(r_1\big[R^{\theta_1}\big], \ldots, r_\ell\big[R^{\theta_\ell}\big],\ \lambda_1, \ldots, \lambda_m\Big) \quad \text{and}$$
$$Q = \big[\, \mathrm{Re}(\mathbf{v}_1) \mid -\mathrm{Im}(\mathbf{v}_1) \mid \cdots \mid \mathrm{Re}(\mathbf{v}_\ell) \mid -\mathrm{Im}(\mathbf{v}_\ell) \mid \mathbf{w}_1 \mid \cdots \mid \mathbf{w}_m \,\big].$$

As another way of thinking about this theorem, recall that any real matrix $A \in \mathcal{M}_n(\mathbb{R})$ with distinct complex (non-real) eigenvalues can be diagonalized as $A = PDP^{-1}$ via some $P, D \in \mathcal{M}_n(\mathbb{C})$, but not via a *real* P and D (in other words, A can be diagonalized over $\mathbb{C}$ but not over $\mathbb{R}$). For example, the matrix

$$A = \frac{1}{4}\begin{bmatrix} 3 & 2 & -1 \\ -5 & 2 & 3 \\ 5 & 2 & 7 \end{bmatrix}$$

has distinct eigenvalues 2 and $(1 \pm i)/2$ and is thus diagonalizable over $\mathbb{C}$, but not over $\mathbb{R}$.

While we cannot *diagonalize* matrices like this one via a real change of basis (i.e., a real similarity), it seems natural to ask how close we can get if we restrict our attention to real matrices. The above theorem answers this question and shows that we can make it *block* diagonal with blocks of size no larger than 2×2. Furthermore, its proof provides a simple method of transforming a complex diagonalization into a real 2×2 block diagonalization, and vice-versa.

Proof of Theorem 3.C.2. While this theorem looks like a bit of a beast at first, we did all of the hard work when proving Theorem 3.C.1, which we can now just repeatedly apply to prove this theorem. In particular, we recall the 2×2 matrix H from Equation (3.C.1) and then we start by defining the various two-dimensional "chunks" of this problem. That is, we let

$$D_j = \begin{bmatrix} r_j e^{i\theta_j} & 0 \\ 0 & r_j e^{-i\theta_j} \end{bmatrix}, \quad P_j = \begin{bmatrix} \mathbf{v}_j \mid \overline{\mathbf{v}_j} \end{bmatrix}, \quad \text{and} \quad Q_j = \begin{bmatrix} \operatorname{Re}(\mathbf{v}_j) \mid -\operatorname{Im}(\mathbf{v}_j) \end{bmatrix}$$

for all $1 \leq j \leq \ell$, and we also define

$$\widetilde{H} = \operatorname{diag}\big(\underbrace{H, H, \ldots, H}_{\ell \text{ copies}}, \underbrace{1, 1, \ldots, 1}_{m \text{ copies}}\big).$$

We already showed in Equation 3.C.2 of the proof of Theorem 3.C.1 that $Q_j H = P_j$ for all $1 \leq j \leq \ell$, so block matrix multiplication then shows that

All of these block
matrix multiplications
work out so cleanly
simply because $\tilde{H}$ is
block diagonal.

$$\begin{aligned} Q\widetilde{H} &= \begin{bmatrix} Q_1 \mid Q_2 \mid \cdots \mid Q_\ell \mid \mathbf{w}_1 \mid \mathbf{w}_2 \mid \cdots \mid \mathbf{w}_m \end{bmatrix} \operatorname{diag}(H, H, \ldots, H, 1, 1, \ldots, 1) \\ &= \begin{bmatrix} Q_1 H \mid Q_2 H \mid \cdots \mid Q_\ell H \mid \mathbf{w}_1 \mid \mathbf{w}_2 \mid \cdots \mid \mathbf{w}_m \end{bmatrix} \\ &= \begin{bmatrix} P_1 \mid P_2 \mid \cdots \mid P_\ell \mid \mathbf{w}_1 \mid \mathbf{w}_2 \mid \cdots \mid \mathbf{w}_m \end{bmatrix} = P. \end{aligned}$$

Similarly, we also showed in the proof of Theorem 3.C.1 that $HD_j = r_j \big[R^{\theta_j}\big] H$ for all $1 \leq j \leq \ell$, so block matrix multiplication shows that

$$\begin{aligned} \tilde{H}D &= \operatorname{diag}(H, H, \ldots, H, 1, 1, \ldots, 1) \operatorname{diag}(D_1, D_2, \ldots, D_\ell, \ \lambda_1, \ldots, \lambda_m) \\ &= \operatorname{diag}(HD_1, HD_2, \ldots, HD_\ell, \ \lambda_1, \ldots, \lambda_m) \\ &= \operatorname{diag}\big(r_1\big[R^{\theta_1}\big]H, r_2\big[R^{\theta_2}\big]H, \ldots, r_\ell\big[R^{\theta_\ell}\big]H, \ \lambda_1, \ldots, \lambda_m\big) \\ &= B\tilde{H}. \end{aligned}$$

Note that $\tilde{H}$ is
invertible by
Exercise 2.2.20, since
all of its diagonal
blocks are invertible.

It follows that $\tilde{H}D\tilde{H}^{-1} = B$, so

$$QBQ^{-1} = Q(\tilde{H}D\tilde{H}^{-1})Q^{-1} = (Q\tilde{H})D(Q\tilde{H})^{-1} = PDP^{-1} = A,$$

as claimed, which completes the proof.   ∎

Example 3.C.3

**A Complex
Diagonalization
of a Real Matrix**

Diagonalize the following matrix over $\mathbb{C}$:

$$A = \frac{1}{4}\begin{bmatrix} 3 & 2 & -1 \\ -5 & 2 & 3 \\ -5 & 2 & 7 \end{bmatrix}.$$

Solution:

so we just give a brief rundown of the key properties of A that we must compute:

- Its characteristic polynomial is $p_A(\lambda) = -\lambda^3 + 3\lambda^2 - (5/2)\lambda + 1$.

- Its eigenvalues are $(1+i)/2$, $(1-i)/2$, and 2.

- Eigenvectors corresponding to those eigenvalues, in the same order, are $\mathbf{v} = (1, i, 1)$, $\overline{\mathbf{v}} = (1, -i, 1)$, and $\mathbf{w} = (0, 1, 2)$.

By just placing these eigenvalues along the diagonal of a diagonal matrix D and the eigenvectors as columns (in the same order) into a matrix P, we see that $A = PDP^{-1}$ is a diagonalization of A over $\mathbb{C}$ when

$$D = \begin{bmatrix} (1+i)/2 & 0 & 0 \\ 0 & (1-i)/2 & 0 \\ 0 & 0 & 2 \end{bmatrix} \quad \text{and} \quad P = \begin{bmatrix} 1 & 1 & 0 \\ i & -i & 1 \\ 1 & 1 & 2 \end{bmatrix}.$$

We list the real eigenvalue of A last so as to be consistent with Theorem 3.C.2.

While the complex diagonalization that we found in the above example is fine for algebraic purposes (for example, we could use it to compute large powers of that matrix very quickly), it does not help us visualize A as a linear transformation. To fill in this gap, we now construct its real block diagonalization described by Theorem 3.C.2.

Example 3.C.4

A Real Block Diagonalization of a Real Matrix

Block diagonalize (in the sense of Theorem 3.C.2) the matrix A from Example 3.C.3 and then illustrate how it acts as a linear transformation on $\mathbb{R}^3$.

Solution:

We first (arbitrarily) pick one of the complex (non-real) eigenvalues of A and write it in polar form. We choose the eigenvalue $(1+i)/2$, so $r = \sqrt{(1/2)^2 + (1/2)^2} = 1/\sqrt{2}$ and $\theta = \arctan((1/2)/(1/2)) = \arctan(1) = \pi/4$. The polar form of this eigenvalue is thus $(1+i)/2 = \frac{1}{\sqrt{2}} e^{i\pi/4}$.

If we denote the eigenvectors corresponding to the eigenvalues $(1+i)/2$ and 2 by $\mathbf{v} = (1, i, 1)$ and $\mathbf{w} = (0, 1, 2)$, respectively, it then follows from Theorem 3.C.2 that

$$B = \left[\begin{array}{c|c} \frac{1}{\sqrt{2}}[R^{\pi/4}] & \mathbf{0} \\ \hline \mathbf{0}^T & 2 \end{array} \right] = \begin{bmatrix} 1/2 & -1/2 & 0 \\ 1/2 & 1/2 & 0 \\ 0 & 0 & 2 \end{bmatrix} \quad \text{and}$$

$$Q = \left[\operatorname{Re}(\mathbf{v}) \mid -\operatorname{Im}(\mathbf{v}) \mid \mathbf{w} \right] = \begin{bmatrix} 1 & 0 & 0 \\ 0 & -1 & 1 \\ 1 & 0 & 2 \end{bmatrix}.$$

Geometrically, this means that on the line spanned by the eigenvector $(0, 1, 2)$, A stretches by a factor of 2 (this is nothing new—this is exactly the geometric intuition of eigenvalues that we have had from the start). Furthermore, if we define a plane $\mathcal{S} \subset \mathbb{R}^3$ by

When constructing real block diagonalizations in this way, the rotation in $\mathcal{S}$ is in the direction from $\operatorname{Re}(\mathbf{v})$ toward $-\operatorname{Im}(\mathbf{v})$.

$$\mathcal{S} = \operatorname{span}(\operatorname{Re}(\mathbf{v}), -\operatorname{Im}(\mathbf{v})) = \operatorname{span}((1, 0, 1), (0, -1, 0))$$

then A acts on $\mathcal{S}$ like a rotation by an angle of $\theta = \pi/4$ composed with a scaling by a factor of $r = 1/\sqrt{2}$, as displayed below:

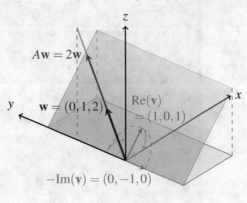

Just like diagonalizations can be used to quickly compute matrix powers, so can the block diagonalizations provided by Theorem 3.C.2. The trick is to recall that $\left[R^\theta\right]^k = \left[R^{k\theta}\right]$ since a rotating $\mathbb{R}^2$ by an angle of $k\theta$ is equivalent to rotating it by an angle of θ a total of k times. It follows that if A is block diagonalized as $A = QBQ^{-1}$ then we can compute its powers quickly via

> It's probably a good idea to have a look back at Example 1.4.12 and Remark 1.4.2 at this point.

$$A^k = (QBQ^{-1})^k = QB^kQ^{-1} = Q\,\text{diag}\left(r_1^k\left[R^{k\theta_1}\right], \ldots, r_\ell^k\left[R^{k\theta_\ell}\right], \ \lambda_1^k, \ldots, \lambda_m^k\right)Q^{-1}.$$

While this method might seem somewhat more involved than just using diagonalization itself at first, it has the advantage over that method of not requiring us to compute large powers of complex numbers. For example, if we tried to compute A^{40} for the matrix A from Example 3.C.3 via its diagonalization, we would have to compute complex numbers like $((1+i)/2)^{40}$, which is perhaps not so easy depending on how comfortable with complex numbers we are (in this particular case, the answer is in fact real: $((1+i)/2)^{40} = 1/2^{20}$).

Example 3.C.5

Large Matrix Powers by Block Diagonalization

Use the block diagonalization of the matrix A from Example 3.C.4 to compute A^{40}.

Solution:
   Recall that $A = QBQ^{-1}$, where

$$B = \left[\begin{array}{c|c} \frac{1}{\sqrt{2}}\left[R^{\pi/4}\right] & \mathbf{0} \\ \hline \mathbf{0}^T & 2 \end{array}\right] \quad \text{and} \quad Q = \begin{bmatrix} 1 & 0 & 0 \\ 0 & -1 & 1 \\ 1 & 0 & 2 \end{bmatrix}.$$

We thus conclude that

$$A^{40} = (QBQ^{-1})^{40} = QB^{40}Q^{-1}$$

$$= \frac{1}{2}\begin{bmatrix} 1 & 0 & 0 \\ 0 & -1 & 1 \\ 1 & 0 & 2 \end{bmatrix} \left[\begin{array}{c|c} \frac{1}{2^{20}}\left[R^{40\pi/4}\right] & \mathbf{0} \\ \hline \mathbf{0}^T & 2^{40} \end{array} \right] \begin{bmatrix} 2 & 0 & 0 \\ -1 & -2 & 1 \\ -1 & 0 & 1 \end{bmatrix}$$

$$= \frac{1}{2}\begin{bmatrix} 1 & 0 & 0 \\ 0 & -1 & 1 \\ 1 & 0 & 2 \end{bmatrix} \begin{bmatrix} 1/2^{20} & 0 & 0 \\ 0 & 1/2^{20} & 0 \\ 0 & 0 & 2^{40} \end{bmatrix} \begin{bmatrix} 2 & 0 & 0 \\ -1 & -2 & 1 \\ -1 & 0 & 1 \end{bmatrix}$$

$$= \frac{1}{2^{21}}\begin{bmatrix} 2 & 0 & 0 \\ 1 - 2^{60} & 2 & 2^{60}-1 \\ 2 - 2^{61} & 0 & 2^{61} \end{bmatrix}.$$

The rotation matrix $\left[R^{40\pi/4}\right]$ is simply the identity matrix, since a rotation by an angle of $\theta = 40\pi/4 = 10\pi$ is equivalent to no rotation at all.

Exercises

solutions to starred exercises on page 473

3.C.1 Construct a decomposition of the form described by Theorem 3.C.1 for each of the following 2×2 matrices. That is, find values for $r, \theta \in \mathbb{R}$ and a matrix $Q \in \mathcal{M}_2(\mathbb{R})$ such that the given matrix has the form $Q\left(r\left[R^{\theta}\right]\right)Q^{-1}$.

*(a) $\begin{bmatrix} 0 & -1 \\ 1 & 0 \end{bmatrix}$ (b) $\begin{bmatrix} 0 & 1 \\ -2 & 0 \end{bmatrix}$

*(c) $\begin{bmatrix} 2 & 1 \\ -2 & 0 \end{bmatrix}$ (d) $\begin{bmatrix} 1 & -\sqrt{3} \\ \sqrt{3} & 1 \end{bmatrix}$

3.C.2 Block diagonalize (in the sense of Theorem 3.C.2) each of the following real matrices.

*(a) $\begin{bmatrix} 1 & 0 & 0 \\ -1 & 0 & 1 \\ 1 & -1 & 0 \end{bmatrix}$ (b) $\begin{bmatrix} 1 & -\sqrt{3} & \sqrt{3} \\ 0 & 3 & 0 \\ -\sqrt{3} & 2 & 1 \end{bmatrix}$

*(c) $\begin{bmatrix} 1 & -2 & 2 & 1 \\ 0 & 1 & 0 & 0 \\ -1 & -1 & 2 & 1 \\ 1 & 0 & 0 & 1 \end{bmatrix}$ (d) $\begin{bmatrix} 4 & -3 & -1 & 2 \\ 6 & -5 & 0 & 1 \\ 4 & -5 & 1 & 0 \\ 0 & -1 & 3 & -2 \end{bmatrix}$

⌨ **3.C.3** Use computer software to diagonalize (in the sense of Theorem 3.C.2) each of the following real matrices.

*(a) $\begin{bmatrix} 2 & 0 & 0 & 0 & -2 \\ 0 & 2 & 1 & 2 & -1 \\ 2 & 0 & 2 & 0 & 0 \\ 0 & -2 & -1 & 2 & 1 \\ 2 & 0 & -2 & 0 & 4 \end{bmatrix}$

(b) $\begin{bmatrix} -2 & -3 & 7 & -1 & -7 & 3 \\ 2 & -9 & -9 & 1 & 13 & -3 \\ -9 & 16 & 31 & -5 & -39 & 9 \\ -13 & 38 & 50 & -8 & -67 & 14 \\ -2 & -1 & 5 & -1 & -5 & 1 \\ 8 & -20 & -29 & 5 & 38 & -9 \end{bmatrix}$

3.C.4 Determine which of the following statements are true and which are false.

*(a) Every matrix $A \in \mathcal{M}_3(\mathbb{R})$ has a real eigenvalue.
(b) Every matrix $A \in \mathcal{M}_4(\mathbb{R})$ has a real eigenvalue.
*(c) Every matrix $A \in \mathcal{M}_n(\mathbb{R})$ that is diagonalizable over $\mathbb{R}$ is diagonalizable over $\mathbb{C}$.
(d) Every matrix $A \in \mathcal{M}_n(\mathbb{R})$ that is diagonalizable over $\mathbb{C}$ is diagonalizable over $\mathbb{R}$.
*(e) Every matrix $A \in \mathcal{M}_n(\mathbb{R})$ can be block diagonalized in the sense of Theorem 3.C.2.
(f) Every rotation matrix $\left[R^{\theta}\right] \in \mathcal{M}_2(\mathbb{R})$ has complex (non-real) eigenvalues.

3.C.5 Compute A^{40} for the matrix A from Example 3.C.4 via complex diagonalization rather than via real block diagonalization as in Example 3.C.5.

*3.C.6 Consider the matrix

$$A = \begin{bmatrix} \sqrt{3}-1 & 1 \\ -2 & \sqrt{3}+1 \end{bmatrix}$$

(a) Diagonalize A and use that diagonalization to find a formula for A^k.
(b) Use block diagonalization to find a formula for A^k that only makes use of real numbers.

3.D   Extra Topic: Linear Recurrence Relations

One of the most useful aspects of diagonalization and matrix functions is their ability to swiftly solve problems that do not even seem to be linear algebraic in nature at first glance. We already saw a hint of how this works in Example 3.4.8, where we developed a closed-form formula for the Fibonacci numbers. In this section, we extend this idea to a larger family of sequences of numbers $x_1, x_2, x_3, \ldots$ that are defined in a similar linear way.

3.D.1   Solving via Matrix Techniques

More specifically, these are called **homogeneous linear recurrence relations with constant coefficients**, but that is a bit of a mouthful.

A **linear recurrence relation** is a formula that represents a term x_n in a sequence of numbers as a linear combination of the immediately previous terms in the sequence. That is, it is a formula for which there exist scalars $a_0, a_1, \ldots, a_{k-1}$ so that

$$x_n = a_{k-1}x_{n-1} + a_{k-2}x_{n-2} + \cdots + a_0 x_{n-k} \quad \text{for all} \quad n \geq k. \qquad (3.D.1)$$

The number of terms in the linear combination is called the **degree** of the recurrence relation (so the recurrence relation (3.D.1) has degree k). In order for the sequence to be completely defined, we also need some **initial conditions** that specify the first term(s) of the sequence. In particular, for a degree-k recurrence relation we require the first k terms to be specified.

For example, the Fibonacci numbers $F_0, F_1, F_2, \ldots$ from Example 3.4.8 satisfy the degree-2 linear recurrence relation $F_n = F_{n-1} + F_{n-2}$, and the initial conditions were the requirements that $F_0 = 0$ and $F_1 = 1$.

Example 3.D.1

Computing Terms via a Linear Recurrence Relation

Compute the first 10 terms of the sequence defined by the linear recurrence relation

$$x_n = 4x_{n-1} - x_{n-2} - 6x_{n-3} \quad \text{for all} \quad n \geq 3,$$

if $x_0 = 3$, $x_1 = -2$, and $x_2 = 8$.

Solution:

We just compute the next terms in the sequence by repeatedly using the given recurrence relation. For example,

$$x_3 = 4x_2 - x_1 - 6x_0 = 4(8) - (-2) - 6(3) = 16,$$
$$x_4 = 4x_3 - x_2 - 6x_1 = 4(16) - 8 - 6(-2) = 68, \quad \text{and}$$
$$x_5 = 4x_4 - x_3 - 6x_2 = 4(68) - 16 - 6(8) = 208.$$

By using this same procedure to compute x_6, x_7, and so on, we see that the first 10 terms of this sequence are

we typically start our sequences at x_0 instead of x_1. We just use this convention to make some results work out more cleanly—it is not actually important.

$$3, -2, 8, 16, 68, 208, 668, 2056, 6308, \text{ and } 19168.$$

It is often desirable to convert a recurrence relation into an explicit formula for the n-th term of the sequence that does not depend on the computation of the previous terms in the sequence. In order to do this, we represent linear recurrence relations via matrices-vector multiplication, just like we did for the Fibonacci numbers in Example 3.4.8. The following definition says how to construct the matrix in this matrix-vector multiplication.

| Definition 3.D.1 | The **companion matrix** of the linear recurrence relation |

$$x_n = a_{k-1}x_{n-1} + a_{k-2}x_{n-2} + \cdots + a_0x_{n-k}$$

Companion Matrix

is the $k \times k$ matrix

We first saw
companion matrices
back in
Exercises 3.3.23
and 3.4.23.

$$C = \begin{bmatrix} 0 & 1 & 0 & \cdots & 0 \\ 0 & 0 & 1 & \cdots & 0 \\ \vdots & \vdots & \vdots & \ddots & \vdots \\ 0 & 0 & 0 & \cdots & 1 \\ a_0 & a_1 & a_2 & \cdots & a_{k-1} \end{bmatrix}.$$

Example 3.D.2

Constructing Companion Matrices

Construct the companion matrices of the following linear recurrence relations:

a) $x_n = x_{n-1} + x_{n-2}$,
b) $x_n = 4x_{n-1} - x_{n-2} - 6x_{n-3}$, and
c) $x_n = x_{n-1} + 4x_{n-3} + 2x_{n-4}$.

Solutions:

The recurrence from
part (a) is the same
as the one from
Example 3.4.8. In
that example, we
used a slightly
different companion
matrix just to make
the algebra work out
slightly cleaner.

a) This is a degree-2 linear recurrence relation with $a_1 = a_0 = 1$, so its companion matrix is

$$C = \begin{bmatrix} 0 & 1 \\ 1 & 1 \end{bmatrix}.$$

b) This is a degree-3 linear recurrence relation with $a_2 = 4$, $a_1 = -1$, and $a_0 = -6$, so its companion matrix is

$$C = \begin{bmatrix} 0 & 1 & 0 \\ 0 & 0 & 1 \\ -6 & -1 & 4 \end{bmatrix}.$$

c) Be careful—this is a degree-4 (*not* degree-3) linear recurrence relation with $a_3 = 1$, $a_2 = 0$, $a_1 = 4$, and $a_0 = 2$, so its companion matrix is

$$C = \begin{bmatrix} 0 & 1 & 0 & 0 \\ 0 & 0 & 1 & 0 \\ 0 & 0 & 0 & 1 \\ 2 & 4 & 0 & 1 \end{bmatrix}.$$

The first main result of this section shows how we can represent the terms in any sequence that is defined by a linear recurrence relation via powers of that relation's companion matrix.

Theorem 3.D.1

Solving Linear Recurrence Relations via Matrix Multiplication

Let $x_0, x_1, x_2, \ldots$ be a sequence satisfying a linear recurrence relation with companion matrix C. For each $n = 0, 1, 2, \ldots$, define $\mathbf{x}_n = (x_n, x_{n+1}, \ldots, x_{n+k-1}) \in \mathbb{R}^k$. Then

$$\mathbf{x}_n = C^n \mathbf{x}_0 \quad \text{for all} \quad n \geq 0.$$

Proof of Theorem 3.D.1. We start by computing

$$C\mathbf{x}_0 = \begin{bmatrix} 0 & 1 & 0 & \cdots & 0 \\ 0 & 0 & 1 & \cdots & 0 \\ \vdots & \vdots & \vdots & \ddots & \vdots \\ 0 & 0 & 0 & \cdots & 1 \\ a_0 & a_1 & a_2 & \cdots & a_{k-1} \end{bmatrix} \begin{bmatrix} x_0 \\ x_1 \\ \vdots \\ x_{k-2} \\ x_{k-1} \end{bmatrix}$$

$$= \begin{bmatrix} x_1 \\ x_2 \\ \vdots \\ x_{k-1} \\ a_0 x_0 + a_1 x_1 + \cdots + a_{k-1} x_{k-1} \end{bmatrix} = \begin{bmatrix} x_1 \\ x_2 \\ \vdots \\ x_{k-1} \\ x_k \end{bmatrix} = \mathbf{x}_1,$$

where we used the given recurrence relation (with $n = k$) to simplify the bottom entry at the right. By multiplying by this matrix n times and repeating this argument, we similarly see that

$$C^n\mathbf{x}_0 = C^{n-1}(C\mathbf{x}_0) = C^{n-1}\mathbf{x}_1 = C^{n-2}(C\mathbf{x}_1) = C^{n-2}\mathbf{x}_2 = \cdots = C\mathbf{x}_{n-1} = \mathbf{x}_n,$$

as desired. ∎

On its own, the above theorem might not seem terribly useful, since computing x_n by repeatedly multiplying a vector by a matrix is more work than just computing it via the original recurrence relation. However, it becomes extremely useful when combined with diagonalization (i.e., the technique of Section 3.4), which lets us derive explicit formulas for powers of matrices.

Notice that $\mathbf{x}_0 = (x_0, x_1, \ldots, x_{k-1})$ contains exactly the initial conditions of the sequence.

In particular, since x_n equals the first entry of $C^n\mathbf{x}_0$ (or the last entry of $C^{n-k+1}\mathbf{x}_0$), any explicit formula for the powers of C immediately gives us an explicit formula for x_n as well. This technique is exactly how we found an explicit formula for the Fibonacci numbers in Example 3.4.8, and we illustrate this method again here with a degree-3 recurrence relation.

Example 3.D.3

Finding a Formula for a Linear Recurrence Relation

This is the same sequence that we investigated in Examples 3.D.1 and 3.D.2(b).

Find an explicit formula for the n-th term of the sequence defined by the recurrence relation

$$x_n = 4x_{n-1} - x_{n-2} - 6x_{n-3} \quad \text{for all} \quad n \geq 3,$$

if $x_0 = 3$, $x_1 = -2$, and $x_2 = 8$.

Solution:

We start by using Theorem 3.D.1, which tells us that x_n equals the top entry of

$$\begin{bmatrix} 0 & 1 & 0 \\ 0 & 0 & 1 \\ -6 & -1 & 4 \end{bmatrix}^n \begin{bmatrix} 3 \\ -2 \\ 8 \end{bmatrix}.$$

To compute the top entry of this vector explicitly, we diagonalize the companion matrix. Its eigenvalues are 3, 2, and -1, with corresponding eigenvectors $(1, 3, 9)$, $(1, 2, 4)$, and $(1, -1, 1)$, respectively. This matrix

P is the transpose of a Vandermonde matrix, which we showed always happens when diagonalizing companion matrices in Exercise 3.4.23.

can thus be diagonalized via

$$D = \begin{bmatrix} 3 & 0 & 0 \\ 0 & 2 & 0 \\ 0 & 0 & -1 \end{bmatrix}, \quad P = \begin{bmatrix} 1 & 1 & 1 \\ 3 & 2 & -1 \\ 9 & 4 & 1 \end{bmatrix}, \quad P^{-1} = \frac{1}{12}\begin{bmatrix} -6 & -3 & 3 \\ 12 & 8 & -4 \\ 6 & -5 & 1 \end{bmatrix}.$$

It follows that

$$\begin{bmatrix} 0 & 1 & 0 \\ 0 & 0 & 1 \\ -6 & -1 & 4 \end{bmatrix}^n \begin{bmatrix} 3 \\ -2 \\ 8 \end{bmatrix} = PD^nP^{-1}\begin{bmatrix} 3 \\ -2 \\ 8 \end{bmatrix}$$

$$= \begin{bmatrix} 1 & 1 & 1 \\ 3 & 2 & -1 \\ 9 & 4 & 1 \end{bmatrix}\begin{bmatrix} 3^n & 0 & 0 \\ 0 & 2^n & 0 \\ 0 & 0 & (-1)^n \end{bmatrix}\left(\frac{1}{12}\begin{bmatrix} -6 & -3 & 3 \\ 12 & 8 & -4 \\ 6 & -5 & 1 \end{bmatrix}\right)\begin{bmatrix} 3 \\ -2 \\ 8 \end{bmatrix}$$

$$= \begin{bmatrix} 1 & 1 & 1 \\ 3 & 2 & -1 \\ 9 & 4 & 1 \end{bmatrix}\begin{bmatrix} 3^n & 0 & 0 \\ 0 & 2^n & 0 \\ 0 & 0 & (-1)^n \end{bmatrix}\begin{bmatrix} 1 \\ -1 \\ 3 \end{bmatrix}$$

$$= \begin{bmatrix} 1 & 1 & 1 \\ 3 & 2 & -1 \\ 9 & 4 & 1 \end{bmatrix}\begin{bmatrix} 3^n \\ -2^n \\ 3(-1)^n \end{bmatrix},$$

and the top entry of this matrix-vector product is $3^n - 2^n + 3(-1)^n$. We thus conclude that $x_n = 3^n - 2^n + 3(-1)^n$ for all n.

3.D.2 Directly Solving When Roots are Distinct

While we could always work through the procedure of Example 3.D.3 in order to find an explicit formula for the terms in a linear recurrence relation (as long as its companion matrix is diagonalizable), there are some observations that we can make to speed up the process. First, we learned in Exercise 3.3.23 that the characteristic polynomial of the companion matrix

$$\begin{bmatrix} 0 & 1 & 0 & \cdots & 0 \\ 0 & 0 & 1 & \cdots & 0 \\ \vdots & \vdots & \vdots & \ddots & \vdots \\ 0 & 0 & 0 & \cdots & 1 \\ a_0 & a_1 & a_2 & \cdots & a_{k-1} \end{bmatrix}$$

is simply $p(\lambda) = (-1)^k(\lambda^k - a_{k-1}\lambda^{k-1} - \cdots - a_1\lambda - a_0)$. Since the leading coefficient of $(-1)^k$ does not affect the roots of this polynomial (i.e., the eigenvalues of the companion matrix), this leads naturally to the following definition:

Definition 3.D.2

Characteristic Polynomial of a Linear Recurrence Relation

The **characteristic polynomial** of the degree-k linear recurrence relation

$$x_n = a_{k-1}x_{n-1} + a_{k-2}x_{n-2} + \cdots + a_0 x_{n-k}$$

is the degree-k polynomial

$$p(\lambda) = \lambda^k - a_{k-1}\lambda^{k-1} - a_{k-2}\lambda^{k-2} - \cdots - a_1\lambda - a_0.$$

A recurrence relation's characteristic polynomial is obtained by moving all terms to one side and replacing each x_{n-j} with λ^{k-j}.

If the roots of this polynomial (i.e., the eigenvalues of the companion matrix) are distinct, then we know from Corollary 3.4.4 that the companion matrix is diagonalizable. We can thus mimic Example 3.D.3 to solve the corresponding recurrence relation in terms of powers of those roots. However, doing so still requires us to compute k eigenvectors, and thus solve k linear systems. The following theorem shows that we can instead solve the recurrence relation by solving just a single linear system instead.

Theorem 3.D.2

Solving Linear Recurrence Relations

Suppose $x_0, x_1, x_2, \ldots$ is a sequence satisfying a linear recurrence relation whose characteristic polynomial has distinct roots $\lambda_0, \lambda_1, \ldots, \lambda_{k-1}$. Let $c_0, c_1, \ldots, c_{k-1}$ be the (necessarily unique) scalars such that

$$x_n = c_0\lambda_0^n + c_1\lambda_1^n + \cdots + c_{k-1}\lambda_{k-1}^n \quad \text{when} \quad 0 \le n < k. \quad (3.D.2)$$

Then the formula (3.D.2) holds for all $n \ge 0$.

In other words, we choose $c_0, c_1, \ldots, c_{k-1}$ so that this formula satisfies the given initial conditions.

Before proving the above theorem, it is worth noting that $\lambda = 0$ is never a root of the characteristic polynomial $p(\lambda) = \lambda^k - a_{k-1}\lambda^{k-1} - a_{k-2}\lambda^{k-2} - \cdots - a_1\lambda - a_0$, so there is never a term of the form 0^n in the explicit formula (3.D.2) (not that it would contribute anything even if it were there). The reason for this is simply that it would imply $a_0 = 0$, which means that the recurrence relation

$$x_n = a_{k-1}x_{n-1} + a_{k-2}x_{n-2} + \cdots + a_0 x_{n-k}$$

actually has degree $k - 1$ (or less), so p is not actually its characteristic polynomial in the first place—a characteristic polynomial cannot have higher degree than the recurrence relation it came from.

The fact that we can find scalars $c_0, c_1, \ldots, c_{k-1}$ satisfying Equation (3.D.2) can be made clearer by writing it out as a linear system. Explicitly, this linear system has the form

For example, plugging $n = 0$ into Equation (3.D.2) gives the first equation in this linear system, plugging in $n = 1$ gives the second equation, and so on.

$$\begin{bmatrix} 1 & 1 & \cdots & 1 \\ \lambda_0 & \lambda_1 & \cdots & \lambda_{k-1} \\ \lambda_0^2 & \lambda_1^2 & \cdots & \lambda_{k-1}^2 \\ \vdots & \vdots & \ddots & \vdots \\ \lambda_0^{k-1} & \lambda_1^{k-1} & \cdots & \lambda_{k-1}^{k-1} \end{bmatrix} \begin{bmatrix} c_0 \\ c_1 \\ c_2 \\ \vdots \\ c_{k-1} \end{bmatrix} = \begin{bmatrix} x_0 \\ x_1 \\ x_2 \\ \vdots \\ x_{k-1} \end{bmatrix}.$$

The fact that this linear system has a unique solution follows from the fact that the coefficient matrix (i.e., the matrix containing the powers of $\lambda_0, \lambda_1, \ldots, \lambda_{k-1}$) is the transpose of a Vandermonde matrix, and we showed that all such matrices are invertible in Example 2.3.8.

Proof of Theorem 3.D.2. We just discussed why $c_0, c_1, \ldots, c_{k-1}$ exist and are

To this end, we recall that we specifically chose $c_0, c_1, \ldots, c_{k-1}$ so that

$$x_n = c_0\lambda_0^n + c_1\lambda_1^n + \cdots + c_{k-1}\lambda_{k-1}^n \quad \text{for} \quad 0 \leq n < k,$$

so we just need to show that this same formula holds for higher values of n. For $n = k$, we use the original recurrence relation to see that

$$
\begin{aligned}
x_k &= \sum_{n=0}^{k-1} a_n x_n && \text{(recurrence that defines } x_k) \\
&= \sum_{n=0}^{k-1} a_n \left(\sum_{j=0}^{k-1} c_j \lambda_j^n \right) && \text{(since } x_n = \sum_{j=0}^{k-1} c_j \lambda_j^n \text{ when } i \leq k-1) \\
&= \sum_{j=0}^{k-1} c_j \left(\sum_{n=0}^{k-1} a_n \lambda_j^n \right) && \text{(swap the order of the sums)} \\
&= \sum_{j=0}^{k-1} c_j \lambda_j^k, && \text{(characteristic polynomial says } \lambda_j^k = \sum_{n=0}^{k-1} a_n \lambda_j^n)
\end{aligned}
$$

This proof can be made a bit more rigorous via induction, if desired.

which is exactly the claimed formula for x_k. Repeating this argument for x_{k+1}, x_{k+2}, and so on shows that the result holds for all n. ∎

The above theorem has a couple of advantages over our previous method based on diagonalization: we only need to solve one linear system (instead of n linear systems for an $n \times n$ matrix), and it can be used even by people who have never heard of eigenvalues, eigenvectors, or diagonalization.

Example 3.D.4

Simpler Method for Solving Linear Recurrence Relations

Find an explicit formula for the n-th term of the sequence defined by the recurrence relation

$$x_n = 8x_{n-1} - 17x_{n-2} + 10x_{n-3} \quad \text{for all} \quad n \geq 3,$$

if $x_0 = 1$, $x_1 = 4$, and $x_2 = 22$.

Solution:

The characteristic polynomial of this recurrence relation is

$$p(\lambda) = \lambda^3 - 8\lambda^2 + 17\lambda - 10 = (\lambda - 5)(\lambda - 2)(\lambda - 1),$$

which has roots $\lambda_0 = 1$, $\lambda_1 = 2$, and $\lambda_2 = 5$. Since these roots are distinct, Theorem 3.D.2 applies, so our goal is to solve the linear system

The matrix on the left contains powers of $\lambda_0 = 1$, $\lambda_1 = 2$, and $\lambda_2 = 5$, while the right-hand side vector contains the initial conditions $x_0 = 1$, $x_1 = 4$, and $x_2 = 22$.

$$
\begin{bmatrix} 1 & 1 & 1 \\ 1 & 2 & 5 \\ 1 & 4 & 25 \end{bmatrix} \begin{bmatrix} c_0 \\ c_1 \\ c_2 \end{bmatrix} = \begin{bmatrix} 1 \\ 4 \\ 22 \end{bmatrix}.
$$

The unique solution is $(c_0, c_1, c_2) = (1, -1, 1)$, so we conclude that

$$x_n = c_0\lambda_0^n + c_1\lambda_1^n + c_2\lambda_2^n = 1 - 2^n + 5^n \quad \text{for all} \quad n \geq 0.$$

Remark 3.D.1

Proving Theorem 3.D.2 via Diagonalization

Another way to demonstrate that the formula provided by Theorem 3.D.2 holds is to represent the sequence via the matrix-vector multiplication $x_n = C^n x_0$, where C is the companion matrix, and then diagonalize C. In particular, we showed in Exercise 3.4.23 that companion matrices with distinct eigenvalues can be diagonalized by the transpose of a Vandermonde matrix. That is, $C = VDV^{-1}$, where

$$V = \begin{bmatrix} 1 & 1 & \cdots & 1 \\ \lambda_0 & \lambda_1 & \cdots & \lambda_{k-1} \\ \lambda_0^2 & \lambda_1^2 & \cdots & \lambda_{k-1}^2 \\ \vdots & \vdots & \ddots & \vdots \\ \lambda_0^{k-1} & \lambda_1^{k-1} & \cdots & \lambda_{k-1}^{k-1} \end{bmatrix} \quad \text{and} \quad D = \begin{bmatrix} \lambda_0 & 0 & \cdots & 0 \\ 0 & \lambda_1 & \cdots & 0 \\ \vdots & \vdots & \ddots & \vdots \\ 0 & 0 & \cdots & \lambda_{k-1} \end{bmatrix}.$$

If we let $\mathbf{c} = (c_0, c_1, \ldots, c_{k-1})$ be a vector for which $V\mathbf{c} = \mathbf{x}_0$, then Theorem 3.D.1 tells us that x_n equals the top entry of

Here we use the fact that if $V\mathbf{c} = \mathbf{x}_0$ then $V^{-1}\mathbf{x}_0 = \mathbf{c}$.

$$\mathbf{x}_n = C^n \mathbf{x}_0 = VD^n V^{-1} \mathbf{x}_0 = VD^n \mathbf{c} \quad \text{for all} \quad n \geq 0.$$

It is straightforward to explicitly compute the top entry of $VD^n\mathbf{c}$ to then see that

$$x_n = c_0 \lambda_0^n + c_1 \lambda_1^n + \cdots + c_{k-1}\lambda_{k-1}^n \quad \text{for all} \quad n \geq 0,$$

as claimed.

It is perhaps worth working through an example to illustrate what happens when the characteristic polynomial of a linear recurrence relation has non-real roots.

Example 3.D.5

Solving a Linear Recurrence Relation with Complex Roots

Find an explicit formula for the n-th term of the sequence defined by the recurrence relation

$$x_n = 2x_{n-1} - x_{n-2} + 2x_{n-3} \quad \text{for all} \quad n \geq 3,$$

if $x_0 = 3$, $x_1 = 8$, and $x_2 = 2$.

Solution:

The characteristic polynomial of this recurrence relation is

$$p(\lambda) = \lambda^3 - 2\lambda^2 + \lambda - 2 = (\lambda - 2)(\lambda^2 + 1),$$

which has roots $\lambda_0 = 2$, $\lambda_1 = i$, and $\lambda_2 = -i$. Since these roots are distinct, Theorem 3.D.2 applies, so our goal is to solve the linear system

Complex roots of real polynomials always come in complex conjugate pairs ($\lambda_1 = i$ and $\lambda_2 = \bar{i} = -i$ here), as do the associated coefficients in the formula for x_n ($c_1 = 1 - 3i$ and $c_2 = \overline{1 - 3i} = 1 + 3i$ here).

$$\begin{bmatrix} 1 & 1 & 1 \\ 2 & i & -i \\ 4 & -1 & -1 \end{bmatrix} \begin{bmatrix} c_0 \\ c_1 \\ c_2 \end{bmatrix} = \begin{bmatrix} 3 \\ 8 \\ 2 \end{bmatrix}.$$

The unique solution is $(c_0, c_1, c_2) = (1, 1 - 3i, 1 + 3i)$, so we conclude that

$$x_n = c_0 \lambda_0^n + c_1 \lambda_1^n + c_2 \lambda_2^n = 2^n + (1 - 3i)i^n + (1 + 3i)(-i)^n \quad \text{for all} \quad n \geq 0.$$

The formula above perhaps seems nonsensical at first, since x_n must be real for all n, but its formula involves complex numbers. However, the imaginary parts of these complex numbers cancel out in such a way that, no matter what value of n is plugged into the formula, the result is always real. For example,

$$x_3 = 2^3 + (1 - 3i)i^3 + (1 + 3i)(-i)^3 = 8 + (-3 - i) + (-3 + i) = 2.$$

If we wish to obtain a formula for x_n consisting entirely of real numbers, we can write the complex roots λ_1 and λ_2 in their polar forms $\lambda_1 = i = e^{i\pi/2}$ and $\lambda_2 = -i = e^{-i\pi/2}$ and simplify:

$$\begin{aligned} x_n &= 2^n + (1 - 3i)i^n + (1 + 3i)(-i)^n \\ &= 2^n + (1 - 3i)e^{i\pi n/2} + (1 + 3i)e^{-i\pi n/2} \\ &= 2^n + (e^{i\pi n/2} + e^{-i\pi n/2}) - 3i(e^{i\pi n/2} - e^{-i\pi n/2}) \\ &= 2^n + 2\cos(\pi n/2) + 6\sin(\pi n/2) \quad \text{for all} \quad n \geq 0, \end{aligned}$$

with the final equality coming from the fact that $e^{i\theta} = \cos(\theta) + i\sin(\theta)$, so $e^{i\theta} + e^{-i\theta} = 2\cos(\theta)$ and $e^{i\theta} - e^{-i\theta} = 2i\sin(\theta)$ (see Appendix A.1).

Alternatively, this same formula could have been obtained by applying the 2×2 block diagonalization techniques of Section 3.C to the companion matrix of this linear recurrence relation.

3.D.3 Directly Solving When Roots are Repeated

If the roots of a recurrence relation's characteristic polynomial are not distinct, then Theorem 3.D.2 does not apply. We might think that we could instead go back to our method based on diagonalization and hope that the companion matrix is diagonalizable, but it turns out that companion matrices are *never* diagonalizable if their eigenvalues are not distinct (this is not obvious, but we proved it in Exercise 3.4.23(b)), so this will not work. Instead, we can make use of the following generalization of Theorem 3.D.2.

Theorem 3.D.3

Solving Linear Recurrence Relations with Repeated Roots

Suppose $x_0, x_1, x_2, \ldots$ is a sequence satisfying a linear recurrence relation whose characteristic polynomial has roots $\lambda_0, \lambda_1, \ldots, \lambda_{m-1}$, with multiplicities $r_0, r_1, \ldots, r_{m-1}$, respectively. Let $q_0, q_1, \ldots, q_{m-1}$ be the (necessarily unique) polynomials with degrees $r_0 - 1, r_1 - 1, \ldots, r_{m-1} - 1$, respectively, such that

$$x_n = \sum_{j=0}^{m-1} q_j(n)\lambda_j^n \quad \text{when} \quad 0 \leq n < k. \quad (3.D.3)$$

Then the formula (3.D.3) holds for all $n \geq 0$.

A polynomial of degree 0 is a constant, so this theorem simplifies to exactly Theorem 3.D.2 if $r_0 = \ldots = r_{m-1} = 1$.

The above theorem seems like quite a mouthful at first glance, but its general structure is very similar to that of Theorem 3.D.2—all that has changed is that the powers of roots are now multiplied by polynomials rather than just scalars. Before proving this result, we work through an example to clarify what it says.

Example 3.D.6

Solving a Linear Recurrence Relation with Repeated Roots

Find an explicit formula for the n-th term of the sequence defined by the recurrence relation

$$x_n = 3x_{n-2} + 2x_{n-3} \quad \text{for all} \quad n \geq 3,$$

if $x_0 = 6$, $x_1 = 0$, and $x_2 = 3$.

Solution:

The characteristic polynomial of this recurrence relation is

$$p(\lambda) = \lambda^3 - 3\lambda - 2 = (\lambda - 2)(\lambda + 1)^2,$$

which has roots $\lambda_0 = 2$ and $\lambda_1 = -1$, with multiplicities 1 and 2, respectively. The polynomials q_0 and q_1 in Theorem 3.D.3 then have degrees 0 and 1, respectively, so they are of the form $q_0(x) = c_0$ and $q_1(x) = c_1 + c_2 x$. Our goal is thus to find c_0, c_1, c_2 such that

$$c_0 2^n + (c_1 + c_2 n)(-1)^n = x_n \quad \text{when} \quad 0 \leq n < 3.$$

By plugging $n = 0$, 1, and 2 into this equation, we arrive at the linear system

$$\begin{bmatrix} 1 & 1 & 0 \\ 2 & -1 & -1 \\ 4 & 1 & 2 \end{bmatrix} \begin{bmatrix} c_0 \\ c_1 \\ c_2 \end{bmatrix} = \begin{bmatrix} 6 \\ 0 \\ 3 \end{bmatrix},$$

which has unique solution $(c_0, c_1, c_2) = (1, 5, -3)$. It follows that

$$x_n = c_0 2^n + (c_1 + c_2 n)(-1)^n = 2^n + (5 - 3n)(-1)^n \quad \text{for all} \quad n \geq 0.$$

> Read the recurrence relation above carefully—it has degree 3, not 2 (the coefficient of x_{n-1} is 0).

> More specifically, the first equation in this linear system comes from plugging in $n = 0$, the second from $n = 1$, and the third from $n = 2$.

Proof of Theorem 3.D.3. We start by showing that $q_0, q_1, \ldots, q_{k-1}$ exist and are unique. To this end, we start by writing the polynomials $q_0, q_1, \ldots, q_{m-1}$ more explicitly as $q_j(n) = c_{j,0} + c_{j,1} n + \cdots + c_{j,r_j-1} n^{r_j-1}$ for $0 \leq j < m$. Then the system of equations (3.D.3) has coefficient matrix

$$V = \begin{bmatrix} V_0 \mid V_1 \mid \cdots \mid V_{m-1} \end{bmatrix},$$

where V_j has the following form for each $0 \leq j < m$:

> Notice that each V_j has size $k \times r_j$, so V has size $k \times (r_0 + \cdots + r_{m-1}) = k \times k$.

$$V_j = \begin{bmatrix} 1 & 0 & 0 & \cdots & 0 \\ \lambda_j & \lambda_j & \lambda_j & \cdots & \lambda_j \\ \lambda_j^2 & 2\lambda_j^2 & 4\lambda_j^2 & \cdots & 2^{r_j-1}\lambda_j^2 \\ \lambda_j^3 & 3\lambda_j^3 & 9\lambda_j^3 & \cdots & 3^{r_j-1}\lambda_j^3 \\ \vdots & \vdots & \vdots & \ddots & \vdots \\ \lambda_j^{k-1} & (k-1)\lambda_j^{k-1} & (k-1)^2\lambda_j^{k-1} & \cdots & (k-1)^{r_j-1}\lambda_j^{k-1} \end{bmatrix}.$$

> The matrix V is the transpose of a Vandermonde matrix if $r_j = 1$ (i.e., V_j is $k \times 1$) for all j. This invertibility proof generalizes Example 2.3.8.

In particular, if we recall that every column in the coefficient matrix corresponds to a variable in the linear system, then the ℓ-th column of V_j (starting counting from $\ell = 0$) corresponds to $c_{j,\ell}$.

To show that $q_0, q_1, \ldots, q_{k-1}$ exist and are unique, it suffices to show that V is invertible (since we can then row-reduce V to I and solve for the $c_{j,\ell}$ coefficients). To this end, we will show that the rows of V form a linearly independent set (which is equivalent to invertibility via Theorem 2.3.5). Well suppose that some linear combination (with coefficients $d_0, d_1, \ldots, d_{k-1}$) of the rows of V equals $\mathbf{0}$. Then

> Keep in mind that
> $0^0 = 1$, so $d_0 0^0 \lambda_j^0 = d_0$,
> but $d_0 0^\ell \lambda_j^0 = 0$ when
> $\ell > 0$.

$$\sum_{n=0}^{k-1} d_n n^\ell \lambda_j^n = 0 \quad \text{for all} \quad 0 \le j < m, \ 0 \le \ell < r_j.$$

We now claim that the above equation implies that the polynomial

$$p(\lambda) = d_0 + d_1 \lambda + \cdots + d_{k-1} \lambda^{k-1}$$

has each of $\lambda_0, \lambda_1, \ldots, \lambda_{k-1}$ as roots with multiplicities $r_0, r_1, \ldots, r_{m-1}$, respectively. We prove this fact in Appendix B.6 (since it is quite long and technical, and the proof we are currently working through is already long enough). Counting multiplicity, we have thus found $r_0 + r_1 + \cdots + r_{m-1} = k$ roots of p, which has degree $k - 1$. It follows that p is the zero polynomial, so $d_0 = d_1 = \cdots = d_{k-1} = 0$. The rows of V are thus linearly independent, so it is invertible, as claimed.

Now that we know that $q_0, q_1, \ldots, q_{m-1}$ exist and are unique, we recall that we specifically chose them so that

$$x_n = q_0(n)\lambda_0^n + q_1(n)\lambda_1^n + \cdots + q_{m-1}(n)\lambda_{m-1}^n \quad \text{when} \quad 0 \le n < k.$$

We thus just need to show that this same formula holds for higher values of n. To do so, we use the original recurrence relation to see that

> The structure of this
> proof is very similar
> to that of
> Theorem 3.D.2, but
> with a couple of
> extra layers of
> ugliness added
> on top.

$$x_k = \sum_{n=0}^{k-1} a_n x_n \qquad \text{(recurrence that defines } x_k)$$

$$= \sum_{n=0}^{k-1} a_n \left(\sum_{j=0}^{k-1} q_j(n)\lambda_j^n \right) \qquad \left(\text{since } x_n = \sum_{j=0}^{k-1} q_j(n)\lambda_j^n \text{ when } n < k\right)$$

$$= \sum_{n=0}^{k-1} a_n \sum_{j=0}^{k-1} \left(\sum_{\ell=0}^{r_j-1} c_{j,\ell} n^\ell \right) \lambda_j^n \qquad \left(\text{since } q_j(n) = \sum_{\ell=0}^{r_j-1} c_{j,\ell} n^\ell\right)$$

$$= \sum_{j=0}^{k-1} \sum_{\ell=0}^{r_j-1} c_{j,\ell} \left(\sum_{n=0}^{k-1} a_n n^\ell \lambda_j^n \right) \qquad \text{(swap the order of the sums)}$$

$$= \sum_{j=0}^{k-1} \sum_{\ell=0}^{r_j-1} c_{j,\ell} \left(k^\ell \lambda_j^k \right) \qquad \text{(Theorem B.6.2: } \lambda_j \text{ is root of char. poly.)}$$

> Again, this argument
> could be made a bit
> more rigorous via
> induction, but the
> proof is already long
> and messy enough
> without it.

$$= \sum_{j=0}^{k-1} q_j(k)\lambda_j^k, \qquad \left(\text{since } q_j(k) = \sum_{\ell=0}^{r_j-1} c_{j,\ell} k^\ell\right)$$

which is exactly the claimed formula for x_k. Repeating this argument for x_{k+1}, x_{k+2}, and so on shows that the result holds for all n. ∎

That proof was a bit rough to get through, so we cleanse our palates by working through another example, this time with a root whose multiplicity is greater than 2.

Example 3.D.7 **Solving a Linear Recurrence Relation with Repeated Roots**	Find an explicit formula for the n-th term of the sequence defined by the recurrence relation $$x_n = 6x_{n-1} - 12x_{n-2} + 10x_{n-3} - 3x_{n-4} \quad \text{for all} \quad n \ge 4,$$ if $x_0 = 0$, $x_1 = 3$, $x_2 = 8$, and $x_3 = 7$.

Solution:

The characteristic polynomial of this recurrence relation is

$$p(\lambda) = \lambda^4 - 6\lambda^3 + 12\lambda^2 - 10\lambda + 3 = (\lambda - 1)^3(\lambda - 3),$$

which has roots $\lambda_0 = 1$ and $\lambda_1 = 3$, with multiplicities 3 and 1, respectively. The polynomials q_0 and q_1 in Theorem 3.D.3 then have degrees 2 and 0, respectively, so they are of the form $q_0(x) = c_0 + c_1 x + c_2 x^2$ and $q_1(x) = c_3$. Our goal is thus to find c_0, c_1, c_2, c_3 such that

$$c_0 + c_1 n + c_2 n^2 + c_3 3^n = x_n \quad \text{when} \quad 0 \le n < 4.$$

By plugging $n = 0$, 1, 2, and 3 into this equation, we arrive at the linear system

$$\begin{bmatrix} 1 & 0 & 0 & 1 \\ 1 & 1 & 1 & 3 \\ 1 & 2 & 4 & 9 \\ 1 & 3 & 9 & 27 \end{bmatrix} \begin{bmatrix} c_0 \\ c_1 \\ c_2 \\ c_3 \end{bmatrix} = \begin{bmatrix} 0 \\ 3 \\ 8 \\ 7 \end{bmatrix},$$

which has unique solution $(c_0, c_1, c_2, c_3) = (1, 2, 3, -1)$. It follows that

$$x_n = c_0 + c_1 n + c_2 n^2 + c_3 3^n = 1 + 2n + 3n^2 - 3^n \quad \text{for all} \quad n \ge 0.$$

Since $1^n = 1$ for all n, the term $(c_0 + c_1 n + c_2 n^2)1^n$ simplifies to just $c_0 + c_1 n + c_2 n^2$ here.

In the explicit formulas for sequences defined by linear recurrence relations that are provided by Theorems 3.D.2 and 3.D.3, if n is large then the term involving largest root of the characteristic polynomial has by far the biggest effect on the overall size of x_n. For example, the following table lists the first 18 Fibonacci numbers (recall from Example 3.4.8 that the Fibonacci numbers satisfy $F_0 = 0$, $F_1 = 1$, and $F_n = F_{n-1} + F_{n-2}$ when $n \ge 2$), as well as the ratios between consecutive terms in the sequence:

n	F_n	F_n/F_{n-1}	n	F_n	F_n/F_{n-1}	n	F_n	F_n/F_{n-1}
0	0	–	6	8	1.6000	12	144	1.6180
1	1	–	7	13	1.6250	13	233	1.6181
2	1	1.0000	8	21	1.6154	14	377	1.6180
3	2	2.0000	9	34	1.6190	15	610	1.6180
4	3	1.5000	10	55	1.6176	16	987	1.6180
5	5	1.6667	11	89	1.6182	17	1597	1.6180

It appears that the ratio between terms converges to some number around 1.6180. To see why this is the case, recall the explicit formula for the Fibonacci numbers:

Since $1 - \phi \approx -0.6180$, which is smaller than 1 in absolute value, $(1 - \phi)^n \to 0$ as $n \to \infty$.

$$F_n = \frac{1}{\sqrt{5}} (\phi^n - (1-\phi)^n), \quad \text{where} \quad \phi = \frac{1 + \sqrt{5}}{2} \approx 1.6180.$$

In particular, since ϕ^n is significantly larger than $(1-\phi)^n$ when n is large, F_n grows in roughly the same way that $\phi^n/\sqrt{5}$ grows, so the Fibonacci sequence grows by roughly a factor of ϕ from term to term. The following result makes this observation precise and generalizes it to other linear recurrence relations.

Corollary 3.D.4

Growth Rate of Linear Recurrence Relations

Suppose $x_0, x_1, x_2, \ldots$ is a sequence satisfying a linear recurrence relation whose characteristic polynomial has roots $\lambda_0, \lambda_1, \ldots, \lambda_{m-1}$ (which may have multiplicities greater than 1, but they are listed here just once). Suppose that one of these roots, λ_0, has strictly larger absolute value than the others:

$$|\lambda_0| > |\lambda_j| \quad \text{for all} \quad 0 < j < m.$$

If the above inequality is not strict, then this limit may not even exist (see Example 3.D.8).

Then

$$\lim_{n \to \infty} \frac{x_{n+1}}{x_n} = \lambda_0.$$

Proof. We start by using Theorem 3.D.3 to write

$$x_n = q_0(n)\lambda_0^n + q_1(n)\lambda_1^n + \cdots + q_{m-1}(n)\lambda_{m-1}^n,$$

where $q_0, q_1, \ldots, q_{m-1}$ are the polynomials described by that theorem. Then

$$\lim_{n \to \infty} \frac{x_{n+1}}{x_n} = \lim_{n \to \infty} \frac{q_0(n+1)\lambda_0^{n+1} + \cdots + q_{m-1}(n+1)\lambda_{m-1}^{n+1}}{q_0(n)\lambda_0^n + \cdots + q_{m-1}(n)\lambda_{m-1}^n}$$

Dividing the numerator and denominator by $q_0(n)\lambda_0^n$ here is a standard technique for evaluating limits of fractions—divide by the fastest-growing term in the denominator.

$$= \lim_{n \to \infty} \frac{q_0(n+1)\lambda_0^{n+1} + \cdots + q_{m-1}(n+1)\lambda_{m-1}^{n+1}}{q_0(n)\lambda_0^n + \cdots + q_{m-1}(n)\lambda_{m-1}^n} \cdot \frac{\left(\frac{1}{q_0(n)\lambda_0^n}\right)}{\left(\frac{1}{q_0(n)\lambda_0^n}\right)}$$

$$= \lim_{n \to \infty} \frac{\frac{q_0(n+1)}{q_0(n)}\lambda_0 + \cdots + \lambda_{m-1}\frac{q_{m-1}(n+1)}{q_0(n)}\left(\frac{\lambda_{m-1}}{\lambda_0}\right)^n}{1 + \cdots + \frac{q_{m-1}(n)}{q_0(n)}\left(\frac{\lambda_{m-1}}{\lambda_0}\right)^n}$$

$$= \frac{\lambda_0 + 0 + \cdots + 0}{1 + 0 + \cdots + 0}$$

$$= \lambda_0,$$

where the second-to-last equality comes from the facts that, for any polynomial q and any number c with $|c| < 1$, we have

$$\lim_{n \to \infty} \frac{q(n+1)}{q(n)} = 1 \quad \text{and} \quad \lim_{n \to \infty} q(n)c^n = 0.$$

The two limits above are hopefully intuitive enough ($q(n+1)$ and $q(n)$ have the same leading terms and the exponential c^n goes to 0 faster than the polynomial $q(n)$ goes to infinity), so we leave their proofs to Appendix B.7. ∎

For example, if we go way back to the sequence defined by $x_0 = 3$, $x_1 = -2$, $x_2 = 8$, and $x_n = 4x_{n-1} - x_{n-2} - 6x_{n-3}$ for $n \geq 3$ from Example 3.D.1, it seems that every term is roughly 3 times as large as the previous one (e.g., $x_8 = 6308$, which is roughly 3 times as large as $x_7 = 2056$). This now makes sense, since we showed in Example 3.D.3 that the roots of its characteristic polynomial are 3, 2, and -1.

Similarly, the sequences from Examples 3.D.4, 3.D.5, 3.D.6, and 3.D.7 have limiting growth rates of 5, 2, 2, and 3, respectively. It is perhaps worth working through a couple of examples for which Corollary 3.D.4 does not apply.

Example 3.D.8

Sequences Without a Limiting Ratio

Explain why Corollary 3.D.4 does not apply to each of the following sequences:

a) $x_0 = 0$, $x_1 = 0$, $x_2 = 1$, $x_n = x_{n-1} - 4x_{n-2} + 4x_{n-3}$ when $n \geq 3$, and
b) $x_0 = 0$, $x_1 = 1$, $x_2 = 0$, $x_n = 2x_{n-1} + 9x_{n-2} - 18x_{n-3}$ when $n \geq 3$.

Solutions:

a) The characteristic polynomial of this linear recurrence relation is

$$p(\lambda) = \lambda^3 - \lambda^2 + 4\lambda - 4 = (\lambda - 1)(\lambda^2 + 4),$$

If the largest root (in absolute value) λ_0 is not real then Corollary 3.D.4 does not apply, since $\overline{\lambda_0}$ is another root with the same absolute value.

which has roots 1, $2i$, and $-2i$. Since the absolute values of these roots are 1, 2, and 2, respectively, there is no one root that has *strictly* larger absolute value than the rest, so Corollary 3.D.4 does not apply.

To get a bit of a better feel for what is happening in this sequence, we list out its first several terms explicitly:

$$0, \ 0, \ 1, \ 1, \ -3, \ -3, \ 13, \ 13, \ -51, \ -51, \ 205, \ 205, \ \ldots$$

We *prove* the claim that the terms in this sequence repeat in Exercise 3.D.8.

It is now a bit more apparent that the limiting ratio discussed by Corollary 3.D.4 indeed does not exist, since this sequence alternates back and forth between repeating the previous term (so $x_n/x_{n-1} = 1$) and growing by roughly a factor of -4 (so $x_n/x_{n-1} \approx -4$). The fact that this sequence grows by a factor of -4 every *two* terms is explained by the fact that the two largest roots of its characteristic polynomial satisfy $(2i)^2 = (-2i)^2 = -4$.

b) The characteristic polynomial of this linear recurrence relation is

$$p(\lambda) = \lambda^3 - 2\lambda^2 - 9\lambda + 18 = (\lambda - 2)(\lambda - 3)(\lambda + 3),$$

which has roots 2, 3, and -3. Since the absolute values of these roots are 2, 3, and 3, respectively, there is no one root that has *strictly* larger absolute value than the rest, so Corollary 3.D.4 does not apply.

Again, it is perhaps helpful to write out the first several terms of the sequence explicitly to see what is happening here:

$$0, \ 1, \ 0, \ 9, \ 0, \ 81, \ 0, \ 729, \ 0, \ 6561, \ 0, \ 59049, \ \ldots$$

This sequence just alternates back and forth between zeros and the powers of 9, so the limiting ratio of Corollary 3.D.4 indeed does not exist—the ratio x_n/x_{n-1} alternates back and forth between 0 and being undefined. The fact that this sequence grows by a factor of 9 every *two* terms is explained by the fact that the two largest roots of its characteristic polynomial satisfy $3^2 = (-3)^2 = 9$.

Remark 3.D.2

The Look-and-Say Sequence

As one particularly interesting and surprising application of the techniques from this section, consider the sequence that begins

$$1, \ 11, \ 21, \ 1211, \ 111221, \ 312211, \ 13112221, \ 1113213211, \ \ldots$$

This is called the **look-and-say sequence**, since each of its terms after the initial 1 is generated by "reading" the digits of the previous term. For example, the term 111221 can be read as "three 1s, two 2s, one 1", so the next term in the sequence is 312211. This term can be read as "one 3, one 1, two 2s, two 1s", so the next term in the sequence is 13112221, and so on.

Although this sequence seems somewhat silly and "non-mathematical", the number of digits in the terms of this sequence behave in a very interesting way. Let x_n denote the number of digits in the n-th term of the look-and-say sequence—for example, $x_5 = 6$ since the fifth terms in the look-and-say sequence is 111221, which has 6 digits. Remarkably, the sequence $x_1, x_2, x_3, \ldots$ satisfies the following degree-74(!!) recurrence relation:

The fact that x_n satisfies this recurrence relation (or any linear recurrence relation at all) is very not obvious—it was proved in (Con86).

$$
\begin{aligned}
x_n = \ & x_{n-1} + 2x_{n-2} - 2x_{n-4} - 4x_{n-5} + x_{n-6} + 4x_{n-7} + 2x_{n-8} \\
& - x_{n-9} - 2x_{n-10} - 3x_{n-13} - 3x_{n-14} + 5x_{n-15} + 8x_{n-16} + 6x_{n-17} \\
& - 12x_{n-18} - 9x_{n-19} - x_{n-20} + x_{n-21} + 13x_{n-22} - 8x_{n-23} + 7x_{n-24} \\
& + 12x_{n-25} - 11x_{n-26} - 4x_{n-27} - 19x_{n-28} + 4x_{n-29} + 16x_{n-30} + 9x_{n-31} \\
& + 9x_{n-32} - 32x_{n-33} + 7x_{n-34} + 5x_{n-35} - 8x_{n-36} + 18x_{n-37} - 19x_{n-38} \\
& + 5x_{n-39} + 20x_{n-40} - 13x_{n-41} + x_{n-42} - 3x_{n-43} - 13x_{n-44} + 17x_{n-46} \\
& - 10x_{n-47} + 10x_{n-48} + 9x_{n-49} - 15x_{n-50} - 8x_{n-51} - x_{n-52} + 23x_{n-53} \\
& - 10x_{n-55} - 25x_{n-56} + 8x_{n-57} + 24x_{n-58} + 9x_{n-59} - 13x_{n-60} - 16x_{n-61} \\
& + x_{n-62} + 7x_{n-63} + 6x_{n-64} + 3x_{n-65} - 13x_{n-66} + 3x_{n-67} - 2x_{n-68} \\
& + 7x_{n-69} + 7x_{n-70} - 9x_{n-71} + 3x_{n-72} - 9x_{n-73} + 6x_{n-74}
\end{aligned}
$$

Be careful—the coefficient of each of $x_{n-3}, x_{n-11}, x_{n-12}, x_{n-45}, x_{n-54}$ is 0.

The characteristic polynomial of the recurrence relation above factors as $p(\lambda) = (\lambda + 1)(\lambda - 1)^2 q(\lambda)$, where $q(\lambda)$ is the following degree-71 polynomial:

This degree-71 polynomial does not have any rational roots, so we cannot really factor it any further.

$$
\begin{aligned}
q(\lambda) = \ & \lambda^{71} - \lambda^{69} - 2\lambda^{68} - \lambda^{67} + 2\lambda^{66} + 2\lambda^{65} + \lambda^{64} - \lambda^{63} \\
& - \lambda^{62} - \lambda^{61} - \lambda^{60} - \lambda^{59} + 2\lambda^{58} + 5\lambda^{57} + 3\lambda^{56} - 2\lambda^{55} \\
& - 10\lambda^{54} - 3\lambda^{53} - 2\lambda^{52} + 6\lambda^{51} + 6\lambda^{50} + \lambda^{49} + 9\lambda^{48} - 3\lambda^{47} \\
& - 7\lambda^{46} - 8\lambda^{45} - 8\lambda^{44} + 10\lambda^{43} + 6\lambda^{42} + 8\lambda^{41} - 5\lambda^{40} - 12\lambda^{39} \\
& + 7\lambda^{38} - 7\lambda^{37} + 7\lambda^{36} + \lambda^{35} - 3\lambda^{34} + 10\lambda^{33} + \lambda^{32} - 6\lambda^{31} \\
& - 2\lambda^{30} - 10\lambda^{29} - 3\lambda^{28} + 2\lambda^{27} + 9\lambda^{26} - 3\lambda^{25} + 14\lambda^{24} - 8\lambda^{23} \\
& - 7\lambda^{21} + 9\lambda^{20} + 3\lambda^{19} - 4\lambda^{18} - 10\lambda^{17} - 7\lambda^{16} + 12\lambda^{15} + 7\lambda^{14} \\
& + 2\lambda^{13} - 12\lambda^{12} - 4\lambda^{11} - 2\lambda^{10} + 5\lambda^9 + \lambda^7 - 7\lambda^6 + 7\lambda^5 \\
& - 4\lambda^4 + 12\lambda^3 - 6\lambda^2 + 3\lambda - 6
\end{aligned}
$$

We could thus use Theorem 3.D.3 to come up with an explicit formula for x_n in terms of powers of 1, -1, and the 71 (distinct) roots of the polynomial q. It is not realistic to write down this entire formula, but we can at least plot those 71 roots to see where they are located in the complex plane:

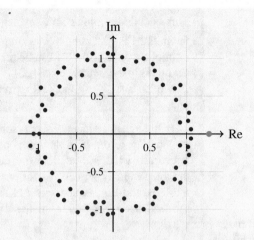

Of these 71 roots,
only 3 are real. The 3
real roots are
approximately
$-1.0882, -1.0112,$
and $1.3036.$

Of particular note is the root approximately equal to 1.3036 (high-lighted in orange above), which is larger in absolute value than any of the other roots. In light of Corollary 3.D.4, this tells us that

$$\lim_{n\to\infty} \frac{x_{n+1}}{x_n} \approx 1.3036.$$

In words, this means that when n is large, the number of digits in the terms of the look-and-say sequence increases by approximately 30.36% from one term to the next.

This limiting ratio, approximately equal to 1.3036, is interesting for the fact that it is a root of a degree-71 polynomial with integer coefficients, but not of any lower-degree polynomial with integer coefficients. Many numbers like $\sqrt{2}$ or $\phi = (1+\sqrt{5})/2$ are roots of low-degree polynomials with integer coefficients (degree 2 in those two cases), and many numbers like π and e are not roots of any polynomial with integer coefficients at all. This limiting ratio is one of the few examples of a naturally-arising number that is the root of a polynomial with integer coefficients, but only a very high-degree one.

Numbers like π and e
that are not roots of
any polynomial with
integer coefficients
are called
transcendental.

Exercises

solutions to starred exercises on page 474

3.D.1 Compute the (potentially complex) roots, and their multiplicities, of the characteristic polynomials of the each of the following linear recurrence relations.

*(a) $x_n = x_{n-1} + 6x_{n-2}$

(b) $x_n = 6x_{n-1} - 9x_{n-2}$

*(c) $x_n = 6x_{n-1} - 4x_{n-2}$

(d) $x_n = 2x_{n-1} - 2x_{n-2}$

*(e) $x_n = 6x_{n-1} - 11x_{n-2} + 6x_{n-3}$

(f) $x_n = 2x_{n-2} + x_{n-3}$

*(g) $x_n = 6x_{n-1} - 12x_{n-2} + 8x_{n-3}$

(h) $x_n = 7x_{n-1} - 8x_{n-2} - 16x_{n-3}$

*(i) $x_n = 5x_{n-1} - 4x_{n-2} - 6x_{n-3}$

(j) $x_n = 7x_{n-1} - 17x_{n-2} + 15x_{n-3}$

3.D.2 Write the companion matrix for each of the linear recurrence relations given in Exercise 3.D.1.

3.D.3 Solve each of the following linear recurrence relations. That is, find an explicit formula for the n-th term of each of the following sequences.

*(a) $x_0 = 4, x_1 = -3,$
    $x_n = x_{n-1} + 6x_{n-2}$ when $n \geq 2$

(b) $x_0 = 1, x_1 = 0,$
    $x_n = 6x_{n-1} - 9x_{n-2}$ when $n \geq 2$

*(c) $x_0 = 2, x_1 = 0,$
    $x_n = -x_{n-2}$ when $n \geq 2$

(d) $x_0 = 2, x_1 = 3,$
    $x_n = 2x_{n-1} - x_{n-2}$ when $n \geq 2$

*(e) $x_0 = 0, x_1 = 2, x_2 = 8,$
    $x_n = 6x_{n-1} - 11x_{n-2} + 6x_{n-3}$ when $n \geq 3$

(f) $x_0 = 0, x_1 = 2, x_2 = 16,$
    $x_n = 6x_{n-1} - 12x_{n-2} + 8x_{n-3}$ when $n \geq 3$

*(g) $x_0 = -2, x_1 = 1, x_2 = 5,$
    $x_n = 4x_{n-1} - 5x_{n-2} + 2x_{n-3}$ when $n \geq 3$

(h) $x_0 = 7$, $x_1 = 8$, $x_2 = 4$,
$x_n = 4x_{n-1} - 6x_{n-2} + 4x_{n-3}$ when $n \geq 3$

3.D.4 Determine which of the following statements are true and which are false.

*(a) The roots of the characteristic polynomial of a linear recurrence relation must all be real.

(b) The sequence defined by the formula $x_n = 2^n - 3^n$ satisfies a linear recurrence relation.

*(c) The sequence defined by the formula $x_n = n^2 - n^3$ satisfies a linear recurrence relation.

(d) The sequence defined by the formula $x_n = \sqrt{2n} - \sqrt{3n}$ satisfies a linear recurrence relation.

*(e) If the characteristic polynomial of a linear recurrence relation has roots 4, 3, and 2, each with multiplicity 2, then $\lim_{n\to\infty} x_{n+1}/x_n = 4$.

3.D.5 For each of the following formulas, find a linear recurrence relation that the formula satisfies.

*(a) $x_n = 2^n + 3^n$

(b) $x_n = 2^n - 3^{n+1}$

*(c) $x_n = 2^n + 3^n + 4^n$

(d) $x_n = (1+n)2^n + 3^n$

*(e) $x_n = (1+n+n^2)2^n + 3^n$

(f) $x_n = 7 + 2^n$

*(g) $x_n = n^2$

(h) $x_n = 2(1+i)^n + 2(1-i)^n$

*(i) $x_n = (n-5)3^n + (3+i)(1-2i)^n + (3-i)(1+2i)^n$

(j) $x_n = 2^n \cos(\pi n/4) - 2^n \sin(\pi n/4)$

3.D.6 Find an explicit formula for the n-th term of each of the following sequences that only involves real numbers and real-valued functions (even though their characteristic polynomials all have non-real roots).

[Hint: Mimic the procedure of Example 3.D.5.]

*(a) $x_0 = 2$, $x_1 = 0$,
$x_n = -x_{n-2}$ when $n \geq 2$

(b) $x_0 = 4$, $x_1 = 4$,
$x_n = 2x_{n-1} - 2x_{n-2}$ when $n \geq 2$

*(c) $x_0 = 3$, $x_1 = 1$, $x_2 = -7$,
$x_n = x_{n-1} - 4x_{n-2} + 4x_{n-3}$ when $n \geq 3$

(d) $x_0 = -1$, $x_1 = 0$, $x_2 = 8$,
$x_n = 4x_{n-1} - 8x_{n-2} + 8x_{n-3}$ when $n \geq 3$

3.D.7 Consider the sequence defined by the formula $x_n = \lfloor (4+\sqrt{11})^n \rfloor$, where $\lfloor \cdot \rfloor$ denotes the "floor" function that rounds down the input, discarding its decimal part. Show that x_n satisfies a linear recurrence relation.

[Hint: The characteristic polynomial of the linear recurrence relation should have $4+\sqrt{11}$ as a root.]

**3.D.8 In Example 3.D.8(a), we claimed that the sequence defined by the linear recurrence relation $x_n = x_{n-1} - 4x_{n-2} + 4x_{n-3}$, with initial conditions $x_0 = 0$, $x_1 = 0$ and $x_2 = 1$, repeats every term (for example, $x_0 = x_1, x_2 = x_3$, and so on). Prove this claim.

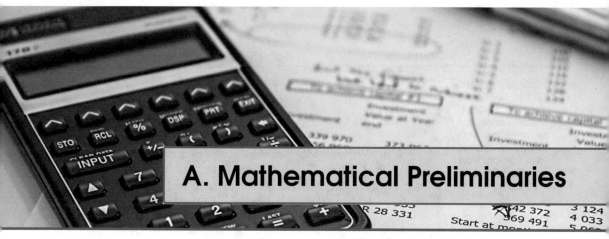

A. Mathematical Preliminaries

In this appendix, we present some of the miscellaneous bits of mathematical knowledge that are not topics of linear algebra themselves, but are nevertheless useful and might be missing from the reader's toolbox.

A.1 Complex Numbers

Throughout this book, the vectors, matrices, and other linear algebraic objects that we work with almost always have entries that come from the set of real numbers (denoted by $\mathbb{R}$), though we sometimes make use of the set of complex numbers (denoted by $\mathbb{C}$) in order to simplify our work. Since complex numbers make linear algebra work so nicely (especially once we start computing eigenvalues in Section 3.3), we now introduce some of their basic properties.

The core idea behind complex numbers is to extend the real numbers in a way that lets us take square roots of negative numbers. To this end, we let i be an object with the property that $i^2 = -1$. It is clear that i cannot be a real number, but we nonetheless think of it like a number anyway, as we will see that we can manipulate it much like we manipulate real numbers. We call any real scalar multiple of i like $2i$ or $-(7/3)i$ an **imaginary number**, and such numbers obey the same laws of arithmetic that we would expect them to (e.g., $2i + 3i = 5i$ and $(3i)^2 = 3^2 i^2 = -9$).

We then let $\mathbb{C}$, the set of **complex numbers**, be the set

$$\mathbb{C} \stackrel{\text{def}}{=} \{a + bi : a, b \in \mathbb{R}\}$$

in which addition and multiplication work exactly as they do for real numbers, as long as we keep in mind that $i^2 = -1$. We call a the **real part** of $a + bi$, and it is sometimes convenient to denote it by $\mathrm{Re}(a + bi) = a$. We similarly call b its **imaginary part** and denote it by $\mathrm{Im}(a + bi) = b$.

Remark A.1.1

Yes, We Can Do That

It might seem extremely strange at first that we can just define a new number i and start doing arithmetic with it. However, this is perfectly fine, and we do this type of thing all the time—one of the beautiful things about mathematics is that we can define whatever we like. However, for that definition to actually be *useful*, it should mesh well with other definitions and objects that we use.

The reason that complex numbers are so useful is that they extend the set of real numbers in such a way that they allow us to solve any polynomial equation (see the upcoming fundamental theorem of algebra—

© Springer Nature Switzerland AG 2021
N. Johnston, *Introduction to Linear and Matrix Algebra*,
https://doi.org/10.1007/978-3-030-52811-9

We actually do lose one thing when we go from $\mathbb{R}$ to $\mathbb{C}$: we cannot order the members of $\mathbb{C}$ in a way that meshes well with addition and multiplication like we can on $\mathbb{R}$.

Theorem A.2.3), yet they do not break any of the usual laws of arithmetic. That is, we still have all of the properties like $ab = ba$, $a+b = b+a$, and $a(b+c) = ab+ac$ for all $a,b,c \in \mathbb{C}$ that we would expect "numbers" to have.

By way of contrast, suppose that we tried to similarly add a new number that lets us divide by zero. If we let ε be a number with the property that $\varepsilon \times 0 = 1$ (i.e., we are thinking of ε as $1/0$ much like we think of i as $\sqrt{-1}$), then we have

$$1 = \varepsilon \times 0 = \varepsilon \times (0+0) = (\varepsilon \times 0) + (\varepsilon \times 0) = 1+1 = 2.$$

We thus cannot work with such a number without breaking at least one of the usual laws of arithmetic.

A.1.1   Basic Arithmetic and Geometry

For the most part, arithmetic involving complex numbers works simply how we might expect it to. For example, to add two complex numbers together we just add up their real and imaginary parts: $(a+bi)+(c+di) = (a+c)+(b+d)i$, which is hopefully not surprising (it is completely analogous to how we can group and add real numbers and vectors). For example,

$$(3+7i)+(2-4i) = (3+2)+(7-4)i = 5+3i.$$

If you are familiar with the acronym "FOIL" for multiplying binomials together, that is what we are using here.

Similarly, to multiply two complex numbers together we just distribute parentheses like we usually do when we multiply numbers together, and we make use of the fact that $i^2 = -1$:

$$(a+bi)(c+di) = ac+bci+adi+bdi^2 = (ac-bd)+(ad+bc)i.$$

For example,

$$(3+7i)(4+2i) = (12-14)+(6+28)i = -2+34i.$$

You can find the solutions of a polynomial equation like this one via the quadratic formula, which we cover in the next section.

One of the primary uses of complex numbers is that they can be used to solve arbitrary polynomial equations—even ones that do not have real solutions. For example, the equation $x^2 - 2x + 5 = 0$ has no real solutions, but it has $x = 1+2i$ as a complex solution, since

$$\begin{aligned}
(1+2i)^2 - 2(1+2i) + 5 &= (1+2i)(1+2i) - (2+4i) + 5 \\
&= (-3+4i) - (2+4i) + 5 \\
&= (-5+5) + (4-4)i \\
&= 0.
\end{aligned}$$

Once we have the (potentially complex) solutions of polynomial equations like this one, we can use them to factor and simplify that polynomial, or better understand the underlying problem that led to that polynomial in the first place.

Much like we think of $\mathbb{R}$ as a line, we think of $\mathbb{C}$ as a plane, which we call the **complex plane**. The set of real numbers takes the place of the x-axis (which we call the real axis) and the set of imaginary numbers takes the place of the y-axis (which we called the imaginary axis). The complex number $a+bi$ thus has coordinates (a,b) on that plane, as in Figure A.1.

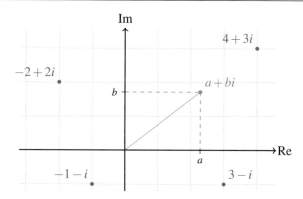

Figure A.1: The complex plane is a representation of the set $\mathbb{C}$ of complex numbers.

We thus think of $\mathbb{C}$ much like we think of $\mathbb{R}^2$ (i.e., we can think of the complex number $a + bi \in \mathbb{C}$ as the vector $(a, b) \in \mathbb{R}^2$), but with a multiplication operation that we do not have on $\mathbb{R}^2$. With this in mind, we define the **magnitude** of a complex number $a + bi$ to be the quantity

The magnitude of a real number is simply its absolute value: $|a| = \sqrt{a^2}$.

$$|a + bi| \overset{\text{def}}{=} \sqrt{a^2 + b^2},$$

which is simply the length of the associated vector (a, b) (i.e., it is the distance between $a + bi$ and the origin in the complex plane).

A.1.2 The Complex Conjugate

One particularly important operation on complex numbers that does *not* have any natural analog on the set of real numbers is **complex conjugation**, which negates the imaginary part of a complex number and leaves its real part alone. We denote this operation by putting a horizontal bar over the complex number it is being applied to so that, for example,

$$\overline{3 + 4i} = 3 - 4i, \qquad \overline{5 - 2i} = 5 + 2i, \qquad \overline{3i} = -3i, \qquad \text{and} \qquad \overline{7} = 7.$$

Geometrically, complex conjugation corresponds to reflecting a number in the complex plane through the real axis, as in Figure A.2.

Applying the complex conjugate twice simply undoes it:
$$\overline{\overline{a + bi}} = \overline{a - bi} = a + bi.$$

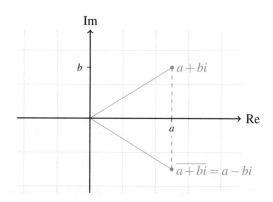

Figure A.2: Complex conjugation reflects a complex number through the real axis.

Algebraically, complex conjugation is useful since many other common operations involving complex numbers can be expressed in terms of it. For example:

- The magnitude of a complex number $z = a + bi$ can be written in terms of the product of z with its complex conjugate: $|z|^2 = z\bar{z}$, since

The product $(a+bi)(a-bi)$ here simplifies a lot because its imaginary part is $(ab-ab)i = 0$.

$$z\bar{z} = (a+bi)\overline{(a+bi)} = (a+bi)(a-bi)$$
$$= a^2 + b^2 = |a+bi|^2 = |z|^2.$$

- The previous point tells us that we can multiply any complex number by another one (its complex conjugate) to get a real number. We can make use of this fact to come up with a method of dividing by complex numbers:

In the first step here, we just cleverly multiply by 1 so as to make the denominator real.

$$\frac{a+bi}{c+di} = \left(\frac{a+bi}{c+di}\right)\left(\frac{c-di}{c-di}\right)$$
$$= \frac{(ac+bd)+(bc-ad)i}{c^2+d^2} = \left(\frac{ac+bd}{c^2+d^2}\right) + \left(\frac{bc-ad}{c^2+d^2}\right)i.$$

- The real and imaginary parts of a complex number $z = a + bi$ can be computed via

$$\operatorname{Re}(z) = \frac{z+\bar{z}}{2} \quad \text{and} \quad \operatorname{Im}(z) = \frac{z-\bar{z}}{2i},$$

since

$$\frac{z+\bar{z}}{2} = \frac{(a+bi)+(a-bi)}{2} = \frac{2a}{2} = a = \operatorname{Re}(z) \quad \text{and}$$
$$\frac{z-\bar{z}}{2i} = \frac{(a+bi)-(a-bi)}{2i} = \frac{2bi}{2i} = b = \operatorname{Im}(z).$$

A.1.3   Euler's Formula and Polar Form

Since we can think of complex numbers as points in the complex plane, we can specify them via their length and direction rather than via their real and imaginary parts. In particular, we can write every complex number in the form $z = |z|u$, where $|z|$ is the magnitude of z and u is a number on the unit circle in the complex plane.

The notation $\theta \in [0, 2\pi)$ means that θ is between 0 (inclusive) and 2π (non-inclusive).

By recalling that every point on the unit circle in $\mathbb{R}^2$ has coordinates of the form $(\cos(\theta), \sin(\theta))$ for some $\theta \in [0, 2\pi)$, we see that every point on the unit circle in the complex plane can be written in the form $\cos(\theta) + i\sin(\theta)$, as illustrated in Figure A.3.

It follows that we can write every complex number in the form $z = |z|(\cos(\theta) + i\sin(\theta))$. However, we can simplify this expression somewhat by using the remarkable fact, called **Euler's formula**, that

In other words, $\operatorname{Re}(e^{i\theta}) = \cos(\theta)$ and $\operatorname{Im}(e^{i\theta}) = \sin(\theta)$.

$$e^{i\theta} = \cos(\theta) + i\sin(\theta) \quad \text{for all} \quad \theta \in [0, 2\pi).$$

If you are not familiar with the exponential function e^x, then you can think of this formula simply as a definition—just think of $e^{i\theta}$ as a shorthand for the expression $\cos(\theta) + i\sin(\theta)$. However, if you have seen the function e^x and are

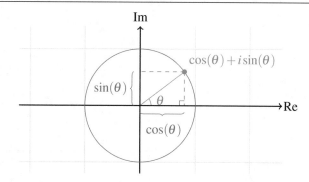

Figure A.3: Every number on the unit circle in the complex plane can be written in the form $\cos(\theta) + i\sin(\theta)$ for some $\theta \in [0, 2\pi)$.

It is worth noting that $|\cos(\theta) + i\sin(\theta)|^2 = \cos^2(\theta) + \sin^2(\theta) = 1$, so these numbers really are on the unit circle.

familiar with Taylor series then you can justify this formula by recalling that e^x, $\cos(x)$, and $\sin(x)$ have Taylor series

$$e^x = 1 + x + \frac{x^2}{2} + \frac{x^3}{3!} + \frac{x^4}{4!} + \frac{x^5}{5!} + \cdots,$$

$$\cos(x) = 1 \quad -\frac{x^2}{2!} \quad +\frac{x^4}{4!} \quad -\cdots, \quad \text{and}$$

$$\sin(x) = \quad x \quad -\frac{x^3}{3!} \quad +\frac{x^5}{5!} \quad \cdots,$$

respectively. Plugging $x = i\theta$ into the Taylor series for e^x then gives

We do not worry about things like convergence here—these details are covered in textbooks on complex analysis.

$$e^{i\theta} = 1 + i\theta - \frac{\theta^2}{2} - i\frac{\theta^3}{3!} + \frac{\theta^4}{4!} + i\frac{\theta^5}{5!} - \cdots$$

$$= \left(1 - \frac{\theta^2}{2} + \frac{\theta^4}{4!} - \cdots\right) + i\left(\theta - \frac{\theta^3}{3!} + \frac{\theta^5}{5!} + \cdots\right),$$

which equals $\cos(\theta) + i\sin(\theta)$, as claimed.

By making use of Euler's formula, we see that we can write every complex number $z \in \mathbb{C}$ in the form $z = re^{i\theta}$, where r is the magnitude of z (i.e., $r = |z|$) and θ is the angle that z makes with the positive real axis (see Figure A.4). This is called the **polar form** of z, and we can convert back and forth between the polar form $z = re^{i\theta}$ and its **Cartesian form** $z = a + bi$ via the formulas

In the formula for θ, $\text{sign}(b) = \pm 1$, depending on whether b is positive or negative. If $b < 0$ then we get $-\pi < \theta < 0$, which we can put in the interval $[0, 2\pi)$ by adding 2π to it.

$$a = r\cos(\theta) \qquad\qquad r = \sqrt{a^2 + b^2}$$

$$b = r\sin(\theta) \qquad\qquad \theta = \text{sign}(b)\arccos\left(\frac{a}{\sqrt{a^2 + b^2}}\right).$$

There is no simple way to "directly" add two complex numbers that are in polar form, but multiplication is quite straightforward: $(r_1 e^{i\theta_1})(r_2 e^{i\theta_2}) = (r_1 r_2)e^{i(\theta_1 + \theta_2)}$. We can thus think of complex numbers as stretched rotations—multiplying by $re^{i\theta}$ stretches numbers in the complex plane by a factor of r and rotates them counter-clockwise by an angle of θ.

Because the polar form of complex numbers works so well with multiplication, we can use it to easily compute powers and roots. Indeed, repeatedly multiplying a complex number in polar form by itself gives $(re^{i\theta})^n = r^n e^{in\theta}$ for all positive integers $n \geq 1$. We thus see that every non-zero complex number

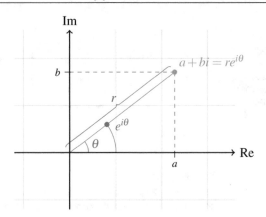

Figure A.4: Every complex number can be written in Cartesian form $a + bi$ and also in polar form $re^{i\theta}$.

You should try to convince yourself that raising any of these numbers to the n-th power results in $re^{i\theta}$. Use the fact that $e^{2\pi i} = e^{0i} = 1$.

has at least n distinct n-th roots (and in fact, *exactly* n distinct n-th roots). In particular, the n roots of $z = re^{i\theta}$ are

$$r^{1/n}e^{i\theta/n}, \quad r^{1/n}e^{i(\theta+2\pi)/n}, \quad r^{1/n}e^{i(\theta+4\pi)/n}, \quad \ldots, \quad r^{1/n}e^{i(\theta+2(n-1)\pi)/n}.$$

Example A.1.1

Computing Complex Roots

Compute all 3 cube roots of the complex number $z = i$.

Solution:

We start by writing z in its polar form $z = e^{\pi i/2}$, from which we see that its 3 cube roots are

$$e^{\pi i/6}, \quad e^{5\pi i/6}, \quad \text{and} \quad e^{9\pi i/6}.$$

We can convert these three cube roots into their Cartesian forms if we want to, though this is perhaps not necessary:

$$e^{\pi i/6} = (\sqrt{3}+i)/2, \quad e^{5\pi i/6} = (-\sqrt{3}+i)/2, \quad \text{and} \quad e^{9\pi i/6} = -i.$$

Geometrically, the cube root $e^{\pi i/6} = (\sqrt{3}+i)/2$ of i lies on the unit circle and has angle one-third as large as that of i, and the other two cube roots are evenly spaced around the unit circle:

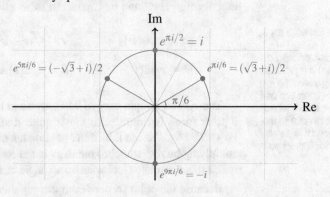

This definition of principal roots requires us to take $\theta \in [0, 2\pi)$.

Among the n distinct n-th roots of a complex number $z = re^{i\theta}$, we call $r^{1/n}e^{i\theta/n}$ its **principal n-th root**, which we denote by $z^{1/n}$. The principal root of a complex number is the one with the smallest angle so that, for example, if

z is a positive real number (i.e., $\theta = 0$) then its principal roots are positive real numbers as well. Similarly, the principal square root of a complex number is the one in the upper half of the complex plane (for example, the principal square root of -1 is i, not $-i$), and we showed in Example A.1.1 that the principal cube root of $z = e^{\pi i/2} = i$ is $z^{1/3} = e^{\pi i/6} = (\sqrt{3}+i)/2$.

A.2 Polynomials

A **polynomial** is a function $p : \mathbb{R} \to \mathbb{R}$ of the form

It is also OK to consider polynomials $p : \mathbb{C} \to \mathbb{C}$ with coefficients in $\mathbb{C}$.

$$p(x) = a_n x^n + a_{n-1} x^{n-1} + \cdots + a_2 x^2 + a_1 x + a_0,$$

where $a_0, a_1, a_2, \ldots, a_{n-1}, a_n \in \mathbb{R}$ are constants (called the **coefficients** of p). The highest power of x appearing in the polynomial is called its **degree** (so, for example, the polynomial displayed above has degree n, as long as $a_n \neq 0$).

A degree-2 polynomial is called a **quadratic**, and it is typically written in the form

$$p(x) = ax^2 + bx + c,$$

where a, b, and c are constants. The graph of a quadratic function is a parabola (see Figure A.5). A degree-3 polynomial is called a **cubic**, and a degree-4 polynomial is called a **quartic**, but quadratics typically receive quite a bit more attention than their higher-degree counterparts since they are so much simpler to work with. The graph of a cubic function typically looks much like a parabola, but with an extra "bend" in it, and the graph of a quartic typically has another bend still (and this pattern continues for higher-degree polynomials).

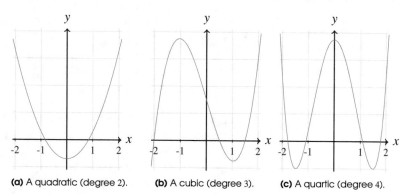

(a) A quadratic (degree 2).   (b) A cubic (degree 3).   (c) A quartic (degree 4).

Figure A.5: Graphs of some low-degree polynomials. Increasing the degree of a polynomial typically adds an additional "bend" to its graph.

A.2.1 Roots of Polynomials

One of the most typical tasks involving polynomials is to find their **roots**, which are the solutions to the equation $p(x) = 0$. Geometrically, the roots of a polynomial are the x values of the points where its graph crosses the x-axis. For example, the roots of the cubic in Figure A.5(b) are approximately $x = -2$, $x = 0.5$, and $x = 1.5$.

We prove the quadratic formula a bit later, in Theorem A.3.3.

If p is a quadratic then one method of finding its roots is to use the **quadratic formula**:

$$x = \frac{-b \pm \sqrt{b^2 - 4ac}}{2a}.$$

We illustrate how to use this formula via some examples.

Example A.2.1

Using the Quadratic Formula

Find all roots of the following quadratics:

a) $x^2 - 6x + 8$,
b) $2x^2 + 8x + 8$, and
c) $x^2 + 2x + 3$.

Solutions:

a) This quadratic has $a = 1$, $b = -6$, and $c = 8$, so the quadratic formula tells us that its roots are

$$x = \frac{6 \pm \sqrt{36 - 32}}{2} = \frac{6 \pm 2}{2} = 3 \pm 1.$$

Its roots are thus $x = 2$ and $x = 4$. This quadratic is displayed below (notice that it crosses the x-axis at $x = 2$ and $x = 4$):

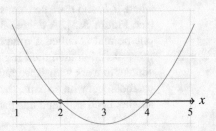

The quadratics in this example show that they can have 2, 1, or 0 distinct roots.

b) This quadratic has $a = 2$, $b = 8$, and $c = 8$, so the quadratic formula tells us that its roots are

$$x = \frac{-8 \pm \sqrt{64 - 64}}{4} = \frac{-8 \pm 0}{4} = -2.$$

It thus has just one root: $x = -2$. This quadratic is displayed below (notice that it just touches the x-axis once, at $x = -2$):

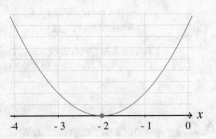

c) This quadratic has $a = 1$, $b = 2$, and $c = 3$, so the quadratic formula tells us that its roots are

$$x = \frac{-2 \pm \sqrt{4 - 12}}{2} = -1 \pm \sqrt{-2}.$$

If you are uncomfortable or unfamiliar with complex numbers like $-1+i\sqrt{2}$, go back to Appendix A.1.

In this situation, context matters: if we demand that roots must be real (i.e., we are considering this quadratic as a function on $\mathbb{R}$) then we conclude that it has no roots. However, if we allow complex roots (i.e., we are considering it as a function on $\mathbb{C}$) then its roots are $-1 \pm i\sqrt{2}$. This quadratic is displayed below (notice that it does not cross the x-axis at all):

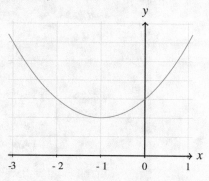

To find the roots of a polynomial of degree higher than 2, we have to be a bit more clever than we were with quadratics, since we typically cannot just plug values into a formula to find the roots. The following theorem provides the main tool that is used for analytically finding roots of polynomials in general.

Theorem A.2.1

Rational Root Theorem

For integers x and y, x is a "divisor" of y if y is an integer multiple of x.

Let p be a polynomial of the form

$$p(x) = a_n x^n + a_{n-1} x^{n-1} + \cdots + a_2 x^2 + a_1 x + a_0,$$

where $a_0, a_1, a_2, \ldots, a_{n-1}, a_n \in \mathbb{R}$ are integers and $a_0, a_n \neq 0$. Then every rational root of p has the form b/c, where b is a divisor of a_0 and c is a divisor of a_n.

We do not prove the above theorem, but instead jump right into working through a few examples to illustrate its usage.

Example A.2.2

Using the Rational Root theorem

Find all rational roots of the following polynomials:
a) $x^2 - 2x - 3$,
b) $2x^3 - 9x^2 - 6x + 5$, and
c) $3x^3 - 14x + 4$.

Solutions:
a) Since $a_2 = 1$ and $a_0 = 3$, the rational root theorem tells us that the only possible rational roots of this polynomial are ± 1 and ± 3 (i.e., the divisors of 3). To check which of these are *actually* roots, we just plug them into the polynomial:

A **rational** number is one that can be written in the form b/c, where b and c are both integers (and $c \neq 0$). For example, 3, 5/7, and $-13/2$ are rational numbers.

$$(1)^2 - 2(1) - 3 = -4 \qquad (-1)^2 - 2(-1) - 3 = 0$$
$$(3)^2 - 2(3) - 3 = 0 \qquad (-3)^2 - 2(-3) - 3 = 12.$$

The rational roots of this polynomial are thus $x = -1$ and $x = 3$ (which we could have also found via the quadratic formula).

b) Since $a_3 = 2$ and $a_0 = 5$, the rational root theorem tells us that the only possible rational roots of this polynomial are ± 5, ± 1, $\pm 5/2$, and $\pm 1/2$. To check which of these are *actually* roots, we just plug them into the polynomial:

$$2(5)^3 - 9(5)^2 - 6(5) + 5 = 0$$
$$2(-5)^3 - 9(-5)^2 - 6(-5) + 5 = -440$$
$$2(1)^3 - 9(1)^2 - 6(1) + 5 = -8$$
$$2(-1)^3 - 9(-1)^2 - 6(-1) + 5 = 0$$
$$2(5/2)^3 - 9(5/2)^2 - 6(5/2) + 5 = -35$$
$$2(-5/2)^3 - 9(-5/2)^2 - 6(-5/2) + 5 = -135/2$$
$$2(1/2)^3 - 9(1/2)^2 - 6(1/2) + 5 = 0$$
$$2(-1/2)^3 - 9(-1/2)^2 - 6(-1/2) + 5 = 11/2.$$

It follows that the rational roots of this polynomial are $x = 5$, $x = 1/2$, and $x = -1$. In fact, we will see shortly that every cubic polynomial has at most 3 roots, so these must be *all* of the (not necessarily rational) roots of this polynomial.

c) Since $a_3 = 3$ and $a_0 = 4$, the rational root theorem tells us that the only possible rational roots of this polynomial are ± 4, ± 2, ± 1, $\pm 4/3$, $\pm 2/3$, and $\pm 1/3$. Plugging these values into the polynomial (just like we did in parts (a) and (b)) shows that $x = 2$ is the only one of these quantities that is *actually* a root.

A.2.2　　Polynomial Long Division and the Factor Theorem

In Example A.2.2(c), we used the rational root theorem to find one root of the cubic $p(x) = 3x^3 - 14x + 4$. However, this cubic actually has two other (necessarily irrational) roots as well. To find them, we can use the fact that $x = 2$ is one of its roots to try to factor $p(x)$ as

> A number is called **irrational** if it is not rational. Examples of irrational numbers include $\sqrt{2}$, π, and e.

$$p(x) = (x - 2)q(x), \tag{A.2.1}$$

where q is a quadratic. If we can find such a factorization of p, then we can use the quadratic formula to find the roots of q (which are the two remaining roots of p).

To find the quadratic q, we use a method called **polynomial long division**, which is a method of "dividing" polynomials by each other that is completely analogous to how we can use long division to divide numbers by each other. This method is best illustrated via an example, so we now work through the polynomial long division that produces the quadratic q in Equation (A.2.1).

Step 1: Divide the highest-degree term in the dividend ($p(x) = 3x^3 - 14x + 4$) by the highest-degree term in the divisor ($x - 2$). In this case, we divide $3x^3$ by x, getting $3x^2$ as our result:

$$
\begin{array}{r}
3x^2 \\
x - 2 \overline{)\, 3x^3 + 0x^2 - 14x + 4}
\end{array}
$$

Step 2: Multiply the divisor $(x-2)$ by the quotient from Step 1 $(3x^2)$ and subtract the result from the dividend $(p(x) = 3x^3 - 14x + 4)$:

Be careful with double negatives: $0x^2 - (-6x^2) = 6x^2$.

$$
\begin{array}{r}
3x^2 \phantom{{}+0x^2 - 14x+4} \\
x-2 \overline{)\,3x^3 + 0x^2 - 14x + 4} \\
\underline{3x^3 - 6x^2 \phantom{{}- 14x+4}} \\
6x^2 - 14x + 4
\end{array}
$$

Step 3: Repeat this procedure, but with the polynomial that resulted from the subtraction in Step 2 $(6x^2 - 14x + 4)$ as the new dividend:

$$
\begin{array}{r}
3x^2 + 6x \phantom{{}-14x+4} \\
x-2 \overline{)\,3x^3 + 0x^2 - 14x + 4} \\
\underline{3x^3 - 6x^2 \phantom{{}- 14x+4}} \\
6x^2 - 14x + 4 \\
\underline{6x^2 - 12x \phantom{{}+4}} \\
-2x + 4
\end{array}
$$

Step 4: Keep on repeating until the degree of the dividend is strictly smaller than the degree of the divisor. If the divisor is in fact a factor of the original polynomial p, then the final remainder will be 0:

$$
\begin{array}{r}
3x^2 + 6x - 2 \\
x-2 \overline{)\,3x^3 + 0x^2 - 14x + 4} \\
\underline{3x^3 - 6x^2 \phantom{{}- 14x+4}} \\
6x^2 - 14x + 4 \\
\underline{6x^2 - 12x \phantom{{}+4}} \\
-2x + 4 \\
\underline{-2x + 4} \\
0
\end{array}
$$

The quotient (i.e., the result of the polynomial long division) is the polynomial written at the top of this calculation $(q(x) = 3x^2 + 6x - 2)$.

The calculation above showed that

$$ p(x) = 3x^3 - 14x + 4 = (x-2)q(x) = (x-2)(3x^2 + 6x - 2). $$

We can now find the remaining roots of p by using the quadratic formula to find the roots of $q(x) = 3x^2 + 6x - 2$, which are

$$ x = \frac{-6 \pm \sqrt{36 + 24}}{6} = -1 \pm \sqrt{\frac{5}{3}}. $$

We (finally!) have shown that the roots of p are 2, $-1 + \sqrt{5/3}$, and $-1 - \sqrt{5/3}$. We now work through another example to make sure that we understand this procedure.

Example A.2.3

Finding Roots of Higher-Degree Polynomials

Find all roots of the polynomial $p(x) = 3x^4 + 4x^3 - 14x^2 - 11x - 2$.

Solution:

Since $a_4 = 3$ and $a_0 = -2$, the rational root theorem tells us that the only possible rational roots of this polynomial are ± 2, ± 1, $\pm 2/3$, and $\pm 1/3$. Plugging these values into the polynomial shows that $x = 2$ and $x = -1/3$ are the only ones that are *actually* roots of p.

To find the remaining (necessarily irrational) roots of p, we divide

Alternatively, we
could have divided
$p(x)$ by

$(x-2)(x+\frac{1}{3})$

$= x^2 - \frac{5}{3}x - \frac{2}{3},$

but multiplying this
divisor through by 3
avoids some ugly
fractions.

$p(x)$ by $3(x-2)(x+1/3) = 3x^2 - 5x - 2$:

$$
\begin{array}{r}
x^2 + 3x + 1 \\
3x^2 - 5x - 2 \overline{\smash{\big)}\ 3x^4 + 4x^3 - 14x^2 - 11x - 2} \\
\underline{3x^4 - 5x^3 - 2x^2\phantom{{}-11x-2}} \\
9x^3 - 12x^2 - 11x - 2 \\
\underline{9x^3 - 15x^2 - 6x\phantom{{}-2}} \\
3x^2 - 5x - 2 \\
\underline{3x^2 - 5x - 2} \\
0
\end{array}
$$

It follows that

$$p(x) = 3x^4 + 4x^3 - 14x^2 - 11x - 2 = 3(x-2)(x+1/3)(x^2+3x+1),$$

so we can find the remaining roots of p by applying the quadratic formula to $x^2 + 3x + 1$:

$$x = \frac{-3 \pm \sqrt{9-4}}{2} = \frac{1}{2}(-3 \pm \sqrt{5}).$$

It follows that the roots of p are 2, $\frac{-1}{3}$, $\frac{1}{2}(-3+\sqrt{5})$, and $\frac{1}{2}(-3-\sqrt{5})$.

If r is a root of a polynomial p and we divide $p(x)$ by $x-r$ via polynomial long division, the remainder of that division always equals 0. It follows that polynomial long division in these cases really does always lead to a factorization of p. We now state (but do not prove) this result more formally:

Theorem A.2.2

Factor Theorem

Let p be a polynomial of degree $n \geq 1$. Then a scalar r is a root of p if and only if there exists a polynomial q with degree $n-1$ such that $p(x) = (x-r)q(x)$.

The (degree-0)
polynomial $p(x) = 0$
has infinitely many
roots.

One of the consequences of the factor theorem is that every polynomial of degree $n \geq 1$ has at most n distinct roots. After all, each root r of a non-constant polynomial corresponds to a linear factor $x-r$ of that polynomial, and each of these factors contributes 1 to its degree.

Remark A.2.1

**The Cubic Formula
and Beyond**

Just like there is a quadratic formula for finding roots of quadratics, there is indeed a **cubic formula** that can be used to explicitly compute the roots of cubics. Specifically, if we want to find the roots of the cubic

$$p(x) = ax^3 + bx^2 + cx + d,$$

we introduce the intermediate variables

$$q = \frac{-b}{3a}, \quad r = q^3 + \frac{bc - 3ad}{6a^2}, \quad \text{and} \quad s = \frac{c}{3a}$$

and then compute one of its roots via

The other 2 roots of
this cubic can be
found by carefully
choosing different
complex cube roots
in this formula.

$$x = q + \sqrt[3]{r + \sqrt{r^2 + (s-q^2)^3}} + \sqrt[3]{r - \sqrt{r^2 + (s-q^2)^3}}.$$

However, this formula is typically not used in practice due to it being cumbersome, and the fact that complex numbers may be required in intermediate steps of the calculation even when the cubic's coefficients and roots are all real. For these reasons, the methods discussed earlier in this section based on the rational root theorem and polynomial long division are preferred.

There is also a **quartic formula** for explicitly computing the roots of quartics, but it is even nastier than its cubic counterpart. However, this is where the formulas end: a remarkable result called the Abel–Ruffini theorem says that there is no explicit formula for the roots of degree-5 (or higher) polynomials in terms of "standard" operations like addition, subtraction, multiplication, division, and roots.

A.2.3 The Fundamental Theorem of Algebra

Some polynomials like $p(x) = x^2 + 1$ do not have any real roots at all, so the upper bound of a degree-n polynomial having at most n roots is sometimes quite loose. One of the most remarkable theorems concerning polynomials says that this never happens as long as we allow the roots to be complex numbers. We now state this theorem, but proving it is far outside of the scope of this book—the interested reader is directed to a book like [FR97] for a more thorough treatment.

Theorem A.2.3

Fundamental Theorem of Algebra

Every non-constant polynomial has at least one complex root.

By combining the fundamental theorem of algebra with the factor theorem, we immediately see that every polynomial of degree n has *exactly* n complex roots and can be factored as

$$p(x) = a(x - r_1)(x - r_2) \cdots (x - r_n),$$

where a is the coefficient of x^n in p and $r_1, r_2, \ldots, r_n$ are the (potentially complex, and not necessarily distinct) roots of p.

To clarify what we mean when we say that a polynomial has exactly n not-necessarily-distinct roots, we need to introduce the **multiplicity** of a root r, which is the exponent of the term $x - r$ in the factorization of the polynomial. We think of a multiplicity-k root as one that "occurs k times", or as k roots that are equal to each other. With this in mind, the fundamental theorem of algebra says that every degree-n polynomial has exactly n complex roots, counting multiplicity (that is, if we add up the multiplicities of each root).

Roots with multiplicity 2 or greater are sometimes called "repeated roots".

For example, the root -2 of the polynomial $2x^2 + 8x + 8 = 2(x + 2)^2$ has multiplicity 2, since the term $x + 2$ is squared, while all of the other roots that we have discussed so far have multiplicity 1. The 2 roots of this polynomial are thus both equal to -2.

A.3 Proof Techniques

While proofs are at the center of modern mathematics, this book serves as one of the first places where the intended reader will encounter them with any regularity. The goal of a proof is to use facts that the reader is already comfortable with to establish new facts (called "theorems") that they can then add to their mathematical toolbox.

At the start of this book, we assume that the reader is familiar with the basic properties of the real numbers, like the fact that $xy = yx$ for all real numbers x and y (this property is called **commutativity**). These properties of real numbers are then used to prove similar properties of vectors, matrices, and other linear algebraic objects.

The first proofs that we see, and the simplest type of proofs that exist, are **direct proofs**: proofs where we just use the definition of whatever object we are working with, together with other properties that we are familiar with, to demonstrate that the proposed theorem is true. We illustrate this type of proof with an example.

Theorem A.3.1

The Squares of Even Integers are Even

If m is an even integer then so is m^2.

Proof. If m is an even integer then there is some integer k such that $m = 2k$. Then

$$m^2 = (2k)^2 = 4k^2 = 2(2k^2),$$

which is also even since it is 2 times the integer $2k^2$. ∎

One of the most difficult hurdles to get over when first learning how to write proofs is how to make use of the information that has been given to you. In Theorem A.3.1, we were told that m is even, so to make use of that information we write m as $m = 2k$. If we were instead trying to prove a theorem about an odd integer m we would write $m = 2k + 1$ for some integer k, and if we were trying to prove a theorem about multiples of 7 then we would write $m = 7k$.

Many proofs just make use of all of the relevant definitions and proceed in pretty much the only way they could. For example, to prove Theorem A.3.1, our hypothesis tells us that m is an even integer, so we start off with the relevant definition (i.e., $m = 2k$ for some integer k). We want to show that m^2 is even based on that, so we then simply compute $m^2 = (2k)^2$, and we finish the proof by again using the definition of "even integer": we write $(2k)^2 = 2(2k^2)$, which is even.

However, proofs are not always this straightforward, and there may be a trick or two that we have to perform along the way to get from the hypothesis to the conclusion. We illustrate this with an example.

Theorem A.3.2

AM–GM Inequality

If x and y are real numbers then $x^2 + y^2 \geq 2xy$.

Proof. Recall that the square of every real number is non-negative, so $(x - y)^2 \geq 0$. However, using some algebra to expand $(x - y)^2$ shows that

$$0 \leq (x - y)^2 = x^2 - 2xy + y^2.$$

Adding $2xy$ to both sides of this inequality shows that $2xy \leq x^2 + y^2$, as desired. ∎

If you are new to proofs, it likely seems quite strange that we considered the quantity $(x - y)^2$ at all in the above proof—nothing in the *statement* of the theorem had anything to do with that quantity, so why bring it up? The answer is simply that it is a trick that got us where we needed to go. Even if you do not believe that you could have come up with the trick yourself, you should be able to follow along with the proof and agree that every step in it is logically correct.

As you read (and write!) more and more proofs, you will see a greater number of tricks like this one, and they will start to become more familiar. Furthermore, you will notice that many of the same tricks get applied over and over in similar situations. For example, if you are trying to prove some arithmetic expression involving quadratic terms (like in Theorem A.3.2), then it is probably a good idea to try to factor or complete the square (which is really all we did in that proof—we factored $x^2 + y^2 - 2xy$ as $(x-y)^2$). This same trick can be used to prove the quadratic formula:

Theorem A.3.3	If a, b, c, and x are real or complex numbers such that $ax^2 + bx + c = 0$ and $a \neq 0$, then
Quadratic Formula	$$x = \frac{-b \pm \sqrt{b^2 - 4ac}}{2a}.$$

Proof. We start by rewriting the quantity $ax^2 + bx + c$ in a different way that groups that linear bx term and quadratic ax^2 term together:

The "$\implies$" arrows here are called "implication arrows", and they mean that the current line follows logically from the one before it.

$$ax^2 + bx + c = 0 \qquad \text{(hypothesis)}$$
$$\implies \quad x^2 + \frac{b}{a}x = -\frac{c}{a} \qquad \text{(divide both sides by } a)$$
$$\implies \quad x^2 + \frac{b}{a}x + \frac{b^2}{4a^2} = \frac{b^2}{4a^2} - \frac{c}{a} \qquad \text{(add } b^2/(4a^2) \text{ to both sides)}$$
$$\implies \quad \left(x + \frac{b}{2a}\right)^2 = \frac{b^2}{4a^2} - \frac{c}{a} \qquad \text{(factor the left-hand side)}$$

We abbreviate the terms "left-hand side" and "right-hand side" as "LHS" and "RHS", respectively.

From here we can just apply standard algebraic manipulations to solve for x:

$$\left(x + \frac{b}{2a}\right)^2 = \frac{b^2}{4a^2} - \frac{c}{a} \qquad \text{(derived above)}$$
$$\implies \quad \left(x + \frac{b}{2a}\right)^2 = \frac{b^2 - 4ac}{4a^2} \qquad \text{(common denom. on RHS)}$$
$$\implies \quad x + \frac{b}{2a} = \pm\frac{\sqrt{b^2 - 4ac}}{2a} \qquad \text{(square root both sides)}$$
$$\implies \quad x = \frac{-b \pm \sqrt{b^2 - 4ac}}{2a} \qquad \text{(subtract } b/(2a))$$

which is exactly what we wanted to prove. ∎

A.3.1 The Contrapositive

Sometimes it is difficult to prove a statement directly, so it is convenient to rearrange the statement into some other (equivalent) form first. For example, every statement of the form "if P then Q" is logically equivalent to its **contrapositive**, which is the statement "if not Q then not P". That is, if the original statement is true then so is its contrapositive, and vice-versa.

The contrapositive is actually quite a bit more intuitive than it seems at first. For example, if the statements P and Q are

P: "I am running", and

Q: "I am sweaty", then

the statement "if P then Q" reads "if I am running then I am sweaty". This statement is logically equivalent to its contrapositive "if not Q then not P", which in this case is "if I am not sweaty then I am not running" (indeed, if I *were* running then I would be sweaty, so if I'm *not* sweaty then I must not be running). The following example illustrates a case where the contrapositive of a statement might be easier to prove than the original statement itself.

Theorem A.3.4 **Even Squares are the Squares of Even Integers** Seriously, try proving this directly for a couple of minutes. You can start by writing $m^2 = 2k$ for some integer k, but you likely will not get much farther than that—it's hard!	If m^2 is an even integer then so is m. *Proof.* This statement is rather difficult to prove directly, so we instead consider its contrapositive: "if m is not even then m^2 is not even", which can be phrased a bit more naturally as "if m is odd then so is m^2". To prove this claim, we proceed just like we did in the proof of Theorem A.3.1. If m is an odd integer then there is some integer k such that $m = 2k + 1$. Then $$m^2 = (2k+1)^2 = 4k^2 + 4k + 1 = 2(2k^2 + 2k) + 1,$$ which is also odd since it is 1 larger than the even integer $2(2k^2 + 2k)$. ∎

Be careful not to confuse the contrapositive with the **converse** "if Q then P", which is not equivalent to "if P then Q". For example, the converse of "if I am running then I am sweaty" is "if I am sweaty then I am running", which is *not* a true statement—maybe I got sweaty from riding my bike.

A.3.2   Bi-Directional Proofs

The statement "P only if Q" is synonymous with "if P then Q".

Sometimes we wish to state a fact that not only does one property imply another one, but actually they imply each other (i.e., they are equivalent to each other). In these situations, we usually say that the first property holds "if and only if" the second property holds. To prove such a statement, care needs to be taken to ensure that we really are showing that each property implies the other, and not simply that the second property follows from the first.

Theorem A.3.5 **Integers Have the Same Parity as Their Squares**	A integer m is even if and only if m^2 is even. *Proof.* This theorem is simply a combination of Theorems A.3.1 and A.3.4. ∎

By "adjacent squares", we mean squares that share a side, not just touch at a corner.

To better illustrate why it really is necessary to prove both directions of an "if and only if" statement, consider the problem of tiling a standard 8×8 chessboard by 2×1 dominoes. That is, we lay 2×1 dominoes on top of the chessboard so that each domino covers exactly two of its adjacent squares, and none of the dominoes overlap. There are numerous ways to do this, with one of them illustrated in Figure A.6.

While a chessboard itself can of course be tiled by dominoes, if we change the shape of the board then this may no longer be the case. For example, if we

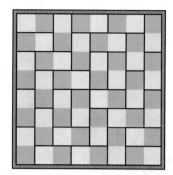

(a) A standard 8×8 chessboard.   **(b)** A domino tiling of a chessboard.

Figure A.6: An illustration of one of the many possible domino tilings of a standard 8×8 chessboard.

remove one of the $8 \times 8 = 64$ squares from the chessboard, then the remaining configuration of 63 squares cannot possibly have a domino tiling, since each domino covers 2 squares and thus no collection of dominoes can cover an odd number of squares.

However, if we remove *two* squares from a chessboard, things become more interesting (see Figure A.7)—the resulting board may or may not have a domino tiling. For example, if we remove two opposite corners from a chessboard then it cannot be tiled by dominoes (this fact is not obvious—try to figure out why), but if we remove two adjacent squares from one corner then it can be tiled by dominoes (after all, we are essentially just removing one domino from a domino tiling of the full chessboard).

If you cannot figure out why the board in Figure A.7(b) cannot be tiled by dominoes, that's OK—the upcoming proof will explain it. In fact, that's the *point* of proofs—to explain and illuminate things.

(a) Any chessboard with one square removed cannot be tiled by dominoes.   **(b)** This board with two opposite corners removed cannot be tiled by dominoes.

Figure A.7: Some chessboards, once they have some squares removed, can no longer be tiled by dominoes.

The following theorem solves this problem and tells us exactly which boards can still be tiled by dominoes after two squares are removed.

Theorem A.3.6

Tiling Chessboards Minus Two Squares

> An 8×8 chessboard with two squares removed can be tiled by dominoes if and only if the two squares that were removed had different colors.

For example, this theorem tells us that we cannot tile the board in Figure A.7(b), since the two corners that we removed were both black. Instead, the only way to be able to tile the resulting 62-square board is if we removed one black square and one white one (so that we have 31 squares of each color remaining).

Proof of Theorem A.3.6. For the "only if" direction, we must show that if the board with two squares removed can be tiled by dominoes, then the two squares that were removed had opposite colors. By the contrapositive, this is equivalent to showing that if the two squares that were removed had the same color then the resulting board has no domino tiling.

To this end, simply notice that each domino in a domino tiling covers exactly one square of each color (since all of the neighbors of a white square are black, and all of the neighbors of a black square are white). However, if two squares of the same color are removed from the chessboard, then the resulting board has 30 squares of one color and 32 squares of the other color, so no arrangement of dominoes can cover them all (the best we can do is use 30 dominoes to cover 60 squares, and then have 2 squares of the same color left over).

We must also prove the "if" direction, that every chessboard with two squares of different colors removed has a domino tiling. To do so, suppose we placed "walls" along certain lines of the chessboard as in Figure A.8(a). Then no matter which two squares of opposite color are removed, there is exactly one way to tile the board with dominoes without crossing those walls, since they form two paths that start and end with opposite-colored squares, as in Figure A.8(b).

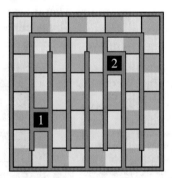

(a) An arrangement of "walls" on a chessboard that will help us tile it.

(b) A tiling of a board with (opposite-colored) squares "1" and "2" removed.

Figure A.8: An illustration of the fact that a chessboard with two squares removed can be tiled by dominoes as long as the two squares that were removed had different colors.

Since we have now demonstrated both implications of the theorem (i.e., both the "if" and "only if" directions), the proof is now complete.   ∎

When starting out with proofs, it is typically best to be careful and explicitly prove both directions in an "if and only statement" separately. However, sometimes one direction of the proof is identical to the other direction read backwards, and in these situations it is okay to just say so or re-word the proof to make it clear that it can be read both forward and backward.

For example, in the proof of the quadratic formula (Theorem A.3.3), we showed that *if* $ax^2 + bx + c = 0$ *then*

$$x = \frac{-b \pm \sqrt{b^2 - 4ac}}{2a}.$$

(A.3.1)

However, the converse of this statement is also true—*if* x is given by the quadratic formula (A.3.1) *then* $ax^2 + bx + c = 0$. To prove this additional implication, we could either substitute the quadratic formula (A.3.1) into the expression $ax^2 + bx + c$ and do a bunch of messy algebra to see that it does simplify to 0, or alternatively we could just notice that every step in the proof of Theorem A.3.3 works backwards as well as forwards. That is, we could change the "if" in the statement of that theorem to "if and only if" just by replacing the implication arrows "$\implies$" with double-sided implication arrows "$\iff$".

Many of the most useful theorems state that numerous different properties are all equivalent to each other (i.e., the first property holds if and only if the second property holds, if and only if the third property holds, and so on). A theorem of this form could be proved by explicitly proving each and every implication among those properties, but this is much more work than is necessary.

Typically, the simplest way to prove a theorem of this type is to prove a "loop" of implications. For example, to prove a theorem that says that properties A, B, C, and D are equivalent to each other, it is enough to show that A implies B, B implies C, C implies D, and D implies A (since then we can follow this loop of implications to see that any one of the four properties implies each of the others). We first make use of this technique in Theorem 2.2.4, and we typically include a schematic in the margin that illustrates which implications we explicitly prove.

A.3.3 Proof by Contradiction

The contrapositive of Section A.3.1 provided us with our first method of **indirect proof**—it let us show that one statement follows logically from another even when there was no clear path from the former to the latter. Another technique along these lines is **proof by contradiction**, in which we assume that the desired statement P is *not* true and then use that assumption to demonstrate two contradictory facts. Since those two contradictory facts cannot simultaneously be true, we then conclude that we must have made a mistake at some point. That is, we conclude that our original assumption (that P is not true) was incorrect (i.e., P *is* true).

This method of proof is particularly useful when proving that certain objects do not exist, or that it is not possible to perform a certain task. The following proof is a standard example of this technique.

Theorem A.3.7

Irrationality of the Square Root of Two

There do not exist integers a and b such that $\sqrt{2} = a/b$. In other words, $\sqrt{2}$ is irrational.

Proof. Suppose (for the sake of establishing a contradiction) that there *did* exist integers a and b such that $\sqrt{2} = a/b$. Since we can cancel common divisors of a and b in this fraction, it follows that we can choose a and b so that at least one of them is odd (just divide each of them by 2 until this happens—doing so does not change the value of the fraction a/b).

A real number that can be written in the form a/b, where a and b are integers, is called **rational**, and all other real numbers are called **irrational**.

Multiplying the equation $\sqrt{2} = a/b$ through by b shows that $\sqrt{2}b = a$, and then squaring shows that $2b^2 = a^2$, so a^2 is even. It follows from Theorem A.3.5 that a is even and can be written in the form $a = 2k$ for some integer k. Since a is even, b must be odd (recall that we chose a and b so that at least one of them is odd).

Plugging $a = 2k$ into the equation $2b^2 = a^2$ shows that $2b^2 = (2k)^2 = 4k^2$, so $b^2 = 2k^2$. It follows that b^2 is even, so using Theorem A.3.5 again shows that b is also even. We have thus reached a contradiction: we have shown that b must be both odd and even. This is of course not possible, so we must have made a mistake somewhere. In particular, our original assumption that a and b exist must be false.   ∎

A.3.4   Proof by Induction

Oftentimes, we observe a pattern and expect that it is true, but it is not immediately clear how to actually *prove* that it is true. For example, if we start adding up consecutive odd natural numbers then we find that

$$1 = 1$$
$$1 + 3 = 4$$
$$1 + 3 + 5 = 9$$
$$1 + 3 + 5 + 7 = 16$$
$$1 + 3 + 5 + 7 + 9 = 25,$$

and so on. In particular, we notice that the sums on the right-hand side are perfect squares (i.e., we have shown that $1 + 3 + 5 + \cdots + (2n - 1) = n^2$, at least for $n \leq 5$). We expect that this pattern continues forever, but how could we *prove* it?

One particularly useful proof technique for showing that a pattern really does continue forever is **proof by induction**. If we have a family of statements $P(1), P(2), P(3), \ldots$, indexed by the natural numbers $1, 2, 3, \ldots$, then proof by induction consists of two pieces:

- The **base case**: we start by showing that $P(1)$ is true.

> The assumption that $P(n)$ holds for a particular value of n is sometimes called the **inductive hypothesis**.

- The **inductive step**: we show that *if* $P(n)$ is true for a particular value of n, *then* $P(n + 1)$ must also be true.

If we can carry out both of the steps described above, it then follows that $P(n)$ must be true for *all* natural numbers n. To see why this is the case, just consider each $n \geq 1$ in order:

$P(1)$: True because of the base case.

$P(2)$: True by the inductive step: if $P(1)$ is true (it is!) then $P(2)$ is true too.

$P(3)$: True by the inductive step: if $P(2)$ is true (it is!) then $P(3)$ is true too.

$P(4)$: True because $P(3)$ is true. You get the idea...

To see how proof by induction works a bit more explicitly, we now use it to prove our conjecture from earlier.

> **Theorem A.3.8**
> **Sum of Odd Natural Numbers**

If $n \geq 1$ is a natural number then $1 + 3 + 5 + \cdots + (2n - 1) = n^2$.

> The base case being trivial (like it was here) is very common, but it is still important to make sure that it holds.

Proof. We prove this result by induction. To be explicit, the statements that we will prove are

$$P(n) : \text{``}1 + 3 + 5 + \cdots + (2n - 1) = n^2\text{''}$$

For the base case, we must show that $P(1)$ is true. Since $P(1)$ simply says that $1 = 1$, this step is complete.

For the inductive step, we suppose that $P(n)$ is true for a particular value of n, and our goal is to use that information to prove that $P(n+1)$ is true as well. The trick that lets us do this is to add $2n+1$ to both sides of $P(n)$:

$$1+3+5+\cdots+(2n-1) = n^2 \qquad (P(n) \text{ is true})$$
$$\implies \quad 1+3+5+\cdots+(2n-1)+(2n+1) = n^2+(2n+1) \qquad (\text{add } 2n+1)$$
$$\implies \quad 1+3+5+\cdots+(2(n+1)-1) = n^2+2n+1 \qquad (\text{rewrite slightly})$$
$$\implies \quad 1+3+5+\cdots+(2(n+1)-1) = (n+1)^2 \qquad (\text{factor RHS})$$

The final line above is exactly the statement $P(n+1)$, so we have shown that $P(n)$ implies $P(n+1)$. This completes the inductive step and the proof. ∎

You can think of induction much like dominoes knocking each other over: the inductive step says that if a particular domino falls over then so does the one after it, and the base case says that the first domino is falling over. Based on these two facts, we conclude that *all* dominoes will fall over.

While proof by induction is a very effective proof technique, such proofs are also often quite un-illuminating. That is, they show *that* a statement is true, but they often are not great at really explaining *why* that statement is true. We thus consider proofs by induction to be a necessary evil that we make use of when we must, but we avoid when we can. For example, we now present another proof of Theorem A.3.8 that we feel is much more illuminating and better explains *why* that theorem is true.

Alternate proof of Theorem A.3.8. We notice that an $n \times n$ square grid can be partitioned into "L"-shaped regions, each consisting of an odd number of squares, as in Figure A.9.

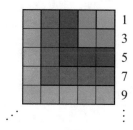

Figure A.9: A demonstration of the fact that adding the first n odd natural numbers gives n^2.

Since the grid overall consists of n^2 squares, and the "L"-shaped regions consist of $1, 3, 5, \ldots, 2n-1$ squares, it follows that $1+3+5+\cdots+(2n-1) = n^2$, as claimed. ∎

B. Additional Proofs

In this appendix, we prove some of the technical results that we made use of throughout the main body of the textbook. Their proofs are messy enough (or unenlightening enough, or simply not "linear algebra-y" enough) that they are hidden away here.

B.1 Block Matrix Multiplication

In Section 1.3.4, we claimed that performing block matrix multiplication gives the same result as performing standard matrix multiplication, as long as the matrices are partitioned so that the block matrix multiplication makes sense. We now prove this claim.

Theorem B.1.1

Block Matrix Multiplication

If two matrices A and B are partitioned as block matrices

$$A = \begin{bmatrix} A_{1,1} & A_{1,2} & \cdots & A_{1,n} \\ A_{2,1} & A_{2,2} & \cdots & A_{2,n} \\ \vdots & \vdots & \ddots & \vdots \\ A_{m,1} & A_{m,2} & \cdots & A_{m,n} \end{bmatrix} \quad \text{and} \quad B = \begin{bmatrix} B_{1,1} & B_{1,2} & \cdots & B_{1,p} \\ B_{2,1} & B_{2,2} & \cdots & B_{2,p} \\ \vdots & \vdots & \ddots & \vdots \\ B_{n,1} & B_{n,2} & \cdots & B_{n,p} \end{bmatrix}$$

then AB can be computed by multiplying their blocks together via the usual matrix multiplication rule. That is, for each $1 \leq i \leq m$ and $1 \leq j \leq p$, the (i, j)-block of AB equals

$$A_{i,1}B_{1,j} + A_{i,2}B_{2,j} + \cdots + A_{i,n}B_{n,j},$$

as long as each of these matrix products make sense.

This proof is ugly no matter how you cut it (that's why it's tucked away back here in the appendix).

Proof. This fact follows almost immediately from the definition of matrix multiplication, but we first have to get our head around the horrendous notation that we must use. Since we are going to be working with individual entries of a block matrix, we need four subscripts—two to tell us which block of the block matrix the entry is in, and two to tell us which row and column of that block it is. Specifically, we use $[A_{i,j}]_{k,\ell}$ to refer to the (k, ℓ)-entry of the matrix $A_{i,j}$, which itself is the matrix that makes up the i-th block row and j-th block column of A.

First, we compute the (k, ℓ)-entry of the (i, j)-block of AB directly from the definition of matrix multiplication. This is somewhat difficult to visualize due to the block structure of A and B, but recall that we want to take the dot product of one row of A with one column of B. In particular, we take the dot product of the k-th row of the i-th block row of A with the ℓ-th column of the j-th block column of B:

The way we have written the block matrices here, each block in the i-th block row of A has the same number of columns (c), but this need not be the case in general.

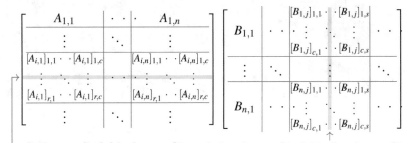

k- th row of i- th block row of A   ℓ- th column of j- th block column of B

To compute this quantity, we sum over each column and block column of A, or equivalently we sum over each row and block row of B, which gives us

$$\sum_{a=1}^{n}\sum_{b}[A_{i,a}]_{k,b}[B_{a,j}]_{b,\ell} \tag{B.1.1}$$

as the (k,ℓ)-entry of the (i,j)-block of AB. We note that the inner sum over b has as many terms in it as there are columns in the a-th block column of A (or as many rows as in the a-th block row of B).

On the other hand, if we compute this entry via the method suggested by the statement of the theorem, we get

The quantity in square brackets on the left here is the same as the sum in the statement of the theorem.

$$\left[\sum_{a=1}^{n}A_{i,a}B_{a,j}\right]_{k,\ell} = \sum_{a=1}^{n}[A_{i,a}B_{a,j}]_{k,\ell} = \sum_{a=1}^{n}\left(\sum_{b}[A_{i,a}]_{k,b}[B_{a,j}]_{b,\ell}\right),$$

which is the same as the direct formula from Equation (B.1.1) and thus completes the proof. ∎

B.2 Uniqueness of Reduced Row Echelon Form

In Section 2.1.3, we claimed that the reduced row echelon form of a matrix is unique. That is, every matrix can be row-reduced to one and only one matrix in reduced row echelon form. We now prove this claim.

Theorem B.2.1

Uniqueness of Reduced Row Echelon Form

Suppose $A \in \mathcal{M}_{m,n}$. There exists exactly one matrix in reduced row echelon form that is row equivalent to A.

Proof. The Gauss–Jordan elimination algorithm shows that every matrix is row equivalent to one that is in reduced row echelon form, so all that we need to show is uniqueness. That is, we need to show that if A is row equivalent to two matrices $R, S \in \mathcal{M}_{m,n}$, both of which are in reduced row echelon form, then $R = S$.

To this end, suppose for the sake of contradiction that $R \neq S$. Let $\widetilde{R}$ and $\widetilde{S}$ be the matrices that are obtained from R and S, respectively, by keeping the leftmost column in which they differ, together with all leading columns to their left, and discarding all others. For example, if

$$R = \begin{bmatrix} 0 & 1 & 3 & 0 & 2 & 0 \\ 0 & 0 & 0 & 1 & -1 & 0 \\ 0 & 0 & 0 & 0 & 0 & 1 \\ 0 & 0 & 0 & 0 & 0 & 0 \end{bmatrix} \quad \text{and} \quad S = \begin{bmatrix} 0 & 1 & 3 & 0 & 4 & 0 \\ 0 & 0 & 0 & 1 & 2 & 0 \\ 0 & 0 & 0 & 0 & 0 & 1 \\ 0 & 0 & 0 & 0 & 0 & 0 \end{bmatrix}$$

then

$$\widetilde{R} = \left[\begin{array}{cc|c} 1 & 0 & 2 \\ 0 & 1 & -1 \\ 0 & 0 & 0 \\ 0 & 0 & 0 \end{array} \right] \quad \text{and} \quad \widetilde{S} = \left[\begin{array}{cc|c} 1 & 0 & 4 \\ 0 & 1 & 2 \\ 0 & 0 & 0 \\ 0 & 0 & 0 \end{array} \right].$$

Since R and S are each row equivalent to A, we know from Exercise 2.1.13(a) that they are row equivalent to each other. It follows that $\widetilde{R}$ and $\widetilde{S}$ are row equivalent as well, since removing columns from a pair of matrices does not affect row equivalence.

If we interpret $\widetilde{R}$ and $\widetilde{S}$ as augmented matrices that represent linear systems (which must have the same solution sets since they are row equivalent), then they must have one of the forms

$\widetilde{R}$ and/or $\widetilde{S}$ have the form on the right if the first in which they differ is leading, and they have the form on the left otherwise.

$$\widetilde{R} = \left[\begin{array}{c|c} I & \mathbf{b} \\ O & \mathbf{0} \end{array} \right] \quad \text{or} \quad \widetilde{R} = \left[\begin{array}{c|c} I & \mathbf{0} \\ O & \mathbf{e}_1 \end{array} \right], \quad \text{and}$$

$$\widetilde{S} = \left[\begin{array}{c|c} I & \mathbf{c} \\ O & \mathbf{0} \end{array} \right] \quad \text{or} \quad \widetilde{S} = \left[\begin{array}{c|c} I & \mathbf{0} \\ O & \mathbf{e}_1 \end{array} \right].$$

We now split this into two cases depending on the form of $\widetilde{R}$.

Case 1: $\widetilde{R}$ **has the first form.** Then the linear system associated with $\widetilde{R}$ has the unique solution $\mathbf{b}$, so the linear system associated with $\widetilde{S}$ must also have the unique solution $\mathbf{b}$, so it must also have the first form with $\mathbf{c} = \mathbf{b}$ (and thus $\widetilde{S} = \widetilde{R}$).

Case 2: $\widetilde{R}$ **has the second form.** Then the linear system associated with $\widetilde{R}$ has no solutions, so the linear system associated with $\widetilde{S}$ must also have no solutions, so it must also have the second form (and thus $\widetilde{S} = \widetilde{R}$).

We have $\widetilde{R} = \widetilde{S}$ in both cases, which contradicts the fact that we chose them so that their last columns differ. It follows that $R = S$, which completes the proof. ∎

B.3 Multiplication by an Elementary Matrix

In Section 2.2, we claimed that multiplying a matrix on the left by an elementary matrix gave the same result as applying the corresponding row operation to that matrix directly. We now prove this claim.

Theorem B.3.1

Multiplication on the Left by an Elementary Matrix

Suppose $A \in \mathcal{M}_{m,n}$. If applying a single elementary row operation to A results in a matrix B, and applying that same elementary row operation to I_m results in a matrix E, then $EA = B$.

Proof. We consider the three types of elementary row operations one at a time. The "multiplication" row operation cR_j is the easiest to work with since in this case, E is diagonal with all diagonal entries equal to 1, except for the (j, j)-entry, which is equal to c. We already noted in Section 1.4.2 that multiplying on the left by a diagonal matrix has the same effect as multiplying row-wise by its diagonal entries, which is exactly the claim here.

For the "addition" row operation $R_i + cR_j$, we notice that the associated elementary matrix E is

$$E = [\, \mathbf{e}_1 \mid \cdots \mid \mathbf{e}_j + c\mathbf{e}_i \mid \cdots \mid \mathbf{e}_m \,],$$

> Yes, the j-th column of E is $\mathbf{e}_j + c\mathbf{e}_i$, not $\mathbf{e}_i + c\mathbf{e}_j$. The reason for the apparent subscript swap is that we are listing the *columns* of E here, not its rows.

where the single "weird" column $\mathbf{e}_j + c\mathbf{e}_i$ occurs in the j-th column of E. If we multiply this matrix by A, which we represent in terms of its rows, then we get

$$EA = [\, \mathbf{e}_1 \mid \cdots \mid \mathbf{e}_j + c\mathbf{e}_i \mid \cdots \mid \mathbf{e}_m \,] \begin{bmatrix} \mathbf{a}_1^T \\ \vdots \\ \mathbf{a}_m^T \end{bmatrix}$$

$$= \mathbf{e}_1\mathbf{a}_1^T + \mathbf{e}_2\mathbf{a}_2^T + \cdots + (\mathbf{e}_j + c\mathbf{e}_i)\mathbf{a}_j^T + \cdots + \mathbf{e}_m\mathbf{a}_m^T$$

$$= \left(\mathbf{e}_1\mathbf{a}_1^T + \mathbf{e}_2\mathbf{a}_2^T + \cdots + \mathbf{e}_m\mathbf{a}_m^T \right) + c\mathbf{e}_i\mathbf{a}_j^T$$

$$= A + c\mathbf{e}_i\mathbf{a}_j^T.$$

> Keep in mind that these products like $\mathbf{e}_1\mathbf{a}_1^T$ are *matrices*, not scalars, since $\mathbf{e}_1$ is $m \times 1$ and $\mathbf{a}_1^T$ is $1 \times n$.

Since $\mathbf{e}_i\mathbf{a}_j^T$ is the matrix that contains the entries from the j-th row of A in its i-th row, it follows that $A + c\mathbf{e}_i\mathbf{a}_j^T$ is exactly the matrix B that is obtained by applying the row operation $R_i + cR_j$ to A, as claimed.

Finally, for the "swap" row operation $R_i \leftrightarrow R_j$, the associated elementary matrix has the form

$$[\, \mathbf{e}_1 \mid \cdots \mid \mathbf{e}_j \mid \cdots \mid \mathbf{e}_i \mid \cdots \mid \mathbf{e}_m \,],$$

where $\mathbf{e}_j$ appears in the i-th column and $\mathbf{e}_i$ appears in the j-th column. By mimicking the block matrix multiplication argument that we made earlier for "addition" row operations, we see that

$$EA = [\, \mathbf{e}_1 \mid \cdots \mid \mathbf{e}_j \mid \cdots \mid \mathbf{e}_i \mid \cdots \mid \mathbf{e}_m \,] \begin{bmatrix} \mathbf{a}_1^T \\ \vdots \\ \mathbf{a}_m^T \end{bmatrix}$$

$$= \mathbf{e}_1\mathbf{a}_1^T + \mathbf{e}_2\mathbf{a}_2^T + \cdots + \mathbf{e}_j\mathbf{a}_i^T + \cdots + \mathbf{e}_i\mathbf{a}_j^T + \cdots + \mathbf{e}_m\mathbf{a}_m^T,$$

which is exactly the matrix that is obtained by swapping the i-th and j-th rows of A. ∎

B.4 Existence of the Determinant

> Throughout this section, we use the same notation as in Section 3.A.3. In particular, S_n is the set of all permutations on $\{1, 2, \ldots, n\}$, σ is a permutation, and $\mathrm{sgn}(\sigma)$ is the sign of that permutation.

In Section 3.2, we defined the determinant as the unique function satisfying the three properties given in Definition 3.2.1. Uniqueness of the determinant followed from the fact that we came up with explicit formulas for computing it (i.e., we showed that any function satisfying those three properties must be given by, for example, the cofactor expansions of Theorem 3.2.8).

However, we never showed that the determinant *exists*. That is, we never showed that any of the formulas we came up with satisfy all three of the defining properties of the determinant (see Remark 3.2.2). We now fill in this gap by showing that the function defined by the formula given in Theorem 3.A.4 satisfies its three defining properties.

Theorem B.4.1 **Permutation Sums** **Satisfy Determinant** **Properties**	Let $A \in \mathcal{M}_n$ and define a function $f : \mathcal{M}_n \to \mathbb{R}$ via $$f(A) = \sum_{\sigma \in S_n} \mathrm{sgn}(\sigma) a_{\sigma(1),1} a_{\sigma(2),2} \cdots a_{\sigma(n),n}.$$ Then f satisfies the three defining properties of the determinant from Definition 3.2.1.

Proof. The fact that $f(I) = 1$ follows simply from the fact that if $A = I$ and σ is the identity permutation (i.e., $\sigma(j) = j$ for all $1 \leq j \leq n$) then

$$\mathrm{sgn}(\sigma) a_{\sigma(1),1} a_{\sigma(2),2} \cdots a_{\sigma(n),n} = a_{1,1} a_{2,2} \cdots a_{n,n} = 1,$$

whereas if σ is any other permutation then the above product contains an off-diagonal entry of I and thus equals 0. By adding up many terms equal to 0 and one term equal to 1, we see that $\det(I) = 1$.

To see that $f(AB) = f(A)f(B)$ for all $A, B \in \mathcal{M}_n$, we recall the Cauchy–Binet formula from Theorem 3.A.5. In particular, in the $m = n$ case this formula tells us that $\det(AB) = \det(A)\det(B)$, and we proved that theorem only using the sum-of-permutations formula for f given in the statement of this theorem (not any of the three defining properties of the determinant or other formulas for it), so we conclude that $f(AB) = f(A)f(B)$ too.

Finally, to see that f is multilinear in the columns of the input matrix A, we just compute

Here we are letting j denote the index of the column in which the $v + cw$ column occurs.

$$
\begin{aligned}
f\big(&[\, \mathbf{a}_1 \mid \cdots \mid \mathbf{v} + c\mathbf{w} \mid \cdots \mid \mathbf{a}_n \,]\big) \\
&= \sum_{\sigma \in S_n} \mathrm{sgn}(\sigma) a_{\sigma(1),1} \cdots [\mathbf{v} + c\mathbf{w}]_{\sigma(j)} \cdots a_{\sigma(n),n} \\
&= \sum_{\sigma \in S_n} \mathrm{sgn}(\sigma) a_{\sigma(1),1} \cdots v_{\sigma(j)} \cdots a_{\sigma(n),n} \\
&\quad + c \sum_{\sigma \in S_n} \mathrm{sgn}(\sigma) a_{\sigma(1),1} \cdots w_{\sigma(j)} \cdots a_{\sigma(n),n} \\
&= f\big([\, \mathbf{a}_1 \mid \cdots \mid \mathbf{v} \mid \cdots \mid \mathbf{a}_n \,]\big) + c f\big([\, \mathbf{a}_1 \mid \cdots \mid \mathbf{w} \mid \cdots \mid \mathbf{a}_n \,]\big),
\end{aligned}
$$

which completes the proof. $\blacksquare$

B.5   Multiplicity in the Perron–Frobenius Theorem

About halfway through the proof of the Perron–Frobenius theorem (Theorem 3.B.2), we claimed that the eigenvalue μ of A had algebraic multiplicity 1. We now prove this claim. Throughout this subsection, we use the same notation as in the proof of Theorem 3.B.2, so L is the function defined in Equation (3.B.1) and $\mu \in \mathbb{R}$ is the maximal value of L on S^n.

Before proving anything about μ, we claim that if $B \in \mathcal{M}_n(\mathbb{R})$ is such that $O \leq B \leq A$ and λ is any eigenvalue of B, then $|\lambda| = \mu$ implies $B = A$. To provide a bit of context, recall that we already showed in Claim 4 of the proof of Theorem 3.B.2 that $|\lambda| \leq \mu$ (regardless of whether or not $B = A$).

To see why this claim holds, let $\mathbf{w}$ be a unit eigenvector corresponding to the eigenvalue λ of B. We showed in the proof of Claim 4 that $L(|\mathbf{w}|) \geq |\lambda|$, so if $|\lambda| = \mu$ then in fact we have $L(|\mathbf{w}|) = \mu$. Claims 1 and 3 then tell us that $A|\mathbf{w}| = \mu|\mathbf{w}|$ and $|\mathbf{w}| > \mathbf{0}$. It follows that

$$(A - B)|\mathbf{w}| \le \mu|\mathbf{w}| - \mu|\mathbf{w}| = \mathbf{0},$$

but since $A - B$ is entrywise non-negative, this is only possible if $(A - B)|\mathbf{w}| = \mathbf{0}$. Since $|\mathbf{w}| > \mathbf{0}$, this then implies $A - B = O$, so $A = B$, which completes the proof of this first claim.

With the above detail out of the way, we can now prove that μ has algebraic multiplicity 1 as an eigenvalue of A. To this end, we consider what happens when we take the derivative of the characteristic polynomial $p_A(\lambda) = \det(A - \lambda I)$ of A. To show that the algebraic multiplicity of μ is 1, it suffices to show that $p'_A(\mu) \ne 0$ (via Theorem B.6.1, for example). To see why this is the case, consider the somewhat uglier multivariable function $f : \mathbb{R}^n \to \mathbb{R}$ defined by

$$f(\lambda_1, \lambda_2, \ldots, \lambda_n) = \det(A - \Lambda),$$

where Λ is the diagonal matrix whose diagonal entries are $\lambda_1, \lambda_2, \ldots, \lambda_n$. If we let A_i denote the $(n-1) \times (n-1)$ matrix that is obtained by removing the i-th row and column from A (and similarly for Λ_i from Λ), and we let $c_{i,j}$ denote the (i, j)-cofactor of $A - \Lambda$, then applying a cofactor expansion along the i-th row of $A - \Lambda$ shows that

$$\det(A - \Lambda) = a_{i,1}c_{i,1} + \cdots + (a_{i,i} - \lambda_i)c_{i,i} + \cdots + a_{i,n}c_{i,n}.$$

Since none of $a_{i,1}, \ldots, a_{i,n}$ or $c_{i,1}, \ldots, c_{i,n}$ depend on λ_i, it follows that

$$\frac{\partial}{\partial \lambda_i} \det(A - \Lambda) = -c_{i,i} = -\det(A_i - \Lambda_i). \tag{B.5.1}$$

By the chain rule from multivariable calculus, we then have

$$
\begin{aligned}
p'_A(\lambda) &= \frac{d}{d\lambda} f(\lambda, \lambda, \ldots, \lambda) \\
&= \frac{\partial}{\partial \lambda_1} f(\lambda_1, \lambda, \ldots, \lambda) + \frac{\partial}{\partial \lambda_2} f(\lambda, \lambda_2, \ldots, \lambda) + \cdots + \frac{\partial}{\partial \lambda_n} f(\lambda, \lambda, \ldots, \lambda_n) \\
&= -\det(A_1 - \lambda I) - \det(A_2 - \lambda I) - \cdots - \det(A_n - \lambda I) \\
&= -p_{A_1}(\lambda) - p_{A_2}(\lambda) - \cdots - p_{A_n}(\lambda),
\end{aligned}
$$

with the second-to-last equality following from Equation (B.5.1).

If we now let B_i be the matrix obtained by replacing the i-th row and column of A by zeros (or equivalently, B_i is obtained from A_i by inserting a row and column of zeros at the i-th index), then $p_{B_i}(\lambda) = -\lambda p_A(\lambda)$ for all i, so

$$p'_A(\lambda) = \frac{1}{\lambda}\left(p_{B_1}(\lambda) + p_{B_2}(\lambda) + \cdots + p_{B_n}(\lambda)\right) \quad \text{for all} \quad \lambda \ne 0.$$

Since $O \le B_i \le A$, we know from our earlier claim that the absolute values of the eigenvalues of B_i are strictly smaller than μ. This tells us that either $p_{B_i}(\mu) > 0$ for all i (if n is even) or $p_{B_i}(\mu) < 0$ for all i (if n is odd). Either way, it follows that $p'_A(\mu) \ne 0$, so we are done.

B.6 Multiple Roots of Polynomials

When proving Theorem 3.D.3, which gives the general form for the solution of a linear recurrence relation, we needed a technical result that relates the multiplicity of a root of a polynomial to another closely-related family of polynomials. We now pin down this result, but first we need to establish the following connection between the multiplicity of a root of a polynomial and the derivatives of that polynomial.

Theorem B.6.1
Multiplicity of Roots and Derivatives

> Suppose p is a polynomial. Then r is a root of p with multiplicity at least m if and only if r is a root of each of p and its first $m-1$ derivatives:
>
> $$p(r) = p'(r) = p''(r) = \cdots = p^{(m-1)}(r) = 0.$$

Proof. We instead prove the (equivalent) statement that if r has multiplicity *exactly* m then is a root of p, p', ..., $p^{(m-1)}$, but not of $p^{(m)}$.

Recall that the factor theorem (Theorem A.2.2) lets us factor p as $p(x) = (x-r)^m q(x)$, where q is some polynomial with $q(r) \neq 0$. We begin by showing that, for every integer $k \geq 0$, the k-th derivative of p has the form

The notation $p^{(k)}$ refers to the k-th derivative of p.

$$p^{(k)}(x) = (x-r)^{m-k}(c_k q(x) + s_k(x)) \quad \text{for all} \quad 0 \leq k \leq m, \qquad \text{(B.6.1)}$$

where $c_k \neq 0$ is a scalar and $s_k(x)$ is a polynomial depending on k with the property that $s_k(r) = 0$ (note that in the $k = 0$ case, we simply have $c_0 = 1$ and $s_k(x) = 0$).

For the $k = 1$ case, we simply use the product rule for derivatives to see that

Recall that the product rule says that $(fg)'(x) = f'(x)g(x) + f(x)g'(x)$.

$$p'(x) = m(x-r)^{m-1}q(x) + (x-r)^m q'(x) = (x-r)^{m-1}(mq(x) + (x-r)q'(x)),$$

so the derivative of p does have the claimed form, with $c_1 = m$ and $s_1(x) = (x-r)q'(x)$.

From here, we proceed by induction: if $p^{(k)}$ has the form (B.6.1) for some value of k then taking its derivative via the product rule shows that

$$p^{(k+1)}(x) = (m-k)(x-r)^{m-(k+1)}\left(c_k q(x) + s_k(x)\right)$$
$$+ (x-r)^{m-k}(c_k q'(x) + s_k'(x))$$
$$= (x-r)^{m-(k+1)}\left((m-k)c_k q(x) + (m-k)s_k(x)\right.$$
$$\left. + (x-r)(c_k q'(x) + s_k'(x))\right),$$

so $p^{(k+1)}$ also has the form (B.6.1), with

$$c_{k+1} = (m-k)c_k \quad \text{and} \quad s_{k+1}(x) = (m-k)s_k(x) + (x-r)(c_k q'(x) + s_k'(x)).$$

This completes the inductive step and thus the proof of the claim that $p^{(k)}$ has the form (B.6.1) for all $0 \leq k \leq m$.

It follows by plugging in $x = r$ that $p^{(k)}(r) = 0$ for all $0 \leq k < m$. Furthermore, since $s_m(r) = 0$ but $c_m, q(r) \neq 0$, it follows that

$$p^{(m)}(r) = c_m q(r) + s_m(r) = c_m q(r) \neq 0,$$

which completes the proof. ∎

With the above result taken care of, we can now prove the result that we actually need.

Theorem B.6.2	Consider polynomials $p_0, p_1, \ldots, p_m$ of the form
Multiple Roots of	
Polynomials	

$$
\begin{aligned}
p_0(x) &= c_0 + c_1 x + c_2 x^2 + \cdots + c_n x^n \\
p_1(x) &= c_1 x + 2c_2 x^2 + \cdots + nc_n x^n \\
p_2(x) &= c_1 x + 2^2 c_2 x^2 + \cdots + n^2 c_n x^n \\
&\vdots \\
p_m(x) &= c_1 x + 2^m c_2 x^2 + \cdots + n^m c_n x^n.
\end{aligned}
$$

Then $r \neq 0$ is a root of p_0 with multiplicity at least $m+1$ if and only if it is a root of each of p_0, p_1, ..., p_m.

Proof. If $m = 0$ then the result is trivial (it just says that r is a root of p_0 if and only if it is a root of p_0), so we start by proving the $m = 1$ case. If r is a root of p_0 with multiplicity $m + 1 = 2$ then Theorem B.6.1 tells us that it is also a root of its derivative, p_0', and thus also a root of

Recall that the
derivative of x^n is
nx^{n-1}.

$$
x p_0'(x) = x(c_1 + 2c_2 x + 3c_3 x^2 + \cdots + nc_n x^{n-1}) = p_1(x).
$$

In the opposite direction, we can just follow this argument backwards: if $r \neq 0$ is a root of $p_1(x) = x p_0'(x)$ then it is a root of $p_0'(x)$, so Theorem B.6.1 tells us that it is a root of p_0 with multiplicity at least 2.

If $m > 1$ then we can just iterate the argument above, using the facts that $p_2(x) = x p_1'(x)$, $p_3(x) = x p_2'(x)$, and so on. ∎

B.7 Limits of Ratios of Polynomials and Exponentials

In the proof of Corollary 3.D.4, we needed to make use of the facts that if q is any polynomial and c is a number with $|c| < 1$ then

$$
\lim_{x \to \infty} \frac{q(x+1)}{q(x)} = 1 \quad \text{and} \quad \lim_{x \to \infty} q(x) c^x = 0. \tag{B.7.1}
$$

We now prove these statements by making use of L'Hôpital's rule from calculus, which says that if

$$
\lim_{x \to \infty} \frac{f(x)}{g(x)}
$$

exists and is an indeterminate form of type "$0/0$" or "∞/∞" (along with some other technical requirements that we won't get into here), then it equals

$$
\lim_{x \to \infty} \frac{f'(x)}{g'(x)}.
$$

For the first of the limits (B.7.1) that we want to evaluate, we write q explicitly as $q(x) = c_k x^k + \cdots + c_1 x + c_0$. Then using L'Hôpital's rule a total of

k times shows that

$$\lim_{x\to\infty}\frac{q(x+1)}{q(x)}=\lim_{x\to\infty}\frac{c_k(x+1)^k+\cdots+c_1(x+1)+c_0}{c_kx^k+\cdots+c_1x+c_0}$$

$$\overset{\text{L'Hôp}}{=}\lim_{x\to\infty}\frac{c_kk(x+1)^{k-1}+\cdots+c_1}{c_kkx^{k-1}+\cdots+c_1}$$

$$\vdots$$

$$\overset{\text{L'Hôp}}{=}\lim_{x\to\infty}\frac{c_kk!(x+1)+c_{k-1}(k-1)!}{c_kk!x+c_{k-1}(k-1)!}$$

$$\overset{\text{L'Hôp}}{=}\lim_{x\to\infty}\frac{c_kk!}{c_kk!}$$

$$=1,$$

as claimed.

To evaluate the other limit (B.7.1), we define $b=1/c$ and notice that $|b|>1$ since $|c|<1$. Then using L'Hôpital's rule a total of k times (again!) shows that

$$\lim_{x\to\infty}q(x)c^x=\lim_{x\to\infty}\frac{c_k(x+1)^k+\cdots+c_1(x+1)+c_0}{b^x}$$

Recall that the derivative of b^x is $b^x\ln(b)$.

$$\overset{\text{L'Hôp}}{=}\lim_{x\to\infty}\frac{c_kk(x+1)^{k-1}+\cdots+c_1}{b^x\ln(b)}$$

$$\vdots$$

This final fraction goes to 0 because the numerator is constant (it equals $c_kk!$), whereas the denominator grows without bound.

$$\overset{\text{L'Hôp}}{=}\lim_{x\to\infty}\frac{c_kk!(x+1)+c_{k-1}(k-1)!}{b^x(\ln(b))^{k-1}}$$

$$\overset{\text{L'Hôp}}{=}\lim_{x\to\infty}\frac{c_kk!}{b^x(\ln(b))^k}$$

$$=0,$$

as claimed.

C. Selected Exercise Solutions

Section 1.1: Vectors and Vector Operations

1.1.1 (a)

(c)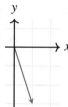

1.1.2 (a)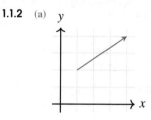

(c)

1.1.3 (a) $(0,1)$ (c) $(2,6)$

1.1.4 (a) (c)

1.1.5 (a) $(2.5,5)$ (c) $(-4,5)$

1.1.6 (a) $(3,2,1)$ (c) $(6,4,-2)$

1.1.7 (a) $\mathbf{v} = 2\mathbf{e}_3$ (c) $\mathbf{x} = \mathbf{e}_1 + 2\mathbf{e}_2$

1.1.9 We want vectors $\mathbf{v}$ and $\mathbf{w}$ such that $\mathbf{v} + \mathbf{w} = \mathbf{x}$ and $\mathbf{v} - \mathbf{w} = \mathbf{y}$. Adding these two equations together tells us that $2\mathbf{v} = \mathbf{x} + \mathbf{y}$, and subtracting the second equation from the first tells us that $2\mathbf{w} = \mathbf{x} - \mathbf{y}$, so $\mathbf{v} = (\mathbf{x} + \mathbf{y})/2 = (2,1)$ and $\mathbf{w} = (\mathbf{x} - \mathbf{y})/2 = (1,-3)$. The negatives of these answers are fine too.

1.1.10 (a) $\mathbf{x} = (2,2)$ (c) $\mathbf{x} = \mathbf{0}$

1.1.11 (a) $\mathbf{x} = \frac{1}{2}(\mathbf{v} - \mathbf{w})$ (c) $\mathbf{x} = 2\mathbf{w} - 4\mathbf{v}$

1.1.12 Algebraically, we note that $c(1,2) = (3,4)$ implies $c = 3$ and $2c = 4$, so $c = 2$. Since c cannot simultaneously be 2 and 3, no such c exists. Geometrically, we notice that $(1,2)$ and $(3,4)$ point in different directions, so no such c exists:

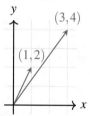

1.1.14 We just use the definition of vector addition, together with associativity of real number addition:

$$
\begin{aligned}
(\mathbf{v} + \mathbf{w}) + \mathbf{x} &= ((v_1,\ldots,v_n) + (w_1,\ldots,w_n)) \\
&\quad + (x_1,\ldots,x_n) \\
&= (v_1 + w_1,\ldots,v_n + w_n) + (x_1,\ldots,x_n) \\
&= (v_1 + w_1 + x_1,\ldots,v_n + w_n + x_n) \\
&= (v_1,\ldots,v_n) + (w_1 + x_1,\ldots,w_n + x_n) \\
&= (v_1,\ldots,v_n) \\
&\quad + ((w_1,\ldots,w_n) + (x_1,\ldots,x_n)) \\
&= \mathbf{v} + (\mathbf{w} + \mathbf{x}).
\end{aligned}
$$

1.1.15 (a) We just use the definition of vector addition and scalar multiplication, together with properties of real numbers:

$$
\begin{aligned}
(c+d)\mathbf{v} &= ((c+d)v_1,\ldots,(c+d)v_n) \\
&= (cv_1 + dv_1,\ldots,cv_n + dv_n) \\
&= (cv_1,\ldots,cv_n) + (dv_1,\ldots,dv_n) \\
&= c\mathbf{v} + d\mathbf{v}.
\end{aligned}
$$

(b) Similarly,

$$
\begin{aligned}
c(d\mathbf{v}) &= c(dv_1,\ldots,dv_n) \\
&= (cdv_1,\ldots,cdv_n) \\
&= ((cd)v_1,\ldots,(cd)v_n) \\
&= (cd)\mathbf{v}.
\end{aligned}
$$

Section 1.2: Lengths, Angles, and the Dot Product

1.2.1 (a) 0 (c) 1 (e) 0

1.2.2 (a) $\|\mathbf{v}\| = 5, \mathbf{u} = (3/5, 4/5)$
(c) $\|\mathbf{v}\| = 6, \mathbf{u} = (-\sqrt{2}/3, -1/2, \sqrt{10}/6, 1/2)$

1.2.3 (a) $\arccos(\sqrt{3}/2) = \pi/6$
(c) $\arccos(-1/\sqrt{2}) = 3\pi/4$
(e) $\mathbf{v} \cdot \mathbf{w} = 0$, so the angle is $\pi/2$

1.2.4 (a) False. A counter-example is given by $\mathbf{v} = (1,0)$, $\mathbf{w} = (1,1)$, and $\mathbf{x} = (1,-1)$. In this case, $\mathbf{v} \cdot \mathbf{w} = \mathbf{v} \cdot \mathbf{x} = 1$, but $\mathbf{w} \neq \mathbf{x}$.
(c) True. The triangle inequality tells us that $\|\mathbf{v} + \mathbf{w}\| \leq \|\mathbf{v}\| + \|\mathbf{w}\| \leq 2$.
(e) True. As long as $|\mathbf{v} \cdot \mathbf{w}| \leq \|\mathbf{v}\|\|\mathbf{w}\|$ (i.e., the Cauchy–Schwarz inequality) is satisfied, we can find vectors with the given lengths and dot product. As an explicit example, let $\mathbf{v} = (1,0,0)$ and $\mathbf{w} = (-1, \sqrt{3}, 0)$.
(g) False. A counter-example is given by $\mathbf{v} = (2,0)$ and $\mathbf{w} = (0,2)$.

1.2.6 Since $\mathbf{v} \cdot \mathbf{w} = \cos(\theta)\|\mathbf{v}\|\|\mathbf{w}\| = \cos(\pi/3)(2\sqrt{3})(2) = 2\sqrt{3}$, we find that $3w_1 + \sqrt{3}w_2 = 2\sqrt{3}$, so $w_2 = 2 - \sqrt{3}w_1$. On the other hand, $\|\mathbf{w}\| = 2$, so $w_1^2 + w_2^2 = 4$. Plugging in our formula for w_2 gives $w_1^2 + (2 - \sqrt{3}w_1)^2 = 4$, and simplifying gives $w_1(w_1 - \sqrt{3}) = 0$, so $w_1 = 0$ or $w_1 = \sqrt{3}$. If we plug these values back into the equation $3w_1 + \sqrt{3}w_2 = 2\sqrt{3}$, we find that $w_1 = 0$ implies $w_2 = 2$, and $w_1 = \sqrt{3}$ implies $w_2 = -1$. There are thus two possible vectors: $\mathbf{w} = (0,2)$ and $\mathbf{w} = (\sqrt{3}, -1)$.

1.2.7 (a) Makes sense.
(c) Does not make sense—we cannot add a vector and a scalar.
(e) Does not make sense—raising a vector to an exponent is not standard notation. Does $\mathbf{v}^2$ mean the vector whose entries are the squares of the entries of $\mathbf{v}$, or the scalar $\|\mathbf{v}\|^2 = \mathbf{v} \cdot \mathbf{v}$, or something else altogether?

1.2.9 (a) There are many possibilities: we require that $\mathbf{w} = (w_1, w_2) \in \mathbb{R}^2$ satisfies $w_1 + 2w_2 = 0$, so we can choose w_2 arbitrarily and set $w_1 = -2w_2$. One possibility is $\mathbf{w} = (-2, 1)$.
(b) No, this is not possible, since we showed in part (a) that the only solutions $\mathbf{y}$ to $\mathbf{v} \cdot \mathbf{y} = 0$ are of the form $\mathbf{y} = y_2(-2, 1)$. However, $\mathbf{y}$ then cannot be orthogonal to $\mathbf{w}$, since $\mathbf{w} \cdot \mathbf{y} = y_2(4+1)$, which equals 0 if and only if $\mathbf{y} = \mathbf{0}$. However, we were asked to find a *non-zero* $\mathbf{y}$.

1.2.11 Start by computing the differences of the three points that are given to us:

$$(1,0,-1) - (2,2,2) = (-1,-2,-3)$$
$$(1,0,-1) - (-1,2,3) = (2,-2,-4)$$
$$(-1,2,3) - (2,2,2) = (-3,0,1).$$

We know that two of these vectors are sides of the rectangle (the other one is a diagonal of the rectangle), but we do not know which two yet. To determine which are which, we use the dot product to find which of these vectors are orthogonal:

$$(-1,-2,-3) \cdot (2,-2,-4) = -2+4+12 = 14$$
$$(-1,-2,-3) \cdot (-3,0,1) = 3+0-3 = 0$$
$$(-3,0,1) \cdot (2,-2,-4) = -6+0-4 = -10.$$

We thus see that $(-1,-2,-3)$ and $(-3,0,1)$ are sides of the rectangle, and $(2,2,2)$ is the vertex that joins these two sides. If we add both of the side vectors to $(2,2,2)$ then we will get the opposite corner of the rectangle, which is $(2,2,2) + (-1,-2,-3) + (-3,0,1) = (-2,0,0)$.

1.2.13 (a) We just use the definitions of the dot product and vector addition, together with properties of the real numbers:

$$\mathbf{v} \cdot (\mathbf{w} + \mathbf{x}) = v_1(w_1 + x_1) + \cdots + v_n(w_n + x_n)$$
$$= (v_1 w_1 + v_1 x_1) + \cdots + (v_n w_n + v_n x_n)$$
$$= \mathbf{v} \cdot \mathbf{w} + \mathbf{v} \cdot \mathbf{x}.$$

(b) We just use the definitions of the dot product and scalar multiplication, together with properties of the real numbers:

$$\mathbf{v} \cdot (c\mathbf{w}) = v_1(cw_1) + \cdots + v_n(cw_n)$$
$$= c(v_1 w_1) + \cdots + c(v_n w_n)$$
$$= c(\mathbf{v} \cdot \mathbf{w}).$$

1.2.15 (a) To prove the desired equality, we compute

$$\overline{\mathbf{w} \cdot \mathbf{v}} = \overline{w_1 v_1} + \cdots + \overline{w_n v_n}$$
$$= w_1 \overline{v_1} + \cdots + w_n \overline{v_n}$$
$$= \overline{v_1} w_1 + \cdots + \overline{v_n} w_n$$
$$= \mathbf{v} \cdot \mathbf{w},$$

where we used the facts that $\overline{\overline{x}} = x$ and $\overline{x \cdot y} = \overline{x} \cdot \overline{y}$ for all $x, y \in \mathbb{C}$.
(b) For this one, we similarly compute

$$\mathbf{v} \cdot (c\mathbf{w}) = \overline{v_1}(cw_1) + \cdots + \overline{v_n}(cw_n)$$
$$= c(\overline{v_1} w_1 + \cdots + \overline{v_n} w_n)$$
$$= c(\mathbf{v} \cdot \mathbf{w}).$$

By combining the above fact with the one that we proved in part (a), we see that

$$(c\mathbf{v}) \cdot \mathbf{w} = \overline{\mathbf{w} \cdot (c\mathbf{v})} = \overline{c}(\overline{\mathbf{w} \cdot \mathbf{v}}) = \overline{c}\mathbf{v} \cdot \mathbf{w}.$$

1.2.17 (a) For the "if" direction: if $\mathbf{w} = \mathbf{0}$ then $\mathbf{v} \cdot \mathbf{w} = 0$ and $\|\mathbf{v}\|\|\mathbf{w}\| = 0$, so equality holds in the Cauchy–Schwarz inequality, and if $\mathbf{v} = c\mathbf{w}$ then $|\mathbf{v} \cdot \mathbf{w}| = |(c\mathbf{w}) \cdot \mathbf{w}| = |c|\|\mathbf{w}\|^2$, whereas $\|\mathbf{v}\|\|\mathbf{w}\| = \|c\mathbf{w}\|\|\mathbf{w}\| = |c|\|\mathbf{w}\|^2$, so equality holds in this case too.

For the "only if" direction, we notice that $\mathbf{v} \cdot \mathbf{w} = \|\mathbf{v}\|\|\mathbf{w}\|$ implies (by tracing the proof of Theorem 1.2.3 backwards) that the length of $\|\mathbf{w}\|\mathbf{v} - \|\mathbf{v}\|\mathbf{w}$ is 0. However, we saw in Theorem 1.2.2 that the only vector with 0 length is $\mathbf{0}$, so it follows that $\|\mathbf{w}\|\mathbf{v} = \|\mathbf{v}\|\mathbf{w}$, so $\mathbf{v}$ and $\mathbf{w}$ are scalar multiples of each other.

(b) For the "if" direction: if $\mathbf{w} = \mathbf{0}$ then the equality just says that $\|\mathbf{v}\| = \|\mathbf{v}\|$, which is true, and if $\mathbf{v} = c\mathbf{w}$ then $\|\mathbf{v} + \mathbf{w}\| = \|(c\mathbf{w}) + \mathbf{w}\| = \|(c+1)\mathbf{w}\| = (c+1)\|\mathbf{w}\| = c\|\mathbf{w}\| + \|\mathbf{w}\| = \|c\mathbf{w}\| + \|\mathbf{w}\| = \|\mathbf{v}\| + \|\mathbf{w}\|$.

For the "only if" direction, recall from the proof of Theorem 1.2.4 that if $\|\mathbf{v} + \mathbf{w}\| = \|\mathbf{v}\| + \|\mathbf{w}\|$ then $2(\mathbf{v} \cdot \mathbf{w}) = 2\|\mathbf{v}\|\|\mathbf{w}\|$. It then follows from part (a) that either $\mathbf{w} = \mathbf{0}$ or there exists $c \in \mathbb{R}$ such that $\mathbf{v} = c\mathbf{w}$; all that remains is to show that $c \geq 0$. To this end, just observe that if it were the case that $\mathbf{v} = c\mathbf{w}$ with $c < 0$ then $2(\mathbf{v} \cdot \mathbf{w}) = 2c\|\mathbf{w}\|^2$, whereas $2\|\mathbf{v}\|\|\mathbf{w}\| = 2|c|\|\mathbf{w}\|^2 = -2c\|\mathbf{w}\|^2$, which contradicts the fact that $2(\mathbf{v} \cdot \mathbf{w}) = 2\|\mathbf{v}\|\|\mathbf{w}\|$. It follows that $c \geq 0$ and we are done.

1.2.19 **(a)** Expand $\|\mathbf{v} + \mathbf{w}\|^2$ and $\|\mathbf{v} - \mathbf{w}\|^2$ using the dot product:

$$\|\mathbf{v} + \mathbf{w}\|^2 + \|\mathbf{v} - \mathbf{w}\|^2$$
$$= (\mathbf{v} + \mathbf{w}) \cdot (\mathbf{v} + \mathbf{w}) + (\mathbf{v} - \mathbf{w}) \cdot (\mathbf{v} - \mathbf{w})$$
$$= (\mathbf{v} \cdot \mathbf{v}) + 2(\mathbf{v} \cdot \mathbf{w}) + (\mathbf{w} \cdot \mathbf{w})$$
$$+ (\mathbf{v} \cdot \mathbf{v}) - 2(\mathbf{v} \cdot \mathbf{w}) + (\mathbf{w} \cdot \mathbf{w})$$
$$= 2\|\mathbf{v}\|^2 + 2\|\mathbf{w}\|^2.$$

(b) The vectors $\mathbf{v} + \mathbf{w}$ and $\mathbf{v} - \mathbf{w}$ are the diagonals of the parallelogram, so part (a) says that the sum of the squares of the two diagonal lengths of a parallelogram is equal to the sum of the squares of the lengths of its four sides. In the special case where the parallelogram is a rectangle, this is exactly the Pythagorean theorem.

1.2.21 The triangle inequality tells us that $\|\mathbf{x} + \mathbf{y}\| \leq \|\mathbf{x}\| + \|\mathbf{y}\|$ for all $\mathbf{x}, \mathbf{y} \in \mathbb{R}^n$. If we choose $\mathbf{x} = \mathbf{v} - \mathbf{w}$ and $\mathbf{y} = \mathbf{w}$, then this says that

$$\|(\mathbf{v} - \mathbf{w}) + \mathbf{w}\| \leq \|\mathbf{v} - \mathbf{w}\| + \|\mathbf{w}\|.$$

After simplifying and rearranging, this becomes

$$\|\mathbf{v}\| - \|\mathbf{w}\| \leq \|\mathbf{v} - \mathbf{w}\|.$$

1.2.23 **(a)** $f(x) = \|\mathbf{v} - x\mathbf{w}\|^2 = (\mathbf{v} - x\mathbf{w}) \cdot (\mathbf{v} - x\mathbf{w}) = x^2\|\mathbf{w}\|^2 - 2x(\mathbf{v} \cdot \mathbf{w}) + \|\mathbf{v}\|^2$

(b) $4(\mathbf{v} \cdot \mathbf{w})^2 - 4\|\mathbf{v}\|^2\|\mathbf{w}\|^2$

(c) Since $f(x) \geq 0$ for all $x \in \mathbb{R}$, we see that f has at most one real root, so the discriminant is non-positive.

(d) By parts (b) and (c), we know that $4(\mathbf{v} \cdot \mathbf{w})^2 - 4\|\mathbf{v}\|^2\|\mathbf{w}\|^2 \leq 0$. By adding $4\|\mathbf{v}\|^2\|\mathbf{w}\|^2$ to both sides, dividing by 4, and then taking square roots, we get the Cauchy–Schwarz inequality: $|\mathbf{v} \cdot \mathbf{w}| \leq \|\mathbf{v}\|\|\mathbf{w}\|$.

Section 1.3: Matrices and Matrix Operations

1.3.1 **(a)** $\begin{bmatrix} 1 & 0 \\ -2 & 3 \end{bmatrix}$ **(c)** $\begin{bmatrix} 3 & -2 \\ 2 & 3 \end{bmatrix}$

1.3.2 **(a)** $\begin{bmatrix} 0 & 2 \\ -6 & 0 \end{bmatrix}$ **(c)** $\begin{bmatrix} 1 & 4 & -3 \\ 3 & 0 & 3 \end{bmatrix}$

(e) $\begin{bmatrix} 4 & -5 \\ 0 & 9 \end{bmatrix}$ **(g)** $\begin{bmatrix} 2 & 4 & -1 \\ 0 & 4 & -2 \end{bmatrix}$

(i) $\begin{bmatrix} 1 & 0 & 0 \\ 7 & 6 & 3 \end{bmatrix}$

1.3.3 **(a)** False. For example, if A is 2×3 and B is 3×2 then AB is 2×2.

(c) True. (Same reasoning as in part (b).)

(e) False. Again, non-commutativity of matrix multiplication gets in the way. $(AB)^2 = (AB)(AB)$, but in general there is no way to swap the As and Bs past each other in the middle. As a specific counter-example, consider

$$A = \begin{bmatrix} 1 & 0 \\ 0 & 0 \end{bmatrix}, B = \begin{bmatrix} 1 & 1 \\ 0 & 1 \end{bmatrix}.$$

(g) False. One counter-example is given by

$$\begin{bmatrix} 1 & 0 \\ 0 & -1 \end{bmatrix}.$$

(i) False. It is a linear combination of the **columns** of A, not its rows.

1.3.4 **(a)** Yes, this makes sense. AB is 2×5.

(c) No, the inner dimensions of A and C are 2 and 5, so we cannot multiply them.

(e) No, we can only subtract matrices that have the same size.

(g) Yes, this makes sense. ABC is 2×2.

(i) Yes, this makes sense. $A + BC$ is 2×2.

1.3.5 Start by observing that

$$A^1 = \begin{bmatrix} 0 & 1 \\ -1 & 0 \end{bmatrix} \quad A^2 = \begin{bmatrix} -1 & 0 \\ 0 & -1 \end{bmatrix}$$

$$A^3 = \begin{bmatrix} 0 & -1 \\ 1 & 0 \end{bmatrix} \quad A^4 = \begin{bmatrix} 1 & 0 \\ 0 & 1 \end{bmatrix}.$$

It follows that $A^4 = I$, so the powers of A cycle from this point on (i.e., $A^5 = A^1$, $A^6 = A^2$, and so on). We thus conclude that

$$A^{1000} = A^{996} = A^{992} = \cdots = A^4 = I.$$

1.3.7 **(a)** $A^2 = \begin{bmatrix} 1 & 2 \\ 0 & 1 \end{bmatrix}$

(b) $A^3 = \begin{bmatrix} 1 & 3 \\ 0 & 1 \end{bmatrix}$

(c) Based on parts (a) and (b), we guess that

$$A^k = \begin{bmatrix} 1 & k \\ 0 & 1 \end{bmatrix} \quad \text{for all integers } k \geq 0.$$

We prove this by induction: the base cases $k = 0, 1$ are clear, and we also already showed the $k = 2, 3$ cases. For the inductive step, assume that $A^\ell = \begin{bmatrix} 1 & \ell \\ 0 & 1 \end{bmatrix}$ for some particular positive integer ℓ. Then

$$A^{\ell+1} = AA^\ell = \begin{bmatrix} 1 & 1 \\ 0 & 1 \end{bmatrix}\begin{bmatrix} 1 & \ell \\ 0 & 1 \end{bmatrix} = \begin{bmatrix} 1 & \ell+1 \\ 0 & 1 \end{bmatrix},$$

which completes the inductive step and thus the proof.

1.3.11 The (i, j)-entry of J_n^2 is defined to be the dot product of the i-th row of J_n with the j-th column of J_n. That is, $[J_n]_{i,j} = (1, 1, \ldots, 1) \cdot (1, 1, \ldots, 1) = n$ for all i, j, so $J_n^2 = nJ_n$.

1.3.12   (a) Yes, this makes sense:

$$\begin{bmatrix} A & A \\ A & A \end{bmatrix}^2 = \begin{bmatrix} 2A^2 & 2A^2 \\ 2A^2 & 2A^2 \end{bmatrix}.$$

  (c) No, this makes no sense. The block matrix multiplication would gives us

$$\begin{bmatrix} A & B \\ B^T & I_3 \end{bmatrix}\begin{bmatrix} O & C^T \\ B^T & I_3 \end{bmatrix}$$

$$= \begin{bmatrix} AO + BB^T & AC^T + BI_3 \\ B^T O + I_3 B^T & B^T C^T + I_3 I_3 \end{bmatrix},$$

but some of these products (like AC^T) do not make sense.

1.3.13   (a) $AB = \left[\begin{array}{cc|cc} 3 & 3 & 3 & 3 \\ 3 & 3 & 3 & 3 \end{array}\right]$

  (c) $AB = \begin{bmatrix} 2 & 4 & 6 \\ 3 & 5 & 7 \\ 4 & 6 & 8 \\ 2 & 4 & 6 \end{bmatrix}$

1.3.15   (a) For AB to exist, we need $r = n$, and for BA to exist, we need $p = m$. For both products to exist, we thus need both of these conditions to be true (i.e., if $A \in \mathcal{M}_{m,n}$, then $B \in \mathcal{M}_{n,m}$).

  (b) Under the restrictions from part (a), AB will have size $m \times m$ and BA will have size $n \times n$. For them to have the same size, we need $m = n$ (and thus $m = n = r = p$).

1.3.17   (a) We use the fact that if $\mathbf{v}$ and $\mathbf{w}$ are column vectors then $\mathbf{v} \cdot \mathbf{w} = \mathbf{v}^T \mathbf{w}$. It follows that $\mathbf{x} \cdot (A\mathbf{y}) = \mathbf{x}^T A\mathbf{y}$ and $(A^T\mathbf{x}) \cdot \mathbf{y} = (A^T\mathbf{x})^T \mathbf{y} = \mathbf{x}^T A\mathbf{y}$. Since both quantities equal $\mathbf{x}^T A\mathbf{y}$, we indeed have $\mathbf{x} \cdot (A\mathbf{y}) = (A^T\mathbf{x}) \cdot \mathbf{y}$.

  (b) Just like in part (a), $\mathbf{x} \cdot (A\mathbf{y}) = \mathbf{x}^T A\mathbf{y}$ and $(B\mathbf{x}) \cdot \mathbf{y} = \mathbf{x}^T B^T \mathbf{y}$. We thus know that $\mathbf{x}^T A\mathbf{y} = \mathbf{x}^T B^T \mathbf{y}$ for all $\mathbf{x} \in \mathbb{R}^m$ and $\mathbf{y} \in \mathbb{R}^n$. However, if we choose $\mathbf{x} = \mathbf{e}_i$ and $\mathbf{y} = \mathbf{e}_j$ then $\mathbf{x}^T A\mathbf{y} = a_{i,j}$ and $\mathbf{x}^T B^T \mathbf{y} = b_{j,i}$, so we see that $a_{i,j} = b_{j,i}$ for all i, j. In other words, $B = A^T$.

1.3.18   (a) Recall from the definition of matrix multiplication that the j-th entry of $A\mathbf{e}_i$ is

$$a_{j,1}[\mathbf{e}_i]_1 + a_{j,2}[\mathbf{e}_i]_2 + \cdots + a_{j,n}[\mathbf{e}_i]_n = a_{j,i},$$

where the equality comes from the fact that $[\mathbf{e}_i]_k = 0$ if $k \neq i$ and $[\mathbf{e}_i]_i = 1$. It follows that

$$A\mathbf{e}_i = \begin{bmatrix} a_{1,i} \\ a_{2,i} \\ \vdots \\ a_{m,i} \end{bmatrix},$$

which is the i-th column of A.

  (b) We again directly use the definition of matrix multiplication: the j-th entry of $\mathbf{e}_i^T A$ is

$$[\mathbf{e}_i]_1 a_{1,j} + [\mathbf{e}_i]_2 a_{2,j} + \cdots + [\mathbf{e}_i]_m a_{n,j} = a_{i,j},$$

where the equality comes from the fact that $[\mathbf{e}_i]_k = 0$ if $k \neq i$ and $[\mathbf{e}_i]_i = 1$. It follows that

$$\mathbf{e}_i^T A = \begin{bmatrix} a_{i,1} & a_{i,2} & \cdots & a_{i,n} \end{bmatrix},$$

which is the i-th row of A.

  (c) We let $\mathbf{a}_i$ denote the i-th column of A and then simply compute

$$AI_n = \begin{bmatrix} A\mathbf{e}_1 & | & A\mathbf{e}_2 & | & \cdots & | & A\mathbf{e}_n \end{bmatrix}$$

$$= \begin{bmatrix} \mathbf{a}_1 & | & \mathbf{a}_2 & | & \cdots & | & \mathbf{a}_n \end{bmatrix}$$

$$= A,$$

where the first equality comes from Theorem 1.3.6 and the second equality comes from part (a) of this exercise.

On the other hand, if we instead let $\mathbf{a}_i$ denote the i-th row of A then we have

$$I_m A = \begin{bmatrix} \mathbf{e}_1^T A \\ \hline \mathbf{e}_2^T A \\ \hline \vdots \\ \hline \mathbf{e}_m^T A \end{bmatrix} = \begin{bmatrix} \mathbf{a}_1 \\ \hline \mathbf{a}_2 \\ \hline \vdots \\ \hline \mathbf{a}_m \end{bmatrix} = A.$$

  (d) The "if" direction of the proof is trivial: if $A = B$ then $A\mathbf{v} = B\mathbf{v}$ for all $\mathbf{v} \in \mathbb{R}^n$. For the "only if" direction, notice that if $A\mathbf{v} = B\mathbf{v}$ then $(A - B)\mathbf{v} = \mathbf{0}$ for all $\mathbf{v} \in \mathbb{R}^n$. In particular then, $(A - B)\mathbf{e}_i = \mathbf{0}$ for all i, so part (a) tells us that the i-th column of $A - B$ is $\mathbf{0}$ for all i. In other words, every entry of $A - B$ is 0, so $A - B = O$ and thus $A = B$.

1.3.19   (a) We compute the (i, j)-entry of both $(A + B) + C$ and $A + (B + C)$:

$$[(A + B) + C]_{i,j} = (a_{i,j} + b_{i,j}) + c_{i,j}$$
$$= a_{i,j} + (b_{i,j} + c_{i,j})$$
$$= [A + (B + C)]_{i,j}.$$

It follows that the entries of $(A + B) + C$ and $A + (B + C)$ are all the same, so $(A + B) + C = A + (B + C)$.

  (b) We compute the (i, j)-entry of both $c(A + B)$ and $cA + cB$:

$$[c(A + B)]_{i,j} = c(a_{i,j} + b_{i,j})$$
$$= ca_{i,j} + cb_{i,j}$$
$$= [cA + cB]_{i,j}.$$

It follows that the entries of $c(A + B)$ and $cA + cB$ are all the same, so $c(A + B) = cA + cB$.

(c) We compute the (i,j)-entry of both $(c+d)A$ and $cA + dA$:

$$[(c+d)A]_{i,j} = (c+d)a_{i,j}$$
$$= ca_{i,j} + da_{i,j}$$
$$= [cA + dA]_{i,j}.$$

It follows that the entries of $(c+d)A$ and $cA + dA$ are all the same, so $(c+d)A = cA + dA$.

(d) We compute the (i,j)-entry of both $c(dA)$ and $(cd)A$:

$$[c(dA)]_{i,j} = c(da_{i,j}) = (cd)a_{i,j} = [(cd)A]_{i,j}.$$

It follows that the entries of $c(dA)$ and $(cd)A$ are all the same, so $c(dA) = (cd)A$.

1.3.20 (a) We compute the (i,j)-entry of both $(AB)C$ and $A(BC)$:

$$[A(BC)]_{i,j} = \sum_{k=1}^{n} a_{i,k}\left(\sum_{\ell=1}^{n} b_{k,\ell}c_{\ell,j}\right)$$
$$= \sum_{\ell=1}^{n}\left(\sum_{k=1}^{n} a_{i,k}b_{k,\ell}\right) c_{\ell,j}$$
$$= [(AB)C]_{i,j}.$$

It follows that the entries of $(AB)C$ and $A(BC)$ are all the same, so $(AB)C = A(BC)$.

(b) We compute the (i,j)-entry of both $A(B+C)$ and $AC + BC$:

$$[A(B+C)]_{i,j} = \sum_{k=1}^{n} a_{i,k}(b_{k,j} + c_{k,j})$$
$$= \sum_{k=1}^{n} a_{i,k}b_{k,j} + \sum_{k=1}^{n} a_{i,k}c_{k,j}$$
$$= [AB + AC]_{i,j}.$$

It follows that the entries of $A(B+C)$ and $AB + AC$ are all the same, so $A(B+C) = AB + AC$.

(c) We compute the (i,j)-entry of both $c(AB)$ and $(cA)B$:

$$[c(AB)]_{i,j} = c\sum_{k=1}^{n} a_{i,k}b_{k,j}$$
$$= \sum_{k=1}^{n}(ca_{i,k})b_{k,j} = [(cA)B]_{i,j}.$$

It follows that the entries of $c(AB)$ and $(cA)B$ are all the same, so $c(AB) = (cA)B$.

1.3.21 We recall that the (i,j)-entry of I_mA is the dot product of the i-th row of I_m (which is $\mathbf{e}_i$) with the j-th column of A. It follows from Example 1.2.1(c) that the (i,j)-entry of I_mA is the i-th entry of the j-th column of A: $a_{i,j}$. All entries of I_mA and A thus coincide, so $I_mA = A$.

1.3.22 (a) Just note that $A^kA^\ell = (AA\cdots A)(AA\cdots A)$, where the first bracket has k copies of A and the second has ℓ copies of A, for a total of $k+\ell$ copies of A (and is thus equal to $A^{k+\ell}$).

(b) Just note that

$$(A^k)^\ell = (AA\cdots A)(AA\cdots A)\cdots(AA\cdots A),$$

where each parenthesis has k copies of A, and there are ℓ parentheses, for a total of $k\ell$ copies of A (which is thus equal to $A^{k\ell}$).

1.3.23 (a) Since the transpose operation swaps columns and rows, performing it twice will swap them back to their original orientations. In symbols, $[(A^T)^T]_{i,j} = [A^T]_{j,i} = [A]_{i,j}$, so $(A^T)^T = A$.

(b) Compute the (i,j)-entry of both sides: $[(A+B)^T]_{i,j} = [A+B]_{j,i} = [A]_{j,i} + [B]_{j,i} = [A^T]_{i,j} + [B^T]_{i,j} = [A^T + B^T]_{i,j}$, so $(A+B)^T = A^T + B^T$.

(c) Compute the (i,j)-entry of both sides: $[(cA)^T]_{i,j} = [cA]_{j,i} = c[A]_{j,i} = c[A^T]_{i,j} = [cA^T]_{i,j}$, so $(cA)^T = cA^T$.

1.3.26 To prove this equality, we perform block matrix multiplication:

$$AB = \begin{bmatrix} \mathbf{a}_1^T \\ \mathbf{a}_2^T \\ \vdots \\ \mathbf{a}_m^T \end{bmatrix} B = \begin{bmatrix} \mathbf{a}_1^T B \\ \mathbf{a}_2^T B \\ \vdots \\ \mathbf{a}_m^T B \end{bmatrix},$$

where the final inequality comes from thinking of A as a $p \times 1$ block matrix and B as a 1×1 block matrix, and performing block matrix multiplication.

1.3.27 (a) Recall from Theorem 1.3.6 that

$$AB = \begin{bmatrix} A\mathbf{b}_1 \mid A\mathbf{b}_2 \mid \cdots \mid A\mathbf{b}_p \end{bmatrix}.$$

Well, Theorem 1.3.5 tells us that

$$A\mathbf{b}_j = b_{1,j}\mathbf{a}_1 + b_{2,j}\mathbf{a}_2 + \cdots + b_{n,j}\mathbf{a}_n,$$

where $\mathbf{a}_1, \mathbf{a}_2, \ldots \mathbf{a}_n$ are the columns of A. Combining these two facts shows that the columns of AB are linear combinations of the columns of A.

(b) We could prove this by mimicking the proof of part (a), and making use of Exercise 1.3.26, but perhaps an easier way is to use the transpose. The rows of AB are the columns of $(AB)^T = B^TA^T$. By part (a), the columns of B^TA^T are linear combinations of the columns of B^T, which are the rows of B. Putting these facts together gives us what we want: the rows of AB are linear combinations of the rows of B.

Section 1.4: Linear Transformations

1.4.1 (a) $\begin{bmatrix} 1 & 2 \\ 3 & -1 \end{bmatrix}$ (c) $\begin{bmatrix} 1 & 1 & 0 \\ 1 & 1 & -1 \end{bmatrix}$

1.4.2 (a) Just place the output vectors as columns into the matrix $[T]$:

$$[T] = \begin{bmatrix} 3 & 1 \\ -1 & 2 \end{bmatrix}.$$

(c) Again, we just place the output vectors as columns in a matrix:

$$[T] = \begin{bmatrix} -1 & 2 \\ 0 & 3 \\ 1 & 0 \end{bmatrix}.$$

(e) This time, we must be more clever. $T(1,0,0) = (1,2,3)$ is the first column of $[T]$. Its second column is $T(0,1,0) = T(1,1,0) - T(1,0,0) = (0,1,2) - (1,2,3) = (-1,-1,-1)$. Its third column is $T(0,0,1) = T(1,1,1) - T(1,1,0) = (0,0,1) - (0,1,2) = (0,-1,-1)$. It follows that

$$[T] = \begin{bmatrix} 1 & -1 & 0 \\ 2 & -1 & -1 \\ 3 & -1 & -1 \end{bmatrix}.$$

1.4.3 (a) True. They are functions with the special properties that $T(\mathbf{v}+\mathbf{w}) = T(\mathbf{v}) + T(\mathbf{w})$ and $T(c\mathbf{v}) = cT(\mathbf{v})$.

(c) True. In particular, the standard matrix of T (which completely determines T) is $[T] = \big[\, T(\mathbf{e}_1) \mid T(\mathbf{e}_2) \,\big]$.

(e) False. If $T(\mathbf{e}_1) = (2,1)$ and $T(\mathbf{e}_2) = (1,3)$ then $T(1,1) = T(\mathbf{e}_1+\mathbf{e}_2) = T(\mathbf{e}_1) + T(\mathbf{e}_2) = (2,1) + (1,3) = (3,4) \neq (3,3)$.

(g) False. For example, suppose $\theta = \pi/2$ and $\mathbf{v} = (0,1,0)$. Then $R_{xz}^{\theta}(\mathbf{v}) = \mathbf{v}$, but $(R_{xy}^{\theta} \circ R_{yz}^{\theta})(\mathbf{v}) = R_{xy}^{\theta}(0,0,1) = (0,0,1) \neq \mathbf{v}$. It follows that $R_{xy}^{\theta} \circ R_{yz}^{\theta} \neq R_{xz}^{\theta}$.

1.4.4 (a) Is not a linear transformation: squaring the entries of $\mathbf{v}$ is not allowed. Specifically, $T(2(1,1)) = (4,2)$, but $2T(1,1) = (2,2)$.

(c) Is not a linear transformation: neither sin nor cos can be applied to the entries of $\mathbf{v}$. Specifically, $T(0,0) = (0,-1) \neq \mathbf{0}$.

(e) Is not a linear transformation: taking the square root of entries of $\mathbf{v}$ is not allowed. Specifically, $T(2(1,1)) = (2\sqrt{2},2)$, but $2T(1,1) = (4,2\sqrt{2})$.

(g) Is not a linear transformation: absolute values of the entries of the vector are not allowed. Specifically, $T(1,1) + T(-1,-1) = (2,2)$, but $T((1,1) + (-1,-1)) = (0,0)$.

1.4.5 (a) Is a linear transformation. Its standard matrix is $\dfrac{1}{10}\begin{bmatrix} 1 & 3 \\ 3 & 9 \end{bmatrix}$.

(c) Is a linear transformation. Its standard matrix is $\dfrac{1}{5}\begin{bmatrix} -3 & 4 \\ 4 & 3 \end{bmatrix}$.

(e) Is a linear transformation. Its standard matrix is $\begin{bmatrix} \cos(\pi/5) & -\sin(\pi/5) \\ \sin(\pi/5) & \cos(\pi/5) \end{bmatrix}$.

1.4.6 (a) The standard matrices of S and T are

$$[S] = \begin{bmatrix} 0 & 2 \\ 1 & 1 \end{bmatrix}, \quad [T] = \begin{bmatrix} 1 & 2 \\ 3 & -1 \end{bmatrix},$$

so the standard matrix of $S \circ T$ is given by the matrix product

$$[S \circ T] = [S][T] = \begin{bmatrix} 6 & -2 \\ 4 & 1 \end{bmatrix}.$$

(c) The standard matrices of S and T are

$$[S] = \begin{bmatrix} 1 & 0 & 0 \\ 1 & 1 & 0 \\ 1 & 1 & 1 \end{bmatrix}, \quad [T] = \begin{bmatrix} 1 & 1 \\ 2 & -1 \\ -1 & 3 \end{bmatrix},$$

so the standard matrix of $S \circ T$ is given by the matrix product

$$[S \circ T] = [S][T] = \begin{bmatrix} 1 & 1 \\ 3 & 0 \\ 2 & 3 \end{bmatrix}.$$

1.4.7 (b) The standard matrices of the projection in the direction of $\mathbf{u} = \frac{1}{\sqrt{2}}(1,1)$ and rotation are

$$[P_{\mathbf{u}}] = \frac{1}{2}\begin{bmatrix} 1 & 1 \\ 1 & 1 \end{bmatrix}, \quad [R^{-\pi/4}] = \frac{1}{\sqrt{2}}\begin{bmatrix} 1 & 1 \\ -1 & 1 \end{bmatrix}.$$

The standard matrix of the composite linear transformations thus is their product:

$$[R^{-\pi/4}][P_{\mathbf{u}}] = \frac{1}{2\sqrt{2}}\begin{bmatrix} 1 & 1 \\ -1 & 1 \end{bmatrix}\begin{bmatrix} 1 & 1 \\ 1 & 1 \end{bmatrix}$$

$$= \frac{1}{\sqrt{2}}\begin{bmatrix} 1 & 1 \\ 0 & 0 \end{bmatrix}.$$

1.4.8 Since T is a linear transformation, we know that $T(c\mathbf{v}) = cT(\mathbf{v})$ for all $c \in \mathbb{R}$ and $\mathbf{v} \in \mathbb{R}^n$. If we choose $c = 0$ and $\mathbf{v} = \mathbf{0}$ then this says that $T(\mathbf{0}) = 0T(\mathbf{0}) = \mathbf{0}$, as desired.

1.4.9 For the "if" direction, assume that $T(c_1\mathbf{v}_1 + \cdots + c_k\mathbf{v}_k) = c_1T(\mathbf{v}_1) + \cdots + c_kT(\mathbf{v}_k)$ for all $\mathbf{v}_1, \ldots, \mathbf{v}_k \in \mathbb{R}^n$ and all $c_1, \ldots, c_k \in \mathbb{R}$. If we choose $k = 1$ then this says that $T(c_1\mathbf{v}_1) = c_1T(\mathbf{v}_1)$, and if we choose $k = 2$ and $c_1 = c_2 = 1$ then it says that $T(\mathbf{v}_1 + \mathbf{v}_2) = T(\mathbf{v}_1) + T(\mathbf{v}_2)$. These are the two defining properties of linear transformations, so T is a linear transformation.

For the "only if" direction, assume that T is a linear transformation. Then we can repeatedly use the two defining properties of linear transformations to see that $T(c_1\mathbf{v}_1 + \cdots + c_k\mathbf{v}_k) = T(c_1\mathbf{v}_1) + T(c_2\mathbf{v}_2 + \cdots + c_k\mathbf{v}_k) = c_1T(\mathbf{v}_1) + T(c_2\mathbf{v}_2 + \cdots + c_k\mathbf{v}_k) = \cdots = c_1T(\mathbf{v}_1) + \cdots + c_kT(\mathbf{v}_k)$ for all $\mathbf{v}_1, \ldots, \mathbf{v}_k \in \mathbb{R}^n$ and all $c_1, \ldots, c_k \in \mathbb{R}$.

1.4.10 (a) We partition A as a block matrix based on its rows, which we call $\mathbf{a}_1, \mathbf{a}_2, \ldots, \mathbf{a}_n$ and then perform block matrix multiplication:

$$DA = \begin{bmatrix} d_1 & 0 & \cdots & 0 \\ 0 & d_2 & \cdots & 0 \\ \vdots & \vdots & \ddots & \vdots \\ 0 & 0 & \cdots & d_n \end{bmatrix}\begin{bmatrix} \underline{\mathbf{a}_1} \\ \underline{\mathbf{a}_2} \\ \vdots \\ \underline{\mathbf{a}_n} \end{bmatrix} = \begin{bmatrix} \underline{d_1\mathbf{a}_1} \\ \underline{d_2\mathbf{a}_2} \\ \vdots \\ \underline{d_n\mathbf{a}_n} \end{bmatrix}.$$

(b) We partition A as a block matrix based on its columns, which we call $\mathbf{a}_1, \mathbf{a}_2, \ldots, \mathbf{a}_n$ and then perform block matrix multiplication:

$$AD = \big[\, \mathbf{a}_1 \mid \mathbf{a}_2 \mid \cdots \mid \mathbf{a}_n \,\big]\begin{bmatrix} d_1 & 0 & \cdots & 0 \\ 0 & d_2 & \cdots & 0 \\ \vdots & \vdots & \ddots & \vdots \\ 0 & 0 & \cdots & d_n \end{bmatrix}$$

$$= \big[\, d_1\mathbf{a}_1 \mid d_2\mathbf{a}_2 \mid \cdots \mid d_n\mathbf{a}_n \,\big].$$

1.4.11 (a) We just string together the relevant definitions: for all $\mathbf{v} \in \mathbb{R}^n$ it is the case that $[S+T]\mathbf{v} = (S+T)(\mathbf{v}) = S(\mathbf{v}) + T(\mathbf{v}) = [S]\mathbf{v} + [T]\mathbf{v} = ([S] + [T])\mathbf{v}$. Since $[S+T]\mathbf{v} = ([S] + [T])\mathbf{v}$ for all $\mathbf{v} \in \mathbb{R}^n$, it follows from Exercise 1.3.18(d) that $[S+T] = [S] + [T]$.

(b) Similar to part (a), for all $\mathbf{v} \in \mathbb{R}^n$ it is the case that $[cS]\mathbf{v} = (cS)(\mathbf{v}) = cS(\mathbf{v}) = c[S]\mathbf{v} = (c[S])\mathbf{v}$. Since $[cS]\mathbf{v} = (c[S])\mathbf{v}$ for all $\mathbf{v} \in \mathbb{R}^n$, it follows from Exercise 1.3.18(d) that $[cS] = c[S]$.

1.4.14 The matrix A is the standard matrix of the rotation counter-clockwise by an angle of $\pi/4$ about the origin, and thus A^{160} is the matrix that implements this transformation 160 times. In other words, it rotates vectors counter-clockwise by an angle of $160\pi/4 = 40\pi$. Since 40π is a multiple of 2π, this rotation does not actually move any vectors at all, so $A^{160} = I$.

1.4.15 We directly compute the product

$$A^T A = \begin{bmatrix} \cos(\theta) & \sin(\theta) \\ -\sin(\theta) & \cos(\theta) \end{bmatrix} \begin{bmatrix} \cos(\theta) & -\sin(\theta) \\ \sin(\theta) & \cos(\theta) \end{bmatrix}.$$

The top-left and bottom-right entries of this product are both $\sin^2(\theta) + \cos^2(\theta) = 1$, while the off-diagonal entries of this product are both $\cos(\theta)\sin(\theta) - \cos(\theta)\sin(\theta) = 0$. We thus conclude that $A^T A = I$, as claimed.

1.4.17 We directly compute A^2:

$$\begin{aligned} A^2 &= (2\mathbf{u}\mathbf{u}^T - I)^2 \\ &= (2\mathbf{u}\mathbf{u}^T - I)(2\mathbf{u}\mathbf{u}^T - I) \\ &= 4\mathbf{u}\mathbf{u}^T \mathbf{u}\mathbf{u}^T - 4\mathbf{u}\mathbf{u}^T + I \\ &= 4\mathbf{u}\mathbf{u}^T - 4\mathbf{u}\mathbf{u}^T + I = I, \end{aligned}$$

as desired. Geometrically, this makes sense since A is the standard matrix of a reflection across a line, and A^2 is the standard matrix of the same reflection across the line twice. However, applying a reflection twice returns each vector to where it started, so it has the same effect as doing nothing at all (i.e., it acts as the identity transformation).

1.4.19 One vector that points in the direction of the line $y = mx$ is $\mathbf{u} = (1, m)$, so we know that the standard matrix of this reflection is

$$\begin{aligned} [F_{\mathbf{u}}] &= 2\mathbf{u}\mathbf{u}^T / \|\mathbf{u}\|^2 - I \\ &= 2 \begin{bmatrix} 1 \\ m \end{bmatrix} \begin{bmatrix} 1 & m \end{bmatrix} / (1 + m^2) - \begin{bmatrix} 1 & 0 \\ 0 & 1 \end{bmatrix} \\ &= \frac{1}{1 + m^2} \begin{bmatrix} 1 - m^2 & 2m \\ 2m & m^2 - 1 \end{bmatrix}. \end{aligned}$$

1.4.23 (a) The transformation $S_{1,2}^c$ has the following effect on the standard unit grid on $\mathbb{R}^2$, with the value of c determining the severity of the shear:

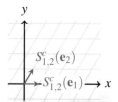

Similarly, the transformation $S_{2,1}^c$ has the following effect on the standard unit grid on $\mathbb{R}^2$:

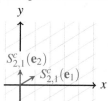

(b) First, we notice that $E_{i,j} = \mathbf{e}_i \mathbf{e}_j^T$, so $E_{i,j}^2 = \mathbf{e}_i \mathbf{e}_j^T \mathbf{e}_i \mathbf{e}_j^T = \mathbf{e}_i (\mathbf{e}_j \cdot \mathbf{e}_i) \mathbf{e}_j^T = \mathbf{0}$, since $i \neq j$. Since $S_{i,j}^c = I + cE_{i,j}$, it follows that

$$\begin{aligned} (S_{i,j}^c)^2 &= (I + cE_{i,j})^2 \\ &= I + cE_{i,j} + cE_{i,j} + c^2 E_{i,j}^2 \\ &= I + 2cE_{i,j} = S_{i,j}^{2c}. \end{aligned}$$

By repeating this argument and multiplying by $S_{i,j}^c$ again and again, we similarly see that $(S_{i,j}^c)^n = S_{i,j}^{nc}$ for all integers $n \geq 1$ (this argument can be made rigorous via induction, if desired).

1.4.24 (a) This matrix is

$$[R_{\mathbf{u}}^\theta] = \begin{bmatrix} 1 & 0 & 0 \\ 0 & \cos(\theta) & -\sin(\theta) \\ 0 & \sin(\theta) & \cos(\theta) \end{bmatrix} = [R_{yz}^\theta].$$

(b) In this case, the standard matrix is

$$[R_{\mathbf{u}}^{\pi/4}] = \frac{1}{9} \begin{bmatrix} 1 + 4\sqrt{2} & 2 - 4\sqrt{2} & 2 + 2\sqrt{2} \\ 2 + 2\sqrt{2} & 4 + 5/\sqrt{2} & 4 - 7/\sqrt{2} \\ 2 - 4\sqrt{2} & 4 - 1/\sqrt{2} & 4 + 5/\sqrt{2} \end{bmatrix}.$$

Multiplying this matrix by $\mathbf{v} = (3, 2, 1)$ gives

$$R_{\mathbf{u}}^{\pi/4}(3, 2, 1) =$$
$$(3 + 2\sqrt{2}, 6 + 5/\sqrt{2}, 6 - 7/\sqrt{2})/3.$$

(c) Multiplying their standard matrices together shows that

$$[R_{yz}^{\pi/3} \circ R_{xy}^{\pi/3}] = \begin{bmatrix} 1/2 & -\sqrt{3}/2 & 0 \\ \sqrt{3}/4 & 1/4 & -\sqrt{3}/2 \\ 3/4 & \sqrt{3}/4 & 1/2 \end{bmatrix}.$$

To see that this is a rotation matrix, we set this matrix equal to $[R_{\mathbf{u}}^\theta]$ from the statement of the question and try to solve for $\mathbf{u}$ and θ. There are many ways to do this, but if we focus on setting the diagonal entries equal to each other first, we find that $u_1^2 = u_3^2 = 3/7$, $u_2^2 = 1/7$, and $\cos(\theta) = 1/8$. Then focusing on the off-diagonal entries tells us that $\mathbf{u} = (\sqrt{3}, 1, \sqrt{3})/\sqrt{7}$ and $\theta = -\arccos(1/8) \approx -1.4455$ (be careful when finding θ—if you choose $\theta = \arccos(1/8) \approx 1.4455$ then the $\sin(\theta)$ terms will have the wrong sign, so you'll get the transpose of the desired matrix).

Section 1.5: Summary and Review

1.5.1   (a) True. If AB is $n \times n$ then A and B must have sizes of the form $n \times m$ and $m \times n$, respectively, so BA is $m \times m$.

     (c) False. $(AB)^3 = (AB)(AB)(AB)$ and in general there is no way to commute the A's and B's past each other in the middle, so it does not necessarily equal $A^3B^3 = AAABBB$.

     (e) False. A simple counter-example is given by

$$A = \begin{bmatrix} 0 & 1 \\ 0 & 0 \end{bmatrix}.$$

     (g) True. This is called "associativity" of composition, and it follows from the fact that applying either linear transformation to a vector $\mathbf{v}$ results in $R(S(T(\mathbf{v})))$.

1.5.3   (a) Projection onto the line in the direction of $\mathbf{u} = (1,0)$.

     (c) Rotation counter-clockwise by an angle of $\theta = \pi/4$.

     (e) None of these.

     (g) Projection onto the line in the direction of $\mathbf{u} = (1,2,2)/3$.

1.5.5 The matrix

$$\begin{bmatrix} \cos(\theta) & -\sin(\theta) \\ \sin(\theta) & \cos(\theta) \end{bmatrix}$$

is the standard matrix of a rotation counter-clockwise by an angle of θ (i.e., it is $[R^\theta]$). Composing n copies of R^θ together results in a rotation by an angle of $n\theta$, so we have

$$[R^\theta]^n = [R^\theta \circ R^\theta \circ \cdots \circ R^\theta] = [R^{n\theta}],$$

which is exactly what we wanted to show.

Section 1.A: Extra Topic: Areas, Volumes, and the Cross Product

1.A.1   (a) $(-4,8,-4)$ (c) $(-4,-2,7)$

1.A.2   (a) $\|(2,1,0) \times (-2,3,0)\| = \|(0,0,8)\| = 8$

     (c) $\|(1,2,3) \times (3,-1,2)\| = \|(7,7,-7)\| = 7\sqrt{3}$

1.A.3   (a) $\frac{1}{2}\|(0,4,0) \times (1,1,0)\| = \frac{1}{2}\|(0,0,-4)\| = 2$

     (c) $\frac{1}{2}\|(-1,1,-1) \times (3,2,1)\| = \frac{1}{2}\|(3,-2,-5)\| = \frac{\sqrt{38}}{2}$

1.A.4   (a) This is a cuboid (rectangular prism) with side lengths 1, 2, and 3, so its volume is $1 \cdot 2 \cdot 3 = 6$.

     (c) $|\mathbf{v} \cdot (\mathbf{w} \times \mathbf{x})| = |(1,1,1) \cdot (3,4,-2)| = |3+4-2| = 5$

1.A.5   (a) In all cases, we compute the area of the parallelogram or parallelepiped with sides given by the columns of the matrix. In this case, the sides are $(1,0)$ and $(0,1)$, which is a square of area 1, so that is our answer.

     (c) $\|(0,2,0) \times (-2,3,0)\| = \|(0,0,4)\| = 4$.

     (e) Now we use Theorem 1.A.3: $|(1,0,0) \cdot ((-3,2,0) \times (2,-7,3))| = 6$.

1.A.6   (a) False. For example, the cross product of two unit vectors can be $\mathbf{0}$ if they are the *same* unit vector.

     (c) False. The notation $\mathbf{v}^2$ does not even mean anything. The correct statement is $\mathbf{v} \times \mathbf{v} = \mathbf{0}$ (via Theorem 1.A.1(c)).

     (e) False. This would be true if $\mathbf{v}, \mathbf{w} \in \mathbb{R}^3$, but since they are in $\mathbb{R}^4$ the cross product does not even make sense.

1.A.8 Recall that $\|\mathbf{v} \times \mathbf{w}\| = \|\mathbf{v}\|\|\mathbf{w}\|\sin(\theta)$. We are told in the question that $\mathbf{v}$ and $\mathbf{w}$ are unit vectors, so $\|\mathbf{v}\| = \|\mathbf{w}\| = 1$. Similarly, we can compute $\|\mathbf{v} \times \mathbf{w}\| = \sqrt{(1/3)^2 + (2/3)^2 + (2/3)^2} = 1$. It follows that $1 = \sin(\theta)$, so $\theta = \pi/2$.

1.A.10 We just mimic the calculation given earlier in the text:

$$\begin{aligned}
\mathbf{w} \cdot (\mathbf{v} \times \mathbf{w}) &= w_1(v_2w_3 - v_3w_2) + w_2(v_3w_1 - v_1w_3) \\
&\quad + w_3(v_1w_2 - v_2w_1) \\
&= w_1v_2w_3 - w_1v_3w_2 + w_2v_3w_1 - w_2v_1w_3 \\
&\quad + w_3v_1w_2 - w_3v_2w_1 \\
&= 0.
\end{aligned}$$

1.A.11   (a) We use the definitions of the cross product and vector addition:

$$\mathbf{v} \times (\mathbf{w} + \mathbf{x}) = \begin{bmatrix} v_2(w_3 + x_3) - v_3(w_2 + x_2) \\ v_3(w_1 + x_1) - v_1(w_3 + x_3) \\ v_1(w_2 + x_2) - v_2(w_1 + x_1) \end{bmatrix}.$$

On the other hand,

$$\mathbf{v} \times \mathbf{w} + \mathbf{v} \times \mathbf{x} = \begin{bmatrix} v_2w_3 - v_3w_2 \\ v_3w_1 - v_1w_3 \\ v_1w_2 - v_2w_1 \end{bmatrix} + \begin{bmatrix} v_2x_3 - v_3x_2 \\ v_3x_1 - v_1x_3 \\ v_1x_2 - v_2x_1 \end{bmatrix}$$

$$= \begin{bmatrix} v_2w_3 - v_3w_2 + v_2x_3 - v_3x_2 \\ v_3w_1 - v_1w_3 + v_3x_1 - v_1x_3 \\ v_1w_2 - v_2w_1 + v_1x_2 - v_2x_1 \end{bmatrix}$$

It is straightforward to compare these two vectors entry-by-entry and see that they are the same.

     (b) Again, we just use the relevant definitions:

$$(c\mathbf{v}) \times \mathbf{w} = \begin{bmatrix} (cv_2)w_3 - (cv_3)w_2 \\ (cv_3)w_1 - (cv_1)w_3 \\ (cv_1)w_2 - (cv_2)w_1 \end{bmatrix}$$

$$= c\begin{bmatrix} v_2w_3 - v_3w_2 \\ v_3w_1 - v_1w_3 \\ v_1w_2 - v_2w_1 \end{bmatrix} = c(\mathbf{v} \times \mathbf{w}).$$

1.A.13 No, it is not associative. To see this, we can choose $\mathbf{v} = \mathbf{e}_1$, $\mathbf{w} = \mathbf{e}_2$, and $\mathbf{x} = \mathbf{e}_2$. Then $(\mathbf{v} \times \mathbf{w}) \times \mathbf{x} = (\mathbf{e}_1 \times \mathbf{e}_2) \times \mathbf{e}_2 = \mathbf{e}_3 \times \mathbf{e}_2 = -\mathbf{e}_1$, but $\mathbf{v} \times (\mathbf{w} \times \mathbf{x}) = \mathbf{e}_1 \times (\mathbf{e}_2 \times \mathbf{e}_2) = \mathbf{e}_1 \times \mathbf{0} = \mathbf{0}$.

1.A.16 (a) To show that $\mathbf{v} \cdot (\mathbf{w} \times \mathbf{x}) = \mathbf{w} \cdot (\mathbf{x} \times \mathbf{v})$ we explicitly compute both sides in terms of the vectors' entries:

$$\mathbf{v} \cdot (\mathbf{w} \times \mathbf{x})$$
$$= \mathbf{v} \cdot (w_2 x_3 - w_3 x_2, w_3 x_1 - w_1 x_3, w_1 x_2 - w_2 x_1)$$
$$= v_1(w_2 x_3 - w_3 x_2) + v_2(w_3 x_1 - w_1 x_3)$$
$$\quad + v_3(w_1 x_2 - w_2 x_1)$$
$$= v_1 w_2 x_3 - v_1 w_3 x_2 + v_2 w_3 x_1 - v_2 w_1 x_3$$
$$\quad + v_3 w_1 x_2 - v_3 w_2 x_1.$$

Similarly,

$$\mathbf{w} \cdot (\mathbf{x} \times \mathbf{v})$$
$$= \mathbf{w} \cdot (x_2 v_3 - x_3 v_2, x_3 v_1 - x_1 v_3, x_1 v_2 - x_2 v_1)$$
$$= w_1(x_2 v_3 - x_3 v_2) + w_2(x_3 v_1 - x_1 v_3)$$
$$\quad + w_3(x_1 v_2 - x_2 v_1)$$
$$= w_1 x_2 v_3 - w_1 x_3 v_2 + w_2 x_3 v_1 - w_2 x_1 v_3$$
$$\quad + w_3 x_1 v_2 - w_3 x_2 v_1.$$

We can directly compare these two expressions to see that they are equal, so $\mathbf{v} \cdot (\mathbf{w} \times \mathbf{x}) = \mathbf{w} \cdot (\mathbf{x} \times \mathbf{v})$. The fact that $\mathbf{w} \cdot (\mathbf{x} \times \mathbf{v}) = \mathbf{x} \cdot (\mathbf{v} \times \mathbf{w})$ can be proved in the exact same way.

(b) Recall from Theorem 1.A.1 that the cross product is anticommutative: $\mathbf{v} \times \mathbf{w} = -(\mathbf{w} \times \mathbf{v})$. By combining this fact with part (a) of this question, we get exactly what we want:

$$\mathbf{v} \cdot (\mathbf{w} \times \mathbf{x}) = -\mathbf{v}(\mathbf{x} \times \mathbf{w}),$$
$$\mathbf{v} \cdot (\mathbf{w} \times \mathbf{x}) = \mathbf{w} \cdot (\mathbf{x} \times \mathbf{v}) = -\mathbf{w} \cdot (\mathbf{v} \times \mathbf{x}), \quad \text{and}$$
$$\mathbf{v} \cdot (\mathbf{w} \times \mathbf{x}) = \mathbf{x} \cdot (\mathbf{v} \times \mathbf{w}) = -\mathbf{x} \cdot (\mathbf{w} \times \mathbf{v}).$$

1.A.18 We could verify this directly by computing the expression in terms of the entries of $\mathbf{v}$, $\mathbf{w}$, and $\mathbf{x}$, but an easier way is to use the result of Exercise 1.A.17:

$$\mathbf{v} \times (\mathbf{w} \times \mathbf{x}) + \mathbf{w} \times (\mathbf{x} \times \mathbf{v}) + \mathbf{x} \times (\mathbf{v} \times \mathbf{w})$$
$$= \big((\mathbf{v} \cdot \mathbf{x})\mathbf{w} - (\mathbf{v} \cdot \mathbf{w})\mathbf{x}\big) + \big((\mathbf{w} \cdot \mathbf{v})\mathbf{x} - (\mathbf{w} \cdot \mathbf{x})\mathbf{v}\big)$$
$$\quad + \big((\mathbf{x} \cdot \mathbf{w})\mathbf{v} - (\mathbf{x} \cdot \mathbf{v})\mathbf{w}\big)$$
$$= \big((\mathbf{v} \cdot \mathbf{x})\mathbf{w} - (\mathbf{x} \cdot \mathbf{v})\mathbf{w}\big) + \big((\mathbf{w} \cdot \mathbf{v})\mathbf{x} - (\mathbf{v} \cdot \mathbf{w})\mathbf{x}\big)$$
$$\quad + \big((\mathbf{x} \cdot \mathbf{w})\mathbf{v} - (\mathbf{w} \cdot \mathbf{x})\mathbf{v}\big)$$
$$= 0.$$

Section 1.B: Extra Topic: Paths in Graphs

1.B.1 (a) The adjacency matrix of this graph is

$$\begin{bmatrix} 0 & 1 & 0 & 1 \\ 1 & 0 & 1 & 1 \\ 0 & 1 & 0 & 0 \\ 1 & 1 & 0 & 0 \end{bmatrix}.$$

(c) The adjacency matrix of this graph is

$$\begin{bmatrix} 0 & 1 & 1 & 1 & 0 & 0 \\ 1 & 0 & 1 & 0 & 0 & 0 \\ 1 & 1 & 0 & 0 & 0 & 1 \\ 1 & 0 & 0 & 0 & 1 & 1 \\ 0 & 0 & 0 & 1 & 0 & 1 \\ 0 & 0 & 1 & 1 & 1 & 0 \end{bmatrix}.$$

1.B.2 (a) The adjacency matrix of this graph is

$$\begin{bmatrix} 0 & 1 & 0 & 1 \\ 0 & 0 & 1 & 1 \\ 0 & 0 & 0 & 0 \\ 0 & 0 & 0 & 0 \end{bmatrix}.$$

(c) The adjacency matrix of this graph is

$$\begin{bmatrix} 0 & 1 & 1 & 1 & 0 & 0 \\ 0 & 0 & 1 & 0 & 0 & 0 \\ 0 & 0 & 0 & 0 & 0 & 1 \\ 1 & 0 & 0 & 0 & 1 & 1 \\ 0 & 0 & 0 & 0 & 0 & 1 \\ 0 & 0 & 1 & 0 & 0 & 0 \end{bmatrix}.$$

1.B.3 (a) The adjacency matrix of this multigraph is

$$\begin{bmatrix} 0 & 1 & 0 & 1 \\ 1 & 0 & 5 & 2 \\ 0 & 5 & 0 & 0 \\ 1 & 2 & 0 & 0 \end{bmatrix}.$$

(c) The adjacency matrix of this graph is

$$\begin{bmatrix} 0 & 1 & 0 & 1 & 0 & 0 \\ 0 & 0 & 2 & 0 & 0 & 0 \\ 0 & 0 & 0 & 0 & 0 & 1 \\ 1 & 0 & 0 & 0 & 2 & 1 \\ 0 & 0 & 0 & 0 & 2 & 1 \\ 0 & 0 & 1 & 0 & 0 & 0 \end{bmatrix}.$$

1.B.4 (a) True. This follows from the fact that there is an edge from vertex i to j if and only if there is an edge from vertex j to i.

(c) False. A simple counter-example is given by

which has a path of length 2 from A to C, but no such path of length 3.

1.B.5 (a)

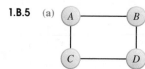

(c)

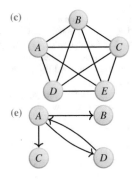

(e)

1.B.6   (a) 1   (c) 1   (e) 2
           (g) 10   (i) 6   (k) 298

1.B.7   (a) The adjacency matrix of this graph is

$$A = \begin{bmatrix} 0 & 1 & 0 & 1 \\ 1 & 0 & 1 & 1 \\ 0 & 1 & 0 & 1 \\ 1 & 1 & 1 & 0 \end{bmatrix}.$$

To compute paths of length 2, we compute A^2:

$$A^2 = \begin{bmatrix} 2 & 1 & 2 & 1 \\ 1 & 3 & 1 & 2 \\ 2 & 1 & 2 & 1 \\ 1 & 2 & 1 & 3 \end{bmatrix}.$$

There is thus $[A^2]_{1,2} = 1$ path of length 2 from vertex A to vertex B.

(c) We start by computing A^4:

$$A^4 = \begin{bmatrix} 10 & 9 & 10 & 9 \\ 9 & 15 & 9 & 14 \\ 10 & 9 & 10 & 9 \\ 9 & 14 & 9 & 15 \end{bmatrix}.$$

There are thus $[A^4]_{1,4} = 9$ paths of length 4 from vertex A to vertex D.

1.B.9   (a)

B

(b) From vertex A to A, there is just one path. From B to B, there is just one path. From B to A there are no paths. From A to B there are n paths of length n: for any $0 \le k \le n-1$, the self-loop at A can be used k times, followed by the edge from A to B, followed by the self-loop at B another $n-k-1$ times.

(c) Part (b) tells us that

$$A^n = \begin{bmatrix} 1 & n \\ 0 & 1 \end{bmatrix},$$

which is exactly the same as the formula for A^n that we found in Exercise 1.3.7(c).

1.B.11 As suggested by the hint, we proceed by induction. For the base case, we note that the $k = 1$ and $k = 2$ cases were already proved for Theorem 1.B.1. For the inductive step, suppose that, for some fixed integer $k \ge 1$, $[A^k]_{i,j}$ counts the number of paths of length k between vertices i and j. Then

$$[A^{k+1}]_{i,j} = [A^k A]_{i,j}$$
$$= [A^k]_{i,1}a_{1,j} + [A^k]_{i,2}a_{2,j} + \cdots + [A^k]_{i,n}a_{n,j}.$$

The first term in the sum above ($[A^k]_{i,1}a_{1,j}$) equals $[A^k]_{i,1}$ exactly if there is a path of length k edge between vertices i and 1 and also an edge between vertices 1 and j. In other words, the quantity $[A^k]_{i,1}a_{1,j}$ counts the number of paths of length $k+1$ between vertices i and j, with vertex 1 as the second-to-last vertex visited.
Similarly, $[A^k]_{i,2}a_{2,j}$ counts the number of paths of length $k+1$ between vertices i and j, with vertex 2 as the second-to-last vertex visited, and so on. By adding these terms up, we see that $[A^{k+1}]_{i,j}$ counts the number of paths of length $k+1$ between vertices i and j, with *any* vertex as the second-to-last vertex visited, completing the inductive step.

Section 2.1: Systems of Linear Equations

2.1.1   (a) Reduced row echelon form.
           (c) Not in row echelon form.
           (e) Not in row echelon form.
           (g) Reduced row echelon form.

2.1.2   (a) $\begin{bmatrix} 1 & 0 \\ 0 & 1 \end{bmatrix}$ (c) $\begin{bmatrix} 1 & 2 \\ 0 & 0 \\ 0 & 0 \end{bmatrix}$

(e) $\begin{bmatrix} 1 & 0 & 0 \\ 0 & 1 & 0 \\ 0 & 0 & 1 \end{bmatrix}$ (g) $\begin{bmatrix} 1 & 0 & 2 \\ 0 & 1 & 3 \\ 0 & 0 & 0 \\ 0 & 0 & 0 \end{bmatrix}$

2.1.3   (a) $\begin{bmatrix} 1 & 0 & 0 & -1 \\ 0 & 1 & 0 & -1/2 \\ 0 & 0 & 1 & -1/2 \\ 0 & 0 & 0 & 0 \end{bmatrix}$

(c) $\begin{bmatrix} 1 & 0 & 0 & 0 & -4/3 & 2/5 \\ 0 & 1 & 0 & 0 & -4/3 & -4/5 \\ 0 & 0 & 1 & 0 & 7/3 & 9/20 \\ 0 & 0 & 0 & 1 & 2 & 1/10 \end{bmatrix}$

2.1.4   (a) No solutions (the last row says $0 = 1$).
           (c) Infinitely many solutions (there are two variables x and y: x is free and $y = 2$).
           (e) No solutions.
           (g) No solutions.

2.1.5   (a) $(x,y) = (1,1)$.

(c) No solutions.

(e) No solutions.

(g) Infinitely many solutions: The variables v, w, and y are leading, while x and z are free. The solutions are of the form

$$(v, w, x, y, z) = (4 + x + 2z, 1/2 + z/2, x, 1, z).$$

2.1.6 (a) Infinitely many solutions: v, w, x, and y are leading variables, and z is free. The solutions are of the form $(v, w, x, y, z) = (-30, 3, 60, 7, 0) + z(-51, 3, 106, 5, 6)/6$.

2.1.7 (a) True. Each of the variables (x, y, and z) are just multiplied by a scalar and added together. The fact that the scalars ($\sin(1)$, $\sqrt[3]{5}$) are ugly does not matter.

(c) False. Lines in $\mathbb{R}^3$ can be "skew": non-parallel and non-intersecting. For example, the line on the x-axis does not intersect the line going through the points $(0, 0, 1)$ and $(0, 1, 1)$ (i.e., the line 1 unit above the y-axis).

(e) False. Every linear system has exactly 0, 1, or infinitely many solutions.

(g) False. For example, here is a system of linear equations with more equations than variables:

$$\begin{aligned} x + y &= 1 \\ 2x + 2y &= 2 \\ 3x + 3y &= 3 \end{aligned}$$

that has infinitely many solutions. One solution is $x = 1$, $y = 0$.

(i) False. It could also have no solutions, as demonstrated by the following system:

$$\begin{aligned} x + y + z &= 1 \\ x + y + z &= 2 \end{aligned}$$

2.1.8 One row echelon form of this matrix is

$$\left[\begin{array}{cc|c} 1 & -1 & 0 \\ 0 & 1 & 1/2 \\ 0 & 0 & t - 1/2 \end{array} \right].$$

We see that the first two rows are always non-zero with leading variables, and the final row is non-zero if and only if $t \neq 1/2$. This linear system thus has (i) no solutions if $t \neq 1/2$, (ii) a unique solution if $t = 1/2$, and (iii) never has infinitely many solutions.

2.1.10 The unique solutions when $h = 0.95$, $h = 1.00$, and $h = 1.05$ are $(w, x, y, z) = (3, -24, 30, 0)$, $(w, x, y, z) = (-4, 60, -180, 140)$, and $(w, x, y, z) = (-11, 144, -390, 280)$, respectively. What is surprising about these solutions is how drastically they change when h is changed just a tiny amount. The reason that these small changes in h lead to such large changes in the solution is that the equations in the linear system are so close to being multiples of each other.

2.1.11 We start by finding a row echelon form for the matrix:

$$\left[\begin{array}{cc|c} a & b & 1 \\ c & d & 1 \end{array} \right] \xrightarrow{aR_2 - cR_1} \left[\begin{array}{cc|c} a & b & 1 \\ 0 & ad - bc & a - c \end{array} \right].$$

If $ad - bc \neq 0$ then the row echelon form has no all-zero rows, so the system has a unique solution.

2.1.13 (a) Since A and R are row equivalent, we can apply a sequence of elementary row operations to turn A into R. Similarly, since B and R are row equivalent, we can apply a sequence of elementary row operations to turn R into B. By stringing these two sequences of row operations together, we can convert A into R into B, so A and B are row equivalent.

(b) If A and B have the same RREF (which we will call R) then A and R are row equivalent, as are B and R, so part (a) tells us that A and B are row equivalent too. In the other direction, if A and B are row equivalent then we can row-reduce A to B and then to the RREF of B. Since every matrix can be row-reduced to exactly one RREF (this is Theorem B.2.1), A and B must have the same RREF.

2.1.15 (a) The system of linear equations that we must solve is

$$\begin{aligned} v_1 + 2v_2 + 3v_3 &= 0 \\ 3v_1 + 2v_2 + v_3 &= 0 \end{aligned}$$

The reduced row echelon form of the associated augmented matrix is

$$\left[\begin{array}{cc c|c} 1 & 0 & -1 & 0 \\ 0 & 1 & 2 & 0 \end{array} \right],$$

so v_3 is a free variable. Arbitrarily choosing $v_3 = 1$ then gives us $v_2 = -2$ and $v_1 = 1$, so the vector $\mathbf{v} = (1, -2, 1)$ is orthogonal to both of the given vectors.

(b) The system of linear equations that we must solve is

$$\begin{aligned} v_1 + 2v_2 \phantom{{}+ v_3} + v_4 &= 0 \\ -2v_1 + v_2 + v_3 + 3v_4 &= 0 \\ -v_1 - v_2 + 2v_3 + v_4 &= 0 \end{aligned}$$

The reduced row echelon form of the associated augmented matrix is

$$\left[\begin{array}{ccc c|c} 1 & 0 & 0 & -7/9 & 0 \\ 0 & 1 & 0 & 8/9 & 0 \\ 0 & 0 & 1 & 5/9 & 0 \end{array} \right],$$

so v_4 is a free variable. Arbitrarily choosing $v_4 = 9$ then gives us $v_3 = -5$, $v_2 = -8$, and $v_1 = 7$, so the vector $\mathbf{v} = (7, -8, -5, 9)$ is orthogonal to all three of the given vectors.

2.1.16 (a) The augmented matrix becomes

$$\left[\begin{array}{cc|c} 1 & 0 & 1 \\ 0 & 1 & 2 \end{array} \right],$$

so the unique solution is $(x, y) = (1, 2)$.

(b) The augmented matrix becomes

$$\left[\begin{array}{cc|c} 0 & -1 & -2 \\ 0 & 1 & 2 \end{array} \right].$$

There are infinitely many solutions of this system of equations: $y = 2$ and x can be anything.

2.1.17 (a) Yes, $\mathbf{b} = (1/2)\mathbf{v}_1$.

(c) Yes, $\mathbf{b} = \mathbf{v}_1 + 5\mathbf{v}_2 - 6\mathbf{v}_3$.

(e) Yes, $\mathbf{b} = \mathbf{v}_1 + 3\mathbf{v}_2$.

(g) Yes, $\mathbf{b} = -\mathbf{v}_1 + \mathbf{v}_2 - 2\mathbf{v}_3$.

2.1.18 The augmented matrix of the system of linear equations that we need to solve is

$$\left[\begin{array}{ccc|c} v_1 & v_2 & v_3 & 0 \\ w_1 & w_2 & w_3 & 0 \end{array}\right]$$

By replacing R_1 with $w_1 R_1$ and then replacing R_2 with $R_1 - v_1 R_2$, the system becomes

$$\left[\begin{array}{ccc|c} w_1 v_1 & w_1 v_2 & w_1 v_3 & 0 \\ 0 & w_1 v_2 - v_1 w_2 & w_1 v_3 - v_1 w_3 & 0 \end{array}\right]$$

(Note that we have to be slightly careful here, as we can only replace R_1 by $w_1 R_1$ if $w_1 \neq 0$. However, since $\mathbf{w} \neq \mathbf{0}$, we can always find *some* non-zero entry of $\mathbf{w}$, so this is just a technicality.)
This augmented matrix corresponds to the linear system

$$(w_1 v_1) x_1 + (w_1 v_2) x_2 + (w_1 v_3) x_3 = 0$$
$$(w_1 v_2 - v_1 w_2) x_2 + (w_1 v_3 - v_1 w_3) x_3 = 0$$

where x_1, x_2, x_3 are the variables we are trying to solve for. In particular, x_1 and x_2 are leading variables, while x_3 is a free variable. Rearranging the above equations in terms of x_3 and then simplifying gives

$$(w_1 v_2 - v_1 w_2) x_1 = (v_3 w_2 - v_2 w_3) x_3$$
$$(w_1 v_2 - v_1 w_2) x_2 = (v_1 w_3 - w_1 v_3) x_3$$

If $c \in \mathbb{R}$ is any scalar, and we choose $x_3 = c(v_1 w_2 - v_2 w_1)$ then we find that $x_1 = c(v_2 w_3 - v_3 w_2)$ and $x_2 = c(v_3 w_1 - v_1 w_3)$, so

$$(x_1, x_2, x_3) = c(v_2 w_3 - v_3 w_2, v_3 w_1 - v_1 w_3, v_1 w_2 - v_2 w_1),$$

as claimed.

2.1.19 If we mix x and y kilograms of the 13% and 5% salt water mixtures, respectively, then the resulting linear system is

$$x + y = 120$$
$$0.13x + 0.05y = 9.6$$

Solving this linear system shows that we should mix $x = 45$ kilograms of the 13% salt water and $y = 75$ kilograms of the 5% salt water.

2.1.21 If we let w, x, y, and z denote how many copies of each molecule are present in the reaction (in the order listed in the exercise), then we get the linear system

$$2x - y - 2z = 0$$
$$w - y = 0$$
$$w - z = 0.$$

Solving this linear system shows that z is free, $w = z$, $x = 3z/2$, and $y = z$. Choosing $z = 2$ so that all of these variables are integers gives the balanced equation

$$2\text{ZnS} + 3\text{O}_2 \rightarrow 2\text{ZnO} + 2\text{SO}_2.$$

2.1.24 The standard matrix of $P_\mathbf{u}$ is

$$[P_\mathbf{u}] = \mathbf{u}\mathbf{u}^T = \frac{1}{9}\begin{bmatrix} 1 & 2 & 2 \\ 2 & 4 & 4 \\ 2 & 4 & 4 \end{bmatrix}.$$

We are thus trying to find vectors $\mathbf{v} = (v_1, v_2, v_3)$ with the property that

$$\frac{1}{9}\begin{bmatrix} 1 & 2 & 2 \\ 2 & 4 & 4 \\ 2 & 4 & 4 \end{bmatrix}\begin{bmatrix} v_1 \\ v_2 \\ v_3 \end{bmatrix} = \begin{bmatrix} 2 \\ 4 \\ 4 \end{bmatrix}.$$

This is a linear system with augmented matrix

$$\left[\begin{array}{ccc|c} 1/9 & 2/9 & 2/9 & 2 \\ 2/9 & 4/9 & 4/9 & 4 \\ 2/9 & 4/9 & 4/9 & 4 \end{array}\right],$$

which has reduced row echelon form

$$\left[\begin{array}{ccc|c} 1 & 2 & 2 & 18 \\ 0 & 0 & 0 & 0 \\ 0 & 0 & 0 & 0 \end{array}\right].$$

It follows that v_2 and v_3 are free variables and v_1 is a leading variable with $v_1 = 18 - 2v_2 - 2v_3$. The vectors $\mathbf{v}$ for which $P_\mathbf{u}(\mathbf{v}) = (2, 4, 4)$ are thus those of the form $\mathbf{v} = (18 - 2v_2 - 2v_3, v_2, v_3)$, where $v_2, v_3 \in \mathbb{R}$ are arbitrary.

2.1.25 (a) This exercise is asking whether or not there exist $c_1, c_2 \in \mathbb{R}$ such that $(3, -2, 1) = c_1(1, 4, 2) + c_2(2, -1, 1)$. This is equivalent to the system of linear equations

$$c_1 + 2c_2 = 3$$
$$4c_1 - c_2 = -2$$
$$2c_1 + c_2 = 1$$

One row echelon form of the associated augmented matrix is

$$\left[\begin{array}{cc|c} 1 & 2 & 3 \\ 0 & -9 & -14 \\ 0 & 0 & -1/3 \end{array}\right].$$

Since the last line of this matrix corresponds to the equation $0 = -1/3$, we see that there are no solutions, so $(3, -2, 1)$ is not a linear combination of $(1, 4, 2)$ and $(2, -1, 1)$.

2.1.26 (a) If we can find $T(1, 0)$ and $T(0, 1)$ then we can just stick these as columns in a matrix to get the standard matrix of T. To find these vectors, we start by writing $(1, 0)$ and $(0, 1)$ as linear combinations of $(1, 1)$ and $(1, -1)$:

$$(1, 0) = c_1(1, 1) + c_2(1, -1)$$
$$(0, 1) = c_3(1, 1) + c_4(1, -1).$$

These are linear systems of equations that are easily solved to get $c_1 = c_2 = c_3 = 1/2$ and $c_4 = -1/2$. It follows that

$$T(1, 0) = T(\tfrac{1}{2}(1, 1) + \tfrac{1}{2}(1, -1))$$
$$= \tfrac{1}{2}T(1, 1) + \tfrac{1}{2}T(1, -1)$$
$$= \tfrac{1}{2}(3, 7) + \tfrac{1}{2}(-1, -1)$$
$$= (1, 3).$$

A similar calculation shows that $T(0, 1) = (2, 4)$, so

$$[T] = \begin{bmatrix} 1 & 2 \\ 3 & 4 \end{bmatrix}.$$

(c) By working similarly as in part (a), we find

$$[T] = \begin{bmatrix} 1 & 2 & 1 \\ 1 & 3 & 2 \\ -1 & 2 & 0 \end{bmatrix}.$$

2.1.27 (a) Multiplying the top equation by xy turns the system into

$$2x + y = 3$$
$$x + y = 2$$

which has unique solution $(x,y) = (1,1)$.

(c) Let $u = 1/x$, $v = 1/y$, and $w = 1/z$. Then the system of equations becomes

$$u \quad\;\; - 2w = -4$$
$$u + v + w = 4$$
$$2u - 6v - 2w = 4$$

Solving this system via row operations gives us the unique solution $(u,v,w) = (2,-1,3)$. Using the fact that $u = 1/x$, $v = 1/y$, and $w = 1/z$ then gives us $(x,y,z) = (1/2,-1,1/3)$ as the unique solution of the original system of equations.

2.1.28 (a) We need to find all matrices $\begin{bmatrix} a & b \\ c & d \end{bmatrix}$ such that

$$\begin{bmatrix} 0 & 1 \\ 1 & 0 \end{bmatrix}\begin{bmatrix} a & b \\ c & d \end{bmatrix} = \begin{bmatrix} a & b \\ c & d \end{bmatrix}\begin{bmatrix} 0 & 1 \\ 1 & 0 \end{bmatrix}.$$

By multiplying out these matrices, we get

$$\begin{bmatrix} c & d \\ a & b \end{bmatrix} = \begin{bmatrix} b & a \\ d & c \end{bmatrix}.$$

This is a system of 4 linear equations in 4 variables, and it is straightforward to solve it to get $c = b$, $d = a$. The matrices that commute with A are thus those of the form $\begin{bmatrix} a & b \\ b & a \end{bmatrix}$.

(b) We need to find all matrices $\begin{bmatrix} a & b \\ c & d \end{bmatrix}$ such that

$$\begin{bmatrix} 1 & 1 \\ 1 & 0 \end{bmatrix}\begin{bmatrix} a & b \\ c & d \end{bmatrix} = \begin{bmatrix} a & b \\ c & d \end{bmatrix}\begin{bmatrix} 1 & 1 \\ 1 & 0 \end{bmatrix}.$$

By multiplying out these matrices, we get

$$\begin{bmatrix} a+c & b+d \\ a & b \end{bmatrix} = \begin{bmatrix} a+b & a \\ c+d & c \end{bmatrix}.$$

This is a system of 4 linear equations in 4 variables, and it is straightforward to solve it to get $c = b$, $d = a - b$. The matrices that commute with C are thus those of the form $\begin{bmatrix} a & b \\ b & a-b \end{bmatrix}$.

(c) If B commutes with everything in $\mathcal{M}_2$ then in particular $AB = BA$ and $BC = CB$, where A and C are defined in parts (a) and (b) of this question. It follows that B must have the form described by our solutions to parts (a) and (b) of this question, so $B = \begin{bmatrix} a & b \\ b & a-b \end{bmatrix}$ (by part (b)), and $a - b = a$ (by part (a)), so $b = 0$, so $B = \begin{bmatrix} a & 0 \\ 0 & a \end{bmatrix} = aI$. On the other hand, it is clear that the matrix aI commutes with everything in $\mathcal{M}_2$ since $(aI)A = aA = A(aI)$.

2.1.29 (a) The system of equations

$$b + m = 2$$
$$b + 3m = 8$$

has unique solution $(b,m) = (-1,3)$, so the unique line that works is $y = 3x - 1$.

(b) The system of equations

$$a + b + c = 3$$
$$4a + 2b + c = 6$$
$$9a + 3b + c = 13$$

has unique solution $(a,b,c) = (2,-3,4)$, so the unique parabola that works is $y = 2x^2 - 3x + 4$.

(c) The relevant system of equations in this case is

$$b + m = 3$$
$$b + 2m = 6$$
$$b + 3m = 13$$

which has no solution. There thus does not exist a line that goes through all 3 points.

(d) The relevant system of equations in this case is

$$a + b + c + d = 3$$
$$8a + 4b + 2c + d = 6$$
$$27a + 9b + 3c + d = 13$$

Representing this system by an augmented matrix and row-reducing yields the reduced row echelon form

$$\begin{bmatrix} 1 & 0 & 0 & 1/6 & \bigm| & 2/3 \\ 0 & 1 & 0 & -1 & \bigm| & -2 \\ 0 & 0 & 1 & 11/6 & \bigm| & 13/3 \end{bmatrix}.$$

Since this system has a free variable, there are infinitely many solutions. In particular, no matter what value of d we choose, the cubic

$$y = (2/3 - d/6)x^3 + (d-2)x^2$$
$$+ (13/3 - 11/6d)x + d$$

goes through the 3 points. [Side note: If we choose $d = 4$, we get exactly the parabola that we found in part (b). We will discuss polynomial interpolation a bit more thoroughly in Remark 2.3.1.]

Section 2.2: Elementary Matrices and Matrix Inverses

2.2.1 (a) $\begin{bmatrix} 0 & 0 & 1 \\ 0 & 1 & 0 \\ 1 & 0 & 0 \end{bmatrix}$ (c) $\begin{bmatrix} 1 & 3 & 0 \\ 0 & 1 & 0 \\ 0 & 0 & 1 \end{bmatrix}$

     (e) $\begin{bmatrix} 3 & 0 & 0 \\ 0 & 1 & 0 \\ 0 & 0 & 1 \end{bmatrix}$ (g) $\begin{bmatrix} 1 & 0 & 0 \\ 0 & 1 & -2 \\ 0 & 0 & 1 \end{bmatrix}$

2.2.2 (a) The identity matrix is its own inverse:
$$\begin{bmatrix} 1 & 0 \\ 0 & 1 \end{bmatrix}.$$

     (c) We can compute the inverse of this matrix either via the Gauss–Jordan (row-reduction) method or via the explicit formula for 2×2 matrices. Its inverse is
$$\frac{1}{5}\begin{bmatrix} -2 & 3 \\ 3 & -2 \end{bmatrix}.$$

     (e) We perform row operations on the augmented system $[A \mid I]$ (where A is the matrix whose inverse we are trying to compute) to find the inverse:
$$\begin{bmatrix} 2 & 4 & 0 & | & 1 & 0 & 0 \\ 1 & -2 & 0 & | & 0 & 1 & 0 \\ 2 & 0 & -1 & | & 0 & 0 & 1 \end{bmatrix}$$
$$\xrightarrow{\frac{1}{2}R_1} \begin{bmatrix} 1 & 2 & 0 & | & 1/2 & 0 & 0 \\ 1 & -2 & 0 & | & 0 & 1 & 0 \\ 2 & 0 & -1 & | & 0 & 0 & 1 \end{bmatrix}$$
$$\xrightarrow[R_3-2R_1]{R_2-R_1} \begin{bmatrix} 1 & 2 & 0 & | & 1/2 & 0 & 0 \\ 0 & -4 & 0 & | & -1/2 & 1 & 0 \\ 0 & -4 & -1 & | & -1 & 0 & 1 \end{bmatrix}$$
$$\xrightarrow{R_3-R_2} \begin{bmatrix} 1 & 2 & 0 & | & 1/2 & 0 & 0 \\ 0 & -4 & 0 & | & -1/2 & 1 & 0 \\ 0 & 0 & -1 & | & -1/2 & -1 & 1 \end{bmatrix}$$
$$\xrightarrow[-R_3]{\frac{-1}{4}R_2} \begin{bmatrix} 1 & 2 & 0 & | & 1/2 & 0 & 0 \\ 0 & 1 & 0 & | & 1/8 & -1/4 & 0 \\ 0 & 0 & 1 & | & 1/2 & 1 & -1 \end{bmatrix}$$
$$\xrightarrow{R_1-2R_2} \begin{bmatrix} 1 & 0 & 0 & | & 1/4 & 1/2 & 0 \\ 0 & 1 & 0 & | & 1/8 & -1/4 & 0 \\ 0 & 0 & 1 & | & 1/2 & 1 & -1 \end{bmatrix}.$$

We can now simply read the inverse off from the right-hand side of the above augmented matrix:
$$\begin{bmatrix} 1/4 & 1/2 & 0 \\ 1/8 & -1/4 & 0 \\ 1/2 & 1 & -1 \end{bmatrix}.$$

     (g) This matrix is not invertible, since it is not even square.

2.2.3 (a) This matrix has inverse
$$\begin{bmatrix} 36 & -7 & -22 & 19 \\ -3 & 1 & 2 & -2 \\ -46 & 9 & 28 & -24 \\ 25 & -5 & -15 & 13 \end{bmatrix}.$$

     (c) This matrix has inverse
$$\frac{1}{10}\begin{bmatrix} -11 & 17 & 26 & -36 & -1 \\ -162 & 264 & 382 & -562 & -22 \\ -80 & 130 & 190 & -280 & -10 \end{bmatrix}.$$

2.2.4 (a) The RREF is I_2 and
$$E = \begin{bmatrix} -2 & 1 \\ 3/2 & -1/2 \end{bmatrix}.$$

     (c) $\begin{bmatrix} -5/3 & 2/3 \\ 4/3 & -1/3 \end{bmatrix}$

     (e) $\begin{bmatrix} 0 & -1/3 & 2/3 \\ 0 & 1/3 & 1/3 \\ 1 & 1/3 & -5/3 \end{bmatrix}$ (not unique)

     (g) $\begin{bmatrix} 0 & 0 & 1 \\ 0 & -1/2 & 3/2 \\ 1 & 1/2 & -3/2 \end{bmatrix}$ (not unique)

2.2.5 (a) True. In fact, the inverse of an elementary matrix is an elementary matrix of the same "type" (i.e., corresponding to the same type of row operation—swap, multiplication, or addition).

     (c) False. The product of elementary matrices is always invertible, so no non-invertible matrix can be written as such a product.

     (e) True. In fact, the identity matrix is its own inverse.

     (g) False. For example, if $A = I$ and $B = -I$ then $A + B = O$, which is not invertible.

     (i) True. If A *were* invertible then we could multiply the equation $A^7 = O$ on the left by A^{-1} 7 times to get
$$(A^{-1})^7 A^7 = (A^{-1})^7 O,$$
which simplifies to $I = O$, which is false. It follows that A is not invertible.

     (k) False. Multiplying that equation on the right by A^{-1} gives $X = BA^{-1}$, which in general is not the same as $A^{-1}B$ since matrix multiplication is not commutative. A specific counter-example is given by
$$X = \begin{bmatrix} 0 & 1 \\ 1 & 0 \end{bmatrix}, \quad A = \begin{bmatrix} 1 & 0 \\ 0 & -1 \end{bmatrix}, \quad B = \begin{bmatrix} 0 & -1 \\ 1 & 0 \end{bmatrix}$$

2.2.7 (a) If $a = b$ then every entry in this matrix is a, so its reduced row echelon form is
$$\begin{bmatrix} 1 & 1 & 1 & \cdots & 1 \\ 0 & 0 & 0 & \cdots & 0 \\ 0 & 0 & 0 & \cdots & 0 \\ \vdots & \vdots & \vdots & \ddots & \vdots \\ 0 & 0 & 0 & \cdots & 0 \end{bmatrix},$$
which is not the identity matrix, so A is not invertible.

     (b) If we let J_n denote the $n \times n$ matrix with every entry equal to 1, then $A = (a - b)I_n + bJ_n$. Based on testing out some small examples, we might conjecture that the inverse of A is a matrix of the form $B = cI_n + dJ_n$ for some $c, d \in \mathbb{R}$ as well, and we just need to carefully choose c and d. Well, we use the fact that $J_n^2 = nJ_n$ (which we proved way back in Exercise 1.3.11) to compute

Since we want this product to equal I_n, we choose $c = 1/(a-b)$ and then solve $bc + d(a-b) + nbd = 0$ to get $d = -b(nd+1/(a-b))/(a-b)$. It follows that A is invertible, and its inverse is

$$A^{-1} = \frac{1}{a-b}(I_n - b(nd+1/(a-b))J_n).$$

2.2.9 (a) $A^2 = \begin{bmatrix} 7 & 4 \\ 12 & 7 \end{bmatrix}$

(c) $A^{-2} = \begin{bmatrix} 7 & -4 \\ -12 & 7 \end{bmatrix}$

2.2.11 (a) Since A is $m \times n$, $A^T A$ is $n \times n$, so $(A^T A)^{-1}$ is also $n \times n$. It follows that $P = A(A^T A)^{-1} A^T$ is $m \times m$.

(b) $P^T = \left(A(A^T A)^{-1} A^T\right)^T = (A^T)^T \left((A^T A)^{-1}\right)^T A^T = A\left((A^T A)^T\right)^{-1} A^T = A(A^T A)^{-1} A^T = P$.

(c) Similar to part (b), this follows from some messy algebra:

$$P^2 = \left(A(A^T A)^{-1} A^T\right)\left(A(A^T A)^{-1} A^T\right)$$
$$= A(A^T A)^{-1}(A^T A)(A^T A)^{-1} A^T$$
$$= A(A^T A)^{-1} A^T$$
$$= P.$$

2.2.12 We use a similar argument to the one used to prove Theorem 2.2.1: row-reducing $[\,P\,|\,Q\,]$ is equivalent to multiplying it on the left by a sequence of elementary matrices $E_1, E_2, \ldots, E_k$:

$$E_k \cdots E_2 E_1 [\,P\,|\,Q\,] = [\,E_k \cdots E_2 E_1 P\,|\,E_k \cdots E_2 E_1 Q\,].$$

Well, if the matrix on the right equals $[\,I\,|\,P^{-1}Q\,]$ then $E_k \cdots E_2 E_1 P = I$, so $E_k \cdots E_2 E_1 = P^{-1}$, so the matrix on the right is $P^{-1}Q$, as claimed.

2.2.13 (a) Multiplying on the left by A^{-1} gives $X = A^{-1}B$.

(c) Multiplying on the left by A^{-1} and on the right by B^{-1} gives $X = A^{-1} I B^{-1} = A^{-1} B^{-1}$.

2.2.16 (a) We can solve the linear system $A\mathbf{x} = \mathbf{0}$ via back substitution. The final equation says $a_{n,n} x_n = 0$, which implies $x_n = 0$ as long as $a_{n,n} \neq 0$. Plugging this into the next equation then gives $a_{n-1,n-1} x_{n-1} = 0$, which implies $x_{n-1} = 0$ as long as $a_{n-1,n-1} \neq 0$. Continuing in this way gives $\mathbf{x} = \mathbf{0}$ as the unique solution of the linear system if all of the diagonal entries of A are non-zero. However, if one of the diagonal entries equals 0 (let's say $a_{j,j} = 0$) then the equation $a_{j,j} x_j = 0$ gives no restriction on x_j, so it is free and thus the linear system has infinitely many solutions. The desired result then follows from Theorem 2.2.4(d).

(b) As suggested by the hint, we first show that if the last k entries of $\mathbf{b}$ equal 0 then the solution $\mathbf{x}$ to $A\mathbf{x} = \mathbf{b}$ also has its last k entries equal to 0. The proof of this fact is almost identical to the back substitution argument from part (a). The final equation in this linear system says $a_{n,n} x_n = 0$, so $x_n = 0$ too, which then gives $a_{n-1,n-1} x_{n-1} = 0$ via back substitution, so $x_{n-1} = 0$, and so on.
To then prove that A^{-1} must be upper triangular, we give names to its columns:

$$A^{-1} = [\,\mathbf{b}_1\,|\,\mathbf{b}_2\,|\,\cdots\,|\,\mathbf{b}_n\,].$$

Then

$$AA^{-1} = A[\,\mathbf{x}_1\,|\,\mathbf{x}_2\,|\,\cdots\,|\,\mathbf{x}_n\,]$$
$$= [\,A\mathbf{x}_1\,|\,A\mathbf{x}_2\,|\,\cdots\,|\,A\mathbf{x}_n\,]$$
$$= [\,\mathbf{e}_1\,|\,\mathbf{e}_2\,|\,\cdots\,|\,\mathbf{e}_n\,] = I,$$

so $A\mathbf{x}_j = \mathbf{e}_j$ for all $1 \leq j \leq n$. Since the last $n - j$ entries of $\mathbf{e}_j$ all equal 0, the last $n - j$ entries of $\mathbf{x}_j$ must equal 0 as well, so A^{-1} is upper triangular.

(c) Since A and A^{-1} are both upper triangular, we can compute the diagonal entries of their product reasonably directly. If we denote the diagonal entries of A by $a_1, a_2, \ldots, a_n$, the diagonal entries of A^{-1} by $b_1, b_2, \ldots, b_n$, and use asterisks ($*$) to denote entries whose values we do not care about, then

$$AA^{-1} = \begin{bmatrix} a_1 & * & \cdots & * \\ 0 & a_2 & \cdots & * \\ \vdots & \vdots & \ddots & \vdots \\ 0 & 0 & \cdots & a_n \end{bmatrix} \begin{bmatrix} b_1 & * & \cdots & * \\ 0 & b_2 & \cdots & * \\ \vdots & \vdots & \ddots & \vdots \\ 0 & 0 & \cdots & b_n \end{bmatrix}$$

$$= \begin{bmatrix} a_1 b_1 & * & \cdots & * \\ 0 & a_2 b_2 & \cdots & * \\ \vdots & \vdots & \ddots & \vdots \\ 0 & 0 & \cdots & a_n b_n \end{bmatrix},$$

which can only equal the identity matrix if $b_1 = 1/a_1$, $b_2 = 1/a_2$, and so on.

(d) Part (a) does not change at all (the proof just uses forward substitution instead of backward substitution). In part (b), A^{-1} is lower triangular whenever A is invertible and lower triangular (which can be proved just by applying the transpose to the result from part (b)). Part (c) also does not change at all.

2.2.17 (a) An inverse of A^{-1} is a matrix $(A^{-1})^{-1}$ such that $A^{-1}(A^{-1})^{-1} = (A^{-1})^{-1} A^{-1} = I$. Replacing $(A^{-1})^{-1}$ in this equation by A makes it true, so it must be the case that $(A^{-1})^{-1} = A$, since matrix inverses are unique.

(b) Again, we just need to show that the claimed inverse, when multiplied by the original matrix, gives the identity matrix. That is, we want to show that $(A^{-1})^T A^T = A^T (A^{-1})^T = I$. Well, $I = I^T = (AA^{-1})^T = (A^{-1})^T A^T$, and similarly $I = I^T = (A^{-1}A)^T = A^T (A^{-1})^T$, as desired.

(c) Notice that if we multiply A^k by A^{-1} k times then we get

$$A^k (A^{-1})^k = (AA \cdots A)(A^{-1} A^{-1} \cdots A^{-1}).$$

The central AA^{-1} cancels to give an identity matrix, leaving $k - 1$ copies of each of A and A^{-1}. We then cancel the central terms again, repeating until there is nothing but an identity matrix left. It follows that $A^k (A^{-1})^k = I$ (and the equation $(A^{-1})^k A^k = I$ is proved similarly), so $(A^k)^{-1} = (A^{-1})^k$.

2.2.18   (a)   We already proved this claim when $k, \ell \geq 0$ in Exercise 1.3.22, and it follows when $k, \ell < 0$ simply by replacing A by A^{-1}, so we just prove the case when $k \geq 0$ and $\ell < 0$. In this case, we notice that $A^k A^\ell = (AA \cdots A)(A^{-1}A^{-1} \cdots A^{-1})$, where the first bracket has k copies of A and the second has $|\ell| = -\ell$ copies of A^{-1}. The central AA^{-1} terms repeatedly cancel out, leaving just $k - |\ell| = k + \ell$ copies of A (if $k \geq |\ell|$) or $|\ell| - k = -(k + \ell)$ copies of A^{-1} (if $|\ell| > k$). In either case, the product thus equals $A^{k+\ell}$, as claimed.

      (b)   Again, we just explicitly prove the $k \geq 0$, $\ell < 0$ case, and note that the other cases are handled similarly. In this case, notice that $(A^k)^\ell = (AA \cdots A)^{-1}(AA \cdots A)^{-1} \cdots (AA \cdots A)^{-1}$, where each bracket has k copies of A, and there are $|\ell| = -\ell$ brackets. Since $(AA \cdots A)^{-1} = A^{-1}A^{-1} \cdots A^{-1}$, it follows that there are a total of $k|\ell| = -k\ell$ copies of A^{-1}, so this product equals $A^{-|k\ell|} = A^{k\ell}$.

2.2.19   If $AB = I$ then taking the transpose of both sides gives $B^T A^T = (AB)^T = I^T = I$. It follows from the half of Theorem 2.2.7 that was proved in the text that A^T is invertible and $(A^T)^{-1} = B^T$. Using the fact that $(A^T)^{-1} = (A^{-1})^T$ and taking the transpose of both sides then gives $A^{-1} = B$, as desired.

2.2.20   (a)   We just compute the product of the two matrices (using block matrix multiplication) and see that we get the identity:

$$\begin{bmatrix} A & O \\ O & D \end{bmatrix} \begin{bmatrix} A^{-1} & O \\ O & D^{-1} \end{bmatrix} = \begin{bmatrix} AA^{-1} & O \\ O & DD^{-1} \end{bmatrix}$$
$$= \begin{bmatrix} I & O \\ O & I \end{bmatrix}.$$

      (b)   We first note that if we multiply the block diagonal matrix by any other matrix, we get

$$\begin{bmatrix} A_1 & O & \cdots & O \\ O & A_2 & \cdots & O \\ \vdots & \vdots & \ddots & \vdots \\ O & O & \cdots & A_n \end{bmatrix} \begin{bmatrix} B_{1,1} & B_{1,2} & \cdots & B_{1,n} \\ B_{2,1} & B_{2,2} & \cdots & B_{2,n} \\ \vdots & \vdots & \ddots & \vdots \\ B_{n,1} & B_{n,2} & \cdots & B_{n,n} \end{bmatrix}$$

$$= \begin{bmatrix} A_1 B_{1,1} & A_1 B_{1,2} & \cdots & A_1 B_{1,n} \\ A_2 B_{2,1} & A_2 B_{2,2} & \cdots & A_2 B_{2,n} \\ \vdots & \vdots & \ddots & \vdots \\ A_n B_{n,1} & A_n B_{n,2} & \cdots & A_n B_{n,n} \end{bmatrix}.$$

       Well, if any of $A_1, A_2, \ldots, A_n$ are not invertible, then the products $A_1 B_{1,1}, A_2 B_{2,2}, \ldots, A_n B_{n,n}$ cannot all equal I, so the above block matrix product cannot equal I either (since its diagonal blocks do not all equal I). The block diagonal matrix is thus not invertible, which completes the "only if" direction of the proof.

On the other hand, if $A_1, A_2, \ldots, A_n$ *are* invertible, then it is straightforward to check that

$$\begin{bmatrix} A_1 & O & \cdots & O \\ O & A_2 & \cdots & O \\ \vdots & \vdots & \ddots & \vdots \\ O & O & \cdots & A_n \end{bmatrix} \begin{bmatrix} A_1^{-1} & O & \cdots & O \\ O & A_2^{-1} & \cdots & O \\ \vdots & \vdots & \ddots & \vdots \\ O & O & \cdots & A_n^{-1} \end{bmatrix}$$
$$= \begin{bmatrix} I & O & \cdots & O \\ O & I & \cdots & O \\ \vdots & \vdots & \ddots & \vdots \\ O & O & \cdots & I \end{bmatrix},$$

so the matrix above on the right is the inverse of the block diagonal matrix above on the left, which completes the "if" direction of the proof.

2.2.21   In all parts of this question, we show that the claimed inverse is correct simply by performing block matrix multiplication and seeing that we get the identity matrix.

      (a) $\begin{bmatrix} A & I \\ I & O \end{bmatrix} \begin{bmatrix} O & I \\ I & -A \end{bmatrix} = \begin{bmatrix} I & A-A \\ O & I \end{bmatrix} =$
$\begin{bmatrix} I & O \\ O & I \end{bmatrix}$

      (c)   This one is a bit uglier:

$$\begin{bmatrix} A & B \\ O & D \end{bmatrix} \begin{bmatrix} A^{-1} & -A^{-1}BD^{-1} \\ O & D^{-1} \end{bmatrix}$$
$$= \begin{bmatrix} AA^{-1} & -AA^{-1}BD^{-1} + BD^{-1} \\ O & DD^{-1} \end{bmatrix}$$
$$= \begin{bmatrix} I & -BD^{-1} + BD^{-1} \\ O & I \end{bmatrix} = \begin{bmatrix} I & O \\ O & I \end{bmatrix}$$

2.2.22   (a)   This inverse can be found by using the formula from Exercise 2.2.21(a), partitioning this matrix via

$$A = \begin{bmatrix} 1 & 2 \\ 3 & 4 \end{bmatrix}.$$

       The inverse is

$$\begin{bmatrix} 0 & 0 & 1 & 0 \\ 0 & 0 & 0 & 1 \\ 1 & 0 & -1 & -2 \\ 0 & 1 & -2 & -3 \end{bmatrix}.$$

      (c)   This inverse can be found by using the formula from Exercise 2.2.21(c), partitioning this matrix via

$$A = \begin{bmatrix} 1 & 1 \\ 1 & 2 \end{bmatrix}, \quad B = \begin{bmatrix} 2 & -1 & 0 \\ 2 & 0 & 1 \end{bmatrix}, \text{ and}$$
$$D = \begin{bmatrix} 1 & 0 & 0 \\ 0 & 2 & 0 \\ 0 & 0 & 3 \end{bmatrix}.$$

       The inverse is

$$\frac{1}{6} \begin{bmatrix} 12 & -6 & -12 & 6 & 2 \\ -6 & 6 & 0 & -3 & -2 \\ 0 & 0 & 6 & 0 & 0 \\ 0 & 0 & 0 & 3 & 0 \\ 0 & 0 & 0 & 0 & 2 \end{bmatrix}.$$

Section 2.3: Subspaces, Spans, and Linear Independence

2.3.1 (a) Not a subspace, since (for example) it does not contain the zero vector.

(c) Not a subspace, since (for example) $(\pi, 0)$ is in the set but $\frac{1}{2}(\pi, 0) = (\pi/2, 0)$ is not in the set.

(e) Not a subspace, since (for example) it does not contain the zero vector.

(g) This is a subspace since it is a line going through the origin. A bit more formally, it is span$\{(-2, 1)\}$, and spans are always subspaces.

(i) Not a subspace, since (for example) $(1, 0)$ and $(0, -1)$ are in the set, but $(1, 0) + (0, -1) = (1, -1)$ is not in the set.

2.3.2 (a) All four vectors are multiples of $(1, 1)$, so their span is a line pointing in that direction. This line has equation $y = x$, or $x - y = 0$.

(c) We want to determine which vectors (x, y) are in the span of these two vectors. We thus want to determine when the following system of linear equations has a solution:

$$\left[\begin{array}{cc|c} 1 & 2 & x \\ 2 & 1 & y \end{array}\right]$$

$$\xrightarrow{R_2 - 2R_1} \left[\begin{array}{cc|c} 1 & 2 & x \\ 0 & -3 & y - 2x \end{array}\right]$$

This system of equations always has a solution, regardless of x and y, so the span of the two original vectors is all of $\mathbb{R}^2$.

2.3.3 (a) All three vectors are multiples of $(1, 1, 1)$, so their span is a line pointing in that direction.

(c) We want to determine which vectors (x, y, z) are in the span of these three vectors. We thus want to determine when the following system of linear equations has a solution:

$$\left[\begin{array}{ccc|c} 1 & 0 & 2 & x \\ 2 & 1 & 5 & y \\ 1 & -1 & 1 & z \end{array}\right]$$

$$\xrightarrow[R_3 - R_1]{R_2 - 2R_1} \left[\begin{array}{ccc|c} 1 & 0 & 2 & x \\ 0 & 1 & 1 & y - 2x \\ 0 & -1 & -1 & z - x \end{array}\right]$$

$$\xrightarrow{R_3 + R_2} \left[\begin{array}{ccc|c} 1 & 0 & 2 & x \\ 0 & 1 & 1 & y - 2x \\ 0 & 0 & 0 & -3x + y + z \end{array}\right].$$

This system has a solution if and only if $-3x + y + z = 0$, which is the equation of a plane, and this plane is the span of the 3 given vectors.

2.3.4 (a) We mimic Example 2.3.3 and find the range of this matrix is the line through the origin and $(1, 1)$ (i.e., the line with equation $y = x$), while its null space is the line through the origin and $(1, -1)$ (i.e., the line with equation $y = -x$).

(c) The range is all of $\mathbb{R}^2$, while the null space is $\{\mathbf{0}\}$.

2.3.5 (a) We mimic Example 2.3.3 and find the range of this matrix is the line through the origin and $(1, 1, 2)$. To find its null space, we row-reduce

$$\left[\begin{array}{ccc|c} 1 & 1 & 2 & 0 \\ 1 & 1 & 2 & 0 \\ 2 & 2 & 4 & 0 \end{array}\right]$$

to

$$\left[\begin{array}{ccc|c} 1 & 1 & 2 & 0 \\ 0 & 0 & 0 & 0 \\ 0 & 0 & 0 & 0 \end{array}\right].$$

This linear system has two free variables (y and z) and one leading variable x with $x + y + 2z = 0$. The null space is thus a plane, and that is its equation.

(c) The range is all of $\mathbb{R}^3$, while the null space is $\{\mathbf{0}\}$.

2.3.6 (a) Linearly independent, since this set consists of two vectors that are not multiples of each other.

(c) Linearly independent. To see this, we place these vectors as columns in a matrix A and solve the linear system $A\mathbf{x} = \mathbf{0}$:

$$\left[\begin{array}{ccc|c} 1 & 1 & 1 & 0 \\ 0 & 1 & 2 & 0 \\ -1 & 1 & -1 & 0 \end{array}\right] \xrightarrow{R_3 + R_1} \left[\begin{array}{ccc|c} 1 & 1 & 1 & 0 \\ 0 & 1 & 2 & 0 \\ 0 & 2 & 0 & 0 \end{array}\right].$$

We could go further and get a row echelon form, but we can see from here that the system has a unique solution $\mathbf{x} = \mathbf{0}$, so the set is linearly independent by Theorem 2.3.4.

(e) Linearly dependent (any set with more vectors than dimensions is linearly dependent—see Exercise 2.3.26).

(g) Linearly independent. To see this, use Theorem 2.3.4.

2.3.7 (a) No, does not span all of $\mathbb{R}^4$.

(c) Yes, spans all of $\mathbb{R}^4$.

2.3.8 (a) Linearly dependent, since

$$(4, 3, 2, 1) = (3, 1, 4, 2) + (2, 4, 1, 3) - (1, 2, 3, 4).$$

(c) Linearly independent.

2.3.9 (a) False. A counter-example is provided by $\mathbf{v}_1 = (1, 0)$, $\mathbf{v}_2 = (0, 1)$, and $\mathbf{v}_3 = (1, 1)$.

(c) True. Any scalar multiple of $\mathbf{0}$ is $\mathbf{0}$, and the sum of $\mathbf{0}$ with itself is $\mathbf{0}$.

(e) False. It could be a line (if $\mathbf{v}$ and $\mathbf{w}$ are multiples of each other) or even the subspace $\{\mathbf{0}\}$ if $\mathbf{v} = \mathbf{w} = \mathbf{0}$.

2.3.11 (a) We want to determine which vectors (x, y, z) are in the span of these three vectors. We thus want to determine when the following system of linear equations has a solution:

$$\left[\begin{array}{ccc|c} 0 & 1 & k & x \\ 1 & 2 & -1 & y \\ -1 & 1 & 4 & z \end{array}\right]$$

$$\xrightarrow{R_1 \leftrightarrow R_2} \left[\begin{array}{ccc|c} 1 & 2 & -1 & y \\ 0 & 1 & k & x \\ -1 & 1 & 4 & z \end{array}\right]$$

$$\xrightarrow{R_3 + R_1} \left[\begin{array}{ccc|c} 1 & 2 & -1 & y \\ 0 & 1 & k & x \\ 0 & 3 & 3 & y + z \end{array}\right]$$

$$\xrightarrow{R_3 - 3R_2} \left[\begin{array}{ccc|c} 1 & 2 & -1 & y \\ 0 & 1 & k & x \\ 0 & 0 & 3 - 3k & y + z - 3x \end{array}\right].$$

If $3 - 3k \neq 0$ (i.e., $k \neq 1$), then this system of equations has a solution for all x, y, z, so the span is (iii) all of $\mathbb{R}^3$ when $k \neq 1$. If $k = 1$, then the bottom row of the matrix is all zeros, so the only way for the system to have a solution is if $y + z - 3x = 0$. This is the equation of a plane, so (ii) the span is a plane when $k = 1$. (i) The span of these three vectors is never a line.

2.3.12   (a)   By Remark 2.3.1, we want to solve the linear system
$$\begin{bmatrix} 1 & 2 \\ 1 & 5 \end{bmatrix} \begin{bmatrix} c_0 \\ c_1 \end{bmatrix} = \begin{bmatrix} 1 \\ 7 \end{bmatrix}.$$
The unique solution to this linear system is $(c_0, c_1) = (-3, 2)$, so the interpolating polynomial is $p(x) = 2x - 3$.

   (c)   Again, we want to solve the linear system
$$\begin{bmatrix} 1 & 1 & 1 \\ 1 & 2 & 4 \\ 1 & 4 & 16 \end{bmatrix} \begin{bmatrix} c_0 \\ c_1 \\ c_2 \end{bmatrix} = \begin{bmatrix} 1 \\ 2 \\ 10 \end{bmatrix}.$$
The unique solution is $(c_0, c_1, c_2) = (2, -2, 1)$, so the interpolating polynomial is $p(x) = x^2 - 2x + 2$.

2.3.13   Since $A\mathbf{0} = \mathbf{0} \neq \mathbf{b}$, the solution set does not contain the zero vector and is thus not a subspace.

2.3.14   It is straightforward to check that $A_1 = 1$ so $A_1^{-1} = 1$, and
$$A_2 = \begin{bmatrix} 1 & 2 \\ 3 & 4 \end{bmatrix} \quad \text{so} \quad A_2^{-1} = \frac{1}{2} \begin{bmatrix} -4 & 2 \\ 3 & -1 \end{bmatrix}.$$
To see that A_n is not invertible when $n \geq 3$, we notice that the first row of A_n is the vector $(1, 2, \ldots, n)$ and its second row is the vector $(n+1, n+2, \ldots, 2n)$. Then the linear combination
$$2(n+1, n+2, \ldots, 2n) - (1, 2, \ldots, n)$$
$$= (2n+1, 2n+2, \ldots, 3n)$$
gives its third row. The rows of A_n thus form a linearly dependent set, so Theorem 2.3.5 tells us that A_n is not invertible.

2.3.15   If $\mathcal{S}$ is a set such that
$$c_1 \mathbf{v}_1 + \cdots + c_k \mathbf{v}_k \in \mathcal{S}$$
for all $\mathbf{v}_1, \ldots, \mathbf{v}_k \in \mathcal{S}$ and all $c_1, \ldots, c_k \in \mathbb{R}$, then we can choose $k = 2$ and $c_1 = c_2 = 1$ to see that $\mathbf{v}_1 + \mathbf{v}_2 \in \mathcal{S}$ whenever $\mathbf{v}_1, \mathbf{v}_2 \in \mathcal{S}$. Similarly, if we choose $k = 1$ then we see that $c_1 \mathbf{v}_1 \in \mathcal{S}$ whenever $\mathbf{v}_1 \in \mathcal{S}$, so both of the defining properties of subspaces hold.

In the other direction, if $\mathcal{S}$ is a subspace of $\mathbb{R}^n$ then we can repeatedly use the two defining properties of subspaces to see that $c_1 \mathbf{v}_1 \in \mathcal{S}$, $c_2 \mathbf{v}_2 \in \mathcal{S}$, ..., $c_k \mathbf{v}_k \in \mathcal{S}$ and thus $c_1 \mathbf{v}_1 + c_2 \mathbf{v}_2 \in \mathcal{S}$, so $(c_1 \mathbf{v}_1 + c_2 \mathbf{v}_2) + c_3 \mathbf{v}_3 \in \mathcal{S}$, and so on until we get $c_1 \mathbf{v}_1 + \cdots + c_k \mathbf{v}_k \in \mathcal{S}$.

2.3.17   (a)   Not in the range.

2.3.18   We could prove this fact directly by using the two defining properties of subspaces, but perhaps a simpler way is to notice that $A\mathbf{v} = \mathbf{v}$ if and only if $(A - I)\mathbf{v} = \mathbf{0}$. In other words, the fixed-point subspace of A is equal to $\text{null}(A - I)$, and we already know that the null space of any matrix is a subspace.

2.3.20   By definition, if the set $\{\mathbf{v}, \mathbf{w}\}$ is linearly dependent then there exist (not both zero) constants $c_1, c_2 \in \mathbb{R}$ such that $c_1 \mathbf{v} + c_2 \mathbf{w} = \mathbf{0}$. If $c_1 = 0$ then $c_2 \neq 0$ then $\mathbf{w} = \mathbf{0}$. On the other hand, if $c_1 \neq 0$ then $\mathbf{v} = (-c_2/c_1)\mathbf{w}$.

In the other direction, if $\mathbf{w} = \mathbf{0}$ then $0\mathbf{v} + 1\mathbf{w} = \mathbf{0}$, so $\{\mathbf{v}, \mathbf{w}\}$ is linearly dependent, and if $\mathbf{v} = c\mathbf{w}$ then $1\mathbf{v} + (-c)\mathbf{w} = \mathbf{0}$, so $\{\mathbf{v}, \mathbf{w}\}$ is linearly dependent.

2.3.21   Let $S = \{\mathbf{v}_1, \mathbf{v}_2, \ldots, \mathbf{v}_k\}$ be the set of vectors. If S is linearly dependent, then there exist (not all 0) scalars $c_1, c_2, \ldots, c_k$ such that
$$c_1 \mathbf{v}_1 + c_2 \mathbf{v}_2 + \cdots + c_k \mathbf{v}_k = \mathbf{0}.$$
Suppose that $c_i \neq 0$ (which we can do since not all of the coefficients are 0). Then we can rearrange the above equation to get
$$\mathbf{v}_i = \frac{-c_1}{c_i}\mathbf{v}_1 + \ldots + \frac{-c_{i-1}}{c_i}\mathbf{v}_{i-1} + \frac{-c_{i+1}}{c_i}\mathbf{v}_{i+1} + \ldots + \frac{-c_k}{c_i}\mathbf{v}_k.$$
It follows that $\mathbf{v}_i$ is a linear combination of the other vectors in the set.

In the other direction, suppose that $\mathbf{v}_i$ is a linear combination of the other vectors in S:
$$\mathbf{v}_i = c_1 \mathbf{v}_1 + \ldots + c_{i-1}\mathbf{v}_{i-1} + c_{i+1}\mathbf{v}_{i+1} + \ldots + c_k \mathbf{v}_k.$$
Rearranging this equation gives
$$c_1 \mathbf{v}_1 + \ldots + c_{i-1}\mathbf{v}_{i-1} - \mathbf{v}_i + c_{i+1}\mathbf{v}_{i+1} + \ldots + c_k \mathbf{v}_k = \mathbf{0},$$
so S is linearly dependent.

2.3.23   Suppose $B = \{\mathbf{0}, \mathbf{v}_1, \mathbf{v}_2, \ldots, \mathbf{v}_k\}$. Then
$$\mathbf{0} + 0\mathbf{v}_1 + 0\mathbf{v}_2 + \cdots + 0\mathbf{v}_k = \mathbf{0}.$$
Since not all of the coefficients in the linear combination above are 0 (notice that the first coefficient in the linear combination is 1), this shows that B is linearly dependent.

2.3.25   (a)   Suppose $B = \{\mathbf{v}_1, \mathbf{v}_2, \ldots, \mathbf{v}_k\}$ is linearly dependent. Then there exist (not all 0) scalars $c_1, c_2, \ldots, c_k$ such that
$$c_1 \mathbf{v}_1 + c_2 \mathbf{v}_2 + \cdots + c_k \mathbf{v}_k = \mathbf{0}.$$
If $C = \{\mathbf{v}_1, \mathbf{v}_2, \ldots, \mathbf{v}_k, \mathbf{w}_1, \mathbf{w}_2, \ldots, \mathbf{w}_m\}$ then we thus have
$$c_1 \mathbf{v}_1 + \cdots + c_k \mathbf{v}_k + 0\mathbf{w}_1 + \ldots + 0\mathbf{w}_m = \mathbf{0}.$$
Since at least one of the coefficients in this linear combination is non-zero, C is thus linearly dependent too.

   (b)   This statement is logically equivalent to the one from part (a). If C is linearly independent, then B cannot possibly be linearly dependent, because by (a) that would mean that C is linearly dependent. So B must be linearly independent as well.

2.3.27   (a)   If $\mathbf{x} \in \text{range}(AB)$ then there exists $\mathbf{y} \in \mathbb{R}^p$ such that $\mathbf{x} = (AB)\mathbf{y} = A(B\mathbf{y})$. It follows that $\mathbf{x} \in \text{range}(A)$ as well (since it can be written as A times a vector), so $\text{range}(AB) \subseteq \text{range}(A)$.

   (b)   If $\mathbf{x} \in \text{null}(B)$ then $B\mathbf{x} = \mathbf{0}$, so $AB\mathbf{x} = A\mathbf{0} = \mathbf{0}$, so $\mathbf{x} \in \text{null}(AB)$. It follows that $\text{null}(B) \subseteq \text{null}(AB)$.

(c) We choose

$$A = \begin{bmatrix} 0 & 1 \\ 0 & 0 \end{bmatrix} \quad \text{and} \quad B = \begin{bmatrix} 0 & 0 \\ 1 & 0 \end{bmatrix}.$$

Then

$$AB = \begin{bmatrix} 1 & 0 \\ 0 & 0 \end{bmatrix},$$

which has range$(AB) = \text{span}\big((1,0)\big)$, which is not contained within range$(B) = \text{span}\big((0,1)\big)$. Similarly, null$(A) = \text{span}\big((1,0)\big)$, which is not contained within null$(AB) = \text{span}\big((0,1)\big)$.

2.3.28 Suppose $d_1, d_2, \ldots, d_n \in \mathbb{R}$ are such that

$$d_1(c_1\mathbf{v}_1) + d_2(c_2\mathbf{v}_2) + \cdots + d_n(c_n\mathbf{v}_n) = \mathbf{0}.$$

If $\{\mathbf{v}_1, \mathbf{v}_2, \ldots, \mathbf{v}_n\}$ is linearly independent then this equation implies $d_1 c_1 = \cdots = d_n c_n = 0$. Since $c_1, \ldots, c_n$ are non-zero, this implies $d_1 = \cdots = d_n = 0$, so $\{c_1\mathbf{v}_1, c_2\mathbf{v}_2, \ldots, c_n\mathbf{v}_n\}$ is linearly independent too. The reverse implication follows from writing $\mathbf{v}_j = (1/c_j)(c_j\mathbf{v}_j)$ for each $1 \le j \le n$ and then using the fact that multiplying vectors by scalars preserves linear independence (which we just proved).

Section 2.4: Bases and Rank

2.4.1 (a) Not a basis, since B consists of 3 vectors in a 2-dimensional space ($\mathbb{R}^2$).

(c) Not a basis, since B contains 2 vectors but $\mathcal{S}$ is 3-dimensional.

(e) Is a basis. Since B contains 2 vectors and $\mathcal{S}$ is 2-dimensional, we just need to check that B is linearly independent:

$$\begin{bmatrix} 1 & 2 & | & 0 \\ 0 & 1 & | & 0 \\ 1 & 3 & | & 0 \end{bmatrix} \xrightarrow{R_3 - R_1} \begin{bmatrix} 1 & 2 & | & 0 \\ 0 & 1 & | & 0 \\ 0 & 1 & | & 0 \end{bmatrix}$$

$$\xrightarrow{R_3 - R_2} \begin{bmatrix} 1 & 2 & | & 0 \\ 0 & 1 & | & 0 \\ 0 & 0 & | & 0 \end{bmatrix}.$$

This linear system thus has a unique solution (the zero vector), so B is linearly independent and thus a basis.

2.4.2 (a) Since a line is 1-dimensional, we just need to find any non-zero vector on the line. $(x,y) = (1,3)$ works, so $B = \{(1,3)\}$ is a basis of this subspace.

(c) A plane is 2-dimensional, so we need 2 linearly independent basis vectors. We thus need any two non-zero vectors in this plane that are not multiples of each other. By picking values of x,y and solving for z, we can find many such vectors, such as $(1,1,-3)$ and $(1,0,-2)$. Since these are not multiples of each other, $B = \{(1,1,-3),(1,0,-2)\}$ is a basis of this plane.

(e) To find what direction this line points in, we solve the linear system defined by the two given planes:

$$\begin{bmatrix} 1 & 1 & -1 & | & 0 \\ 2 & -1 & 2 & | & 0 \end{bmatrix}$$

$$\xrightarrow{R_2 - 2R_1} \begin{bmatrix} 1 & 1 & -1 & | & 0 \\ 0 & -3 & 4 & | & 0 \end{bmatrix}$$

It follows that z is a free variable. Arbitrarily choosing $z = 3$ and solving by back substitution gives $y = 4$ and $x = -1$. The line points thus in the direction of the vector $(x,y,z) = (-1,4,3)$, so $B = \{(-1,4,3)\}$ is a basis of this line.

2.4.3 (a) $\{(1,2,3,4),(3,5,7,9)\}$

(c) $\{(1,1,4,5,1),(2,-7,4,2,2),(5,2,4,5,3)\}$

2.4.4 There are many possible solutions to this exercise.

(c) $\{(2,4,3,4),(3,2,-1,-1),(1,0,0,0),(0,1,0,0)\}$

2.4.5 (a) This matrix is already in RREF (and so is its transpose), so we can read bases for the four fundamental subspaces directly from the columns of A.
Basis of range(A): $\{(1,0),(0,1)\}$
Basis of null(A): $\{\}$ (null$(A) = \{\mathbf{0}\}$)
Basis of range(A^T): $\{(1,0),(0,1)\}$
Basis of null(A^T): $\{\}$ (null$(A^T) = \{\mathbf{0}\}$)

(c) This matrix will be in RREF after swapping its first two rows (and the RREF of A^T is similarly simple).
Basis of range(A): $\{(0,1,0)\}$
Basis of null(A): $\{(1,0,0),(0,-2,1)\}$
Basis of range(A^T): $\{(0,1,2)\}$
Basis of null(A^T): $\{(1,0,0),(0,0,1)\}$

(e) Basis of range(A): $\{(1,0,0),(0,2,1)\}$
Basis of null(A): $\{\}$ (null$(A) = \{\mathbf{0}\}$)
Basis of range(A^T): $\{(1,0),(0,1)\}$
Basis of null(A^T): $\{(0,-1,2)\}$

(g) This matrix is already in RREF, and we can directly see the first 3 of these bases.
Basis of range(A): $\{(1,0,0),(0,1,0),(0,0,1)\}$
Basis of null(A): $\{(-2,1,0,0,0),(-3,0,1,1,0)\}$
Basis of range(A^T): $\{(1,2,0,3,0),(0,0,1,-1,0), (0,0,0,0,1)\}$
Basis of null(A^T): $\{\}$ (null$(A^T) = \{\mathbf{0}\}$)

(i) The RREF of this matrix is

$$\begin{bmatrix} 1 & 0 & -1 & -3 & -3/2 \\ 0 & 1 & 0 & -1/2 & -1/4 \\ 0 & 0 & 0 & 0 & 0 \end{bmatrix}.$$

Basis of range(A): $\{(0,-1,-2),(-4,2,0)\}$
Basis of null(A): $\{(1,0,1,0,0),(3,1/2,0,1,0), (3/2,1/4,0,0,1)\}$
Basis of range(A^T): $\{(1,0,-1,-3,-3/2), (0,1,0,-1/2,-1/4)\}$
Basis of null(A^T): $\{(-1,-2,1)\}$

2.4.6 To find the rank of these matrices, we compute a row echelon form of them and count the number of non-zero rows.

(a) rank 1, nullity 1
(e) rank 3, nullity 0
(c) rank 2, nullity 0
(g) rank 2, nullity 0

2.4.7 (a) If we call this matrix A then $\text{rank}(A) = 4$, nullity$(A) = 4$, and we have:
Basis of range(A): $\{(0, -6, 4, -1), (2, -1, -2, 0), (-1, 1, 1, 0), (-1, 6, -2, 1)\}$ (or the standard basis of $\mathbb{R}^4$)
Basis of null(A): $\{(2, -1, 0, 0, 0, 0, 0, 0), (1, 0, 1, -1, 0, 0, 0, 0), (-1, 0, 0, 0, 2, -1, 0, 0), (1, 0, 2, 0, 1, 0, -1, 0)\}$
Basis of range(A^T): $\{(1, 2, 0, 1, 0, -1, 1, 0), (0, 0, 1, 1, 0, 0, 2, 0), (0, 0, 0, 0, 1, 2, 1, 0), (0, 0, 0, 0, 0, 0, 0, 1)\}$
Basis of null(A^T): $\{\}$ (null$(A^T) = \{\mathbf{0}\}$)

2.4.8 (a) True. Every basis of a k-dimensional subspace must consist of k vectors.

(c) False. The vectors $\mathbf{v}_1, \mathbf{v}_2, \ldots, \mathbf{v}_k$ might be linearly dependent.

(e) False. A basis is made up of the non-zero rows of the row echelon form itself—not the corresponding rows of A. For example, the reduced row echelon form of

$$A = \begin{bmatrix} 1 & 1 \\ 2 & 2 \\ 1 & 2 \end{bmatrix} \text{ is } \begin{bmatrix} 1 & 0 \\ 0 & 1 \\ 0 & 0 \end{bmatrix}.$$

It follows that range(A^T) is 2-dimensional, with basis $\{(1, 0), (0, 1)\}$. However, if we took the first two rows of A itself, we would get the set $\{(1, 1), (2, 2)\}$, which is not even linearly independent (indeed, its span is 1-dimensional and thus cannot be range(A^T)).

(g) True. The identity matrix is its own row echelon form, and it has n non-zero rows, so its rank is n.

(i) False. All that we can say is that $\text{rank}(A + B) \leq \text{rank}(A) + \text{rank}(B)$, thanks to Theorem 2.4.11. For a specific counter-example, notice that if $A = B = I_n$ then $\text{rank}(A + B) = \text{rank}(2I_n) = n$, but $\text{rank}(A) + \text{rank}(B) = n + n = 2n$.

(k) False. All that we can say is that $\text{rank}(AB) \leq \min\{\text{rank}(A), \text{rank}(B)\}$, thanks to Theorem 2.4.11. Again, $A = B = I_n$ serves as a counter-example.

(m) True. This is the statement of Theorem 2.4.8(c).

(o) False. These quantities are the same as each other if $m = n$, but otherwise they are necessarily different, since the rank-nullity theorem (Theorem 2.4.10) tells us that

$$\text{nullity}(A) = n - \text{rank}(A) \quad \text{and}$$
$$\text{nullity}(A^T) = m - \text{rank}(A^T) = m - \text{rank}(A).$$

2.4.10 Since every column of this matrix is the same, its range (i.e., span of its columns) is 1-dimensional, so its rank is 1.

2.4.12 It is straightforward to check that $\text{rank}(A_2) = 2$, so we assume from here on that $n \geq 3$. Well, the first row of A_n is the vector $(1, 2, \ldots, n)$ and its second row is the vector $(n+1, n+2, \ldots, 2n)$. Then each of its remaining rows can be written as a linear combination of those two rows. In particular, if $k \geq 3$ then the k-th row can be written as

$$(k-1)(n+1, n+2, \ldots, 2n) - (k-2)(1, 2, \ldots, n)$$
$$= ((k-1)n+1, (k-1)n+2, \ldots, kn).$$

The span of the rows of A_n thus equals the span of its first two rows (and those two rows are not multiples of each other), so A_n has rank 2.

2.4.13 (a) By performing the row operation $R_2 - 2R_1$, we get to the following row echelon form of this matrix:

$$\begin{bmatrix} 1 & x \\ 0 & 4 - 2x \end{bmatrix}.$$

The rank of this matrix thus equals 1 whenever the bottom row of this REF is all-zero (i.e., when $x = 2$) and the rank equals 2 otherwise (i.e., when $x \neq 2$).

(c) This matrix has row echelon form

$$\begin{bmatrix} 1 & 2 & x \\ 0 & -1-x & x+1 \\ 0 & 0 & x^2 - 1 \end{bmatrix}.$$

If $x = -1$ then the bottom two rows are both zero, so the rank is 1. If $x = 1$ then the bottom row is zero but the other two rows are not, so the rank is 2. If $x \neq \pm 1$ then the rank is 3.

2.4.14 (a) If A is diagonal then simply swapping some of its rows (so that its non-zero entries are in the topmost rows) produces one of its row echelon forms. One basis of its range is thus simply the set of its columns that contain a non-zero diagonal entry. Since its rank is the dimension of the range, which is the number of elements in a basis of the range, we conclude that the rank of A equals the number of non-zero entries on its diagonal.

(b) The simplest example is

$$A = \begin{bmatrix} 0 & 1 \\ 0 & 0 \end{bmatrix},$$

which has 0 non-zero diagonal entries but rank 1.

2.4.16 (a) If $x = 1$ then this is the matrix J_n from Exercise 2.4.10, which we already showed has rank 1.

(b) This is somewhat tricky to show directly (e.g., by computing a row echelon form of A), so we just recall from Exercise 2.2.7 that this matrix is invertible whenever $x \neq 1$, so it has rank n.

2.4.17 (a) Theorem 2.4.11 tells us that $\text{rank}(AB) \leq \text{rank}(B)$, so we just need to prove the opposite inequality. Since A is invertible, we know that $A^{-1}(AB) = B$, so using Theorem 2.4.11 again tells us that $\text{rank}(B) = \text{rank}(A^{-1}(AB)) \leq \text{rank}(AB)$, as desired.

(b) This is proved almost identically to part (a), but instead of multiplying on the left by A^{-1}, we multiply on the right by B^{-1}.

2.4.19 Recall from Theorem 2.3.5 that A is invertible if and only if its columns form a linearly independent set. Since there are n columns, Theorem 2.4.4(b) tells us that these columns being linearly independent is equivalent to them forming a basis of $\mathbb{R}^n$. It follows that A is invertible if and only if its columns form a basis of $\mathbb{R}^n$ (i.e., properties (a) and (d) of the theorem are equivalent). The equivalence of (a) and (e) is almost identical.

The equivalence of (b) and (d) follows immediately from the fact that range(A) equals the span of the columns of A, and the equivalence of (a) and (c) follows from the equivalence of conditions (a) and (d) in Theorem 2.2.4.

2.4.22 If $\{\mathbf{v}_1, \mathbf{v}_2, \ldots, \mathbf{v}_k\}$ is a linearly independent set, then we are done: $\{\mathbf{v}_1, \mathbf{v}_2, \ldots, \mathbf{v}_k\}$ is a basis of $\mathcal{S}$ since it also spans $\mathcal{S}$. If it is a linearly dependent set, then one of its members (let's call it $\mathbf{v}_i$) is a linear combination of the other members and thus (by Exercise 2.3.24)

$$\text{span}(\mathbf{v}_1, \ldots, \mathbf{v}_k) = \text{span}(\mathbf{v}_1, \ldots, \mathbf{v}_{i-1}, \mathbf{v}_{i+1}, \ldots, \mathbf{v}_k).$$

We then repeat: if $\{\mathbf{v}_1, \ldots, \mathbf{v}_{i-1}, \mathbf{v}_{i+1}, \ldots, \mathbf{v}_k\}$ is linearly independent then it must be a basis, whereas if it is linearly dependent we can remove another vector without changing the span. By repeatedly removing vectors we must eventually reach a linearly independent set (since any set containing exactly one non-zero vector is linearly independent) and thus a basis.

2.4.23 (a) To show linear dependence, we find a non-zero linear combination of these vectors that adds to 0. We could do this by solving a linear system, but in this case it is easy enough to "eyeball" a solution:

$$2(1, -1, 2) + 0(-1, 2, 3) - (2, -2, 4) = (0, 0, 0).$$

This set of vectors is thus linearly dependent. Since they are not all multiples of each other, their span must be 2-dimensional, so we know from Exercise 2.4.22 that there is a 2-vector subset of B that is a basis of $\mathcal{S}$.

(b) If we remove $(-1, 2, 3)$ from B, we are left with the set $\{(1, -1, 2), (2, -2, 4)\}$, and the span of this set is just the line (not a plane!) through the origin and the point $(1, -1, 2)$. Phrased another way, $(-1, 2, 3)$ is not a linear combination of $(1, -1, 2)$ and $(2, -2, 4)$, so we must instead remove either of those two other vectors.

2.4.24 We can use Theorem 2.4.3(b) to remove 0 or more vectors from B to create a basis of $\mathcal{S}$. However, we know that all bases of $\mathcal{S}$ contain $\dim(\mathcal{S})$ vectors, and B already contains exactly this many vectors. We thus conclude that the only possibility is that B becomes a basis when we remove 0 vectors from it—i.e., B itself is already a basis of $\mathcal{S}$.

2.4.26 (a) If $\mathbf{v} \in \text{range}(A)$ then we can write it as $\mathbf{v} = A\mathbf{x}$ for some $\mathbf{x} \in \mathbb{R}^n$. Then

$$\mathbf{v} \cdot \mathbf{w} = (A\mathbf{x}) \cdot \mathbf{w} = \mathbf{x} \cdot (A^T \mathbf{w}) = \mathbf{x} \cdot \mathbf{0} = 0.$$

(b) If $\mathbf{w} \in \text{range}(A^T)$ then we can write it as $\mathbf{w} = A^T \mathbf{x}$ for some $\mathbf{x} \in \mathbb{R}^m$. Then

$$\mathbf{v} \cdot \mathbf{w} = \mathbf{v} \cdot (A^T \mathbf{x}) = (A\mathbf{v}) \cdot \mathbf{x} = \mathbf{0} \cdot \mathbf{x} = 0.$$

2.4.27 (a) We observe that if A and B have the same RREF R then the augmented matrices $[\, A \mid \mathbf{0} \,]$ and $[\, B \mid \mathbf{0} \,]$ have the same RREF $[\, R \mid \mathbf{0} \,]$, so the linear systems $A\mathbf{x} = \mathbf{0}$ and $B\mathbf{x} = \mathbf{0}$ have the same solution set (in particular, the same solution set as the equation $R\mathbf{x} = \mathbf{0}$), so $\text{null}(A) = \text{null}(B)$.

For the converse, we can proceed in much the same way as the proof of Theorem B.2.1. If A and B have different RREFs R and S, respectively, then they have a first column where they differ. Then consider two cases: that column is leading in one of the RREFs and not the other, or that column is not leading in either RREF. In either case, we can construct a vector $\mathbf{x}$ such that $R\mathbf{x} = \mathbf{0}$ and $S\mathbf{x} \neq \mathbf{0}$, so $\text{null}(A) \neq \text{null}(B)$.

(b) Recall that one way to construct a basis of range(A^T) is to simply put the non-zero rows of the reduced row echelon form of A into a set. If A and B have the same RREF then these bases are the same, so $\text{range}(A^T) = \text{range}(B^T)$. For the converse, we argue much like we did in part (a). If A and B have different RREFs R and S, respectively, then there is a first (i.e., topmost) row where they differ. Then consider two cases: the leading entries in that row of R and S are or are not in the same position. In either case, we can show that row in R is not a linear combination of the rows of S (and that row of S is not a linear combination of the rows of R). It follows that $\text{range}(A^T) \neq \text{range}(B^T)$.

2.4.29 (a) Recall that there is a subset of the columns of A that form a basis of range(A). This set will be linearly independent and consist of rank(A) vectors. On the other hand, there is no linearly independent set of columns of A with more than this many vectors, since the range of A contains the span of these vectors, and would thus be larger than rank(A)-dimensional, which contradicts the definition of rank(A) as the dimension of range(A).

(b) This follows from part (a) and the fact that $\text{rank}(A) = \text{rank}(A^T)$.

2.4.30 For the "if" direction, notice that if we think of $\mathbf{v}$ and $\mathbf{w}^T$ as $m \times 1$ and $1 \times n$ matrices, respectively, then $\text{rank}(\mathbf{v}) = \text{rank}(\mathbf{w}^T) = 1$. Then Theorem 2.4.11(b) tells us that $\text{rank}(A) = \text{rank}(\mathbf{vw}^T) \leq 1$. Since $\mathbf{v}$ and $\mathbf{w}$ are non-zero, $A \neq O$ so $\text{rank}(A) \neq 0$. It follows that $\text{rank}(A) = 1$.

To prove the opposite direction, we note that $\text{rank}(A) = 1$ implies that range(A) has dimension 1. In other words, the columns of A are all multiples of each other. In particular, if the first column of A is $\mathbf{v}$ then there are constants $w_2, w_3, \ldots, w_n$ such that the 2nd column is $w_2\mathbf{v}$, the 3rd column is $w_3\mathbf{v}$, and so on. In other words,

$$A = \begin{bmatrix} \mathbf{v} \mid w_2\mathbf{v} \mid \cdots \mid w_n\mathbf{v} \end{bmatrix}.$$

If $\mathbf{w}^T = (1, w_2, w_3, \ldots, w_n)$ then the above expression for A is equivalent to $A = \mathbf{vw}^T$, so we are done. Note that if this first column of A is $\mathbf{0}$ then this argument does not quite work, but it can be patched up by letting $\mathbf{v}$ be the first non-zero column of A.

Section 2.5: Summary and Review

2.5.1 (a) Not row equivalent, since they have RREFs

$$\begin{bmatrix} 1 & 1 \\ 0 & 0 \end{bmatrix} \quad \text{and} \quad \begin{bmatrix} 1 & 0 \\ 0 & 1 \end{bmatrix},$$

respectively.

(c) Row equivalent, since they both have RREF

$$\begin{bmatrix} 1 & 0 & 1/3 \\ 0 & 1 & 1/3 \\ 0 & 0 & 0 \end{bmatrix}.$$

(b) Row equivalent, since they both have RREF

$$\begin{bmatrix} 1 & 0 & 0 \\ 0 & 1 & 0 \\ 0 & 0 & 1 \\ 0 & 0 & 0 \end{bmatrix}.$$

2.5.2 (a) True. This is the statement of Theorem B.2.1.

(c) False. Row-reducing $[A \mid \mathbf{b}]$ produces a matrix of the form $[R \mid \mathbf{c}]$, where $\mathbf{c}$ in general may be different from $\mathbf{b}$. It follows that $A\mathbf{x} = \mathbf{b}$ has the same solution set as $R\mathbf{x} = \mathbf{c}$ for *some* vector $\mathbf{c}$, but $\mathbf{c} \neq \mathbf{b}$ in general.

(e) True. This is exactly the RREF decomposition (Theorem 2.5.3).

(g) False. If range$(A) = \{\mathbf{0}\}$ then $A = O$, which is not invertible (in fact, the zero matrix is, in a sense, the "least invertible" matrix that exists).

(i) True. If A and B are row equivalent then they have the same RREF. Since the rank of a matrix equals the number of non-zero rows in its RREF, they must have the same rank too.

2.5.3 The "only if" implication is trivial (if $A = B$ then $A\mathbf{v}_j = B\mathbf{v}_j$ for all $1 \leq j \leq n$). For the "if" direction, suppose $\mathbf{v} \in \mathbb{R}^n$ and write $\mathbf{v}$ as a linear combination of the basis vectors:

$$\mathbf{v} = c_1\mathbf{v}_1 + c_2\mathbf{v}_2 + \cdots + c_n\mathbf{v}_n.$$

Then the fact that $A\mathbf{v}_j = B\mathbf{v}_j$ for all $1 \leq j \leq n$ tells us that

$$\begin{aligned} A\mathbf{v} &= A(c_1\mathbf{v}_1 + c_2\mathbf{v}_2 + \cdots + c_n\mathbf{v}_n) \\ &= c_1 A\mathbf{v}_1 + c_2 A\mathbf{v}_2 + \cdots + c_n A\mathbf{v}_n \\ &= c_1 B\mathbf{v}_1 + c_2 B\mathbf{v}_2 + \cdots + c_n B\mathbf{v}_n \\ &= B(c_1\mathbf{v}_1 + c_2\mathbf{v}_2 + \cdots + c_n\mathbf{v}_n) = B\mathbf{v}. \end{aligned}$$

It then follows from Exercise 1.3.18(d) that $A = B$.

2.5.5 As noted in the hint, we just have to explain why P can be chosen to be invertible. The reason for this is that if we row-reduce

$$[A \mid I] \quad \text{to} \quad [R \mid E]$$

via a sequence of row operations encoded by elementary matrices $E_1, E_2, \ldots, E_k$ then

$$\begin{aligned} [R \mid E] &= E_k \cdots E_2 E_1 [A \mid I] \\ &= [E_k \cdots E_2 E_1 A \mid E_k \cdots E_2 E_1]. \end{aligned}$$

In particular, $R = E_k \cdots E_2 E_1 A$, so $A = E_1^{-1} E_2^{-1} \cdots E_k^{-1} R$ (since elementary matrices are invertible). We can thus choose $P = E_1^{-1} E_2^{-1} \cdots E_k^{-1}$, which will also be invertible, since the product of invertible matrices is invertible.

2.5.6 For the "only if" direction, recall that if A and B are row equivalent then they have the same RREF R. By Theorem 2.5.3, we know that there exist invertible matrices $P_1, P_2 \in \mathcal{M}_m$ such that $A = P_1 R$ and $B = P_2 R$. Then multiplying on the left by $P_1 P_2^{-1}$ shows that

$$P_1 P_2^{-1} B = (P_1 P_2^{-1}) P_2 R = P_1 R = A,$$

so we can choose $P = P_1 P_2^{-1}$.

For the "if" direction, we note that if $A = PB$ and P is invertible, then we can write P as a product of elementary matrices: $P = E_1 E_2 \cdots E_k$. It follows that $A = E_1 E_2 \cdots E_k B$, so A can be reached from B via the sequence of row operations corresponding to these elementary matrices. In other words, A and B are row equivalent.

Section 2.A: Extra Topic: Linear Algebra Over Finite Fields

2.A.1 (a) $(x, y, z) = (0, 1, 0)$.

(c) $(w, x, y, z) = (0, 1, 0, 0)$ and $(w, x, y, z) = (1, 0, 1, 0)$.

(e) $(v, w, x, y, z) = (1, 1, 0, 0, 0)$.

2.A.2 (a) $(x, y, z) = (1, 0, 1)$.

2.A.3 (a) There are no solutions.

2.A.4 (a) (c)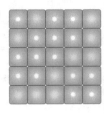

2.A.5 (a) (c)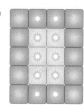

2.A.6 (a) False. They must have 0 solutions or a number of solutions that is a non-negative integer power of 2.

    (c) False. We showed right after Example 2.A.3 that only 1 out of every 16 starting configurations has a solution on a 4×4 grid.

    (e) True. This follows from the fact that, for each i and j, button i is a neighbor of button j if and only if button j is a neighbor of button i.

2.A.7 Since we only care about the evenness/oddness of these equations, they are really mod-2 equations that we can solve by representing it as a linear system over $\mathbb{Z}_2$. The augmented system looks like

$$\left[\begin{array}{ccccc|c} 1 & 1 & 1 & 1 & 1 & 0 \\ 0 & 1 & 0 & 1 & 0 & 1 \\ 1 & 1 & 0 & 1 & 1 & 1 \\ 0 & 1 & 1 & 0 & 1 & 0 \\ 1 & 1 & 1 & 1 & 0 & 0 \end{array}\right].$$

The unique solution to this linear system is $(v,w,x,y,z) = (0,1,1,0,0)$, though the original even/odd linear system can differ from this one by any even number in any entry (for example, $(v,w,x,y,z) = (4,-1,3,2,8)$ is another solution).

2.A.9 (a) If we order the buttons in standard reading order as usual then the linear system that we need to solve is $A\mathbf{x} = \mathbf{v}_e - \mathbf{v}_s$, where $\mathbf{v}_e - \mathbf{v}_s = (1,1,1,1,1,1,1,1,1)$ and

$$A = \begin{bmatrix} 1 & 1 & 1 & 0 & 0 & 0 & 0 & 1 & 0 \\ 1 & 1 & 1 & 0 & 0 & 1 & 0 & 0 & 1 \\ 1 & 1 & 1 & 0 & 1 & 0 & 0 & 0 & 0 \\ 0 & 0 & 0 & 1 & 1 & 0 & 0 & 1 & 0 \\ 0 & 0 & 1 & 1 & 1 & 1 & 1 & 0 & 0 \\ 0 & 1 & 0 & 0 & 1 & 1 & 0 & 0 & 0 \\ 0 & 0 & 0 & 0 & 1 & 0 & 1 & 0 & 0 \\ 1 & 0 & 0 & 1 & 0 & 0 & 0 & 1 & 1 \\ 0 & 1 & 0 & 0 & 0 & 0 & 0 & 1 & 1 \end{bmatrix}.$$

    (b) The unique solution of the linear system from part (a) is $\mathbf{x} = (1,0,0,1,0,1,1,0,1)$, which means we win the game by pressing buttons 1, 4, 6, 7, and 9:

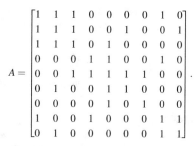

2.A.10 (a) First notice that B must have an even number of ones in it since its lower triangular portion is identical to its upper triangular portion. On the other hand, if the j-th row has m_j ones in it, and each m_j is odd, then in total there are

$$m_1 + m_2 + \cdots + m_k$$

ones in B. This sum is even if and only if k is even.

    (b) Just apply part (a) to the matrix $B - I$.

Section 2.B: Extra Topic: Linear Programming

2.B.1 (a) The optimal value is 3, which is attained at $(x_1,x_2) = (3,0)$.

    (c) The optimal value is 4, which is attained at $(x_1,x_2) = (1,1)$.

2.B.2 (a) The optimal value is $85/8$, which is attained at $(x_1,x_2) = (21,1)/8$.

    (c) The optimal value is $37/18$, which is attained at $(x_1,x_2,x_3) = (3,11,23)/18$.

    (e) The optimal value is $31/30$, which is attained at $(x_1,x_2,x_3,x_4) = (6,2,0,23)/30$.

2.B.3 (a) The optimal value is $1/2$, which is attained at $(x_1,x_2) = (3,1)/2$.

    (c) The optimal value is 0, which is attained at $(x_1,x_2,x_3) = (1,0,1)$.

    (e) The optimal value is $7/4$, which is attained at $(x_1,x_2,x_3,x_4) = (8,2,0,1)/8$.

2.B.4 (a)
minimize: $2y_1 + 3y_2$
subject to: $2y_1 + \ y_2 \geq 0$
           $-2y_1 + 3y_2 \geq 1$
           $-\ y_1, \quad y_2 \geq 0$

    (c)
maximize: $3y_1 + \ y_2 + 3y_3$
subject to: $2y_2 + \ y_3 \geq 1$
           $y_1 + 2y_2 + 2y_3 \geq 2$
           $-\ y_1 - \ y_2 + 2y_3 \geq -1$
           $-\ y_1, \qquad\quad y_3 \geq 0$

    (e)
minimize: $3y_1 + \ y_2 + 3y_3$
subject to: $2y_1 + \qquad 3y_3 \geq 1$
           $4y_1 + 3y_2 - \ y_3 = 2$
           $-\ y_1 + \ y_2 + 2y_3 \geq 1$
           $2y_2 + 2y_3 = 2$
           $y_1, \qquad\quad y_3 \geq 0$

2.B.5 (a) False. For example, if $\mathbf{x} = (1,0)$, $\mathbf{b} = (2,2)$, and

$$A = \begin{bmatrix} 1 & 2 \\ 2 & 3 \end{bmatrix},$$

then $A\mathbf{x} \leq \mathbf{b}$ but

$$A^{-1}\mathbf{b} = \begin{bmatrix} -3 & 2 \\ 2 & -1 \end{bmatrix}\begin{bmatrix} 2 \\ 2 \end{bmatrix} = \begin{bmatrix} -2 \\ 2 \end{bmatrix} \not\geq \begin{bmatrix} 1 \\ 0 \end{bmatrix} = \mathbf{x}.$$

(c) True. If $\mathbf{x} \geq \mathbf{y}$ then $x_j \geq y_j$ for each $1 \leq j \leq n$, so $c_j x_j \geq c_j y_j$ for each j, so

$$\mathbf{c} \cdot \mathbf{x} = \sum_{j=1}^{n} c_j x_j \geq \sum_{j=1}^{n} c_j y_j = \mathbf{c} \cdot \mathbf{y}.$$

(e) False. See the margin note beside the linear program (2.B.4).

2.B.6 The optimal value of this linear program is 1.4, which is attained at $(x_1, x_2) = (0.6, 0.8)$. To see that its feasible region contains no points with integer coordinates, we note that adding -1 times the first constraint to the second constraint shows that $x_2 \geq 1/2$, whereas adding the first constraint to -2 times the third constraint shows that $x_2 \leq 4/5$. In particular, this means that x_2 cannot be an integer. Alternatively, x_1 can similarly be bounded between $3/7$ and $3/4$, or the feasible region can just be plotted.

2.B.8 (a) Adding the two constraints shows that $2x_2 \leq c$, so $x_2 \leq c/2$, so certainly the optimal value of the linear program cannot exceed $c/2$. On the other hand, $(1/2, c/2)$ is a feasible point of the linear program attaining this value in the objective function.

(b) The first constraint tells us that $x_1 \leq 1 - x_2/c$ and the second constraint tells us that $x_1 \geq x_2/c$. Since $x_2 \geq 0$, it follows that $0 \leq x_1 \leq x_2$, and the only way that either equality can hold is if $x_2 = 0$, so the optimal value of the integer linear program cannot exceed 0. On the other hand, $(0,0)$ is a feasible point of this integer linear program attaining the value of 0 in the objective function.

2.B.9 The dual of a linear program in standard form looks like

$$\begin{aligned} \text{minimize:} \quad & \mathbf{b} \cdot \mathbf{y} \\ \text{subject to:} \quad & A^T \mathbf{y} \geq \mathbf{c} \\ & \mathbf{y} \geq \mathbf{0} \end{aligned}$$

which we can write in the form

$$\begin{aligned} -\text{maximize:} \quad & (-\mathbf{b}) \cdot \mathbf{y} \\ \text{subject to:} \quad & (-A^T)\mathbf{y} \leq -\mathbf{c} \\ & \mathbf{y} \geq \mathbf{0} \end{aligned}$$

Then the dual of this linear program is

$$\begin{aligned} -\text{minimize:} \quad & (-\mathbf{c}) \cdot \mathbf{x} \\ \text{subject to:} \quad & (-A^T)^T \mathbf{x} \geq -\mathbf{b} \\ & \mathbf{x} \geq \mathbf{0} \end{aligned}$$

which can be written in the form

$$\begin{aligned} \text{maximize:} \quad & \mathbf{c} \cdot \mathbf{x} \\ \text{subject to:} \quad & A\mathbf{x} \leq \mathbf{b} \\ & \mathbf{x} \geq \mathbf{0} \end{aligned}$$

which is exactly the original linear program.

2.B.12 Consider the linear program

$$\begin{aligned} \text{maximize:} \quad & 0 \\ \text{subject to:} \quad & (A - I)\mathbf{x} = \mathbf{0} \\ & \mathbf{x} \geq \mathbf{0} \end{aligned}$$

and the associated dual problem

$$\begin{aligned} \text{minimize:} \quad & 0 \\ \text{subject to:} \quad & (A^T - I)\mathbf{y} = \mathbf{0} \\ & \mathbf{y} \geq \mathbf{0} \end{aligned}$$

Notice that the rows of A^T sum to 1 so $A^T \mathbf{1} = \mathbf{1}$ (where $\mathbf{1} = (1, 1, \ldots, 1)$). It follows that the latter problem is feasible, since we can choose $\mathbf{y} = \mathbf{1}$ in it. By strong duality, it follows that the original primal problem is feasible as well, so there exists a vector $\mathbf{x} \in \mathbb{R}^n$ such that $A\mathbf{x} = \mathbf{x}$ and $\mathbf{x} \geq \mathbf{0}$.

Section 2.C: Extra Topic: More About the Rank

2.C.1 In this question, we compute a rank decomposition by mimicking the method of Example 2.C.1. However, keep in mind that rank decompositions are very non-unique, so your answers might look different from the answers provided here.

(a) $\begin{bmatrix} 1 \\ -1 \end{bmatrix}\begin{bmatrix} 1 & -1 \end{bmatrix}$

(c) $\begin{bmatrix} 2 & 4 & 0 \\ 1 & -2 & 0 \\ 2 & 0 & -1 \end{bmatrix}\begin{bmatrix} 1 & 0 & 0 \\ 0 & 1 & 0 \\ 0 & 0 & 1 \end{bmatrix}$

(e) $\begin{bmatrix} 3 \\ 1 \end{bmatrix}\begin{bmatrix} 2 & 1 \end{bmatrix}$

(g) $\begin{bmatrix} 2 & 6 \\ 5 & 4 \\ 3 & -2 \end{bmatrix}\begin{bmatrix} 1 & 0 & 2 & 5/11 \\ 0 & 1 & -1/2 & -7/22 \end{bmatrix}$

2.C.2 (a) The rank is 3, and

$$C = \begin{bmatrix} 2 & -1 & 0 \\ 1 & 0 & 0 \\ -1 & 0 & 2 \\ 2 & 0 & 1 \end{bmatrix} \quad \text{and}$$

$$R = \begin{bmatrix} 2 & 2 & 1 & 2 & 1 & 1 \\ 0 & -1 & 2 & 0 & 1 & -1 \\ -1 & 1 & 2 & 0 & -1 & 1 \end{bmatrix}.$$

2.C.3 (a) Rank 1, invertible 1×1 submatrix is any of the matrix's individual entries.

(c) Rank 2, invertible 2×2 submatrix

$$\begin{bmatrix} 1 & 2 \\ 1 & 3 \end{bmatrix}.$$

(e) Rank 2, invertible 2×2 submatrix

$$\begin{bmatrix} 1 & 2 \\ 4 & 5 \end{bmatrix}.$$

2.C.4 (a) The rank is 3 and its top-left 3×3 submatrix works:

$$\begin{bmatrix} 4 & 5 & 0 \\ 2 & 2 & 1 \\ -4 & 0 & 3 \end{bmatrix}$$

is invertible.

2.C.5 (a) True. This follows from the rank-one sum decomposition (Theorem 2.C.2).

(c) False. The rank of a matrix is determined by its largest invertible submatrix, not its largest non-invertible submatrix. For example, the 4×4 identity matrix has rank 4, but has a non-invertible 3×3 submatrix (its top-right 3×3 submatrix, for example).

(e) False. The following matrix has rank 2, for example:

$$\begin{bmatrix} 1 & 1 & 1 \\ 1 & 0 & 0 \\ 1 & 0 & 0 \end{bmatrix}.$$

2.C.6 It is clear that $\tilde{C} \in \mathcal{M}_{m,r}$ and $\tilde{R} \in \mathcal{M}_{r,n}$, so we just need to show that $A = \tilde{C}\tilde{R}$. To see this, we just compute $\tilde{C}\tilde{R} = CP^{-1}PR = CR = A$.

2.C.8 (a) Recall the RREF decomposition (Theorem 2.5.3), which says that we can write $A = PR$, where $P \in \mathcal{M}_m$ is invertible and $R \in \mathcal{M}_{m,n}$ is the RREF of A. We thus just have to show that we can write R in the form

$$R = \begin{bmatrix} I_{\text{rank}(A)} & O \\ O & O \end{bmatrix} Q,$$

where $Q \in \mathcal{M}_n$ is invertible. To this end, notice that the non-zero rows of R form a linearly independent set (they form a basis of range(A^T), after all), so by Theorem 2.4.3(a) we can extend them to a basis of $\mathbb{R}^n$. We can choose Q to be the matrix whose first rank(A) rows are the non-zero rows of R, and whose remaining rows are the vectors that we added to create a basis (thus making Q invertible).

(b) If $A = PBQ$ for some invertible P and Q, then the fact that rank(A) = rank(B) follows immediately from Exercise 2.4.17.

On the other hand, if rank(A) = rank(B) (and we denote this common rank by r) then we can find invertible matrices $P_1, P_2, Q_1,$ and Q_2 such that

$$A = P_1 \begin{bmatrix} I_r & O \\ O & O \end{bmatrix} Q_1 \quad \text{and} \quad B = P_2 \begin{bmatrix} I_r & O \\ O & O \end{bmatrix} Q_2.$$

Multiplying on the left by P_2^{-1} and on the right by Q_2^{-1} shows that

$$A = P_1 P_2^{-1} B Q_2^{-1} Q_1,$$

so we can choose $P = P_1 P_2^{-1}$ and $Q = Q_2^{-1} Q_1$ to get $A = PBQ$.

Section 2.D: Extra Topic: The LU Decomposition

2.D.1 (a) $L = \begin{bmatrix} 1 & 0 \\ 2 & 1 \end{bmatrix}, U = \begin{bmatrix} 1 & 2 \\ 0 & 1 \end{bmatrix}.$

(c) $L = \begin{bmatrix} 1 & 0 \\ -1 & 1 \end{bmatrix}, U = \begin{bmatrix} 3 & 1 & 2 \\ 0 & -2 & 1 \end{bmatrix}.$

(e) $L = \begin{bmatrix} 1 & 0 & 0 \\ 2 & 1 & 0 \\ 1 & 3 & 1 \end{bmatrix}, U = \begin{bmatrix} 1 & 2 \\ 0 & -1 \\ 0 & 0 \end{bmatrix}.$

(g) $L = \begin{bmatrix} 1 & 0 & 0 \\ 3 & 1 & 0 \\ -2 & -1 & 1 \end{bmatrix}, U = \begin{bmatrix} 1 & -4 & 5 \\ 0 & 3 & -7 \\ 0 & 0 & 1 \end{bmatrix}.$

2.D.2 (a) $P = \begin{bmatrix} 0 & 1 \\ 1 & 0 \end{bmatrix}, L = \begin{bmatrix} 1 & 0 \\ 0 & 1 \end{bmatrix}, U = \begin{bmatrix} 1 & 3 \\ 0 & 2 \end{bmatrix}.$

(c) $P = \begin{bmatrix} 1 & 0 & 0 \\ 0 & 0 & 1 \\ 0 & 1 & 0 \end{bmatrix}, L = \begin{bmatrix} 1 & 0 & 0 \\ 1 & 1 & 0 \\ 1 & 0 & 1 \end{bmatrix}, U = \begin{bmatrix} 1 & 1 & 1 \\ 0 & 1 & 2 \\ 0 & 0 & 1 \end{bmatrix}.$

2.D.3 (a) $(x, y) = (5, -3).$

(c) $(x, y, z) = (-3, 6, -2).$

2.D.4 (a) False. See Exercise 2.D.13 for an explicit counter-example. Also note that this answer being "false" is the entire reason that we introduced the PLU decomposition.

(c) True. This is simply the definition of a permutation matrix.

(e) True. This follows immediately from the definition of row echelon form.

2.D.5 Let B be the top-left $r \times r$ submatrix of A. If the top-left $k \times k$ submatrices of A are invertible for all $1 \leq k \leq r$ then the same is true of B, which is square, so Theorem 2.D.5 applies to it and shows that B has a unit LU decomposition $B = LU$.

Suppose now that $A \in \mathcal{M}_{m,n}$ with $m < n$ (i.e., A is wider than it is tall). If we write $A = \begin{bmatrix} B \mid C \end{bmatrix}$ and let $\tilde{U} = \begin{bmatrix} U \mid L^{-1}C \end{bmatrix}$ (note that L is invertible since its diagonal entries equal 1) then block matrix multiplication shows that

$$L\tilde{U} = \begin{bmatrix} LU \mid LL^{-1}C \end{bmatrix} = \begin{bmatrix} B \mid C \end{bmatrix} = A,$$

so A has a unit LU decomposition too. The case when $m > n$ is similar.

2.D.6 (a) Suppose that A had two unit LU decompositions $A = L_1U_1 = L_2U_2$. Since A is invertible, we know from Exercise 2.4.25 that each of L_1, L_2, U_1, and U_2 must be invertible as well, so doing some matrix algebra shows that $L_1^{-1}L_2 = U_1U_2^{-1}$.

Since the inverse of an upper (lower) triangular matrix is also upper (lower) triangular, and the product of two upper (lower) triangular matrices is again upper (lower) triangular, it follows that $L_1^{-1}L_2$ and $U_1U_2^{-1}$ must in fact be diagonal. However, since the diagonal entries of L_1^{-1} and L_2 all equal 1, it follows that $L_1^{-1}L_2 = I$, so $L_1 = L_2$. Multiplying the equation $L_1U_1 = L_2U_2$ on the left by L_1^{-1} then shows that $U_1 = U_2$ as well, so these two unit LU decompositions are in fact the same.

(b) Here are two different unit LU decompositions $A = L_1U_1 = L_2U_2$:

$$L_1 = \begin{bmatrix} 1 & 0 \\ 0 & 1 \end{bmatrix}, \quad U_1 = \begin{bmatrix} 0 & 0 \\ 0 & 1 \end{bmatrix}, \quad \text{and}$$

$$L_1 = \begin{bmatrix} 1 & 0 \\ 1 & 1 \end{bmatrix}, \quad U_1 = \begin{bmatrix} 0 & 0 \\ 0 & 1 \end{bmatrix}.$$

2.D.7 If A has a unit LU decomposition $A = LU$ then block matrix multiplication shows that the top-left $k \times k$ submatrix of A has the form $\widetilde{L}\widetilde{U}$, where $\widetilde{L}$ and $\widetilde{U}$ are the top-left $k \times k$ submatrices of L and U, respectively.

Since A is invertible, so are L and U (by Exercise 2.4.25), so their diagonal entries are non-zero, so the diagonal entries of $\widetilde{L}$ and $\widetilde{U}$ are also non-zero. It follows that $\widetilde{L}$ and $\widetilde{U}$ are invertible, so $\widetilde{L}\widetilde{U}$ is also invertible (i.e., the top-left $k \times k$ submatrix of A is invertible).

2.D.8 (a) Has a unit LU decomposition since its top-left square submatrices are

$$[1] \quad \text{and} \quad \begin{bmatrix} 1 & 3 \\ 3 & 2 \end{bmatrix},$$

both of which are invertible.

(c) Has a unit LU decomposition since its top-left square submatrices are

$$[3], \quad \begin{bmatrix} 3 & 1 \\ -1 & 1 \end{bmatrix}, \quad \text{and} \quad \begin{bmatrix} 3 & 1 & -1 \\ -1 & 1 & 3 \\ 2 & -2 & 1 \end{bmatrix},$$

all of which are invertible.

2.D.9 (a) We have $A = LU$, where

$$L = \begin{bmatrix} 1 & 0 & 0 & 0 & 0 & 0 \\ \frac{-1}{2} & 1 & 0 & 0 & 0 & 0 \\ 0 & \frac{-2}{5} & 1 & 0 & 0 & 0 \\ 0 & 0 & \frac{-5}{18} & 1 & 0 & 0 \\ 0 & 0 & 0 & \frac{-18}{67} & 1 & 0 \\ 0 & \frac{2}{5} & \frac{1}{9} & \frac{2}{67} & \frac{-19}{37} & 1 \end{bmatrix}, \quad \text{and}$$

$$U = \begin{bmatrix} 2 & -1 & 0 & 0 & 0 & 0 \\ 0 & \frac{5}{2} & -1 & 0 & 1 & 0 \\ 0 & 0 & \frac{18}{5} & -1 & \frac{2}{5} & 0 \\ 0 & 0 & 0 & \frac{67}{18} & \frac{-8}{9} & 0 \\ 0 & 0 & 0 & 0 & \frac{185}{67} & -1 \\ 0 & 0 & 0 & 0 & 0 & \frac{55}{37} \end{bmatrix}.$$

(b) $\dfrac{1}{275} \begin{bmatrix} 171 & 67 & 16 & -3 & -28 & -14 \\ 67 & 134 & 32 & -6 & -56 & -28 \\ 17 & 34 & 82 & 19 & -6 & -3 \\ 1 & 2 & 21 & 82 & 32 & 16 \\ -13 & -26 & 2 & 34 & 134 & 67 \\ -40 & -80 & -15 & 20 & 95 & 185 \end{bmatrix}$

2.D.10 If $A = LDU$ then setting $\widetilde{U} = DU$ gives $A = L\widetilde{U}$ as a unit LU decomposition of A, since $\widetilde{U}$ is also upper triangular.

In the opposite direction, if $A = LU$ is a unit LU decomposition of A then we can let D be the matrix with the same diagonal entries as U and set $\widetilde{U} = D^{-1}U$ to get $A = LD\widetilde{U}$ as an LDU decomposition of A (since $LD\widetilde{U} = LDD^{-1}U = LU = A$). This argument relies on D being invertible (i.e., U having non-zero diagonal entries), which is guaranteed by A being invertible.

2.D.11 (a) $L = \begin{bmatrix} 1 & 0 \\ 2 & 1 \end{bmatrix}, D = \begin{bmatrix} 1 & 0 \\ 0 & 1 \end{bmatrix}, U = \begin{bmatrix} 1 & 2 \\ 0 & 1 \end{bmatrix}.$

(c) $L = \begin{bmatrix} 1 & 0 & 0 \\ -1 & 1 & 0 \\ 3 & 1 & 1 \end{bmatrix}, D = \begin{bmatrix} 1 & 0 & 0 \\ 0 & -1 & 0 \\ 0 & 0 & -2 \end{bmatrix},$

$U = \begin{bmatrix} 1 & 2 & -1 \\ 0 & 1 & 3 \\ 0 & 0 & 1 \end{bmatrix}.$

2.D.13 (a) This matrix is already lower triangular and is thus essentially its own LU decomposition:

$$A = \begin{bmatrix} 0 & 0 \\ 1 & 1 \end{bmatrix}\begin{bmatrix} 1 & 0 \\ 0 & 1 \end{bmatrix}.$$

(b) The only elementary row operations that could be applied to A to put a leading entry in its top-left corner are $R_1 \leftrightarrow R_2$ or $R_1 + cR_2$, so Corollary 2.D.3 tells us that it does not have a unit LU decomposition. Alternatively, suppose for a moment that A did have a unit LU decomposition:

$$\begin{bmatrix} 0 & 0 \\ 1 & 1 \end{bmatrix} = \begin{bmatrix} 1 & 0 \\ \ell_{2,1} & 1 \end{bmatrix}\begin{bmatrix} u_{1,1} & u_{1,2} \\ 0 & u_{2,2} \end{bmatrix}$$

$$= \begin{bmatrix} u_{1,1} & u_{1,2} \\ \ell_{2,1}u_{1,1} & \ell_{2,1}u_{1,2} + u_{2,2} \end{bmatrix}.$$

Comparing the $(1,1)$-entries of these matrices shows that $u_{1,1} = 0$. However, this then implies that the $(2,1)$-entry also equals 0, which contradicts the fact that the $(2,1)$-entry of A is 1. It follows that A does not have a unit LU decomposition.

(c) It is straightforward to eyeball the PLU decomposition $A = PLU$ with

$$P = \begin{bmatrix} 0 & 1 \\ 1 & 0 \end{bmatrix}, \quad L = \begin{bmatrix} 1 & 0 \\ 0 & 1 \end{bmatrix}, \quad U = \begin{bmatrix} 1 & 1 \\ 0 & 0 \end{bmatrix}.$$

2.D.14 (a) We get one unit PLU decomposition if $P = I$:

$$A = \begin{bmatrix} 1 & 0 \\ 2 & 1 \end{bmatrix} \begin{bmatrix} 2 & -1 \\ 0 & 1 \end{bmatrix},$$

and we can get another one by choosing P to be the other 2×2 permutation matrix:

$$A = \begin{bmatrix} 0 & 1 \\ 1 & 0 \end{bmatrix} \begin{bmatrix} 1 & 0 \\ 1/2 & 1 \end{bmatrix} \begin{bmatrix} 4 & -1 \\ 0 & -1/2 \end{bmatrix}.$$

(b) If $A \in \mathcal{M}_n$ then we can usually find a different unit PLU decomposition for each of the $n!$ different permutation matrices P (i.e., no matter which permutation matrix P we choose, we can usually find an L and U such that $A = PLU$). We thus expect to be able to find $n!$ different unit PLU decompositions for most $n \times n$ matrices. In particular, we expect to be able to find 6 unit PLU decompositions of 3×3 matrices and 24 such decompositions of 4×4 matrices.

2.D.15 Recall that if the j-th column of P is $\mathbf{p}_j$ then the (i,j)-entry of $P^T P$ is

$$[P^T P]_{i,j} = \mathbf{p}_i \cdot \mathbf{p}_j.$$

Since each column of P contains exactly one 1 and all other entries equal to 0, we conclude that $\mathbf{p}_i \cdot \mathbf{p}_j = 1$ if $i = j$, so the diagonal entries of $P^T P$ equal 1. On the other hand, if $i \neq j$ then the columns $\mathbf{p}_i$ and $\mathbf{p}_j$ have their 1s in different positions (otherwise there would be a *row* of P with two 1s in it), so $\mathbf{p}_i \cdot \mathbf{p}_j = 0$ if $i \neq j$. In other words, the off-diagonal entries of $P^T P$ equal 0. It follows that $P^T P = I$, so $P^T = P^{-1}$.

Section 3.1: Coordinate Systems

3.1.1 (a) $[\mathbf{v}]_B = (2, 1/2)$, since $2(3,0) + \frac{1}{2}(0,2) = (6,1)$. This linear combination can be found by "eyeballing" or by solving the linear system $(6,1) = c_1(3,0) + c_2(0,2)$.

(c) $[\mathbf{v}]_B = (1/2, 3/2)$ since $(2,3,1) = \frac{1}{2}(1,0,-1) + \frac{3}{2}(1,2,1)$.

3.1.2 (a) $B = \{(2,3), (2,-1)\}$ works.

3.1.4 (a) $\begin{bmatrix} 1 & 3 \\ 2 & 4 \end{bmatrix}$ (c) $\begin{bmatrix} 11/10 & 5/2 \\ 3/10 & 1/2 \end{bmatrix}$

3.1.5 (a) $\begin{bmatrix} 2 & 0 \\ -5 & 2 \end{bmatrix}$ (c) $\begin{bmatrix} 2 & 0 \\ 0 & -3 \end{bmatrix}$

3.1.6 (a) $\begin{bmatrix} 10 & -5 & 0 \\ 5 & -6 & -4 \\ -11 & 5 & -1 \end{bmatrix}$

3.1.7 (a) $\text{rank}(A) = 1$ and $\text{rank}(B) = 2$, so they are not similar (the trace cannot be used, since $\text{tr}(A) = \text{tr}(B) = 2$).

(c) $\text{tr}(A) = 6$ and $\text{tr}(B) = 3$, so they are not similar (but their ranks both equal 2).

3.1.8 (a) $\text{rank}(A) = \text{rank}(B) = 2$ and $\text{tr}(A) = \text{tr}(B) = 3$, so we need to work harder to determine whether or not A and B are similar. Recall that A and B are similar if there exists an invertible matrix $P \in \mathcal{M}_2$ such that

$$A = PBP^{-1}, \quad \text{and} \quad AP = PB.$$

Multiplying these matrices out gives

$$\begin{bmatrix} p_{1,1} & p_{1,2} \\ 2p_{2,1} & 2p_{2,2} \end{bmatrix} = \begin{bmatrix} 3p_{1,1} + p_{1,2} & -2p_{1,1} \\ 3p_{2,1} + p_{2,2} & -2p_{2,1} \end{bmatrix}.$$

This is a linear system with 4 variables, and its solutions have $p_{1,2}$ and $p_{2,2}$ free, with $p_{1,1} = -p_{1,2}/2$ and $p_{2,1} = -p_{2,2}$. We just need to find any invertible matrix of this form, and one possibility is

$$P = \begin{bmatrix} 1 & -2 \\ -1 & 1 \end{bmatrix}.$$

Then $A = PBP^{-1}$, so A and B are similar.

3.1.9 (a) True. Recall that if $B = \{\mathbf{v}_1, \mathbf{v}_2, \ldots, \mathbf{v}_k\}$ then $[\mathbf{v}_j]_B = \mathbf{e}_j$ for each $1 \leq j \leq k$. The j-th column of $P_{B \leftarrow B}$ is thus $\mathbf{e}_j$, so $P_{B \leftarrow B} = I$.

(c) True. This follows from the fact that $PIP^{-1} = PP^{-1} = I$ for all invertible matrices P.

(e) False. A counter-example is given by

$$A = \begin{bmatrix} 1 & 0 \\ 0 & 1 \end{bmatrix}, \quad B = \begin{bmatrix} 1 & 1 \\ 0 & 1 \end{bmatrix},$$

both of which have rank and trace equal to 2. However, A and B are not similar: $A = I$, so $PAP^{-1} = A$ for all invertible P (i.e., A is only similar to itself and thus cannot be similar to B).

(g) False. For example, if $A = B = I \in \mathcal{M}_2$ then $\text{tr}(AB) = 2$ but $\text{tr}(A)\text{tr}(B) = 4$.

3.1.10 If every vector in S can be written as a linear combination of the members of B, then B spans S by definition. We thus only need to show linear independence of B. To this end, suppose that $\mathbf{v} \in S$ can be written as a linear combination of the members of $B = \{\mathbf{v}_1, \mathbf{v}_2, \ldots, \mathbf{v}_k\}$ in exactly one way:

$$\mathbf{v} = c_1\mathbf{v}_1 + c_2\mathbf{v}_2 + \cdots + c_k\mathbf{v}_k.$$

If B were linearly dependent, then there must be a non-zero linear combination of the form

$$\mathbf{0} = d_1\mathbf{v}_1 + d_2\mathbf{v}_2 + \cdots + d_k\mathbf{v}_k.$$

By adding these two linear combinations, we see that

$$\mathbf{v} = (c_1 + d_1)\mathbf{v}_1 + (c_2 + d_2)\mathbf{v}_2 + \cdots + (c_k + d_k)\mathbf{v}_k.$$

Since not all of the d_j's are zero, this is a different linear combination that gives $\mathbf{v}$, which contradicts uniqueness. It follows that B must in fact be linearly independent.

3.1.11 (a) Write $B = \{\mathbf{v}_1, \mathbf{v}_2, \ldots, \mathbf{v}_k\}$ and give names to the entries of $[\mathbf{v}]_B$ and $[\mathbf{w}]_B$: $[\mathbf{v}]_B = (c_1, c_2, \ldots, c_k)$ and $[\mathbf{w}]_B = (d_1, d_2, \ldots, d_k)$. This means that

$$\mathbf{v} = c_1\mathbf{v}_1 + c_2\mathbf{v}_2 + \cdots + c_k\mathbf{v}_k$$
$$\mathbf{w} = d_1\mathbf{v}_1 + d_2\mathbf{v}_2 + \cdots + d_k\mathbf{v}_k.$$

By adding these equations we see that

$$\mathbf{v} + \mathbf{w} = (c_1 + d_1)\mathbf{v}_1 + \cdots + (c_k + d_k)\mathbf{v}_k,$$

which means that $[\mathbf{v} + \mathbf{w}]_B = (c_1 + d_1, c_2 + d_2, \ldots, c_k + d_k)$, which is the same as $[\mathbf{v}]_B + [\mathbf{w}]_B$.

(b) Write $B = \{\mathbf{v}_1, \mathbf{v}_2, \ldots, \mathbf{v}_k\}$ and give names to the entries of $[\mathbf{v}]_B$ similar to those in part (a): $[\mathbf{v}]_B = (d_1, d_2, \ldots, d_k)$. This means that

$$\mathbf{v} = d_1\mathbf{v}_1 + d_2\mathbf{v}_2 + \cdots + d_k\mathbf{v}_k.$$

It follows that

$$c\mathbf{v} = (cd_1)\mathbf{v}_1 + (cd_2)\mathbf{v}_2 + \cdots + (cd_k)\mathbf{v}_k,$$

which means that $[c\mathbf{v}]_B = (cd_1, cd_2, \ldots, cd_k)$, which is the same as $c[\mathbf{v}]_B$.

3.1.12 (a) This follows fairly quickly from Exercise 3.1.11. We just notice that $[\mathbf{v}]_B = \mathbf{0}$ if and only if $\mathbf{v} = \mathbf{0}$, so

$$c_1\mathbf{w}_1 + \cdots + c_m\mathbf{w}_m = \mathbf{0}$$

if and only if

$$c_1[\mathbf{w}_1]_B + \cdots + c_m[\mathbf{w}_m]_B$$
$$= [c_1\mathbf{w}_1 + \cdots + c_m\mathbf{w}_m]_B = [\mathbf{0}]_B = \mathbf{0}.$$

In particular, $c_1 = \cdots = c_m = 0$ is the only solution to the former equation if and only if it is the only solution to the latter equation, so C is linearly independent if and only if D is linearly independent.

(b) Not only does every vector $\mathbf{v} \in S$ have a coordinate vector $[\mathbf{v}]_B \in \mathbb{R}^k$, but conversely every vector $\mathbf{x} \in \mathbb{R}^k$ can be written as the coordinate vector of some vector from S: $\mathbf{x} = [\mathbf{v}]_B$ for some $\mathbf{v} \in S$. We thus conclude that the linear system

$$c_1\mathbf{w}_1 + \cdots + c_m\mathbf{w}_m = \mathbf{v}$$

has a solution for all $\mathbf{v} \in S$ (i.e., C spans S) if and only if the linear system

$$c_1[\mathbf{w}_1]_B + \cdots + c_m[\mathbf{w}_m]_B$$
$$= [c_1\mathbf{w}_1 + \cdots + c_m\mathbf{w}_m]_B = [\mathbf{v}]_B$$

(c) This follows immediately from combining parts (a) and (b).

3.1.13 Since P is invertible, its columns (which are the coordinate vectors with respect to B of some vectors in S) form a basis of $\mathbb{R}^k$. That is,

$$P \stackrel{\text{def}}{=} [\,[\mathbf{v}_1]_B \mid [\mathbf{v}_2]_B \mid \cdots \mid [\mathbf{v}_k]_B\,]$$

for some vectors $\mathbf{v}_1, \ldots, \mathbf{v}_k \in S$ with the property that

$$\{[\mathbf{v}_1]_B, [\mathbf{v}_2]_B, \ldots, [\mathbf{v}_k]_B\}$$

is a basis of $\mathbb{R}^k$. It follows from Exercise 3.1.12 that $C = \{\mathbf{v}_1, \mathbf{v}_2, \ldots, \mathbf{v}_k\}$ is a basis of S, and (by definition) we have $P = P_{B \leftarrow C}$.

3.1.14 (a) Let $\mathbf{v} \in S$ be any vector. Then

$$(P_{C \leftarrow E} P_{E \leftarrow B})[\mathbf{v}]_B = P_{C \leftarrow E}(P_{E \leftarrow B}[\mathbf{v}]_B)$$
$$= P_{C \leftarrow E}[\mathbf{v}]_E = [\mathbf{v}]_C.$$

By uniqueness of change-of-basis matrices (Theorem 3.1.3), it thus follows that $P_{C \leftarrow E} P_{E \leftarrow B} = P_{C \leftarrow B}$.

(b) By using Exercise 2.2.12, we see that row-reducing $[\,P_{E \leftarrow C} \mid P_{E \leftarrow B}\,]$ produces the matrix

$$[\,I \mid P_{E \leftarrow C}^{-1} P_{E \leftarrow B}\,] = [\,I \mid P_{C \leftarrow E} P_{E \leftarrow B}\,],$$

which we know from part (a) equals $[\,I \mid P_{C \leftarrow B}\,]$.

3.1.15 (a) $\begin{bmatrix} 1 & 1 & 5/6 & 3/2 \\ -2 & -1 & -2/3 & 1 \\ 1 & 2 & 4/3 & 3 \\ 1 & 1 & 1/12 & -1/4 \end{bmatrix}$

3.1.16 We just mimic the proof of Theorem 1.4.1. To see that $[T]_B[\mathbf{v}]_B = [T(\mathbf{v})]_B$, first write $[\mathbf{v}]_B = (c_1, \ldots, c_n)$ and then use block matrix multiplication:

$$[T]_B[\mathbf{v}]_B = [\,[T(\mathbf{v}_1)]_B \mid \cdots \mid [T(\mathbf{v}_n)]_B\,] \begin{bmatrix} c_1 \\ \vdots \\ c_n \end{bmatrix}$$
$$= c_1[T(\mathbf{v}_1)]_B + \cdots + c_n[T(\mathbf{v}_n)]_B$$
$$= [T(c_1\mathbf{v}_1 + \cdots + c_n\mathbf{v}_n)]_B$$
$$= [T(\mathbf{v})]_B.$$

To verify that $[T]_B$ is unique, suppose that $A \in \mathcal{M}_n$ is *any* matrix such that $[T(\mathbf{v})]_B = A[\mathbf{v}]_B$ for all $\mathbf{v} \in \mathbb{R}^n$. Then $[T]_B[\mathbf{v}]_B = A[\mathbf{v}]_B$ for all $\mathbf{v} \in \mathbb{R}^n$. It follows from Exercise 1.3.18(d) that $A = [T]B$, which completes the proof.

3.1.17 Recall from the rank-nullity theorem (Theorem 2.4.10) that for every matrix $A \in \mathcal{M}_n$, we have $\text{nullity}(A) = n - \text{rank}(A)$. Since we already know that A and B being similar implies $\text{rank}(A) = \text{rank}(B)$, it follows that A and B being similar implies

$$n - \text{nullity}(A) = n - \text{nullity}(B),$$

so $\text{nullity}(A) = \text{nullity}(B)$.

3.1.19 (a) We compute $\text{tr}(A+B)$ directly from the definition of the trace:

$$\text{tr}(A+B) = \sum_{j=1}^{n} [A+B]_{j,j} = \sum_{j=1}^{n} a_{j,j} + b_{j,j}$$

$$= \sum_{j=1}^{n} a_{j,j} + \sum_{j=1}^{n} b_{j,j} = \text{tr}(A) + \text{tr}(B).$$

(b) Similarly, we now compute $\text{tr}(cA)$ directly from the definition of the trace:

$$\text{tr}(cA) = \sum_{j=1}^{n} [cA]_{j,j} = \sum_{j=1}^{n} ca_{j,j}$$

$$= c\sum_{j=1}^{n} a_{j,j} = c\,\text{tr}(A).$$

3.1.20 We start by seeing what happens if we choose $B = \mathbf{e}_i\mathbf{e}_j^T$ and $C = \mathbf{e}_j\mathbf{e}_k^T$ for some i,j,k with $i \neq k$:

$$f(ACB) = f(A\mathbf{e}_j\mathbf{e}_k^T\mathbf{e}_i\mathbf{e}_j^T) = 0f(A\mathbf{e}_j\mathbf{e}_j^T) = 0,$$

$$f(ABC) = f(A\mathbf{e}_i\mathbf{e}_j^T\mathbf{e}_j\mathbf{e}_k^T) = f(A\mathbf{e}_i\mathbf{e}_k^T).$$

Since $f(ACB) = f(ABC)$, it follows that $f(A\mathbf{e}_i\mathbf{e}_k^T) = 0$. By now choosing $A = \mathbf{e}_j\mathbf{e}_i^T$, we see that

$$f(A\mathbf{e}_i\mathbf{e}_k^T) = f(\mathbf{e}_j\mathbf{e}_i^T\mathbf{e}_i\mathbf{e}_k^T) = f(\mathbf{e}_j\mathbf{e}_k^T) = 0$$

for all j,k. Since f is linear and every matrix $A \in \mathcal{M}_n$ can be written in the form $A = \sum_{j,k=1}^{n} a_{j,k}\mathbf{e}_j\mathbf{e}_k^T$, it follows that

$$f(A) = f\left(\sum_{j,k=1}^{n} a_{j,k}\mathbf{e}_j\mathbf{e}_k^T\right) = \sum_{j,k=1}^{n} a_{j,k}f(\mathbf{e}_j\mathbf{e}_k^T) = 0$$

for all $A \in \mathcal{M}_n$.

3.1.21 (a) To give names to all of the vectors and scalars we will need to work with, suppose $B = \{\mathbf{v}_1, \mathbf{v}_2, \ldots, \mathbf{v}_k\}$, $[\mathbf{v}]_B = (c_1, c_2, \ldots, c_k)$ and $[\mathbf{w}]_B = (d_1, d_2, \ldots, d_k)$. Then

$$\mathbf{v} \cdot \mathbf{w} = (c_1\mathbf{v}_1 + \cdots + c_k\mathbf{v}_k) \cdot (d_1\mathbf{v}_1 + \cdots + d_k\mathbf{v}_k)$$

$$= c_1d_1(\mathbf{v}_1 \cdot \mathbf{v}_1) + c_1d_2(\mathbf{v}_1 \cdot \mathbf{v}_2) + \cdots$$

$$+ c_kd_{k-1}(\mathbf{v}_k \cdot \mathbf{v}_{k-1}) + c_kd_k(\mathbf{v}_k \cdot \mathbf{v}_k)$$

$$= c_1d_1 + c_2d_2 + \cdots + c_kd_k,$$

where the final equality comes from the fact that $\mathbf{v}_i \cdot \mathbf{v}_j = 1$ if $i = j$ and $\mathbf{v}_i \cdot \mathbf{v}_j = 0$ if $i \neq j$ (since B is an orthonormal basis). But

$$[\mathbf{v}]_B \cdot [\mathbf{w}]_B = c_1d_1 + c_2d_2 + \cdots + c_nd_n$$

too, directly from the definition of the dot product, so we are done.

(b) We can use part (a):

$$\big\|[\mathbf{v}]_B\big\| = \sqrt{[\mathbf{v}]_B \cdot [\mathbf{v}]_B} = \sqrt{\mathbf{v} \cdot \mathbf{v}} = \|\mathbf{v}\|.$$

(c) There are lots of possible counter-examples. For example, if $B = \{(1,1), (0,1)\}$ and $\mathbf{v} = (1,0)$, $\mathbf{w} = (0,1)$ then $[\mathbf{v}]_B = (1,-1)$ and $[\mathbf{w}]_B = (0,1)$. It follows that

$$\mathbf{v} \cdot \mathbf{w} = 0 \quad \text{but} \quad [\mathbf{v}]_B \cdot [\mathbf{w}]_B = -1, \text{ and}$$

$$\|\mathbf{v}\| = 1 \quad \text{but} \quad \big\|[\mathbf{v}]_B\big\| = \sqrt{2}$$

3.1.22 We start by supposing that $\mathbf{v}_1, \mathbf{v}_2, \ldots, \mathbf{v}_k \in B$ and $c_1, c_2, \ldots, c_k \in \mathbb{R}$ are such that

$$c_1\mathbf{v}_1 + c_2\mathbf{v}_2 + \cdots + c_k\mathbf{v}_k = \mathbf{0}.$$

Our goal is to show that $c_1 = c_2 = \cdots = c_k = 0$. To this end, we start by computing $\mathbf{v}_1 \cdot \mathbf{0}$ in two different ways:

$$0 = \mathbf{v}_1 \cdot \mathbf{0}$$

$$= \mathbf{v}_1 \cdot (c_1\mathbf{v}_1 + c_2\mathbf{v}_2 + \cdots + c_k\mathbf{v}_k)$$

$$= c_1(\mathbf{v}_1 \cdot \mathbf{v}_1) + c_2(\mathbf{v}_1 \cdot \mathbf{v}_2) + \cdots + c_k(\mathbf{v}_1 \cdot \mathbf{v}_k)$$

$$= c_1\|\mathbf{v}_1\|^2 + 0 + \cdots + 0.$$

Since all of the vectors in B are non-zero we know that $\|\mathbf{v}_1\| \neq 0$, so this implies $c_1 = 0$. A similar computation involving $\mathbf{v}_2 \cdot \mathbf{0}$ shows that $c_2 = 0$, and so on up to $\mathbf{v}_k \cdot \mathbf{0}$ showing that $c_k = 0$, so we conclude $c_1 = c_2 = \cdots = c_k = 0$ and thus B is linearly independent.

3.1.23 We use the interpretation of matrix multiplication as the matrix whose entries are all possible dot products of the first matrix with the second matrix:

$$A^TA = \begin{bmatrix} \mathbf{v}_1^T \\ \mathbf{v}_2^T \\ \vdots \\ \mathbf{v}_n^T \end{bmatrix} \begin{bmatrix} \mathbf{v}_1 \mid \mathbf{v}_2 \mid \cdots \mid \mathbf{v}_n \end{bmatrix}$$

$$= \begin{bmatrix} \mathbf{v}_1^T\mathbf{v}_1 & \mathbf{v}_1^T\mathbf{v}_2 & \cdots & \mathbf{v}_1^T\mathbf{v}_n \\ \mathbf{v}_2^T\mathbf{v}_1 & \mathbf{v}_2^T\mathbf{v}_2 & \cdots & \mathbf{v}_2^T\mathbf{v}_n \\ \vdots & \vdots & \ddots & \vdots \\ \mathbf{v}_n^T\mathbf{v}_1 & \mathbf{v}_n^T\mathbf{v}_2 & \cdots & \mathbf{v}_n^T\mathbf{v}_n \end{bmatrix}$$

$$= \begin{bmatrix} 1 & 0 & \cdots & 0 \\ 0 & 1 & \cdots & 0 \\ \vdots & \vdots & \ddots & \vdots \\ 0 & 0 & \cdots & 1 \end{bmatrix}$$

$$= I.$$

Side note: Matrices $A \in \mathcal{M}_n$ with the property that $A^TA = I$ are called **unitary matrices**.

3.1.24 (a) Since $\{R^\theta(\mathbf{e}_1), R^\theta(\mathbf{e}_2)\}$ consists of 2 vectors in $\mathbb{R}^2$, we just need to show that they are orthogonal unit vectors. Well, $R^\theta(\mathbf{e}_1) = (\cos(\theta), \sin(\theta))$, which has length $\sqrt{\sin^2(\theta) + \cos^2(\theta)} = \sqrt{1} = 1$, and similarly $R^\theta(\mathbf{e}_2) = (-\sin(\theta), \cos(\theta))$, which has length 1. (Another way to see that they have length 1 is to simply notice that rotating a unit vector does not change its length.) To see that they are orthogonal to each other, we compute

$$R^\theta(\mathbf{e}_1) \cdot R^\theta(\mathbf{e}_2)$$

$$= (\cos(\theta), \sin(\theta)) \cdot (-\sin(\theta), \cos(\theta))$$

$$= -\cos(\theta)\sin(\theta) + \sin(\theta)\cos(\theta) = 0.$$

(b) Suppose that $\{\mathbf{u}_1, \mathbf{u}_2\}$ is an orthonormal basis of $\mathbb{R}^2$. Since $\mathbf{u}_1 = (a,b)$ is a unit vector, we know that $|a| \leq 1$, so we can find θ such that $\cos(\theta) = a$. Then $b = \pm\sqrt{1-\cos^2(\theta)} = \pm|\sin(\theta)|$. If b is positive, we have written $\mathbf{u}_1 = (\cos(\theta), \sin(\theta))$. If b is negative, then replace θ by $-\theta$ so that we have $\mathbf{u}_1 =$

Next, since $\mathbf{u}_2$ is orthogonal to $\mathbf{u}_1$, we know that it is on the line perpendicular to the line in the direction of $(\cos(\theta), \sin(\theta))$. It follows that $\mathbf{u}_2$ is a multiple of $(-\sin(\theta), \cos(\theta))$. Since $\mathbf{u}_2$ is a unit vector, the only possible multiples are 1 and -1, so we know that either $\mathbf{u}_2 =$

$(-\sin(\theta), \cos(\theta))$ or $\mathbf{u}_2 = (\sin(\theta), -\cos(\theta))$. In the former case, we are done: $\mathbf{u}_1 = R^\theta(\mathbf{e}_1)$ and $\mathbf{u}_2 = R^\theta(\mathbf{e}_2)$. In the latter case, we use the facts that $\sin(\theta) = \cos(\theta - \pi/2)$ and $-\cos(\theta) = \sin(\theta - \pi/2)$ to see that $\mathbf{u}_1 = R_{\theta-\pi/2}(\mathbf{e}_2)$ and $\mathbf{u}_2 = R^{\theta-\pi/2}(\mathbf{e}_1)$, so again we are done.

Section 3.2: Determinants

3.2.1 (a) 0
       (e) 44

(c) 8
(g) 18

3.2.2 (a) -545

(c) -4008

3.2.3 (a) 6
       (e) $27/2$

(c) 12
(g) $2^7 3^3 = 3456$

3.2.4 (a) 6
       (e) -12

(c) -6
(g) 144

3.2.5 (a) -4

(c) 8

3.2.6 (a) False. $\det(-A) = (-1)^n \det(A)$, so the given property is true when n (the size of the matrix) is odd, but false when it is even.

(c) False. For example, if $A = I_2$ and $B = -I_2$ then $\det(A + B) = \det(O) = 0$, but $\det(A) + \det(B) = 1 + 1 = 2$.

(e) False. All that we can say is that if the RREF of A is I then it is invertible, so $\det(A) \neq 0$. Row operations can change determinants, so the determinant might not equal $\det(I) = 1$.

(g) True. If the columns of A are linearly dependent then it is not invertible, so $\det(A) = 0$.

3.2.8 Let

$$A = \begin{bmatrix} 1 & 0 \\ 0 & 1 \\ 0 & 0 \end{bmatrix} \text{ and } B = \begin{bmatrix} 1 & 0 & 0 \\ 0 & 1 & 0 \end{bmatrix}.$$

Then

$$AB = \begin{bmatrix} 1 & 0 & 0 \\ 0 & 1 & 0 \\ 0 & 0 & 0 \end{bmatrix} \text{ and } BA = \begin{bmatrix} 1 & 0 \\ 0 & 1 \end{bmatrix},$$

which have determinants 0 and 1, respectively.

3.2.10 Since the range of $[P_{\mathbf{u}}]$ is 1-dimensional, it is not invertible (as long as $[P_{\mathbf{u}}]$ is 2×2 or larger), so $\det([P_{\mathbf{u}}]) = 0$.

3.2.13 (a) We just use multilinearity of the determinant (i.e., its defining property (c)) repeatedly:

$$\det(cA) = \det\left(\left[\, c\mathbf{a}_1 \mid c\mathbf{a}_2 \mid \cdots \mid c\mathbf{a}_n \,\right]\right)$$
$$= c \det\left(\left[\, \mathbf{a}_1 \mid c\mathbf{a}_2 \mid \cdots \mid c\mathbf{a}_n \,\right]\right)$$
$$= c^2 \det\left(\left[\, \mathbf{a}_1 \mid \mathbf{a}_2 \mid \cdots \mid c\mathbf{a}_n \,\right]\right)$$
$$\vdots$$
$$= c^n \det\left(\left[\, \mathbf{a}_1 \mid \mathbf{a}_2 \mid \cdots \mid \mathbf{a}_n \,\right]\right) = c^n \det(A).$$

(b) If A is not invertible then $\det(A) = 0$ and A^T must also not be invertible, so $\det(A^T) = 0$ as well.

On the other hand, if A is invertible then Theorem 2.2.4 tells us that we can write A as a product of elementary matrices: $A = E_1 E_2 \cdots E_k$. Then $\det(A) = \det(E_1 E_2 \cdots E_k) = \det(E_1) \cdots \det(E_k)$. Also, $A^T = E_k^T \cdots E_2^T E_1^T$, so $\det(A^T) = \det(E_k^T \cdots E_2^T E_1^T) = \det(E_k^T) \cdots \det(E_2^T) \det(E_1^T)$, so it suffices to show that $\det(E) = \det(E^T)$ whenever E is an elementary matrix.

To this end, we note that if E is an elementary matrix corresponding to a "swap" row operation $R_i \leftrightarrow R_j$ or a "multiplication" row operation cR_i then $E^T = E$, so it is trivially the case that $\det(E^T) = \det(E)$. For the one remaining case (i.e., the "addition" row operation $R_i + cR_j$), we just note that each of E and E^T are triangular (one of them is lower triangular and the other is upper triangular), so Theorem 3.2.4 tells us that their determinants equal the product of their diagonal entries. But E and E^T have the *same* diagonal entries (which are all 1), so $\det(E) = \det(E^T)$, which completes the proof.

3.2.16 (a) First notice that

$$\begin{bmatrix} A & B \\ O & C \end{bmatrix} = \begin{bmatrix} I & B \\ O & C \end{bmatrix} \begin{bmatrix} A & O \\ O & I \end{bmatrix}.$$

The determinant of the matrix on the left is the product of the determinants of the matrices on the right, so it suffices to show that the matrices on the right have determinants equal to $\det(C)$ and $\det(A)$, respectively.

Suppose A is $k \times k$. For the matrix

$$\begin{bmatrix} I_k & B \\ O & C \end{bmatrix},$$

notice that computing its determinant via a cofactor expansion down its leftmost column shows that

$$\det\left(\begin{bmatrix} I_k & B \\ O & C \end{bmatrix}\right) = \det\left(\begin{bmatrix} I_{k-1} & \widetilde{B} \\ O & C \end{bmatrix}\right),$$

where $\widetilde{B}$ is the matrix obtained from B by removing its topmost row. Repeating this argument again a total of k times similarly shows that

$$\det\left(\begin{bmatrix} I_k & B \\ O & C \end{bmatrix}\right) = \det(C),$$

as desired. The fact that

$$\det\left(\begin{bmatrix} A & O \\ O & I \end{bmatrix}\right) = \det(A)$$

can be proved similarly by using cofactor expansions along the *last* column of the matrix.

(b) This fact is trivial if $n = 1$ and we already proved it in the $n = 2$ case in part (a). To prove it in general, we use induction and start with the inductive step since we have already proved the base case.

Suppose that the claim holds for a particular value of n. We can partition an $(n+1) \times (n+1)$ block triangular matrix as a 2×2 block triangular matrix as follows:

$$\begin{bmatrix} A_1 & * & \cdots & * & * \\ O & A_2 & \cdots & * & * \\ \vdots & \vdots & \ddots & \vdots & \vdots \\ O & O & \cdots & A_n & * \\ O & O & \cdots & O & A_{n+1} \end{bmatrix}$$

$$= \begin{bmatrix} \begin{array}{cccc|c} A_1 & * & \cdots & * & * \\ O & A_2 & \cdots & * & * \\ \vdots & \vdots & \ddots & \vdots & \vdots \\ O & O & \cdots & A_n & * \\ \hline O & O & \cdots & O & A_{n+1} \end{array} \end{bmatrix}.$$

By using the inductive hypothesis and part (a), we then see that

$$\det\left(\begin{bmatrix} \begin{array}{cccc|c} A_1 & * & \cdots & * & * \\ O & A_2 & \cdots & * & * \\ \vdots & \vdots & \ddots & \vdots & \vdots \\ O & O & \cdots & A_n & * \\ \hline O & O & \cdots & O & A_{n+1} \end{array} \end{bmatrix}\right)$$

$$= \det\left(\begin{bmatrix} A_1 & * & \cdots & * \\ O & A_2 & \cdots & * \\ \vdots & \vdots & \ddots & \vdots \\ O & O & \cdots & A_n \end{bmatrix}\right) \det(A_{n+1})$$

$$= \left(\prod_{j=1}^{n} \det(A_j)\right) \det(A_{n+1}) = \prod_{j=1}^{n+1} \det(A_j),$$

which completes the inductive step and the proof.

3.2.18 Following the hint, we compute

$$\begin{bmatrix} I_m & -A \\ B & I_n \end{bmatrix} \begin{bmatrix} I_m & A \\ O & I_n \end{bmatrix} = \begin{bmatrix} I_m & O \\ B & I_n + BA \end{bmatrix}.$$

Since this matrix is block triangular, Exercise 3.2.16 (together with the fact that taking the transpose of a matrix does not change its determinant) tells us that the determinant of this product equals

$$\det\left(\begin{bmatrix} I_m & O \\ B & I_n + BA \end{bmatrix}\right) = \det(I_m)\det(I_n + BA)$$

$$= \det(I_n + BA).$$

However, we can swap the order of the matrix product without changing the determinant, so this also equals

$$\det\left(\begin{bmatrix} I_m & A \\ O & I_n \end{bmatrix} \begin{bmatrix} I_m & -A \\ B & I_n \end{bmatrix}\right) = \det\left(\begin{bmatrix} I_m + AB & O \\ B & I_n \end{bmatrix}\right)$$

$$= \det(I_m + AB)\det(I_n)$$

$$= \det(I_m + AB).$$

We thus conclude that $\det(I_m + AB) = \det(I_n + BA)$, as desired.

3.2.20 (a) We just multiply

$$\begin{bmatrix} I_n & 0 \\ \mathbf{w}^T & 1 \end{bmatrix} \begin{bmatrix} I_n + \mathbf{v}\mathbf{w}^T & \mathbf{v} \\ \mathbf{0}^T & 1 \end{bmatrix} \begin{bmatrix} I_n & 0 \\ -\mathbf{w}^T & 1 \end{bmatrix}$$

$$= \begin{bmatrix} I_n & \mathbf{v} \\ \mathbf{0}^T & \mathbf{w}^T\mathbf{v} + 1 \end{bmatrix}$$

(b) By Exercise 3.2.16, we know that the determinant of a block triangular matrix is the product of the determinants of its diagonal blocks. It follows that

$$\det\left(\begin{bmatrix} I_n & 0 \\ \mathbf{w}^T & 1 \end{bmatrix} \begin{bmatrix} I_n + \mathbf{v}\mathbf{w}^T & \mathbf{v} \\ \mathbf{0}^T & 1 \end{bmatrix} \begin{bmatrix} I_n & 0 \\ -\mathbf{w}^T & 1 \end{bmatrix}\right)$$

$$= \det\left(\begin{bmatrix} I_n + \mathbf{v}\mathbf{w}^T & \mathbf{v} \\ \mathbf{0}^T & 1 \end{bmatrix}\right) = \det(I_n + \mathbf{v}\mathbf{w}^T).$$

However, using the result of part (a) shows that this determinant also equals

$$\det\left(\begin{bmatrix} I_n & \mathbf{v} \\ \mathbf{0}^T & \mathbf{w}^T\mathbf{v} + 1 \end{bmatrix}\right) = \mathbf{w}^T\mathbf{v} + 1,$$

so $\det(I_n + \mathbf{v}\mathbf{w}^T) = 1 + \mathbf{w}^T\mathbf{v}$.

(c) By the hint, $A + \mathbf{v}\mathbf{w}^T = A(I_n + A^{-1}\mathbf{v}\mathbf{w}^T)$. Applying the result of part (b) then shows that

$$\det(A + \mathbf{v}\mathbf{w}^T) = \det(A)\det(I_n + A^{-1}\mathbf{v}\mathbf{w}^T)$$

$$= \det(A)\det(I_n + (A^{-1}\mathbf{v})\mathbf{w}^T)$$

$$= \det(A)(1 + \mathbf{w}^T A^{-1}\mathbf{v}),$$

as desired.

Section 3.3: Eigenvalues and Eigenvectors

3.3.1 (a) 6　　　　(c) 0　　　　(e) -4

3.3.2 (a) A corresponding eigenvector is $\mathbf{v} = (5, -4)$. Note that any scalar multiple of this eigenvector also works (and the same remark applies to all parts of this exercise).

(c) $\mathbf{v} = (3, 2, 0)$

(e) $\mathbf{v} = (-1, 1, 0, 1)$

3.3.3 (a) We start by finding the eigenvalues:

$$\det\left(\begin{bmatrix} 1-\lambda & 2 \\ -1 & -2-\lambda \end{bmatrix}\right)$$
$$= (1-\lambda)(-2-\lambda) + 2 = \lambda^2 + \lambda = \lambda(\lambda+1),$$

so the eigenvalues are $\lambda = 0$ and $\lambda = -1$, each of which have algebraic multiplicity 1 (and thus geometric multiplicity 1 as well).
To find a basis of the eigenspace corresponding to $\lambda = 0$, we solve the linear system $(A - 0I)\mathbf{v} = \mathbf{0}$:

$$\begin{bmatrix} 1 & 2 & | & 0 \\ -1 & -2 & | & 0 \end{bmatrix} \xrightarrow{R_2 + R_1} \begin{bmatrix} 1 & 2 & | & 0 \\ 0 & 0 & | & 0 \end{bmatrix}.$$

From here we see that v_2 is a free variable and $v_1 = -2v_2$, so all eigenvectors are of the form $\mathbf{v} = v_2(-2, 1)$. It follows that $\{(-2, 1)\}$ is a basis of this eigenspace.
Similarly, to find a basis of the eigenspace corresponding to $\lambda = -1$, we solve the linear system $(A + I)\mathbf{v} = \mathbf{0}$:

$$\begin{bmatrix} 2 & 2 & | & 0 \\ -1 & -1 & | & 0 \end{bmatrix} \xrightarrow{R_2 + \frac{1}{2}R_1} \begin{bmatrix} 2 & 2 & | & 0 \\ 0 & 0 & | & 0 \end{bmatrix}.$$

From here we see that v_2 is a free variable and $v_1 = -v_2$, so all eigenvectors are of the form $\mathbf{v} = v_2(-1, 1)$. It follows that $\{(-1, 1)\}$ is a basis of this eigenspace.

(c) The only eigenvalue is 0 with algebraic multiplicity 2. A basis for the corresponding eigenspace is $\{(1, 0)\}$, so this eigenvalue has geometric multiplicity 1.

(e) The eigenvalues of a triangular matrix (and thus of a diagonal matrix) are its diagonal entries: 3, -2, and 7, each with algebraic multiplicity 1 (and thus geometric multiplicity 1 as well). Bases for the corresponding eigenspaces are $\{\mathbf{e}_1\}$, $\{\mathbf{e}_2\}$, and $\{\mathbf{e}_3\}$, respectively.

(g) This matrix is triangular, so its eigenvalues are its diagonal entries 2, -3, and 1, with algebraic multiplicities 2, 1, and 1, respectively. Bases for the corresponding eigenspaces are $\{(1, 0, 0, 0)\}$, $\{(1, -5, 0, 0)\}$, and $\{(1, -1, -2, 0)\}$, respectively, so all of the eigenvalues have geometric multiplicity 1.

3.3.4 (a) The eigenvalues are 2 and 8 (with algebraic and geometric multiplicity 1 each), and 7 (with algebraic and geometric multiplicity 2). Bases for their eigenspaces are $\{(1, 1, -1, -1)\}$, $\{(2, -1, 2, -1)\}$, and $\{(1, 1, 0, -1), (0, 1, -2, 0)\}$, respectively.

(c) The eigenvalues are 6 and 10 (with algebraic and geometric multiplicity 2 each), and 4 (with algebraic and geometric multiplicity 1). Bases for their eigenspaces are $\{(0, 0, 1, 0, -1), (1, 0, 0, 2, 0)\}$, $\{(1, -1, -1, 0, 1), (2, -2, 0, 1, 0)\}$, and $\{(0, 2, 2, -1, 0)\}$, respectively.

3.3.5 (a) False. We saw a counter-example to this claim in Example 3.3.8.

(c) False. For example, the following two matrices are row equivalent

$$\begin{bmatrix} 1 & 0 \\ 0 & 2 \end{bmatrix}, \quad \begin{bmatrix} 3 & 0 \\ 0 & 4 \end{bmatrix},$$

but the first matrix has eigenvalues 1 and 2, while the second matrix has eigenvalues 3 and 4.

(e) False. The closest true statement that can be made is $\det(A - \lambda I) = (-1)^n \det(\lambda I - A)$.

(g) True. If a matrix A is diagonal then $a_{i,j} = 0$ whenever $i \neq j$, so $a_{i,j} = a_{j,i}$ for all i, j.

(i) True (by definition).

(k) True. Since geometric multiplicity cannot exceed algebraic multiplicity, and geometric multiplicity cannot equal 0, the geometric multiplicity must equal 1.

(m) True. If two matrices have the same characteristic polynomial then they have the same eigenvalues, and the trace of a matrix equals the sum of its eigenvalues (by Theorem 3.3.1), so their trace must also be the same.

3.3.6 We start by computing the eigenvalues of A:

$$\det(A - \lambda I) = \det\left(\begin{bmatrix} k-\lambda & 1 \\ -1 & 1-\lambda \end{bmatrix}\right)$$
$$= (k-\lambda)(1-\lambda) + 1$$
$$= \lambda^2 - (1+k)\lambda + (k+1)$$
$$= 0.$$

Using the quadratic formula to solve this equation for λ, we get $\lambda = \frac{1}{2}(1+k) \pm \frac{1}{2}\sqrt{(1+k)^2 - 4k - 4} = \frac{1}{2}(1+k) \pm \frac{1}{2}\sqrt{k^2 - 2k - 3}$. We get two distinct eigenvalues exactly if the discriminant is positive: $k^2 - 2k - 3 > 0$. Similarly, we get only one real eigenvalue if the discriminant is 0, and we get no real eigenvalues if it is < 0.
The discriminant $k^2 - 2k - 3$ factors as $(k-3)(k+1)$, so we find that (i) we get two distinct real eigenvalues when $k > 3$ or $k < -1$, (ii) we get exactly one distinct eigenvalue when $k = 3$ or $k = -1$, and (iii) we get no real eigenvalues when $-1 < k < 3$.

3.3.8 If $A\mathbf{v} = \lambda\mathbf{v}$ then

$$A\overline{\mathbf{v}} = \overline{A}\overline{\mathbf{v}} = \overline{A\mathbf{v}} = \overline{\lambda\mathbf{v}} = \overline{\lambda}\overline{\mathbf{v}},$$

as desired.

3.3.10 Recall that the characteristic polynomial of a 2×2 matrix A is $p_A(\lambda) = \lambda^2 - \text{tr}(A)\lambda + \det(A)$. Applying the quadratic formula to this polynomial gives the eigenvalues of A as

$$\lambda = \frac{\text{tr}(A) \pm \sqrt{\text{tr}(A)^2 - 4\det(A)}}{2}.$$

3.3.12 The only possible eigenvalues are $\lambda = \pm 1$ and $\lambda = \pm i$. The fact that these eigenvalues are all possible can be seen simply by placing them on the diagonal of a diagonal matrix. Conversely, to see that no other eigenvalues are possible, let λ be an eigenvalue of A corresponding to an eigenvector $\mathbf{v}$, so $A\mathbf{v} = \lambda\mathbf{v}$. Multiplying by A on the left three times shows that $A^4\mathbf{v} = \lambda^4\mathbf{v}$. On the other hand, $A^4 = I$, so $A^4\mathbf{v} = \mathbf{v}$, so $\lambda^4\mathbf{v} = \mathbf{v}$, so $\lambda^4 = 1$.

3.3.13 (a) Setting $\det\left(\left[R^\theta\right] - \lambda I\right) = 0$ gives

$$\det\left(\begin{bmatrix} \cos(\theta) - \lambda & -\sin(\theta) \\ \sin(\theta) & \cos(\theta) - \lambda \end{bmatrix}\right)$$
$$= (\cos(\theta) - \lambda)^2 + \sin^2(\theta)$$
$$= \lambda^2 - 2\cos(\theta)\lambda + 1$$
$$= 0.$$

Solving this equation (via the quadratic formula) gives $\lambda = \cos(\theta) \pm \sqrt{\cos^2(\theta) - 1}$.

(b) We found the eigenvalues to be $\lambda = \cos(\theta) \pm \sqrt{\cos^2(\theta) - 1}$, which are real if and only if $\cos^2(\theta) - 1 \geq 0$, i.e., $\cos(\theta) = \pm 1$. In other words, $\theta = k\pi$ for some integer k.

This makes sense geometrically because if $\theta = 2k\pi$ then all vectors are being rotated back around to themselves, so every vector is an eigenvector of $\left[R^\theta\right]$ in this case (with eigenvalue 1). If $\theta = (2k+1)\pi$ then every vector is rotated around so that it is pointing in the exact opposite direction, so again every vector is an eigenvector (with eigenvalue -1). However, if $\theta \neq k\pi$, then every vector is rotated so that it is not parallel to where it started, so there are no (real) eigenvectors and hence no (real) eigenvalues.

As a side note, we show a bit later (in Theorem 3.C.1) that *every* 2×2 matrix with complex (non-real) eigenvalues looks like a scaling of a rotation in some basis.

3.3.15 (a) We use the fact that if $\mathbf{v}$ and $\mathbf{w}$ are column vectors then $\mathbf{v} \cdot \mathbf{w} = \mathbf{v}^*\mathbf{w}$. It follows that $\mathbf{x} \cdot (A\mathbf{y}) = \mathbf{x}^*A\mathbf{y}$ and $(A^*\mathbf{x}) \cdot \mathbf{y} = (A^*\mathbf{x})^*\mathbf{y} = \mathbf{x}^*A\mathbf{y}$. Since both quantities equal $\mathbf{x}^*A\mathbf{y}$, we indeed have $\mathbf{x} \cdot (A\mathbf{y}) = (A^*\mathbf{x}) \cdot \mathbf{y}$.

(b) Just like in part (a), $\mathbf{x} \cdot (A\mathbf{y}) = \mathbf{x}^*A\mathbf{y}$ and $(B\mathbf{x}) \cdot \mathbf{y} = \mathbf{x}^*B^*\mathbf{y}$. We thus know that $\mathbf{x}^*A\mathbf{y} = \mathbf{x}^*B^*\mathbf{y}$ for all $\mathbf{x} \in \mathbb{C}^m$ and $\mathbf{y} \in \mathbb{C}^n$. However, if we choose $\mathbf{x} = \mathbf{e}_i$ and $\mathbf{y} = \mathbf{e}_j$ then $\mathbf{x}^*A\mathbf{y} = a_{i,j}$ and $\mathbf{x}^*B^*\mathbf{y} = \overline{b_{j,i}}$, so we see that $a_{i,j} = \overline{b_{j,i}}$ for all i, j. In other words, $B = A^*$.

3.3.16 Note that 0 is an eigenvalue of A if and only if there exists a non-zero vector $\mathbf{v}$ such that $A\mathbf{v} = 0\mathbf{v} = \mathbf{0}$, which is equivalent to A not being invertible via Theorem 2.2.4.

3.3.18 Recall from Theorem 3.2.4 that the determinant of a triangular matrix is the product of its diagonal entries. It follows that

$$p_A(\lambda) = \det(A - \lambda I)$$
$$= (a_{1,1} - \lambda)(a_{2,2} - \lambda)\cdots(a_{n,n} - \lambda),$$

which has roots $a_{1,1}, a_{2,2}, \ldots, a_{n,n}$. In other words, the eigenvalues of A are the diagonal entries of A.

3.3.21 (a) Recall from Exercise 3.3.20 that A and A^T have the same eigenvalues. If λ is an eigenvalue of A^T (and thus of A as well) with corresponding eigenvector $\mathbf{v}$ then

$$A^*\overline{\mathbf{v}} = \overline{A^T\mathbf{v}} = \overline{\lambda\mathbf{v}} = \overline{\lambda}\,\overline{\mathbf{v}},$$

so $\overline{\lambda}$ is an eigenvalue of A^* (with corresponding eigenvector $\overline{\mathbf{v}}$). The reverse implication follows from the fact that $(A^*)^* = A$ so if $\overline{\lambda}$ is an eigenvalue of A^* then $\overline{\overline{\lambda}} = \lambda$ is an eigenvalue of A.

(b) We compute $\mathbf{w} \cdot (A\mathbf{v})$ in two different ways:

$$\mathbf{w} \cdot (A\mathbf{v}) = \mathbf{w} \cdot (\lambda\mathbf{v}) = \overline{\lambda}(\mathbf{w} \cdot \mathbf{v}) \quad \text{and}$$
$$\mathbf{w} \cdot (A\mathbf{v}) = (A^*\mathbf{w}) \cdot \mathbf{v} = (\mu\mathbf{w}) \cdot \mathbf{v} = \mu(\mathbf{w} \cdot \mathbf{v}).$$

Since $\mu \neq \overline{\lambda}$, the only way that this can happen is if $\mathbf{w} \cdot \mathbf{v} = 0$ (or equivalently, $\mathbf{v} \cdot \mathbf{w} = 0$).

3.3.22 (a) We just compute

$$B\mathbf{v}_1 = (A - \lambda_1\mathbf{v}_1\mathbf{v}_1^*)\mathbf{v}_1$$
$$= A\mathbf{v}_1 - \lambda_1\mathbf{v}_1 = \lambda_1\mathbf{v}_1 - \lambda_1\mathbf{v}_1 = \mathbf{0}.$$

(b) We already showed in part (a) that B has 0 as an eigenvalue, so we just need to show that its other eigenvalues are $\lambda_2, \lambda_3, \ldots, \lambda_n$. To this end, let $\mathbf{w}_j$ be an eigenvector of A^* corresponding to the eigenvalue $\overline{\lambda_j}$ for $2 \leq j \leq n$. Then

$$B^*\mathbf{w}_j = (A - \lambda_1\mathbf{v}_1\mathbf{v}_1^*)^*\mathbf{w}_j$$
$$= (A^* - \overline{\lambda_1}\mathbf{v}_1\mathbf{v}_1^*)\mathbf{w}_j = A^*\mathbf{w}_j = \overline{\lambda_j}\mathbf{w}_j,$$

where the second-to-last equality follows from Exercise 3.3.21 and the assumption that $\lambda_1 \neq \lambda_j$. The statement is still true even if $\lambda_1 = \lambda_j$, but we have to be careful to *choose* $\mathbf{w}_j$ to be orthogonal to $\mathbf{v}_1$ (it does not come for free). It follows that B^* has $\overline{\lambda_2}, \overline{\lambda_3}, \ldots, \overline{\lambda_n}$ as eigenvalues, so B has $\lambda_2, \lambda_3, \ldots, \lambda_n$ as eigenvalues.

(c) Consider the matrix

$$A = \begin{bmatrix} 1 & 1 \\ 0 & 2 \end{bmatrix},$$

which has eigenvalues $\lambda_1 = 1$ and $\lambda_2 = 2$ with corresponding unit eigenvectors $\mathbf{v}_1 = (1,0)$ and $v_2 = (1,1)/\sqrt{2}$, respectively. Then

$$B = A - \lambda_1\mathbf{v}_1\mathbf{v}_1^* = \begin{bmatrix} 0 & 1 \\ 0 & 2 \end{bmatrix}$$

still has $\lambda_2 = 2$ as an eigenvalue, but its corresponding eigenvectors are now the non-zero multiples of $\mathbf{v}_2 = (1,2)$.

In general, almost any triangular matrix can be used to show that the eigenvectors might change here.

3.3.23 (a) We just compute this characteristic polynomial via the formula for the determinant of a 3×3 matrix:

$$\det(C - \lambda I) = \det\left(\begin{bmatrix} -\lambda & 1 & 0 \\ 0 & -\lambda & 1 \\ -a_0 & -a_1 & -a_2 - \lambda \end{bmatrix}\right)$$
$$= -\lambda^2(a_2 + \lambda) - a_0 - a_1\lambda$$
$$= -(\lambda^3 + a^2\lambda^2 + a_1\lambda + a_0),$$

(b) We already proved the $n = 3$ base case (and the $n = 1$ and $n = 2$ cases are even simpler), so we jump right to the inductive step. We induct on n (the size of the matrix) and assume that the result holds for $(n-1) \times (n-1)$ companion matrices. By taking a cofactor expansion along the first column of C, we see that

$\det(C - \lambda I)$

$$= \det \left(\begin{bmatrix} -\lambda & 1 & 0 & \cdots & 0 \\ 0 & -\lambda & 1 & \cdots & 0 \\ \vdots & \vdots & \vdots & \ddots & \vdots \\ 0 & 0 & 0 & \cdots & 1 \\ -a_0 & -a_1 & -a_2 & \cdots & -a_{n-1} - \lambda \end{bmatrix} \right)$$

$$= -\lambda \det \left(\begin{bmatrix} -\lambda & 1 & \cdots & 0 \\ \vdots & \vdots & \ddots & \vdots \\ 0 & 0 & \cdots & 1 \\ -a_1 & -a_2 & \cdots & -a_{n-1} - \lambda \end{bmatrix} \right)$$

$$+ (-1)^n a_0$$

$$= -\lambda \left((-1)^{n-1} (\lambda^{n-1} + a_{n-1} \lambda^{n-2} + \cdots + a_2 \lambda + a_1) \right) + (-1)^n a_0$$

$$= (-1)^n (\lambda^n + a_{n-1} \lambda^{n-1} + \cdots + a_1 \lambda + a_0),$$

where the second-to-last equality follows from the inductive hypothesis and the fact that we are taking the determinant of an $(n-1) \times (n-1)$ companion matrix.

3.3.24 (a) We simply note that if $\mathbf{1} = (1, 1, \ldots, 1)$ then $A\mathbf{1} = \mathbf{1}$, since $A\mathbf{1}$ is simply the vector whose entries are the sums of the rows of A. It follows that $\mathbf{1}$ is an eigenvector of A with corresponding eigenvalue 1.

(b) Let $\mathbf{v}$ be an eigenvector of A corresponding to an eigenvalue λ, scaled so that its largest entry is $v_i = 1$, as suggested by the hint. Since $A\mathbf{v} = \lambda \mathbf{v}$, we conclude then that

$$\sum_{j=1}^{n} a_{i,j} v_j = \lambda v_i = \lambda.$$

Taking the absolute value of both sides of this equation then shows that

$$|\lambda| = \left| \sum_{j=1}^{n} a_{i,j} v_j \right| \leq \sum_{j=1}^{n} |a_{i,j} v_j| \leq \sum_{j=1}^{n} |a_{i,j}| = 1,$$

where the first inequality is the triangle inequality and the second inequality follows from the fact that we scaled $\mathbf{v}$ so that $|v_j| \leq 1$ for all j.

(c) We simply note that A is column stochastic if and only if A^T is row stochastic, and recall that the eigenvalues of A and A^T coincide.

Section 3.4: Diagonalization

3.4.1 In all of these cases, your diagonalization might look slightly different than the solution presented here due to choosing different eigenvectors for the columns of P or a different ordering of the eigenvalues/eigenvectors.

(a) $D = \begin{bmatrix} 0 & 0 \\ 0 & 2 \end{bmatrix}$, $P = \begin{bmatrix} -1 & 1 \\ 1 & 1 \end{bmatrix}$, $P^{-1} = \frac{1}{2} \begin{bmatrix} -1 & 1 \\ 1 & 1 \end{bmatrix}$.

(c) Not diagonalizable, since its eigenvalue $\lambda = 1$ has algebraic multiplicity 2 but geometric multiplicity 1.

(e) $D = \begin{bmatrix} 2 & 0 & 0 \\ 0 & 1 & 0 \\ 0 & 0 & -1 \end{bmatrix}$, $P = \begin{bmatrix} 1 & 0 & 0 \\ 0 & 1 & 1 \\ 0 & 1 & -1 \end{bmatrix}$,

$P^{-1} = \frac{1}{2} \begin{bmatrix} 2 & 0 & 0 \\ 0 & 1 & 1 \\ 0 & 1 & -1 \end{bmatrix}$.

(g) $D = \begin{bmatrix} 3 & 0 & 0 \\ 0 & -1 & 0 \\ 0 & 0 & 2 \end{bmatrix}$, $P = \begin{bmatrix} 1 & 0 & 3 \\ 0 & 1 & -2 \\ 0 & 0 & -3 \end{bmatrix}$,

$P^{-1} = \frac{1}{3} \begin{bmatrix} 3 & 0 & 3 \\ 0 & 3 & -2 \\ 0 & 0 & -1 \end{bmatrix}$.

3.4.2 (a) This matrix can be diagonalized in the same way as in Exercise 3.4.1(a). Any real diagonalization of a matrix is also a complex diagonalization of it.

(c) $D = \begin{bmatrix} 2+i & 0 \\ 0 & 2-i \end{bmatrix}$, $P = \begin{bmatrix} 1 & 1 \\ i & -i \end{bmatrix}$,

$P^{-1} = \frac{1}{2} \begin{bmatrix} 1 & -i \\ 1 & i \end{bmatrix}$.

(e) $D = \begin{bmatrix} 2 & 0 & 0 \\ 0 & i & 0 \\ 0 & 0 & -i \end{bmatrix}$, $P = \begin{bmatrix} 1 & 0 & 0 \\ 0 & 1 & i \\ 0 & 1 & -i \end{bmatrix}$,

$P^{-1} = \frac{1}{2} \begin{bmatrix} 2 & 0 & 0 \\ 0 & 1 & -i \\ 0 & 1 & i \end{bmatrix}$.

(g) Not diagonalizable, since its eigenvalue $\lambda = i$ has algebraic multiplicity 3 but geometric multiplicity 2.

3.4.3 (a) This matrix A can be diagonalized over $\mathbb{R}$ (and thus $\mathbb{C}$) as $A = PDP^{-1}$, where D is diagonal with diagonal entries 1, 2, 3, 3, and 4 (in that order), and

$$P = \begin{bmatrix} -1 & -1 & 1 & -1 & -1 \\ 0 & 0 & 0 & 1 & 1 \\ -1 & 0 & 1 & 0 & 0 \\ -1 & 0 & 1 & -1 & 0 \\ 1 & 1 & 0 & 0 & 1 \end{bmatrix}.$$

(c) This matrix A can be diagonalized over $\mathbb{C}$ (but not $\mathbb{R}$) as $A = PDP^{-1}$, where D is diagonal with diagonal entries $2+i$, $2+i$, $2-i$, $2-i$, and 0 (in that order), and

$$P = \begin{bmatrix} 2 & 2+i & 2 & 2-i & 2 \\ 1-i & -1 & 1+i & -1 & 0 \\ 2-i & -1-i & 2+i & -1+i & 2 \\ 2-i & 1 & 2+i & 1 & 2 \\ -i & 2 & i & 2 & 0 \end{bmatrix}.$$

3.4.4 (a) $\begin{bmatrix} (e^2+1)/2 & (e^2-1)/2 \\ (e^2-1)/2 & (e^2+1)/2 \end{bmatrix}$

(b) $\begin{bmatrix} \sin(2)/2 & \sin(2)/2 \\ \sin(2)/2 & \sin(2)/2 \end{bmatrix}$

3.4.5 (a) $\begin{bmatrix} 5 & 1 & 1 & 2 & -2 \\ -4 & 0 & -1 & -2 & 2 \\ 5 & 2 & 2 & 3 & -2 \\ 1 & 1 & 0 & 2 & 0 \\ 5 & 2 & 1 & 3 & -1 \end{bmatrix}$

3.4.6 (a) False. We can *always* find such a P and D: just choose $P = D = O$. To make this statement true, we must add the requirement that P is invertible.

(c) False. Every complex matrix has n eigenvalues counting algebraic multiplicity (as noted in the text, this follows from the Fundamental Theorem of Algebra).

(e) False. It can have repeated eigenvalues, as long as the geometric multiplicity of each eigenvalue equals its algebraic multiplicity.

(g) False. Diagonalizability and invertibility are not at all related. For example, the matrix

$$\begin{bmatrix} 1 & 0 \\ 0 & 0 \end{bmatrix}$$

is diagonal (and thus diagonalizable via $P = I$) but not invertible.

(i) True. If that single eigenvalue is λ then $A = PDP^{-1} = P(\lambda I)P^{-1} = \lambda PP^{-1} = \lambda I$.

3.4.8 We start by diagonalizing A. Its eigenvalues are -1 and 8, with corresponding eigenvectors $(1,1)$ and $(2,-1)$, respectively. It follows that $A = PDP^{-1}$, where

$$D = \begin{bmatrix} -1 & 0 \\ 0 & 8 \end{bmatrix}, P = \begin{bmatrix} 1 & 2 \\ 1 & -1 \end{bmatrix}, P^{-1} = \frac{1}{3}\begin{bmatrix} 1 & 2 \\ 1 & -1 \end{bmatrix}.$$

One cube root of A is thus

$$B = PD^{1/3}P^{-1}$$

$$= \frac{1}{3}\begin{bmatrix} 1 & 2 \\ 1 & -1 \end{bmatrix}\begin{bmatrix} -1 & 0 \\ 0 & 2 \end{bmatrix}\begin{bmatrix} 1 & 2 \\ 1 & -1 \end{bmatrix}$$

$$= \begin{bmatrix} 1 & -2 \\ -1 & 0 \end{bmatrix}.$$

3.4.9 Everything in Example 3.4.8 is the same with the Lucas numbers as it was with the Fibonacci numbers, except for the initial condition $L_0 = 2$ instead of $F_0 = 0$. We can thus copy down the diagonalization from that example and change that initial condition to get

$$\begin{bmatrix} L_{n+1} \\ L_n \end{bmatrix} = \begin{bmatrix} 1 & 1 \\ 1 & 0 \end{bmatrix}^n \begin{bmatrix} 1 \\ 2 \end{bmatrix}$$

$$= PD^nP^{-1}\begin{bmatrix} 1 \\ 2 \end{bmatrix}$$

$$= \frac{1}{\sqrt{5}}PD^n\begin{bmatrix} 1 & \phi-1 \\ -1 & \phi \end{bmatrix}\begin{bmatrix} 1 \\ 2 \end{bmatrix}$$

$$= P\begin{bmatrix} \phi^n & 0 \\ 0 & (1-\phi)^n \end{bmatrix}\begin{bmatrix} 1 \\ 1 \end{bmatrix}$$

$$= \begin{bmatrix} \phi & 1-\phi \\ 1 & 1 \end{bmatrix}\begin{bmatrix} \phi^n \\ (1-\phi)^n \end{bmatrix}$$

$$= \begin{bmatrix} \phi^{n+1} + (1-\phi)^{n+1} \\ \phi^n + (1-\phi)^n \end{bmatrix},$$

where we used the fact that $2\phi - 1 = \sqrt{5}$ in the fourth equality above. It follows that

$$L_n = \phi^n + (1-\phi)^n.$$

3.4.12 (a) Since A is diagonalizable, we can find an invertible matrix $P \in \mathcal{M}_n$ and a diagonal matrix $D \in \mathcal{M}_n$ such that $A = PDP^{-1}$. Since P and P^{-1} are invertible, Exercise 2.4.17 implies $\text{rank}(A) = \text{rank}(PDP^{-1}) = \text{rank}(DP^{-1}) = \text{rank}(D)$. Since D is diagonal, its rank is the number of non-zero diagonal entries that it has (see Exercise 2.4.14), which is the number of non-zero eigenvalues of A.

(b) For example,

$$A = \begin{bmatrix} 0 & 1 \\ 0 & 0 \end{bmatrix}$$

has eigenvalue 0 with algebraic multiplicity 2 (and thus no non-zero eigenvalues), but its rank is 1.

3.4.14 (a) Theorem 3.4.3 tells us that if A is diagonalizable then it has a set of eigenvectors that forms a basis of $\mathbb{R}^n$ (or $\mathbb{C}^n$, as appropriate). Theorem 3.4.1 then tells us that $A = PDP^{-1}$, where D is diagonal with the eigenvalues of A as its diagonal entries and P has the corresponding eigenvectors as its columns. Since B has the same eigenvalues and corresponding eigenvectors, the same theorem also implies that B this is a valid diagonalization of B. That is, $B = PDP^{-1}$ too, so $A = B$.

(b) For example, the matrices

$$A = \begin{bmatrix} 0 & 1 \\ 0 & 0 \end{bmatrix} \quad \text{and} \quad B = \begin{bmatrix} 0 & 2 \\ 0 & 0 \end{bmatrix}$$

have the same eigenvalues ($\lambda = 0$ with algebraic multiplicity 2) and the same eigenvectors ($\{(1,0)\}$ is a basis of the corresponding eigenspace for both matrices), but $A \neq B$.

3.4.15 (a) Recall from Appendix A.1.3 that every complex number has exactly k distinct k-th roots. It follows that if $A = PDP^{-1}$ then A has at least k^n distinct k-th roots of the form $P\widetilde{D}P^{-1}$, where $\widetilde{D}$ is one of the k^n diagonal matrices that can be obtained from D by taking a k-th root of each of its n diagonal entries.

To see that there are no other k-th roots of A (so there are *exactly* k^n k-th roots), notice that if $B^k = A$ then each eigenvalue λ of B is such that λ^k is an eigenvalue of A, and the corresponding eigenvectors for B and A are the same. It then follows from Exercise 3.4.14 that B must be one of the k^n matrices that are diagonalized by the same matrix P as A described above.

(b) Suppose $A = PDP^{-1}$ where the top two diagonal entries of D are the same value, which we will call λ, and we partition P as $P = \begin{bmatrix} \mathbf{v}_1 \mid \mathbf{v}_2 \mid P_2 \end{bmatrix}$. If we let

$$Q = \begin{bmatrix} c_1\mathbf{v}_1 + c_2\mathbf{v}_2 \mid d_1\mathbf{v}_1 + d_2\mathbf{v}_2 \mid P_2 \end{bmatrix}$$

be such that $\{c_1\mathbf{v}_1 + c_2\mathbf{v}_2, d_1\mathbf{v}_1 + d_2\mathbf{v}_2\}$ is linearly independent (so Q is invertible) then $A = QDQ^{-1}$ as well. If we let $\widetilde{D}$ be a k-th root of D with its top-left two entries different from each other (i.e., they are different k-th roots of λ) then there are infinitely many different matrices of the form $Q\widetilde{D}Q^{-1}$ (by varying the coefficients c_1, c_2, d_1, d_2 in the definition of Q), each of which is a k-th root of A.

3.4.16 This follows immediately from the fact that every diagonalization is produced via eigenvalues and eigenvectors (Theorem 3.4.1) and every diagonalizable matrix is completely determined by its eigenvalues an eigenvectors (Exercise 3.4.14).

3.4.17 See the solution to Exercise 3.4.16.

3.4.18 (a) If $A\mathbf{v} = \lambda\mathbf{v}$, multiplying on the left by A shows that $A^2\mathbf{v} = \lambda A\mathbf{v} = \lambda^2\mathbf{v}$.

(b) If $k \geq 1$ then this follows just by multiplying the equation $A\mathbf{v} = \lambda\mathbf{v}$ on the left by A a total of $k-1$ times. If $k = 0$ then it follows from the fact that every vector $\mathbf{v}$ is an eigenvector of $A^0 = I$ with eigenvalue $\lambda^0 = 1$.

(c) We just make use of part (b) repeatedly. If $p(x) = c_k x^k + \cdots + c_1 x + c_0$ then

$$
\begin{aligned}
p(A)\mathbf{v} &= \left(c_k A^k + \cdots + c_1 A + c_0 I \right)\mathbf{v} \\
&= c_k A^k \mathbf{v} + \cdots + c_1 A\mathbf{v} + c_0\mathbf{v} \\
&= c_k \lambda^k \mathbf{v} + \cdots + c_1 \lambda\mathbf{v} + c_0\mathbf{v} \\
&= \left(c_k \lambda^k + \cdots + c_1\lambda + c_0 \right)\mathbf{v} = p(\lambda)\mathbf{v}.
\end{aligned}
$$

3.4.19 (a) Diagonalize A as $A = PDP^{-1}$. Then

$$
\begin{aligned}
A^r A^s &= (PD^r P^{-1})(PD^s P^{-1}) \\
&= PD^r D^s P^{-1} = PD^{r+s}P^{-1} = A^{r+s},
\end{aligned}
$$

where the second-to-last equality follows from the fact that diagonal matrix multiplication works entrywise and $x^r x^s = x^{r+s}$ for all scalars x.

(b) Diagonalize A as $A = PDP^{-1}$. Then

$$
\begin{aligned}
\left(A^r\right)^s &= \left(PD^r P^{-1}\right)^s = P\left(D^r\right)^s P^{-1} \\
&= PD^{rs}P^{-1} = A^{rs},
\end{aligned}
$$

where the second-to-last equality follows from the fact that diagonal matrix multiplication works entrywise and $(x^r)^s = x^{rs}$ for all scalars x.

3.4.21 Recall from Theorem 3.4.9 that $\det(e^A) = e^{\mathrm{tr}(A)}$, and similarly $\det(e^{A+B}) = e^{\mathrm{tr}(A+B)}$. We thus conclude that

$$
\begin{aligned}
\det\left(e^{A+B}\right) &= e^{\mathrm{tr}(A+B)} = e^{\mathrm{tr}(A)+\mathrm{tr}(B)} = e^{\mathrm{tr}(A)}e^{\mathrm{tr}(B)} \\
&= \det(e^A)\det(e^B) = \det(e^A e^B).
\end{aligned}
$$

3.4.23 (a) Recall from Example 2.3.8 that V is invertible, so we just need to show that its columns are eigenvectors of C with corresponding eigenvalues $\lambda_1, \lambda_2, \ldots, \lambda_n$. Well, let $\mathbf{v}_j$ be the j-th column of V. Then

$$
\begin{aligned}
C\mathbf{v}_j &= \begin{bmatrix} 0 & 1 & \cdots & 0 \\ \vdots & \vdots & \ddots & \vdots \\ 0 & 0 & \cdots & 1 \\ -a_0 & -a_1 & \cdots & -a_{n-1} \end{bmatrix} \begin{bmatrix} 1 \\ \lambda_j \\ \vdots \\ \lambda_j^{n-2} \\ \lambda_j^{n-1} \end{bmatrix} \\
&= \begin{bmatrix} \lambda_j \\ \lambda_j^2 \\ \vdots \\ \lambda_j^{n-1} \\ -a_0 - a_1\lambda_j - \cdots - a_{n-1}\lambda_j^{n-1} \end{bmatrix} = \lambda_j\mathbf{v}_j,
\end{aligned}
$$

where the final entry $-a_0 - a_1\lambda_j - a_2\lambda_j^2 - \cdots - a_{n-1}\lambda_j^{n-1}$ equals λ_j^n, since λ_j is a root of the polynomial p. It follows that $\mathbf{v}_j$ is indeed an eigenvector of C with corresponding eigenvalue λ_j, which completes the proof.

(b) Suppose that λ is an eigenvalue of C that is repeated (i.e., has algebraic multiplicity greater than 1). We will show that it has geometric multiplicity equal to 1 and thus C is not diagonalizable by Theorem 3.4.3.

To this end, suppose $\mathbf{v} = (v_1, v_2, \ldots, v_n)$ is an eigenvector of C with corresponding eigenvalue λ. Then $C\mathbf{v} = \lambda\mathbf{v}$ implies

$$C\mathbf{v} = \begin{bmatrix} 0 & 1 & \cdots & 0 \\ \vdots & \vdots & \ddots & \vdots \\ 0 & 0 & \cdots & 1 \\ -a_0 & -a_1 & \cdots & -a_{n-1} \end{bmatrix} \begin{bmatrix} v_1 \\ v_2 \\ \vdots \\ v_{n-1} \\ v_n \end{bmatrix}$$

$$= \begin{bmatrix} v_2 \\ v_3 \\ \vdots \\ v_n \\ -a_0 v_1 - \cdots - a_{n-1} v_n, \end{bmatrix} = \begin{bmatrix} \lambda v_1 \\ \lambda v_2 \\ \vdots \\ \lambda v_{n-1} \\ \lambda v_n \end{bmatrix}.$$

The top entry of these vectors tell us that $v_2 = \lambda v_1$, the next entry tells us that $v_3 = \lambda v_2$, and so on up to $v_n = \lambda v_{n-1}$. It follows that $v_j = \lambda^{j-1} v_1$ for $1 \leq j \leq n-1$. Then the bottom entry of the vectors above tell us that

$$\lambda v_n = -a_0 v_1 - a_1 v_2 - \cdots - a_{n-1} v_n$$
$$= v_1(-a_0 - a_1\lambda - \cdots - a_{n-1}\lambda^{n-1}) = \lambda^n v_1,$$

so $v_n = \lambda^{n-1} v_1$. We have thus shown that every eigenvector of C corresponding to the eigenvalue λ has the form $\mathbf{v} = v_1(1, \lambda, \lambda^2, \ldots, \lambda^{n-1})$. This set of vectors is 1-dimensional, so the geometric multiplicity of λ is 1.

Section 3.5: Summary and Review

3.5.1 (a) Not similar. For example, $\text{rank}(A) = 1$ but $\text{rank}(B) = 2$.

(c) Not similar. For example, $\text{tr}(A) = 6$ but $\text{tr}(B) = 7$.

(e) Similar. To see this, just note that both are diagonalizable and have the same characteristic polynomials $(\lambda - 2)(\lambda - 3)$. Explicitly, $A = PBP^{-1}$ if

$$P = \frac{1}{5}\begin{bmatrix} 2 & -1 \\ 1 & 2 \end{bmatrix}.$$

(g) Similar. To see this, just note that both are diagonalizable and have the same characteristic polynomials $(\lambda - 1)(\lambda - 3)^2$. Explicitly, $A = PBP^{-1}$ if

$$P = \begin{bmatrix} 1 & 0 & 1 \\ 1 & 1 & 1 \\ 0 & 1 & 1 \end{bmatrix}.$$

3.5.2 (a) True. If $A\mathbf{v} = 3\mathbf{v}$, then multiplying on the left by A gives $A^2\mathbf{v} = 3A\mathbf{v} = 9\mathbf{v}$.

(c) True. This was noted explicitly in the text.

(e) True. If the eigenvalues of A and B are distinct then they are diagonalizable, and if they are diagonalizable with the same eigenvalues (with the same multiplicities) then they are similar by Theorem 3.5.1.

3.5.3 (a) They both have rank 2 (A is already in RREF and B can be put into RREF just by swapping rows).

(b) The characteristic polynomial of each matrix is $p_A(\lambda) = p_B(\lambda) = \lambda^4$.

(c) Both matrices only have 0 as their eigenvalues, corresponding to the eigenspace span$\{(1,0,0,0), (0,1,0,0)\}$. In particular, for each matrix the eigenvalue 0 has geometric multiplicity 2.

(d) Following the hint, let $P \in \mathcal{M}_4$ and write the linear system $AP = PB$ explicitly in terms of the entries of P:

$$\begin{bmatrix} P_{3,1} & P_{3,2} & P_{3,3} & P_{3,4} \\ P_{4,1} & P_{4,2} & P_{4,3} & P_{4,4} \\ 0 & 0 & 0 & 0 \\ 0 & 0 & 0 & 0 \end{bmatrix}$$

$$= \begin{bmatrix} 0 & 0 & P_{1,2} & P_{1,3} \\ 0 & 0 & P_{2,2} & P_{2,3} \\ 0 & 0 & P_{3,2} & P_{3,3} \\ 0 & 0 & P_{4,2} & P_{4,3} \end{bmatrix}.$$

It follows that P has the form

$$P = \begin{bmatrix} P_{1,1} & 0 & P_{1,3} & P_{1,4} \\ P_{2,1} & 0 & P_{2,3} & P_{2,4} \\ 0 & 0 & 0 & P_{1,3} \\ 0 & 0 & 0 & P_{2,3} \end{bmatrix}.$$

Since no matrix P of this form is invertible (if a matrix has a zero column then it is not invertible), we conclude that A and B are not similar.

3.5.4 Suppose that

$$A = \begin{bmatrix} a & b \\ c & d \end{bmatrix}$$

is such that $A^2 = N$. Multiplying A by itself gives

$$\begin{bmatrix} a^2 + bc & b(a+d) \\ c(a+d) & bc+d^2 \end{bmatrix} = \begin{bmatrix} 0 & 1 \\ 0 & 0 \end{bmatrix}.$$

The bottom-left entry of this matrix equation tells us that either $c = 0$ or $a+d = 0$. However, if $a+d = 0$ then the top-right entry must also equal 0 (but it equals 1), so we conclude that $c = 0$. The above matrix equation thus simplifies to

$$\begin{bmatrix} a^2 & b(a+d) \\ 0 & d^2 \end{bmatrix} = \begin{bmatrix} 0 & 1 \\ 0 & 0 \end{bmatrix}.$$

The top-left entry tells us that $a = 0$, the bottom-right entry tells us that $d = 0$, and then the top-right entry gives us a contradiction since it says $b(0+0) = 1$. It follows that the equation $A^2 = N$ has no solution.

Section 3.A: Extra Topic: More About Determinants

3.A.1 (a) $\begin{bmatrix} 1 & 0 \\ 0 & 1 \end{bmatrix}$ (c) $\begin{bmatrix} 2 & -3 \\ -3 & 2 \end{bmatrix}$

(e) $\begin{bmatrix} 2 & 1 & 4 \\ 4 & -2 & 8 \\ 0 & 0 & -8 \end{bmatrix}$ (g) $\begin{bmatrix} -3 & 12 & -8 \\ 5 & -4 & 0 \\ -2 & 0 & 0 \end{bmatrix}$

3.A.2 (a) $\begin{bmatrix} 3 & 27 & -6 & -6 \\ -5 & -31 & 17 & 3 \\ 5 & -11 & 4 & -3 \\ -1 & -2 & -5 & 9 \end{bmatrix}$

3.A.3 Throughout this solution, we give the determinant of A (the coefficient matrix) as well as the matrices $A_1, A_2, \ldots, A_n$ that are used in Cramer's rule.
(a) $\det(A) = -3$, $\det(A_1) = -3$, $\det(A_2) = -3$, $(x,y) = (1,1)$.
(c) $\det(A) = 4$, $\det(A_1) = 2$, $\det(A_2) = 8$, $\det(A_3) = 6$, $(x,y,z) = (1/2, 2, 3/2)$.
(e) $\det(A) = -2$, $\det(A_1) = 1$, $\det(A_2) = -2$, $\det(A_3) = -1$, $(x,y,z) = (-1/2, 1, 1/2)$.

3.A.4 (a) $(1\,2\,3)$ (c) $(1\,3\,2\,4)$
(e) $(5\,4\,3\,2\,1)$ (g) $(3\,5\,1\,2\,4\,6)$

3.A.5 (a) $(3\,2\,1)$ (c) $(4\,3\,2\,1)$
(e) $(3\,1\,5\,2\,4)$ (g) $(5\,1\,4\,2\,6\,3)$

3.A.6 (a) True. In fact, we can rearrange Theorem 3.A.1 to see that
$$(\text{cof}(A))^{-1} = \frac{1}{\det(A)} A^T.$$
(c) True. This can be seen directly from the definition of the cofactor matrix or by rearranging Theorem 3.A.1 to show
$$(\text{cof}(I)) = \det(I)(I^T)^{-1} = I.$$
(e) False. We encountered a counter-example to this claim in Example 3.A.1(b). The matrix in that example has rank 2, but its cofactor matrix has rank 1. Note, however, that if $\text{rank}(A) = n$ then $\text{rank}(\text{cof}(A)) = n$ by part (a) of this exercise.
(g) True. This follows from the fact that $\iota \circ \iota = \iota$.
(i) True. There are $n!$ permutations in S_n. When $n = 5$, there are $5! = 5 \cdot 4 \cdot 3 \cdot 2 \cdot 1 = 120$ permutations.

3.A.8 This follows from repeatedly applying the result of Exercise 3.A.7. For example, that exercise tells us that $\text{cof}(A^2) = \text{cof}(AA) = \text{cof}(A)\text{cof}(A) = (\text{cof}(A))^2$. Applying this result repeatedly similarly gives $\text{cof}(A^k) = (\text{cof}(A))^k$ for all integers $k \geq 1$.
The $k = 0$ case follows from the fact that $\text{cof}(I) = I$. If $k = -1$ then we notice that $\text{cof}(A^{-1}) = A^T / \det(A) = (\text{cof}(A))^{-1}$. For $k < -1$ we just repeatedly use Exercise 3.A.7 and the $k = -1$ case.

3.A.10 We use Theorem 3.A.1 to see that $\text{cof}(A) = \det(A)(A^T)^{-1}$. Taking the determinant of both sides gives us
$$\det(\text{cof}(A)) = \det\left(\det(A)(A^T)^{-1}\right)$$
$$= (\det(A))^n \det((A^T)^{-1})$$
$$= (\det(A))^n / \det(A)$$
$$= (\det(A))^{n-1}$$
as desired.

3.A.13 (a) This follows from multiplicativity of the determinant:
$$\text{sgn}(\sigma \circ \tau) = \det(P_{\sigma \circ \tau}) = \det(P_\sigma P_\tau)$$
$$= \det(P_\sigma)\det(P_\tau) = \text{sgn}(\sigma)\text{sgn}(\tau).$$
(b) Let $\tau = (2\,1\,3\,4\,5\,\cdots\,n)$ be the permutation that swaps 1 and 2 but leaves all other inputs alone. Then $\text{sgn}(\tau) = -1$ and as σ ranges over all permutations in S_n so does $\sigma \circ \tau$. If strictly more than half of the permutations σ had sign 1, it would follow from part (a) that strictly more than half of the permutations $\sigma \circ \tau$ had sign -1 (and similarly if we replaced "more" by "fewer" in this sentence). This does not make sense, so exactly half of the permutations must have each sign.

3.A.14 If $m > n$ then $\text{rank}(A) \leq n < m$ so $\text{rank}(AB) \leq n < m$ by Theorem 2.4.11. However, AB is an $m \times m$ matrix, so it is thus not invertible and $\det(AB) = 0$.

3.A.16 (a) 1 (c) 2
(e) -8 (g) 18

3.A.17 We observe that
$$\text{per}(A^T) = \sum_{\sigma \in S_n} a_{1,\sigma(1)} a_{2,\sigma(2)} \cdots a_{n,\sigma(n)}$$
$$= \sum_{\sigma \in S_n} a_{\sigma^{-1}(1),1} a_{\sigma^{-1}(2),2} \cdots a_{\sigma^{-1}(n),n}$$
$$= \sum_{\sigma \in S_n} a_{\sigma(1),1} a_{\sigma(2),2} \cdots a_{\sigma(n),n}$$
$$= \text{per}(A),$$
where the second-to-last equality follows from the fact that the set of inverses of permutations is exactly the same as the set of permutations themselves.

3.A.19 Almost any matrices work. For example, if
$$A = \begin{bmatrix} 1 & 1 \\ 1 & 1 \end{bmatrix} \quad \text{and} \quad B = \begin{bmatrix} 1 & 2 \\ 3 & 4 \end{bmatrix}$$
then $\text{per}(A) = 2$, $\text{per}(B) = 10$, but $\text{per}(AB) = 48$.

Section 3.B: Extra Topic: Power Iteration

3.B.1 (a) $\lambda \approx 5.37$, $\mathbf{v} \approx (0.42, 0.91)$
(c) $\lambda \approx -3.33$, $\mathbf{v} \approx (0.73, -0.42, -0.55)$
(e) $\lambda \approx 7.92$, $\mathbf{v} \approx (0.73, 0.29, 0.57, 0.24)$

3.B.2 (a) $\lambda_1 \approx 5.37$, $\mathbf{v}_1 \approx (0.42, 0.91)$
$\lambda_2 \approx -0.37$, $\mathbf{v}_2 \approx (0.82, -0.57)$
(c) $\lambda_1 \approx 16.12$, $\mathbf{v}_1 \approx (0.23, 0.53, 0.82)$
$\lambda_2 \approx -1.12$, $\mathbf{v}_2 \approx (0.79, 0.09, -0.61)$
$\lambda_3 \approx 0.00$, $\mathbf{v}_2 \approx (0.41, -0.82, 0.41)$

3.B.3 (a) $\lambda \approx 11.58$,
$\mathbf{v} \approx (0.15, 0.25, 0.01, 0.54, -0.27, 0.46, 0.58)$

3.B.4 (a) False. Diagonalizability has nothing to do with whether or not power iteration converges.
(c) True. This follows from Theorem 3.B.1.
(e) True. This is part of the Perron–Frobenius theorem (Theorem 3.B.2).

3.B.6 Recall that $\mathbf{v}_k$ approaches the eigenspace corresponding to the dominant eigenvalue of A, so $A\mathbf{v}_k \approx \lambda_1 \mathbf{v}_k$ when k is large, so $\|A\mathbf{v}_k\| \approx \|\lambda_1 \mathbf{v}_k\| = \lambda_1$ when k is large.

The above argument is the key idea. If we want to make it rigorous, we note that since $\mathbf{v}_k$ approaches the eigenspace corresponding to the dominant eigenvalue, we have

$$\lim_{k \to \infty} \left(A\mathbf{v}_k - \lambda_1 \mathbf{v}_k \right) = \mathbf{0},$$

so

$$\lim_{k \to \infty} \|A\mathbf{v}_k - \lambda_1 \mathbf{v}_k\| = 0.$$

By the reverse triangle inequality (Exercise 1.2.21), this implies

$$\lim_{k \to \infty} \|A\mathbf{v}_k\| = \lim_{k \to \infty} \|\lambda_1 \mathbf{v}_k\|,$$

as long as at least one of these limits exists. The limit on the right is simply

$$\lim_{k \to \infty} \|\lambda_1 \mathbf{v}_k\| = \lim_{k \to \infty} \lambda_1 = \lambda_1,$$

which proves the claim.

3.B.8 If A^k is positive then the Perron–Frobenius theorem applies to it. Since A has the same eigenvectors and the corresponding eigenvalues of A are simply some k-th roots of the corresponding eigenvalues of A^k, parts (b), (c), and (d) follow immediately. Part (a) just needs the minor adjustment that $|\lambda_1| > |\lambda_j|$ for all $2 \le j \le n$ since a k-th root of a positive number is not necessarily positive.

3.B.9 In all parts of this question, we refer to the original given matrix as A.
(a) $k = 2$: $A^2 = \begin{bmatrix} 1 & 1 \\ 1 & 2 \end{bmatrix}$
(c) $k = 4$: $A^4 = \begin{bmatrix} 8 & 8 \\ 8 & 8 \end{bmatrix}$
(e) $k = 4$: $A^4 = \begin{bmatrix} 5 & 4 & 1 \\ 4 & 6 & 3 \\ 1 & 3 & 2 \end{bmatrix}$

3.B.10 (a) $\lambda \approx 1.62$, $\mathbf{v} \approx (0.53, 0.85)$
(c) $\lambda \approx 2.00i$, $\mathbf{v} \approx (0.71, 0.71)$
(e) $\lambda \approx 1.80$, $\mathbf{v} \approx (0.59, 0.74, 0.33)$

3.B.12 (a) Neither irreducible (irreducible matrices by definition have all entries non-negative) nor primitive.
(c) Both primitive and irreducible.
(e) Neither irreducible nor primitive.

3.B.14 Pages B and C are both linked by page A, whereas pages D and E are both linked by page B. Since page A has a higher rank than page B, it boosts up the ranks of the pages it links to, so pages B and C end up with higher rank than pages D and E.

Section 3.C: Extra Topic: Complex Eigenvalues of Real Matrices

3.C.1 (a) We recognize this as the rotation matrix $[R^{\pi/2}]$, so a decomposition of the desired form comes just from choosing $Q = I$, $r = 1$, and $\theta = \pi/2$.
(c) This matrix has eigenvalues $1 \pm i = \sqrt{2}e^{\pm i\pi/4}$ with corresponding eigenvectors $(1, -1 \pm i)$, so we can choose $r = \sqrt{2}$, $\theta = \pi/4$, and

$$Q = \begin{bmatrix} 1 & 0 \\ -1 & -1 \end{bmatrix}.$$

3.C.2 (a) $B = \text{diag}\left(1, [R^{\pi/2}]\right)$, $Q = \begin{bmatrix} 1 & 0 & 0 \\ 0 & 1 & 0 \\ 1 & 0 & -1 \end{bmatrix}$.
(c) $B = \text{diag}\left(1, 2, \sqrt{2}[R^{\pi/4}]\right)$,
$$Q = \begin{bmatrix} 0 & 1 & 1 & 0 \\ 1 & 0 & 0 & 0 \\ 1 & 0 & 0 & -1 \\ 0 & 1 & 0 & 1 \end{bmatrix}.$$

3.C.3 (a) $B = \text{diag}\left(2\sqrt{2}[R^{\pi/4}], 2\sqrt{2}[R^{\pi/4}], 4\right)$,
$$Q = \begin{bmatrix} 1 & 0 & 0 & 1 & 1 \\ -1 & 1 & 0 & 1 & 0 \\ 0 & 1 & -1 & 0 & 1 \\ 1 & 1 & 1 & 0 & -1 \\ 0 & 1 & -1 & 0 & -1 \end{bmatrix}.$$

3.C.4 (a) True. If n is odd then every matrix $A \in \mathcal{M}_n(\mathbb{R})$ has a real eigenvalue. The reason for this is that the eigenvalues of A come in complex conjugate pairs, so there is necessarily an even number of non-real eigenvalues, but the total number of eigenvalues (counting algebraic multiplicity) is n, which is odd.
(c) True. Since real numbers are complex numbers, any real diagonalization is automatically a complex diagonalization as well.

(e) False. We can only block diagonalize matrices that can be diagonalized over $\mathbb{C}$. For example, we cannot block diagonalize the matrix

$$\begin{bmatrix} 1 & 1 & 1 \\ 0 & 1 & 1 \\ 0 & 0 & 1 \end{bmatrix}$$

in the sense of that theorem, since its eigenvalue 1 has algebraic multiplicity 3 but geometric multiplicity 1.

3.C.6 (a) This matrix can be diagonalized as $A = PDP^{-1}$, where

$$P = \begin{bmatrix} 1 & 1 \\ 1+i & 1-i \end{bmatrix}, \quad D = \begin{bmatrix} \sqrt{3}+i & 0 \\ 0 & \sqrt{3}-i \end{bmatrix}.$$

It follows that
$$A^k = PD^kP^{-1} =$$

$$\frac{1}{2}\begin{bmatrix} 1 & 1 \\ 1+i & 1-i \end{bmatrix}\begin{bmatrix} (\sqrt{3}+i)^k & 0 \\ 0 & (\sqrt{3}-i)^k \end{bmatrix}\begin{bmatrix} 1+i & -i \\ 1-i & i \end{bmatrix},$$

which can be multiplied together if desired (it is an ugly mess).

(b) One block diagonalization of this matrix is $A = QBQ^{-1}$, where

$$Q = \begin{bmatrix} 1 & 0 \\ 1 & -1 \end{bmatrix} \quad \text{and} \quad B = 2\left[R^{\pi/6}\right].$$

It follows that

$$A^k = QB^kQ^{-1}$$

$$= 2^k\begin{bmatrix} 1 & 0 \\ 1 & -1 \end{bmatrix}\left[R^{k\pi/6}\right]\begin{bmatrix} 1 & 0 \\ 1 & -1 \end{bmatrix},$$

which has top-left entry equal to $2^k(\cos(k\pi/6) - \sin(k\pi/6))$, top-right entry $2^k\sin(k\pi/6)$, bottom-left entry $-2^{k+1}\sin(k\pi/6)$, and bottom-right entry $2^k(\cos(k\pi/6) + \sin(k\pi/6))$.

Section 3.D: Extra Topic: Linear Recurrence Relations

3.D.1 (a) The characteristic polynomial of this linear recurrence relation is $\lambda^2 - \lambda - 6$, which has roots $\lambda_0 = 3$ and $\lambda_1 = -2$, each with multiplicity 1.
(c) $3 \pm \sqrt{5}$, each with multiplicity 1.
(e) 1, 2, and 3, each with multiplicity 1.
(g) 2, with multiplicity 3.
(i) 3 and $1 \pm \sqrt{3}$, each with multiplicity 1.

3.D.2 (a) $\begin{bmatrix} 0 & 1 \\ 6 & 1 \end{bmatrix}$ (c) $\begin{bmatrix} 0 & 1 \\ -4 & 6 \end{bmatrix}$

(e) $\begin{bmatrix} 0 & 1 & 0 \\ 0 & 0 & 1 \\ 6 & -11 & 6 \end{bmatrix}$ (g) $\begin{bmatrix} 0 & 1 & 0 \\ 0 & 0 & 1 \\ 8 & -12 & 6 \end{bmatrix}$

(i) $\begin{bmatrix} 0 & 1 & 0 \\ 0 & 0 & 1 \\ -6 & -4 & 5 \end{bmatrix}$

3.D.3 (a) The characteristic polynomial of this linear recurrence relation is $\lambda^2 z - \lambda - 6$, which has roots $\lambda_0 = 3$ and $\lambda_1 = -2$. There thus exist (unique) scalars c_0 and c_1 such that $x_n = c_0 3^n + c_1(-2)^n$ for all n. Using the fact that $x_0 = 4$ and $x_1 = -3$ shows that $c_0 = 1$ and $c_1 = 3$, so $x_n = 3^n + 3(-2)^n$.
(c) $x_n = i^n + (-i)^n$. Alternatively, a formula that does not involve complex numbers is $x_n = 2\cos(\pi n/2)$.
(e) $x_n = 3^n - 1$.
(g) $x_n = (2n-3) + 2^n$.

3.D.4 (a) False. We saw numerous counter-examples to this claim throughout this section, and a simple example is the linear recurrence relation $x_n = -x_{n-2}$, which has characteristic polynomial $\lambda^2 + 1$, which has roots $\pm i$.

(c) True. In light of Theorem 3.D.3, we want to construct a linear recurrence relation with 1 as a root of multiplicity 4 (since the polynomial $n^2 - n^3$ has degree $4 - 1 = 3$). One such linear recurrence relation is $x_n = 4x_{n-1} - 6x_{n-2} + 4x_{n-3} - x_{n-4}$, which has characteristic polynomial $(\lambda - 1)^4$.
(e) True. This follows from Corollary 3.D.4. Note that the root 4 being repeated is not a problem—there would only be a problem if there were two *different* largest roots with the same absolute value (e.g., 4 and -4).

3.D.5 (a) The characteristic polynomial of this linear recurrence relation must have roots 2 and 3, so it is $p(\lambda) = (\lambda-2)(\lambda-3) = \lambda^2 - 5\lambda + 6$, so the recurrence itself must be

$$x_n = 5x_{n-1} - 6x_{n-2}.$$

(c) $x_n = 9x_{n-1} - 26x_{n-2} + 24x_{n-3}$
(e) $x_n = 9x_{n-1} - 30x_{n-2} + 44x_{n-3} - 24x_{n-4}$
(g) $x_n = 3x_{n-1} - 3x_{n-2} + x_{n-3}$
(i) $x_n = 8x_{n-1} - 26x_{n-2} + 48x_{n-3} - 45x_{n-4}$

3.D.6 (a) The characteristic polynomial of this recurrence relation is $\lambda^2 + 1$, which has roots $\pm i = e^{\pm i\pi/2}$. After solving for the coefficients in the solution, we see that $x_n = i^n + (-i)^n$. By substituting in the polar forms $(\pm i)^n = (e^{\pm i\pi/2})^n = e^{\pm in\pi/2}$, we see that

$$x_n = i^n + (-i)^n$$
$$= e^{in\pi/2} + e^{-in\pi/2} = 2\cos(n\pi/2).$$

(c) $x_n = 1 + 2^{n+1}\cos(n\pi/2)$.

3.D.8 We prove this claim by induction. The base case is trivial since we were told that $x_0 = x_1 = 0$. For the inductive step, suppose $x_{n-1} = x_{n-2}$. Then using the recurrence relation gives $x_{n+1} = x_n - 4x_{n-1} + 4x_{n-2} = x_n - 4x_{n-1} + 4x_{n-1} = x_n$, which completes the induc-

Bibliography

[Chv83] V. Chvatal, *Linear Programming* (W. H. Freeman, 1983)

[Con86] J.H. Conway, The weird and wonderful chemistry of audioactive decay. Eureka **46**, 5–16 (1986)

[Fel91] S.L. Feld, Why your friends have more friends than you do. Am. J. Sociol. **96**(6), 1464–1477 (1991)

[FR97] B. Fine, G. Rosenberger, *The Fundamental Theorem of Algebra*. Undergraduate Texts in Mathematics (Springer, New York, 1997)

[Han96] J. Hannah, A geometric approach to determinants. Am. Math. Mon. **103**(5), 401–409 (1996)

[HL10] F.S. Hillier, G.J. Lieberman, *Introduction to Operations Research*, 9th edn. (McGraw–Hill, 2010)

[Joh20] N. Johnston, *Advanced Linear and Matrix Algebra*. Undergraduate Texts in Mathematics (Springer, New York, 2020)

[Mar99] R.K. Martin, *Large Scale Linear and Integer Optimization: A Unified Approach* (Springer, US, 1999)

[Win03] W.L. Winston, *Operations Research: Applications and Algorithms*, 4th edn. (Cengage Learning, 2003)

© Springer Nature Switzerland AG 2021
N. Johnston, *Introduction to Linear and Matrix Algebra*,
https://doi.org/10.1007/978-3-030-52811-9

Index

© Springer Nature Switzerland AG 2021
N. Johnston, *Introduction to Linear and Matrix Algebra*,
https://doi.org/10.1007/978-3-030-52811-9

Symbol Index

Printed in the United States
by Baker & Taylor Publisher Services